Contemporary Polymer Chemistry

THIRD EDITION

Harry R. Allcock
Department of Chemistry
The Pennsylvania State University

Frederick W. Lampe
Formerly, Department of Chemistry
The Pennsylvania State University

James E. Mark
Department of Chemistry
University of Cincinnati

Pearson Education, Inc.
Upper Saddle River, New Jersey 07458

Library of Congress Cataloging-in-Publication Data

Allcock, H. R.
  Contemporary polymer chemistry / Harry R. Allcock, Frederick W. Lampe, James E. Mark.—3rd ed.
       p. cm.
  Includes bibliographical references and index.
  ISBN 0-13-065056-0
  1. Polymers.   2. Polymerization.
  [DNLM: 1. Polymers—chemistry.   2. Macromolecular Systems.   QD 381 A354c 2003]   I. Lampe, Frederick Walter   II. Mark, James E.   III. Title.

QD381.A44 2003
547.7—dc21   2002154591

Acquisitions Editor: *Nicole Folchetti*
Editorial Assistant: *Chris Cella*
Vice President and Director of Production and Manufacturing, ESM: *David W. Riccardi*
Production Editor: *Patty Donovan*
Director of Creative Services: *Carole Anson*
Creative Director: *Paul Belfanti*
Art and Cover Director: *Jayne Conte*
Managing Editor, AV Production & Management: *Patty Burns*
Art Editor: *Jessica Einsig*
Production Manager, Artworks: *Ronda Whitson*
Manager, Production Technologies, Artworks: *Matt Haas*
Illustrator, Artworks: *Kathryn Anderson, Royce Copenheaver, Dan Knopsnyder, Mark Landis, Audrey Simonetti, Stacy Smith*
Quality Assurance, Artworks: *Pamela Taylor, Ken Mooney, Timothy Nguyen*
Manufacturing Manager: *Trudy Pisciotti*
Manufacturing Buyer: *Lynda Castillo*
Marketing Manager: *Christine Henry*

© 2003, 1990, 1981 by Pearson Education, Inc.
Pearson Education, Inc.
Upper Saddle River, New Jersey 07458

Printed in the United States of America

10   9   8   7   6   5   4   3   2   1

ISBN 0-13-065056-0

Pearson Education Ltd., *London*
Pearson Education Australia Pty. Ltd., *Sydney*
Pearson Education Singapore, Pte. Ltd.
Pearson Education North Asia Ltd., *Hong Kong*
Pearson Education Canada, Inc., *Toronto*
Pearson Educación de Mexico, S.A. de C.V.
Pearson Education—Japan, *Tokyo*
Pearson Education Malaysia, Pte. Ltd.

Brief Contents

Contents

Contents

Preface

This third edition of *Contemporary Polymer Chemistry,* like the two preceding editions, is designed as an introduction to polymers for students of chemistry, physics, chemical engineering, materials science, and biomaterials. It assumes a basic knowledge of subjects taught in university undergraduate programs in the above disciplines. Specifically, the book aims to broaden the perspective of specialists in different technical areas to the point where they can appreciate the scope, importance, and future potential of polymer chemistry and technology. Thus, in writing this book we have kept in mind the individual who has a sound knowledge of basic science but needs to know more about polymers for future academic research or teaching or for entry into the polymer industry. For this reason, many topics that are well-known to practicing polymer scientists are handled here from first principles.

More rigorous and more comprehensive treatments exist for nearly all of the topics discussed in this book. However, few attempts have been made to bring together synthetic, structural, kinetic, thermodynamic, and use-oriented material in one volume. Our aim has been to provide a broad, coherent introduction to modern polymer chemistry and to direct the reader to more detailed sources for advanced study.

This edition has undergone an appreciable expansion compared to the earlier volumes. This reflects the widening scope of the field and the need to include recent advances in polymer synthesis as well as to broaden the treatment of polymer characterization methods and the physics and materials science aspects of the subject. Contributions by new co-author James E. Mark have added considerably to this widened perspective. A number of new topics have also been introduced, including insights into rubberlike elasticity, viscoelasticity, biomimicry, and the materials science of structure-property relationships. The list of references for further reading and the study questions, problems, and solutions have been updated extensively throughout the book.

The book is divided into five parts. *Part I* (Chapters 1–9) provides an introduction to the different classes of polymers and the ways in which they are synthesized and modified. Individual chapters deal with condensation, free-radical, and ionic or coordination polymerization, with photolytic, high energy radiation, and electrolytic polymerization, polymerization of cyclic compounds, biological macromolecules, with the ways that synthetic polymers can be modified chemically, and with polymers that contain inorganic elements. Several of the chapters in this section have been revised to reflect recent developments. For example, new sections have been included on molecular weight distributions, dendrimers and telechelic polymers, polymer surface chemistry, organometallic-initiated polymerizations, atom transfer radical reactions, polymerizations in supercritical carbon dioxide, and inorganic polymers. Thus, the emphasis in these chapters is on descriptive chemistry, general principles, and synthetic issues. The material is this section should be understandable to students who have taken undergraduate courses in general chemistry, and organic, inorganic, or biological chemistry. These chapters form the groundwork for the sections that follow.

Part II (Chapters 10–13) deals with thermodynamics, equilibria, and polymerization kinetics. Chapter 10 provides an elementary overview of the underlying principles that determine whether a monomer or a cyclic compound will polymerize or if a polymer will depolymerize. Chapters 11, 12, and 13 deal respectively with the kinetics of condensation, free-radical, and ionic polymerizations. A unique feature of these chapters is the full derivation of the kinetic expressions, with every attempt made to explain the underlying principles for each step. This should enable the treatments to be understood by anyone with a basic background of third year undergraduate physical chemistry.

Part III (Chapters 14–19) covers the physical methods that are employed for the characterization of polymers. Individual chapters cover "absolute" molecular weight measurements by osmometry, light scattering and ultracentrifugation; secondary molecular weight methods such as solution viscosity and gel permeation chromatography; thermodynamics of high polymer solutions; polymer morphology; glass transitions and crystallinity; conformational analysis; and X-ray diffraction techniques. The sections on secondary molecular weight methods, thermodynamics of polymer solutions, solid state characterization, X-ray scattering, and other physical methods have been expanded to provide more detail than in the earlier editions. In all these chapters the reader is introduced to the underlying theory and, where appropriate, to the practical approaches used. Each chapter provides the basic groundwork for elementary experimental work in these areas or for further, more detailed, studies.

In *Part IV* (Chapters 20 and 21) we discuss the engineering aspects of polymer science, including the fabrication of polymers and testing techniques. Here too, new sections have been added that deal with polymer chain orientation and materials reinforcement, with the behavior of polymers within fabrication machinery, and a much expanded section on rubberlike materials.

In *Part V* (Chapters 22–24) the emphasis is on the uses of polymers and the ways in which the polymer scientist can correlate molecular structure with properties and applications. Chapter 22 provides an overview of how the practicing chemist intuitively relates molecular structure to polymer properties as a route to the design of new materials. An addition to this chapter is an overview of how the molecular features of polymers become translated into solid state materials properties. Chapter 22 is one of the most important sections of the book for those who are encountering polymers for the first time. Chapter 23 gives an account of the rapidly expanding field of electroactive and electro-optical polymers, a subject that it likely to be the focus of much research in the future. The treatment in this edition includes electronically conductive materials and their use in devices, ionically conducting polymers and their role in advanced batteries and fuel cells, soft lithography, polymers for lenses and optical waveguides, optically responsive polymers, light-emitting polymers, and polymers for photovoltaic cells. The last chapter (Chapter 24) deals with the biomedical uses of synthetic polymers, a topic that continues to grow in importance year by year, and which accounts for an increasing proportion of the total research effort in polymer science.

Appendix I is a brief review of polymer nomenclature, and Appendix II is a compilation of physical property data and uses for a number of commercially important and research intensive polymers. This latter appendix provides perspective and serves as a reference source as the reader encounters new polymers at different points in the book.

The book may be used in several different ways. We recommend that the reader should follow the sequence outlined above, although specialized topics, such as those discussed in Chapters 10–13, might be absorbed best during a second reading. However, for readers who prefer to approach the subject from the viewpoint of properties and uses, we suggest starting with Chapters 1 and 22–25, and then following the remaining chapters in sequence beginning with Chapter 2.

In writing this new edition we have been helped considerably by a number of individuals. We are especially grateful to the many users of the earlier editions who have made numerous suggestions for improvements and to the reviewers of the manuscript for this edition. Virtually all of their suggestions have been incorporated into this edition, including the availability of a booklet of answers to the numerical questions.

Finally, we are sad to report that Fred Lampe died suddenly while this new edition was being prepared. His death was a serious blow to all who knew him. His engaging personality, good cheer, fairness, common sense, and deep fascination with physical chemistry are severely missed. Most of the sections of this book for which he was mainly responsible have been updated, but have otherwise retained their original character. We believe that his classic and, to a large degree, timeless contributions to this book will constitute a fitting memorial to him.

HARRY R. ALLCOCK
JAMES E. MARK

About the Authors

Harry R. Allcock is Evan Pugh Professor of Chemistry at The Pennsylvania State University. He received his B.Sc. and Ph.D. degrees from the University of London. After holding postdoctoral positions at Purdue University and the National Research Council of Canada, he spent five years as a polymer research scientist in American industry before joining The Pennsylvania State University in 1966. Trained initially as a mechanistic organometallic chemist, his research interests have included the synthesis of new organic and inorganic polymers, the use of inorganic and organometallic compounds as polymerization initiators, radiation-induced polymerizations, ring-chain equilibria, organosilicon compounds, and the structural examination of polymers by X-ray diffraction, NMR, and conformational analysis techniques. A major interest throughout his career has been the design and synthesis of new biomedical materials. He and his coworkers synthesized the first stable polyphosphazenes, and his research group has been responsible for many of the major developments in this field. His pioneering research has been recognized by numerous awards including the American Chemical Society National Awards in Polymer Chemistry and Materials Chemistry, and the A.C.S Polymer Division Herman Mark Award in Polymer Science. He is also a Guggenheim Fellow and a recipient of the American Institute of Chemists Chemical Pioneer Award. He has held a number of visiting lectureships, has published three research monographs, co-authored three other volumes, and co-edited three additional books. Professor Allcock has authored or co-authored more than 440 research papers and reviews, holds 54 patents, and has trained more than 100 graduate students and postdoctoral students in his laboratory.

Frederick W. Lampe (1927–2000) was a Professor of Chemistry at the Pennsylvania State University. Professor Lampe was born in Chicago. He served in the United States Navy from 1944 to 1946, and later received a B.S. degree from Michigan State University and a Ph.D. degree from Columbia University. He then spent seven years as a research scientist with the Humble Oil and Refining Company in Texas before moving to Penn State in 1960. Professor Lampe was a physical chemist whose polymer interests were in the areas of radiation-induced polymerizations, kinetics of polymerization processes, application of mass spectrometry to polymer degradation processes, statistical mechanics, and molecular weight methods. He was also interested in gaseous ion reactions, photochemistry, and the effects of ionizing radiation on materials. He was the author or co-author of over 160 research papers and review articles, and held five patents on polymer chemistry and radiation chemistry. He was also a visiting professor at the University of Freiburg, Germany, and at the Hahn-Meitner Nuclear Research Institute in Berlin. His honors included the Alexander von Humboldt Senior Scientist Award, a National Science Foundation Senior Postdoctoral Fellowship, and a Robert A. Welch Lecturship. Professor Lampe also served for five years as the Head of the Chemistry Department at Penn State.

James E. Mark was born in Wilkes-Barre, Pennsylvania. He received a B.S. degree in Chemistry from Wilkes College and his Ph.D. in Physical Chemistry from the University of Pennsylvania. After serving as a Postdoctoral Fellow at Stanford University with Professor Paul J. Flory, he was an Assistant Professor at the Polytechnic Institute of Brooklyn before moving to the University of Michigan, where he became a Full Professor in 1972. In 1977 he assumed the position of Professor of Chemistry at the University of Cincinnati, where he was Chairman of the Physical Chemistry division and Director of the Polymer Research Center. In 1987 he was named to the first Distinguished Professorship at Cincinnati. Dr. Mark's research interests include the physical chemistry of polymers, elasticity of polymer networks, liquid crystalline polymers, hybrid organic-inorganic composites, and a variety of computer simulations. He has lectured widely on polymer chemistry, is a consultant to industry, and has organized a number of polymer-related short courses. He has published over 600 research papers and has coauthored or coedited eighteen books. Among his awards are the A.C.S Applied Polymer Science Award and the A.C.S. Polymer Division Paul. J. Flory Polymer Education Award, the Whitby Award, and the Charles Goodyear Medal from the A.C.S Rubber Division. He is the founding Editor of the journal *Computational and Theoretical Polymer Science* and serves on the Editorial Boards of a number of other journals. He has also been a Turner Alfrey Visiting Professor and has received the Edward W. Morley Award from the A.C.S. Cleveland Section.

1

The Scope
of Polymer Chemistry

INTRODUCTION

During the past 60 years, polymer chemistry has had a marked and very direct practical impact on the way of life of people in nearly every region of the earth. Before the beginning of World War II, relatively few materials were available for the manufacture of the articles needed for a civilized life. Steel, glass, wood, stone, brick, and concrete accounted for most of the construction and manufacturing needs of the population, while cotton, wool, jute, and a few other agricultural products provided the raw materials for clothing or fabric manufacture.

The rapid increase in the range of manufactured products following World War II resulted directly from the development of a broad range of new fibers, plastics, elastomers, adhesives, and resins. These new materials are polymers, and their impact on our present way of life is almost incalculable. Products made from polymers are all around us—clothing made from synthetic fibers, polystyrene cups, Fiberglas® boats, nylon bearings, plastic bags, polymer-based paints, epoxy glue, polyurethane foam cushions, silicone heart valves, Teflon®-coated cookware—the list is almost endless.

It is not surprising, therefore, that more than 50% of all chemists and chemical engineers, large numbers of physicists and mechanical engineers, and nearly all materials scientists and textile technologists are involved with research or development work with polymers. Add to this the fact that biochemistry, biophysics, and molecular biology are fields in which polymer chemistry plays a paramount role, and it is clear why the study of macromolecules is one of the most important and rapidly growing branches of science.

Polymer chemistry is not a specialized side branch of traditional chemistry. Instead, it is a uniquely broad discipline that *encompasses* the whole of chemistry and several other fields as well. Areas of science have always prospered when research workers trained in one specialized area turn their attention to a related area. This has been and still is especially true in polymer research. The challenge in polymer chemistry is the application of fundamental chemical and physical techniques and ideas to large and complex molecules. This is a demanding task, and it requires the very best approaches that traditional chemistry can provide.

It will become clear that polymer chemistry, perhaps more than any other research area, cuts across the traditional lines of organic, inorganic, physical, and analytical chemistry; physics; engineering; biology; and even medicine. A newcomer to polymer science needs to be able to blend together knowledge from all these fields. It is to assist in that process that this book has been written.

DEFINITIONS

Many of the terms and definitions used in polymer chemistry are not encountered in conventional chemical textbooks, and for this reason the following summary of terminology is given. Some of these definitions will seem fairly obvious, but others will need explanation.

Monomers

A monomer is any substance that can be converted into a polymer. For example, ethylene is a monomer that can be polymerized to polyethylene (reaction 1). An

$$CH_2{=}CH_2 \longrightarrow -CH_2-CH_2-CH_2-CH_2-CH_2-CH_2-etc. \tag{1}$$

amino acid is a monomer which, by loss of water, can polymerize to give polypeptides (reaction 2). The term *monomer* is used very loosely—sometimes it applies to dimers or trimers if they, themselves, can undergo further polymerization.

$$n H_2N - \underset{\underset{H}{|}}{\overset{\overset{R}{|}}{C}} - \overset{\overset{O}{||}}{C} - OH \xrightarrow{-H_2O} \left[N - \underset{\underset{H}{|}}{\overset{\overset{R}{|}}{C}} - \overset{\overset{O}{||}}{C} \right]_n \tag{2}$$

Dimers, Trimers, and Oligomers

The polymerization of a monomer often occurs in a sequential manner. In other words, two monomer molecules first react together to form a *dimer*. The dimer may then react with a third monomer to yield a *trimer*, and so on. Dimers are usually linear molecules, but trimers, tetramers, pentamers, and so on, can be linear or cyclic. The reactions outlined in schemes (3) to (5) illustrate the relationship between monomers, dimers, and trimers for three systems. Low-molecular-weight polymerization products, for example, dimers, trimers, tetramers, pentamers, and so on—cyclic or linear—are known as *oligomers*. Some care should be taken to avoid use of the term "polymer" to describe materials that are really oligomers, because these two types of products have very different properties.

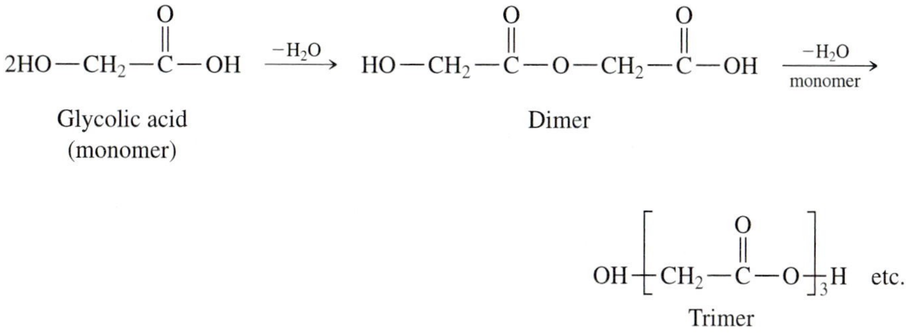

$$2HO-CH_2-\overset{\overset{\displaystyle O}{\|}}{C}-OH \xrightarrow{-H_2O} HO-CH_2-\overset{\overset{\displaystyle O}{\|}}{C}-O-CH_2-\overset{\overset{\displaystyle O}{\|}}{C}-OH \xrightarrow[\text{monomer}]{-H_2O}$$

Glycolic acid
(monomer)   Dimer

(3)

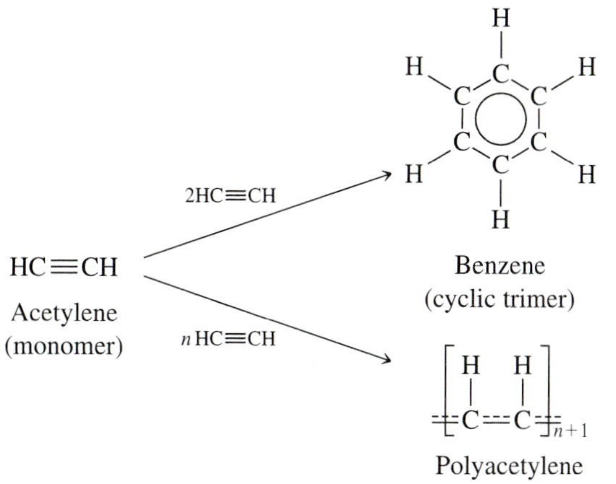

$$OH-\!\!\left[CH_2-\overset{\overset{\displaystyle O}{\|}}{C}-O\right]_{\!3}\!\!-H \quad \text{etc.}$$

Trimer

$HC\!\equiv\!CH$

Acetylene
(monomer)

$2HC\!\equiv\!CH \longrightarrow$ Benzene (cyclic trimer)

(4)

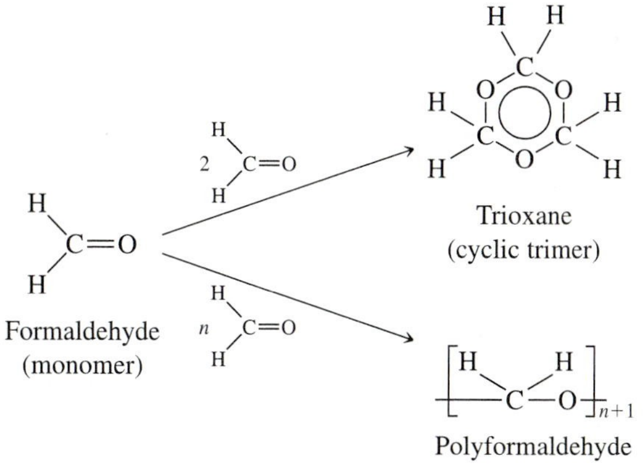

$n\,HC\!\equiv\!CH \longrightarrow$

$$\left[\overset{\overset{\displaystyle H}{|}}{C}\!=\!\overset{\overset{\displaystyle H}{|}}{C}\right]_{n+1}$$

Polyacetylene

$$\underset{H}{\overset{H}{>}}C\!=\!O$$

Formaldehyde
(monomer)

$2\,\underset{H}{\overset{H}{>}}C\!=\!O \longrightarrow$ Trioxane (cyclic trimer)

(5)

$n\,\underset{H}{\overset{H}{>}}C\!=\!O \longrightarrow$

$$\left[\overset{\overset{\displaystyle H}{|}}{C}-O\right]_{n+1}$$

Polyformaldehyde

Polymers

The term *polymer* is used to describe *high-molecular-weight substances*. However, this is a very broad definition, and in practice it is convenient to divide polymers into subcategories according to their molecular weight and structure.

Although there is no general agreement on this point, in this book we will consider *low polymers* to have molecular weights below about 10,000 to 20,000 and *high polymers* to have molecular weights between 20,000 and several million.[1] Obviously, this is a rather arbitrary dividing line, and a better definition might be based on the number of repeating units in the structure. For example, since polymer properties become almost independent of molecular weight when more than 1000 to 2000 repeating units are present, this point could also constitute a satisfactory dividing line between low and high polymers.

Linear Polymers

A *linear polymer* consists of a long chain of skeletal atoms to which are attached the substituent groups. Polyethylene **(1)** is one of the simplest examples. Linear polymers are usually *soluble* in some solvent, and in the solid state at normal temperatures they exist as elastomers, flexible materials, or glasslike thermoplastics. In addition to polyethylene, typical linear-type polymers include poly(vinyl chloride) or PVC **(2)**, poly(methyl methacrylate) (also known as PMMA, Lucite, Plexiglas, or Perspex) **(3)**, polyacrylonitrile.[2] (Orlon or Creslan) **(4)**, and nylon 66 **(5).**

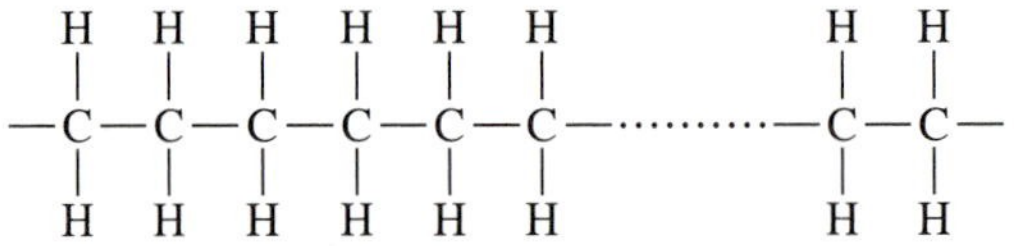

1

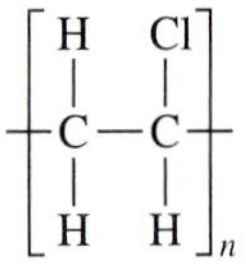

2 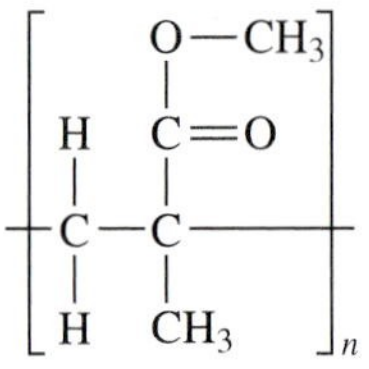   3 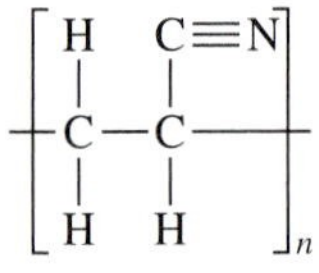   4

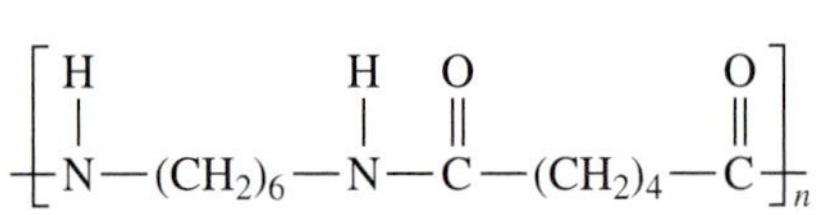

5

[1]The term "molar mass" has been recommended to replace "molecular weight," which still predominates in most of the current literature.
[2]Note that parentheses are not normally used in the name of a polymer if the monomer has a one-word name. The parentheses simply avoid ambiguity.

Branched Polymers

A *branched polymer* can be visualized as a linear polymer with branches of the same basic structure as the main chain. A branched polymer structure is illustrated in Figure 1.1. Branched polymers are often soluble in the same solvents as the corresponding linear polymer. In fact, they resemble linear polymers in many of their properties. However, they can sometimes be distinguished from linear polymers by their lower tendency to crystallize or by their different solution viscosity or light-scattering behavior. Heavily branched polymers may swell in certain liquids without dissolving completely.

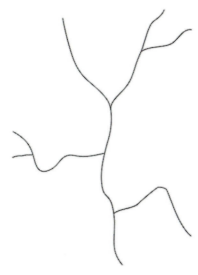

Figure 1.1 Branched polymer.

Crosslinked Polymers

A *crosslinked* or *network polymer* is one in which chemical linkages exist between the chains, as illustrated in Figure 1.2. Such materials are usually swelled by "solvents," but they *do not dissolve*. In fact, this insolubility can be used as a cautious criterion of a crosslinked structure. Actually, the amount by which the polymer is swelled by a liquid depends on the density of crosslinking: the more crosslinks present, the smaller is the amount of swelling. If the degree of crosslinking is high enough, the material may be a rigid, high-melting, unswellable solid, such as diamond. Light crosslinking of chains favors the formation of rubbery elastomeric properties.

Stars and Dendrimers

Star polymers have arms radiating from a common core (Figure 1.3). The number of arms may vary from three to six or more. Such polymers are prepared either by growing the arms by polymerization from a multifunctional core, or by linking

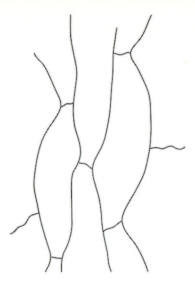

Figure 1.2   Crosslinked macromolecule.

preformed polymer molecules to a core through reactive end groups on the polymer. In principle, there is no limit to the length of the arms in a star polymer.

Dendrimeric polymers have the types of architecture shown in Figure 1.4, except that the structure is usually three-dimensional and the overall final outer shape is that of a sphere. Dendrimers can be produced in two ways. First, they are accessible through the reactions of a multifunctional core with a tri- (or higher) functional monomer—the so-called "core-first" method. The growth of the molecule is usually carried out in successive layers or "generations" moving further and further from the core, with each generation doubling the number of functional sites on the outside of the molecule. As the number of branch points increases, and as the reaction zone moves further and further from the core, less and

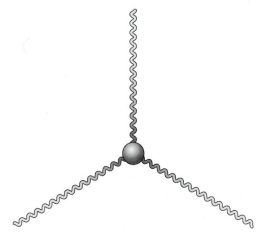

Figure 1.3   Tri-star polymer.

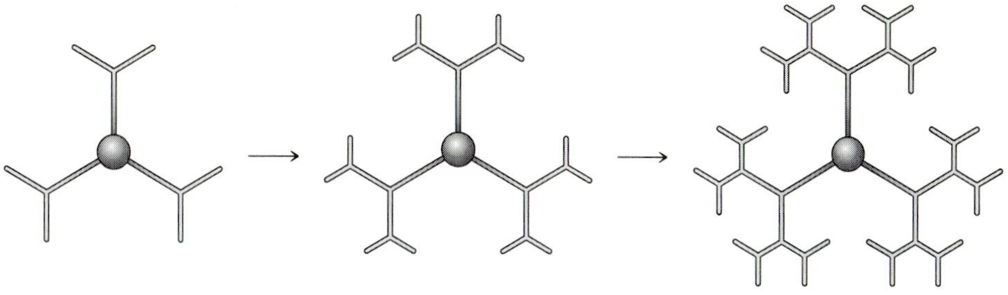

1st generation   2nd generation   3rd generation

Figure 1.4 Stages in the growth of a "starburst" dendrimeric polymer.

less space is available for subsequent reactions. Thus, these molecules reach a size and spherical shape within three or four generations that prevents further growth. A second approach to dendrimer formation is the "arms-first" method in which each highly branched arm is synthesized first, and several of these are then linked in a final step to the core.

The solution- and solid-state properties of star polymers and especially dendrimers usually differ markedly from those of their linear counterparts with the same molecular weight. The spherical shape of a dendrimer restricts intermolecular entanglements. Hence, both the solution- and bulk-viscosities are lower than expected.

Cyclolinear Polymers

Cyclolinear polymers are a special type of linear polymer formed by the linking together of ring systems. Benzene rings are often incorporated into polymers of this type **(6),** but heterocyclic and inorganic rings can also be utilized in the same way. The properties of cyclolinear polymers resemble those of conventional linear polymers, except that the solubility of the cyclolinear species is often low. The tendency for crystallization may be very high.

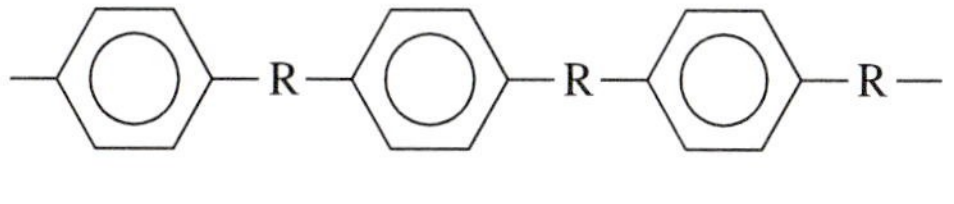

6

Ladder Polymers

As the name suggests, a *ladder polymer* consists of linear molecules in which two skeletal strands are linked together in a *regular* sequence by crosslinking units, as illustrated diagramatically in **7.** In practice, aromatic rings may constitute the linking units **(8)** or, in ladder-type silicone polymers, silicon–oxygen units serve the same function (see Chapter 9). As might be expected, ladder polymers have a more rigid molecular structure than do conventional linear polymers, and they

Definitions   **7**

are often much less soluble. However, they frequently display very good thermal stability, because molecular-weight decreases must be preceded by the cleavage of two bonds at each cleavage site. Spiropolymers of the type shown in Figure 1.5 are sometimes included in the ladder polymer classification.

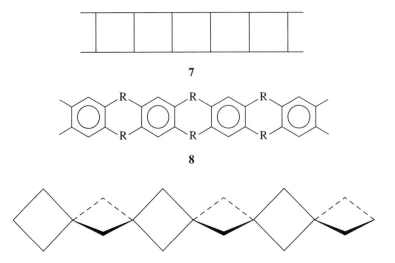

Figure 1.5 Schematic representation of a spiropolymer structure.

Cyclomatrix Polymers

Many polymer systems are known in which ring systems are linked together to form a three-dimensional matrix of connecting units. These materials are known as *cyclomatrix polymers.* Organic and inorganic rings can be incorporated into such systems, and typical examples are provided by the structures found in some silicate minerals and silicone resins **(9)**. Since a three-dimensional network of bonds is formed in these systems, the polymers are highly insoluble, rigid, very high melting, and usually stable at elevated temperatures. Structures of this type are often found in thermosetting resins (see page 47) and in heat-resistant wire coatings.

Graphite is a special example of a cyclomatrix polymer. It has a structure made up of sheets of fused aromatic rings. Individual sheets or layers are sandwiched between the neighboring layers and are held in place by weak van der Waals forces.

Copolymers

A *copolymer* is a polymer made from two or more different monomers. For example, if styrene and acrylonitrile are allowed to polymerize in the same reaction vessel, a copolymer will be formed which contains both styrene and acrylonitrile residues (reaction 6). Many commercial synthetic polymers are copolymers. It should be noted that the sequence of monomer units along a copolymer chain can vary according to the method and mechanism of synthesis. Three different types of sequencing arrangements are commonly found.

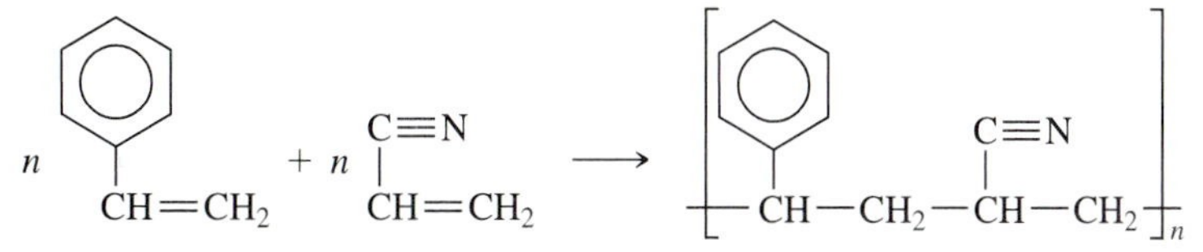

$$ (6) $$

1. *Random copolymers.* In random copolymers, no definite sequence of monomer units exists. A copolymer of monomers A and B might be depicted by the arrangement shown in **10.** Random copolymers are often

$$ -A-B-B-B-A-A-B-A-A-A-A-B-A-B-B-B- $$

10

formed when olefin-type monomers copolymerize by free-radical-type processes (see Chapter 3). The properties of random copolymers are usually quite different from those of the related homopolymers.

2. *Alternating copolymers.* As the name implies, alternating copolymers contain a regular alternating sequence of two monomer units **(11).** Olefin poly-

$$ -A-B-A-B-A-B-A-B- $$

11

merizations that take place through ionic-type mechanisms (see Chapter 4) can yield copolymers of this type. Again, the properties of the copolymer usually differ markedly from those of the two related homopolymers.

3. *Block copolymers.* Block copolymers contain a block of one monomer connected to a block of another, as illustrated in sequence **12.** Block

$$ -A-A-A-A-A-A-A-A-B-B-B-B-B-B-B- $$

12

Definitions   9

copolymers are often formed by ionic polymerization processes. Unlike other copolymers, they retain many of the physical characteristics of the two homopolymers.

Terpolymers

A *terpolymer* contains three different monomer units. These can be sequenced randomly or in blocks.

Graft Copolymers

A graft copolymer has the type of structure shown in **13**. Graft copolymers can be obtained in two ways. First, monomer B can be polymerized from sites along the length of polymer A. Second, two preformed polymers derived from A and B can be induced to react with each other to form a graft structure. For example, if polymer B has end-groups that will react with the side groups on polymer A, a graft copolymer will be formed. Graft copolymers are sometimes produced when two polymers are mixed and either irradiated with X-rays or gamma-rays, or are subjected to mechanical shearing.

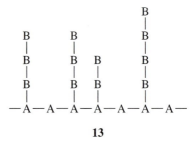

13

The chains of polymer B may be broken and the active chain ends will then be grafted onto polymer A. Graft copolymers often display properties that are derived from the two homopolymers.

Telechelic Polymers

A telechelic polymer is one that bears reactive functional groups at one or both of its chain ends (Figure 1.6). These are frequently produced by living polymerization processes (see Chapters 4 and 9). Telechelic macromolecules are used to prepare block copolymers (by linking the two terminal groups of two different polymers), to make star or dendritic species, or to prepare network structures by end-linking.

Average Molecular Weights and Distributions

A major distinguishing feature of high polymers is their enormous molecular weight. Molecular weights of 20,000 Daltons are routine, and values as high as 2,000,000 Daltons are not uncommon. However, unlike small molecules such as

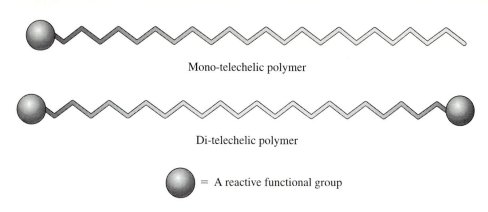

Mono-telechelic polymer

Di-telechelic polymer

= A reactive functional group

Figure 1.6   Telechelic polymers.

benzene or chloroform, or biological polymers like enzymes, a sample of a synthetic polymer has no single, fixed molecular weight. Instead, there is a *distribution* of different molecular weights in the same sample of material (Figure 1.7). For this reason, it is necessary to speak of *average* molecular weights rather than a single defining value.

Several different types of average molecular weights are used in polymer chemistry, the most important of which are known as number average, M_n, and weight average, M_w, values. They are defined as shown in equations (7) and (8):

$$\overline{M}_n = \sum_i X_i M_i = \frac{\sum_i N_i M_i}{\sum_i N_i} \tag{7}$$

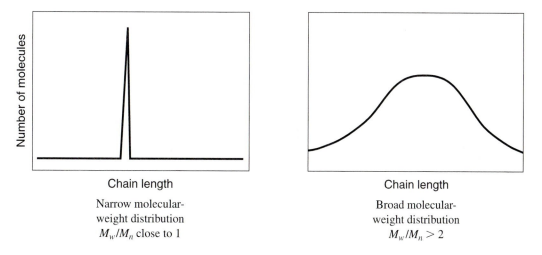

Narrow molecular-weight distribution
M_w/M_n close to 1

Broad molecular-weight distribution
$M_w/M_n > 2$

Figure 1.7   Molecular weight distributions.

$$\overline{M}_w = \sum_i W_i M_i = \frac{\sum_i w_i M_i}{\sum_i w_i} = \frac{\sum_i N_i M_i^2}{\sum_i N_i M_i} \qquad (8)$$

where N_i is the number of molecules of molecular weight M_i, and X_i is the *number fraction* or *mole fraction* of molecules having molecular weight M_i, and where w_i is the weight of molecules of molecular weight M_i, and W_i is the weight fraction of molecules with molecular weight M_i.

As will be seen in Chapter 14, $\overline{M}_n$ can be obtained from the study of the osmotic pressure of polymer solutions, whereas $\overline{M}_w$ values are obtained from light-scattering or ultracentrifugation experiments. The technique known as gel permeation chromatography or size exclusion chromatography (Chapter 15) gives both $\overline{M}_n$ and $\overline{M}_w$ values. In general, $\overline{M}_w$ values are higher than $\overline{M}_n$ because the calculations for $\overline{M}_w$ give more emphasis to the larger molecules, while $\overline{M}_n$ calculations give equal emphasis to all molecules.

The fraction M_w/M_n (called the polydispersity) is a measure of the molecular weight distribution. If the value of M_w/M_n is close to 1 (1.01 or 1.02 for instance), the distribution is very narrow. If it is, say, 2 or higher, the distribution is considered to be very broad. The molecular-weight distribution affects several important polymer properties. For example, polymers with very broad distributions are less prone to crystallize than their narrow distribution counterparts, and they often have lower solidification temperatures. The shorter chains plasticize the bulk material and make it softer. Thus, together with the glass transition temperature, T_g, and the crystalline melting temperature, T_m (see below), M_w/M_n is a crucial characteristic of any synthetic polymer.

Polymer Morphology

Polymer morphology is the study of the solid-state structure and behavior of polymers. It includes investigations of what happens when polymers are mixed in the solid state, and of phase separations. It also involves crystallinity and other phase transitions, and the influence of materials history on strength, elasticity, and polymer-chain orientation. The field also deals with the response of polymers to temperature changes and exposure to solvents. It also includes the membrane behavior of polymers and the transport of liquids, gases, and ions through the solid polymer structure. Ideally, there should be a seamless interaction between research on the synthesis and structure of polymers at the molecular level and morphological investigations, because both aspects are equally important. Chapters 17 and 19 to 21 provide an introduction to polymer morphology.

Thermoplastics

Basically, a *thermoplastic* is any material that softens when it is heated. However, the term is commonly used to describe a substance that passes through a definite sequence of property changes as its temperature is raised. In Figure 1.8 the ther-

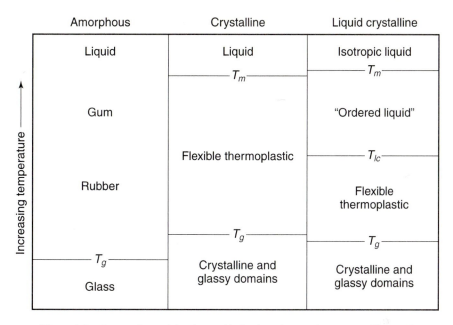

Figure 1.8 Comparison of the thermal behavior of amorphous, crystalline, and liquid crystalline polymers. The main property differences exist in the middle temperature range, where an amorphous polymer is an elastomer or a gum, a crystalline polymer is a tough, flexible material, and a liquid crystalline system shows an ordered liquid state.

moplastic characteristics of amorphous, crystalline, and liquid-crystalline polymers are compared. An amorphous polymeric material contains randomly entangled chains. A microcrystalline (usually abbreviated to "crystalline") material contains domains in which the polymer chains are packed in an ordered array. These "crystalline" domains are embedded in an amorphous polymer matrix.

Both amorphous and crystalline thermoplastics are glasses at low temperatures, and both change from a glass to a rubbery elastomer or flexible plastic as the temperature is raised. This change from glass to elastomer usually takes place over a fairly narrow temperature range (2 to 5°C), and this transition point is known as the *glass transition temperature* (T_g). For many polymers, the glass transition temperature is the most important characterization feature. It can be compared to the characteristic melting point of a low-molecular-weight compound, although care should be taken to remember that T_g is definitely *not* a melting temperature in the accepted sense of the word. It is more a measure of the ease of torsion of the backbone bonds rather than of the ease of separation of the molecules.

At temperatures above T_g, amorphous polymers behave in a different manner from crystalline polymers. As the temperature of an amorphous polymer is raised, the hard rubbery phase *gradually* gives way to a soft, extensible elastomeric phase, then to a gum, and finally to a liquid. No sharp transition occurs from one phase to the other, and only a gradual change in properties is perceptible.

Crystalline polymers, on the other hand, retain their rubbery elastomeric or flexible properties above the glass transition, until the temperature reaches the melting temperature (T_m). At this point, the material liquefies. At the same time, melting is accompanied by a loss of the optical birefringence and crystalline X-ray diffraction effects that are characteristic of the crystalline state.

Some polymers show an additional phase transition between the glass transition temperature and the formation of a true isotropic liquid. These are the so-called *liquid-crystalline* polymers. A liquid-crystalline polymer, when heated, passes through the normal glass and microcrystalline phases but undergoes a quasi melting transition (T_{lc}) at a temperature below the final liquefaction point (T_m). Between T_{lc} and T_m the polymer has some characteristics of a molten material (it flows, for example), but physical techniques reveal the retention of some structural order. Side groups may be loosely stacked in this phase, or skeletal segments may retain some alignment. Only at the highest temperature transition (T_m) is all structure lost, and the material then becomes an isotropic liquid.

The amorphous, crystalline, and liquid crystalline behavior described above is characteristic of linear and branched polymers, copolymers, or cyclolinear polymers. In general, these characteristics are not shown by heavily crosslinked polymers or cyclomatrix materials. These latter substances retain their rigidity when heated. Melting phenomena occur only when the crosslink units or backbone bonds become thermally broken. Lightly crosslinked polymers show many of the conventional thermoplastic properties, with the exception that the true liquid phase may not be formed.

Elastomers

In view of the information just given, it will be clear that an *elastomer* is a flexible polymer that is in the temperature range between its glass transition temperature and its liquefaction temperature. In practice, elastomeric properties become more obvious if the polymer chains are lightly crosslinked. In particular, the liquefaction temperature may be raised by crosslinking, and the polymer may exhibit elastomeric properties over a wider temperature range.

Elastomeric properties appear when the backbone bonds can readily undergo torsional motions to permit uncoiling of the chains when the material is stretched (Figure 1.9). Crosslinks between the chains prevent the macromolecules from slipping past each other and thus prevent the material from becoming permanently elongated when held under tension. An important question connected with elasticity is this: Why do the chains revert to the highly coiled state when the tension on the elastomer is released? The answer lies in the fact that a highly coiled polymer system has a higher degree of disorder and, therefore, a higher entropy than a stretched, oriented sample. Thus, the elastic behavior is a direct consequence of the tendency of the system to assume spontaneously a state of maximum entropy. Since free energy, enthalpy, and entropy are related by the well-known expression, $\Delta G = \Delta H - T\Delta S$, a stretched rubber band im-

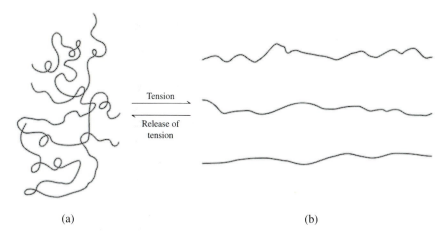

(a)   (b)

Figure 1.9 Rubbery elastomeric properties result from the stress-induced uncoiling and recoiling of polymer chains. (a) Relaxed: high entropy. (b) Stretched:
ordered—low entropy. Note that this is independent of the contribution to elasticity played by cross-links.

mediately held to the lips is warm, and the same material appears cold immediately after contraction.

Plasticizers

Many pure polymers are too rigid for use as flexible films. Poly(vinyl chloride) in
the pure state is a rather rigid material. Only when this polymer is softened by the
addition of low-volatility liquids, such as phthalate esters, can it be used as a
flexible-film or Tygon tubing. Such liquid additives are known as *plasticizers.*
There are also some unusual cases where a low molecular-weight material in
small amounts can act as an *antiplasticizer.* Apparently these molecules interact
so strongly with the polymer chains that they suppress some low-temperature
motions, causing a stiffening effect. Increasing the amount of the material beyond the amount required to saturate this interaction then causes the expected
plasticization and softening of the material.

Thermosetting Resin

The term *thermosetting polymer* refers to a range of systems which exist initially
as liquids but which, on heating, undergo a reaction to form a solid, highly
crosslinked matrix. A typical example is provided by the condensation of methylol melamine to give the hard, tough, crosslinked melamine resin (see Chapter 2).
Partly polymerized systems which are still capable of liquid flow are called
prepolymers. Prepolymers are often preferred as starting materials in technol-

Definitions   **15**

ogy. In practical terms, an uncrosslinked thermoplastic material can be re-formed into a different shape by heating; a *thermosetting* polymer cannot.

Polymer Blends

When two or more polymers are mixed together mechanically, the product is known as a *polymer blend*. Many polymer blends display properties that are different from those of the individual polymers. Polymer blends can be of two main types, miscible or immiscible. Superimposed on these alternatives are three additional classifications: (1) simple cellular domain mixtures of the polymers, (2) interpenetrating coils of macromolecules that are randomly mixed at the molecular level, and (3) secondary block or graft copolymers formed by the physical breaking of bonds, followed by bonding between the different polymeric fragments. The latter type of process can occur when two or more polymers are milled or masticated together. The mechanical shearing can result in the homolytic cleavage of bonds, followed by cross-recombination.

Tacticity

Olefin molecules that contain one unique side group, such as propylene, $CH_2=CH(CH_3)$, or styrene, $CH_2=CH(C_6H_5)$, yield polymers that possess an asymmetric center at each monomer residue. The tacticity of a polymer describes the sequencing of these asymmetric centers along the chain. Three primary possibilities exist, called isotactic, syndiotactic, or atactic (heterotactic) sequencing. This subject is discussed in more detail in Chapters 4 and 17. Here it is sufficient to note that the tacticity of a polymer markedly affects the bulk physical properties. For example, an isotactic polymer may form a microcrystalline solid that has greater strength and rigidity than its atactic counterpart.

DIFFERENT TYPES OF POLYMERS

It is convenient to classify polymers according to the types of reactions involved in their synthesis. The three main polymerization reaction types are (1) condensation reactions, (2) addition reactions, and (3) ring-opening polymerizations. As will be shown later, this classification is an oversimplification, but it provides a useful starting point for an understanding of the field.

Condensation Polymers (Step Reactions)

Condensation processes take place when two or more molecules react with each other with the concurrent loss of water, ammonia, or other small molecules such as methanol. This type of reaction is used as a basis for the synthesis of many important polymers, such as nylon, polyesters, phenol–formaldehyde, and urea–formaldehyde resins. It is also the basis for the laboratory formation of silicates and polyphosphates.

Nylon is a condensation polymer formed by the reaction of a diamine with a dicarboxylic acid. In particular, nylon 66 is made by the condensation of hexamethylenediamine with adipic acid as shown in equation (9).

$$H_2N{+}CH_2{+}_6NH_2 \ + \ HO\overset{\displaystyle O}{\overset{\|}{C}}{-}(CH_2)_4{-}\overset{\displaystyle O}{\overset{\|}{C}}OH \ \longrightarrow \ 1{:}1 \ Salt \ \xrightarrow{-H_2O}$$

(9)

$$H{+}NH(CH_2)_6NH\overset{\displaystyle O}{\overset{\|}{C}}(CH_2)_4\overset{\displaystyle O}{\overset{\|}{C}}{+}_nOH$$

Polyesters, such as Dacron, Terylene, or Mylar, are made by the condensation of a dicarboxylic acid with a diol (reaction 10)(typically, $R = C_6H_4$ and $R' = CH_2CH_2$). The water formed in this reaction is removed by distillation. A variant of this synthesis is the use of the methyl ester of the diacid and the removal of methanol. Typical products are made from terephthalic acid ($R = C_6H_4$) and ethylene glycol (R' is CH_2CH_2). Alkyd paint resins are manufactured from phthalic anhydride and glycerol, and the trifunctionality of the latter reagent ensures the formation of a crosslinked structure (see Chapters 2 and 11).

$$H{-}O{-}\overset{\displaystyle O}{\overset{\|}{C}}{-}R{-}\overset{\displaystyle O}{\overset{\|}{C}}{-}O{-}H + H{-}O{-}R'{-}OH \ \xrightarrow[-H_2O]{catalyst}$$

$$H{+}O{-}\overset{\displaystyle O}{\overset{\|}{C}}{-}R{-}\overset{\displaystyle O}{\overset{\|}{C}}{-}O{-}R'{+}_nOH$$

Phenol–formaldehyde (Bakelite) resins are hard, rigid polymers made by alkylation-condensation reactions between phenols and formaldehyde as discussed in Chapter 2.

Urea–formaldehyde resins are formed by the "methylolation" reaction between urea and formaldehyde, and condensation of the methylol units yields the polymer shown in reaction (11).

$$H_2N{-}\overset{\displaystyle O}{\overset{\|}{C}}{-}NH_2 \ + \ CH_2O \ \longrightarrow \ HOCH_2NH{-}\overset{\displaystyle O}{\overset{\|}{C}}{-}NHCH_2OH \ \xrightarrow{-H_2O}$$

(11)

$$[OCH_2NH{-}\overset{\displaystyle O}{\overset{\|}{C}}{-}NHCH_2]_n$$

Inorganic condensation polymers, such as polysilicate glasses and polyphosphates, are formed by removal of water from the appropriate di-, tri-, or tetrahydroxy monomer, as illustrated by reaction (12).

$$\text{HO}-\overset{\overset{\displaystyle O}{\parallel}}{\underset{\underset{\displaystyle OH}{}}{P}}-\text{OH} \;+\; \text{HO}-\overset{\overset{\displaystyle O}{\parallel}}{\underset{\underset{\displaystyle OH}{}}{P}}-\text{OH} \xrightarrow[-H_2O]{heat} \text{HO}\left[O-\overset{\overset{\displaystyle O}{\parallel}}{\underset{\underset{\displaystyle OM}{}}{P}}\right]_n\text{OH} \qquad (12)$$

(M = An alkali metal)   Polyphosphate

Biological condensation polymerizations, such as those involved in enzyme-catalyzed condensation reactions, are responsible for the polymerization of some amino acids to proteins, for the condensation of sugars to polysaccharides such as starch, cellulose, and glycogen, and for the synthesis of nucleic acids such as DNA and RNA. A separate chapter (Chapter 8) is devoted to a consideration of biological polymers. From the viewpoint of fundamental polymer chemistry it should be remembered that the physiological properties of biological macromolecules are often determined as much by the *molecular conformation* as by the chemical composition. This aspect is discussed in Chapters 8 and 18. It should also be noted that many naturally occurring condensation polymers, particularly proteins and cellulose, are chemically modified before they are used in technology. This subject is covered in Chapter 8.

Addition Polymers

Addition polymers are macromolecules formed by the addition reactions of olefins, acetylenes, aldehydes, or other compounds with "unsaturated" bonds. These reactions are summarized by the scheme shown in (13). Many well-known thermo-

$$m\;\overset{\displaystyle H}{\underset{\displaystyle H}{}}C{=}C\overset{\displaystyle R}{\underset{\displaystyle H}{}} \;+\; n\;\overset{\displaystyle H}{\underset{\displaystyle H}{}}C{=}C\overset{\displaystyle R}{\underset{\displaystyle H}{}} \;\longrightarrow\; \left[\overset{\displaystyle H}{\underset{\displaystyle H}{}}\overset{\displaystyle |}{\underset{\displaystyle |}{C}}-\overset{\displaystyle R}{\underset{\displaystyle H}{}}\overset{\displaystyle |}{\underset{\displaystyle |}{C}}-\overset{\displaystyle H}{\underset{\displaystyle H}{}}\overset{\displaystyle |}{\underset{\displaystyle |}{C}}-\overset{\displaystyle R}{\underset{\displaystyle H}{}}\overset{\displaystyle |}{\underset{\displaystyle |}{C}}\right]_{n-m} \qquad (13)$$

plastics are addition-type polymers, the differences between the various materials being associated primarily with the presence of different substituent groups attached to the main chain. For example, high polymers are known (and manufactured on a large scale) in which Cl, CN, C_6H_5, $CH_3OC(O)$, CH_3, and a variety of other units are present as side groups to the main chain. The equations and names shown in (14) to (23) illustrate some addition polymerization processes.

(a) $n\text{CH}_2{=}\text{CH}_2 \;\longrightarrow\; +\text{CH}_2{-}\text{CH}_2\overset{}{}_n$ Polyethylene   (14)

(b) $n\text{CH}_2{=}\overset{\overset{\displaystyle Cl}{|}}{\text{CH}} \;\longrightarrow\; \left(\text{CH}_2{-}\overset{\overset{\displaystyle Cl}{|}}{\text{CH}}\right)_n$ Poly(vinyl chloride)   (15)
                                                                                 (PVC)

$$(c) \quad nCH_2{=}CH{-}C{\equiv}N \longrightarrow {+}CH_2{-}CH{(}C{\equiv}N{)}{+}_n \qquad \text{Polyacrylonitrile} \qquad (16)$$
(acrylic fiber, Creslan)

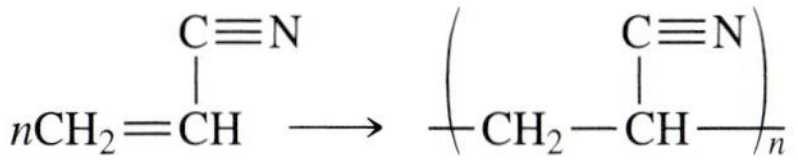

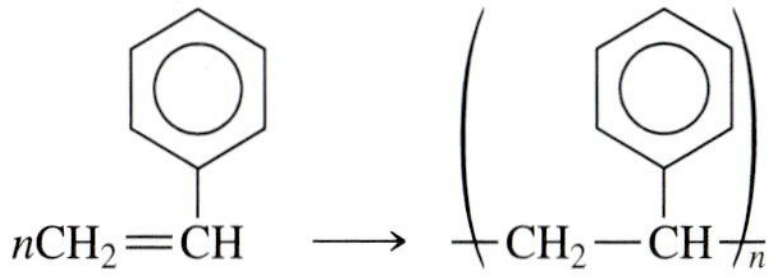

$$(d) \quad nCH_2{=}CH{-}C_6H_5 \longrightarrow {+}CH_2{-}CH{-}{+}_n \qquad \text{Polystyrene} \qquad (17)$$

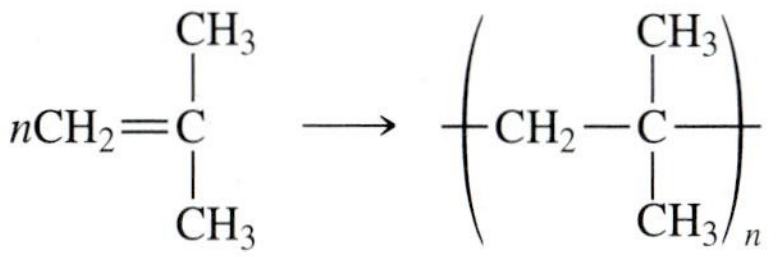

$$(e) \quad nCH_2{=}C(CH_3)_2 \longrightarrow {+}CH_2{-}C(CH_3)_2{+}_n \qquad \text{Polyisobutylene} \qquad (18)$$
(butyl rubber)

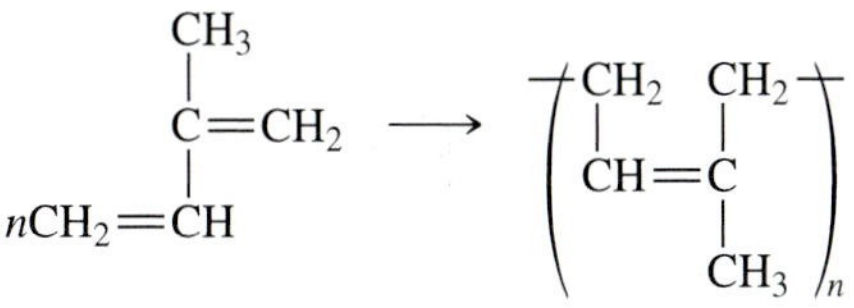

$$(f) \quad nCH_2{=}CH{-}C(CH_3){=}CH_2 \longrightarrow \left(\begin{array}{c} CH_2 \quad CH_2 \\ CH{=}C \\ CH_3 \end{array} \right)_n \qquad \textit{cis}\text{-1,4-Polyisoprene} \qquad (19)$$
(natural rubber)

$$(g) \quad nCH_2{=}CH{-}CH{=}CH_2 \longrightarrow \left(\begin{array}{c} CH{-}CH_2 \\ CH_2{-}CH \end{array} \right)_n \qquad \textit{trans}\text{-1,4-Polybutadiene} \qquad (20)$$

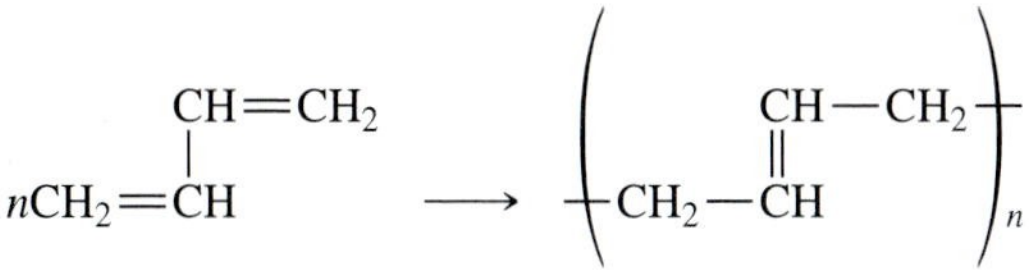

$$(h) \quad nCH_2{=}CH{-}C(Cl){=}CH_2 \longrightarrow \left(\begin{array}{c} Cl \\ C{-}CH_2 \\ CH_2{-}CH \end{array} \right)_n \qquad \textit{trans}\text{-1,4-Polychloroprene} \qquad (21)$$
(Neoprene rubber)

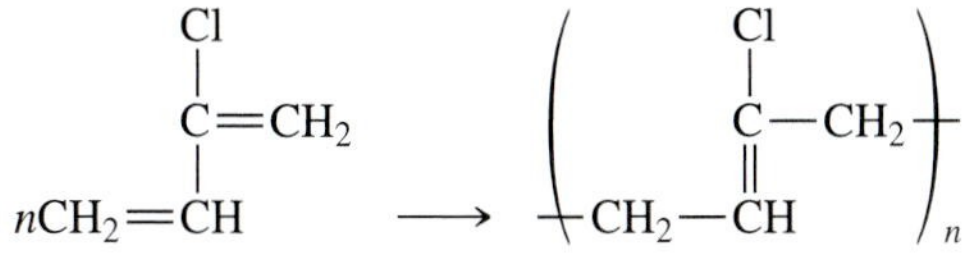

$$(i) \quad nCF_2{=}CF_2 \longrightarrow {+}CF_2{-}CF_2{+}_n \qquad \text{Poly(tetrafluoroethylene)} \qquad (22)$$
(Teflon[1])

$$(j) \quad nCH_2{=}O \longrightarrow {+}CH_2{-}O{+}_n \qquad \text{Polyformaldehyde} \qquad (23)$$
(polyoxymethylene, Delrin[2])

It should be recognized that polymers such as polyacrylonitrile, polybutadiene, and polychloroprene contain additional unsaturation that can be utilized in subsequent high-temperature or crosslinking reactions. Addition polymerizations are considered in more detail in Chapters 3, 4, 5, 12, and 13.

[1]DuPont's registered trademark for its fluorocarbon resins.
[2]DuPont's registered trademark for its acetal resins.

Ring-Opening Polymerizations

The treatment of some cyclic compounds with catalysts brings about cleavage of the ring followed by polymerization to yield high-molecular-weight polymers.

For example, as shown in reaction (24), trioxane polymerizes to yield poly-formaldehyde (polyoxymethylene). Caprolactam polymerizes to nylon 6 (reaction 25), and epoxides undergo ring-opening reactions to yield polyethers (26).

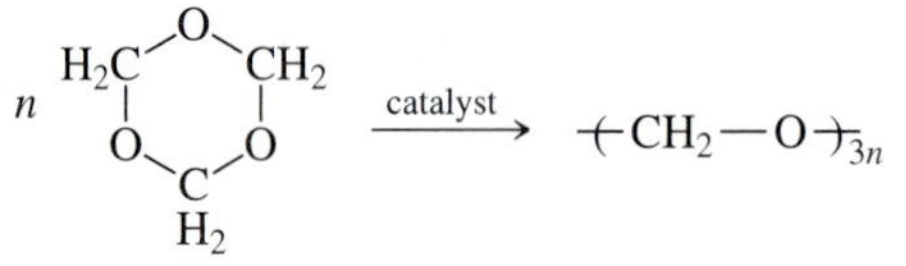

$$n \quad \overset{\displaystyle H_2C \overset{O}{\diagup\diagdown} CH_2}{\underset{\displaystyle O \diagdown \underset{H_2}{C} \diagup O}{}} \quad \xrightarrow{\text{catalyst}} \quad +CH_2-O\,)_{3n} \tag{24}$$

$$\overset{\displaystyle \overbrace{}^{}-NH}{(CH_2)_5-C=O} \quad \longrightarrow \quad \left[NH-(CH_2)_5-\overset{O}{\overset{\|}{C}} \right]_n \tag{25}$$

$$\overset{O}{\underset{CH_2-CHR}{\diagup\diagdown}} \quad \xrightarrow{\text{tertiary amines}} \quad \left[O-CH_2-\overset{R}{\underset{|}{CH}} \right]_n \tag{26}$$

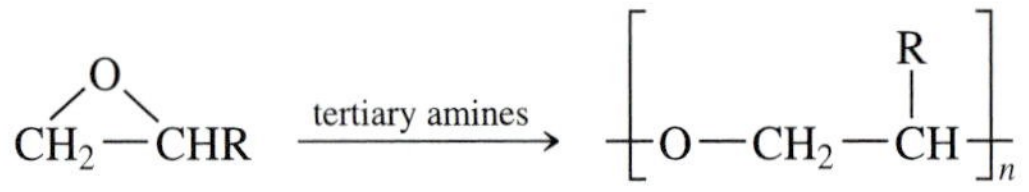

A number of inorganic ring systems also polymerize by ring-opening reactions. Rhombic sulfur, cyclic siloxanes, and cyclic chlorophosphazenes behave in this way (reactions 27 to 29).

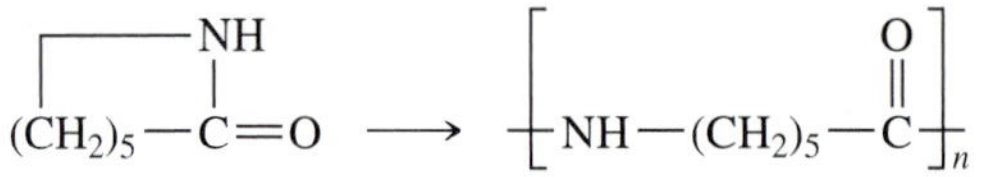

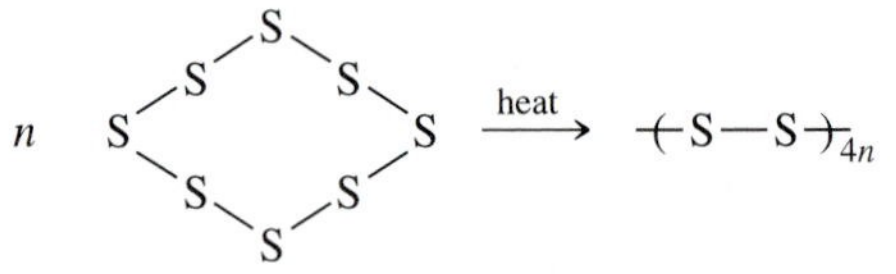

$$n \quad \overset{\displaystyle S}{\underset{\displaystyle S}{S \diagup\diagdown S}} \quad \xrightarrow{\text{heat}} \quad +S-S\,)_{4n} \tag{27}$$

Rhombic sulfur Plastic sulfur

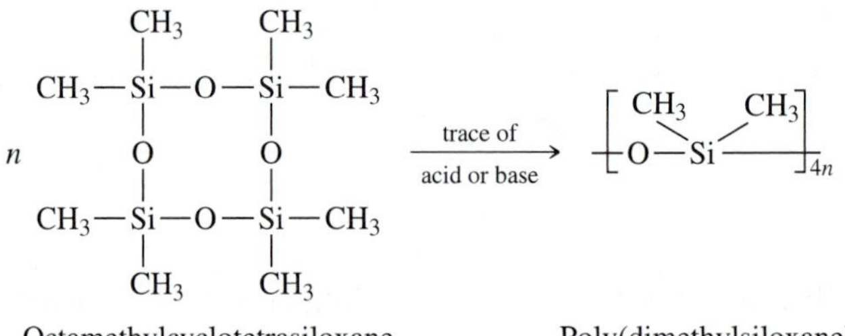

$$n \quad \begin{matrix} CH_3 & CH_3 \\ | & | \\ CH_3-Si-O-Si-CH_3 \\ | & | \\ O & O \\ | & | \\ CH_3-Si-O-Si-CH_3 \\ | & | \\ CH_3 & CH_3 \end{matrix} \quad \xrightarrow[\text{acid or base}]{\text{trace of}} \quad \left[O-Si \overset{CH_3}{\underset{CH_3}{\diagdown\diagup}} \right]_{4n} \tag{28}$$

Octamethylcyclotetrasiloxane Poly(dimethylsiloxane)

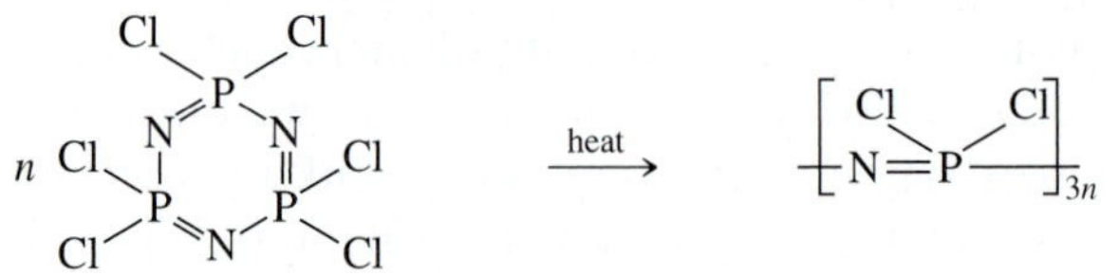

$$n \quad \begin{matrix} Cl & Cl \\ \diagdown & \diagdown \\ Cl & N\!\!=\!\!P\,N \,Cl \\ \diagdown P \diagup \diagdown P \diagup \\ Cl \diagup N \diagdown Cl \end{matrix} \quad \xrightarrow{\text{heat}} \quad \left[N\!\!=\!\!P \overset{Cl}{\underset{Cl}{\diagdown\diagup}} \right]_{3n} \tag{29}$$

Hexachlorocyclotriphosphazene Poly(dichlorophosphazene)
(inorganic rubber)

Ring-opening polymerizations will be considered further in Chapters 6, 9, and 10.

Initiators

A few polymerization reactions take place spontaneously, but most require the addition of small quantities of compounds called *initiators*. As the name suggests, these species start the polymerization process. Many of them are not catalysts in the true sense of that word, because they are changed chemically as they trigger the start of the chain growth process. Different initiators are required for different types of polymerization processes. For example, peroxides, azo compounds, Lewis acids, organometallic species, or high-energy irradiation are used to initiate different addition-type polymerizations. Dehydration or dehydrohalogenation reagents may be needed to induce condensation reactions, and so on. The search for improved initiators has been and still is a major research area in polymer chemistry.

HISTORICAL OVERVIEW

The Macromolecular Hypothesis

Although natural polymers have been used by human beings since antiquity, their structure was not understood, even in elementary terms, until the late-19th century. The problem encountered by the earliest investigators was a general unwillingness on the part of the scientific community to believe that giant covalently bonded molecules could exist. Indeed, the physical tools for the measurement of molecular weights in solution (based on the work of Raoult and van't Hoff) did not exist until the 1800s. In 1888, Brown and Morris used a cryoscopic technique to estimate the molecular weight of a starch hydrosylate at about 30,000. The same technique was used by Gladstone and Hibbert to estimate that the molecular weight of rubber was between 6000 and 12,000, or perhaps even higher.

However, the demonstration that materials such as starch, rubber, or proteins had high molecular weights did not convince the scientific community that these materials had polymeric structures. On the contrary, it was generally assumed that the high values for the molecular weights resulted from defects in the cryoscopic method, or were caused by association of smaller molecules. This view persisted even as the rubber industry prospered after 1839 and synthetic polymers came into limited use. Evidence about the structure of polymers began to accumulate between 1890 and 1919 from the work of Emil Fischer on proteins. However, it was not until 1920 that Staudinger put forward the idea of covalently bonded macromolecular structures for polystyrene, rubber, and polyoxymethylene and began the process of providing convincing evidence in favor of this structure. Even in the following 10 years, this hypothesis was subjected to intense criticism.

Structural Work

The application of physics and physical chemistry to macromolecular systems dates back to the early attempts made in the late-19th century to understand the unusual properties of natural polymers, such as rubber, polysaccharides, and proteins. However, it was not until the 1920 to 1930 period that Meyer and Mark in Germany began to establish the structure of cellulose and rubber with the use of X-ray diffraction techniques. Explanations of rubbery elasticity in terms of polymer conformations were put forward by Kuhn, Guth, and Mark between 1930 and 1934, Kuhn, in particular, was the first to apply statistical methods to the study of macromolecules.

The application of light scattering to macromolecular systems was made by Debye during World War II. It was also during this period that Flory began a series of investigations into the applications of statistical methods, conformational analysis, and other fundamental physicochemical techniques, to polymer science. During the early 1950s, Watson, Crick, Wilkins, Franklin, Kendrew, and Hodgkin successfully applied X-ray diffraction analysis to the structure determination of biological polymers, such as DNA, hemoglobin, and insulin.

Single crystals of polyethylene were first reported by Keller and Till in 1957. Increasingly, during the period 1960 to 1980, nuclear magnetic resonance (NMR) analysis of polymers in solution proved to be a vital physical tool. The introduction of solid-state NMR methods in the 1980s has had a major impact on polymer structural analysis. In addition, the use of lasers in light-scattering experiments, the development of Fourier transform infrared analysis, and the introduction of surface probe analysis techniques have been responsible for major advances in our understanding of polymer structures.

Synthetic Polymers

The synthesis of polymers can be traced to the years 1838 and 1839 when the photochemical polymerization of vinyl chloride was observed, styrene was found to polymerize to a glass, and the sulfur-induced vulcanization (crosslinking) of natural rubber was discovered. This last discovery by MacIntosh and Hancock in Britain and Goodyear in the United States provided a commercial incentive for the study of polymeric materials, since it led to the widespread use of rubber in tires and rainwear.

As shown in Table 1.1 a number of important new polymers were prepared and commercialized in the period between 1890 and 1930, many of which were based on the use of chemical reactions (acetylation, nitration) carried out on cellulose. It is perhaps astonishing to realize that most of these technological developments occurred during a time when the polymeric nature of these products was not recognized or believed.

The 15 years between 1930 and 1945 represent the springboard for the development of modern synthetic polymer chemistry. By 1930, the macromolecular hypothesis was beginning to be accepted. Moreover, in 1929, Carothers at the

TABLE 1.1 APPROXIMATE SEQUENCE OF POLYMER SYNTHESIS DEVELOPMENTS

1838, 1839	Polymerization of vinyl chloride and styrene. Vulcanization of rubber.
1868	"Celluloid" (cellulose nitrate plus camphor).
1893	Rayon (regenerated cellulose).
1910	Styrene-diene copolymers. Phenolic resins.
1914	Cellulose acetate as aircraft "dope."
1920	Cellulose nitrate lacquers (automobiles).
1924	Cellulose acetate fibers.
1927	Cellulose acetate plastics. Poly(vinyl chloride) manufacture.
1929	Urea-formaldehyde resins.
1930	General acceptance of the macromolecular hypothesis.
1931	Poly(methyl methacrylate).
1936	Poly(vinyl acetate) and poly(vinyl butyrate) in laminated safety glass. Nylon 66 manufacture.
1937	Polystyrene manufacture.
1939	Melamine-formaldehyde resins. Neoprene rubber. Polysulfide rubber (Thiokol).
1939–1945	Manufacture of polyethylene (U.K.). Polybutadiene rubber (Germany). Acrylonitrile-butadiene rubber (Germany). Polyurethanes (Germany). Styrene-butadiene rubber (U.S.A.). Polyisobutylene (butyl rubber) (U.S.A.).
1945–1960	Epoxy resins, acrylonitrile-butadiene-styrene (ABS) polymers, polyesters, polyacrylonitrile, polysiloxanes (silicones), linear polyethylene, polypropylene, "living" anionic polymerizations, cationic polymerizations, polyoxymethylene, polycarbonates, polyurethane foams, fluorocarbon polymers.
1960–1980s	cis-Polyisoprene rubber, cis-polybutadiene rubber, ethylene-propylene rubber, polyimides, poly(phenylene oxides), polysulfones, styrene-butadiene block copolymers, aromatic polyamides, aromatic ladder polymers, group-transfer polymerization, cyclopolymerizations, olefin metathesis polymerization, polyphosphazenes, polysilanes.
1980s–2000s	ADMET polymerizations, atom radical polymerizations, polymerizations in super-critical CO_2, poly(phenylenevinylenes), polyaniline, poly(ferrocenophanes), non-linear optical polymers, telechelic polymers, dendrimers, ceramers, living polymerizations to polyphosphazenes, polyester synthesis in bacterial cells, and metallocene catalysts.

DuPont Company began his classical series of attempts to prepare high polymers from well-characterized, low-molecular-weight compounds. The success of these studies provided a final verification of the macromolecular theory. The work also led to the synthesis of polyamides and polyesters. Nylon 66 was, in fact, produced commercially in 1938. Much of the spadework in polymer synthesis that was to form the foundation of the subject as we know it today was done during this period. Moreover, the technological emergencies generated by World War II provided the stimulus for a massive effort by chemists to synthesize new macromolecular materials. For example, the development of polyethylene as an insulator for radar equipment and the large scale synthesis of replacements for natural rubber took place during this time.

The development of polymer synthesis in the period 1945 to 1960 was accelerated by the large-scale availability of new monomers from the petrochemicals industry, coupled with a series of key scientific discoveries. These include the pivotal development by Ziegler and Natta of organometallic compounds as initiators for the polymerization of ethylene and catalysts for the stereoregular polymerization of other olefins. The recognition of nonterminated ("living") polymers by Szwarc also occurred during this period, as did many other synthetic advances (Table 1.1).

Many additional discoveries have emerged since the 1960s (Table 1.1), and most of these have transformed polymer science and its related technologies. However, by the late 1980s it was already evident that polymer science was entering a new phase.

This phase of synthetic polymer chemistry involved a subtle shift in emphasis. Many of the most accessible (and inexpensive) "commodity" polymers have been studied in detail and their applications have been developed extensively. Interest is now focused on the synthesis and study of entirely new polymers that have specialized, high-performance properties, such as very high strength, low-temperature flexibility, electrical conductivity, flame resistance, high-temperature stability, oil and fuel resistance, biomedical compatibility, or suitability for pyrolysis to ceramics. Newer methods of synthesis, such as the macromolecular substitution route to polyphosphazenes, group transfer polymerization, and the use of metathesis-type initiators, are providing access to a wide range of specialty polymers with unique combinations of properties. Aromatic ladder polymers have received considerable attention, as have main-chain liquid crystalline polyamides and polyesters. Polymers are now being used as stationary substrates for the binding of transition metal catalysts or enzymes.

The 1990s and early 2000s saw numerous additional advances. New polymer architectures have been developed such as star and dendritic structures, as well as new types of block and graft copolymers. Novel polymer synthesis methods, such as atom transfer radical-, metallocene-, and acyclic diene metathesis-polymerizations have become evident, as well as a dramatic expansion of work on ring-opening metathesis reactions. Polymerization reactions carried out in supercritical carbon dioxide as a solvent have demonstrated how large-scale reactions can be performed in an environmentally friendly manner. A very broad expansion of the number of different polyphosphazenes with different properties and uses has occurred, and new inorganic–organic polymers such as poly(ferrocenophanes) have appeared. Polymers have also played an important role in the field of nanostructures, with polymer molecules being used to create very small features on a surface and to serve as guest molecules in nano-tunnels that penetrate crystal lattices. Hybrid polymer-ceramic materials have been developed further, as have polymers for use as semiconductor resists, electroluminescent materials, electrolytes in batteries, and proton conducting membranes in fuel cells. The interface with biology and medicine has continued to be strengthened by intensive research on micelles, vesicles, microspheres, hydrogels, and biocompatible surfaces. Thus, it is clear that polymer science continues to be a vibrant and expanding field with an increasing influence on all the other scientific and engineering fields that make use of polymeric materials. In fact, it has become the central science on which nearly every other scientific and technological field depends. This trend is expected to continue for the foreseeable future.

Engineering and Materials Science

Polymer chemistry is a highly developed science in the sense that all the classical perspectives now exist, from synthesis and molecular structural work, through theoretical physics and physical chemistry, to engineering. Polymers have always been used in technology because of their engineering advantages (strength, elasticity, light weight, corrosion resistance, low cost). As will be evident from some of the later chapters, few polymers are used technologically in their chemically pure form. Thus, much of the current research and development work connected with polymers involves studies of the ways in which pure polymers can be modified to make them more suitable for specific applications. Such modifications can involve attempts to physically modify the crystallinity of a polymer, or result from the addition of plasticizers, reinforcement agents, or even the addition of other polymers. This interface between polymer chemistry, engineering, rheology, and solid-state science is often described under the general umbrella term of *materials science*. The period since the mid-1960s has seen a striking growth of interest in the materials science aspects of polymer science as investigators seek to improve the properties of well-known synthetic polymers. This type of work is likely to become even more important as technology continues to demand new materials with improved properties at a time when most of the readily available organic polymers have already been commercialized.

Materials science includes the fields of metals, ceramics, and a wide range of inorganic semiconductors, optical, and electro-optical materials, as well as polymers. Many modern devices and constructs combine and utilize materials from all four of these large areas. It seems clear that the technological future of polymer science will require a close interaction between synthetic chemists on the one hand and physical chemists and materials scientists on the other.

STUDY QUESTIONS

1. Refer to an organic chemistry textbook and suggest synthetic routes that could be used to produce monomers such as vinyl chloride, tetrafluoroethylene, styrene, ethylene glycol, and terephthalic acid from petroleum.

2. What types of "crash programs" can you foresee being implemented if the supply of monomers from oil were to be interrupted for several years as a result of international upheavals?

3. Excluding the specific reactions mentioned in this chapter, suggest examples, based on your own prior experience and study, of an addition polymerization, a condensation polymerization, a ring-opening polymerization, and a biological condensation polymerization.

4. How would you distinguish experimentally between an amorphous, a crystalline, and a crosslinked polymer?

5. Devise several reactions that might lead to the synthesis of cyclolinear, ladder, and spiro polymers. After you have spent 10 minutes on this problem, refer to Chapter 2 and reassess your ideas.

6. Based on your intuitive knowledge of chemistry or physics, suggest why rubber is an elastomer at room temperature but glass is not.

7. Without referring to later chapters, suggest ways in which a graft copolymer might be prepared. Then do the same thing for a block copolymer.

8. Consider three samples of polyisobutylene. All molecules in sample A have a molecular weight of 30,000 g/mol; all those in B have a molecular weight of 70,000 g/mol; and those in C have a molecular weight of 100,000 g/mol. If 10 g of A are mixed with 5 g of B and 1 g of C, calculate $\overline{M_n}$ and $\overline{M_w}$.

9. Before starting to read the next chapter, spend 30 minutes scanning the rest of the book to obtain some perspective of its contents. Glance at the section headings, structural formulas, figures, and the questions at the end of each chapter. Then study individual chapters in detail.

SUGGESTIONS FOR FURTHER READING

FLORY, P. J., *Principles of Polymer Chemistry*. Ithaca, NY: Cornell University Press, **1953**.

FRECHET, J. M., "Functional Polymers and Dendrimers," *Science*," **1994**, *263*, 1710.

GROSBERG, A. Y., and KHOKHLOV, A. R., *Giant Molecules. Here, There, and Everywhere*. San Diego: Academic Press, **1997**.

HERMES, M. E., *Enough for a Lifetime* (a biography of Wallace Carothers), American Chemical Society, Washington, D.C., and the Chemical Heritage Foundation, Philadelphia, **1996**.

MORAWETZ, H., *Polymers: The Origins and Growth of a Science*. New York: Wiley, **1985**.

TOMALIA, D. A., NAYLOR, A. M., and GODDARD, W. A., "Starburst Dendrimers," *Angew. Chem., Int. Ed. Engl.* **1990,** *29*, 138.

TONELLI, A. E., and SRINIVASARAO, M., *Polymers from the Inside Out*, New York: Wiley Interscience, **2001**.

2

Condensation and Other Step-Type Polymerizations

INTRODUCTION

Definitions

As discussed in Chapter 1, polymerization processes can be classified very roughly into three categories: condensation, addition, and ring-opening polymerizations. In condensation reactions, monomer molecules react to release a small molecule such as water. Addition polymers are formed by the polyaddition reactions of olefins or carbonyl compounds. Ring-opening polymerizations take place by cleavage of a ring with concurrent or subsequent addition of the linear product to the end of a growing chain. Chapters 3 to 5 deal with addition polymerizations, and Chapter 6 covers ring-opening processes. Here we are concerned with polymers of the type shown in Table 2.1, formed by condensations and other step reactions.

The polymerization categories just described reflect different monomer structures. However, a related but distinct classification is based on the *general mechanistic pathways* that are involved. This classification divides polymerization processes into step reactions and chain reactions. *Step reactions* are those in which the chain growth occurs in a slow, stepwise manner. Two monomer molecules react to form a dimer. The dimer can then react with another monomer to form a trimer, or with another dimer to form a tetramer. Thus, the average molecular weight of the system increases slowly over a period of time. Condensation polymerizations and some noncondensation reactions, such as Diels–Alder additions, fall into this category. On the other hand, *chain polymerizations* take place by a *rapid* addition of olefin molecules to a growing chain end. Because chain

TABLE 2.1 SUMMARY OF STEP-TYPE POLYMERS DESCRIBED IN THIS CHAPTER

Name	General Repeating Structure
	Polymers Formed by Condensation Reactions

Polyesters

$$-O-R-O-\overset{\overset{\displaystyle O}{\|}}{C}-R'-\overset{\overset{\displaystyle O}{\|}}{C}-$$

Polycarbonates

$$-O-R-O-\overset{\overset{\displaystyle O}{\|}}{C}-$$

Polyanhydrides

$$-O-\overset{\overset{\displaystyle O}{\|}}{C}-R-\overset{\overset{\displaystyle O}{\|}}{C}-$$

Polyamides

$$-\overset{\overset{\displaystyle O}{\|}}{C}-R-\overset{\overset{\displaystyle O}{\|}}{C}-\overset{\overset{\displaystyle H}{|}}{N}-R'-\overset{\overset{\displaystyle H}{|}}{N}-$$

Polyimides

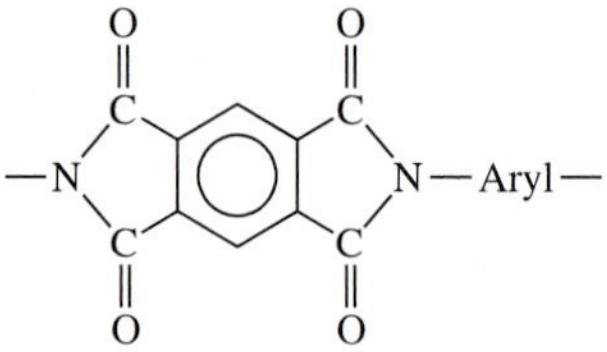

Polybenzoxazoles
(cis-form)

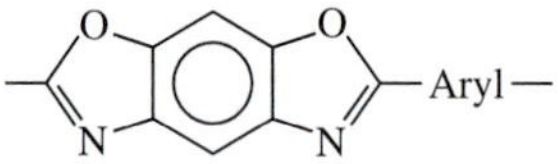

Polybenzthiazoles
(trans-form)

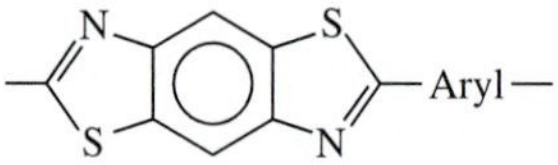

Polybenzimidazoles

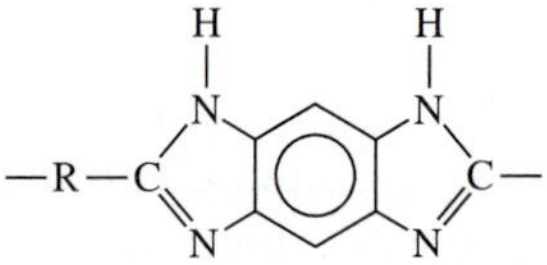

Polyquinoxalines

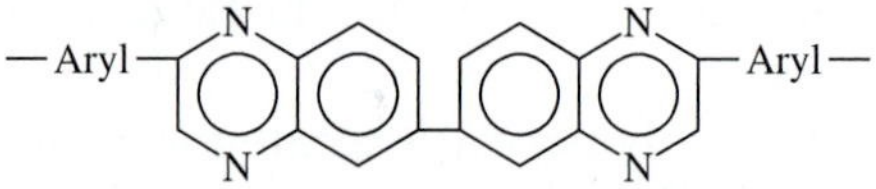

Aromatic ladder polymers

Various structures, often based on

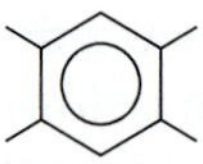

or units

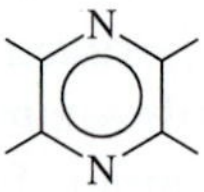

Phenol–formaldehyde polymers

$$-CH_2-Aryl-CH_2-O-$$

Urea–formaldehyde polymers

$$-N(H)-C(O)-N(H)-CH_2-O-$$

(*continued*)

TABLE 2.1 (CONTINUED)

Name	General Repeating Structure
Melamine–formaldehyde polymers	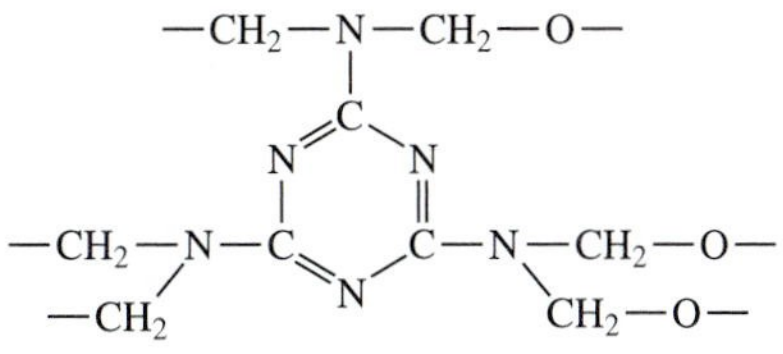
Polyacetals	$-CH_2-O-R-O-$

Polymers Formed by Noncondensation, Step-Type Reactions

Polysulfones (polyethers)	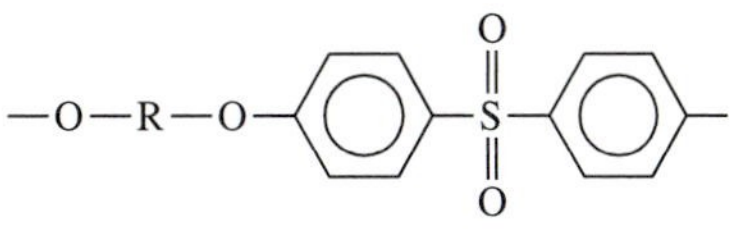
Poly(phenylene oxide) (polyethers)	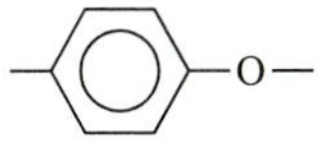
Diels–Alder formed polymers	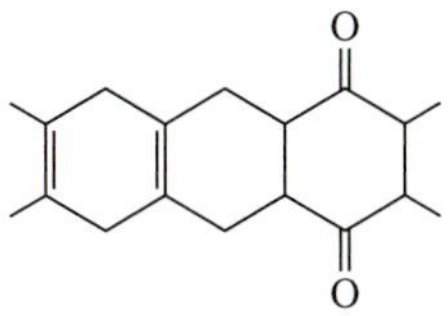
Polyurethanes	$-C(O)-O-R-O-C(O)-N(H)-R'-N(H)-$
Polyarylenes	

growth occurs rapidly, the system usually contains only unreacted monomer and high polymer. Intermediate polymers usually cannot be isolated. A rather clear-cut distinction exists between these two mechanistic types.

Comparison of Step and Chain Polymerizations

The distinction between step and chain polymerizations is an important concept in polymer chemistry. Hence, it is worthwhile at this point to summarize the observable differences between the two types of reactions.

First, in step polymerization, any two molecules in the system can react with each other. Initially, the monomers react to form dimers, dimers can react with dimers, and so on. On the other hand, in chain polymerization, chain growth takes place only at the ends of a few "initiated" chains.

Second, as a result of this difference, step reactions are characterized by a disappearance of the monomer at an early stage in the polymerization, and by the existence of a broad molecular-weight distribution in the later stages of the reaction. With chain polymerizations, the monomer concentration decreases steadily throughout the reaction and, ideally, at any stage the reaction mixture should contain only monomer and high polymer.

Third, the variation in polymer molecular weight at different stages in the reaction provides another distinguishing feature. In step reactions, the polymer molecular weight rises steadily during the reaction. In a chain reaction, high polymer is formed rapidly from each "initiated" monomer. Hence, the molecular weight of each polymer molecule does not increase appreciably after the initial rapid propagation. At longer reaction times there may be an increase in the number of polymer molecules, but not in the molecular weight of those already formed.

Types of Condensation Reactions

Typical condensation polymerizations are those which involve the elimination of a water molecule at each condensation step. The formation of polyesters and polyamides are two examples, as illustrated by the general reactions (1) and (2). Variations include condensation by species which possess both hydroxyl and carboxylic acid or amino and carboxylic acid units on the same molecule (reactions 3 and 4). Also included in the condensation or step-type category are the transesterification-type processes that occur when carboxylic acid esters react with alcohols (5). Acid chlorides can also be allowed to react with diamines in the synthesis of polyamides (6). A cyclization-type condensation may be used to yield polybenzimidazoles, as illustrated by the reaction shown in (7). These and other step-type polymerizations are discussed in later sections.

$$HO-R-OH + HO-\overset{O}{\overset{\|}{C}}-R'-\overset{O}{\overset{\|}{C}}-OH \xrightarrow{-H_2O} H \left[O-R-O-\overset{O}{\overset{\|}{C}}-R'-\overset{O}{\overset{\|}{C}} \right]_n OH \tag{1}$$

$$H_2N-R-NH_2 + HO-\overset{O}{\overset{\|}{C}}-R'-\overset{O}{\overset{\|}{C}}-OH \xrightarrow{-H_2O} H \left[\overset{H}{\overset{|}{N}}-R-\overset{H}{\overset{|}{N}}-\overset{O}{\overset{\|}{C}}-R'-\overset{O}{\overset{\|}{C}} \right]_n OH \tag{2}$$

$$HO-R-\overset{O}{\overset{\|}{C}}-OH \xrightarrow{-H_2O} H \left[O-R-\overset{O}{\overset{\|}{C}} \right]_n OH \tag{3}$$

$$H_2N-R-\overset{O}{\overset{\|}{C}}-OH \xrightarrow{-H_2O} H \left[\overset{O}{\overset{|}{N}}-R-\overset{O}{\overset{\|}{C}} \right]_n OH \tag{4}$$

$$HO-R-OH + R'O-\overset{\overset{\displaystyle O}{\|}}{C}-R''-\overset{\overset{\displaystyle O}{\|}}{C}-OR' \xrightarrow{-R'OH}$$

$$H\left[O-R-O-\overset{\overset{\displaystyle O}{\|}}{C}-R''-\overset{\overset{\displaystyle O}{\|}}{C}\right]_{n}O-R' \tag{5}$$

$$H_2N-R-NH_2 + Cl-\overset{\overset{\displaystyle O}{\|}}{C}-R'-\overset{\overset{\displaystyle O}{\|}}{C}-Cl \xrightarrow{-HCl} H\left[\overset{\overset{\displaystyle H}{|}}{N}-R-\overset{\overset{\displaystyle H}{|}}{N}-\overset{\overset{\displaystyle O}{\|}}{C}-R'-\overset{\overset{\displaystyle O}{\|}}{C}\right]_{n}Cl \tag{6}$$

$$HO-\overset{\overset{\displaystyle O}{\|}}{C}-R-\overset{\overset{\displaystyle O}{\|}}{C}-OH + \begin{matrix} H_2N \\ H_2N \end{matrix}\text{Aryl}\begin{matrix} NH_2 \\ NH_2 \end{matrix} \xrightarrow{-H_2O} \left[R-C\overset{NH}{\underset{N}{\diagdown}}\text{Aryl}\overset{NH}{\underset{N}{\diagdown}}C\right]_{n} \tag{7}$$

Mechanism of Condensation Polymerization

The processes illustrated by reactions (1) to (6) are almost certainly more complex than is implied by these equations. In fact, such reactions fall into a category known as *carbonyl addition-elimination reactions*, illustrated by the following sequence:

$$R-\overset{\overset{\displaystyle O}{\|}}{C}-X + Y^{\ominus} \rightleftharpoons \left[R-\overset{\overset{\displaystyle O^{\ominus}}{|}}{\underset{\underset{\displaystyle X}{|}}{C}}-Y\right] \longrightarrow R-\overset{\overset{\displaystyle O}{\|}}{C}-Y + X^{\ominus}$$

The presence of the carbonyl group is believed to stabilize the tetracoordinate transition state, perhaps to such a degree that it has a finite existence as a transient intermediate. However, the speed of the overall reaction may depend on the ease of elimination of $X^{\ominus}$ rather than $Y^{\ominus}$ from the intermediate as well as on the initial formation of the intermediate. The equilibrium formation of the intermediate is usually favored by the presence of metal cations or protonic or Lewis acids that can coordinate to the $C-O^{\ominus}$ unit. Hence, catalysts such as metal oxides or acetates or sulfonic acids are added, especially to polyesterification reactions, to speed up the process.

Requirements for High Molecular Weight

High-molecular-weight condensation polymers can be obtained if the starting materials are pure and the functional groups (NH_2 and COOH, OH and COOH) are present in exactly equal amounts. If one or the other group exists in excess, that group will remain unreacted at the chain ends of low- or medium-molecular-weight polymers. Thus, the establishment of an exact 1:1 ratio of the two functional groups constitutes a critical practical requirement. Three options exist. First, and most obviously, the two reagents (for example, the diol and dicarboxylic acid) can be added together in as close to a 1:1 ratio as is experimentally feasible. Second, a monomer can be chosen which contains both of the functional groups. Glycolic acid, $HOCH_2COOH$, or an amino acid, $H_2NRCOOH$, are examples. Third, use can be made of the fact that some dicarboxylic acids form 1:1 salts with diamines. Hexamethylenediamine, $H_2N(CH_2)_6NH_2$, forms a 1:1 salt with adipic acid, $HOOC(CH_2)_4COOH$. This salt can be purified and used directly for the preparation of nylon 66.

Sometimes a need exists for the preparation of low- or medium-molecular-weight polymers. This can be accomplished by the addition to a condensation polymerization system of a small amount of a mono-carboxylic acid, alcohol, amine, and so on. For example, if 1 mol % of acetic acid is added to a polyester condensation system, the degree of polymerization will be restricted to values of about 200 or less. Of course, the presence of a slight excess of a difunctional reagent will serve a similar purpose. However, the use of a monofunctional reagent has the advantage that the polymer chain ends are no longer active. Hence scrambling and molecular-weight changes are less likely to occur during subsequent heating of the polymer.

Scrambling Reactions

Condensation polymers that contain active groups may undergo molecular weight changes at elevated temperatures. For example, the carboxylic acid end groups of one polymer could attack the amide linkage of another (8). The total number of molecules remains the same, but the average molecular weight may change.

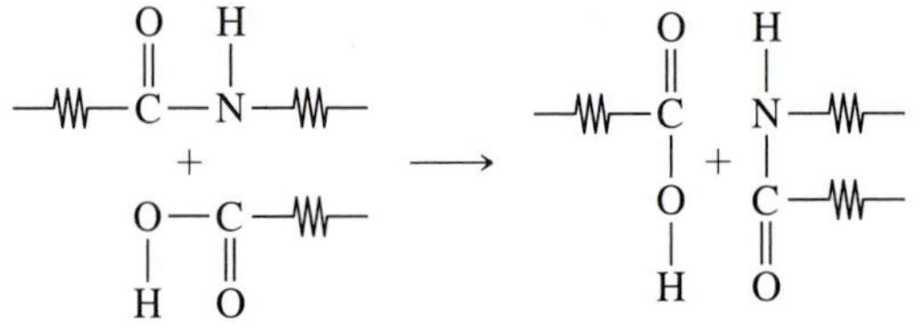

$$\tag{8}$$

Similarly, if two different polymer samples which have different molecular-weight distributions are mixed and heated, the system will scramble to achieve the most probable molecular-weight distribution. This is illustrated in Figure 2.1.

Ring Formation versus Polymerization

At every step in a polymerization reaction the prospect exists that *cyclization* may occur at the expense of linear chain growth. This is especially true with step-type polymerizations, where the chain growth is slow. Consider, for example, the

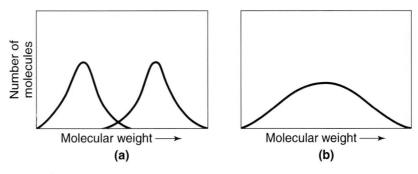

Figure 2.1 Mixture of two polymers with different molecular-weight distributions: (a) will undergo scrambling reactions at elevated temperatures to yield an equilibrium mixture that represents the most probable molecular-weight distribution (b).

initial steps in the polymerization of an amino acid such as glycine **(1)**. Following the initial condensation to give **2,** two courses of action are available to the system. Linear propagation may continue to yield **4.** Alternatively, **2** may cyclize to yield the diketopiperazine **(3).**

Cyclization can yield dimers, trimers, or higher cyclic oligomers. It can occur as a side reaction in the synthesis of polyamides or polyesters. As discussed later in Chapter 10, the ease of cyclization depends on the skeletal bond angles, bond lengths, side-group steric effects, the torsional mobility of the bonds, and ring strain. In general, once the degree of polymerization has exceeded that needed to generate a 12- or 15-membered ring, the probability of cyclization declines. Linear propagation can be facilitated relative to cyclization by the use of high concentrations of the reactants. This effect is a consequence of the fact that linear propagation is a bimolecular process, whereas cyclization is unimolecular.

SPECIFIC CONDENSATION POLYMERIZATIONS

Linear Polyesters

A representative *linear polyester*, poly(ethylene terephthalate), will be discussed here. This polymer is widely used to make plastic soft-drink bottles because it is impermeable to liquids and gases. Poly(ethylene terephthalate) **(7)** can be prepared by a two-step ester interchange reaction between the dimethyl ester of terephthalic acid **(5)** and ethylene glycol **(6)**. The polymer can be prepared in

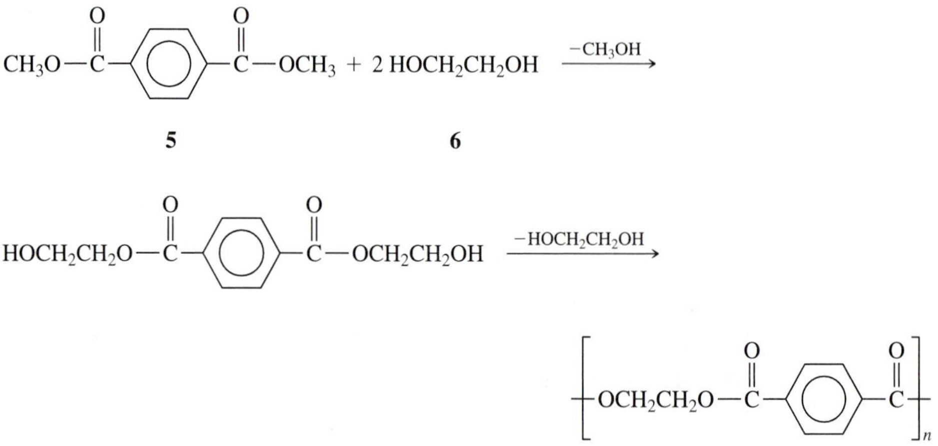

$$CH_3O-\overset{\overset{O}{\|}}{C}-\!\!\!\langle\bigcirc\rangle\!\!\!-\overset{\overset{O}{\|}}{C}-OCH_3 \;+\; 2\,HOCH_2CH_2OH \quad\xrightarrow{-CH_3OH}$$

5   6

$$HOCH_2CH_2O-\overset{\overset{O}{\|}}{C}-\!\!\!\langle\bigcirc\rangle\!\!\!-\overset{\overset{O}{\|}}{C}-OCH_2CH_2OH \quad\xrightarrow{-HOCH_2CH_2OH}$$

$$\left[\!\!-OCH_2CH_2O-\overset{\overset{O}{\|}}{C}-\!\!\!\langle\bigcirc\rangle\!\!\!-\overset{\overset{O}{\|}}{C}-\!\!\right]_n$$

7

the laboratory from a 1:2.4 molar ratio of dimethyl terephthalate to ethylene glycol.[1] The mixture is heated at temperatures up to 197°C for about 3 h in the presence of condensation catalysts such as calcium acetate and antimony trioxide. During this first stage, methanol is evolved. The second step takes place at about 283°C with evolution of ethylene glycol, facilitated by a reduction of the pressure or by a stream of inert gas. The stoichiometric balance between the two monomers is achieved during this step. Failure to remove the last trace of excess ethylene glycol would yield a low-molecular-weight product. This phase (the true polymerization) requires about 3 h of reaction. The product is a tough, flexible polymer that can be melt-pressed into films or drawn from the melt into strong, orientable fibers. The polymer melts at 260 to 270°C.

Poly(ethylene terephthalate) is only one representative of a substantial class of known polyesters. Some are crystalline, others are amorphous. For the crystalline polymers, the melting temperature is often lowered by the asymmetric introduction of substituent groups into the aromatic ring. Methyl groups attached to the diol residue may prevent crystallization.

---

[1] Sorenson, W. R., and Campbell, T. W., *Preparative Methods of Polymer Chemistry*, 2nd ed. (New York: Wiley-Interscience, **1968**), p. 131.

Polyesters can also be prepared by the condensation reactions of aromatic hydroxycarboxylic acids, such as *p*-hydroxybenzoic acid. Such "aromatic polyesters" often show liquid-crystalline character at elevated temperatures, a consequence of the "rigid-rod" character of the macromolecules (see Chapter 22).

Polyesters can also be synthesized by allowing a diol, HO-R-OH, dissolved in alkaline water to react with a di-acid chloride, Cl-C(O)R′C(O)-Cl, dissolved in an organic solvent that is immiscible with water. The reaction mixture is stirred vigorously to create an emulsion and to increase the area of contact between the two liquid phases. Polymerization takes place at the interface between the two liquids, with the HCl liberated being neutralized by the aqueous base. Interfacial polymerizations are important in several areas of polymer synthesis and are mentioned again later in this chapter for the preparation of polycarbonates and polyamides.

Branched and Crosslinked Polyesters

The formation of a linear polyester from *di*functional reagents has just been described. However, if one of the reagents is a tri- or multifunctional species, polymerization will generate a branched polymer. For example, if glycerol **(8)** is allowed to react with a diacid or its anhydride, each glycerol residue will generate one branch point (structure **9**, where G = glycerol residues and A = acid residues). Such mol-

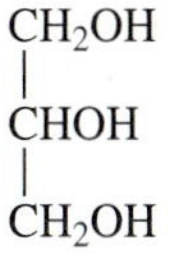

CH_2OH
|
$CHOH$
|
CH_2OH

8 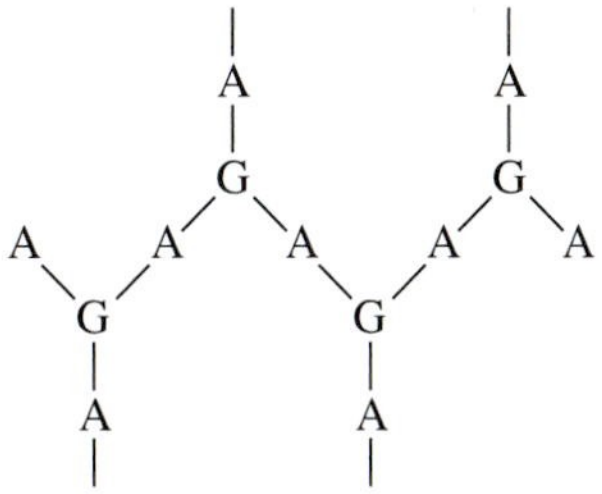   9

ecules can grow to very high molecular weights and an infinitely large polymer network will be formed. If internal coupling occurs (reaction of a hydroxyl group and an acid function from branches of the same or different molecules), the polymer will become crosslinked. In practice, extensive branching and crosslinking cause "gelation" of the polymer. In this state, the polymer is swellable by solvents, but it does not dissolve. Highly crosslinked polymers are totally unaffected by solvents.

Clearly, the degree of branching or crosslinking in a polyester system can be controlled by the amount of triol added relative to diol.

Polycarbonates

Polycarbonates are polymers with the general structure shown in **10.** They are polyesters derived from carbonic acid, $(HO)_2C{=}O$. Of particular importance are the polycarbonates derived from 2,2-bis(4-hydroxyphenyl)propane **(11)**, also

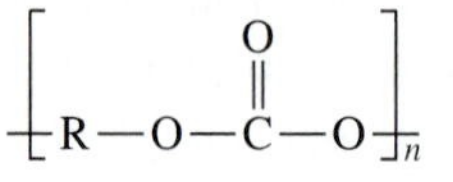

$$
\left[R-O-\underset{\substack{\| \\ O}}{C}-O \right]_n
$$

10

(structure 11: HO—C₆H₄—C(CH₃)₂—C₆H₄—OH)

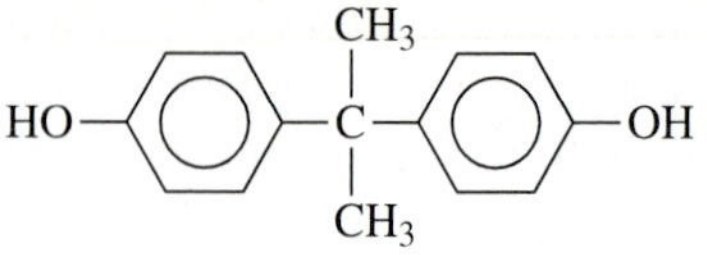

11

known as Bisphenol A. This compound is synthesized on a large scale by the condensation of phenol with acetone. Two convenient reaction routes are available for the preparation of polycarbonates from this monomer. In the first, an ester exchange reaction is performed between molten Bisphenol A and an organic carbonate, such as diphenyl carbonate (reaction 9).

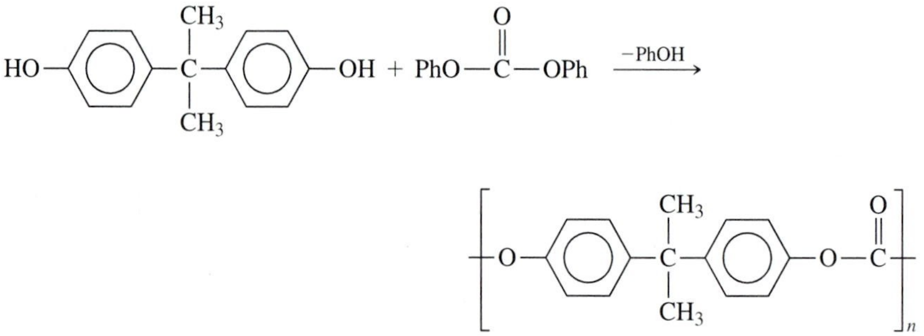

$$
HO-\text{C}_6\text{H}_4-\underset{\substack{CH_3 \\ | \\ | \\ CH_3}}{C}-\text{C}_6\text{H}_4-OH + PhO-\underset{\substack{\| \\ O}}{C}-OPh \xrightarrow{-PhOH}
$$

(9)

$$
\left[O-\text{C}_6\text{H}_4-\underset{\substack{CH_3 \\ | \\ | \\ CH_3}}{C}-\text{C}_6\text{H}_4-O-\underset{\substack{\| \\ O}}{C} \right]_n
$$

Initially, a prepolymer is formed by heating of the mixture at 180 to 220°C in vacuum for 1 to 3 h. The temperature is then raised slowly to 280 to 300°C and the pressure is lowered to remove the last traces of phenol. The residual molten polymer is highly viscous. It sets to a transparent solid when cooled. This can be melt-pressed or solution-cast (see Chapter 20) to fabricate tough films.

A second synthetic route to polycarbonates involves a reaction of Bisphenol A with phosgene (10). This can be accomplished[1] by bubbling phosgene into

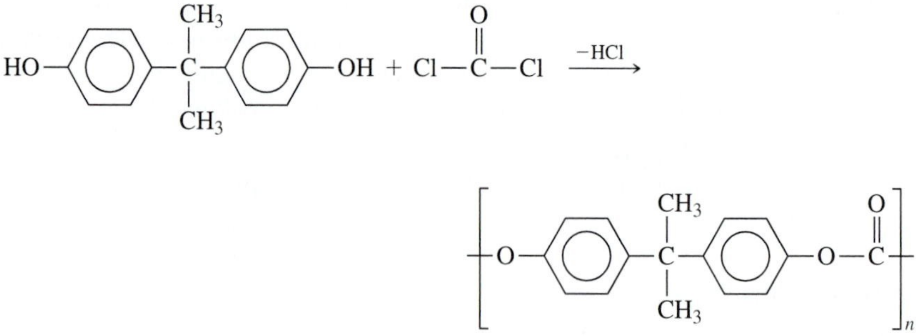

$$
HO-\text{C}_6\text{H}_4-\underset{\substack{CH_3 \\ | \\ | \\ CH_3}}{C}-\text{C}_6\text{H}_4-OH + Cl-\underset{\substack{\| \\ O}}{C}-Cl \xrightarrow{-HCl}
$$

(10)

$$
\left[O-\text{C}_6\text{H}_4-\underset{\substack{CH_3 \\ | \\ | \\ CH_3}}{C}-\text{C}_6\text{H}_4-O-\underset{\substack{\| \\ O}}{C} \right]_n
$$

[1]With suitable precautions. Phosgene is a highly toxic gas.

a solution of Bisphenol A in pyridine at 25 to 30°C. The pyridine functions as a hydrogen chloride acceptor. The polymer can be isolated by precipitation in water or methanol. An alternative procedure makes use of a heterophase emulsion system. Phosgene is passed into a rapidly stirred emulsion of Bisphenol A in methylene chloride, aqueous sodium hydroxide, and quaternary ammonium halide catalyst. Recovery of the polymer is effected by separation of the organic phase and evaporation of the methylene chloride.

The use of highly toxic gaseous phosgene can be avoided by replacing it with trichloromethyl chloroorthoformate, $CCl_3OC(O)Cl$, which decomposes under the polymerization reaction conditions, essentially to yield phosgene in solution. The quaternary ammonium halide, such as tetrabutylammonium bromide, accelerates the transport of molecules across the water-organic solvent interface. Interfacial polymerizations similar to this are also used to prepare polyamides (see pages 38 to 42).

Polyanhydrides

Polyanhydrides are macromolecules formed, in principle, by the dehydrative condensation of dicarboxylic acids. However, in practice, certain embellishments to this direct condensation are employed.[1] These are summarized in reactions (11) to (13).

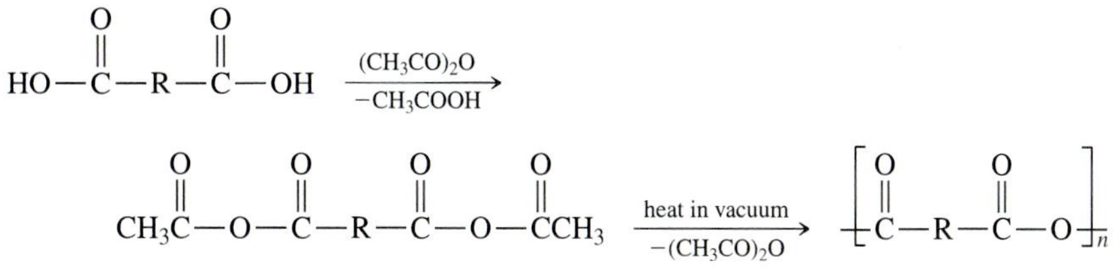

$$
\underset{\substack{\| \\ HO-C-R-C-OH}}{\overset{O \qquad O}{}} \quad \xrightarrow[-CH_3COOH]{(CH_3CO)_2O}
$$

(11)

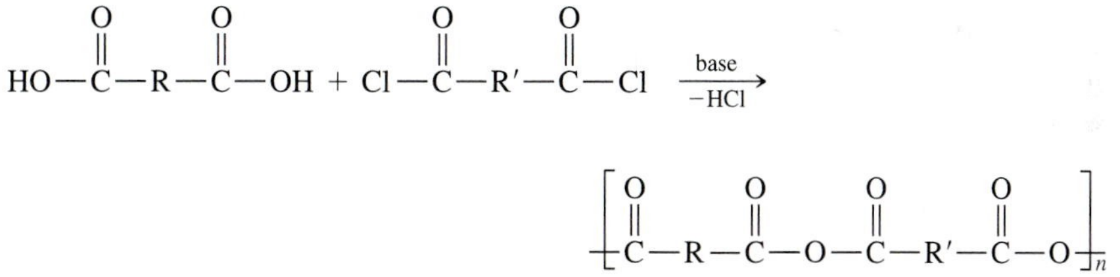

$$
\underset{\substack{CH_3C-O-C-R-C-O-CCH_3}}{\overset{O \quad O \quad O \quad O}{}} \quad \xrightarrow[-(CH_3CO)_2O]{heat\ in\ vacuum} \quad \left[\underset{\substack{C-R-C-O}}{\overset{O \qquad O}{}}\right]_n
$$

$$
\underset{\substack{HO-C-R-C-OH}}{\overset{O \quad O}{}} + \underset{\substack{Cl-C-R'-C-Cl}}{\overset{O \quad O}{}} \quad \xrightarrow[-HCl]{base}
$$

(12)

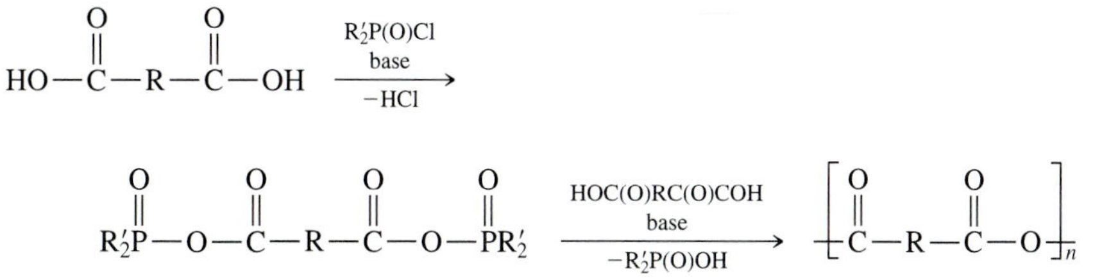

$$
\left[\underset{\substack{C-R-C-O-C-R'-C-O}}{\overset{O \quad O \quad O \quad O}{}}\right]_n
$$

$$
\underset{\substack{HO-C-R-C-OH}}{\overset{O \quad O}{}} \quad \xrightarrow[-HCl]{\substack{R'_2P(O)Cl \\ base}}
$$

(13)

$$
\underset{\substack{R'_2P-O-C-R-C-O-PR'_2}}{\overset{O \quad O \quad O \quad O}{}} \quad \xrightarrow[-R'_2P(O)OH]{\substack{HOC(O)RC(O)COH \\ base}} \quad \left[\underset{\substack{C-R-C-O}}{\overset{O \qquad O}{}}\right]_n
$$

[1]Leong, K. W., Simonte, V., and Langer, R., *Macromolecules*, **1987**, *20*, 705.

Reaction (11) is carried out as a melt polymerization, with acetic anhydride being removed under vacuum. This method, in general, yields the highest-molecular-weight polymers, although careful control of the polymerization temperature is needed to avoid decomposition reactions that occur above 130°C. The second method (12), the so-called Schotten-Baumann condensation, takes place rapidly at room temperature in solution to give medium-molecular-weight polymers ($\overline{M}_n \approx 10^4$). A third method (13) also yields medium-molecular-weight polymers under mild reaction conditions. In addition, polyanhydrides can be prepared by the ring-opening polymerization of cyclic anhydrides (see Chapter 6).

Polyanhydrides are intrinsically unstable to moisture, being converted to small-molecule dicarboxylic acids. At first sight, this might seem to be disadvantageous, but in fact this property allows some (aliphatic) polyanhydrides to be used as "erodable" matrices in living systems for the controlled release of chemotherapeutic drug molecules (see Chapter 24). Somewhat better hydrolytic stability is found for the polyanhydrides that contain aromatic rings in the chain. This is partly a consequence of a high degree of crystallinity, since amorphous polymers hydrolyze more rapidly.

Polyamides

General features Polyamides or "nylons" were among the first synthetic high polymers to be made and used on a large scale. Their discovery by Carothers and co-workers in 1935 served to usher in the new era of synthetic polymers, which has altered technology and everyday life almost beyond recognition.

Four principal methods are available for the synthesis of high-molecular-weight polyamides: (1) the reaction between a dicarboxylic acid and a diamine, (2) the dehydration–condensation of an amino acid, (3) the reaction between a diacid chloride and a diamine, and (4) the ring-opening polymerization of cyclic amides. This fourth route is discussed in Chapter 6. The first three routes will be considered here.

Melt polymerization The direct interaction between a dicarboxylic acid and a diamine is the classical method for the synthesis of polyamides. In practice, it is preferable to ensure the existence of a 1:1 ratio of the two reactants by the prior isolation of a 1:1 salt of the two. The overall procedure is summarized by the reaction scheme shown in (14). The actual polymerization process is known as a *melt polymerization* because the reaction takes place above the melting points of both the reactants and the polymer. This type of reaction should be distinguished from "interfacial polymerizations," discussed earlier.

$$HO-\overset{\overset{\displaystyle O}{\|}}{C}-R-\overset{\overset{\displaystyle O}{\|}}{C}-OH + H_2N-R'-NH_2 \longrightarrow$$

$$\underset{\underset{\oplus}{H_3N}-R'-\underset{\oplus}{NH_3}}{\overset{\ominus}{O}-\overset{\overset{\displaystyle O}{\|}}{C}-R-\overset{\overset{\displaystyle O}{\|}}{C}-O^{\ominus}} \quad \xrightarrow[\text{heat}]{-H_2O} \quad \left[\overset{\overset{\displaystyle O}{\|}}{C}-R-\overset{\overset{\displaystyle O}{\|}}{C}-\overset{\overset{\displaystyle H}{|}}{N}-R'-\overset{\overset{\displaystyle H}{|}}{N}\right]_n \tag{14}$$

Salt

Probably the best known melt polymerization involves the reaction between hexamethylenediamine, $H_2N(CH_2)_6NH_2$ and adipic acid, $HOOC(CH_2)_4COOH$, to yield nylon 66, $-[NH-(CH_2)_6-NH-CO-(CH_2)_4-CO]_n$. The numerals, in the trivial name refer to the numbers of carbon atoms in the two monomers. The first number gives the number of carbon atoms in the diamine. The following description outlines the procedure that can be followed on a laboratory scale.[1]

A 1:1 salt is first prepared by the addition of a warm solution of hexamethylenediamine in dry ethanol to an ethanolic solution of an equimolar amount of adipic acid, followed by cooling to room temperature. The white, crystalline salt melts at about 196 to 197°C. The salt is then introduced into a polymerization tube. Air is displaced from the tube by nitrogen and the tube is sealed. The next step requires that the tube be heated at 215°C for 1.5 to 2 h. However, during this step, the tube is a *potential bomb*. The high internal pressures could cause it to explode violently. Hence, this reaction should never be carried out unless the tube is shielded by an open steel tube container and a steel explosion shield. Even when cooled after this stage of the reaction, it should be handled only behind shielding and with a protective glove. Failure to observe such precautions could have disastrous consequences, and, for this reason, only those who already have substantial synthetic experience in the laboratory should attempt this reaction. After the initial step, the tube is opened (again with adequate shielding), and the contents are heated to 270°C first in a nitrogen atmosphere and later in a vacuum for about 1 h. Even while the tube is cooling, it should be shielded to prevent shattered glass from being hurled out as the tube cracks. On a large scale, the reaction is carried out in an autoclave. The tough, white polymer can be melt-cast into films or melt-drawn into fibers. It has a melting point of 267°C, and it is soluble in formic acid, phenols, and cresols.

Polymerization of amino acids   The polymerization of amino acids, such as 11-aminoundecanoic acid, $H_2N-(CH_2)_{10}-COOH$, can be accomplished by melt polymerization techniques at 220°C. However, an alternative procedure

[1]Sorenson and Campbell, op. cit., p. 74.

often used to prepare "model" polypeptides involves the use of dicyclohexylcarbodiimide **(12)** as a dehydrating agent. In this case, the polymerizations are carried out at moderate temperatures.

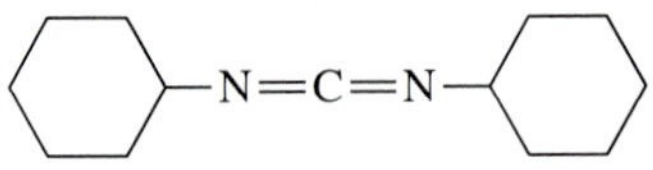

12

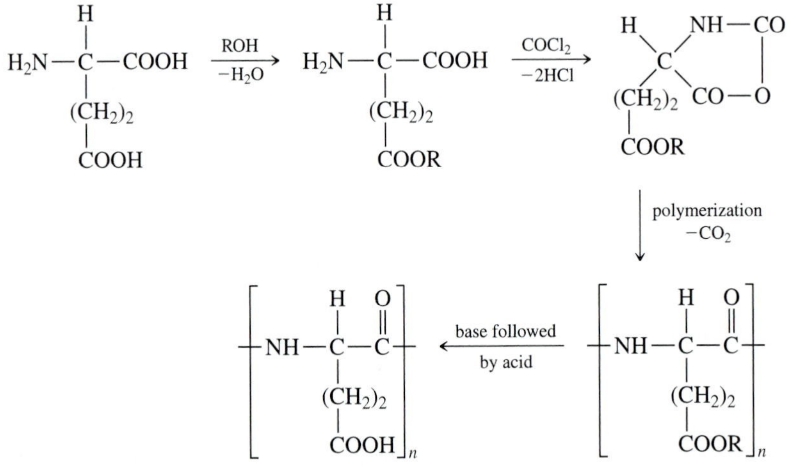

(15)

13

Poly(glutamic acid) **(13)** is prepared by the sequence shown in reaction (15). This polymer can be solution-spun to yield silk-like fibers. These are susceptible to decomposition by microorganisms.

Interfacial polymerization   The process of interfacial polymerization mentioned earlier can best be illustrated by the reaction between a diamine and a diacid chloride. An interfacial polymerization takes place when the two monomers are present in two immiscible solvents. Reaction then occurs at the interface between the two liquids. In practice, the procedure can be carried out effectively only if the polymerization reaction is rapid at moderate temperatures. However, it provides a valuable method for the synthesis of polymers at temperatures far below those required for melt polymerization. The polymer molecular weights obtained by the interfacial procedure are generally higher than those obtained by the melt method. Monomer molecules tend to react more readily with growing polymer molecules than with other monomer molecules because the reaction is too rapid to allow the monomer to diffuse through the layer of polymer. Hence, rigorous step-type polymerization is not maintained. Furthermore, an exact 1:1 balance of the two monomers is not required.

Two examples will serve to illustrate the technique. First, ethylenediamine **(14)** can be allowed to react with terephthaloyl chloride **(15)** to yield poly(ethylene terephthalamide) **(16)** (reaction 16). The diamine is dissolved in a solution of potassium hydroxide (the hydrochloride acceptor) in water, and the diacid chloride is dissolved in methylene chloride. The two layers are emulsified in a kitchen blender or by means of a high-speed laboratory stirrer at room temperature for 10 min. The polymer is then simply removed by filtration, washed, and dried.

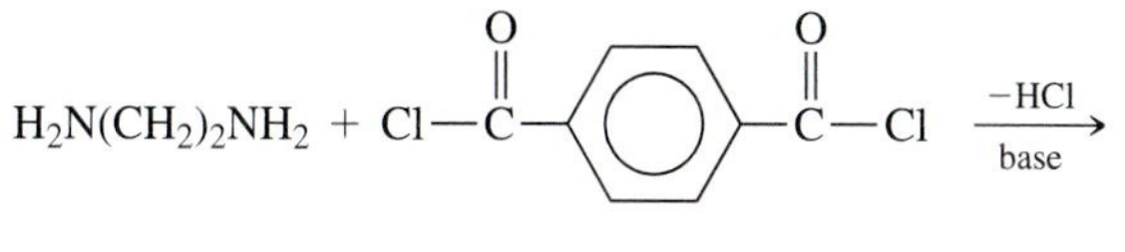

$$H_2N(CH_2)_2NH_2 \ + \ Cl-\underset{\underset{\displaystyle 15}{}}{\overset{\overset{\displaystyle O}{\parallel}}{C}}-\!\!\!\!\bigcirc\!\!\!\!-\overset{\overset{\displaystyle O}{\parallel}}{C}-Cl \ \xrightarrow[\text{base}]{-HCl}$$

14   15

(16)

$$\left[\!\!\!- \overset{\overset{\displaystyle H}{|}}{N}-(CH_2)_2-\overset{\overset{\displaystyle H}{|}}{N}-\overset{\overset{\displaystyle O}{\parallel}}{C}-\!\!\!\!\bigcirc\!\!\!\!-\overset{\overset{\displaystyle O}{\parallel}}{C}-\!\!\!\right]_n$$

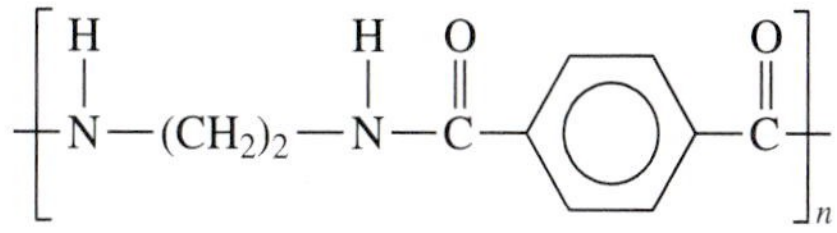

16

Another method, often used in lecture demonstrations, makes use of a combined polymerization and fiber drawing technique. The preparation of poly(hexamethylsebacamide) (nylon 610) by this method is a good example. A solution of sebacoyl chloride, $Cl-C(O)-(CH_2)_8-C(O)-Cl$, in tetrachloroethylene is placed in the bottom of a beaker. Over it is carefully poured a solution of hexamethylenediamine in water. Polymer forms at the interface. The polymer may then be grasped with tweezers or tongs and raised from the beaker. This causes a continuous filament of polymer to be drawn from the mixture. Mechanical devices are sometimes used (Figure 2.2) to pull a substantial length of monofilament from the reaction.

Aromatic polyamides   Some of the most important recent developments in polyamide synthesis have been in the field of so-called "aramid" polymers, macromolecules without aliphatic units in the main chain.[1] An example is Kevlar,[2] a polyamide prepared from *para*-phenylenediamine and terephthaloylchloride. Kevlar is used in bullet proof vests and in composites for motorcycle helmets.

[1]Morgan, P. W., *L. Polym. Sci. Polym. Symp.*, **1985,** *72,* 27.
[2]This is a trade name of the DuPont Company.

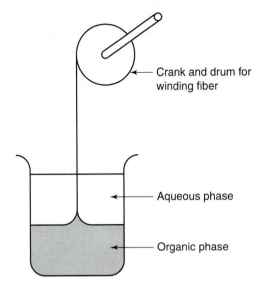

Crank and drum for winding fiber

Aqueous phase

Organic phase

Figure 2.2 Monofilament fiber being drawn directly from an interfacial polymerization experiment.

$$H_2N-\!\!\!\bigcirc\!\!\!-NH_2 \;+\; Cl-\overset{\overset{O}{\|}}{C}-\!\!\!\bigcirc\!\!\!-\overset{\overset{O}{\|}}{C}-Cl \;\xrightarrow{-HCl}$$

$$\left[N-\!\!\!\bigcirc\!\!\!\underset{H}{-}N-\overset{\overset{O}{\|}}{C}-\!\!\!\bigcirc\!\!\!-\overset{\overset{O}{\|}}{C} \right]_n$$

17

Because aliphatic carbon atoms are a common source of thermal instability (see Chapter 7), the absence of aliphatic units in the chain of this polymer **(17)** generates a high thermooxidative stability. However, the most intriguing properties are a result of the rigid-rod nature of this inflexible polymer. The material is highly crystalline due to the facile packing of the extended chains, so much so that one of the few liquids that will dissolve the polymer is sulfuric acid. The rigid-rod structure also gives rise to main-chain liquid crystallinity, a subject that is discussed in Chapter 22. A related polymer prepared from *meta*-phenylenediamine lacks this linear structure and is, therefore, more soluble.

Polyimides

Polyimides are polymers formed by the condensation reactions of dianhydrides with diamines. A typical structure is shown in **19** (reaction 17). Polyimides have been synthesized from aromatic anhydrides and aliphatic diamines, and from aromatic anhydrides and aromatic diamines. The latter products are called *aromatic polyimides*. Aromatic polyimides are often infusible and insoluble. Hence, precipitation occurs during the intermediate stages of polymerization. To circumvent this problem, the condensation is performed in two stages. First, a noncyclized, soluble, linear high polymer **(18)** is generated by a fast reaction in a suitable polar solvent at temperatures below 70°C. Polymer **18** can be fabricated into a suitable shape and then cyclized to **19** by heating at temperatures up to 300°C. Such polymers are rigid, high melting, thermally stable materials. Some flexibility can be introduced into the polymer structure by the use of diamines which contain a flexible linkage unit. An example is 4,4'-diaminodiphenyl ether, $H_2N-C_6H_4-O-C_6H_4-NH_2$. Polyimides are used for the fabrication of machined devices and in laminates for high-temperature applications.

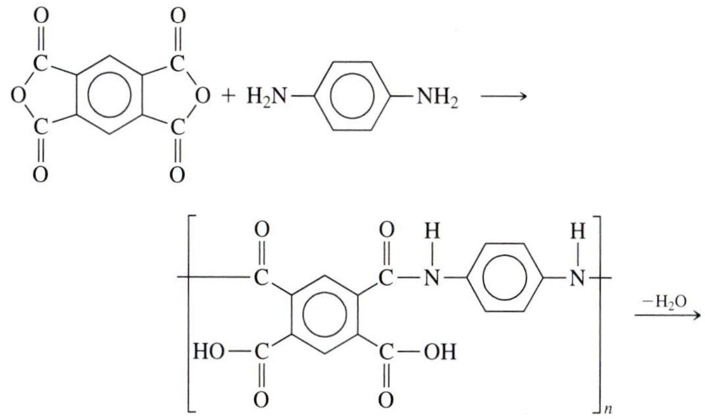

$$(17)$$

18

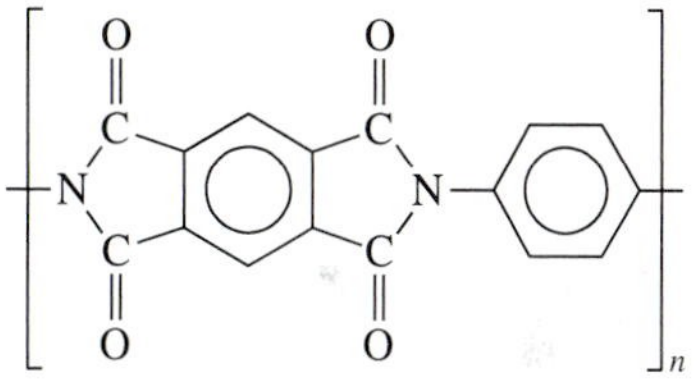

19

Polybenzimidazoles[1]

Polybenzimidazoles **(21)** are polymers formed by the condensation of dicarboxylic acids with aromatic tetramines **(18)**. An example is the polymer obtained by the condensation of a dicarboxylic acid with 3,3'-diaminobenzidine **(20)**. Again a two-stage reaction is preferred, although the reaction between sebacic acid, $HOOC(CH_2)_8COOH$, and 3,3'-diaminobenzidine **(20)** can apparently be carried out at 265°C for 3.5 h in a one-step process. Normally, however, the condensation is not complete until the polymer is heated to 350 to 400°C. Higher-molecular-weight, thermally stable polymers can often be obtained by the use of the diphenyl esters of aromatic diacids. Polybenzimidazoles are used

[1]Marvel, C. S., in *High Temperature Polymers*, C. L. Segal, ed. (New York: Dekker, **1967**), p. 1.

for the manufacture of heat-resistant components and in wire enamels and coatings for high-temperature uses.

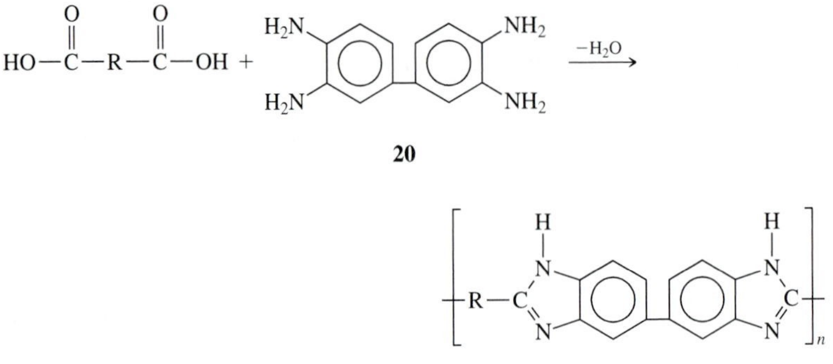

20

(18)

21

Polybenzoxazoles and Polybenzthiazoles[1]

Another class of condensation polymers, known for their rigid-rod character and their thermooxidative stability, are the polybenzoxazoles. A typical example is the polymer shown as **22,** prepared by the reaction of 4,6-diamino-1,3-benzenediol dihydrochloride with terephthalic acid in a medium of polyphosphoric acid (reaction 19).

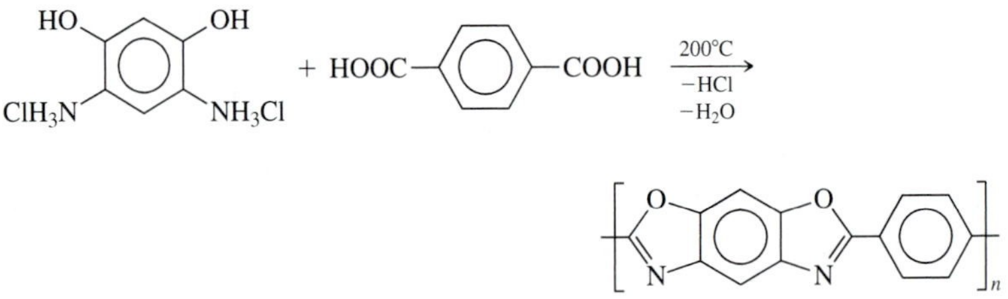

(19)

22

Analogues with sulfur in place of oxygen in the heterocyclic rings are prepared from 2,5-diamino-1,4-benzenedithiol dihydrochloride and terephthalic acid (reaction 20).

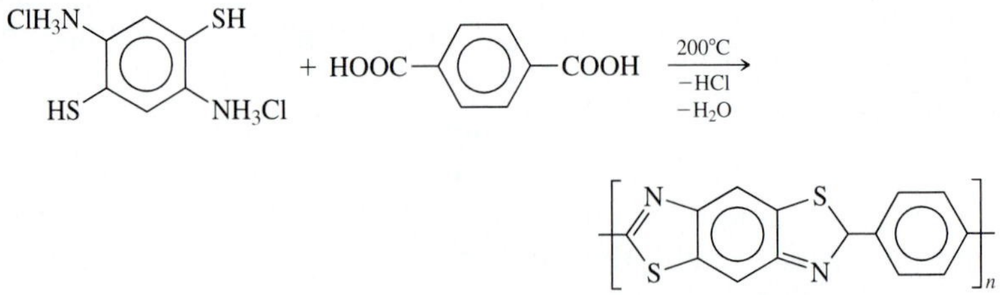

(20)

[1]For more on this subject, see a series of papers by various authors on the subject of polybenzoxazoles and polybenzthiazoles in *Macromolecules*, **1981,** *14,* 909–950.

Polyquinoxalines

Polyquinoxalines are aromatic polymers that are prepared by the condensation of aromatic diglyoxals with aromatic tetramines. A typical example is the reaction shown in scheme (21). This type of polymer has very high thermal stability. Again it is preferable to carry out the reaction in two stages. The first phase comprises a solution polymerization in a solvent such as hexamethylphosphoramide. The second stage is a high-temperature pyrolysis at about 375°C in vacuum. Most poly(phenylquinoxalines) are high-temperature thermoplastic structural materials.

$$(21)$$

Aromatic Ladder Polymers

All the condensation polymers discussed up to this point are single-strand-type polymers. Even though some ladder-type structure may be present (as in the polyquinoxalines, polybenzimidazoles, or polyimides discussed previously), at least one skeletal bond per repeating unit is not a ladder structure. These single-strand linkages allow some limited skeletal flexibility, but they also provide sites for irreversible skeletal cleavage at high temperatures.

Aromatic ladder polymers represent an attempt to prepare macromolecules with no single-strand linkages. Such materials are likely to have very high thermal stabilities. Every backbone cleavage reaction in a single-strand polymer causes a decrease in chain length and an eventual deterioration of the polymer properties. In a ladder polymer *two* bonds must cleave in the same connecting residue before the chain can break. The likelihood that this will happen is low. Moreover, if one bond in a ladder polymer breaks, the two cleaved ends will be held in juxtaposition by the remaining intact strand and "healing" of the broken bond is likely to occur. The fact that ladder structures are generally very stiff also has the advantage of increasing transition temperatures. This stiffening decreases the entropy of melting ΔS_m, and because the melting point T_m is given by $\Delta H_m/\Delta S_m$ (where ΔH_m is the heat of melting), T_m can increase considerably. The stiffness also hinders some of the chain's motions, and this can increase the glass transition temperature T_g (above which these motions make the polymer more flexible). Hence, aromatic ladder polymers are capable of withstanding very high temperatures. Typical aromatic ladder polymers are polybenzimidazoles of the type discussed above, and polyimidazopyrrolones or polyquinoxalines with structures such as those shown in **23** and **24** (schemes 22 and 23). The condensation of quinones with aminophenols leads to the formation of polymers such as **25**, as shown in reaction (24). Similarly, the condensation of tetraketones or their precursors with tetramines leads to the formation of ladder polymers, such as **26** (reaction 25). A number of aromatic ladder polymers are attracting attention as rigid-rod macromolecules because of their unusual rheological properties.

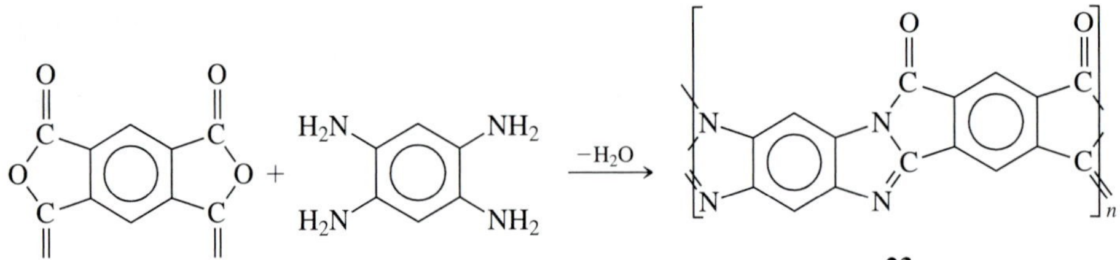

23

(22)

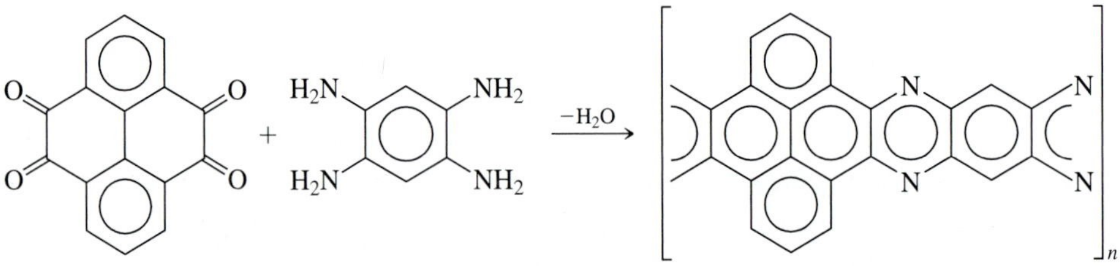

24

(23)

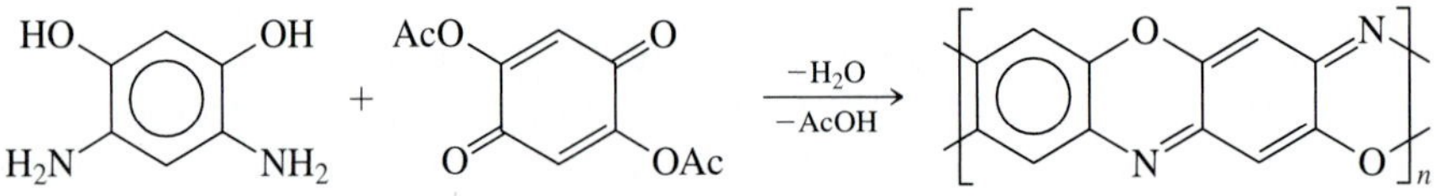

25

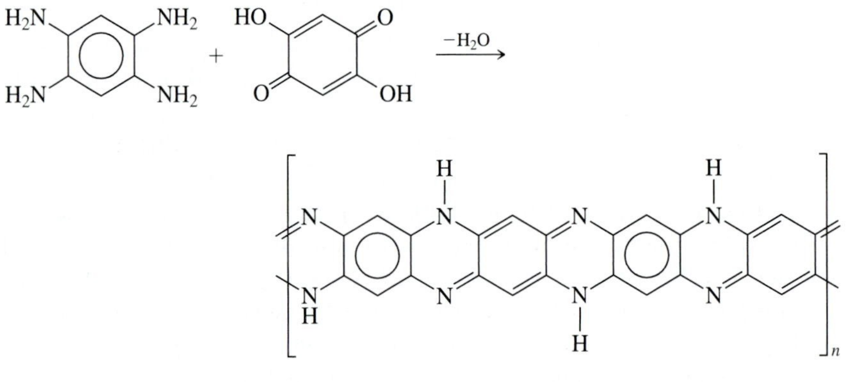

(24)

26

(25)

It should be noted that other ladder polymer systems have been synthesized from siloxane residues and by the pyrolysis of polyacrylonitrile. These reactions are discussed in Chapters 7 and 9, respectively.

Phenol–, Urea– and Melamine–Formaldehyde Resins

The reactions involved in the formation of resins from the reaction of formaldehyde with phenol, urea, or melamine are complex. The initial steps in these processes are not condensations, but the later "curing" stages involve elimination of water. Hence, these reactions are considered here under the classification of condensation polymerizations.

Phenol–formaldehyde polymers Resins based on the reactions of phenol and formaldehyde were known as early as 1872. During the first half of the 20th century, in the form of "Bakelite resins," they provided one of the few synthetic polymeric materials available for fabrication into moldable, thermosetting objects. Three stages can be identified in the preparation of a network polymer from this system.

First, a phenol is allowed to react with formaldehyde to yield a mixture of methylolphenols (reaction 26). This reaction can be carried out under acidic or basic conditions. If acidic conditions are used, spontaneous condensation occurs to form low polymers. The products from the base-catalyzed reactions are soluble prepolymers called "resoles."

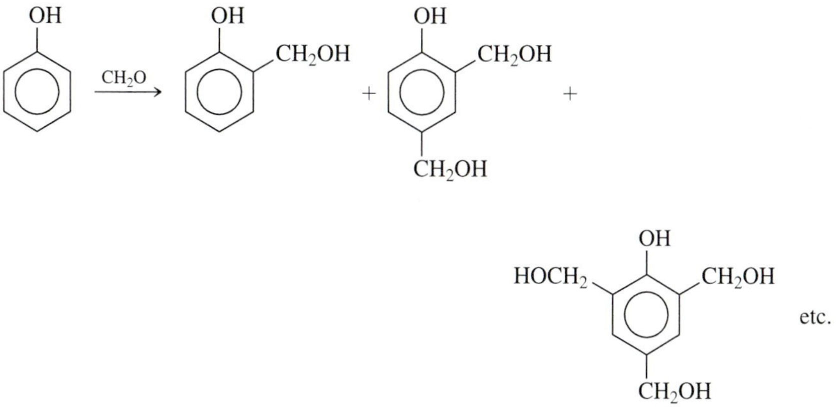

(26)

Heating of the prepolymer at temperatures up to about 105°C under neutral or acidic conditions brings about condensation to generate moldable cyclolinear or branched-type polymers by reactions such as (27). At higher temperatures, methylene bridges are established by reactions of the type shown in (28), which involve the elimination of water and formaldehyde. These structures are believed to predominate in the final crosslinked resin. However, the chemistry of these steps is much more complex than is implied by these reactions.

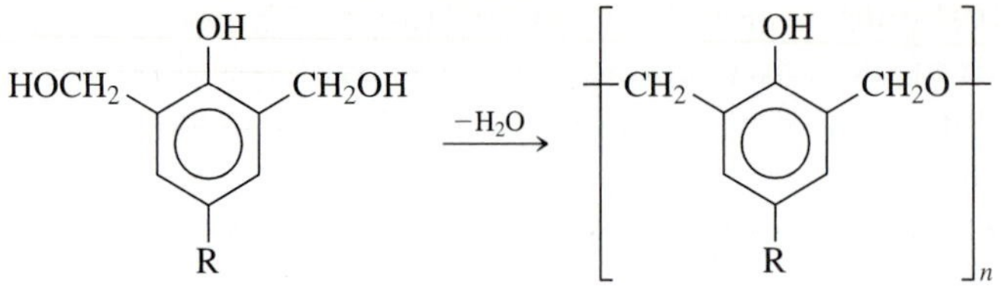

$$\text{(27)}$$

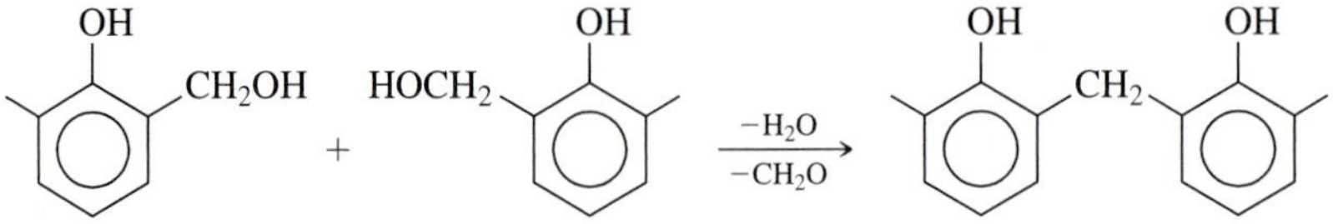

$$\text{(28)}$$

Novolac resins are phenol–formaldehyde polymers made from the acid-catalyzed low polymers discussed earlier. They may be crosslinked by the addition of hexamethylene-tetramine, $(CH_2)_6N_4$, with the concurrent elimination of ammonia.

Urea–formaldehyde resins Formaldehyde reacts with amino compounds to generate methylol derivatives. Condensation of the methylol units can then occur, often at moderate temperatures. Urea and melamine (triamino-s-triazine, $-(N{=}CNH_2)_3$) are amino-compounds that participate in these types of reactions.

Urea reacts with aqueous formaldehyde in alkaline media to yield monomethylol– and dimethylol–urea derivatives (**27** and **28**) (reaction 29). Such products can either be isolated as discrete compounds or the mixture can be used directly for resin formation. Water is removed from the mixture to form a syrup, which may then be acidified and heated near 100°C to bring about condensation and gelation. The exact crosslinking mechanism is uncertain, but condensation could occur between $-CH_2OH$ and NH_2 groups, or between two CH_2OH groups, with liberation of water. Methylene bridges could also be formed by loss of both water and formaldehyde, and numerous other reactions have been postulated. Urea–formaldehyde resins are clear and colorless. They are harder than phenol–formaldehyde resins and they can be reinforced with fillers or can be colored for different uses.

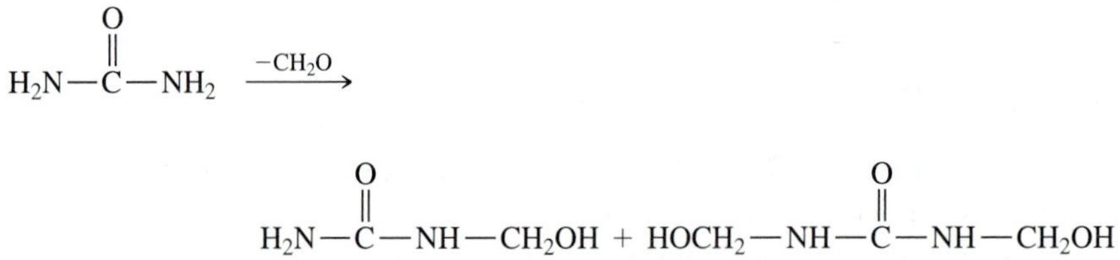

$$H_2N{-}\overset{\overset{\displaystyle O}{\|}}{C}{-}NH_2 \xrightarrow{-CH_2O}$$

$$\text{(29)}$$

$$H_2N{-}\overset{\overset{\displaystyle O}{\|}}{C}{-}NH{-}CH_2OH + HOCH_2{-}NH{-}\overset{\overset{\displaystyle O}{\|}}{C}{-}NH{-}CH_2OH$$

$$\qquad\qquad\textbf{27}\qquad\qquad\qquad\qquad\qquad\qquad\textbf{28}$$

Melamine–formaldehyde polymers   Melamine (triamino-*s*-triazine) **(29)** can also be methylolated (30) to yield a range of products which contain up to six methylol groups **(30)**. These materials are soluble in water and can conveniently be applied to textile fabrics as aqueous solutions. Crosslinked resins are obtained at elevated temperatures via the formation of methylene or methylene–ether linkages. Fabrics containing cured methylol-melamine have "permanent press" characteristics. Bulk polymers made from molding powders constitute the well-known "melamine resins."

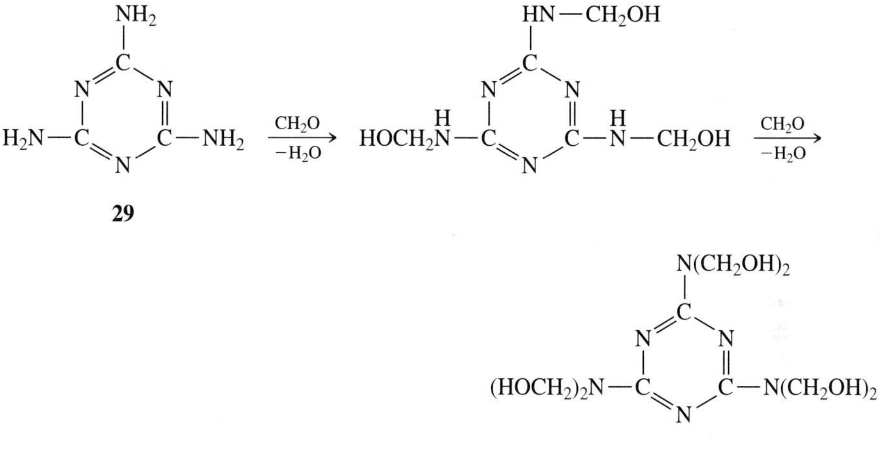

$$\underset{\textbf{29}}{\text{(structure of melamine)}} \xrightarrow[-H_2O]{CH_2O} \text{(methylolated melamine)} \xrightarrow[-H_2O]{CH_2O}$$

(30)

$$\underset{\textbf{30}}{\text{(fully methylolated melamine)}}$$

Polyacetals

Formaldehyde reacts with aliphatic diols to eliminate water and form polyethers known as *polyacetals*. The overall process is shown in reaction (31). This condensation process is, in reality, a carbonyl addition-substitution reaction which takes place in the presence of acid catalysts. Long-chain diols are normally required to reduce the tendency for cyclization. High-melting crystalline polyformals are produced when formaldehyde reacts with cyclic diols. It should be noted that the term "polyacetal" is also sometimes applied to aldehyde *addition* polymers (see Chapter 4).

$$CH_2{=}O + HO{-}R{-}OH \xrightarrow{-H_2O} \pod{CH_2{-}O{-}R{-}O}_n \tag{31}$$

OTHER STEP-TYPE POLYMERIZATIONS

Condensation reactions form only one class of step-type polymerizations. A number of other organic reaction processes have also been employed for the step synthesis of polymers, and a few examples are mentioned briefly in the following sections.

Polyethers by Aromatic Substitution

Although condensation reactions between aromatic diols can yield low-molecular-weight polyethers, high polymers may be prepared by nucleophilic substitution or oxidative coupling routes. The nucleophilic substitution route is used for the syntheses of the so-called *polysulfone* (33) from the disodium salt of Bisphenol A (31) and 4,4'-dichlorodiphenyl sulfone (32) (reaction 32). The sulfone group in (32) serves to activate the chlorine atoms to nucleophilic replacement

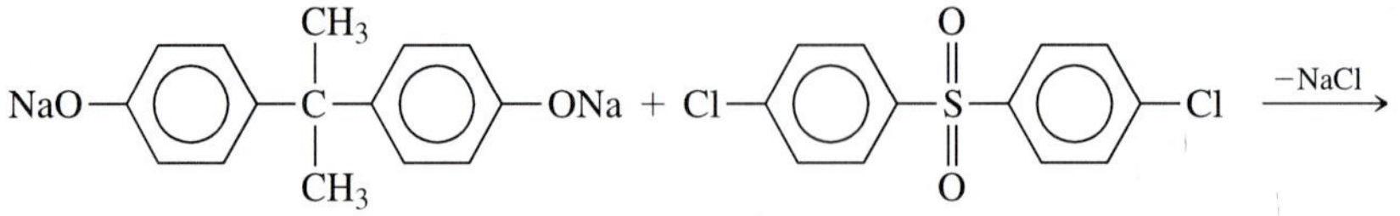

$$\text{NaO}-\text{C}_6\text{H}_4-\underset{\underset{\text{CH}_3}{|}}{\overset{\overset{\text{CH}_3}{|}}{\text{C}}}-\text{C}_6\text{H}_4-\text{ONa} \;+\; \text{Cl}-\text{C}_6\text{H}_4-\underset{\overset{\text{O}}{\|}}{\underset{\underset{\text{O}}{\|}}{\text{S}}}-\text{C}_6\text{H}_4-\text{Cl} \xrightarrow{-\text{NaCl}}$$

31   32

(32)

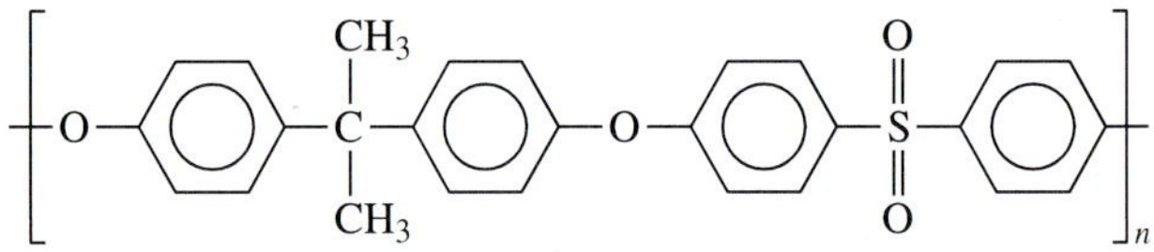

33

in a way that would not be possible in a conventional chloroaromatic compound. Dimethyl sulfoxide is the preferred solvent for such reactions, and the interaction is carried out at 160°C. The polymer softens near 200°C. It can be cast or pressed into films at 280°C and fibers can be obtained by melt techniques. Polysulfones are used for electrical insulating applications and for the fabrication of heat-resistant articles.

Polyethers by Oxidative Coupling Reactions

Aromatic polyethers are also produced when oxygen is bubbled through solutions of 2,6-disubstituted phenols in an organic solvent. A catalyst complex, formed from a cuprous salt and a tertiary amine, is required for the reaction. The overall process is illustrated by the scheme shown in (33). Polymer 34 is known commercially as a "poly(phenylene oxide)." It is used for the manufacture of machined parts, especially when heat stability is needed.

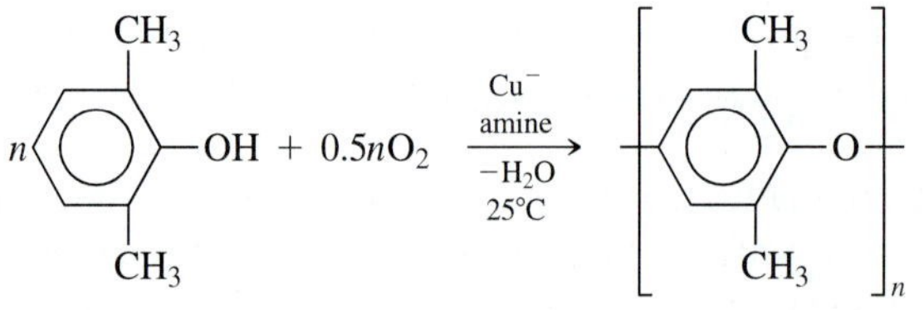

$$n\;\text{(2,6-dimethylphenol)}-\text{OH} + 0.5n\text{O}_2 \xrightarrow[\substack{-\text{H}_2\text{O}\\25°\text{C}}]{\substack{\text{Cu}^-\\\text{amine}}} \left[\text{(dimethylphenylene)}-\text{O} \right]_n$$

(33)

34

Diels–Alder Addition Polymers

The importance of ladder polymers has been discussed earlier in this chapter. Another attractive route to polymers of this type involves Diels–Alder-type additions between, for example, quinones and vinyl compounds, as shown in the formation of **35** in reaction (34). Such reactions occur rapidly at only moderate temperatures. However, side reactions often lead to the formation of structural irregularities.

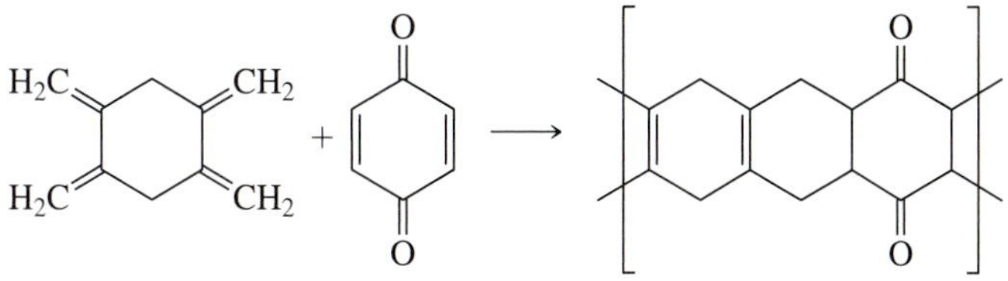

$$(34)$$

35

Polyurethanes and Polyureas

Polyurethanes form one of the most important classes of synthetic polymers. They can be used as elastomers ("Spandex" or foam rubber), as surface coatings (paints and wood varnishes), or as adhesives, and have been widely used as biomedical materials. A problem with polyurethane foam elastomers is that they break down over long periods of time (decades) to form gums, presumably a result of reactions between the urethane linkage (-NH-COO-) and atmospheric moisture. They are also highly flammable.

Polyurethanes are macromolecules formed by the linkage of preformed polymers through their hydroxyl end-groups by reaction with a di-isocyanate. The overall reaction is shown in equation (35).

$$HO-polymer-OH \ + \ O{=}C{=}N-R-N{=}C{=}O \ \longrightarrow$$
$$O{=}C{=}N-R-NH-C(O)-O-polymer \qquad (35)$$
$$-O-C(O)-NH-R-N{=}C{=}O$$

If an excess of the di-isocyanate is employed, the resultant macromolecule is end-terminated by isocyanate groups, which can then react with a different polymeric diol to extend the chain, or with a triol to crosslink the system.

The polymeric diol can be a hydroxyl-terminated poly(ethylene oxide) or a poly(organosiloxane) (Chapter 9) or, in principle, almost any other type of polymer with hydroxyl-group termination. When two different polymeric diols are employed, one may be a hydrophilic block and the other a hydrophobic segment, or one can be a "soft" flexible block, such as poly(ethylene oxide) and the other a "hard" inflexible segment. The di-isocyanate can also form a hard segment as shown in species **36** and **37**.

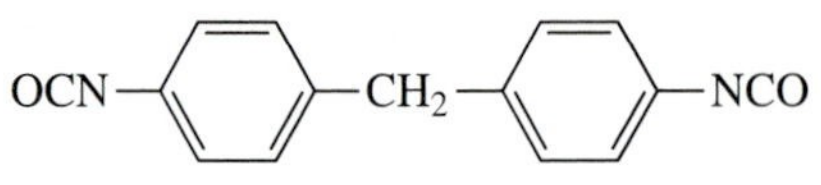

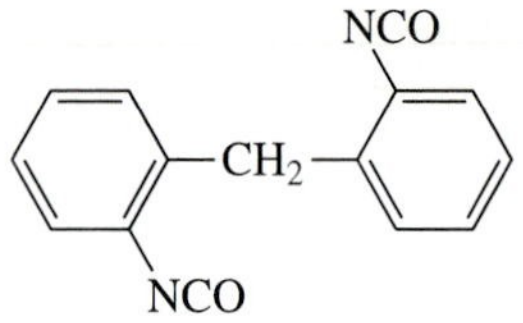

36   **37**

The soft and hard blocks can phase-separate to strengthen an elastomeric material. The incorporation of *short-chain* diols ("chain extenders") increases the ratio of hard urethane units to polymer chains, and increases the tear-strength and stiffness.

Urethane coupling reactions are catalyzed by molecules such as dimethyltin dilaurate, $Me_2Sn[OC(O)(CH_2)_{10}CH_3]_2$, in which the tin atoms coordinate to the terminal oxygen of the isocyanate group and facilitate attack by the oxygen atom of a hydroxyl group on the isocyanate carbon atom. Other catalysts include triethylenediamine ("Dabco") or other Lewis bases. These also catalyze the reactions of isocyanate groups with water to give carbon dioxide which forms the bubbles in foam rubber.

Closely related to polyurethanes are *polyureas*. These are formed by the reaction of a di-isocyanate with an amine-terminated polymer, as shown in equation (36).

$$H_2N-polymer-NH_2 \ + \ O{=}C{=}N-R-N{=}C{=}O \ \longrightarrow$$

$$O{=}C{=}N-R-NH-C(O)-NH-polymer \tag{36}$$

$$-NH-C(O)-NH-R-N{=}C{=}O$$

For example, the precursor polymer can be an amine-terminated poly(ethylene oxide). The urea linkage can give better elastomer properties than polyurethanes. Amino groups react more vigorously with isocyanates and hence give high-molecular-weight polymers faster. Diamines such as **38** serve to stiffen a polyurea.

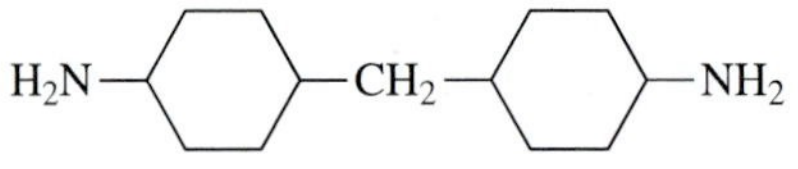

38

Polymers from Electrophilic Aromatic Substitutions

Experimental aromatic cyclolinear polymers have been prepared by Friedel–Crafts-type substitution reactions and related processes. For example, poly(arylene-alkylenes) with molecular weights as high as 12,000 are obtained from the reaction of benzene with 1,2-dichloroethane in the presence of aluminum chloride. Insoluble poly(*p*-phenylenes) (**39**) can be isolated from the reaction of benzene with combined Lewis acid and oxidizing agent systems.

39

CONCLUSIONS

Several step-type systems have been excluded from this chapter because they involve compounds of the inorganic elements. These are more conveniently treated in Chapter 9. However, it should be clear from the multiplicity of reactions mentioned in this chapter that the synthesis of new polymers by condensation and other step-type reactions has been and probably will continue to be a very fertile field for pioneering synthetic research. The current trend is obvious—that future work in this field will almost certainly emphasize the synthesis of new polymers that are stable at high temperatures, that resist burning, or which have unusual electrical properties.

Because of the great variety of (mostly organic) step reactions that have been discovered, this subject is still dominated by descriptive reaction chemistry. Only the best known systems have been analyzed mechanistically and kinetically by detailed physical chemical techniques. This mechanistic aspect is considered in Chapter 11. By contrast, vinyl-addition-type polymerizations involve the synthesis of many different polymers by the use of only one type of reaction—the linear addition of olefins. Hence, greater opportunities have existed in the olefin polymerization field for the analysis of detailed reaction mechanisms and for the application of physical techniques. This will be clear from the treatment of olefin polymerization in the following chapters.

STUDY QUESTIONS

1. Draw diagrams to compare the changes in molecular-weight distribution that occur with time for **(a)** a step-type polymerization, and **(b)** a chain-reaction polymerization.
2. For what reasons might you wish to deliberately limit the molecular weight of a condensation polymerization product by the addition of an excess of, say, one of the reactant species?
3. Draw up a list of different dicarboxylic acids and diols that might be used in polyester formation and attempt to predict advantages or potential problems for each pairwise combination. What property differences do you predict for the final polymers?
4. Give possible reasons why Bisphenol A is a widely used monomer in step-type polymerization syntheses.
5. If branching and crosslinking in a polyester can be induced by the addition of glycerol, what specific reagents might be employed to achieve a similar result in polyamide or polyether synthesis?

6. Suggest a mechanism for the oxidative coupling of phenols to yield polyaromatic ethers. Would phenols that contain more than two methyl groups attached to the phenyl ring be expected to react faster or slower than those with two methyl groups or less?

7. Draw up a list of molecular structural requirements that would be needed for an organic polymer that could withstand temperatures above 350°C.

8. Compile a list of amino compounds other than melamine or urea that might be used for methylolation reactions and subsequent conversion to resins. Speculate on the advantages and disadvantages of each compound for this particular use.

9. Glance through an organic chemistry textbook [e.g., J. D. Roberts and M. C. Caserio, *Basic Principles of Organic Chemistry* (New York: Benjamin, 1977)] and identify any general reactions that are accessible for small-molecule compounds but are not mentioned here as step-type polymerization routes. Suggest possible advantages or problems that might be found if these reactions could be used as polymerization processes.

SUGGESTIONS FOR FURTHER READING

ALBERTSSON, A-C., and LUNDMARK, S., "Synthesis of Poly(adipic anhydride) by Use of Ketene," *Macromol. Sci., Chem.*, **1988**, *A25*, 247.

BEEVER, W. H., and STILLE, J. K., "Synthesis and Thermal Properties of Aromatic Polymers Containing 3,6-Quinoline Units in the Main Chain," *Macromolecules*, **1979**, *12*, 1033.

HERGENROTHER, P. M., "Heat-Resistant Polymers," in *Encyclopedia of Polymer Science and Engineering*, Vol. 7 (H. F. Mark, N. Bikales, C. G. Overberger, G. Menges, and J. I. Kroschwitz, eds.). New York: Wiley, **1987**, p. 639.

HERGENROTHER, P. M., WAKELYN, N. T., and HAVENS, S. J., "Polyimides containing Carbonyl and Ether Connecting Groups," *J. Polymer Sci. (A)*, **1987**, *25*, 1093.

MARVEL, C. S., "Trends in High Temperature Polymer Synthesis," *Macromol. Chem.*, **1975,** *C13* (2), 219.

MORGAN, P. W., "Aromatic Polyamides," *ChemTech.*, **1979**, *9*, 316.

NOREN, G. K., and STILLE, J. K., "Polyphenylenes," *J. Polymer Sci. (D), Macromol. Rev.*, **1971**, *5*, 385.

ODIAN, G., *Principles of Polymerization* (3rd ed.), New York: Wiley-Interscience, **1991**.

OVERBERGER, C. G., and MOORE, J. A., "Ladder Polymers," *Advan. Polymer Sci.*, **1970**, *7*, 113.

STILLE, J. K., "New Developments in High Temperature and High Strength Polyquinolines," *Contemp. Top. Polymer Sci.*, **1984**, *5*, 209.

STILLE, J. K., "Polyquinolines," *Macromolecules*, **1981**, *14*, 870.

TSURUYA, S., NAKAGAWA, K., and MASAI, M., "Catalysis of Copper(II) Chelate-Amine Complexes in the Oxidative Coupling of 2,6-Dialkylphenols." *J. Polymer Sci.*, **1987**, *25*, 995.

3

Free-Radical Polymerization

ADDITION REACTIONS

A great many synthetic polymers are prepared by the polyaddition reactions of unsaturated organic compounds. In general terms, such polymerizations are described by the scheme shown in reaction (1). Such reactions can be induced either by the addition of free-radical-forming reagents or by ionic initiators such as acids or organometallic species. Ionic polymerizations are discussed in Chapter 4. Here we will concentrate on the use of free-radical polymerization processes.

$$
n \; \underset{\overset{|}{H}}{\overset{\overset{|}{H}}{C}} = \underset{\overset{|}{H}}{\overset{\overset{|}{R}}{C}} \; \longrightarrow \; \left[\underset{\overset{|}{H}}{\overset{\overset{|}{H}}{C}} - \underset{\overset{|}{H}}{\overset{\overset{|}{R}}{C}} \right]_n \tag{1}
$$

FREE-RADICAL ADDITION REACTIONS

In a free-radical addition polymerization, the growing chain end bears an unpaired electron **(1)**. Addition of each monomer molecule to the chain end involves an attack by the radical site on the unsaturated monomer. Thus, the unpaired electron is transferred to the new chain end at each addition step.

1

Such polymerizations are very common. In fact, many olefins will undergo "spontaneous" polymerization reactions during storage. In some cases, a major problem is to *prevent* the polymerization of an olefin. This is usually accomplished by the addition of an "inhibitor" to the system. The inhibitor stabilizes the olefin until such time as the laboratory experiment or the manufacturing process can be effected. Free-radical polymerizations can be carried out in the bulk liquid phase (without a solvent), or in solution.

Free-radical polymerization reactions are of enormous importance in technology. The monomers for these reactions are available in large quantities from the petrochemicals industry (e.g., from reaction sequences that start from ethylene, acetylene, or acetone), and the polymers obtained from these monomers form the foundation of much of the polymer industry. Low-density polyethylene, poly(methyl methacrylate), polystyrene, polyacrylonitrile, poly(vinyl chloride), and many other commercially important polymers are manufactured by free-radical processes.

Free-radical polymerizations are, on the whole, much better understood in a fundamental, mechanistic sense than are most of the step-type polymerizations discussed in Chapter 2. This is, in large measure, a consequence of the fact that different free-radical polymerization reactions constitute minor variants of one specific reaction type. Hence, subtle comparisons can be made of the effects of side-group changes, initiator variations, solvent effects, and so on. For this reason, a more detailed mechanistic approach will be taken in this chapter and in Chapter 12.

INITIATORS FOR FREE-RADICAL POLYMERIZATION

Table 3.1 summarizes the types of reagents that can induce the free-radical polymerization of vinyl compounds. The first four categories in Table 3.1 include compounds that dissociate into free radicals when heated or subjected to radiation.

TABLE 3.1   FREE-RADICAL INITIATORS

Initiator Type	Formula or Example
	$$\begin{matrix} O & & O \\ \| & & \| \end{matrix}$$
1. Organic peroxides or hydroperoxides	Benzoyl peroxide, $PhCOOCPh$
2. Redox agents	Persulfates + reducing agents, hydroperoxides + ferrous ion
3. Azo compounds	Azobisisobutyronitrile, $Me_2C(CN)N=NC(CN)Me_2$
4. Organometallic reagents	Silver alkyls
5. Heat, light, ultraviolet-, or high-energy radiation	
6. Electrolytic electron transfer	

Categories 5 and 6 include those *physical* influences that generate free radicals from the monomer itself or the solvent. In general, these initiator systems can be handled without the rigorous removal of atmospheric moisture. (This contrasts with most ionic initiators—see Chapter 4.) However, atmospheric oxygen must be excluded. Some care must be exercised in the use of organic peroxides because a number of these reagents detonate when subjected to shock or high temperatures. Benzoyl peroxide is commonly used because it is among the least shock sensitive of these reagents. The role of the initiator is discussed in more detail in a later section.

MONOMERS FOR FREE-RADICAL POLYMERIZATION

A wide variety of unsaturated organic compounds can be induced to undergo free-radical polymerization. Some of these are listed in Table 3.2. In general, the monomer structure can be represented by the formula $CH_2=CHR$, where the group R is an organic unit, a halogen or pseudo-halogen ligand (such as $C\equiv N$), or even an inorganic residue. A few suitable monomers have the formula $CH_2=C(R)(R')$, in which two substituent groups are attached to the α-carbon atom. Many of these monomers contain electron-withdrawing substituent groups, although the electron-directing effect of the substituent has a less critical influence on a free-radical polymerization than it does on ionic-type polymerizations.

TABLE 3.2 MONOMERS FOR FREE-RADICAL POLYMERIZATION

Styrene	$CH_2=CHPh$	Ethylene	$CH_2=CH_2$
α-Methylstyrene	$CH_2=C(Me)Ph$	Vinyl chloride	$CH_2=CHCl$
1,3-Butadiene	$CH_2=CH-CH=CH_2$	Vinylidene chloride	$CH_2=CCl_2$
Methyl methacrylate	$CH_2=C(Me)COOMe$	Tetrafluoroethylene	$CF_2=CF_2$
Vinyl esters	$CH_2=CHOOCR$	Acrylonitrile	$CH_2=CHC\equiv N$
N-Vinyl pyrrolidone	(structure)	Acrylamide	$CH_2=CHCONH_2$

SOLVENTS AND SYSTEMS

Free-radical polymerizations are often carried out in the bulk phase. In such cases, the monomer itself functions as a solvent in the initial stages of polymerization. This method is useful for the direct polymerization to polymer castings. However, if the polymerization is very exothermic, the reaction may become violent, or bubbles or char may result from the local release of heat.

Polymerizations in solution serve to minimize these problems. However, many organic solvents function as chain transfer agents (see page 69) for free-radical reactions and the polymer molecular weights may be lowered accordingly. Water is a satisfactory solvent for the polymerization of monomers such as acrylonitrile or acrylamide that are soluble or partly soluble in this medium, especially when water-soluble persulfate initiators are used. For example, poly(tetrafluoroethylene) (Teflon) is prepared on a large scale by the polymerization of tetrafluoroethylene in water, with the use of persulfate ions as initiating species.

Water can be used as a reaction medium even if the monomer is insoluble, since the monomer can be suspended as fine droplets in a stirred aqueous-organic heterophase medium. Alternatively, a detergent may be added to the system to generate an emulsion of the monomer in water. A watersoluble initiator is used and this penetrates the emulsion particles to initiate polymerization.

TYPICAL EXPERIMENTAL PROCEDURES

Thermal Polymerization of Styrene (Reaction 2)

Commercial styrene (2) must first be freed from trace impurities and inhibitors. This can be accomplished by vacuum distillation or by passage of the liquid monomer through an alumina chromatography column. Once purified, the monomer is sealed under nitrogen in a glass polymerization tube and heated at 125°C for 1 to 7 days. The tube contents become progressively more viscous as polymerization continues. The product (3) can be purified from monomer and oligomers by dissolving it in benzene, followed by precipitation of the polymer in methanol. The product made in this way often has a molecular weight near 150,000.

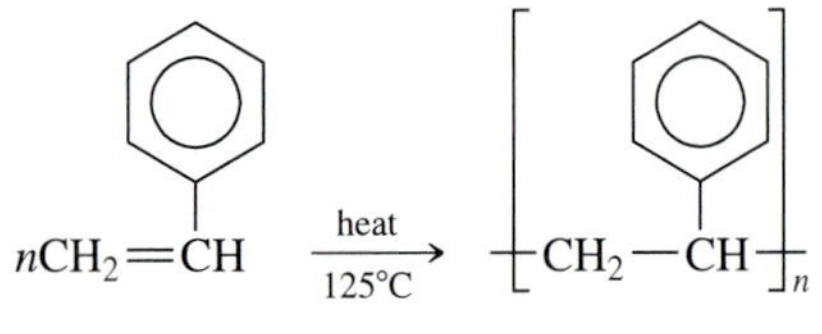

(2)

$$n\mathrm{CH_2}{=}\mathrm{CH} \xrightarrow[125°C]{\text{heat}} {\left[\mathrm{CH_2}{-}\mathrm{CH}\right]}_n$$

2   3

Bulk Polymerization of Methyl Methacrylate Initiated by AIBN (Reaction 3)

Freshly distilled methyl methacrylate monomer (4) is treated with a small amount of azobisisobutyronitrile (in an approximately 200:1 monomer/initiator weight ratio) and a trace of methacrylic acid. The mixture is then allowed to polymerize at about 40°C during one day to give a transparent, glassy polymer (5). The polymerization can be effected in a mold (or the bottom of a glass beaker), in which case poly(methyl methacrylate) may be added to the initial reaction mixture to raise the viscosity. Sheets of poly(methyl methacrylate) glass can be prepared by the use of a mold of two glass plates clamped together and separated by a rubber gasket (see Chapter 20).

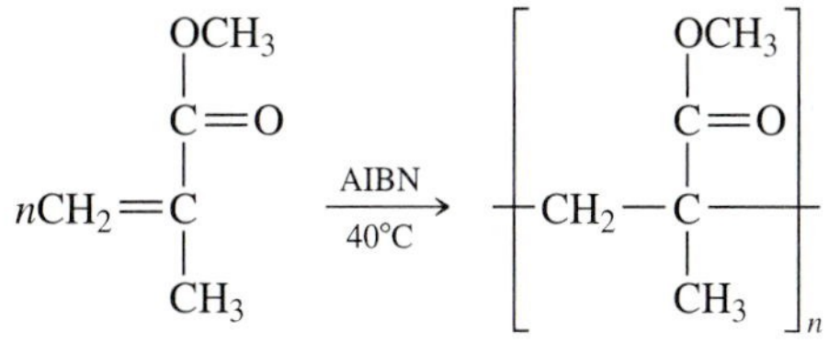

$$n\text{CH}_2\!=\!\underset{\underset{\text{CH}_3}{|}}{\overset{\overset{\text{OCH}_3}{|}}{\overset{|}{\underset{|}{\text{C}\!=\!\text{O}}}}}\overset{}{\underset{\mathbf{4}}{}}\xrightarrow[\text{40°C}]{\text{AIBN}}\left[\!-\text{CH}_2\!-\!\underset{\underset{\text{CH}_3}{|}}{\overset{\overset{\text{OCH}_3}{|}}{\overset{|}{\underset{|}{\text{C}\!-\!}}}}\overset{\overset{\text{C}\!=\!\text{O}}{|}}{}\!-\!\right]_{n}\underset{\mathbf{5}}{} \tag{3}$$

Emulsion Polymerization of Acrylonitrile (Reaction 4)

Acrylonitrile **(6)** must first be freed of inhibitor by passage through a column of silica gel. Polymerization takes place in a stirred mixture of acrylonitrile and water to which has been added a detergent, potassium persulfate, and a trace of sodium bisulfite. Oxygen is excluded by a nitrogen atmosphere. A white emulsion of poly-acrylonitrile **(7)** in water is formed during 2 to 3 h at about 35°C, and the polymer can be isolated by coagulation and filtration.

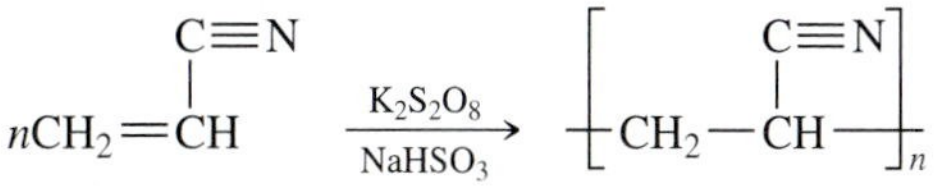

$$n\text{CH}_2\!=\!\underset{\underset{\mathbf{6}}{}}{\overset{\overset{\text{C}\equiv\text{N}}{|}}{\text{CH}}}\xrightarrow[\text{NaHSO}_3]{\text{K}_2\text{S}_2\text{O}_8}\left[\!-\text{CH}_2\!-\!\underset{\underset{\mathbf{7}}{}}{\overset{\overset{\text{C}\equiv\text{N}}{|}}{\text{CH}}}\!-\!\right]_{n} \tag{4}$$

CHAIN REACTIONS

Free-radical polymerizations are chain reactions. The addition of a monomer molecule to an active chain end regenerates the active site at the chain end. Hence, a large number of monomer molecules are "consumed" for each active site introduced into the system. In chain-reaction polymerizations we may recognize four distinct types of processes. They are:

1. *Chain initiation*—a process in which highly reactive transient molecules or active centers are formed.
2. *Chain propagation*—the addition of monomer molecules to the active chain end, accompanied by regeneration of the terminal active site.
3. *Chain transfer*—involving the transfer of the active site to another molecule (e.g., monomer). The molecule that has lost the active site is now "dead" from a chain-propagation point of view. The molecule that has accepted the active site can start a new chain.
4. *Chain termination*—a reaction in which the active chain centers are destroyed. Chain reactions are found in free-radical, anionic, and cationic vinyl-type polymerizations. In free-radical processes all of the four steps listed above can usually be identified.

Schematically, a free-radical polymerization sequence can be represented by the reactions shown in (5) to (11). In these equations M represents a molecule of monomer; $R' \cdot$ is an initiating free radical from the initiator; $R_n \cdot$ is the propagating free radical with a degree of polymerization, n; YZ is a chain transfer agent that may be solvent, monomer, initiator, or polymer molecules; and P_n is the final inactive polymer. It should be noted that in some cases the initiator molecule may be the monomer itself

$$\text{Initiator} \longrightarrow 2R' \cdot \qquad \Big\} \quad \text{Initiation} \qquad (5)$$
$$R' \cdot + M \longrightarrow R_i \cdot \qquad \qquad \qquad \qquad (6)$$

$$R_i \cdot + M \longrightarrow R_2 \cdot \qquad \Big\} \quad \text{Propagation} \qquad (7)$$
$$R_n \cdot + M \longrightarrow R_{n+1} \cdot \qquad \qquad \qquad (8)$$

$$R_n \cdot + YZ \longrightarrow R_n Y + Z \cdot \quad \text{Chain transfer} \qquad (9)$$

$$R_n \cdot + R_m \cdot \longrightarrow P_{n+m} \qquad \Big\} \quad \text{Termination} \qquad (10)$$
$$R_n \cdot + R_m \cdot \longrightarrow P_n + P_m \qquad \qquad \qquad (11)$$

In practical terms, this general chain-reaction sequence can be identified in the polymerization of vinyl chloride. This reaction can be initiated by di-t-butyl peroxide. The sequence for this system is shown in reactions (12) to (17). The chain-transfer steps are not included in this sequence.

$$\text{Me}_3\text{COOCMe}_3 \xrightarrow{\text{heat}} 2\text{Me}_3\text{CO} \cdot \quad \text{Initiator dissociation} \qquad (12)$$

$$\text{Me}_3\text{CO} \cdot + \text{CH}_2{=}\text{CHCl} \longrightarrow \text{Me}_3\text{COCH}_2 - \overset{\displaystyle H}{\underset{\displaystyle Cl}{\overset{|}{\underset{|}{C}}}} \cdot \quad \text{Initiation} \qquad (13)$$

$$\text{Me}_3\text{COCH}_2 - \overset{\displaystyle H}{\underset{\displaystyle Cl}{\overset{|}{\underset{|}{C}}}} \cdot + \text{CH}_2{=}\text{CHCl} \longrightarrow$$

$$(14)$$

$$\text{Me}_3\text{CO} - \text{CH}_2 - \overset{\displaystyle H}{\underset{\displaystyle Cl}{\overset{|}{\underset{|}{C}}}} - \text{CH}_2 - \overset{\displaystyle H}{\underset{\displaystyle Cl}{\overset{|}{\underset{|}{C}}}} \cdot \quad \text{Initial propagation}$$

Free-Radical Polymerization   Chapter 3

$$Me_3CO-\left[CH_2-\underset{\underset{Cl}{|}}{\overset{\overset{H}{|}}{C}}\right]_n CH_2-\underset{\underset{Cl}{|}}{\overset{\overset{H}{|}}{C}}\cdot \ + \ CH_2{=}CHCl \ \longrightarrow$$

(15)

$$Me_3CO-\left[CH_2-\underset{\underset{Cl}{|}}{\overset{\overset{H}{|}}{C}}\right]_{n+1} CH_2-\underset{\underset{Cl}{|}}{\overset{\overset{H}{|}}{C}}\cdot \quad \text{Propagation}$$

$$Me_3CO-\left[CH_2-\underset{\underset{Cl}{|}}{\overset{\overset{H}{|}}{C}}\right]_n CH_2-\underset{\underset{Cl}{|}}{\overset{\overset{H}{|}}{C}}\cdot \ + \ \cdot\underset{\underset{Cl}{|}}{\overset{\overset{H}{|}}{C}}-CH_2-\left[\underset{\underset{Cl}{|}}{\overset{\overset{H}{|}}{C}}-CH_2\right]_m OCMe_3 \ \longrightarrow$$

(16)

$$Me_3CO-\left[CH_2-\underset{\underset{Cl}{|}}{\overset{\overset{H}{|}}{C}}\right]_n CH_2-\underset{\underset{Cl}{|}}{\overset{\overset{H}{|}}{C}}-\underset{\underset{Cl}{|}}{\overset{\overset{H}{|}}{C}}-CH_2-\left[\underset{\underset{Cl}{|}}{\overset{\overset{H}{|}}{C}}-CH_2\right]_m OCMe_3 \quad \text{Termination}$$

$$Me_3CO-\left[CH_2-\underset{\underset{Cl}{|}}{\overset{\overset{H}{|}}{C}}\right]_n CH_2-\underset{\underset{Cl}{|}}{\overset{\overset{H}{|}}{C}}\cdot \ + \ \cdot\underset{\underset{Cl}{|}}{\overset{\overset{H}{|}}{C}}-CH_2-\left[\underset{\underset{Cl}{|}}{\overset{\overset{H}{|}}{C}}-CH_2\right]_m OCMe_3 \ \longrightarrow$$

(17)

$$Me_3CO-\left[CH_2-\underset{\underset{Cl}{|}}{\overset{\overset{H}{|}}{C}}\right]_n CH_2-\underset{\underset{Cl}{|}}{\overset{\overset{H}{|}}{C}}-H \ + \ \underset{\underset{Cl}{|}}{\overset{\overset{H}{|}}{C}}{=}CH-\left[\underset{\underset{Cl}{|}}{\overset{\overset{H}{|}}{C}}-CH_2\right]_m OCMe_3 \quad \text{Termination}$$

FREE-RADICAL INITIATORS

Initiation of a free-radical polymerization requires the production of free radicals in the presence of the unsaturated monomer. The initiating free radicals can be produced directly from the monomer (by irradiation with high-energy radiation, for example), but it is more normal for the radicals to be generated from an added initiator. Such an initiator is usually a molecule that can be decomposed thermally or by irradiation to yield a pair of initiating radicals. The thermal decomposition of di-*t*-butyl peroxide mentioned in the preceding section is an example of this type of reaction. Alternatively, some initiators function by means of redox reactions, in which case one initiator molecule may yield only one initiating radical. The following sections outline the alternative mechanisms that are available for initiation.

Thermal Decomposition of Initiators

Numerous substances decompose to free radicals when heated. If the decomposition temperature corresponds to a convenient temperature range for polymerization, the substance may be useful as an initiator. In fact, it is the dependence of the initiator decomposition rate on the temperature which determines the usefulness of the compound as an initiator. Such thermal decompositions usually yield two free radicals from one initiator molecule by a first-order reaction process.

As will be shown in detail in Chapter 12, the rate of polymerization is proportional to the square root of the rate of initiation. Hence, an increase in the initiator concentration or an elevation of the temperature has a dramatic accelerating effect on the polymerization rate. On the other hand, the average chain length is lowered by an increase in the rate of initiation. Thus, the choice of an initiator and the selection of a temperature for initiation requires a compromise to be made between the need to obtain a reasonably fast polymerization and the need to make high-molecular-weight polymer.

An arbitrary, but useful, rule that can be used to estimate the practical temperature range for a thermal initiator is as follows: *the rate of formation of initiating radicals should be in the range of* 10^{-7} *to* 10^{-6} *mol-liter*$^{-1}$ *s*$^{-1}$ *at an initiator concentration of* 0.1 *M*. In other words, its useful operating range is that temperature range in which the first-order decomposition rate constant is in the region of 10^{-6} to 10^{-5} s^{-1}. Because most free-radical polymerizations must be carried out at temperatures below 150°C to prevent side reactions, the list of useful thermal initiators is largely restricted to organic peroxides, organic hydroperoxides, azo compounds, and metal alkyls. Each of these classes of initiator will now be considered in turn.

Dialkyl peroxides, ROOR, decompose thermally by cleavage of the oxygen-oxygen bond to yield two alkoxy radicals, RO·, as shown by (18). An example of such a decomposition was given earlier for the specific case where R is Me$_3$C.

$$\text{ROOR} \longrightarrow \text{RO·} + \text{RO·} \qquad (18)$$

As shown in the earlier example, the alkoxy radicals may then initiate the polymerization by reaction with the monomer. However, the alkoxy radicals may decompose to alkyl radicals and aldehydes or ketones before initiation can occur. For example, in the case of di-*t*-butyl peroxide, this additional decomposition yields methyl radicals and acetone, as shown in (19). The methyl radical then completes the initiation process by reaction with the monomer. The first-order rate constant for the decomposition of di-t-butyl peroxide is[1] $6.3 \times 10^{15}e^{-37,500\,cal/RT}$ s^{-1}. Application of the rule described above indicates that the predicted useful initiator temperature range for di-t-butyl peroxide is 100 to 120°C.

$$Me_3CO\cdot \longrightarrow (Me)_2C{=}O + CH_3\cdot \qquad (19)$$

Diacylperoxides, RC(O)OO(O)CR, decompose similarly by an initial cleavage of the oxygen–oxygen bond. In general, these compounds provide a somewhat lower useful temperature range for initiation than do the dialkyl peroxides. For example, benzoyl peroxide undergoes the initial bond scission shown in (20) with a first-order rate constant[2] of $1 \times 10^{14}\,e^{-29,900/RT}$ s^{-1}. This rate constant indicates that benzoyl peroxide is a useful thermal initiator in the temperature range of 60 to 80°C. Despite the relatively low temperatures at which this initiator is useful, the benzoyl radicals formed by the initial bond rupture (20) may decompose by the reaction shown in (21) before they can react with the monomer. However, the occurrence of reaction (21) has very little effect on the overall rate of initiation, because the phenyl radicals formed in (21) can themselves add to the monomer to bring about initiation.

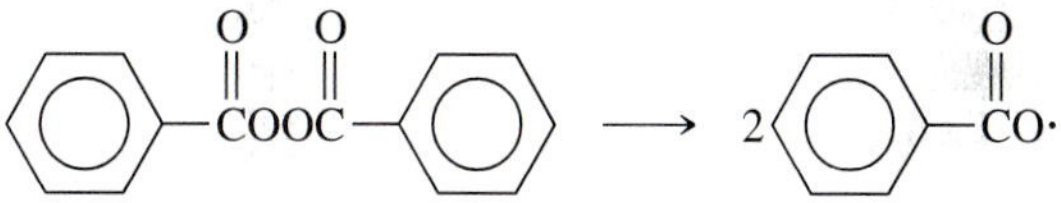

$$(20)$$

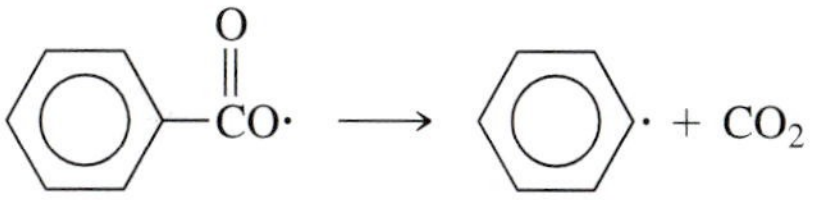

$$(21)$$

Organic hydroperoxides, ROOH, have also been used as thermal initiators. Here also the initial decomposition may involve rupture of the oxygen–oxygen bond.

However, an alternative hydroperoxide decomposition process also exists in which two molecules of hydroperoxide react as shown in equation (22). This is a lower-energy pathway.

$$2\,ROOH \longrightarrow RO\cdot + ROO\cdot + H_2O \qquad (22)$$

Hydroperoxides are susceptible to attack by HO· and RO· (or R·) radicals at the temperatures normally used for initiation. The free-radical products of these radical reactions, as shown in (24) to (26), are peroxy radicals. Often, these

[1]Richardson, W. H., and O'Neal, H. E., in *Comprehensive Chemical Kinetics*, Vol. 5. C. H. Bamford and C. F. H. Tipper, eds. (Amsterdam: Elsevier, **1972**).
[2]Ibid.

are not sufficiently reactive to add to the monomer. Thus, they do not initiate polymerization.

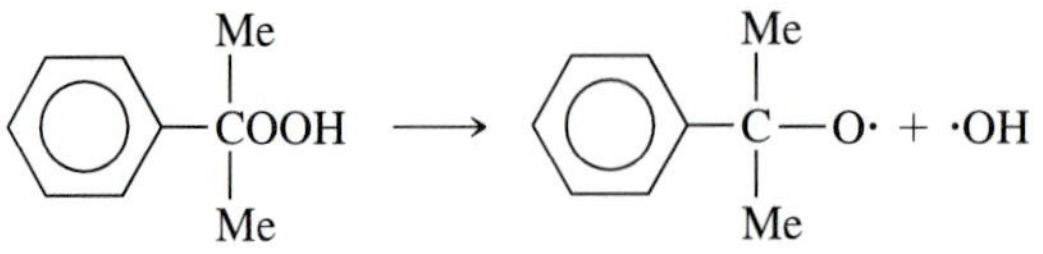

$$\text{Ph}-\underset{\underset{\text{Me}}{|}}{\overset{\overset{\text{Me}}{|}}{C}}-\text{COOH} \longrightarrow \text{Ph}-\underset{\underset{\text{Me}}{|}}{\overset{\overset{\text{Me}}{|}}{C}}-\text{O}\cdot + \cdot\text{OH} \tag{23}$$

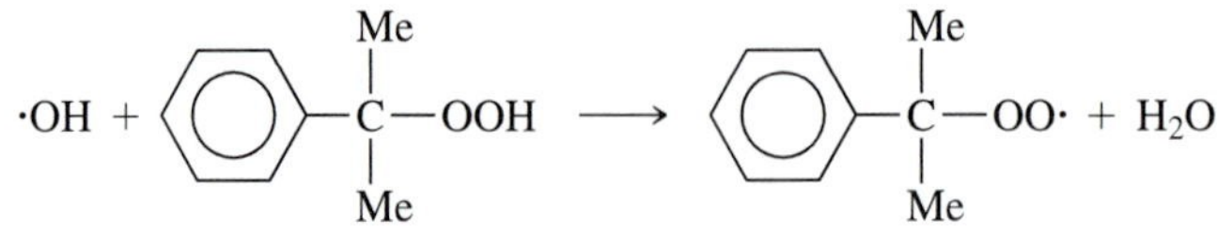

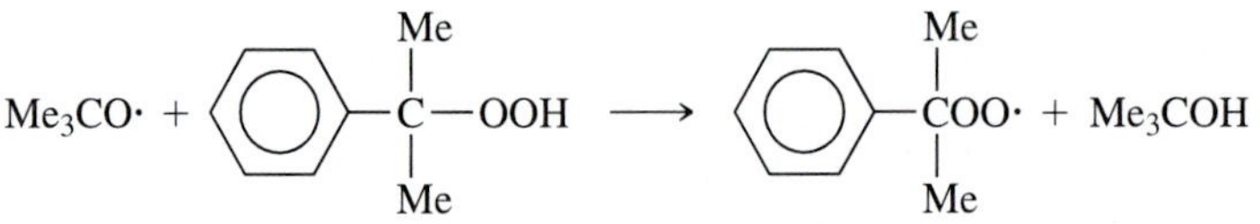

$$\cdot\text{OH} + \text{Ph}-\underset{\underset{\text{Me}}{|}}{\overset{\overset{\text{Me}}{|}}{C}}-\text{OOH} \longrightarrow \text{Ph}-\underset{\underset{\text{Me}}{|}}{\overset{\overset{\text{Me}}{|}}{C}}-\text{OO}\cdot + \text{H}_2\text{O} \tag{24}$$

$$\text{Me}_3\text{CO}\cdot + \text{Ph}-\underset{\underset{\text{Me}}{|}}{\overset{\overset{\text{Me}}{|}}{C}}-\text{OOH} \longrightarrow \text{Ph}-\underset{\underset{\text{Me}}{|}}{\overset{\overset{\text{Me}}{|}}{C}}\text{OO}\cdot + \text{Me}_3\text{COH} \tag{25}$$

$$\text{CH}_3\cdot + \text{Ph}-\underset{\underset{\text{Me}}{|}}{\overset{\overset{\text{Me}}{|}}{C}}-\text{OOH} \longrightarrow \text{Ph}-\underset{\underset{\text{Me}}{|}}{\overset{\overset{\text{Me}}{|}}{C}}\text{OO}\cdot + \text{CH}_4 \tag{26}$$

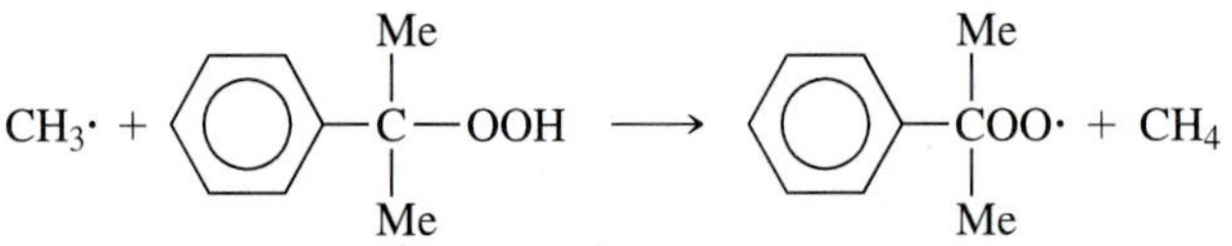

Azo compounds, $RN=NR$, such as azobisisobutyronitrile, decompose thermally to give nitrogen and two alkyl radicals, as shown in reaction (27). Azo compounds are useful and popular thermal initiators because they offer a wide range of useful operating temperatures. Moreover, because nitrogen gas is formed during the decomposition, measurement of the amount of nitrogen produced gives the experimenter a measure of the number of initiating radicals formed. This, in turn, allows an estimate to be made of the efficiency of initiation versus radical recombination (within the solvent cage).

$$\underset{\underset{\text{(CH}_3)_2\text{CN}}{}}{\overset{\overset{\text{CN}}{|}}{}}=\underset{}{\overset{\overset{\text{CN}}{|}}{\text{NC(CH}_3)_2}} \longrightarrow 2\underset{}{\overset{\overset{\text{CN}}{|}}{(\text{CH}_3)_2\text{C}}}\cdot + \text{N}_2 \tag{27}$$

Table 3.3 lists activation energies and useful operating temperature ranges (ΔT) (as defined previously) for a number of azo compounds. Azomethane and azoisopropane are much too stable to be important as thermal initiators at normal polymerization temperatures. The effective temperature range shown for azobisisobutyronitrile (AIBN) illustrates why it is one of the most popular thermal initiators.

TABLE 3.3 ACTIVATION ENERGIES AND USEFUL TEMPERATURE RANGE FOR INITIATION BY R'N=NR' COMPOUNDS

R'	E_{act}(kcal/mol)	Useful Initiation Temperature Range (°C)*
CH_3	50.2	225–250
$(CH_3)_2CH$	40.8	180–200
$C_6H_5(CH_3CH)$	36.5	105–125
$(C_6H_5)_2CH$	26.6	20–35
$(CH_3)_2(CN)C\cdot$	30.8	40–60

Source: Richardson, W. H., and O'Neal, H. E., in *Comprehensive Chemical Kinetics*, Vol. 5, C. H. Bamford and C. F. H. Tipper, eds. (Amsterdam: Elsevier, 1972).

*Defined by the condition that k_1 is in the region 10^{-6} to 10^{-5} s^{-1}.

All the free-radical initiators discussed above are appropriate only for polymerizations carried out at room temperature or above. However, silver alkyls have been used to initiate radical polymerizations at temperatures as low as −20 to −60°C. Ethylsilver, for example, is believed to decompose by the reaction shown in (28). Because silver alkyls decompose below room temperature they must, of course, be prepared at low temperatures and used as soon as possible.

$$Ag-C_2H_5 \longrightarrow Ag + C_2H_5\cdot \qquad (28)$$

Initiation by Redox Reactions

In one of the example polymerizations given earlier, the water-soluble initiator used was a persulfate salt together with bisulfite ion. Initiation by this system falls into a category generally described as *redox reactions*. The bisulfite ion (HSO_3^-) reduces the persulfate ion ($S_2O_8^{2-}$) to yield sulfate (SO_4^{2-}) and the $SO_4^{\cdot-}$ radical ion. This latter species reacts with water to generate the bisulfate ion (HSO_4^-) and a hydroxyl radical ($\cdot OH$). Thiosulfate ion ($S_2O_3^{2-}$) can also be used as a reducing agent. Moreover, traces of ferric ion may also participate. The reactions shown in (29) to (31) summarize these processes. The radicals formed in these reactions then initiate polymerization.

$$S_2O_8^{2-} + HSO_3^- \longrightarrow SO_4^{2-} + SO_4^{\cdot-} + HSO_3\cdot \qquad (29)$$

$$S_2O_8^{2-} + S_2O_3^{2-} \longrightarrow SO_4^{2-} + SO_4^{\cdot-} + S_2O_3^{\cdot-} \qquad (30)$$

$$HSO_3^- + Fe^{3+} \longrightarrow HSO_3\cdot + Fe^{2+} \qquad (31)$$

Related redox reactions can be carried out in aqueous media with the use of alkyl hydroperoxides and a reducing agent, such as ferrous ion. Thus, cumyl hydroperoxide reacts with ferrous ion to yield cumyloxy radicals, as shown in reaction (32). The reaction is sufficiently fast that cumyloxy radicals are formed at temperatures as low as 15 to 50°C, whereas temperatures of 85 to 105°C are needed for the direct thermal cleavage of this peroxide. Moreover, side reactions and further cleavage of the cumyloxy radical are unlikely at this lower temperature. A further advantage of this system is that the peroxide cleavage process can be monitored by the conversion of ferrous to ferric ion.

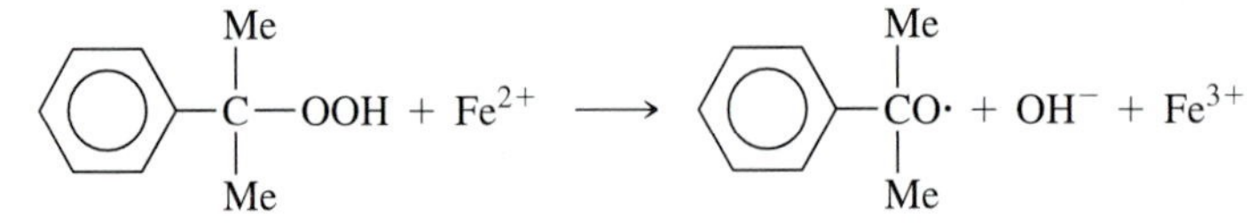

$$
\text{C}_6\text{H}_5\text{C(Me)(Me)}-\text{OOH} + \text{Fe}^{2+} \longrightarrow \text{C}_6\text{H}_5\text{C(Me)(Me)}-\text{CO}\cdot + \text{OH}^- + \text{Fe}^{3+} \qquad (32)
$$

Direct Thermal and Photolytic Initiation

Some monomers undergo free-radical polymerization simply when heated or when exposed to light. Styrene is a monomer that is particularly well-known for this type of behavior. The mechanism of initiation is still not completely understood.

It was originally thought that initiation involves the formation of a triplet state diradical by the collision of two monomer molecules. However, evidence produced later suggested that a monoradical is the chain propagating species. More recently, it has been proposed that the initial reaction is a Diels–Alder cycloaddition of two styrene molecules, followed by a bimolecular radical-forming process (equation 33).[1]

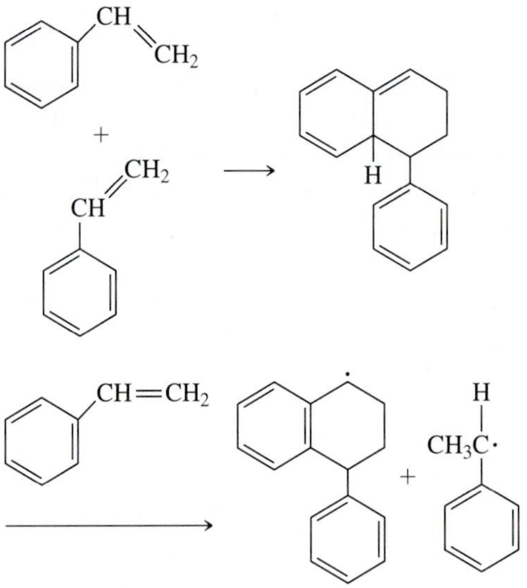

$$(33)$$

[1]Kauffmann, H. F., *Makromol. Chem.*, **1981**, *180*, 2649, 2665, 2681.

It is interesting that the mechanism of this, one of the oldest-known polymerizations[1] is so complex and still somewhat obscure.

The polymerization of many monomers can be induced by irradiation with light if a suitable "photosensitizer" is also present in the system. This method of initiation will be discussed in more detail in Chapter 5.

Initiation by Ionizing Radiation

Free-radical polymerizations can also be induced by irradiation of the monomer with high-energy radiation, such as X-rays, γ-rays, α-particles, high-energy electrons, protons, and so on. The monomer can be in the bulk phase or in solution. The absorption of energy by the monomer is far less selective when ionizing radiation is used than when light is employed. All the species present in the system (monomer and solvent) absorb energy and decompose to yield free radicals. The radicals then initiate polymerization.

However, the process is far more complicated than that found for the absorption of light. Ionizing radiation produces positive and negative ions as well as free radicals. The ions may react with the monomer or other ions to yield secondary ions or free radicals, or they may initiate ionic chain polymerization. Thus, the reaction mechanisms can be extremely complicated. This subject is discussed further in Chapter 5.

REACTIONS OF INITIATOR RADICALS WITH THE MONOMER

Clearly, the second important step in a radical polymerization is the addition of the initiating radical to a monomer molecule. Only a fraction of the initiating radicals formed behave in this way, because it is generally found that the rate of chain initiation is lower than the rate at which the initiating radicals are formed.

This wastage of initiating radicals occurs because of competition from alternative fast reactions. First, it must be recognized that initiator radicals are usually formed in pairs by the cleavage of a peroxide or azo compound. At the instant of dissociation, the two radicals are imprisoned together in a "cage" of monomer or solvent molecules. The mean free path of the radicals is no more than the average diameter of a small molecule. Hence, the two initiating radicals will have excellent opportunities for collision and recombination before they are separated by diffusion. This effect is important even in a cage of monomer molecules because a reaction between the radical and a monomer generally requires many more collisions than does recombination of two free radicals. *Primary* or *geminate recombination* is the name applied to this process.

Second, those initiating radicals that do escape from their original partners in the solvent cage may combine with radicals from other solvent cages. This further lowers the efficiency of initiation. This process is known as *secondary re-*

[1]E. Simon, *Ann. Chem.*, **1839,** *31,* 265.

combination. Its occurrence may be reduced by decreasing the rate of radical formation or by increasing the monomer concentration.

Third, initiator radicals may be wasted by reactions with propagating polymer radicals in such a way as to terminate the chains; and fourth, the initiator radicals may react with the original initiator to generate radicals of low reactivity. Add to this the fact that radicals can abstract chlorine or other radicals from solvents or hydrogen from terminated polymer molecules, and it is clear why the efficiency of initiation is generally low (Table 3.4). The maximum initiation efficiency can usually be obtained by the use of low temperature (to reduce the rate of radical formation) and by the use of low concentrations of initiator relative to the monomer (to improve the probability of a successful initiation).

Quantitative studies[1] indicate that the efficiency of initiation (usually given the symbol *f*) lies most often between 0.1 and 0.8. It correlates roughly with the viscosity of the medium and is approximately independent of temperature.

TABLE 3.4 POSSIBLE FATES OF INITIATOR RADICALS

$R\cdot \rightarrow$

→ Radical recombination within the solvent cage: $R\cdot + \cdot R \rightarrow R-R$
→ Secondary recombination outside the cage: $R\cdot + \cdot R \rightarrow R-R$
→ Reaction with polymer radicals: $R\cdot + \cdot \text{—}\wedge\wedge\text{—} \rightarrow R\text{—}\wedge\wedge\text{—}$
→ Reaction with initiator: $R\cdot + R'-R' \rightarrow R-R' + \cdot R'$
→ Radical abstraction: $R\cdot + H-R' \rightarrow R-H + \cdot R'$
→ Reaction with solvent: $R\cdot + CCl_4 \rightarrow R-Cl + Cl_3C\cdot$
→ Chain initiation: $R\cdot + \text{monomer} \rightarrow R\text{-monomer}\cdot$

RADICAL CHAIN PROPAGATION

Chain propagation involves the addition of a free radical to the double bond of a monomer molecule. The product must itself be a free radical and the process can be repeated. In fact, it is common for thousands of monomer molecules to add successively to the end of the chain.

The most likely form of monomer addition is called *head-to-tail addition.* Reaction (34) illustrates this behavior. Alternatively, the addition may involve a *head-to-head* (reaction 35) or a *tail-to-tail* reaction (reaction 36). Although it might be supposed that species **9, 11**, and **13** might be distributed randomly throughout the molecular chain, it is, in fact, found that head-to-tail linkages (**9**) are in great excess. For example, in poly(vinyl alcohol), $-\!\!\left[\text{CH}_2\text{CH(OH)}\right]_{n}$, it has been found[2] that the percentage of **11** (and, therefore, also of **13**) is only 1.1% in polymer prepared at 25°C, and only 1.8% in polymer prepared at 100°C.

[1]Allen, P. E. M., and Patrick, C. R., *Kinetics and Mechanisms of Polymerization Reactions* (New York: Wiley, **1974**), p. 114.
[2]Flory, P. J., and Leutner, F. S., *J. Polymer Sci.*, **1948**, *3*, 880; **1950**, *5*, 267.

$$R\text{---}\!\wedge\!\text{---}CH_2\text{---}\overset{\displaystyle H}{\underset{\displaystyle X}{C}}\!\cdot\ +\ CH_2\!=\!\overset{\displaystyle H}{\underset{\displaystyle X}{C}}\ \longrightarrow\ R\text{---}\!\wedge\!\text{---}CH_2\text{---}\overset{\displaystyle H}{\underset{\displaystyle X}{C}}\text{---}CH_2\text{---}\overset{\displaystyle H}{\underset{\displaystyle X}{C}}\!\cdot \qquad (34)$$

8   **9**

$$R\text{---}\!\wedge\!\text{---}CH_2\text{---}\overset{\displaystyle H}{\underset{\displaystyle X}{C}}\!\cdot\ +\ \overset{\displaystyle H}{\underset{\displaystyle X}{C}}\!=\!CH_2\ \longrightarrow\ R\text{---}\!\wedge\!\text{---}CH_2\text{---}\overset{\displaystyle H}{\underset{\displaystyle X}{C}}\text{---}\overset{\displaystyle H}{\underset{\displaystyle X}{C}}\text{---}CH_2\!\cdot \qquad (35)$$

10   **11**

$$R\text{---}\!\wedge\!\text{---}\overset{\displaystyle H}{\underset{\displaystyle X}{C}}\text{---}CH_2\!\cdot\ +\ CH_2\!=\!\overset{\displaystyle H}{\underset{\displaystyle X}{C}}\ \longrightarrow\ R\text{---}\!\wedge\!\text{---}\overset{\displaystyle H}{\underset{\displaystyle X}{C}}\text{---}CH_2\text{---}CH_2\text{---}\overset{\displaystyle H}{\underset{\displaystyle X}{C}}\!\cdot \qquad (36)$$

12   **13**

The principal reason for the preference of head-to-tail addition lies in the greater thermodynamic stability of a free radical such as **8** relative to that of one such as **12**, and perhaps also to steric inhibition of steps such as reaction (35). It can, in fact, be shown that the addition of methyl radicals to propylene, $CH_2\!=\!CHMe$, is 3.8 kcal/mol more favorable if the methyl group adds to the CH_2 unit rather than to the CHMe component. Thus, attack on the CH_2 unit should be favored up to a temperature of about 225°C. A similar preference for attack on the CH_2 group exists when butyl radicals are used. However, although these thermodynamic effects would be all-important when equilibrium conditions prevail, it seems more likely that kinetic factors (such as steric effects and activation energies) may exert the strongest influence.

CHAIN TRANSFER REACTIONS

In an "ideal" free-radical polymerization, chains become initiated, they propagate linearly, and they are then terminated. Such ideal circumstances are often found in ionic polymerizations, but only rarely in free-radical processes. The deviation from ideality occurs when a propagating oligomer or polymer radical reacts with another molecule, not by addition, but by abstraction. By "abstraction" we mean a process in which a radical fragment is removed from the second molecule with concurrent generation of a radical residue from that second molecule. This is illustrated by reaction (37), in which YZ represents the second molecule, which could be monomer, solvent, initiator, polymer molecules, or other molecules deliberately or accidentally incorporated into the reaction mixture.

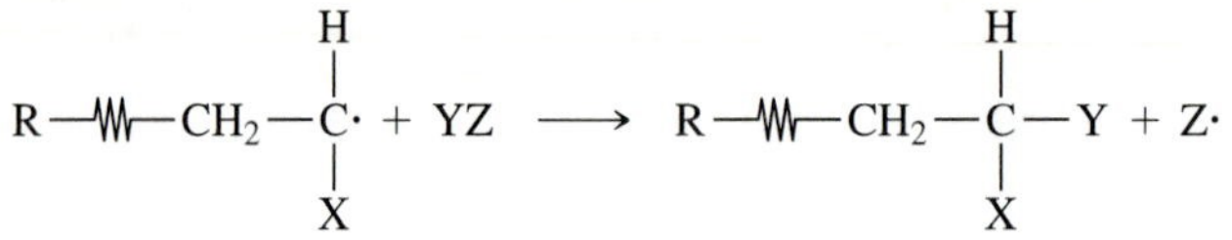

$$R\text{---}\mathcal{W}\text{---}CH_2\text{---}\overset{\overset{\displaystyle H}{|}}{\underset{\underset{\displaystyle X}{|}}{C}}\cdot + YZ \longrightarrow R\text{---}\mathcal{W}\text{---}CH_2\text{---}\overset{\overset{\displaystyle H}{|}}{\underset{\underset{\displaystyle X}{|}}{C}}\text{---}Y + Z\cdot \qquad (37)$$

14

It is important to note that although the polymer chain (**14**) is now effectively terminated, $Z\cdot$ may initiate a new chain if its reactivity is comparable to that of a normal propagating radical. Such a process is called *chain transfer*. The number of radicals growing at any instant is unchanged, but *the average chain length of the* polymer produced *will be reduced*.

When the second molecule is a polymer molecule, the ultimate result is generally the formation of a *branched* polymer. Thus, if hydrogen radical abstraction from a polymer leads to the formation of a residue such as **15**, the radical site on this residue can initiate growth of a branch, such as the one shown in **16**.

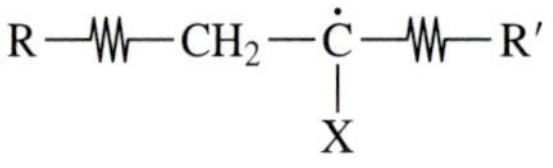

$$R\text{---}\mathcal{W}\text{---}CH_2\text{---}\overset{}{\underset{\underset{\displaystyle X}{|}}{\dot{C}}}\text{---}\mathcal{W}\text{---}R' \qquad\qquad R\text{---}\mathcal{W}\text{---}CH_2\text{---}\overset{}{\underset{\underset{\displaystyle X}{|}}{C}}\text{---}\mathcal{W}\text{---}R'$$

15 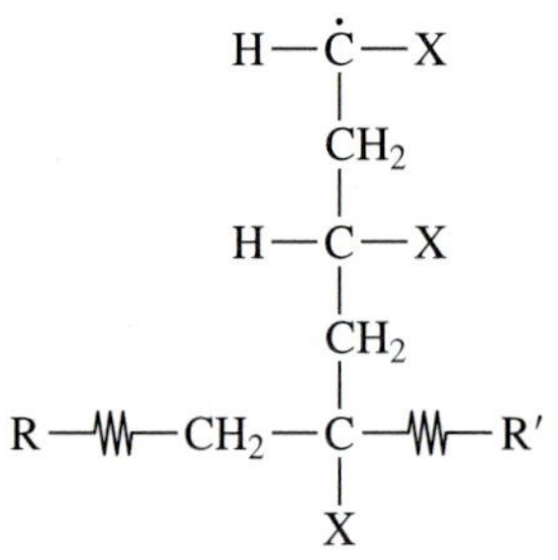   **16**

It should also be noted that if the second molecule, YZ, yields a radical fragment, $Z\cdot$, which is unreactive toward the monomer, the substance YZ is termed an *inhibitor* and the process is termed "inhibition." Such compounds are often added to monomers to inhibit polymerization during storage or transportation. If, on the other hand, $Z\cdot$ is reactive toward the monomer but less reactive than the normal propagating radicals, then YZ is called a *retarder*.

FREE-RADICAL CHAIN TERMINATION

Free-radical chains can be terminated by reaction of a growing polymer radical with some other free radical in the system. First, in the simplest sense, the polymer radical may react with initiator radicals (which will still be produced after the start of the reaction) as shown in (38). However, this is wastage of initiating radicals, and in practice, should be avoided by keeping the rate of initiation sufficiently low.

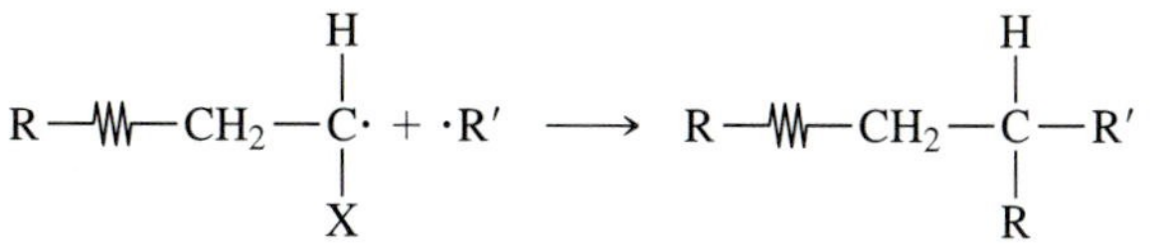

$$R-\text{W}-CH_2-\overset{\overset{\displaystyle H}{|}}{\underset{\underset{\displaystyle X}{|}}{C}}\cdot + \cdot R' \longrightarrow R-\text{W}-CH_2-\overset{\overset{\displaystyle H}{|}}{\underset{\underset{\displaystyle R}{|}}{C}}-R' \qquad (38)$$

Second, and more important, the termination may occur either by combination with another polymer radical (reaction 39), or by transfer of an atom (usually hydrogen) from one polymer radical to another (reaction 40). If reaction (39) predominates, at least one head-to-head linkage must be found in the final terminated polymer. This amounts to 1% of the total linkages for a polymer, with an average degree of polymerization of 100 units. This in itself is sufficient to account for nearly all the head-to-head units in many polymers. Note also that termination by reaction (39) can bring about a considerable increase in the molecular weight of the final polymer.

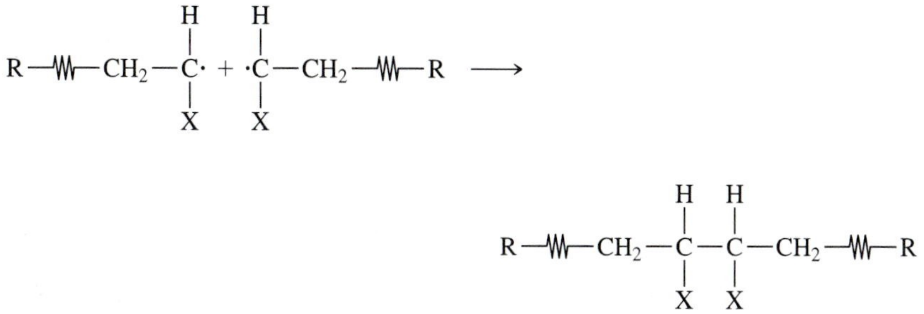

$$R-\text{W}-CH_2-\overset{\overset{\displaystyle H}{|}}{\underset{\underset{\displaystyle X}{|}}{C}}\cdot + \cdot\overset{\overset{\displaystyle H}{|}}{\underset{\underset{\displaystyle X}{|}}{C}}-CH_2-\text{W}-R \longrightarrow$$

$$(39)$$

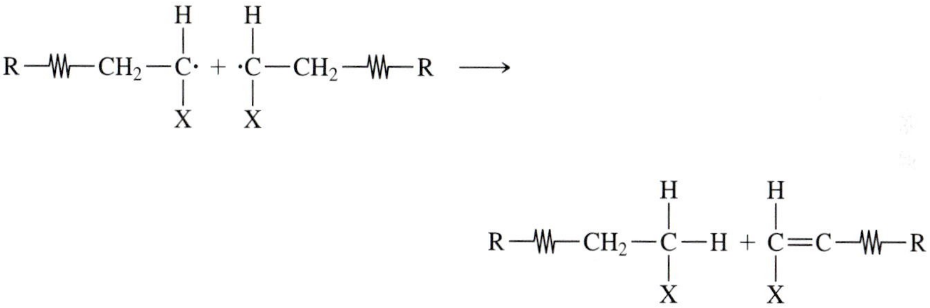

$$R-\text{W}-CH_2-\overset{\overset{\displaystyle H}{|}}{\underset{\underset{\displaystyle X}{|}}{C}}-\overset{\overset{\displaystyle H}{|}}{\underset{\underset{\displaystyle X}{|}}{C}}-CH_2-\text{W}-R$$

$$R-\text{W}-CH_2-\overset{\overset{\displaystyle H}{|}}{\underset{\underset{\displaystyle X}{|}}{C}}\cdot + \cdot\overset{\overset{\displaystyle H}{|}}{\underset{\underset{\displaystyle X}{|}}{C}}-CH_2-\text{W}-R \longrightarrow$$

$$(40)$$

$$R-\text{W}-CH_2-\overset{\overset{\displaystyle H}{|}}{\underset{\underset{\displaystyle X}{|}}{C}}-H + \overset{\overset{\displaystyle H}{|}}{C}=C-\text{W}-R$$

When reaction (40) (known as "disproportionation") predominates, two chemically different types of polymer molecules are produced. The predominance of one of the processes shown in (39) or (40) depends on the nature of the monomer and on the temperature. The dependence on temperature results from the fact that the combination reaction (step 39) usually has a lower activation energy than one which requires the breaking of a chemical bond (step 40). Hence, combination will normally be preferred at low temperatures, but disproportionation (40) should become more significant at high temperatures. This has been demonstrated for methyl methacrylate[1] and styrene[2] polymerization. The relative probabilities of occurrence of reactions (39) and (40) will equal the relative reaction rates of these two processes. Hence, the relative probabilities can be calculated from

$$\frac{\text{probability of combination }(c)}{\text{probability of disproportionation }(d)} = \frac{k_c}{k_d} = \frac{A_c}{A_d}\, e^{(E_d - E_c)/RT} \qquad (41)$$

where k and A are the rate constants and frequency factors, respectively, and E is the activation energy.

In a practical sense, the relative occurrence of combination and disproportionation can, in principle, be measured from the average number of initiator fragments, R', per polymer molecule that are found in the final product. Reaction (39) should yield polymer with two initiator fragments per molecule, whereas reaction (40) should give polymer with only one R' unit per molecule. Although such measurements are difficult to make accurately, a few definitive experiments have been carried out with the use of ^{14}C radioactively labeled initiator molecules. The results for several polymer systems are shown in Table 3.5.

TABLE 3.5 TERMINATION OF FREE-RADICAL POLYMERIZATION AT 60°C

Monomer	Formula	Disproportionation (%)	Combination (%)
Acrylonitrile[*]	$CH_2{=}CH{-}CN$	~0	~100
Methyl methacrylate[†]	$CH_2{=}C{-}C{-}OCH_3$ (with CH_3 above central C and O double-bonded below)	79	21
Styrene[‡]	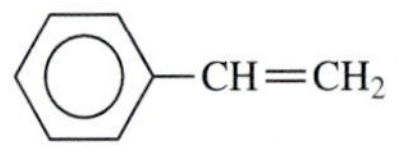 (phenyl)${-}CH{=}CH_2$	23	77
Vinyl acetate[§]	$CH_2{=}CH{-}OCCH_3$ (with O double-bonded above C)	~100	~0

[*]Bamford, C. H., Jenkins, A. D., and Johnston, R, *Trans. Faraday Soc.*, **1959**, *55,* 179.
[†]Bevington, J. C., Melville, H. W., and Taylor, R. P., *J. Polymer Sci.*, *12*; 449; **1954**, *14,* 463.
[‡]Berger, K. C., *Makromo Chem.*, **1975**, *176,* 3575.
[§]At 90°C. C. H. Bamford and A. D. Jenkins, *Nature*, **1955**, *176,* 78.

[1]Bevington, J. C., Melville, H. W., and Taylor, R. P., *J. Polymer Sci.*, **1954**, *13,* 449.
[2]Berger, K. C., *Makromol. Chem.*, **1975**, *176,* 3575.

ATOM TRANSFER RADICAL POLYMERIZATION

Although free-radical polymerizations have generally been assumed to be highly susceptible to chain-transfer processes (because of the high reactivity of carbon free-radicals), it is now known that such side reactions can be suppressed to generate "living" (i.e., nonterminated) polymerizations.[1] In these reactions, depletion of the monomer leads not to an irreversible deactivation of the system, but rather to the formation of quiescent end units that will resume chain propagation when more monomer is added. These are called atom transfer radical polymerizations for the following reasons.

A typical initiator for this type of polymerization is a combination of 1-chloro-1-phenylethane and copper(I) bipyridyl (**17**). These two species generate a radical initiator by the process shown in equation (42).

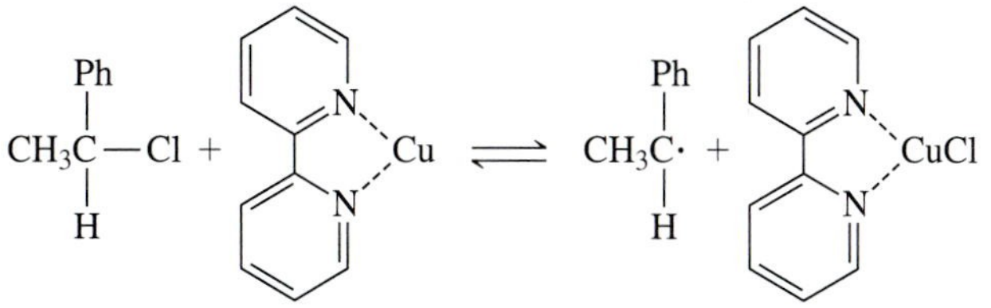

$$\text{(42)}$$

(Cu (I) (bpy)) (Cu (II) (bpy) Cl)

17

Radical initiation of, say, styrene then occurs to give a styryl dimer radical (**19**) which is promptly rechlorinated from the Cu(II) chloride complex. The polymer chain then grows as the process is repeated.

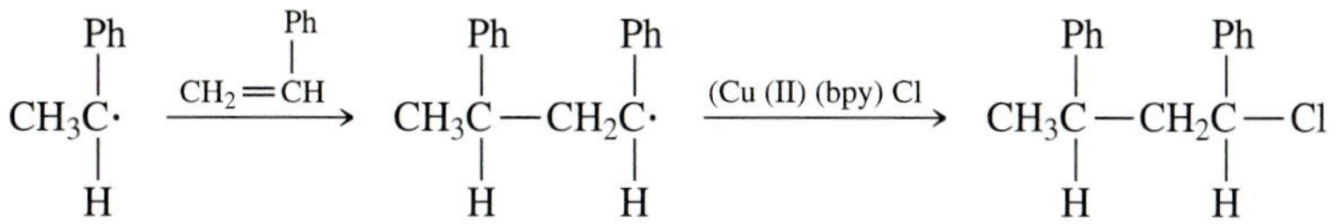

$$\text{(43)}$$

18   **19**

The addition of a stable free radical, such as nitroxide, to the mixture suppresses termination which would otherwise occur by hydrogen abstraction at stages **18** or **19** in equation (43). For example, the nitroxide radical shown in equation (44) reacts *reversibly* with a radical chain end, providing just enough protection to prevent attack by the chain end on a C-H bond elsewhere in the system.

[1]Wang, J. -S., and Matyjaszewski, K., *J. Am. Chem. Soc.*, **1995**, *117*, 5614.

$$\sim\!\!\!\sim\!\!\!\sim\!\!\!\sim\!\!\!\sim\!\!\!CH_2\underset{\underset{H}{|}}{\overset{\overset{Ph}{|}}{C}}\cdot \;+\; \cdot O-N \Big\langle \text{(piperidine ring, Me, Me / Me, Me)} \Big\rangle \;\rightleftharpoons\; \sim\!\!\!\sim\!\!\!\sim\!\!\!\sim\!\!\!\sim\!\!\!CH_2\underset{\underset{H}{|}}{\overset{\overset{Ph}{|}}{C}}-O-N \Big\langle \text{(piperidine ring, Me, Me / Me, Me)} \Big\rangle$$

(44)

Thus, the polymerization is living, and the chain lengths are controlled by the ratio of monomer to initiator. The molecular-weight distribution is now narrow, and the formation of block copolymers is also possible.

POLYMERIZATIONS IN SUPERCRITICAL CARBON DIOXIDE

Supercritical carbon dioxide, which has a critical temperature, T_c, of 31.1°C and a critical pressure, T_p, of 73.8 bar, is a low-viscosity liquid that can be used as a medium for polymerization reactions.[1-3] It is also inert to free radicals and, thus, is an attractive solvent for free-radical polymerizations without the complications of chain transfer reactions that involve the solvent. Fluorinated monomers and polymers are soluble in this medium, and this opens up opportunities for the synthesis of fluoropolymers that are difficult or impossible to prepare in other solvents.

Free radical polymerizations in supercritical CO_2 can be divided into two types. First, there are those in which both the monomer and polymer are soluble in the medium, and these proceed like normal solution-phase polymerizations, but with advantages such as the ease of solvent removal (by evaporation), little or no chain transfer to the solvent, and the avoidance of flammable or toxic solvents. Monomers that are suitable for homogeneous reactions in supercritical CO_2 include tetrafluoroethylene, 1,1-dihydroperfluorooctyl acrylate, or styrene derivatives with fluorinated alkyl groups in the *para*-position of the benzene ring. In addition, fluorinated monomers can be copolymerized with some non-fluorinated species such as methyl methacrylate, ethylene, or styrene without precipitation of the polymer.

Second, and more frequently, the monomer and the initiator are soluble in supercritical CO_2, but the polymer is not. This allows *dispersion* polymerization reactions to take place which yield the polymer in the form of fine colloidal particles with diameters that range from 100 nm to 10 μm, depending on reaction conditions and on the presence of dispersion stabilizers. Perhaps surpris-

[1]Quadir, A., Snook, R., Gilbert, R., DeSimone, J. M., *Macromolecules*, **1997**, *30*, 6015–23.
[2]Canelas, D. A., Betts, D. E., DeSimone, J. M., Yates, M. Z., Macromolecules, **1998**, *31*, 6794–6805.
[3]Hamilton, J. G., Rooney, J. J., DeSimone, J. M., Mistele, C., *Macromolecules*, **1998**, *31*, 4387–89.

ingly, this type of process often yields higher molecular weight polymers than can be obtained by the homogeneous route because of the presence of the stabilizer and plasticization of the precipitated polymer by CO_2. Styrene, methyl methacrylate, and other common organic monomers and their polymers behave in this way.

From an environmental and processing point of view, free-radical vinyl polymerizations in supercritical CO_2 have many advantages over the use of classical organic solvents. Moreover, the method is not restricted to free radical reactions. Cationic polymerization of formaldehyde and vinyl ethers (see Chapter 4) is also possible in CO_2, as is the ring-opening polymerization of epoxides or oxetanes (Chapter 6), and the ring-opening metathesis polymerization (ROMP) of norbornene (Chapter 4). Condensation polymerizations (Chapter 2) can also take place in this medium. Thus, it seems clear that the use of supercritical CO_2 as a polymerization solvent is likely to increase markedly in the coming years.

STUDY QUESTIONS

1. Write the set of elementary reactions analogous to (5) to (11) that comprise the free-radical chain polymerizations of vinyl acetate $(CH_2{=}CHOC(O)CH_3)$ and methyl methacrylate $(CH_2{=}C(CH_3)C(O)OCH_3)$ when the process is initiated by azobisisobutyronitrile. If carbon tetrachloride is added to the systems, describe the effects of chain transfer with carbon tetrachloride on the polymer composition and average molecular weight.

2. Show by equations analogous to those in Problem 1 how branching may arise in the free-radical polymerization of isoprene $(CH_2{=}C(CH_3)CH{=}CH_2)$ initiated by di-t-butyl peroxide even in the absence of chain transfer. How do you think the presence of dissolved oxygen (which may be viewed as a diradical) would affect the polymerization?

3. The first-order rate constant for the decomposition of diethyl peroxide has been reported to be $1.0 \times 10^{14} e^{-35,000 \text{ cal}/RT} \text{ s}^{-1}$ [Pryor, W. A., Huston, D. M., Fiske, T. R., Pickering, T. L., and Ciuffarin, E., *J. Am. Chem. Soc.*, **1964**, *86*, 4237]. Predict the temperature range in which diethyl peroxide would be a useful initiator.

4. J. C. Bevington [*Trans. Faraday Soc.*, **1955**, *51*, 1392] has shown that the fraction of radicals produced in the thermal decomposition of azobisisobutyronitrile that actually initiate the polymerization of styrene in benzene increases with increasing monomer concentration from a value of 0.17 at the lowest monomer concentration used, to an apparent limiting value of 0.65. Discuss possible reasons for the increase in this fraction as the monomer concentration is increased and explain why an efficiency of greater than 0.65 cannot be achieved.

5. The following results [Bevington, J. C., Melville, H. W., and Taylor, R. P., *J. Polymer Sci.*, **1954**, *12*, 449] have been obtained for the temperature dependence of the percentage of chain termination by disproportionation in methyl methacrylate:

$T(°C)$	40	60	80
% Disproportionation	50	59	70

(a) Calculate the difference in activation energy between disproportionation and combination.

(b) At what temperature will termination be 90% by combination?

6. K. C. Berger [*Makromol, Chem* **1975**, *176*, 3575] has reported the following percentages of termination by combination in styrene polymerization:

$t(°C)$	30	52	62	70	80
% Combination	86	80	77	68	60

(a) Evaluate the difference in activation energy between disproportionation and combination in the termination step.

(b) Calculate the temperature at which the rates of combination and termination are equal.

7. Using the chemical literature for the necessary data, evaluate the "useful temperature range" for ethylsilver as an initiator.

8. Discuss the difference between free-radical chain termination and the termination of a growing free radical. How can the latter proceed without the former?

SUGGESTIONS FOR FURTHER READING

ALEXANDER, E. A., and NAPPER, D. H., "Emulsion Polymerization," *Progr. Polymer Sci.* (A. D. Jenkins, ed.), **1971**, *3*, 145.

AMBADE, A. V., and KUMAR, A., "Controlling the Degree of Branching in Vinyl Polymerization," *Progr. Polym. Sci.*, **2000**, *25*(8), 1141.

CANELAS, D. A., and DeSIMONE, J. M., "Polymerization in Liquid and Super-critical Carbon Dioxide", *Adv. Polym. Sci.*, **1997**, *133*, 103.

CANELAS, D. A., BETTS, D. E., DeSIMONE, J. M., YATES, M. Z., and JOHNSTON, K. P., "Poly(vinyl acetate) and Poly(vinyl acetate-co-ethylene) Latexes via Dispersion Polymerization in Carbon Dioxide," *Macromolecules*, **1998**, *31*, 6794.

HAMILTON, J. G., ROONEY, J. J., DeSIMONE, J. M., and MISTELE, C., "Stereo-chemistry of Ring-Opened Metathesis Polymers Prepared in Liquid CO_2 at High Pressure Using $Ru(H_2O)_6(Tos)_2$ as Catalyst," *Macromolecules*, **1998**, *31*, 4387.

KAMACHI, K., "Influence of Solvent on Free Radical Polymerization of Vinyl Compounds," in *Advances in Polymer Science*, Vol. 38. Berlin: Springer-Verlag, **1981**, p. 55.

KUBISA, P., "Radical Polymerization with Reversible Deactivation of Active Species," *Polymery*, **2000**, *45*(11–12), 741.

MATYJASZEWSKI, K. (ed.), *Controlled Radical Polymerization, ACS Symp. Ser.*, **1998**, *685*.

MAYADUNNE, R. T. A., RIZZARDO, E., CHIEFARI, J., CHONG, Y. K., MOAD, G., and THANG, S. H., "Living Radical Polymerization with Reversible Addition-Fragmentation Chain Transfer (RAFT Polymerization) Using Dithiocarbamates as Chain Transfer Agents," *Macromolecules*, **1999**, *32*, 6977.

MISHRA, M. K., and YAGEI, Y., *Handbook of Radical Vinyl Polymerization*, New York: Dekker, **1998**.

MOAD, G., and SOLOMON, D. H., *The Chemistry of Free Radical Polymerization*, Oxford (U.K.) and Tarrytown, N. Y., Pergamon, **1995**.

MOAD, G., CHIEFARI, J., CHONG, Y. K., KRISTINA, J., MAYADUNNE, R. T. A., POSTMA, A., RIZZARDO, E., and THANG, S. H., "Living Free Radical Polymerization with Reversible Addition-Fragmentation Chain Transfer (the Life of RAFT)," *Polym. Int.*, **2000**, *49*(9), 993.

ODIAN, G., *Principles of Polymerization*, New York: Wiley-Interscience, **1991**.

OTSU, T., "Inferter Concept and Living Radical Polymerization," *J. Polym. Sci., Polym. Chem.* **2000**, *38*(12), 2121.

QUADIR, M. A., SNOOK, R., GILBERT, R., and DeSIMONE, J. M., "Emulsion Polymerization in a Hybrid Carbon Dioxide/Aqueous Medium," *Macromolecules*, **1997**, *30*, 6015.

PATTEN, T. E., and MATYJASZEWSKI, K., "Atom Transfer Radical Polymerization and the Synthesis of Polymeric Materials," *Adv. Mater.*, **1998**, *10*, 901.

SCHELLEKENS, M. A. J., and KLUMPERMAN, B. J., "Synthesis of Polyolefin Block and Graft Copolymers," *J. Macromol. Sci. Reviews*, **2000**, *C40*(2–3), 167.

SEYMOUR, R. B., and CARRAHER, C. E., Jr., *Polymer Chemistry*, 2nd. ed. New York: Dekker, **1988**, Ch. 9 and 10.

VEREGIN, R. P. N., GEORGES, M. K., KAZMEIER, P. M., and HAMER, G. K., "Free Radical Polymerizations for Narrow Dispersity Resins, -ESR Studies of the Kinetics and Mechanism," *Macromolecules*, **1993**, *26*, 5316.

WALL, L. A. (ed.), *Fluoropolymers*, New York: Wiley-Interscience, **1971, 1972**.

4

Ionic and Coordination Polymerization

INTRODUCTION

An ionic polymerization is an addition polymerization in which the growing chain ends bear a negative or a positive charge:

$\ominus$ Anionic polymerization

$\oplus$ Cationic polymerization

If the growing chain end bears a negative charge, the process is called an *anionic polymerization*. If the chain end bears a positive charge, the reaction is a *cationic polymerization* process. These two alternative mechanisms are associated with different catalyst systems and different reaction conditions. Some polymerization reactions, particularly those initiated by Ziegler–Natta or transition metal complexes, are best described as "coordination polymerizations," since the reaction mechanisms are believed to involve complexes formed between the transition metal and the π-electrons of the monomer. However, most of the known reactions of this type bear many similarities to anionic polymerizations and they are considered here as part of that classification. Thus, the rest of this chapter is divided into a discussion of first anionic, and then cationic processes. It should be emphasized from the beginning that the reaction conditions for ionic polymerizations differ markedly from those discussed in Chapter 3 for free-radical reactions.

STEREOREGULAR POLYMERS

Ionic and coordination polymerizations can provide access to polymers that have a stereoregular structure. Any monomer molecule that possesses an asymmetric center at a skeletal atom has the capacity to form stereoregular polymers. Three primary possibilities exist with respect to the sequence in which the monomer units enter the chain. These are called *isotactic, syndiotactic*, or *atactic* polymerizations.

Isotactic polymers are characterized by the presence of only one symmetry type of monomer residue in the chain. For example, a representation of isotactic polystyrene is shown in **1.** Or, more symbolically, this structure can be represented by **2.** One of the characteristic features of isotactic polymers is their ability to crystallize readily. This is a consequence of the fact that the regular disposition of substituent groups along the chain permits the molecule to assume a regular helical conformation and allows adjacent chains to pack together in an ordered manner.

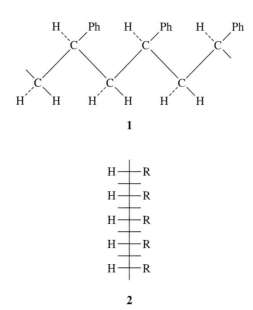

1

2

Syndiotactic polymers are characterized by *alternating* configuration of residues, as depicted in **3.** This structure is represented by the symbolism shown in **4.** Syndiotactic polymers also tend to crystallize readily, again because of the opportunities that exist for the formation of helices and for efficient chain packing.

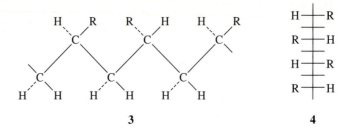

3 **4**

Atactic or heterotactic polymers contain no regular sequence of monomer residues along the chain. Because of this, the polymers are characterized by a low tendency for crystallization.

The simplest stereoregular systems are the ones described above in which the monomer has the structure $CH_2=CHR$. For monomers of structure $CHR=CHR$ and $CHR=CHR'$, the following additional possibilities exist:

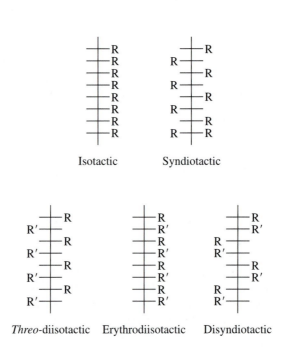

Isotactic   Syndiotactic

Threo-diisotactic   Erythrodiisotactic   Disyndiotactic

And for regular *copolymers* synthesized from ethylene and $CHR-CHR$, the following four alternative structures can be visualized:

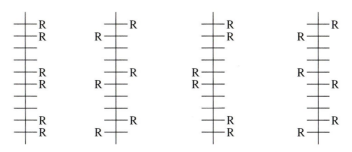

Erythrodiisotactic   Threodiisotactic   Erythrodisyndiotactic   Threodisyndiotactic

One further item of stereoregular nomenclature is necessary when a regular sequence of double bonds is present in the main chain as, for instance, in poly(1,4-butadiene). The configuration at each double bond may be *cis* or *trans*, as follows:

Cis-tactic   *Trans*-tactic

Finally, polymers can be formed that contain blocks of one type of stereoregular sequence followed by another. These are called *stereoblock* polymers.

In general, *ionic* catalysts, especially *anionic* catalysts, favor the formation of stereoregular polymers; thus, they also favor the synthesis of microcrystalline polymers. On the other hand, free-radical catalysts tend to favor the formation of atactic polymers with a corresponding low degree of crystallinity.

ANIONIC AND COORDINATION POLYMERIZATION

Initiators[1] for Anionic Polymerization

Table 4.1 lists different classes of initiators that are known to initiate anionic- or coordination-type polymerizations. Of these initiators the alkali metal suspensions are prepared by dispersion of a molten alkali metal in an inert organic solvent; organolithium reagents are prepared by the reaction of lithium metal with an organic halide; and Grignard reagents are obtained by the reaction of magnesium turnings with an organic halide. These catalysts and the aluminum alkyls can be obtained commercially. However, radical anions such as sodium naphthalenide must be prepared from naphthalene and a sodium mirror in an ether-type solvent (reaction 1). The appearance of an intense green color accompanies

[1]The terms "initiator" and "catalyst" are often used interchangeably in polymer chemistry. However, the initiating species is usually not recoverable at the end of the reaction, and the word "catalyst" is, therefore, something of a misnomer.

TABLE 4.1 CLASSICAL ANIONIC AND COORDINATION INITIATORS

Initiator Class	Formula or Example
1. Alkali metal suspensions	Sodium in tetrahydrofuran or in liquid ammonia
2. Alkyl or aryllithium reagents	nC_4H_9Li
3. Grignard reagents	RMgX (R = alkyl or aryl, X = halogen)
4. Aluminum alkyls	AlR_3
5. Organic radical anions	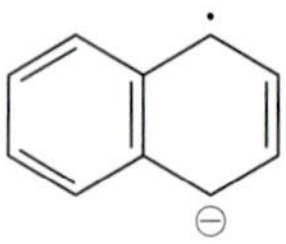 Na^+ (sodium naphthalenide)
6. Some Ziegler-Natta catalysts	$TiCl_4 + AlR_3$
7. Metallocene initiators	$R_2Si(C_9H_6)_2ZrCl_2$
8. ROMP catalysts	$(Cy_3P)_2Ru(Cl_2)CHPh*$
9. ADMET initiators	$(ArO)_2WCl_3$
10. Transition metal π-allyl complexes	$(\pi C_4H_7)_2Ni$
11. Transition metal oxides	Oxides of V, Ti, Cr, Co, Ni, or W
12. Ionizing radiation	X-rays, γ-rays, or electrons

*Cy = cyclohexyl

the formation of this species. Water must be rigorously excluded from the system.

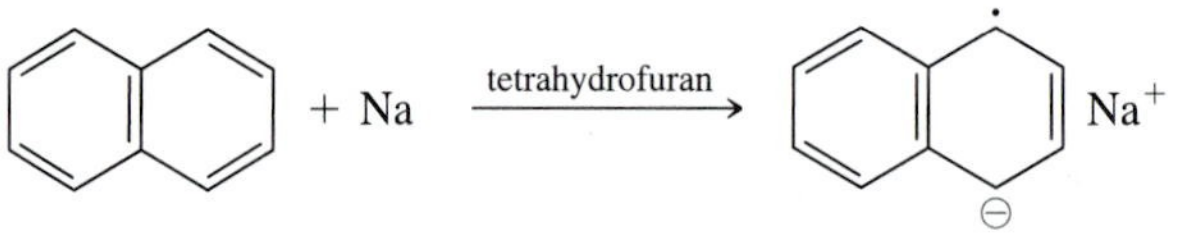

$$(1)$$

As a precautionary measure, note that aluminum alkyls are spontaneously flammable in the atmosphere, and Grignard reagents and organolithium compounds may explode if isolated in the solid state. Hence, these reagents are nearly always handled as solutions in organic solvents under an inert atmosphere.

Monomers for Anionic Polymerization

Typical monomers that can be polymerized by anionic mechanisms include *styrene* (5), *methyl methacrylate* (6), and *acrylonitrile* (7). The important thing to remember is that monomers which are suitable for anionic polymerization generally contain *electron-withdrawing substituent groups*. Some anionic polymerizations are characterized by *highly colored* reaction mixtures.

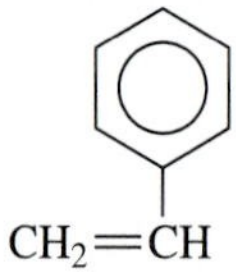

5 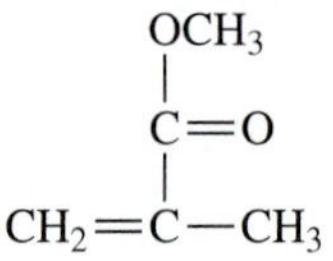   6 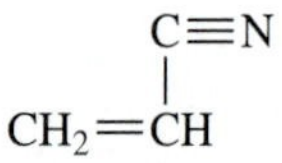   7

A Typical Experimental Procedure

As an example, we will consider the polymerization of styrene using *n*-butyllithium as a catalyst. The reaction takes place in an organic solvent such as ether or tetrahydrofuran. Water and carbon dioxide must be *rigorously* excluded from the solvent, the monomer, the catalyst, and the inside of the reaction apparatus. The monomer itself should be very pure and free from inhibitors. Furthermore, some means must be utilized for the introduction of very small measured quantities of the initiator species.

An apparatus that can be used is shown in Figure 4.1. It consist of a glass three-necked flask fitted with a rubber serum cap stopper, a mercury or silicone oil bubbler, and with the flask connected to a solvent distillation system. A Teflon-covered magnetic stirrer bar is contained in the reaction flask. A stream of dry nitrogen or helium can be used to sweep the air from the system.

The solvent (in this case tetrahydrofuran) is dried by distillation from calcium hydride or lithium aluminum hydride[1] into the reaction flask. Care should be taken that some tetrahydrofuran remains behind in the distillation flask to avoid the possibility that explosive peroxides will be concentrated in the residue. Once the solvent has been distilled into the flask, the condenser system can be removed and replaced by a glass stopper.

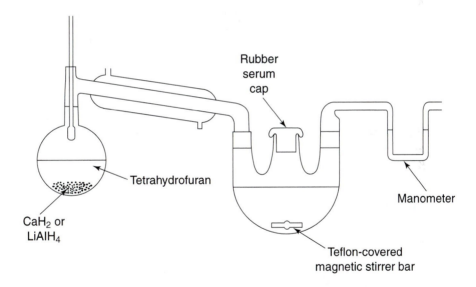

Figure 4.1 Apparatus for the anionic polymerization of styrene with the use of *n*-butyllithium as an initiator.

---

[1]Under no circumstances should these hydrides be allowed to come into contact with water, or a fire or explosion could result. Lithium aluminum hydride in particular should be handled with great care, and distillations of solvent from this drying agent should be carried out behind a shield.

Commercial monomers, such as styrene, always contain small amounts of "inhibitors"—compounds that retard the free-radical polymerization process. Usually, the inhibitor can be removed by the passage of the liquid monomer through a chromatography tube that has been packed with alumina. Vacuum distillation of the monomer can also be used as a purification process. The styrene can be introduced into the reaction vessel from a hypodermic syringe through the serum cap stopper. At this stage, the temperature of the reaction mixture is often lowered.

The catalyst—in this case, n-butyllithium—can be obtained commercially as a solution of known concentration in a hydrocarbon such as pentane. Its concentration should be checked by acid–base titration or from the amount of butane liberated on hydrolysis. A known amount of catalyst solution is then added by hypodermic syringe through the serum cap stopper into the stirred solution of the monomer. A color change occurs rapidly from colorless to orange. After completion of the reaction, the active chain ends are destroyed by the addition of water or dry ice, the solution becomes colorless, and the polymer can be isolated. If dry ice is used for termination, a subsequent treatment with dilute aqueous acid will be needed to convert the lithium carboxylate end groups to free carboxylic acid residues.

It should be noted that, in many anionic polymerizations, the amount of catalyst added determines the chain length of the polymer (see Chapter 13, equation 9). In principle, each catalyst molecule will generate one polymer chain. Thus, the more catalyst that is added, the lower will be the polymer molecular weight. In fact, the polystyrene prepared by this method has a very narrow molecular-weight distribution, and this is a consequence of the small amount of chain transfer in the system.

Similar experimental procedures can be employed when other catalysts are used. Sodium naphthalenide is more sensitive to moisture and carbon dioxide than alkyllithium reagents: consequently, a glass high-vacuum system is frequently employed for the reaction of naphthalene with a sodium mirror. The solvent must be ultra-dry before sodium naphthalenide can be used.

The Mechanism of Anionic Polymerization (General)

The overall reaction can be divided into initiation, propagation, and termination steps.

1. *Initiation* takes place by addition of the initiator across the double bond of the monomer (reaction 2). In this sequence, R' represents an electron-withdrawing group such as phenyl or cyano. Sequence (2) is actually an oversimplification because some organometallic initiators, such as n-butyllithium, exist as aggregates in solution. Thus breakdown of, say, a tetrameric aggregate may be necessary before initiation can occur. Most anionic catalysts have the capacity to react as an organometallic ion pair. For example, although n-butyllithium is generally considered to have a covalent type of

structure, heterolytic cleavage of the lithium-carbon bond to give ions can occur in the presence of a suitable monomer (3). Note that, in undergoing addition to the monomer, the ion pair adds in such a way that the anion from the catalyst becomes attached to the carbon atom farthest from the electron-withdrawing group, R'. This is, of course, the carbon atom that possesses the lower electron density (4). Note also that in initiation and in the subsequent steps, the active site at the end of the chain is accompanied by a counterion. The presence of the counterion explains many of the differences from free-radical polymerization.

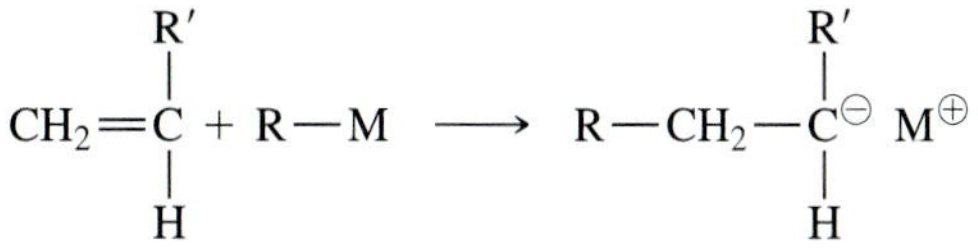

$$\underset{\underset{H}{|}}{\overset{\overset{R'}{|}}{CH_2{=}C}} + R{-}M \longrightarrow R{-}CH_2{-}\underset{\underset{H}{|}}{\overset{\overset{R'}{|}}{C}}^{\ominus}\ M^{\oplus} \tag{2}$$

$$n{-}C_4H_9Li \longrightarrow nC_4H_9^{\ominus}\ Li^{\oplus} \tag{3}$$

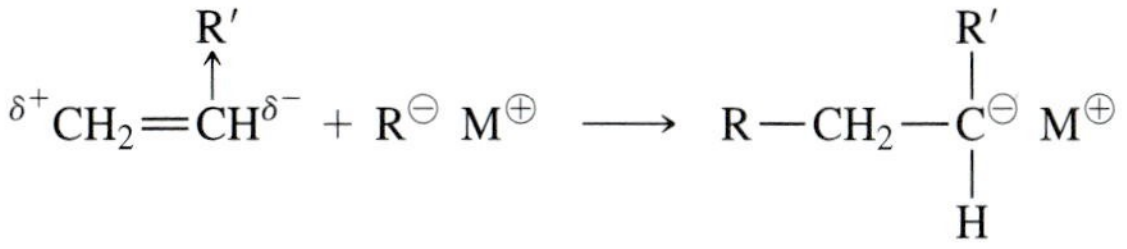

$$\overset{\delta^+}{C}H_2{=}\overset{\overset{R'}{\uparrow}}{C}H^{\delta^-} + R^{\ominus}\ M^{\oplus} \longrightarrow R{-}CH_2{-}\underset{\underset{H}{|}}{\overset{\overset{R'}{|}}{C}}^{\ominus}\ M^{\oplus} \tag{4}$$

2. *Propagation* then involves the successive insertion of monomer molecules into the terminal "ionic" bond (5). This process of chain growth continues until all the monomer has been consumed, or until the reaction is terminated.

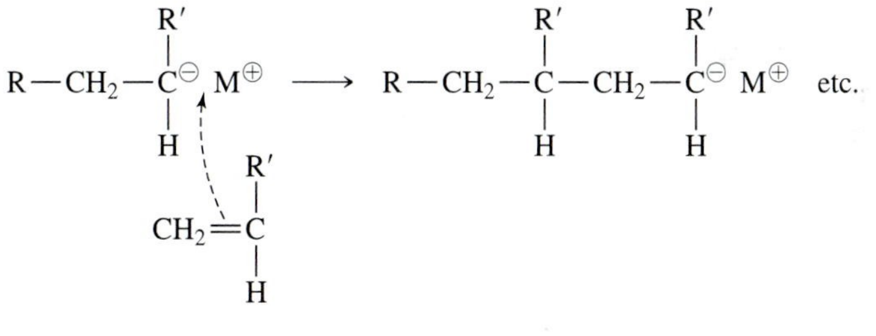

$$R{-}CH_2{-}\underset{\underset{H}{|}}{\overset{\overset{R'}{|}}{C}}^{\ominus}\ M^{\oplus} \longrightarrow R{-}CH_2{-}\underset{\underset{H}{|}}{\overset{\overset{R'}{|}}{C}}{-}CH_2{-}\underset{\underset{H}{|}}{\overset{\overset{R'}{|}}{C}}^{\ominus}\ M^{\oplus}\quad \text{etc.} \tag{5}$$

$$\underset{\underset{H}{|}}{\overset{\overset{R'}{|}}{CH_2{=}C}}$$

3. *Chain transfer* or *chain branching* does not occur to any appreciable extent with anionic systems, and this is especially true if the reaction is carried out at low temperatures.

4. *Termination* of the chains occurs either accidentally or deliberately when the active chain end reacts with a molecule of carbon dioxide or with water, alcohols, or other protonic reagents (6 or 7). However, it is important to note that if termination reagents are absent, the chains could remain active indefinitely. In practice, it is impossible to remove all traces of water molecules from the inside of glass equipment and, even if that were possible, the

Si—OH groups at the glass surface could presumably function as termination agents. However, in a well-dried system the brilliant colors that indicate the integrity of the active end groups may persist for days.

$$R-\!\!\!\!\!\backslash\!\backslash\!\!\!\!\overset{\overset{\displaystyle R'}{|}}{C}{\ominus}\ M^{\oplus} \quad \begin{array}{c} \overset{H_2O}{\nearrow} \\[2pt] \underset{\underset{HCl}{followed\ by}}{\searrow\ CO_2} \end{array} \quad \begin{array}{l} R-\!\!\!\!\!\backslash\!\backslash\!\!\!\!\overset{\overset{\displaystyle R'}{|}}{\underset{\underset{H}{|}}{C}}-H + MOH \qquad (6) \\[30pt] R-\!\!\!\!\!\backslash\!\backslash\!\!\!\!\overset{\overset{\displaystyle R'}{|}}{\underset{\underset{H}{|}}{C}}-COOH + MCl \quad (7) \end{array}$$

"Living" Polymers

The term *"living" polymer* is applied to ionic polymerizations that are not terminated. These systems have the capability of continuing chain growth if more of the same monomer or, indeed, a second monomer is added. Provided that chain terminators are absent, there is no reason in theory why chain growth should stop if more and more monomer is added. In practice, small amounts of terminators are inevitably introduced. Chain growth also slows down eventually because of the high viscosity of the system or because the chains become insoluble.

Initiators that yield some of the most dramatic "living" polymerizations are those which form radical anions during the initiation process. Metallic sodium or sodium naphthalenide function in this way. Consider first the role of sodium metal. The polymerization process can be divided into discrete steps, as follows:

1. *Initiation* occurs when an electron is transferred from sodium to the monomer to generate a radical anion (8).

$$CH_2\!\!=\!\!\overset{\overset{\displaystyle R}{|}}{\underset{\underset{H}{|}}{C}} + Na \longrightarrow \cdot CH_2\!\!=\!\!\overset{\overset{\displaystyle R}{|}}{\underset{\underset{H}{|}}{C}}{\ominus}\ Na^{\oplus} \qquad (8)$$

2. *Dimerization* of the radical anion then takes place to form a *di*anion (9).

$$Na^{\oplus}\ {\ominus}\overset{\overset{\displaystyle R}{|}}{\underset{\underset{H}{|}}{C}}\!-\!CH_2\cdot + \cdot CH_2\!-\!\overset{\overset{\displaystyle R}{|}}{\underset{\underset{H}{|}}{C}}{\ominus}\ Na^{\oplus} \longrightarrow Na^{\oplus}\ {\ominus}\overset{\overset{\displaystyle R}{|}}{\underset{\underset{H}{|}}{C}}\!-\!CH_2\!-\!CH_2\!-\!\overset{\overset{\displaystyle R}{|}}{\underset{\underset{H}{|}}{C}}{\ominus}\ Na^{\oplus} \qquad (9)$$

3. *Propagation* can then occur *at both ends* of the dimer (9) by the insertion of monomer molecules into the ionic bonds. Thus, the growing polymer chain has the structure shown in (10).

$$\text{Na}^{\oplus}\;{}^{\ominus}\underset{\underset{H}{|}}{\overset{\overset{R}{|}}{C}}\text{---}W\text{---}CH_2\text{---}CH_2\text{---}W\text{---}\underset{\underset{H}{|}}{\overset{\overset{R}{|}}{C}}{}^{\ominus}\;\text{Na}^{\oplus} \qquad (10)$$

Sodium naphthalenide functions as a catalyst in a very similar way. The naphthalene radical anion **(8)** either transfers an electron to the monomer to form a monomer radical anion (as described above), or the naphthalene radical anion itself may initiate monomer polymerization directly.

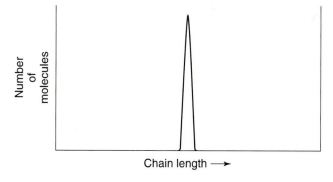

8

One of the characteristics of "living" polymer systems is that they yield polymers with very narrow molecular-weight distributions (Figure 4.2). This phenomenon is discussed in more detail in Chapter 13. However, from a qualitative point of view, it is sufficient to note that the narrow molecular-weight distribution is largely a consequence of the rapidity of the initiation step. All the chains are initiated essentially at the same instant (at the point of catalyst injection), and all of them grow at the same rate until the monomer is consumed. The fact that chain transfer (e.g., proton abstraction) does not occur at low temperatures also prevents a broadening of the molecular-weight distribution. Thus, anionic reactions are much "cleaner" than free-radical polymerizations.

Number of molecules

Chain length ⟶

Figure 4.2 Extremely narrow molecular-weight distribution that is characteristic of anionic and radical-anionic polymerization products.

Copolymerization

Because many anionic chain ends remain "alive," particularly at low temperatures, even though the monomer has been consumed, chain growth can be restarted by the addition of more of the same monomer—or addition of a different monomer. If a different monomer is added, a copolymer is formed. If A represents the first monomer and B the second, a growing copolymer might be represented by the structure shown in **9**. Subsequently, either more of the monomer A, or a third monomer, C, might be added. In this way it is possible to synthesize *block copolymers*.

$$^{\ominus}\text{BBBBBBBBAAAAAAAACH}_2-\text{CH}_2\text{AAAAAAAABBBBBBBB}^{\ominus}$$

9

In practice some restrictions exist with respect to which monomers can be used in a block-copolymerization reaction. The two monomers should have similar electron affinities if mutual reinitiation is to take place. However, if monomer B has a much higher electron affinity (i.e., it contains a more powerful electron-withdrawing substituent) than monomer A, then, although a terminal unit of an A-block will initiate the propagation of a B-block, the terminal unit of the B-block will not initiate the addition of more molecules of A. In other words, terminal residues of type B cannot bond to monomer A. However, they will initiate propagation of an even more polar monomer, C.

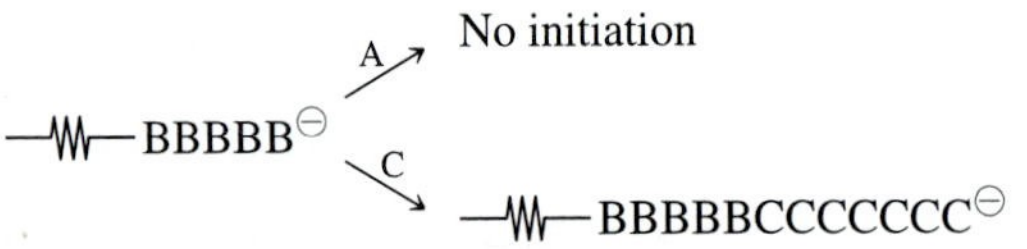

$$-\text{W}-\text{BBBBB}^{\ominus} \quad \begin{array}{c} \xrightarrow{A} \text{No initiation} \\ \xrightarrow{C} -\text{W}-\text{BBBBBCCCCCC}^{\ominus} \end{array}$$

(11)

(12)

This is illustrated schematically in (11) and (12). For example, an active chain made up of methyl methacrylate units (B) will undergo continued growth in the presence of acrylonitrile (C), but not in the presence of styrene (A). The polarographic reduction potential of a monomer provides a good measure of its electron affinity. Thus, a block copolymerization series can be worked out from the relative order of the reduction potentials.

Ziegler–Natta Catalysts

In the early 1950s, Karl Ziegler and his co-workers discovered that ethylene reacts with aluminum alkyls in a high-pressure system to yield organometallic oligomers or polyethylene (reaction 13). It was subsequently found that the addition of transition metal *"cocatalysts,"* such as $TiCl_4$ or VCl_4 to the aluminum alkyl generated a system that would polymerize ethylene at *atmospheric pressure* and *room temperature*. High-molecular-weight polyethylene can, in fact, be made by bubbling ethylene gas into a suspension of the catalyst in a liquid-hydrocarbon medium. Natta and co-workers in Italy found that catalysts of this type induce the formation of crystalline, stereoregular polymers. Since that time an enormous amount of subsequent research has been carried out on these polymerizations, and their industrial utility has been exploited widely. Ziegler–Natta catalysis has such wide ramifications that a separate volume would be needed to do it justice. However, in the following sections, we consider briefly some of the general characteristics of these reactions, the stereoregular nature of the polymerization, the composition of the catalyst, the reaction mechanisms, and an extension of this catalytic pathway to the polymerization of dienes and cycloolefins.

$$AlEt_3 + 3nCH_2{=}CH_2 \longrightarrow Al[(CH_2{-}CH_2)_n{-}Et]_3 \tag{13}$$

General features of the reaction First, it must be emphasized that Ziegler–Natta catalysis has the capability to yield unbranched and stereospecific polymers. The polyethylene produced by this process is linear and has a higher density than that prepared by free-radical techniques. Ziegler–Natta polymerization of propylene can yield an isotactic stereoregular form of polypropylene that has extremely useful technological properties. Many nonpolar unsaturated organic monomers can be polymerized in a similar way. The following example illustrates how such polymerizations are carried out on a laboratory scale.[1]

A catalyst system can be prepared by the successive addition of solutions of titanium tetrachloride and triisobutylaluminum[2] in decahydronaphthalene to decahydronaphthalene diluent under an atmosphere of dry nitrogen. A brown-black suspended precipitate is formed which changes to a deep violet color when the mixture is heated to 185°C for 40 min. This is called the *"aging"* step. The mixture is then cooled, cyclohexane is added as a reaction solvent, and additional triisobutylaluminum is introduced to yield a purplish-black suspension. The polymerization of ethylene can be accomplished simply by allowing ethylene gas to bubble through this mixture. The flask must be cooled by an external water–ice bath because heat is evolved during the polymerization. The polyethylene precipitates for as long as ethylene is introduced into the system. The polymer can be isolated by pouring the slurry into stirred isopropanol, followed by filtration. Isotactic polypropylene is prepared in the same way by allowing propylene gas to bubble into the catalyst suspension.

Stereoregularity As discussed in an earlier section of this chapter, monomers that are asymmetric with respect to the disposition of side groups

---

[1] For specific details, see, for example, Sorenson W. R. and Campbell, T. W., *Preparative Methods of Polymer Chemistry*, 2nd ed. (New York: Wiley-Interscience, **1967**), pp. 289–312.

[2] Aluminum alkyls are dangerously pyrophoric when exposed to atmospheric oxygen or water. They should be handled only in an inert atmosphere and with appropriate safety precautions.

about the double bond can yield polymers with a specific stereochemical arrangement of the side groups. Ziegler–Natta catalysts generate a high degree of stereoregularity. Vinyl monomers usually yield isotactic polymers with these initiator systems, although syndiotactic polypropylene can be formed under certain conditions. The degree of stereoregularity generated depends on a number of factors, including the homogeneous or heterogeneous nature of the catalyst, the detailed composition and history of the catalyst, and the nature of the side group in the monomer. The influence of the catalyst will be discussed in the next section.

Composition of the catalyst A Ziegler–Natta catalyst is made from two components: (1) a transition metal compound from groups 4–10 (formerly groups IVB to VIIIB) of the periodic table, and (2) an organometallic compound, usually derived from a group 1, 2, or 13 (IA to IIIA) metal. The transition metal component employed is usually a halide or oxyhalide of titanium, vanadium, chromium, molybdenum, or zirconium, and the second component often consists of an alkyl, aryl, or hydride of aluminum, lithium, magnesium, or zinc. Perhaps the best known systems are those derived from $TiCl_4$ or $TiCl_3$ and an aluminum trialkyl. The catalyst systems may be heterogeneous (some titanium-based systems) or soluble (most vanadium-containing species). The stereoregularity of the catalytic process can also be altered by the addition of Lewis bases, such as amines.

Changes in the catalyst system affect the yield of polymer, the chain length, and the degree of stereoregularity. The effects of gross catalyst changes on the stereoregularity of polypropylene are shown in Table 4.2. For one particular aluminum alkyl, changes in the transition metal halide affect the percentage stereoregularity, as shown in Table 4.3. The stereoregularity of the polypropylene *decreases* with increasing size of the organic group attached to aluminum.

The nature of Ziegler–Natta catalyst systems is still a subject for debate. The soluble catalysts appear to have well-defined structures. For example, the catalyst system generated for bis(cyclopentadienyl)titanium dichloride and triethylaluminum has a halogen-bridged structure **(10)**.

10

However, $TiCl_4$ and $TiCl_3$ give rise to much more complex initiator systems, the structure of which is still not clear. One fact does appear to be certain, the true catalysts are *not* simple coordination adducts formed from the original metal halide and aluminum alkyl. A critical "aging" period for the catalyst is often needed before it achieves its highest activity, and complex reactions occur during this period. These reactions probably included an initial exchange of substituent groups between the two metals to form transition metal–carbon bonds by inter-

TABLE 4.2 STEREOREGULARITY OF POLYPROPYLENE WITH DIFFERENT CATALYSTS SYSTEMS

Catalyst	Stereoregularity (%)
R_3Al^* + $TiCl_4$	35.2
R_3Al + α-$TiCl_3$	84.7
R_3Al + β-$TiCl_3$	45
R_3Al + $TiCl_4$ + NaF	97
R_3Al + $TiCl_4$ + compounds of P, As, or Sb	98
R_3Al + $TiCl_3$ + amine	81
R_3Al + $Ti(O\text{-iso-Bu})_4^\dagger$	20
R_3Al + $V(acac)_3^\ddagger$	0
R_3Al + $Ti(C_5H_5)Cl_2$	70–90
R_3Al + $Ti(C_5H_5)_2Cl_2$	85
R_2AlX^4 + $TiCl_3$	90–99
$RAlX_2$ + γ-$TiCl_3$ + amine	>99
$RAlX_2$ + $TiCl_3$ + $HPT^\P$	97
RNa + $TiCl_3$	90
RNa + $TiCl_4$	90
RLi + $TiCl_4$	90
R_2Zn + $TiCl_3$	65
R_2Zn + $TiCl_3$ + amine	93

Source: Reprinted by permission. D. O. Jordan, *The Stereochemistry of Macromolecules*, Vol. I, A. D. Ketley, ed. (New York: Marcel Dekker, Inc., **1967**).

*R = alkyl.

† Bu = butyl.

‡ (acac) = acetylacetonate.

§X = halogen.

¶HPT = hexamethyl phosphoric triamide.

TABLE 4.3 INFLUENCE OF THE TRANSITION METAL ON THE STEREOREGULARITY OF POLYPROPYLENE*

Transition Metal Compound	Stereoregularity (%)
$TiCl_4$	48
$TiBr_4$	42
$TiCl_3$, α, γ, or δ	80–92
$TiCl_3$, β	40–50
$ZrCl_4$	55
VCl_3	73
$CrCl_3$	36
VCl_4	48
$VOCl_3$	32

Source: Reprinted by permission. D. O. Jordan, *The Stereochemistry of Macromolecules*, Vol. I, A. D. Ketley, ed. (New York: Marcel Dekker, Inc., **1967**).

* The organometallic compound is $Al(C_2H_5)_3$ in each case.

actions such as those shown in (14) and (15). These organotitanium halides are unstable and can undergo reductive decomposition processes, such as those shown in (16) and (17). Note that $TiCl_3$ can be used as an initial catalyst component in place of $TiCl_4$. Further reduction

$$AlR_3 + TiCl_4 \rightleftharpoons R_2AlCl + RTiCl_3 \tag{14}$$

$$R_2AlCl + TiCl_4 \rightleftharpoons RAlCl_2 + RTiCl_3 \tag{15}$$

$$RTiCl_3 \rightarrow R \cdot + TiCl_3 \tag{16}$$

$$R_2TiCl_2 \rightarrow R \cdot + RTiCl_2 \tag{17}$$

may yield $TiCl_2$, which can itself react with the aluminum trialkyl ligand-exchanged species. For the heterogeneous catalysts, the reactions are more complicated than is implied by these equations. However, the analogous vanadium-containing systems are soluble and may well be represented fairly accurately by reactions such as those shown in (14) to (17).

In all these systems, one of the most important steps is the *reduction* of the transition metal to a low-valency state in which the metal possesses unfilled ligand sites. These low-valency transition metal species are believed to be the real catalysts or precursors of the real catalysts.

The polymerization mechanism Convincing evidence exists that Ziegler-Natta catalysts function by the formation of transient π-complexes between the olefin and the low-valence transition metal species. Stable π-complexes between olefins and transition metals are well known. The complexes are held together by an overlap of the *d*-orbitals of the transition metal with the π-orbitals of the olefin (Figure 4.3). As shown in Figure 4.3a, the d_{xy} (or d_{xz} or d_{yz}) orbital

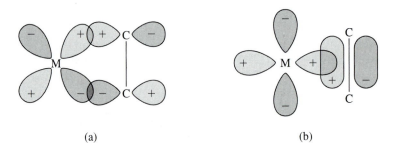

(a)   (b)

Figure 4.3   Orbital overlap scheme for the formation of a π-type interaction between an olefin and a transition metal: (a) use of the π-antibonding orbitals of the olefin; (b) overlap of one lobe of a $d_{x^2-y^2}$ orbital from the metal and a π-bonding orbital of the olefin.

of the metal can overlap the π-antibonding orbitals of the olefin. Overlap is also possible between one lobe of a $d_{x^2-y^2}$ orbital and an olefin π-orbital, as shown in Figure 4.3b. In Ziegler–Natta catalysis, π-complex formation could not occur if the transition metal occupied its highest valence state (in other words, if all the coordination sites were occupied by strongly bound ligands). Hence, the reduction process outlined above is critically important for catalyst activity. However, the exact details of the coordination behavior and the olefin addition mechanism are still obscure.

Two general mechanisms have been proposed, one involving coordination of the olefin to a vacant site on the transition metal (the "monometallic" mechanism), and the other supposing a participation by both the transition metal and the aluminum atom (the "bimetallic" mechanism).

The monometallic mechanism is summarized in Scheme 1. The olefin (in this case, ethylene) becomes coordinated to a vacant site on the transition metal co-

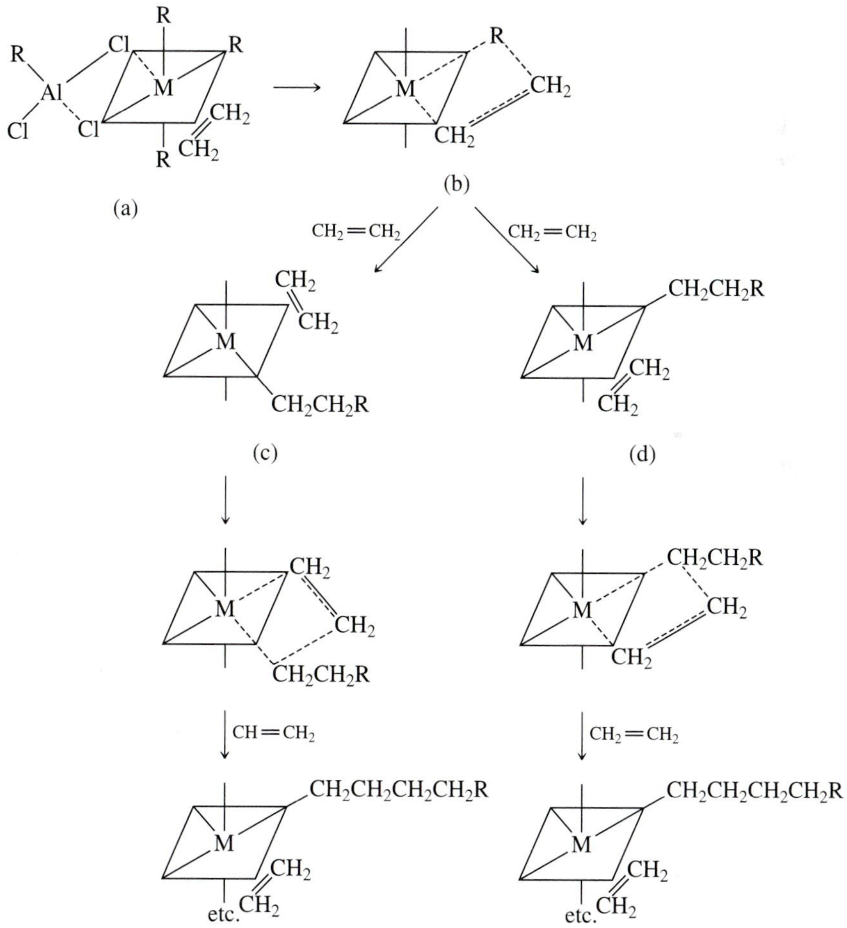

(18)

Scheme 1

ordination sphere (a). A transfer of an organic group, R, then takes place from the metal to the olefin (b). In other words, the olefin becomes inserted into the weak metal–carbon bond. An incoming olefin molecule then coordinates to the new vacant site just formed, ready for insertion into the metal–alkane bond. The mode of insertion will depend on the geometry of the switch to a metal–carbon σ-bond and on the location of the vacant coordination site (c or d). It will be clear that this mechanism allows a partial understanding of the reasons for stereospecific addition. The side groups on the olefin sterically determine the coordination and insertion geometries. The presence of the π-bond would prevent orientational scrambling once the monomer is attached to the catalyst.

The bimetallic mechanism is summarized in Scheme 2. The alkyl groups in the catalyst (a) are presumed to function as bridging units. Such behavior is well documented in other organometallic species. The initial π-coordination of the olefin [as shown in (b)] is followed by an insertion of the olefin into a titanium-carbon bond to yield a new bridged species (d). The coordination and insertion of successive olefin molecules can then take place by the same pathway.

Both mechanisms seem plausible for catalysis in solution. When a heterogeneous-type system is present, defects in the crystal surface may also be responsible for the stereoregular character of a polymerization.

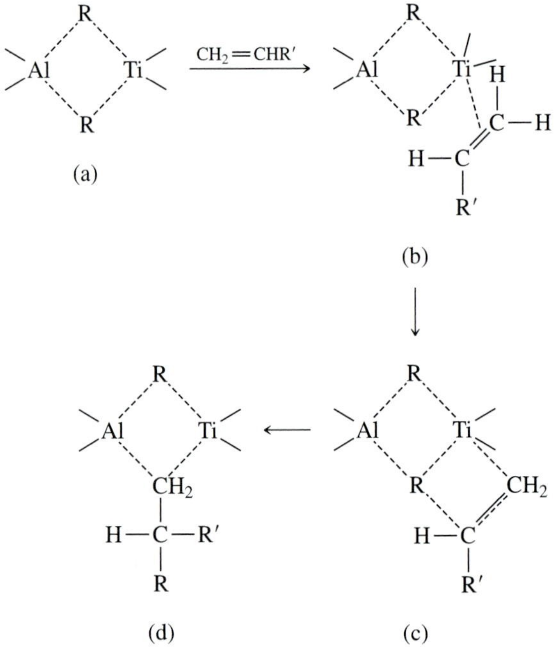

Scheme 2

Diene and cycloolefin polymerization Ziegler–Natta initiators are effective catalysts for the polymerization of dienes as well as monoolefins.

1,3-Butadiene or isoprene undergo stereospecific polymerization with Ziegler–Natta systems to yield *trans*-1,4-, *cis*-1,4-, syndiotactic, or isotactic-1,2-polybutadiene, or *trans*-1,4-, *cis*-1,4-, or 3,4-polyisoprene, depending on the particular catalyst employed. For example, an AlR_3/VCl_4 system gives a 97 to 98% yield of *trans*-1,4-polybutadiene, but an AlR_3/TiI_4 catalyst yields 93 to 94% of the *cis*-1,4-polymer. The reasons for these effects are still a matter for speculation.

Finally, it is known that Ziegler–Natta catalysts can be used for the ring-opening polymerization of cycloolefins such as cyclobutenes, cyclooctadienes, cyclodecadienes, and so on. Cyclobutene itself gives a mixture of *cis*- and *trans*-polybutadiene in the presence of $Al(C_2H_5)_3/TiCl_4$ catalyst systems.

π-Allyl Complexes as Catalysts

The polymerization of diolefins, such as butadiene, is an important technological problem. It is of considerable interest, therefore, that transition metal π-complexes such as $(\pi\text{-}C_4H_7)_2Ni$ **(11)**, can catalyze the conversion of butadiene to high polymers. In fact, a broad range of related complexes, such as $(\pi\text{-}C_4H_7)_3Cr$, $(\pi\text{-}C_4H_7)_3Nb$, $(\pi\text{-}C_4H_7)_4Ti$, or $(\pi\text{-}C_4H_7)_4Zr$ behave in the same way. The yields, rates, and stereospecificity of the polymerizations are comparable to those found in Ziegler–Natta catalysis.

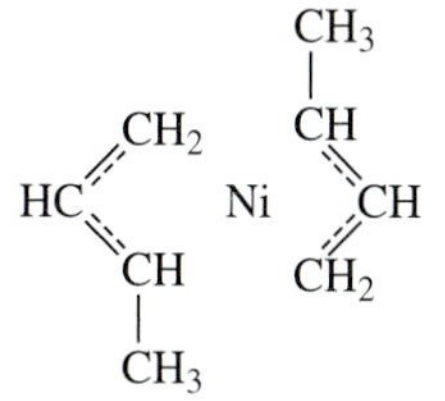

11

However, the mechanism of catalysis by π-allyl complexes is by no means fully established. It is known that the polymer formed in such reactions can be *cis*- or *trans*-polybutadiene and it is believed that the mode of attachment of the diene to the metal determines the stereochemistry of addition. Moreover, it is known

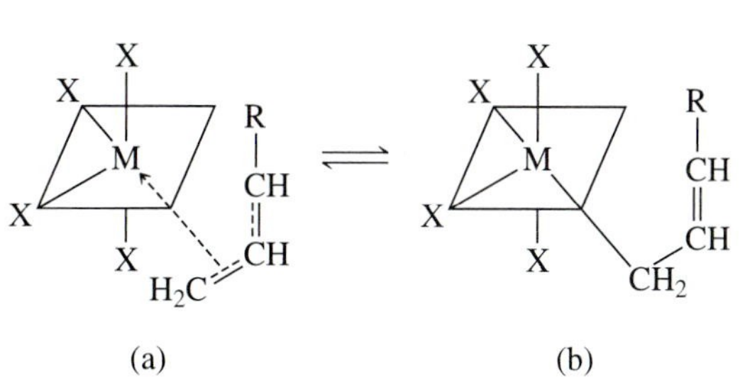

(a)   (b)   (19)

12

that π-allyl complexes can be formed from, and may rearrange to, σ-complexes in which the ligand is covalently bound to the metal through one carbon atom **(12).** A transient rearrangement of this type (19) could strongly influence the pattern of diene insertion into the ligand-metal bond.

Metallocene-Induced Polymerizations

Metallocenes are sandwich-type organometallic molecules in which two unsaturated rings (usually cyclopentadienyl or a related structure) are coordinated to a metal. The first of these molecules discovered was ferrocene **(13),** but many other species are known, including the examples shown as **(14)** and **(15).**

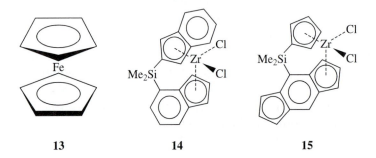

Titanium and hafnium complexes are also employed. Complexes such as **(14)** or **(15)** are often used in the presence of a cocatalyst such as methylalumoxane or diethylaluminum chloride.

The polymerization of ethylene or propylene by metallocene initiators has had a major impact on polyolefin production, and these reactions show several advantages over Ziegler–Natta systems. For example, unlike most Ziegler systems, the metallocenes are single-site initiators that allow a sophisticated control over the structure and properties of the final polymer. Thus, pure isotactic or syndiotactic polypropylene can be produced by the use of catalysts **(14)** or **(15),** respectively. This is the only way to prepare pure syndiotactic polypropylene, which is softer, more flexible, clearer, and more impact-resistant than the isotactic form. The polymer molecular weight, molecular-weight distribution, and melting point can all be controlled. Moreover, because in certain zirconocene catalysts the two organic groups can rotate around the metal above a certain temperature, it is possible to alter the structure by raising or lowering the temperature. Tacticity in the final polymer depends on the asymmetry of the catalyst. Thus, by varying the temperature and the concentration of monomer it is possible to produce polymers with blocks of atactic and isotactic polypropylene, a material that is much more elastic than either the pure atactic or isotactic forms.

The mechanism of catalysis is believed to resemble the one shown for Ziegler–Natta reactions in the sense that an incoming olefin molecule coordinates to the transition metal and is then inserted between the end of the growing polymer chain and the metal. An asymmetric shape of the catalyst controls the mode of insertion and hence the type of structure formed.

Metallocene catalysts have a few disadvantages compared to Ziegler–Natta species. They are more difficult to synthesize and consequently more expensive, and they are difficult to remove from the final product. They are also more sensitive to oxygen and moisture. Nevertheless, they represent a marked improvement for ease of control over the structure of the final macromolecule.

Other Anionic-Type Catalysts

A variety of transition metal oxides supported on alumina, silica, or charcoal induce the stereospecific polymerization of dienes. Typical catalysts include the oxides of vanadium, titanium, chromium, cobalt, nickel, and tungsten. The mechanisms of catalyst action are in most cases only partly understood. However, some type of π-coordination mechanism, similar to Ziegler–Natta catalysis, may exist.

Dienes can also be polymerized stereospecifically by the use of Alfin catalysts. These are heterogeneous catalyst systems formed by the interaction of, for example, allylsodium, sodium isopropoxide, and sodium chloride. The latter component may function as a catalyst support. Alfin catalysts are used for the preparation of high-molecular-weight, stereoregular polyisoprene or poly-(*trans*-1,4-butadiene). The mechanisms of these reactions are not fully understood and both anionic and free-radical mechanisms have been suggested.

Catalysts for Alternating Copolymerization of CO and Ethylene

Certain compounds of palladium, such as $[Pd(PPh_3)_x(MeCN)_{4-x}]^{2+}(BF_4^-)_2$ (where $x = 1$ to 3), have the ability to induce the alternating copolymerization of carbon monoxide and ethylene at room temperature to yield polyketones of structure **16**.

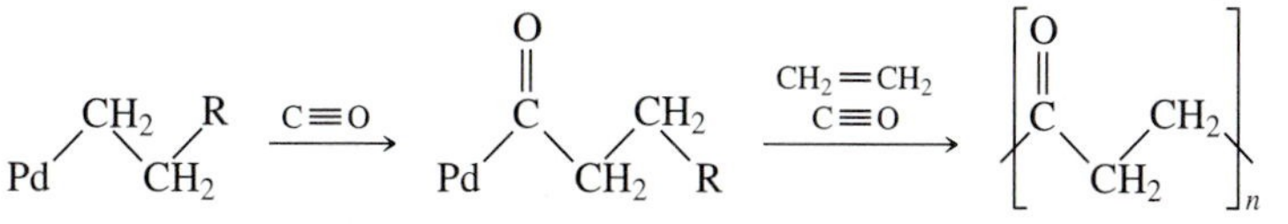

$$(20)$$

16

The mechanism of this reaction is believed to involve an alternating insertion of ethylene and carbon monoxide into a palladium–carbon, bond, as illustrated in equation (20).

Ring-Opening Metathesis Polymerization (ROMP)

The term *olefin metathesis* applies to a number of related reactions that are exemplified by equations (21) to (23). Although reaction (21) appears to be fundamentally different from the other two, in fact all three are induced by similar transition metal catalyst systems, and the reaction mechanisms in each case appear to proceed through related organometallic intermediates—specifically, through a metallocarbene.

$$
\begin{matrix}
R & H \\
\diagdown & \diagup \\
C=C \\
\diagup & \diagdown \\
H & H
\end{matrix}
\quad \rightleftharpoons \quad
\begin{matrix}
R & H \\
\diagdown & \diagup \\
C=C \\
\diagup & \diagdown \\
H & H
\end{matrix}
\quad + \quad
\begin{matrix}
R & H \\
\diagdown & \diagup \\
C=C \\
\diagup & \diagdown \\
H & H
\end{matrix}
\qquad (21)
$$

$$ \text{17} \rightleftharpoons \text{18} \qquad (22) $$

17   **18**

$$ \text{19} \rightleftharpoons \text{[--CH---CH--]}_n \qquad (23) $$

19

The first metathesis reaction was reported in 1955, when norbornene (bi-cyclo-[2.2.1]heptene) was treated with a catalyst derived from titanium tetrachloride and ethylmagnesium bromide. The metathesis polymerization of **19** was commercialized in 1976. Metathesis polymerization of cyclopentene (**17**) yields a polymer (**18**) that has the lowest glass transition temperature ($-114°C$) known for a hydrocarbon backbone polymer, and is a prospective substitute for other elastomers in rubber tires.

A wide range of cyclic olefins polymerize under metathesis conditions. These include monomers that range from cyclobutenes through cyclooctenes and higher ring systems, with the exception of strain-free cyclohexenes. Alkyl substituents at an unsaturated carbon atom retard polymerization. A broad range of bicyclic olefins, such as **19**, polymerize under metathesis conditions. Cyclooctatetraene can be polymerized to polyacetylene (see later).

The catalyst systems that induce metathesis polymerizations include species derived from WCl_6, $WOCl_4$, MoO_3, and ruthenium or rhenium halides. Organometallic compounds are sometimes required as *cocatalysts* or are formed when the metal halide reacts with the olefin.

Typical modern molybdenum or ruthenium catalysts that induce ROMP reactions are shown in **20** to **22**.[1–4]

[1]Schrock, R. R., Murdzek, J. S., Bazan, G. C., Robbins, J., Dimare, M., and O'Regan, M. J., *Am Chem Soc.* **1990,** *112,* 3875–86.

[2]Nguyen, S. T., and Grubbs, R. H., *Organometal. Chem.* **1995,** p. 195.

[3]Schwab, P., Grubbs, R. H., and Ziller, J. W,. *J. Am. Chem. Soc.* **1996,** *118,* 100–10.

[4]Bielawski, C. W., and Grubbs, R. H., *Angew. Chem. Int. Ed.* **2000,** *39,* 2903–5

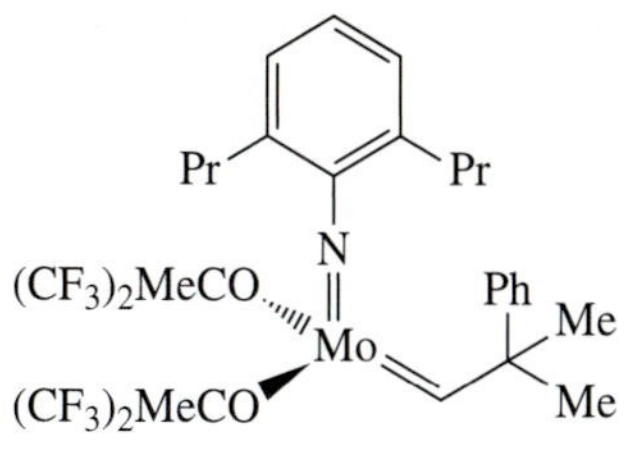

Pr = Propyl   Cy = Cyclohexyl   R = Mesityl (C_6H_2–2,4,6–Me_3)

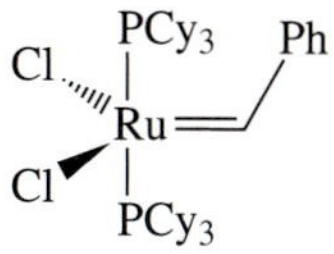

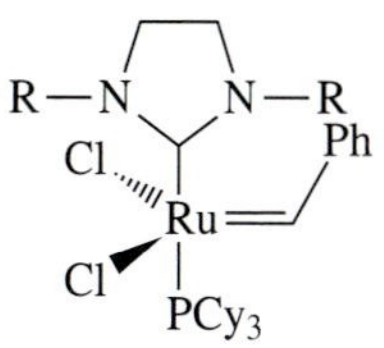

20   **21**   **22**

The molybdenum catalyst (**20**) is more active than the ruthenium species (**21**) and is capable of polymerizing many sterically hindered or electronically deactivated cyclic olefins. However, the ruthenium catalysts are more stable under laboratory conditions, and they tolerate a wider range of functional groups on the monomer, such as hydroxyl, carboxylic acid, and aldehydic units. Ruthenium catalyst (**22**) shows the highest activity but gives polymers with a broad polydispersity, probably because of chain transfer, and a mixed tacticity.

Although no universal agreement exists on the polymerization reaction mechanisms induced by these catalysts, there is strong evidence that metathesis polymerizations proceed through formation of transient metal carbene intermediates and metallocyclobutanes. Such a mechanism is summarized in Scheme 3.

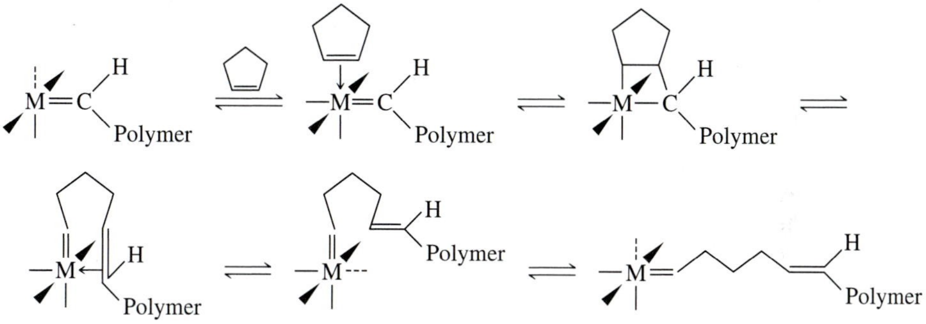

(M = metal, | = bond to ligand, ⦙ = vacant coordination site)

Scheme 3

"Living" metathesis polymerizations that yield block copolymers and narrow molecular-weight distributions are now known.

Monomers that can be polymerized by these catalysts range from norbornene and substituted norbornenes to cyclopentene, 2-methylcyclopentene, cyclooctene, cyclooctadiene, both hydroxy- and acyl-cyclooctenes, and 1,5-dimethyl-1,5-cyclooctadiene.

From the viewpoint of electroactive polymers (Chapter 23), it is interesting that two different metathesis polymerizations provide access to polyacetylene.

These are illustrated in reactions (24) and (25). The *cis*-form then undergoes thermal isomerization to trans polyacetylene.

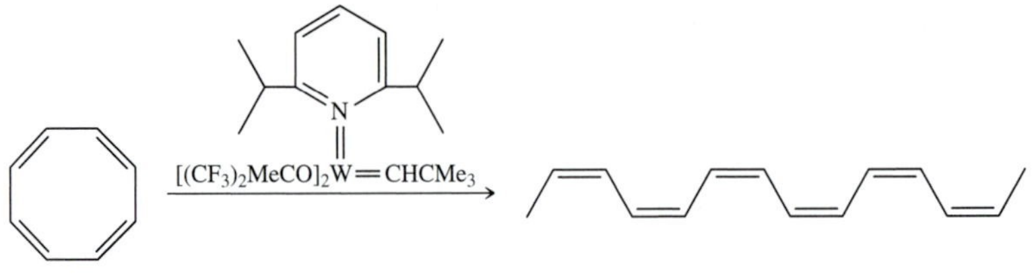

(24)

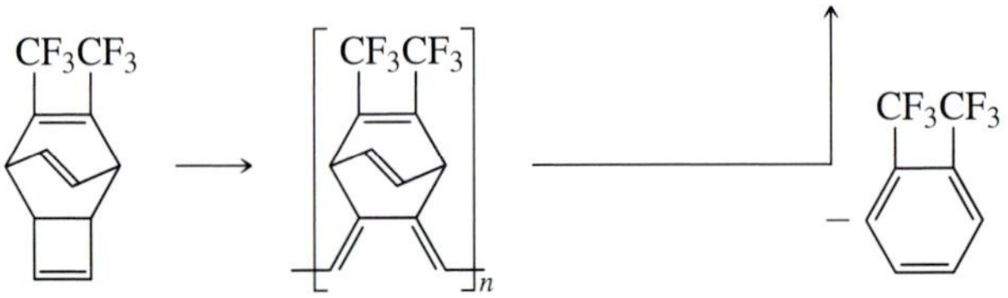

(25)

Finally, it should be noted that nonpolymerization olefin metathesis processes (reaction 21) provide access to a wide variety of olefinic and vinyl monomers that are converted to polymers via other polymerization pathways.

Acyclic Diene Metathesis Polymerization (ADMET)

Another use of metathesis chemistry for polymerization is found in ADMET reactions.[1] In these, a diene is treated with a metathesis catalyst to bring about polymerization, with the concurrent loss of ethylene (equation 26).

(26)

This is an equilibrium process, and it can be driven toward the high polymer only by the removal of ethylene, typically by volatilization under low-pressure conditions.

The catalysts for this process are metal alkylidene-type species of the type shown in (23) and (24) or ROMP-type catalysts such as (25) or (26). The classical ROMP catalysts, (25) and (26), must be activated to form the metal carbenes by treatment with $SnBu_4$, $SnBu_3H$, or $SnMe_4$.

[1]Tindall, D., Pawlow, J. H., and Wagener, K. R., in "Alkene Metathesis in Organic Synthesis," Vol 1, p. 184: in *Topics in Organometallic Chemistry* (A. Furstner, ed.) Springer-Verlag, Bern, Heidelberg, **1998**.

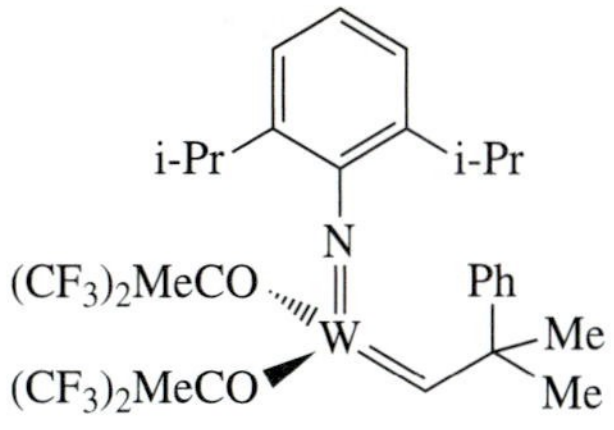

23 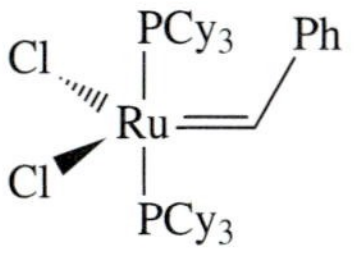 **24** 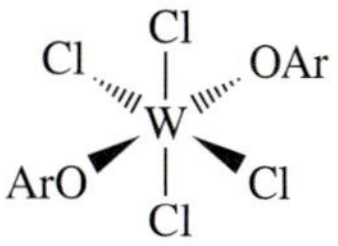 **25** 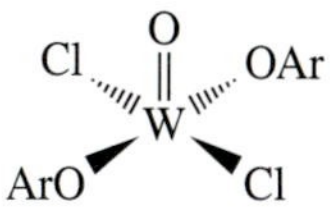 **26**

Ar = 2,6–dibromo– or diphenylbenzene

A large number of dienes have been polymerized by this process, including the examples shown in **27** to **30**.

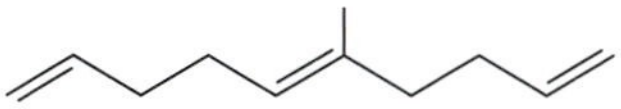

27

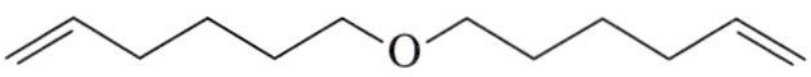

28

29

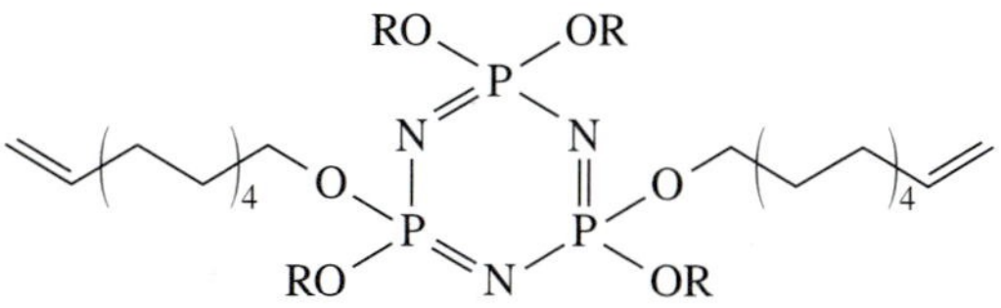

30

OR = OAr or $OCH_2CH_2OCH_2CH_2OCH_3$

A requirement for the incorporation of functional groups into the main chain is that the olefin units in the monomer must be separated from the functional site (-O-, -S-, phosphazene, etc.) by at least two carbon atoms. This is to ensure that the functional site does not coordinate to the catalyst and bring the polymerization to a halt.

Product polymers such as 1,4-polybutadiene can be depolymerized in high yield to the diene monomer using a metathesis catalyst in the presence of ethylene. This provides a route for recycling polymers or producing telechelic oligomers for incorporation into block copolymers.

GROUP TRANSFER POLYMERIZATION

This is a relatively new approach to the living solution polymerization of vinyl monomers that contain carbonyl or nitrile units in the side-group structure. Block copolymers are accessible by the sequential introduction of two different monomers. The process involves the transfer of a trimethylsilyl group from a carbonyl (or nitrile) function at the terminus of a growing polymer chain to the carbonyl or nitrile side unit of an incoming monomer, with subsequent or concurrent attachment of the vinyl group to the end of the chain. The reaction is initiated in the presence of the bifluoride ion (F_2^-). The example shown in Scheme 4 illustrates the polymerization of methyl methacrylate initiated by a silyl ketene acetal in the presence of HF_2^-. The nucleophile (HF_2^-) is assumed to bond to the silicon atom of the initiator, while the carbonyl group of the first entering monomer coordinates to the same silicon and then undergoes vinyl addition to the complex. Subsequent propagation steps then follow the same sequence.

This is the so-called associative mechanism proposed by the team at DuPont which discovered this process.[1] However, others have suggested another interpretation[2-5] in which enolate anions and silyl ketone end-groups are in rapid equilibrium with a 5-coordinate silicon species. This equilibrium generates a low concentration of enolate anionic end units which allow silyl groups to be exchanged between growing polymer chains (equation 27).

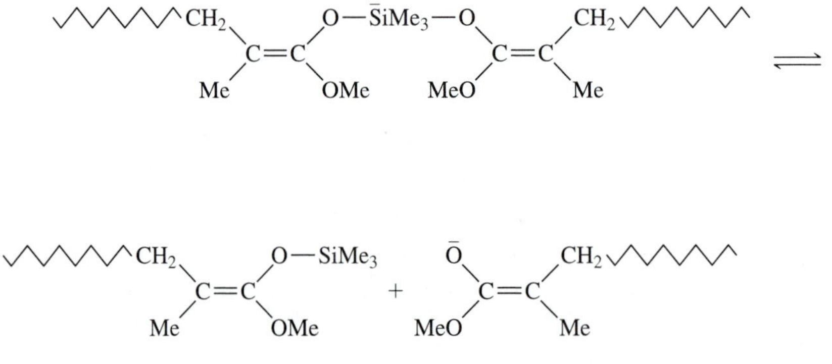

$$(27)$$

[1]Webster, O. W., *Makromol. Chem. Symp.* **1993**, *67*, 365.
[2]Muller, A. H. E., *Makromol. Chem., Macromol. Symp.* **1990**, *87*, 32.
[3]Bywater, S., *ibid*, **1993**, *339*, 67.
[4]Quirk, R. P., Ren, J., Bidinger, G., *ibid.* **1993**, *351*, 67.
[5]Jenkins, A. D., *Eur. Poly. J.*, **1999**, *649*, 27.

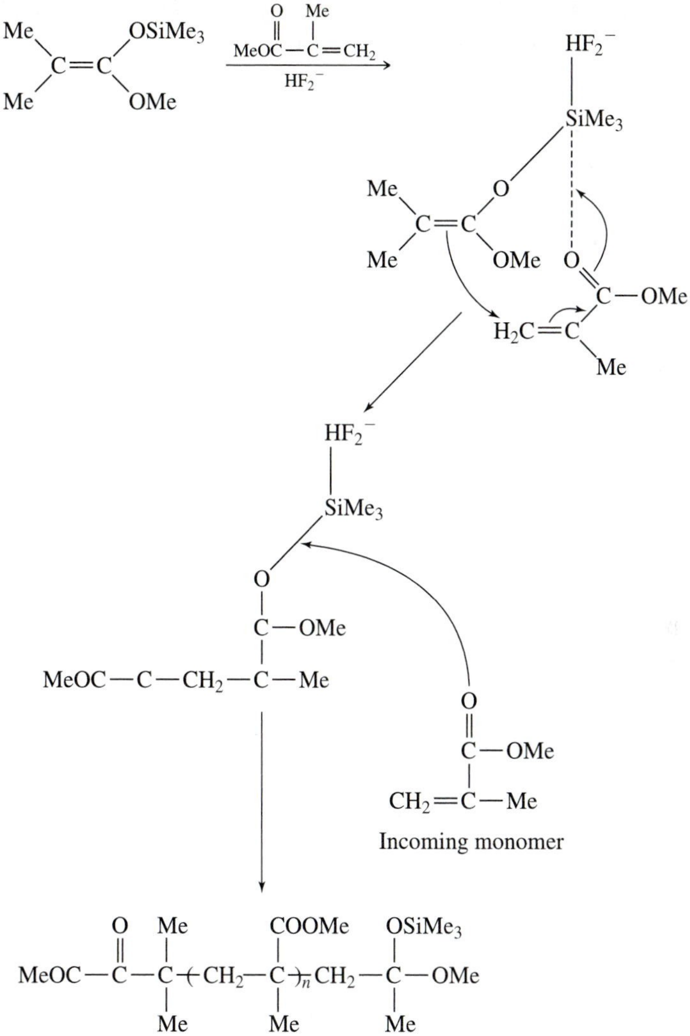

Poly(methyl methacrylate)

Scheme 4

The products formed by group-transfer polymerization are certainly very simi-lar to those formed by classical anionic polymerization techniques, but at higher temperatures. The exact mechanism of this process is still in dispute. Indeed, both pathways may be accessible under different experimental conditions.

From this mechanism it will be clear that monomers that do not possess a side-group carbonyl or nitrile unit will not participate in the reaction, and this is a limitation. However, the process has several striking advantages. It proceeds rapidly at room temperature, gives polymer in quantitative yield, and provides a narrow molecular-weight distribution. In fact, the polymers are living. Moreover, the degree of polymerization can be controlled by the ratio of monomer to silyl ketone acetal initiator. Group transfer polymerization provides access to polymers with side groups that would not survive conventional free-radical polymerization without crosslinking (e.g., polymers with allyl ester side groups). The method can also yield macromolecules with some stereoregularity, although the stereoregularity is in general inferior to that obtained by Ziegler–Natta methods.

CATIONIC POLYMERIZATION

Initiators for Cationic Polymerization

As mentioned previously, cationic polymerizations are those in which the growing chain end bears a positive charge. Table 4.4 lists representative compounds in the two main classes of compounds that are known to initiate cationic polymerizations. These two classes comprise the strong protonic acids and the Lewis acids.

Most or perhaps all of the Lewis acids function as catalysts only if a cocatalyst, such as water or methanol, is present in an equimolar concentration or less. In such cases, the actual catalyst is a proton–anion complex formed by a process such as the one shown in reaction (28). An excess of the cocatalyst destroys the catalytic properties of the system.

$$F_3B{:}OH_2 \rightleftharpoons [F_3BOH]^- + H^+ \qquad (28)$$

Monomers for Cationic Polymerization

Monomers that polymerize under the influence of cationic catalysts usually contain *electron-supplying* substituent groups. Examples include isobutylene (**31**), 1,3-butadiene (**32**), vinyl ethers (**33**), *para*-substituted styrenes (**34**), α-methylstyrenes (**35**), and aldehydes (**36**).

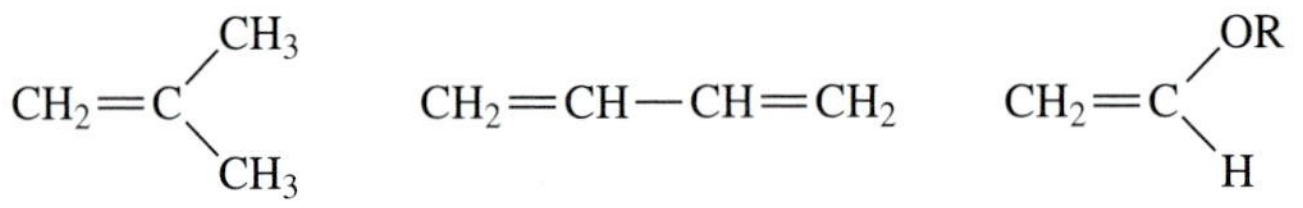

| 31 | 32 | 33 |

TABLE 4.4 CATIONIC CATALYSTS

Catalyst Class	Example Formulas
1. Strong protonic acids	H_2SO_4
	$HClO_4$
	HCl
2. Lewis acids and their complexes	BF_3
	$BF_3{:}O(C_2H_5)_2$
	BCl_3
	$TiCl_4$
	$AlCl_3$
	$SnCl_4$

R or OR

CH_2=C
   |
   H

34

R or OR

CH_2=C
   |
   CH_3

35

O=C with R and H

36

Experimental Conditions for Cationic Polymerizations

Low-temperature polymerization conditions are nearly always needed to suppress unwanted side reactions. Furthermore, the solvents used must be unreactive and very dry.

   A classical example of a cationic reaction is provided by the polymerization of α-methylstyrene in the presence of boron trifluoride hydrate. Toluene may be used as a solvent after it has been dried over calcium hydride. Similarly, the α-methylstyrene should be distilled before use and dried over the same drying agent. The boron trifluoride can conveniently be obtained from a gas cylinder. The water needed as a cocatalyst will already be present in the system in spite of the precautions taken to exclude moisture. Two experimental approaches are available. In one (Figure 4.4), a beer or soft drink bottle is charged with a solution of the monomer, and air is removed with a stream of dry nitrogen. The

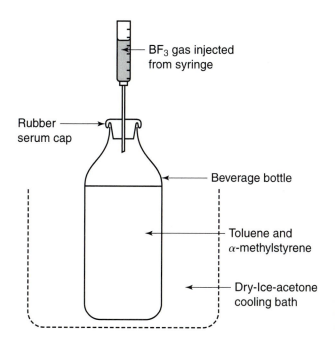

BF$_3$ gas injected from syringe

Rubber serum cap

Beverage bottle

Toluene and α-methylstyrene

Dry-Ice-acetone cooling bath

Figure 4.4 Apparatus for the small-scale cationic polymerization of α-methylstyrene.

bottle is capped with a rubber serum cap-stopper, and the bottle and contents are cooled to −78°C. The catalyst is introduced as a gas through the serum cap by means of a hypodermic syringe, and the "reactor" is then agitated for several hours at −78°C. When polymerization is complete (as indicated by the presence of a viscous reaction mixture), the cap is removed and the reaction products are poured into methanol to destroy the catalyst.

The second method is more suitable for following the rates of polymerization of gaseous monomers. The reaction is carried out in a glass vessel attached to a vacuum line (Figure 4.5). Monomer (e.g., isobutylene), solvent, catalyst, and cocatalyst are stored separately in gas storage bulbs or in flasks attached to the vacuum line, and quantities are measured as gases in bulbs of known volume. Stirring is by means of a magnetic stirrer, and the progress of the reaction is followed by observation of the monomer vapor pressure by means of a manometer. The reaction vessel must be cooled to dissipate the heat evolved. An apparatus such as this can be used for the study of the effects of different catalysts and cocatalysts on the reaction.

The Mechanism of Cationic Polymerization

Again it is convenient to divide the overall mechanism into initiation, propagation, chain transfer, and termination steps.

1. *Initiation* actually involves two sequential steps: the generation of a proton and the addition of that proton to the monomer. Strong protonic acids in

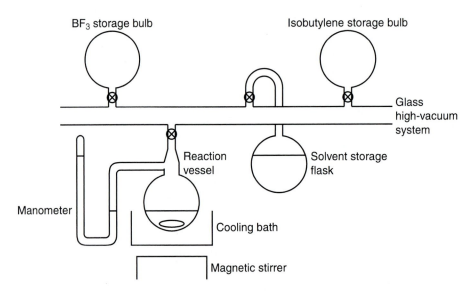

Figure 4.5 Glass high-vacuum system designed for following the rate of polymerization of a volatile monomer under rigorously controlled conditions. (Based on a system by A. M. Eastham.)

nonaqueous media liberate protons by the conventional ionization processes shown in (29), (30), and (31). Lewis acids interact with the cocatalyst to yield a proton donor by reactions such as (32), (33), or (34).

$$H_2SO_4 \rightleftharpoons H^+ + HSO_4^- \tag{29}$$

$$HClO_4 \rightleftharpoons H^+ + ClO_4^- \tag{30}$$

$$HCl \rightleftharpoons H^+ + Cl^- \tag{31}$$

$$BF_3 + H_2O \rightarrow F_3B{:}OH_2 \rightleftharpoons F_3BOH^- + H^+ \tag{32}$$

$$BF_3 + CH_3OH \rightarrow F_3B{:}OCH_3H \rightleftharpoons F_3BOCH_3^- + H^+ \tag{33}$$

$$SnCl_4 + H_2O \rightarrow [Cl_4Sn{:}OH_2] \rightleftharpoons Cl_4SnOH^- + H^+ \tag{34}$$

Boron trifluoride and stannic chloride do not normally function as catalysts in the absence of a cocatalyst such as water or methanol. However, even when such cocatalysts are not deliberately added to the system, enough adsorbed water molecules are present on the inside surface of the apparatus to cocatalyze the reaction. It can, in fact, be shown that an optimum ratio of catalyst to cocatalyst exists (often 1:1) that gives a maximum reaction rate. Larger amounts of cocatalyst decrease the reaction rate by destruction of the catalyst. Lowering of the cocatalyst concentration below the optimum amount lowers the polymerization rate. As shown in Figure 4.6, the very lowest cocatalyst concentrations cannot be attained reproducibly, but

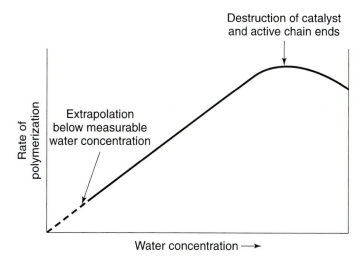

Figure 4.6   Falloff in the rate of a cationic polymerization as the amount of the cocatalyst (water) is reduced below the optimum ratio. See, for example, papers by Evans, A. G., and Meadows, G. W., *Trans. Faraday Soc.*, **1950**, *46*, 327; and A. M. Eastham, *J. Am. Chem. Soc.*, **1956**, *78*, 6040.

extrapolation of the curve strongly suggests that the polymerization rate would be zero in the complete absence of a cocatalyst. Thus, for practical purposes, nearly all cationic catalysts can be considered as species of formula $H^{\oplus}X^{\ominus}$.

Initiation of a monomer molecule, in principle, then involves the addition of the catalyst ion pair across the double bond (35). The mode of addition across the double bond will be such that the proton will add to the

$$H^{\oplus}X^{\ominus} + CH_2{=}\overset{\displaystyle R}{\underset{\displaystyle R}{C}} \longrightarrow HCH_2{-}\overset{\displaystyle R}{\underset{\displaystyle R}{\overset{\oplus}{C}}}X^{\ominus} \tag{35}$$

carbon atom that bears the greatest electron density, thereby forming the most stable carbonium ion. If the side group, R, is an electron-supplying group, then the addition will take place as shown in (35).

2. *Propagation* takes place by successive insertion of monomer molecules into the cation–anion "bond." The first propagation step, to yield a dimer, is illustrated in (36). At low temperatures, chain growth takes place rapidly without appreciable chain transfer.

$$CH_3{-}\overset{\displaystyle R}{\underset{\displaystyle R}{\overset{\oplus}{C}}}\ X^{\ominus} \qquad \longrightarrow \quad CH_3{-}\overset{\displaystyle R}{\underset{\displaystyle R}{C}}{-}CH_2{-}\overset{\displaystyle R}{\underset{\displaystyle R}{\overset{\oplus}{C}}}\ X^{\ominus} \tag{36}$$
$$CH_2{=}\overset{\displaystyle R}{\underset{\displaystyle R}{C}}$$

3. *Chain transfer* becomes important at temperatures near room temperature. An important transfer step involves the donation of a proton from a terminal side group to a monomer molecule (37). The newly initiated monomer molecule can, of course, generate a new chain. If more chains are initiated in this way, the average chain length of polymer in the system will be reduced. Hence, a need exists to maintain low reaction temperatures if high-molecular-weight polymers are desired. (See also Chapter 13, equations 48 and 50.)

$$\left[CH_3{-}\mathcal{W}{-}\overset{\displaystyle CH_3}{\underset{\displaystyle R}{C}}\right]^{\oplus}\ X^{\ominus} + CH_2{=}\overset{\displaystyle R}{\underset{\displaystyle R}{C}} \longrightarrow CH_3{-}\mathcal{W}{-}\overset{\displaystyle CH_2}{\underset{\displaystyle R}{C}} + CH_3{-}\overset{\displaystyle R}{\underset{\displaystyle R}{\overset{\oplus}{C}}}\ X^{\ominus} \tag{37}$$

4. *Termination* of the polymer chain can occur by the transfer mechanism discussed above or by the loss of a proton to the X^- counterion, thereby terminating also the kinetic chain. Termination can also take place by the

reaction of a growing chain end with traces of water or other protonic reagents (38). This mode of termination can be important even in rigorously dried glassware.

$$R-\text{W}-\overset{\overset{\displaystyle R}{|}}{\underset{\underset{\displaystyle R}{|}}{C}}{}^{\oplus}\ \ X^{\ominus} + H_2O \ \longrightarrow \ R-\text{W}-\overset{\overset{\displaystyle R}{|}}{\underset{\underset{\displaystyle R}{|}}{C}}OH + HX \qquad (38)$$

The mechanistic steps described above assume the participation of ions or ion pairs in the polymerization process. However, evidence exists that some cationic-type monomers may polymerize by a nonclassical, "covalent" mechanism. One example of such a mechanism is described later under the heading of "living" carbocationic polymerizations.

Special Characteristics of Classical Cationic Polymerizations

The kinetic characteristics of cationic polymerization are considered in Chapter 13. Here, attention is drawn to the consequences of these kinetic features. Because chain transfer reactions can be important in cationic reactions, a curious relationship exists between the rates of chain transfer, propagation, and termination as the reaction temperature is lowered. *The rate of polymerization will accelerate as the temperature is lowered.* This contrasts with the behavior of free-radical polymerizations. Furthermore, if the activation energy for termination and transfer is greater than that for propagation, the molecular weight also will increase as the temperature is lowered.

Living Carbocationic Polymerization

Until recently it was assumed that cationic polymerization processes could not give rise to "living" polymers comparable to the anionic systems, because the propagating cation in a classical carbocationic polymerization is intrinsically unstable with respect to chain transfer to the monomer. However, it has been found that initiators derived from tertiary esters, such as *t*-butyl acetate, complexed to boron trifluoride are capable of initiating a living polymerization from monomers such as isobutylene.

The mechanism of these reactions is still not clear. However, it has been suggested[1] that an initiating complex of the type shown in **37** is first formed and that this then induces the insertion of an isobutylene molecule, as shown in Scheme 5. Unlike classical cationic polymerizations, these reactions are relatively insensitive to moisture. Moreover, because of their living character they can be used to prepare block copolymers. Other living polymerizations have also been

---

[1]Faust, R., and Kennedy, J. P., *J. Polymer Sci. (A), Polymer Chem. Ed.*, **1987,** *45,* 1847.

discovered based on the use of hydrogen iodide/iodine initiators with monomers such as vinyl ethers or *p*-methoxystyrene.

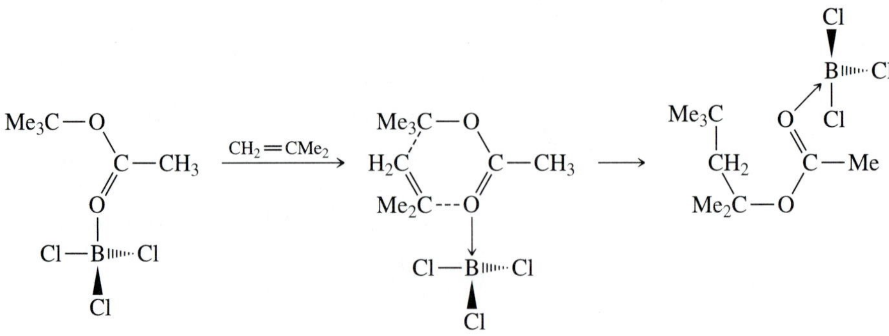

37

Scheme 5

Reversible Termination in Living Cationic Polymerization

There are intriguing ramifications to living cationic polymerizations, as summarized by Szwarc.[1] Consider a polymer of type **38** in equation (39), which is terminated by a carbon-halogen bond (*C-X*). The *C-X* bond is unlikely to ionize to give a $CH_2^+ X^-$ function, and so this end-group is "dead"—it cannot lead to the addition of another monomer molecule. However, if a suitable complexing agent (*Y*) is added, ionization may be assisted, and a series of equilibria established, as shown in equation (39).

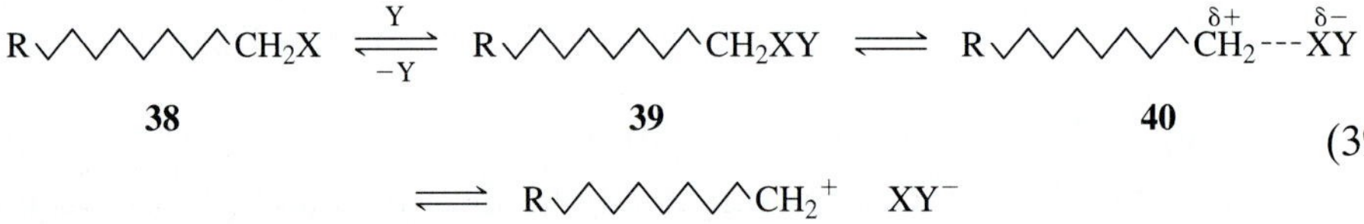

$$\text{(39)}$$

It is assumed that both **38** and **39** are inactive for chain propagation, but **40** and **41** may well be active. Thus, under conditions where **39, 40**, and **41** are participating in an equilibrium, the possibility exists that some chains will be active, **40** and/or **41**, and some will be dead **39**. Thus, a reversible termination can be envisaged. If for any reason active chain ends are destroyed (say, by termination) the equilibrium will be reestablished by the ionization of **39**. Thus, the system is "living" even though the lifetime of the active species, **40** or **41**, is short.

Moreover, because the equilibria will depend on the nature of the Lewis acid, *Y*, and the terminal group, *X*, the investigator can tune the system to control the concentration of the pool of species **39** and hence influence the propa-

---

[1]Szwarc, M., *Ionic Polymerization Fundamentals*, Hanser, New York, **1996**.

gation rate and the degree of living character. Furthermore, the polymerization rate will depend on the (fast) formation of species **39** from **38** and the rapid dissociation of **39** to **40** and **41**. Thus, polymerization rates can be very high.

Cationic Polymerization of Aldehydes

Most of the discussion so far has been directed to the cationic polymerization of olefins. However, a second group of monomers—the aldehydes—are especially prone to cationic polymerization.

For example, formaldehyde can be polymerized in the presence of a boron trifluoride–water catalyst. The reaction mechanism is believed to resemble the one discussed above for olefin monomers. Initiation involves the protonation of the aldehyde oxygen atom (reactions 40 and 41).

$$F_3B{:}OH_2 \longrightarrow F_3^{\ominus}BOH + H^{\oplus} \quad (H^+X^-) \tag{40}$$

$$H^+X^- + O{=}CH_2 \longrightarrow HO-\overset{\overset{\displaystyle H}{|}}{\underset{\underset{\displaystyle H}{|}}{C}}{}^{\oplus}\ \ X^{\ominus} \xrightarrow{\ O{=}CH_2\ }$$

(41)

$$HO-CH_2(\!-O-CH_2\!)_{\!n}\!-O-\overset{\overset{\displaystyle H}{|}}{\underset{\underset{\displaystyle H}{|}}{C}}{}^{\oplus}\ \ X^{\ominus}$$

The prospect that boron trifluoride alone can function as a catalyst in this reaction has also been proposed (42). However, linear propagation by this mechanism would require the progressive separation of the anionic and cationic charges—an unlikely process. Charge separation could only be avoided by a "pseudocyclic" propagation mechanism, such as the one shown in **42**.

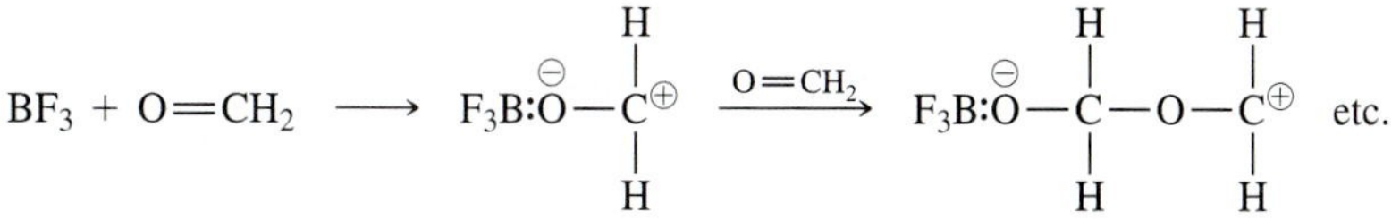

$$BF_3 + O{=}CH_2 \longrightarrow F_3B{:}\overset{\ominus}{O}-\overset{\overset{\displaystyle H}{|}}{\underset{\underset{\displaystyle H}{|}}{C}}{}^{\oplus} \xrightarrow{\ O{=}CH_2\ } F_3B{:}\overset{\ominus}{O}-\overset{\overset{\displaystyle H}{|}}{\underset{\underset{\displaystyle H}{|}}{C}}-O-\overset{\overset{\displaystyle H}{|}}{\underset{\underset{\displaystyle H}{|}}{C}}{}^{\oplus}\quad \text{etc.} \tag{42}$$

$$
\begin{array}{c}
H-\overset{|}{\underset{|}{C}}-H\quad O \\
F_3B{:}\underset{\ominus}{O}\quad \overset{\displaystyle}{\underset{\oplus}{CH_2}} \\
\text{-----}CH_2{=}O
\end{array}
$$

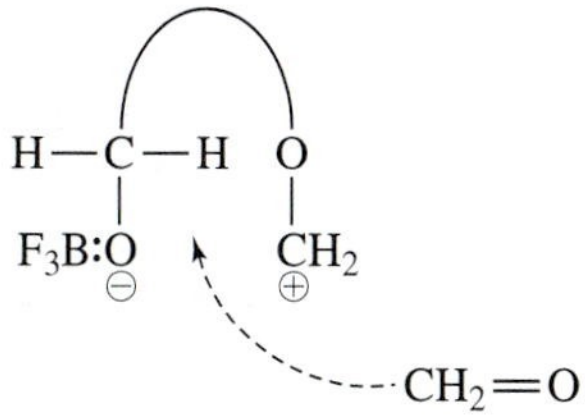

42

Formaldehyde also polymerizes in the absence of added catalyst, especially when cooled. Traces of formic acid impurity could function as catalytic species. Polyformaldehyde is a tough thermoplastic material. It is offered commercially under the trade name Delrin[1] and is particularly useful for the fabrication of gear wheels.

Acetaldehyde also polymerizes "spontaneously" below its freezing point, or when treated with protonic acids or Lewis acids. Amorphous polyacetaldehyde is a rubbery polymer with a molecular weight in the region of approximate 40,000. It readily depolymerizes back to acetaldehyde, particularly if the chain ends remain uncapped and active.

Higher aldehydes such as propionaldehyde or butyraldehyde can be polymerized only at low temperatures and high pressures. They depolymerize to the parent aldehyde when exposed to the atmosphere at ambient pressures. This depolymerization behavior represents a rather profound thermodynamic characteristic of many polymer systems, and it is considered in more detail in Chapter 10.

COMPARISON OF POLYMERS PREPARED BY DIFFERENT ROUTES

This is an appropriate point in the book to summarize some general conclusions regarding the advantages and disadvantages of the polymer synthesis methods discussed up to this point, and others to be mentioned later. The guidelines are given in Table 4.5. These guidelines should be viewed as generalities only, since some of the disadvantages can be minimized in specific instances, for example, by the development of living carbocationic systems or the discovery of new organometallic coordination initiators.

STUDY QUESTIONS

1. Given 100 g of styrene, and the appropriate apparatus and conditions to conduct the anionic polymerization of this monomer with *n*-butyllithium, calculate the expected average molecular weight of the resultant living polymer chains if you could introduce exactly 500 molecules of *n*-butyllithium. Assume no termination and total usage of the monomer and initiator.

2. Make a list of the main differences between anionic, cationic, free-radical, and condensation polymerizations. Which of these reaction types normally gives the lowest proportion of side reactions?

3. Suppose that in an anionic or coordination polymerization, an opportunity exists to use several different initiators, RM, which contain the same R group, but different metals, M^1, M^2, M^3, etc. Speculate on the effects that might be observed if the $R-M$ bond becomes progressively more covalent along the series $R-M^1$, $R-M^2$, etc.

4. What storage conditions would you choose if faced with the need to preserve an unterminated anionic polymer for a period of 10 years?

[1]DuPont's registered trademark for its acetal resins.

TABLE 4.5   ADVANTAGES AND DISADVANTAGES OF DIFFERENT POLYMER SYNTHESIS ROUTES

Polymerization or Synthesis Type	Advantages	Disadvantages
Condensation	Large number of potential monomers	Relatively low molecular weight and broad molecular-weight distribution; no stereo selectivity; slow; requires heat
Free radical	Large number of monomers	Poor selectivity and no stereo control; requires heat or radiation
Classical anionic	Narrow-molecular-weight distribution; some stereo control; block copolymers accessible via "living" polymers	Limited number of monomers; requires low temperatures
Ziegler-Natta	Very high selectivity and stereo specificity; room temperature reactions	Suitable for olefins but not vinyl compounds; molecular weights not especially high
Metathesis	Yields polymers with unsaturation in main chain; can be "living"	Limited to cyclic olefins (or to alkynes)
Group transfer	Living polymers and room-temperature reactions; allows polymerization of monomers with "sensitive" side groups	Limited primarily to monomers with $C=O$ or $C\equiv N$ units in side groups; limited stereo control
Classical cationic	Molecular-weight control possible	Molecular weights limited by chain transfer; limited to olefins; cooling required; sensitive to moisture
"Living" cationic	Control of molecular weight possible; narrow molecular-weight distribution; allows block copolymer synthesis; relatively insensitive to moisture	
Ring opening	Can yield higher molecular weights than condensation; more convenient than most condensation processes	Limited number of monomers; broad molecular-weight distributions; need to heat or irradiate
Macromolecular substitution	Provides access to polymers not available through other routes; efficient reactions with some inorganic polymers; provides access to very wide range of derivative polymers	Inefficient substitution with most organic polymers; limited number of inorganic macromolecular intermediates available

* This table was developed from one suggested by H. F. Mark.

5. Design a laboratory apparatus or a large-scale reaction system that would allow an *n*-butyllithium-initiated vinyl polymerization to be carried out in a continuous-flow reactor. What advantages or disadvantages can you foresee for such a process? Repeat the exercise for the use of an Alfin catalyst or a heterogeneous Ziegler–Natta system.

6. Comment on the reasons why the following monomers are not normally polymerized by ionic processes: acrylic acid, allyl alcohol, acrylamide, vinyl chloride.

7. What effects might you predict in a BF_3-catalyzed polymerization if the solvent was changed from methylene chloride to benzene?

8. What types of experiments can you devise to probe the mechanism of an organometallic reagent-initiated addition polymerization? Why has so much difficulty been encountered by investigators who seek to explain the catalytic mechanisms in these reactions? Why is it important to know the answers to these questions?

SUGGESTIONS FOR FURTHER READING

Anionic Polymerization

ARJUNAN, P., McGRATH, J. E., and HANLON T. L., (Eds.), *Olefin Polymerization: Emerging Frontiers*, Washington, D. C.: American Chemical Society, **2000.**

HALASA, A. F., SCHULZ, D. N., and TATE, D. P., "Organolithium Catalysis of Olefin, and Diene Polymerization," *Adv. Organomet. Chem* (F. G. A. Stone and R. West, Eds.), **1979,** *18.*

HSIEH, H. L., *Anionic Polymerization: Principles and Practical Applications*, New York: Marcel Dekker, **1996.**

MORTON, M., and FETTERS, L. J., "Homogeneous Anionic Polymerization of Unsaturated Monomers," *J. Polymer Sci. (D), Macromol. Rev.*, **1967,** *2,* 71.

ROBBINS, O., *Ionic Reactions and Equilibria*, New York: Macmillan, **1997.**

SZWARC, M., *Living Polymers and Mechanisms of Anionic Polymerization*, Advances in Polymer Science Series. New York: Springer-Verlag, **1983.**

SZWARC, M. *Ionic Polymerization and Living Polymers*, New York: Chapman & Hall, **1993**.

VAN BEYLEN, M., BYWATER, S., SMETS, G., SZWARC, M., and WORSFELD, D. J., "Developments in Anionic Polymerization: A Critical Review," *Advan. Polymer. Sci.*, **1988,** *86,* 87.

Coordination Polymerization

AMASS, A. J., "Metathesis Polymerization," in *Comprehensive Polymer Science*, Vol. 4, Part II (G. C. Eastmond, A. Ledwith, S. Russo, and P. Sigwalt, eds.). Oxford: Pergamon Press, **1989,** p. 1.

BRINTZINGER, H. H., FISCHER, D., MULHAUPT, R., REIGER, B., and WAYMOUTH, R. M., "Stereospecific Olefin Polymerization with Chiral Metallocene Catalysts," *Angew. Chem. Ed. Engl.*, **1995,** *34,* 1143.

CARRICK, W. L., "The Mechanism of Olefin Polymerization by Ziegler-Natta Catalysts," *Fortschr. Hochpolymer-Forsch*, **1973,** *12,* 65.

DRAGUTAN, V., and STRECK, R., *Catalytic Polymerization of Cycloolefins: Ionic, Ziegler–Natta, and Ring-Opening Metathesis Polymerization*. New York: Elsevier, **2000.**

FEAST, W. J., "Metathesis Polymerization Applications," in *Comprehensive Polymer Science*, Vol. 4, Part II (G. C. Eastmond, A. Ledwith, S. Russo, and P. Sigwalt, eds.), Oxford: Pergamon Press, **1989,** p. 135.

HU, Y., CARLSON, E. D., FULLER, G. G., and WAYMOUTH, R. M., "Elastomeric Polypropylenes from Unbridged 2-Phenylindene Zirconocene Catalysts: Temperature Dependence of Crystallinity and Relaxation Properties," *Macromolecules*, **1999,** *32,* 3334.

IVIN, K. J., *Olefin Metathesis*. London: Academic Press, **1983.**

IVIN, K. J., *Olefin Metathesis and Metathesis Polymerization*. San Diego: Academic Press, **1997.**

KLABUNDE, U., MULHAUPT, R., HERSKOVITZ, T., JANOWICZ, A. H., CALABRESE, J., and ITTEL, S. D., "Ethylene Homopolymerization with P,O-Chelated Nickel Catalysts," *J. Polymer Sci. (A), Polymer Chem. Ed.*, **1987,** *25,* 1989.

KLAVETTER, F. L., and GRUBBS, R. H., "Polycyclooctatetraene: Synthesis and Properties," *J. Am. Chem. Soc.*, **1988,** *110* (23), 7807.

LEINO, R., GOMEZ, F. J., COLE, A. P., and WAYMOUTH, R. M., "Syndiospecific Polypropylene Polymerization with C1 Symmetric Group 4 ansa-Metallocene Catalysts," *Macromolecules*, **2001,** *34,* 2072.

MCKNIGHT, A., and WAYMOUTH, R. M., "Group 4 ansa-Cyclopentadienylamido Catalysts for Olefin Polymerization," *Chem. Rev.*, **1998,** *98,* 2587.

SCHROCK, R. R., FELDMAN, J., GRUBBS, R. H., and CANNIZZO, L., "Ring-Opening Polymerization of Norbornene by a Living Tungsten Alkylidene Complex," *Macromolecules,* **1987,** *20,* 1169.

SCHROCK, R. R., "Ring-Opening Metathesis Polymerization," in *Ring-Opening Polymerization* (D. J. Brunelle, Ed.) New York; Hanser, **1993,** Ch. 4.

SEN, A., "The Copolymerization of Carbon Monoxide with Olefins," *Advan. Polymer Sci.*, **1986,** *73/74,* 125.

TAIT, P. J., "Monoalkene Polymerization: Ziegler–Natta and Transition Metal Catalysts," in *Comprehensive Polymer Science*, Vol. 4, Part II (G. C. Eastmond, A. Ledwith, S. Russo, and P. Sigwalt, eds.). Oxford: Pergamon Press, **1989,** p. 1.

TINDALL, D., PAWLOW, J. H., and WAGENER, K. B., "Recent Advances in ADMET Chemistry" in *Alkene Metathesis in Organic Synthesis* (A. Furstner, ed.), Heidelberg: Springer-Verlag, **1998,** pp. 183–196.

WEBSTER, O. W., "Group Transfer Polymerization," in *Encyclopedia of Polymer Science and Engineering*, Vol. 7 (H. F. Mark, N. Bikales, C. G. Overberger, G. Menges, and J. I. Kroschwitz, eds.) New York: Wiley, **1987,** p. 580.

XU, G., CHUNG, T. C., "Borane Chain Transfer in Metallocene-Mediated Oletin Polymerization," *J. Am. Chem. Soc.*, **1999,** *121,* 6763.

ZAMBELLI, A., SACCHI, M. C., and LOCATELLI, P., "Syndiotactic Specific Polymerization of Propene: Some Evidence for Insertion of the Monomer on the Metal-Carbon Bond," *Macromolecules*, **1979,** *12,* 1051.

Cationic Polymerization

FAUST, R., SHAFFER, T. D. (Eds.), *Cationic Polymerization: Fundamentals and Applications*, Washington, D.C.: American Chemical Society, **1997.**

FAUST, R., and KENNEDY, J. P., "Living Carbocationic Polymerization. IV. Living Polymerization of Isobutylene," *J. Polymer Sci. (A), Polymer Chem. Ed.*, **1987,** *45,* 1847.

HIGASHIMURA, T., MIYAMOTO, M., and SAWAMOTO, M., "Mechanism of Living Polymerization of Vinyl Ethers by the Hydrogen Iodide/Iodine Initiating System," *Macromolecules*, **1985,** *18,* 611.

KENNEDY, J. P., *Cationic Polymerization of Olefins*. New York: Wiley, **1975.**

LYONS, A. R., "Polymerization of Vinyl Ketones," *J. Polymer Sci. (D), Macromol. Rev.*, **1972,** *6,* 251.

MATYJASZEWSKI, K., and SIGWALT, P., "Are There Covalent Active Species in the Polymerization of Styrene Initiated by Trifluoromethanesulfonic Acid?" *Makromol. Chem.*, **1986,** *187,* 2299.

SAWAMOTO, M., and AIDA, T., "Living Polymerizations: Cationic Alkenyl" in *Encyclopedia of Polymer Sci. and Eng., Supplement Vol.*, New York, Wiley, **1989,** pp. 412–420.

5

Photolytic, Radiation, and Electrolytic Polymerization

INTRODUCTION

The use of *chemical* catalysts and initiators provides only one method for the induction of olefin polymerization. Polymerization can also be induced by supplying the initiation energy through irradiation with visible or ultraviolet light, high-energy or ionizing radiation, or by the passage of an electric current. The conversion of a monomer to a polymer will occur through the normal propagation, termination, and transfer reactions (see Chapters 3 and 4). Only the initiation processes will be unusual.

PHOTOCHEMICAL POLYMERIZATION

Advantages and Uses

There are several advantages to be gained by the use of light for the initiation of vinyl polymerization. Perhaps the most obvious advantage in laboratory research is the avoidance of chemical contamination by initiator residues. Moreover, there is a marked convenience to photochemical reactions that appeals to many researchers.

From a practical standpoint, photopolymerization and photocross-linking can be used as a photographic process. In one type of system a plate or film is coated with a monomer such as acrylamide [$CH_2\!=\!CHC(O)NH_2$], together with a small amount of a divinyl compound. A photosensitizing dye may also be present. Exposure of parts of the film to light causes photopolymerization of the acrylamide, together with cross-linking through the divinyl residues. After the

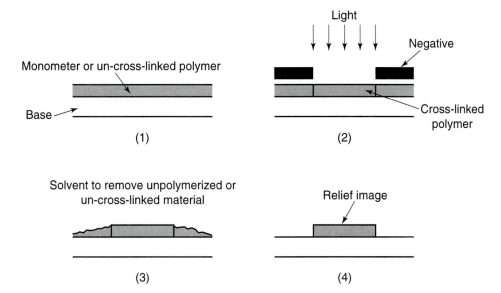

Light

Monometer or un-cross-linked polymer

Base

Negative

Cross-linked
polymer

(1)

(2)

Solvent to remove unpolymerized or
un-cross-linked material

Relief image

(3)

(4)

Figure 5.1 Photopolymerization as a photographic process. Illumination of a monomer or a cross-linkable polymer forms an insoluble relief image.

exposure, the film is "developed" by washing with water. Unpolymerized monomer is removed from those areas that were shielded from light, leaving a relief polymeric image corresponding to the exposed portions (Figure 5.1). A colored image is left if dyestuffs were incorporated into the matrix.

An alternative process makes use of the photoinduced cross-linking of polymers. For example, poly(vinyl cinnamate) or mixtures of cellulose cinnamate and poly(vinyl cinnamate) remain soluble if protected from light. However, exposure of a film of these materials to strong light (usually ultraviolet light) through a negative brings about cross-linking and insolubilization of those areas that lie beneath the transparent sections of the negative. The cross-linking mechanism involves cyclodimerization of cinnamoyl groups on different chains.

Images made either by photopolymerization or photocross-linking are now used extensively in the preparation of letterpress and lithography plates, holography, and (coupled with etching processes) in the manufacture of printed and integrated circuits. The use of photo-sensitive polymers in semiconductor microlithography is discussed in more detail in Chapters 7 and 23. Like conventional silver halide photography, photopolymerization provides a method for *amplification* of the original photochemical event. In this case, the amplification involves the chain reaction of an addition polymerization. However, the fastest photopolymerization processes have only one thousandth of the speed and sensitivity to visible light of medium-speed silver halide systems, but it is anticipated that future developments in this area could bring about the development of faster emulsions. Hopefully, this will occur before the use of silver in photography becomes drastically curtailed.

Monomers that Undergo Photopolymerization

Any monomer that will undergo chain reaction polymerization is susceptible to photopolymerization or photosensitized polymerization. The absorption of light simply produces free radicals or ions. Also, the chain-propagation steps and the termination reactions are generally not affected. The advantage of photopolymerization and photosensitized polymerization is that the initiation process may take place over a wide range of temperatures and with a greater specificity than is found in chemically initiated systems.

Some monomers, such as vinyl alkyl ketones and vinyl bromide, absorb 300-nm or longer wavelength light and dissociate directly to free radicals. These radicals initiate polymerization. Other monomers, such as styrene or methyl methacrylate, are susceptible to direct photopolymerization when exposed to 300-nm or shorter wavelength light. The detailed mechanism of the formation of the propagating radicals in this case is not completely understood, but it appears to involve the conversion of an electronically excited singlet state of the monomer to a long-lived excited triplet state.

Direct photopolymerization is not restricted to the initiation of free-radical chains. For example, the cationic chain polymerization of isobutylene has been initiated by irradiation with light of 117 and 123 nm wavelength (vacuum ultraviolet radiation).[1] The direct absorption of such radiation by isobutylene produces positive ions and electrons, and the former initiate the polymerization. Presumably other monomers that are susceptible to ionic chain polymerization will also undergo direct photopolymerization when irradiated with vacuum ultraviolet radiation.

In spite of these facts, only a few unsaturated monomers are known which absorb light between 250 and 500 nm—the most convenient wavelength range for experimental work. For other monomers, a *photosensitizer* must be added to the system. Photosensitizers are compounds that absorb light in a convenient region of the spectrum and then dissociate into free radicals or transfer energy directly to the monomer. Photosensitizers are considered in more detail later.

Experimental Technique: A Photosensitized Polymerization of Styrene

A simple photopolymerization apparatus, constructed from Pyrex tubing, ordinary laboratory corks, or rubber stoppers, and a 15-W "black light"[2] is shown in Figure 5.2. The light emitted from this lamp has its maximum intensity at a wavelength of about 360 nm, and this matches well the near-ultraviolet absorption maximum of azoalkanes. The useful surface area of the lamp for emission is about 280 cm^2 and the light flux emitted may be taken as 1.0×10^{-8} einstein/cm$^2 \cdot$ s. (An einstein is equivalent to 6.02×10^{23} light quanta.) The lamp A is held in place by holes drilled in the stoppers B, as are a filling tube C and a drain tube D.

[1]Schlag, E. W., and Sparapany, J. J., *J. Am. Chem. Soc.* **1964**, *86*, 1875.
[2]Available commercially from the General Electric Company.

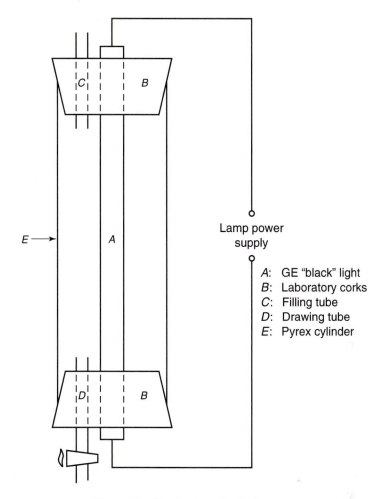

Figure 5.2 Simple photochemical reactor.

Before assembly of the apparatus, the stoppers are thoroughly cleaned and then soaked for several hours in the solvent to be used. A 3-L sample is prepared of a 2 M solution of freshly distilled styrene and 0.2 M azoisopropane in benzene. The azo compound is the photosensitizer. The solution is purged of oxygen by bubbling nitrogen gas through it for an hour. It is then introduced into the apparatus through tube C, and the irradiation is started. After 2 h the irradiation is stopped, the solution is removed from the reactor, and the benzene solvent is removed by evaporation.[1] About 8 g of polystyrene of average molecular weight 20,000 should remain as a residue.

---

[1]Evaporation of the benzene should be carried out in a rotary evaporator or in a fume hood, but *not* on the open bench. Benzene is toxic.

To calculate the rate of initiation using equation (13), to be described later, it should be assumed[1] that the primary quantum yield, ϕ, is 0.3 and that the extinction coefficient, ε, is 10 liters-mol^{-1}-cm^{-1}.

Mechanism of Initiation by Direct Photolysis of the Monomer

Many monomers undergo free-radical chain polymerization when exposed to ultraviolet or visible light. The photopolymerization of a pure monomer occurs by the same chain *propagation* process that was discussed in Chapter 3. However, an all-inclusive mechanism has not yet been established for photolytic *initiation*.

Two different types of direct photoinitiation can be recognized. In the first, light absorption yields an electronically excited monomer molecule which subsequently decomposes to give radical fragments. Examples of monomers in this category include alkyl vinyl ketones and vinyl bromide, both of which dissociate when irradiated with ultraviolet light by the reactions shown in equations (1) to (3). Here $R\cdot$ denotes an alkyl radical. Following photodissociation of the monomer, the resultant monoradicals add to the monomer, and radical chain polymerization (Chapter 3) takes place.

$$R-\overset{\overset{O}{\parallel}}{C}-CH{=}CH_2 \xrightarrow{hv} R-\overset{\overset{O}{\parallel}}{C}\cdot + \cdot CH{=}CH_2 \tag{1}$$

$$R-\overset{\overset{O}{\parallel}}{C}\cdot \longrightarrow R\cdot + CO \tag{2}$$

$$CH_2{=}CHBr \xrightarrow{hv} Br\cdot + CH_2{=}CH\cdot \tag{3}$$

The rate of initiation of the chains and the dependence of this rate on the monomer concentration and temperature is quite different in photochemical initiation from the situation found in thermal initiation. For example, when a thermal initiator, I, is used as a source of radicals, the rate of formation of the initiating radicals is

$$\frac{d[R\cdot]}{dt} = 2k_d[I] \tag{4}$$

where k_d is the rate constant for the first-order decomposition of the initiator and [I] is the initiator concentration. For photoinitiation using a pure monomer, the rate of radical formation averaged over the system may be written as

$$\frac{d[R\cdot]}{dt} = \frac{2\phi I_0 A}{V}(1 - e^{-\alpha[M]L}) = \frac{2\phi I_0 A}{V}\left(\alpha[M]L - \frac{1}{2}(\alpha[M]L)^2 + \cdots\right) \tag{5}$$

[1]Calvert, J. G., and Pitts, J. N., *Photochemistry* (New York: Wiley, **1966**), p. 464.

where I_0 is the number of einsteins per cm^2 incident on the system per second; α the absorption coefficient (an average, if the light is not monochromatic) in liters-mol^{-1} cm^{-1}; [M] the monomer concentration in mol/liter; L the length of the light path in cm; A the area of the system illuminated in cm^2; V the volume in cm^3; and ϕ the fraction of the light quanta absorbed that result in the formation of a pair of radicals. If the system is to be irradiated uniformly, the fraction of the incident light that is absorbed must be kept small. This condition may be ensured by requiring that the term [M]L in (5) is kept sufficiently small that, in the expanded form of the exponential expression, all terms after the first may, with negligible error be discarded. Under such conditions, and with replacement of α by the more familiar extinction coefficient, ε, (5) is transformed into (6). The geometric factor (LA/V) is unity, or nearly so, for most photolytic systems employed.

$$\frac{d[\text{R} \cdot ']}{dt} = 4.6\phi I_0 \varepsilon [\text{M}] \frac{LA}{V} \tag{6}$$

Comparison of (5) and (6) with (4) illustrates the main differences between initiation by a thermal dissociation of an initiator and by the photodissociation of a monomer. The formation of the initiating radicals by a *thermal* dissociation process is strongly dependent on the temperature and is independent of the monomer concentration. On the other hand, the terms $\phi I_0 \alpha$ and $\phi I_0 \varepsilon$ in (5) and (6) are almost independent of temperature. Hence *the rate of initiating radical formation in a photoinitiation process is almost independent of the temperature but is proportional to the monomer concentration.*

The second type of initiation mechanism is exemplified by the photopolymerization of styrene or methyl methacrylate. Absorption of light in this case does not result in decomposition of monomer molecules. Instead, it has been suggested[1] that the absorption of light produces an excited singlet state of the monomer which may either fluoresce (7) or be converted to an excited (and long-lived) triplet state (8). The latter may be regarded as a diradical, that is, $\cdot \text{CH}_2 - \dot{\text{C}}(\text{H})\text{R}$. Attack on the monomer by this diradical (9) ultimately yields two monoradicals (11), which, in turn, initiate polymerization. An example of this complex initiation mechanism is illustrated for styrene photopolymerization by equations (7) to (11).

$$\text{Ph}-\text{CH}=\text{CH}_2 \xrightarrow{h\nu} (\text{PhCH}=\text{CH}_2)^* \begin{cases} \rightarrow \text{PhCH}=\text{CH}_2 + h\nu' & (7) \\ \rightarrow \text{Ph}\dot{\text{C}}\text{H}-\text{CH}_2\cdot & (8) \end{cases}$$

$$\text{Ph}\dot{\text{C}}\text{H}-\text{CH}_2\cdot + \text{PhCH}=\text{CH}_2 \longrightarrow \text{Ph}\dot{\text{C}}\text{H}-\text{CH}_2-\text{CH}_2-\dot{\text{C}}\text{HPh} \tag{9}$$

$$\text{Ph}\dot{\text{C}}\text{HCH}_2\text{CH}_2\dot{\text{C}}\text{HPh} \longrightarrow 2\,\text{PhCH}=\text{CH}_2 \tag{10}$$

$$\text{Ph}\dot{\text{C}}\text{HCH}_2\text{CH}_2\dot{\text{C}}\text{HPh} + \text{PhCH}=\text{CH}_2 \longrightarrow \text{CH}_3\dot{\text{C}}\text{HPh} + \text{PhCH}=\text{CHCH}_2\dot{\text{C}}\text{HPh} \tag{11}$$

[1]Norrish, R. G., and Simons, J. P., *Proc. Roy. Soc.* (London), **1959**, *A251,* 4.

It can be shown by the steady-state methods discussed in Chapter 11 that, at high monomer concentrations, essentially all the diradicals react by step (11) rather than by (10), and the rate of formation of the initiating radicals will be proportional to [M], the monomer concentration. On the other hand, if the monomer concentration is so low that the diradicals mainly revert to monomer by step (10), the rate of initiator formation will be proportional to $[M]^2$, one power of [M] coming from (7) and the second from (9). In both cases these rates are proportional to the light intensity and to the extinction coefficient of the monomer, as required by expression (6).

Photosensitized Polymerizations

In order for *direct* photopolymerization to occur it is essential that the monomer should absorb some of the light impinging on the system. However, even if such direct light absorption does not occur, polymerization can still be initiated if photosensitizers are present. Photosensitizers are substances that produce free radicals when they absorb ultraviolet or visible light. The same substances that are used for thermal initiation are often used for photosensitization. For example, azo compounds and peroxides are photosensitizers, and the photoinitiation reaction is the same as is the thermal initiation process. However, much lower reaction temperatures can be employed when light absorption is used to initiate radical formation. Moreover, many initiators can be used as photosensitizers even though they do not dissociate *thermally* at convenient rates or temperatures. Examples of some photosensitizers are given in Table 5.1.

For example, azoisopropane does not dissociate sufficiently rapidly below 180°C (Table 3.1) to be a useful thermal initiator. However, it photodissociates even at low temperatures when irradiated with near-ultraviolet light:

$$Me_2CHN=NCHMe_2 + h\nu(300\ nm < \lambda < 400\ nm) \longrightarrow 2Me_2\overset{\overset{\displaystyle H}{\displaystyle |}}{C}\cdot + N_2 \qquad (12)$$

As might be expected from the previous discussion, the rate of radical formation from such photosensitizers is given by

$$\frac{d[R\cdot']}{dt} = 4.6\phi I_o \varepsilon_s[S]\frac{LA}{V} \qquad (13)$$

where ε_s is the extinction coefficient of the photosensitizer, [S] the concentration of photosensitizer, and the other symbols are as defined earlier.

Condensed ring aromatic compounds, such as anthracene give rise to a more complex type of photosensitization process. Photoirradiation generates the excited triplet state of anthracene. This excited molecule then interacts with the monomer ultimately to produce monoradicals. The monoradicals initiate the polymerization. With anthracene, denoted by A, and a monomer, $RCH=CH_2$, this initiation mechanism is depicted in steps (14) to (19).

TABLE 5.1 PHOTOSENSITIZERS FOR PHOTOPOLYMERIZATION

Type of Compound	Example	Mechanism of Polymerization
Carbonyl compounds	Acetone	Radical formation
	Biacetyl	Radical formation
	Benzophenone	Radical formation
	Benzoin	Radical formation
	α-Chloroacetone	Radical formation
Condensed ring aromatics	Anthracene	Energy transfer
Peroxides	t-Butyl peroxide	Radical formation
	Hydrogen peroxide	Radical formation
Organic sulfides	Diphenyl disulfide	Radical formation
	Dibenzoyl disulfide	Radical formation
Azo compounds	Azoisopropane	Radical formation
	Azobisisobutyronitrile	Radical formation
	Aryldiazonium salts	Cationic (epoxides)
Halogen-containing compounds	Chlorine	Radical formation
	Chloroform	Radical formation
	Carbon tetrachloride	Radical formation
	Bromotrichloromethane	Radical formation
	Bromoform	Radical formation
	Bromine	Radical formation
Metal carbonyls	Manganese pentacarbonyl and carbon tetrachloride	Radical formation
	Rhenium pentacarbonyl and carbon tetrachloride	Radical formation
Inorganic ions	$FeOH^{2+}$	Radical formation
	$FeCl_4^-$	Radical formation

$$A + h\nu \longrightarrow A^* \tag{14}$$

$$A^* \longrightarrow A + h\nu \text{ (fluorescence)} \tag{15}$$

$$A^* \longrightarrow {}^3A \quad \text{(triplet state)} \tag{16}$$

$${}^3A + M \longrightarrow (AM)^* \text{ (intermediate diradical)} \tag{17}$$

$$(AM)^* \longrightarrow A + M \tag{18}$$

$$(AM)^* + A \longrightarrow RCH{=}CH{-}A\cdot + HA\cdot \tag{19}$$

A complicated mathematical expression is needed to describe the rate of formation of the initiating monoradicals. However, if all the triplet-state anthracene molecules react with the monomer, the rate will be independent of the monomer concentration. In all cases, the rate of formation of the initiating radicals will be proportional to the intensity of the incident light. At sufficiently high sensitizer concentrations the rate will be proportional to [A], while at very low sensitizer concentration it will be proportional to $[A]^2$.

A somewhat different type of photosensitizer is exemplified by the metal carbonyl/carbon tetrachloride systems.[1] For example, photolysis of Mn_2CO_{10} or

[1]Bamford, C. H., Crowe, P. A., and Wayne, R. P., *Proc. Roy. Soc.* (London), **1965,** *A284, 455.*

$Re_2(CO)_{10}$ in the presence of small amounts of CCl_4 initiates the free-radical polymerization of vinyl monomers. Although the mechanism is not completely understood, the actual initiating radical is thought to be $\cdot CCl_3$. The radical is produced by the reaction of the metal carbonyl photolysis products with carbon tetrachloride.

Quantum Yields

The quantum yield of a simple photochemical reaction is defined as the number of molecules of the product formed, or reactant consumed, per quantum of light absorbed. In photopolymerization the quantum yield for initiation is defined by the number of *chains* initiated per quantum of light absorbed. This may be written in terms of rates as

$$\phi_i = \frac{\text{rate of chain initiation}}{\text{rate of light absorption}} \tag{20}$$

When a simple photoinitiation occurs and the absorbing compound dissociates directly to two monoradicals, ϕ_i must lie between zero and 2.

The *overall* quantum yield of photopolymerization refers to the number of molecules of *monomer consumed* per quantum absorbed. This may be written as shown in (21). Since the formation of a high polymer requires the existence of long-chain processes, ϕ_p is generally very large, typically being of the order of several hundred to several thousand.

$$\phi_p = \frac{\text{rate of monomer consumption}}{\text{rate of light absorption}} \tag{21}$$

RADIATION-INDUCED POLYMERIZATION

General Characteristics

Chain reaction polymerizations can also be induced by irradiation of a pure monomer (or a solution of the monomer) with high-energy or ionizing radiation. Alpha particles, beta rays (electrons), gamma rays, or high-velocity particles from an accelerator can all be used. A wide variety of monomers may be polymerized in this way—styrene, acrylonitrile, methyl methacrylate, vinyl chloride, butadiene, isoprene, or practically any other monomer that polymerizes by a free-radical mechanism. In addition, monomers such as isobutylene can be induced to undergo low-temperature cationic polymerization during irradiation.

In a practical sense, the rapid polymerization of monomers and the cross-linking of polymers by high-energy radiation have been considered as possible manufacturing processes. However, for reasons that will become clear from the following sections, radiation-induced reactions are less convenient to carry out on a large scale than are chemically induced reactions, although future developments may change this picture.

Experimental Methods

Monomers and solvents for radiation-induced polymerization must be purified in the same way as for conventional free-radical or ionic polymerization. Oxygen is usually detrimental to the reaction, and the monomer is normally sealed in an evacuated reaction vessel or protected from oxygen by an atmosphere of nitrogen.

A variety of reaction vessels have been used, depending on the monomer, the type of radiation, and the scale of polymerization. Figure 5.3 shows a simple reaction cell that is convenient for small-scale laboratory experiments when gamma- or X-rays are employed. It consists of a small, shallow glass cylinder to which is attached a glass side arm. The monomer is introduced into the cell through the side arm, air is removed by evacuation, and the side arm is sealed. The cell is mounted in a heating or cooling bath with the upper flat cell face above the level of the heating or cooling fluid. The cell is then positioned below the X-ray or gamma-ray source and the contents are irradiated from above. The polymer can be removed either by dissolving it in a suitable solvent or by breaking the cell.

Alternatively, if the experimenter has access to a γ-ray nuclear irradiation facility, other procedures can be used. A particularly simple nuclear irradiation arrangement allows the sample to be lowered into position close to a ^{60}Co source. In a typical ^{60}Co irradiation facility, the ^{60}Co is encapsulated in many stainless steel "pencils" which are themselves maintained beneath the surface of a "swimming pool" of water. The pencils can be arranged around a vertical aluminum tube (ca. 7.5 cm diameter) that extends up to the surface of the pool. Various configurations of the pencils are used to supply differing dose rates to the sample. The sample to be irradiated is simply lowered into position and kept in place until the predetermined dose (total energy input) has been delivered. It is then removed for analysis. Typical γ-ray dose rates in a facility of this kind may be in the range of 0.1 to 1 Mrad h^{-1} (10^7 to 10^8 ergs-g^{-1}-h^{-1}).

More compact nuclear irradiation sources generally use lead instead of water for shielding purposes. An example of this type is the ^{137}Cs irradiation unit shown diagrammatically in Figure 5.4. The ^{137}Cs is contained in fixed tubes surrounding a larger main irradiation chamber. Auxiliary irradiation chambers are

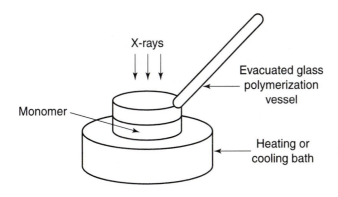

X-rays

Monomer

Evacuated glass polymerization vessel

Heating or cooling bath

Figure 5.3 Radiation-induced polymerization of a monomer in an evacuated glass ampoule.

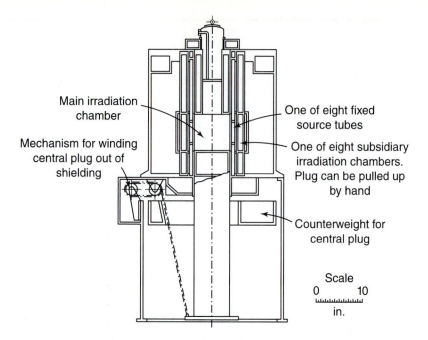

Main irradiation chamber

Mechanism for winding central plug out of shielding

One of eight fixed source tubes

One of eight subsidiary irradiation chambers. Plug can be pulled up by hand

Counterweight for central plug

Scale

0   10

in.

Figure 5.4 Equipment for the irradiation of monomer or polymer samples with the use of a shielded cesium 137 source. [Reproduced from A. J. Swallow, *Radiation Chemistry of Organic Compounds* (New York: Pergamon Press, 1960); with permission from Pergamon Press.]

located adjacent to each tube of ^{137}Cs. Again, samples are placed in the irradiation chambers for predetermined periods of time, and are then removed for examination.

The Absorption of Radiation

The initiation of polymerization by high-energy radiation is a far less selective process than is initiation by light. A wide variety of monomers can be induced to polymerize by almost all forms of high-energy radiation. In fact, the initiation of polymerization seems to result more from the gross "damage" sustained by the system than from the selective induction of specific chemical reactions. Because of this lack of selectivity, it is not possible to calculate a meaningful quantum yield. The energy yield in radiation-induced reactions is generally described by the G-value. This is defined by (22) for any substance A.

$$G(\pm A) = 100\left(\frac{\text{number of molecules of A formed or consumed}}{\text{number of electron-volts of energy absorbed}}\right) \quad (22)$$

The high-energy irradiation of either a pure monomer or a solution of a monomer generates both free radicals and ions. Thus, either free-radical or ionic chain polymerizations may be induced. The preponderance of one or the other mechanism depends on the nature of the monomer, as well as on the temperature and the purity of the reagents. Some monomers, such as acrylic esters, vinyl esters, and vinyl fluoride, polymerize only by free-radical mechanisms. Isobutylene polymerizes only by a cationic chain mechanism. Others, such as styrene, acrylonitrile, or isoprene, apparently polymerize by both radical and ionic types of chain mechanisms. Ionic polymerizations are generally favored at low temperatures. Both types of mechanism are sensitive, in different degrees, to the presence of common impurities such as oxygen or water. Examples of both types of mechanism are discussed in later sections.

Gamma radiation is the most convenient type of high-energy radiation for the initiation of polymerization because the high penetrating power permits *uniform* irradiation of the system. Moreover, because gamma rays are absorbed to the same extent by solids as by liquids, solid monomers can be polymerized readily. This permits low-temperature polymerizations to be performed with many monomers.

The irradiation of a pure monomer will give rise to the simplest initiation reactions, because only one chemical species absorbs the radiation. However, systems that contain both monomer and a solvent undergo more complicated initial reactions because both species can absorb radiation energy to produce ions and free radicals. The details of these energy absorption processes are outside the scope of this book. However, as a very useful approximation, it is possible to assume that the fraction of the total energy absorbed by a particular species in solution is proportional to the electron fraction of that species. The *electron fraction* is defined as the number of electrons contained in that component divided by the total number of electrons in the whole solution.

For example, consider a 0.1 M solution of styrene in carbon tetrachloride. The concentration of carbon tetrachloride is about 10 M, and the numbers of electrons in styrene and carbon tetrachloride molecules are 56 and 74, respectively. Therefore, the electron fraction of carbon tetrachloride is calculated by equation (23):

$$\varepsilon CCl_4 = \frac{74\ M_{CCl_4}}{74\ M_{CCl_4} + 56\ M_{C_8H_8}} = \frac{740}{740 + 5.6} = 0.993 \qquad (23)$$

Thus, during the irradiation of this solution, 99.3% of the total energy absorbed will be absorbed by the solvent, and hence most of the initiating ions and free radicals will be derived from the solvent. In this example, the styrene polymerization will be initiated predominantly by species such as CCl_2^+, CCl_3^+, $Cl\cdot$, $CCl_3\cdot$, etc.

Three distinct phases may be identified in a radiation-induced polymerization. The interaction of the radiation with monomer and solvent molecules occurs within 10^{-16} to 10^{-15} s. In this brief interval the only significant motion is that of electrons, and the products of this first phase are electronically excited molecules, ions, and electrons. In the second phase, which takes place some 10^{-14} to 10^{-10} s after the initial interaction, the excited molecules, ions, and electrons dissociate or react with monomer to yield a set of initiating free radicals and ions. The third phase consists of the normal elementary reactions of initiation, propagation, transfer, and termination characteristic of chain reaction polymerization. This third phase occurs between 10^{-10} to 10^{-1} s after the initial interaction.

Free-Radical Chain Initiation

As discussed earlier, a radiation-catalyzed free-radical polymerization differs from a chemically induced one mainly in terms of the initiation mechanism. The propagation, chain transfer, and termination processes are similar to those described in Chapter 3. Hence, the present comments will focus only on the initiation process.

The initiation step in a radiation-induced free-radical polymerization of a pure monomer can be symbolized by equation (24). The symbol $\rightsquigarrow$ [1] means "under high-energy irradiation." M represents the monomer, and $R \cdot'$ denotes the initiating radicals. If a radiation-induced polymerization is carried out in solution, the initiation step depicted in (25) must also be taken into account. In (25), S denotes the solvent and $S \cdot$ represents initiating radicals derived from the solvent. Thus, even in these simple terms, two types of radicals can be generated. However, even in a well-defined initiation process, $R \cdot'$ and $S \cdot$ represent a whole *set* of initiating radicals and not just single species. Moreover, it should be kept in mind that the simple equations (24) and (25) are not to be considered as discrete chemical reactions. Rather, they represent the overall production of free radicals by a very fast *sequence* of elementary processes that follow the initial energy absorption. These processes include ion and excited molecule dissociations, ion and electron recombinations, and very fast collision reactions of ions and translationally "hot" atoms and radicals. For a discussion of the rapid elementary processes that follow radiation absorption the reader should consult texts and monographs on radiation chemistry such as those listed at the end of the chapter.

$$M \quad \rightsquigarrow \quad 2R \cdot' \qquad (24)$$

$$S \quad \rightsquigarrow \quad 2S \cdot \qquad (25)$$

The most important characteristic of the initiation step is the *yield* of radicals generated by the absorption of a given radiation dose. This characteristic is defined by the G-value or 100-eV yield of initiating radicals. This is referred to as $G_M(R \cdot')$ or $G_s(S \cdot)$, depending on whether the radiation energy is being absorbed by the monomer or by the solvent. A number of 100-eV yields of initiating radicals for some typical monomers and solvents have been determined by

---

[1]This should not be confused with the similar symbolism that represents a polymer chain (see page 32).

TABLE 5.2 100-EV YIELDS ON INITIATING RADICALS IN MONOMERS AND SOLVENTS

Monomer or Solvent	Formula	$G(\text{R}\cdot')$ (radicals/100 eV)
Acrylonitrile	$CH_2{=}CHCN$	5.0
Isobutylene	$CH_2{=}C(CH_3)_2$	3.9
Methyl acrylate	$CH_2{=}CH{-}\overset{\overset{\textstyle O}{\|}}{C}{-}OCH_3$	6.3
Methyl methacrylate	$CH_2{=}C(CH_3)\overset{\overset{\textstyle O}{\|}}{C}{-}OCH_3$	6.1
Styrene	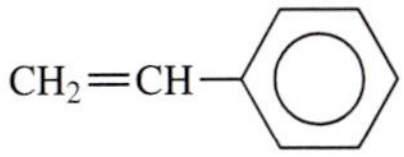 $CH_2{=}CH{-}\hexagon$	0.66
Vinyl acetate	$CH_2{=}CH{-}O{-}\overset{\overset{\textstyle O}{\|}}{C}{-}CH_3$	9.6
Benzene	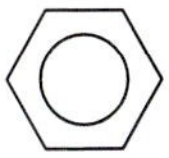 $\hexagon$	0.66
n-Hexane	$CH_3(CH_2)_4CH_3$	5.8
Toluene	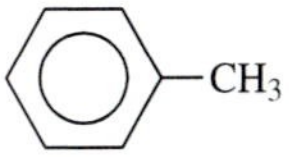 $\hexagon{-}CH_3$	2.4

radical scavenging techniques. Some typical values are shown in Table 5.2. A knowledge of such values permits a calculation to be made of the rate of formation of the radiation-initiated radicals in the same way as was described earlier for thermal and photochemical initiation. Thus, for the radiation-induced polymerization of a pure monomer or a monomer in solution, the rate of formation of initiating radicals is described by (26) and (27), respectively, where the ε's are electron fractions and Q_A is the energy absorbed per unit volume.

$$\frac{d[\text{R}\cdot']}{dt} = \frac{G(\text{R}\cdot')}{100}\left(\frac{dQ_A}{dt}\right) \qquad \text{(pure monomer)} \qquad (26)$$

$$\frac{d[\text{R}\cdot']}{dt} = \left(\frac{\varepsilon_M G_M(\text{R}\cdot')}{100} + \frac{\varepsilon_s G_s(\text{S}\cdot)}{100}\right)\left(\frac{dQ_A}{dt}\right) \quad \text{(solution)} \qquad (27)$$

The energy absorption per unit volume, Q_A, and the rate of energy absorption per unit volume, dQ_A/dt, are related to the dose and dose rates[1] by the density of the system.

[1]These terms are generally used by radiation chemists to express energy absorption per unit mass.

Radiation-Induced Polymerization

Radiation chemists often express doses in a unit called the rad, which is defined as 100 ergs/g. A typical dose rate from a ^{60}Co γ-ray source would be 10^6 rads/h or 1 Mrad/h. The following example illustrates a calculation of the rate of formation of initiating radicals for a comparison with thermal and photochemical initiation.

Consider the irradiation of pure acrylonitrile (density = 0.81 g/cm^3) at 20°C with γ-rays, with a dose rate of 1 Mrad/h. It is possible to calculate both the energy absorption and the rate of initiating radical formation. The energy absorption rate per unit volume, $dQ_A/dt = 1 \times 10^6$ rads/h $\times 10^2$ ergs/g-rad $\times$ 0.81 g/cm^3 $\times$ 6.24 $\times 10^{11}$ eV/erg $\times$ 2.78 $\times 10^{-4}$ h/s = 1.4 $\times 10^{16}$ eV/cm^3-s. The rate of initiating radical formation may be derived from this value, equation (26), and the value listed in Table 5.2. Thus, $d[R\cdot']/dt = 0.050 \times 1.4 \times 10^{16} = 7.0 \times 10^{14}$ radicals/cm^3-s. This is equivalent to the rate of radical formation from 0.1 M benzoyl peroxide solution at 60°C.

Ionic Chain Initiation

The formation of ions from the monomer and solvent can be symbolized by (28) and (29). Again, the processes shown in (28) and (29) are not elementary reactions. They represent the overall formation of initiating ions by very fast sequences of reactions that follow an initial ionization event. Thus, M^+, S^+, and e^- formed in the initial event are converted by very rapid reactions to the initiating ions denoted by R^+ and R^- in (28) and (29). The reader is again referred to texts and monographs on radiation chemistry for a detailed discussion.

$$M \xrightarrow{} R^+ + R^- \tag{28}$$

$$S \xrightarrow{} R'^+ + R'^- \tag{29}$$

The G-values for the *initial* formation of ion pairs in liquids are in the range 3 to 4. Although this is comparable to the G-values found for free-radical formation, the efficiency of ionic initiation is much lower than that of free-radical initiation. Most of the "gegenions" formed initially do not separate from each other, but instead undergo mutual charge neutralization. The radiation yield of "free" ions (ions capable of initiating polymerization) depends on the dielectric constant of the medium. In hydrocarbons, which are solvents of low dielectric constant (i.e., values of 2 to 4), the G-value for "free" ions is only 0.1, which means that only 1 out of every 30 or 40 ions that are formed goes on to initiate an ionic polymerization. Alcohols, which have dielectric constants in the range 20 to 40 show free-ion G-values of 0.6 to 1.5. Water, which has a dielectric constant of 78, has a free ion G-value of about 2.5.

By analogy with free-radical polymerization, it is possible to use equation (30) to describe the radiation-induced rate of formation of initiating positive ions in the pure monomer, and equation (31) for a monomer in solution. As before, the ε's are

$$\frac{d[\mathrm{R}^+]}{dt} = \frac{G(\mathrm{R}^+)}{100}\left(\frac{dQ_A}{dt}\right) \tag{30}$$

$$\frac{d[\mathrm{R}^+]}{dt} = \left(\frac{\varepsilon_M G_M(\mathrm{R}^+)}{100} + \frac{\varepsilon_S G_S(\mathrm{R}^+)}{100}\right)\frac{dQ_A}{dt} \tag{31}$$

electron fractions, Q_A is the energy absorbed per unit volume, and M and S refer to the monomer and the solvent, respectively. Identical expressions may be written for the rates of formation of negative ions. The rates of formation of the initiating ions can be estimated from the radiation dose rate using an analogous method to the one describe above.

Ionic chain polymerizations are especially sensitive to traces of impurities, and these impurities can exert a strong inhibiting effect. For example, in the cationic polymerization of styrene, the propagating chains are effectively broken by proton transfer to a water molecule, as shown in reaction (32). It is very difficult to remove the last traces of water from a chemical system, and this experimental problem considerably delayed the recognition that radiation-induced polymerization can proceed by cationic chain mechanisms. Thus, styrene that has been dried by distillation from sodium-potassium alloy polymerizes under irradiation about 200 times faster than styrene that has merely been subjected to a single conventional distillation. Ionic polymerizations are generally much faster than free-radical polymerizations. If water is present, only the slower radical component of a radiation-induced polymerization can be observed because the cationic component is effectively inhibited.

$$\mathrm{R}\!-\!\!\left[\mathrm{CH_2}\!-\!\mathrm{CH}\right]\!\!-\!\!\mathrm{CH_2}\!-\!\overset{\overset{\textstyle H}{|}}{\mathrm{C}}{}^+ + \mathrm{H_2O} \longrightarrow \mathrm{R}\!-\!\!\left[\mathrm{CH_2}\!-\!\mathrm{CH}\right]\!\!-\!\!\mathrm{CH}\!=\!\mathrm{CH} + \mathrm{H_3O}^+ \tag{32}$$

Solid-State Radiation-Induced Polymerization

General features Some crystalline monomers or crystalline cyclic compounds can be induced to polymerize in the solid state. Typically, crystals of the monomer are irradiated with electrons, gamma rays, or X-rays, often at low temperatures. The source of radiation may be an accelerator, radioactive isotope, or an X-ray generator. Polymerization may occur either at the temperature of irradiation or during subsequent warming of the crystals to room temperature. The latter type of process is called *postpolymerization*. After polymerization is complete, the unchanged monomer can be dissolved away to leave the polymer behind, or the monomer can be removed by volatilization in vacuum or by reprecipitation into a nonsolvent for the polymer.

Monomers that have been polymerized in this way include acrylamide, acrylonitrile, styrenes, isobutylene, isoprene, butadiene, acrylic and methacrylic acids and their salts, formaldehyde, acetaldehyde, and acetone. Cyclic compounds that undergo solid-state radiation-catalyzed polymerization include trioxane, hexamethylcyclotrisiloxane, β-propiolactone, diketene, 3,3-bischloromethylcyclooxabutane, and hexachlorocyclotriphosphazene. The following examples illustrate the behavior of specific unsaturated monomers or cyclic trimers during solid-state polymerization.

Diacetylenes   A number of diacetylenes crystallize in a stacked crystalline arrangement that allows solid-state polymerization. The process is depicted in Figure 5.5. Polymerization can be induced by irradiation of the crystals with X-rays, γ-rays, or ultraviolet light, or simply by heating at temperatures below the melting point. Individual chains grow independently of each other starting from random points in the lattice. As polymerization proceeds, the color of each crystal changes to deep red with a metallic luster as the span of polyconjugation expands with the growth of each chain. The polymerization mechanism involves carbenes (i.e., diradicals) associated with the terminal carbon atoms.

Acrylamide   Acrylamide is a solid at room temperature (m.p. 84°C). The γ-ray-induced polymerization of this monomer was first achieved in 1954, an event which stimulated the initial development of the solid-state polymerization field. Irradiation at −78°C results in no appreciable polymerization.

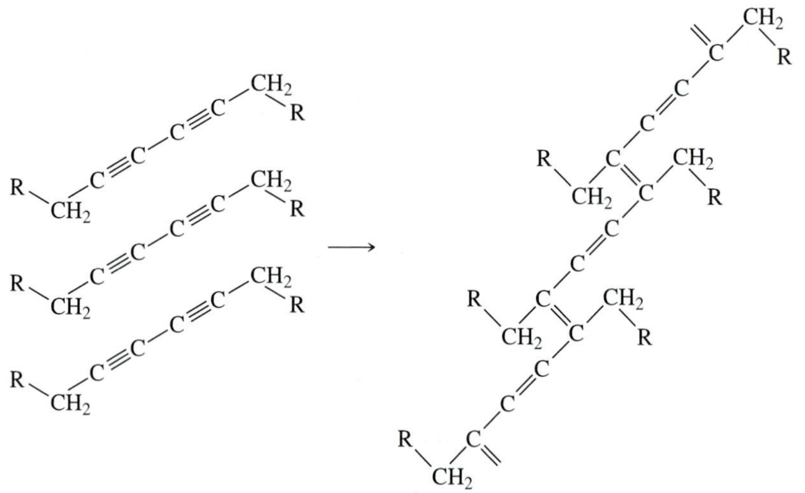

Figure 5.5   Radiation polymerization of diacetylenes. R can be various groups, including — OSO_2 — ⌬ — CH_3 or *N*-carbazolyl.

However, rapid warming of the preirradiated polymer results in a violent polymerization process. A gradual increase in the temperature brings about a slow polymerization which may continue over a period of several months. Total conversion of monomer to polymer can be achieved during postpolymerization at 27°C.

It has been shown by electron spin resonance techniques that free radicals are formed during the irradiation process and that the radical concentration remains constant as the postpolymerization proceeds. Each radical is associated with one chain. Thus, if the polymerization is a free-radical process, it would appear that a radical termination step is absent, presumably because the radicals are trapped in a polymer matrix. However, the presence of free radicals does not prove a free-radical mechanism. Radical anions or radical cations can generate *ionic* polymerizations, even though radicals can be detected.

The polyacrylamide formed by this process is armorphous. Apparently, polymerization takes place at defects in the crystal lattice. If the crystal is scratched, polymerization proceeds rapidly along the scratch. Hence, it is assumed that defects and strains are generated at the crystal–polymer interfaces. Trapped radicals are perhaps released at these locations, thus favoring further polymerization.

Acrylonitrile   Continuous irradiation of solid acrylonitrile at −196°C gives polyacrylonitrile, but the conversion to polymer never exceeds about 5%. However, if the 5% polymer–95% monomer mixture is allowed to melt and is then cooled and reirradiated, a higher conversion to polymer takes place. This effect may result from the crystallization of pure monomer crystals following melting, or it could be a consequence of the introduction of additional polymer–monomer interface defects. It has also been reported that irradiation of acrylonitrile at −196°C with 7.8×10^5 rads, followed by subsequent heating to room temperature, results in an explosive polymerization before the melting point is reached. Large quantities of polymer are apparently formed in this process.

It is believed that the polymerization at −196°C follows an ionic mechanism, but that free-radical processes predominate at temperatures near the melting point.

Vinyl carboxylates and derivatives   Salts of acrylic and methacrylic acids, $M^+ {}^-OOC-CH=CH_2$ or $M^+ {}^-OOC-C(CH_3)=CH_2$, can be polymerized by γ-irradiation in the solid state. However, changes in the cation, M^+, affect the rate of polymerization. At −78°C the potassium salt polymerizes faster than the sodium salt, which in turn polymerizes faster than the lithium salt. These changes have been ascribed to differences in the crystal structures of the different salts. Similarly, different hydrates of acrylate salts polymerize at different rates. Methyl methacrylate resists polymerization in the solid state.

Aldehydes   Irradiation of solid formaldehyde with ionizing radiation at −196°C can generate an almost explosive polymerization process. The violence of the reaction increases if irradiated solid formaldehyde is warmed or subjected to mechanical shock. Another curious feature of this reaction is the high G-value (5.4×10^6) for a 52% conversion and an average degree of polymerization of 10^4. This fact suggests that conventional ionic or free-radical initiation mechanisms are not operative. It has been proposed that irradiation generates a trapped formaldehyde excited state, such as $H_2C^+ \!\!-\!\! O^-$, which initiates polymerization when the temperature is raised. The polymer formed by this technique has a fibrous appearance.

Trioxane   Crystals of trioxane, $(O\!\!-\!\!CH_2)_3$, undergo solid-state polymerization when irradiated with γ-rays or α-particles (see Chapter 6). The temperature during irradiation may be from $-78°C$ to $+55°C$. Radiation-catalyzed polymerization does not take place in the molten state or in solution. There is no induction period before polymerization begins, and the polymerization rate is apparently not influenced by free-radical scavengers or by air. The reaction rates are proportional to the radiation dose rate. For these reasons an ionic mechanism has been proposed, although this viewpoint is not universally accepted. The "in-source" polymerization of trioxane can be completely inhibited by the application of high pressures.

At 55°C the conversion of trioxane to polyoxymethylene does not exceed about 35%. It is believed that this limit corresponds to the point at which the irradiation process begins to cause decomposition of the polymer. The polymer matrix has the outward appearance of large crystals and, indeed, the polymer fibers obtained are the same length as the single crystals of the trimer from which they were formed. The crystallinity of these fibers is appreciable and the orientation is excellent. The polyoxymethylene formed by γ-irradiation of trioxane has a higher melting point and better heat stability than polymer formed by the conventional polymerization of formaldehyde.

X-ray diffraction analyses of both trioxane crystals and the polymer formed within them show that the trioxane crystal structure influences the route of the propagation reaction. Trioxane crystals are trigonal, with the molecular packing arrangement shown in Figure 5.6. The polymer crystals are twinned in such a way that two different helical chain axes can be discerned for the polyoxymethylene. These two chain orientations arise either by propagation down the c-axis of the trioxane crystal or by the linkage of molecules oriented along a diagonal axis oriented 76.1° to the c-axis. These two possibilities are illustrated in Figure 5.6, with the atomic separations (in angstroms) that must be bridged during polymerization indicated on the diagram.

α-Particle irradiation of trioxane crystals apparently generates a different twinned modification. In this case, polymerization originates from the damage streaks left by the α-particle tracks. Tetroxane, $(O\!\!-\!\!CH_2)_4$, also polymerizes under the influence of γ-rays to give a twinned polymer crystal. However, this twinned structure differs from that of trioxane because the crystal structure of tetroxane is different (monoclinic).

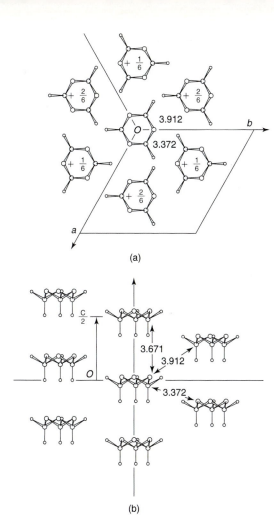

(a)

(b)

Figure 5.6 (a) Packing of trioxane molecules between the planes $z = 0$ and $z = \frac{1}{2}$, viewed down the c-axis. The z-height of each molecule is indicated. (b) Packing of the molecules along the contiguous threefold axis lying in the (110) plane. Polymerization may occur between molecules stacked above each other on one threefold axis or between molecules on different axes, displaced by 1/6 of c. [The separations are indicated by the 3.671 Å and 3.372 Å distances shown in (b).] [From Busetti, V., Mammi, M., and Carazzolo, G., *Z. Krist.*, **1963**, *119*, 310.]

Other cyclic compounds Hexachlorocyclotriphosphazene, $(NPCl_2)_3$, (see Chapter 7) polymerizes in the crystalline state when irradiated with X-rays. The maximum yield of polymer is about 10%. In fact, the rate of polymerization increases with temperature until the melting point (114°C) is reached, at which point the polymerization rate falls to zero. Presumably, the active sites are stabilized within the crystalline lattice but are destroyed rapidly in the liquid state.

Radiation polymerization of vinyl monomers trapped in clathrate crystals Some small molecules crystallize in such a way that cavities exist within the crystal structure. These cavities are normally occupied by molecules of the solvent used for crystallization. Such systems are called *clathrates*. This phenomenon often occurs when the matrix molecules (the "host") have an awkward shape that prevents efficient packing within one of the allowed crystal systems. In specific cases the voids have an elongated shape, and may extend completely through the crystal in the form of "channels" or "tunnels." The guest molecules that may occupy the tunnels can include olefins or vinyl monomers. Gamma-irradiation of such clathrates often brings about polymerization of the monomer molecules within each tunnel. The architecture of the tunnel leads to the formation of one polymer molecule in each tunnel, which provides a unique opportunity for the study of polymers under conditions where they are isolated from each other. Furthermore, the packing of monomer molecules in each tunnel may induce stereoregular polymerization. The polymer molecules may be isolated after irradiation either by dissolving the host in a suitable solvent and precipitating the polymer, or simply by volatilizing the host molecules away from the polymer.

A number of host systems have been studied as matrices for this type of polymerization. These include urea and thiourea, deoxycholic acid, perhydrotriphenylene (1), and certain spirocyclophosphazenes (2 and 3).[1,2] The crystal packing pattern for host 2 is depicted in Figure 5.7. Host 2 gives rise to 5 Å diameter tunnels, whereas those of 3 are 10 Å wide because of the longer side arms.

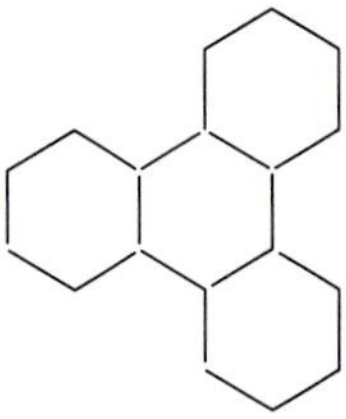

| 1 | 2 | 3 |

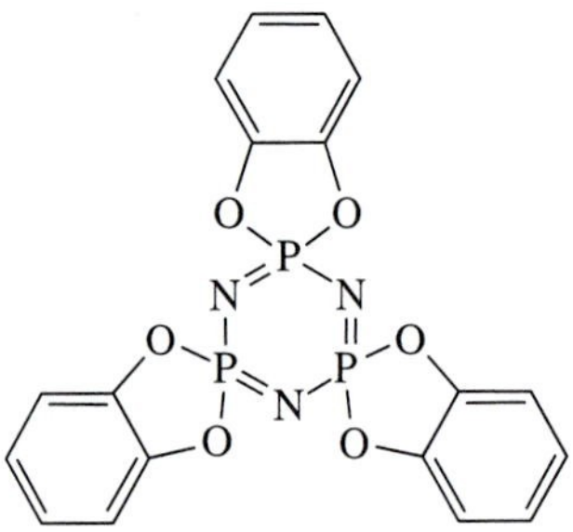

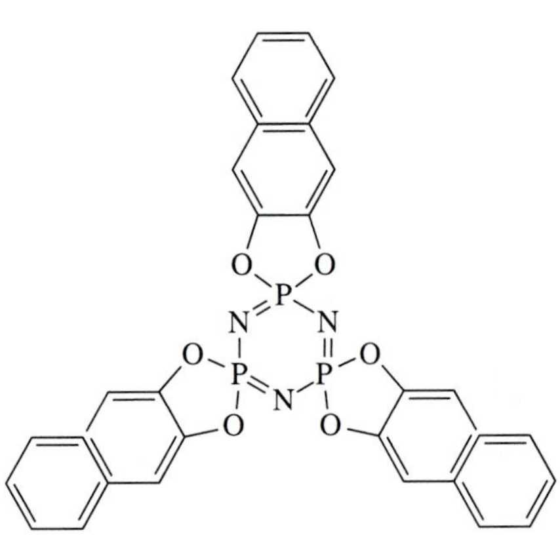

Monomer molecules that have been polymerized in tunnel clathrates include butadiene, 1,3-pentadiene, isoprene, 2,3-dimethylbutadiene, vinyl chloride, acrylic acid, acrylic anhydride, acrylonitrile, methyl methacrylate, and divinylbenzene. An illustration of the utility of the tunnel clathrate method is that polymerization of divinylbenzene yields an uncross-linked macromolecule, whereas polymerization outside the tunnel allows extreme cross-linking. Copolymerization of two different monomers has also been accomplished via the tunnel clathrate method. Perhaps the most intriguing aspect of this field is the relation-

[1]Allcock, H. R., Dudley, G. K., and Silverberg, E. R., *Macromolecules*, **1994,** *27,* 1039–1044.
[2]Allcock, H. R., Silverberg, E. R., Dudley, G. K., and Pucher, S. R., *Macromolecules*, **1994,** *27,* 7550–7555.

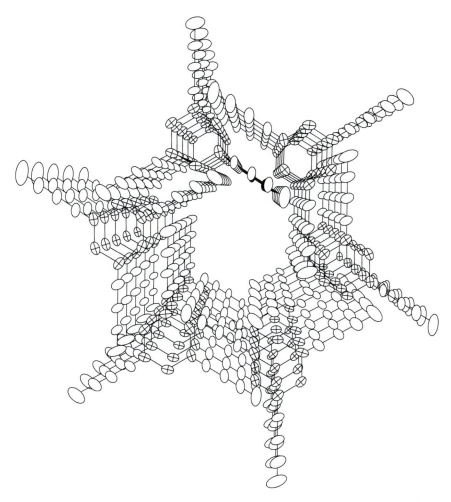

Figure 5.7 Representation of a perspective view down one of the tunnels that penetrate the hexagonal crystalline lattice of tris(*o*-phenylenedioxy)cyclotriphosphazene (I). The 5-Å-diameter tunnels accommodate the organic monomer molecules.

ship between the crystal structure of the monomer or clathration host and the polymerization pathway. This subject provides a fertile area for future fundamental research.

Conclusions   A number of practical advantages are inherent in the radiation-catalyzed solid-state process. First, and most obviously, the method allows the polymerization of compounds which have low ceiling temperatures[1] and which

_____

[1]The ceiling temperature is the temperature above which the polymer cannot exist (see Chapter 10).

cannot be polymerized thermally by conventional means. Second, solid-state reactions provide a method for polymerization in the absence of a potentially reactive solvent. Third, polymer order and crystallinity can be introduced by the use of the monomer crystal structure as a "template." Fourth, contamination by catalyst molecules or residues is avoided. Fifth, some "monomers" yield different polymers in the liquid and solid states. The conversion of diketene $(CH_2=\overset{\quad}{C}-O-C(O)-CH_2)$ to the polymer $-\!\!\left[C(=CH_2)-CH_2-\right.$ $\left.C(O)-O\right]_n$ occurs only in the solid state.

PLASMA POLYMERIZATION

A plasma is a system of gaseous ions and radicals formed by radio-frequency (megahertz) induction across a gas at low pressure. The plasma "activates" molecules in the gas, ionizes and fragments them, and produces larger molecules from recombination of the fragments. The polymers formed are deposited on the walls of the reaction chamber or on any object placed in the gas flow downstream of the plasma region.

It should be obvious from this description that plasma polymerizations involve more complex intermediates and mechanisms than are normally encountered in polymerization reactions. Hence, the deposited products may have unexpected compositions. Moreover, almost any organic molecule can be polymerized by this process. The method is used to prepare coherent, pinhole-free coatings. Some of the materials formed in this way represent a half-way stage between classical polymers and ceramics (see Chapter 9).

ELECTROCHEMICALLY-INITIATED POLYMERIZATIONS

General Principles

The passage of an electric current through a liquid system takes place by the transport of electrons from the cathode to the anode. During conventional electrolysis, the current is carried through the solution by ions, as shown in Figure 5.8.

If an unsaturated monomer is present in solution, an electron can be transferred from the cathode to the monomer to generate a radical anion. At the anode, an electron can be removed from the unsaturated compound to generate a radical cation. These processes are illustrated by (33) and (34).

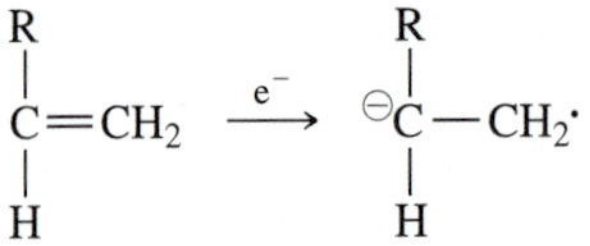

$$\underset{\overset{|}{H}}{\overset{\overset{R}{|}}{C}}=CH_2 \quad \xrightarrow{e^-} \quad \underset{\overset{|}{H}}{\overset{\overset{R}{|}}{\overset{\ominus}{C}}}-CH_2\text{·} \qquad \text{at the cathode} \tag{33}$$

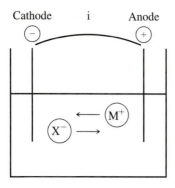

Figure 5.8 Conventional electrolysis.

$$\underset{\underset{H}{|}}{\overset{\overset{R'}{|}}{C}}=CH_2 \quad \xrightarrow{-e^-} \quad \underset{\underset{H}{|}}{\overset{\overset{R'}{|}}{^\oplus C}}-CH_2^{\cdot} \qquad \text{at the anode} \qquad (34)$$

In practice each vinyl monomer has its own specific reduction potential below which electron transfer from the electrode to the monomer does not occur. Electron-withdrawing groups in the monomer facilitate reduction and thereby lower the reduction potential. Conversely, electron-supplying groups raise the reduction potential of a monomer but lower the oxidation potential at the anode.

It will be clear that electroinitiated monomers, such as those shown in (33) and (34), could generate anionic, cationic, or free-radical chain reactions. In fact, difficulty is often experienced in the elucidation of the exact polymerization mechanism. The following examples will illustrate a few of the possibilities and problems.

Acrylonitrile, methyl methacrylate, and styrene polymerize in dimethyl formamide solvent under electrolytic conditions. Sodium nitrate is added as the supporting electrolyte. Apparently, acrylonitrile polymerizes by an anionic mechanism initiated by direct cathodic reduction of the monomer and dimerization of the radical anion (35). More than three polymer chains are formed for each electron transferred to the monomer, a result which suggests the existence of a chain transfer process. The polymer molecular weight is independent of the current density or the monomer concentration.

$$\underset{\underset{H}{|}}{\overset{\overset{C\equiv N}{|}}{CH_2=C}} \quad \xrightarrow{e^-} \quad \underset{\underset{H}{|}}{\overset{\overset{C\equiv N}{|}}{\cdot CH_2-C^\ominus}} \quad \longrightarrow \quad \underset{\underset{H}{|}}{\overset{\overset{C\equiv N}{|}}{^\ominus C}}-CH_2-CH_2-\underset{\underset{H}{|}}{\overset{\overset{C\equiv N}{|}}{C^\ominus}} \qquad (35)$$

Electrolysis of solutions of styrene or methyl methacrylate in dimethylformamide in the presence of tetramethylammonium chloride leads to the formation of polystyrene or poly(methyl methacrylate) by an anionic mechanism. Copolymerizations are also possible. For example, copolymers of styrene and methyl methacrylate can be formed by the electroinitiation of mixtures of the monomers in dimethyl formamide, and this could indicate either an anionic or a free-radical process. However, in tetrahydrofuran, poly(methyl methacrylate) homopolymer is the principal product.

Electrochemical initiation of vinyl polymerization has also been attempted in heterophase monomer-aqueous systems. Polymers can be obtained from styrene, vinyl chloride, methyl methacrylate, and other monomers, but free-radical mechanisms appear to operate under these conditions. Cationic chain processes are believed to take place in nitrobenzene solvent when perchlorate salts are used as the supporting electrolyte.

Polymerization on an Electrode Surface

The electropolymerization method has been used with great success to generate insoluble films of electronically conducting polymers at an electrode surface. Indeed, this is a principal method for the synthesis of polypyrrole and polyaniline, two polymers that have received widespread attention as "organic metals" (see Chapter 23).

$$
\begin{array}{ccccc}
& & & & \\
\text{pyrrole} & \xrightarrow{-e} & \text{pyrrole radical cation} & \longrightarrow & \text{polypyrrole} \\
& & & & \mathbf{3}
\end{array}
\tag{36}
$$

Polypyrrole **(3)** is formed by electrochemical oxidation of pyrrole. A solvent such as acetonitrile and a supporting electrolyte are needed. Oxidation of pyrrole at the anode generates radical cations. These combine to form polymer *on the surface of the anode*. Because the polymer is an electronic conductor of electricity (see Chapter 23), the deposition of polymer does not bring the process to a halt. Instead, polymer continues to be deposited until a relatively thick, free-standing film is formed. The film can then be separated from the electrode by the use of a sharp blade. Large-scale continuous deposition processes have been developed that make use of a polished rotating drum as the anode, with deposited polymer being removed as a continuous film.

N-substituted pyrroles have been polymerized in the same way, including species in which a "dopant" constitutes the substituent. Thiophene and its derivatives also polymerize to polythiophenes on electrochemical oxidation. Polyaniline is a more complex structure generated by the electrochemical reactions of aniline. It, too, is of interest as an electronically conducting material. These polymers are discussed in more detail in Chapter 23. Finally, the electropolymerization of vinyl monomers that bear organometallic units in the side groups has been used to deposit polymer films at the reducing electrode of an electrochemical cell. The transition metals in the polymer matrix function as electron-transfer, electrode mediator sites, and these systems are of considerable interest for new catalytic processes (see Chapter 7).

STUDY QUESTIONS

1. What practical advantages exist for the use of radiation, photochemically, or electrolytically induced polymerizations compared with those initiated by the addition of chemical reagents? What are the disadvantages?

2. Under what circumstances would the high dielectric constant of water be an advantage if this solvent were to be used as a polymerization solvent?

3. Survey the X-ray crystallographic literature (start with *Chemical Abstracts*) and examine the crystal structures of as many vinyl monomers or cyclic compounds as you can find. In each case, speculate on whether or not the molecular arrangement in the crystal would be suitable for a solid-state polymerization.

4. A parallel beam of light of wavelength 366 nm and having an intensity of 3.0×10^{15} photons/cm^2-s is incident on the window of a cylindrical cell that is 0.500 in. in diameter and 1.50 in. in length. If the cell contains 0.0100 mol/L of azoethane dissolved in benzene and the extinction coefficient of azoethane at 366 nm is 9.00 L/mol-cm, calculate the rate of absorption of photons per unit volume.

5. When the photolysis in Problem 4 was carried out for 15 min, it was found that the N_2 evolved from the solution exerted a pressure of 2.59 torr when contained in a vessel of 10-cm^3 volume at 27°C.
   (a) Calculate the primary quantum yield of decomposition of the azoethane.
   (b) Calculate the rate of formation of C_2H_5 radicals in the solution.
   (c) Assuming an efficiency of initiation of 0.4, calculate the rate of initiation of polymerization in a similar solution containing 3 mol/L of styrene.

6. Given a light source that emits 10^{16} quanta/cm^2-s at 366 nm and some azobisisobutyronitrile ($\varepsilon = 9.5$ at 366 nm with a primary quantum yield of 0.43), design a system for the photosensitized polymerization of methyl methacrylate in benzene. The total volume of solution should be 50 cm^3 and the initiation rate should be 1×10^{-7} mol/L-s. Assume the efficiency of initiation to be 0.50.

7. In a 15-min photopolymerization of methyl methacrylate using the system of Problem 6, 0.25 g of a polymer having an average molecular weight of 50,000 is obtained. Calculate the overall quantum yield of the polymerization, that is, the quantum yield for disappearance of monomer.

8. For γ-ray-induced polymerization of the following systems, calculate the fraction of energy absorbed by the monomer(s): (a) 5 *M* methyl acrylate in cyclohexane ($\rho = 0.779$ g/cm^3), (b) 0.1 *M* vinyl bromide in toluene ($\rho = 0.867$ g/cm^3); (c) an equimolar mixture of styrene ($\rho = 0.906$ g/cm^3) and (d) methyl acrylate ($\rho = 0.950$ g/cm^3).

9. A certain ^{60}Co irradiator produces a dose rate of 1.5 Mrads/h. Using data in this chapter and assuming an initiation efficiency of 0.5, calculate the rate of initiation of free radical polymerization in the following liquid monomers: (a) styrene ($\rho = 0.906$ g/cm^3);

(b) vinyl acetate ($\rho = 0.93$ g/cm^3); **(c)** acrylonitrile ($\rho = 0.806$ g/cm^3); **(d)** methyl acrylate ($\rho = 0.950$ g/cm^3).

10. A solution of 1 M acrylonitrile ($\rho = 0.806$ g/cm^3) in benzene ($\rho = 0.879$ g/cm^3) is irradiated with ^{60}Co γ-rays at a dose rate of 1.0 Mrads/h. Assuming that all radicals derived from either solvent or monomer initiate polymerization with an efficiency of 0.7, calculate the rate of initiation of free radical polymerization.

11. In Problem 10, the G-value for depletion of monomer is found to be 1500 molecules/100 eV. Calculate the rate of polymerization of acrylonitrile.

12. Suppose that you have available to you a ^{60}Co source with a dose rate of 0.50 Mrads/h and you wish to polymerize methyl acrylate ($\rho = 0.95$ g/cm^3) in benzene solution ($\rho = 0.879$ g/cm^3) using an initiation rate of 10^{-7} mol/L-s. Assuming an initiation efficiency of 0.5, calculate the weight of methyl acrylate that must be added to 100 cm^3 of benzene to prepare the solution.

SUGGESTIONS FOR FURTHER READING

Photochemical polymerization

COHEN, A. B., "Photopolymer Images," *Ind. Res.*, December, 39 **1976**.

CURTIS, H., IRVING, E., and JOHNSON, B. F. G., "Organometallic Photoinitiated Polymerizations," *Chem. Britain*, (April **1986**) *22*, 4, 327.

LABANA, S. S., "Photopolymerization," *J. Macromol. Sci., Rev. Macromol. Chem.*, **1974** *C11*(2), 229.

RANBY, B., "Photoinitiated Reactions of Organic Polymers," in *Polymer Science in the Next Decade* (O. Vogl and E. H. Immergut, eds.). New York: Wiley, **1987**, pp. 121–133.

Radiation-induced polymerization

ALLCOCK, H. R., FERRAR, W. T., and LEVIN, M. L., "Polymerization in Clathrate Tunnel Systems: Stereocontrolled Polymerization of Unsaturated Monomers in Crystals of Tris(*o*-phenylenedioxy)cyclotriphosphazene," *Macromolecules*, **1982**, *15*, 697.

ALLCOCK, H. R., and LEVIN, M. L., "Stereocontrolled Polymerization of Acrylic Monomers within a Tris(*o*-phenylenedioxy)cyclotriphosphazene Tunnel Clathrate," *Macromolecules*, **1985**, *18*, 1324.

CARAZZOLO, G., LEGHISSA, S., and MAMMI, M., "Polyoxymethylene from Trioxane by Solid State Polymerization," *Makromol. Chem.*, **1963**, *60*, 171.

CHATANI, Y., UCHIDA, T., TADOKORO, H., HAYASHI, K., NISHII, M., and OKAMURA, S., "X-Ray Crystallographic Study of Solid State Polymerization of Trioxane and Tetraoxymethylene," *J. Macromol. Sci., Phys.*, **1968**, *B2*(4), 567.

EASTMOND, G. C., "Solid State Polymerization," *Progr. Polymer Sci.* (A. D. Jenkins, ed.), **1970**, *2*, 1.

FARINA, M., "Inclusion Compounds of Perhydrotriphenylene," in *Inclusion Compounds*, Vol. 2 (J. L. Atwood, J. E. D. Davies, and D. D. MacNichol, eds.). London: Academic Press, **1984**, Chap. 3, p. 69.

FARINA, M., "Inclusion Polymerization," in *Inclusion Compounds*, Vol. 2 (J. L. Atwood, J. E. D. Davies, and D. D. MacNichol, eds.). London: Academic Press, **1984**, Chap. 10, p. 297.

GARRATT, P. G., "Radiation-Induced Solid State Polymerization," *Polymer*, **1962**, *3*, 323.

HENGLEIN, A., SCHNABEL, W., and WENDENBURG, J., Einführung in die Strahlenchemie. Weinheim, West Germany: Verlag Chemie, **1969**, pp. 299–354.

HERZ, J. E., and STANNETT, V., "Copolymerization in the Crystalline Solid State," *J. Polymer Sci. (D), (Macromol. Rev.)*, **1968**, *3*, 1.

STANNETT, V. T., and DEFFIEUX, A., "Cationic Polymerization and Grafting Initiated by High Energy Radiation," in *Cationic Polymerization and Related Processes*. New York: Academic Press, **1984**.

TABATA, Y., "Solid State Polymerization," *Advan. Macromol. Chem.*, **1968**, *1*, 283.

WEGNER, G., "The Mechanism of Solid-State Polymerizations," *Mol. Cryst. Liq. Cryst.*, **1979**, *52*(1–4), 535.

WEGNER, G., in *Molecular Metals* (W. E. Hatfield, ed.). New York: Plenum Press, **1979**.

WILSON, J. E., *Radiation Chemistry of Monomers, Polymers and Plastics*. New York: Dekker, **1974**.

Plasma polymerization

SHEN, M., and BELL, A. T., (eds.), *Plasma Polymerization, ACS Symp. Ser.*, **1979**, *108*, 344.

YASUDA, H., *Plasma Polymerization*. New York: Academic Press, **1985**.

Electrochemically initiated polymerizations

BRETTENBACH, J. W., OLAJ, O. F., and SOMMER, F., "Polymerisationsanregung durch Elektrolyse," *Fortschr. Hochpolymer.-Forsch.*, **1972**, *9*, 47.

DENISEVICH, P. D., ABRUNA, H. D., LEIDNER, C. R., MEYER, T. J., and MURRAY, R. W., "Electropolymerization of Vinylpyridine and Vinylbipyridine Complexes of Iron and Ruthenium: Homopolymers, Copolymers, Reactive Polymers," *Inorg. Chem.*, **1982**, *21*, 2153.

FUNT, B. L., "Electrolytically Controlled Polymerizations," *J. Polymer Sci. (D), Macromol. Rev.*, **1967**, *1*, 35.

NAARMANN, H., "Electrochemical Polymerization:-An Interesting Principle for Synthesis of Electrically Conductive Polymers," *Angew. Makromol. Chem.*, **1988**, *162*, 1.

NAARMANN, H., "New Aspects on Intrinsically Conducting Organic Systems and Their Synthesis," *Makromol. Chem., Macromol. Symp.*, **1987**, *8*, 1.

REYNOLDS, J. R., "Advances in the Chemistry of Conducting Organic Polymers: A Review," *J. Mol. Electron.*, **1986**, *2*, 1.

REYNOLDS, J. R., "Electrically Conductive Polymers," *ChemTech.*, **1988**, *18*, 440.

STREET, G. B., LINDSAY, S. E., NAZZAI, A. I., and WYNNE, K. J., "The Structure and Properties of Polypyrrole," *Mol. Cryst. Liq. Cryst.*, **1985**, *118*, 137.

YAMAZAKI, N., "Electrolytically Initiated Polymerization," *Advan. Polymer Sci.*, **1969**, *6*, 377.

6

Polymerization
of Cyclic
Organic Compounds

INTRODUCTION

Two of the three general classes of polymerization processes—condensation and olefin or vinyl polymerizations—have been discussed in Chapters 2 to 5. Here, we introduce the third category, the ring-opening polymerization of cyclic compounds, (**1**), by the overall process shown schematically in reaction (1).

$$\begin{matrix} A-B \\ | \quad | \\ (A-B)_n \end{matrix} \quad \longrightarrow \quad \underset{}{+}A-B\underset{n+1}{)} \tag{1}$$

1

Two main differences from the other two polymerization types can be emphasized. First, in contrast to condensation reactions, polymerization does not result in the loss of a small molecule. Second, ring-opening polymerization does not involve a loss of multiple-bonding enthalpy, whereas the loss of unsaturation is a powerful driving force for vinyl polymerization. In fact, the cyclic compounds and the polymers formed from them are often remarkably similar in enthalpy per repeating segment. The consequences of this fact will be explored more fully in Chapter 10.

It should be noted that ring-opening polymerization is the principal method for the synthesis of inorganic polymers, a topic that is covered in Chapter 9. In this chapter, we are mainly concerned with outlining the scope, experimental techniques, and general mechanisms for the polymerization of cyclic compounds, with a special emphasis on organic derivatives.

Cyclic organic compounds that have been polymerized include cyclic ethers, lactones (cyclic esters), lactams (cyclic amides), and imines (cyclic amines). In addition to these classical ring-opening polymerizations, a number of nonclassical reactions have been developed that fall outside the established categories. These include zwitterion polymerizations and "no catalyst" reactions, various cyclopolymerizations (polymerization with concurrent cyclization), polymerization of bicyclic monomers, and the unusual polymerization of *para*-xylene. Although some of these polymer synthesis reactions are not strictly "ring-opening polymerizations," they do bear a conceptual relationship to the other syntheses and are discussed at the end of this chapter. A summary of cyclic systems that have been polymerized is given in Table 6.1.

CYCLIC COMPOUNDS THAT RESIST POLYMERIZATION

Benzene **(2)**, *s*-triazine **(3)**, borazines **(4)**, cyclohexane **(5)**, tetrahydropyran **(6)**, and 1,4-dioxane **(7)**, have so far resisted all attempts to induce ring-opening polymerization. Both mechanistic and thermodynamic reasons can be put forward to rationalize this behavior. First, it is possible that suitable catalysts have not yet been found for the cleavage of skeletal bonds and the initiation of an ionic propagation. However, a more serious reason is apparently connected with the absence of ring strain in these compounds. Moreover, benzene, *s*-triazine, and borazine are especially stabilized by aromatic or pseudoaromatic π-bonding. Thus these compounds constitute an "energy trap" in the polymeric series. Polymers can be made from the monomers (acetylene, nitriles, ethylene, etc.) provided that the "trap" can be avoided. However, once cyclotrimerization has occurred, further polymerization is essentially blocked. These and other influences on ring-opening polymerizations are discussed further in Chapter 10.

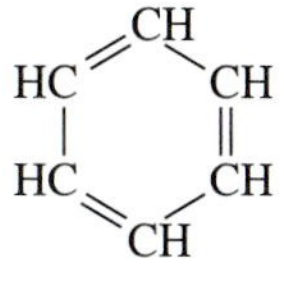

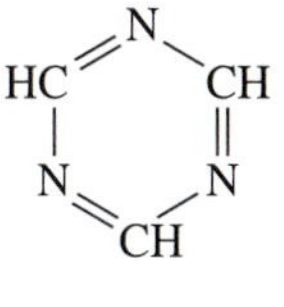

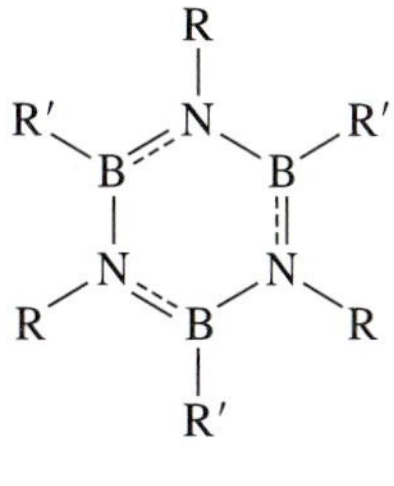

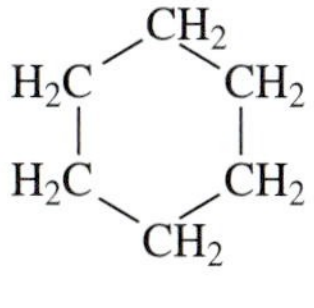

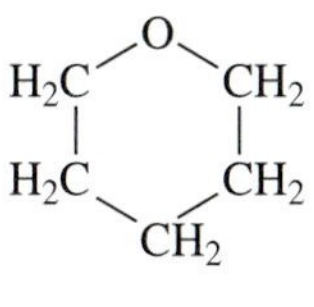

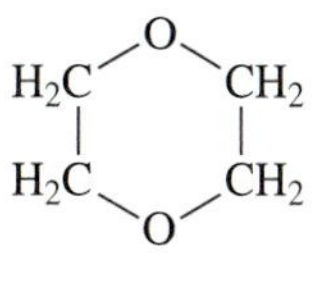

TABLE 6.1 MONOMERS, CATALYSTS, AND POLYMERIZABILITY IN RING-OPENING POLYMERIZATION

Class	Structure	Ring sizes [a]	Catalyst type [b]
Olefin		4, 5, 8	ROMP (W, Mo, Ru, Re, Ti, Ta)
Ether	O	3, 4, 5, 7	Cationic, anionic, covalent nucleophilic
Thioether	S	3, 4	Cationic, anionic, covalent nucleophilic
Amine	 NR	3, 4, 7	Cationic, covalent nucleophilic
Lactone	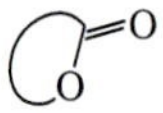	4, 6, 7, 8	Anionic, cationic, covalent nucleophilic
Thiolactone	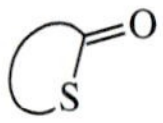	4–8	Anionic, covalent nucleophilic, cationic
Lactam	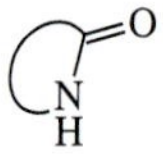	4–8, higher	Anionic, cationic
Disulfide	 S—S	4–8, higher	Radical
Anhydride	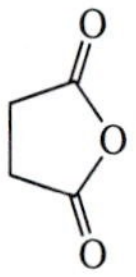	5, 7, 8, higher	Anionic
Carbonate	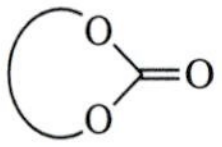	6, 7, 8, 20, higher	Anionic, covalent nucleophilic
Formal	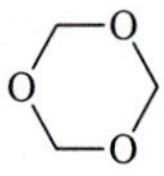	5, 7, 8, higher	Cationic
Silicone	 SiR_2	6, 8, 10, higher	Anionic, cationic
Phosphazene	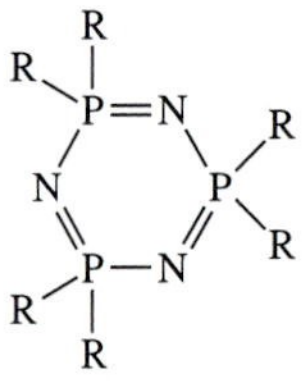	6	Cationic
Phosphonite	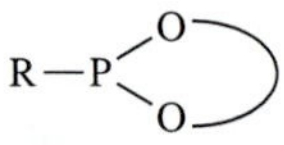 $R-P$	3, 5, 6, 7	Covalent nucleophilic, redox
Isoxazole	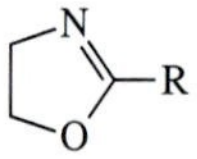 —R	5	Cationic, covalent nucleophilic

[a] Ring sizes affording high-molecular-weight polymers.

[b] Covalent nucleophilic mechanisms include alkylating agents (e.g., benzyl chloride), Lewis acids (e.g., BF_3), and organometallics (e.g., $R-Sn-X$).

Reproduced with permission from D. J. Brunelle, *Ring-Opening Polymerization*, New York, Hanser, **1993**, p. 3.

GENERAL MECHANISMS OF CLASSICAL RING-OPENING POLYMERIZATIONS

Two general types of mechanisms have been proposed for ring-opening polymerizations. In the first, the catalyst is presumed to attack the ring initially, with concurrent or subsequent cleavage. The resultant ionic or zwitterionic end group then attacks another ring with concurrent ring cleavage, and so on. This overall process is illustrated in reaction (2).

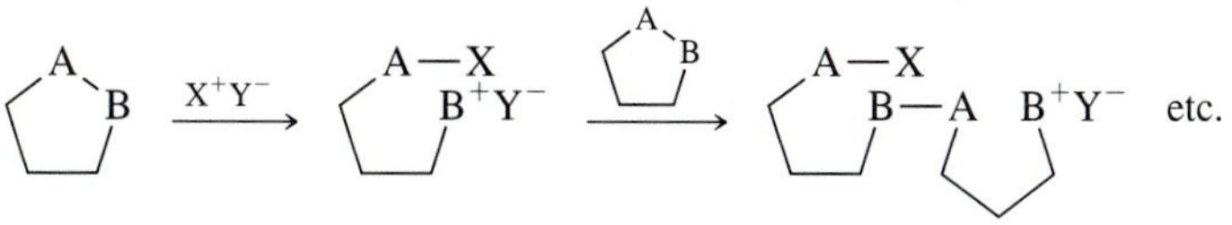

$$\text{(2)}$$

The alternative mechanism supposes that an initial ring cleavage does not occur. Instead, the primary interaction of the catalyst with the cyclic monomer generates a coordination intermediate (usually an oxonium ion), which then functions as the true initiating species. This is illustrated in reaction (3). In many cases, the distinction between these two pathways is difficult to establish.

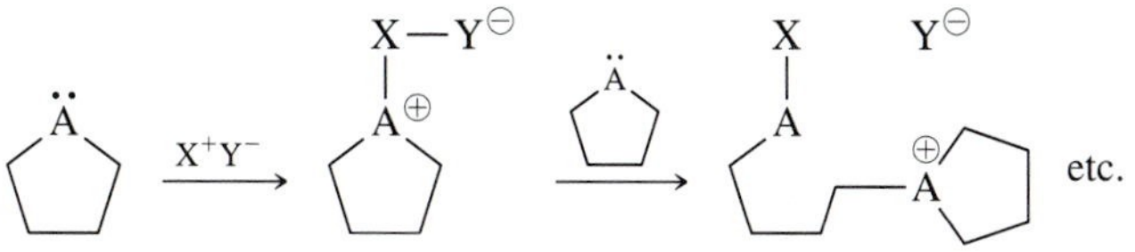

$$\text{(3)}$$

CYCLIC ETHERS

Trioxane

Trioxane **(8)** polymerizes under a variety of reaction conditions to yield polyoxymethylene (polyformaldehyde) **(9)**, as shown in reaction (4). The polymerization takes place (a) in the presence of boron trifluoride or other Lewis acid catalysts, (b) during sublimation of the cyclic trimer, or (c) during γ-irradiation of the crystalline trimer.

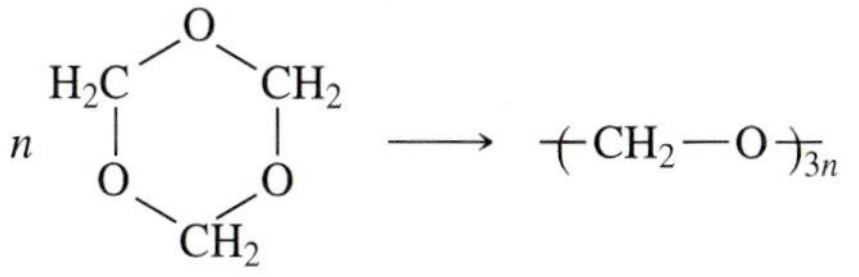

$$n \quad \begin{array}{c} H_2C \diagdown O \diagup CH_2 \\ | \quad \quad | \\ O \diagdown \quad \diagup O \\ CH_2 \end{array} \longrightarrow \quad \left(CH_2 - O \right)_{3n} \qquad \text{(4)}$$

$$\quad \quad \quad \quad 8 \quad \quad \quad \quad \quad \quad \quad 9$$

The experimental procedure used for the Lewis-acid-catalyzed polymerization of trioxane is quite straightforward. A concentrated solution of trioxane in cyclohexane in a glass reaction vessel is purged of air by bubbling a stream of dry nitrogen through it. Boron trifluoride etherate catalyst is then added and the mixture is heated at 55 to 60°C and stirred out of contact with the air for several hours. Polyoxymethylene separates from the solution as a white powder. The polymer can be compression-molded at 180 to 220°C to give tough, translucent films that can be oriented by stretching. Polyoxymethylene has a marked tendency to depolymerize to formaldehyde at moderate temperatures. This process can be retarded by "end capping" of the chains by acylation or etherification, or by copolymerization with a small amount of an epoxide. The partial polymerization of trioxane to polyoxymethylene during *sublimation* has been ascribed to the presence of traces of free formaldehyde which polymerizes, or to formic acid impurity which functions as a cationic catalyst.

The radiation-induced solid-state polymerization of trioxane has already been mentioned in Chapter 5. However, a solid-state polymerization can also be achieved by allowing solutions of Lewis acid catalysts to come into contact with trioxane crystals. Polymerization then proceeds inward from the crystal outer surfaces. Tetroxane, $(O-CH_2)_4$, also polymerizes in the crystalline state, especially when irradiated with γ-rays.

Different reaction conditions apparently generate different polymerization mechanisms for trioxane. The general acid-catalyzed reactions are believed to involve the initiation and propagation steps depicted in (5) and (6). Thus propagation probably takes place by the insertion of the trimer molecules into the $CH_2^+ \cdots X^-$ ionic bond of **10**, **11**, and so on. Polymerization can also be accompanied by depolymerization and equilibration. The latter processes may regenerate trioxane or yield the monomer (formaldehyde). Moreover, tetroxane, $(O-CH_2)_4$, may also be generated by a "back-biting" process. Such processes are discussed in a more general sense in Chapter 10. Chain termination can occur either by reaction of the chain ends with anions, or by hydride abstraction by the terminal carbonium ion.

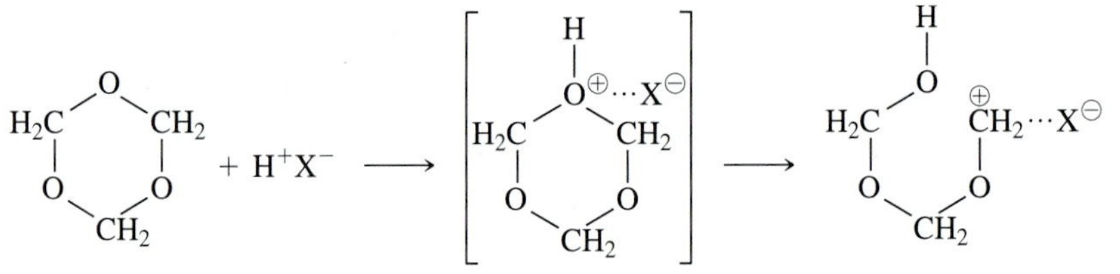

$$(5)$$

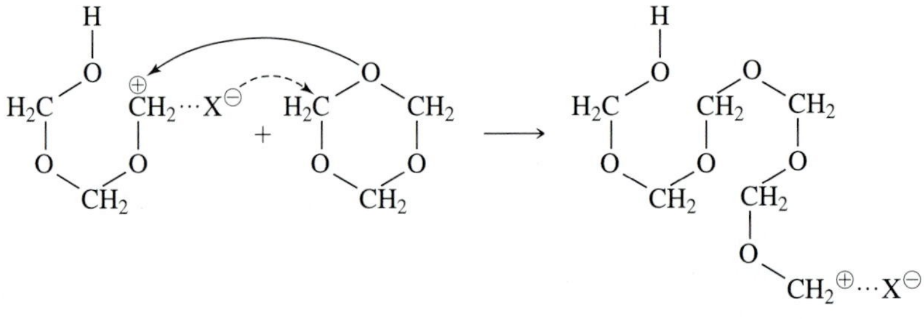

$$(6)$$

The polymerization of trioxane in methylene chloride solution under the influence of *anhydrous* Lewis acid catalysts, such as boron trifluoride, may follow a different mechanism. Apparently no cocatalyst (water) is required, and it has been proposed that polymerization involves the formation of a zwitterion (12). The sequence is illustrated in reactions (7) and (8). Because continued propagation would progressively increase the number of skeletal atoms that separate the charges, it must be assumed that the chain ends remain in close proximity. If this is true, the polymerization can be viewed as the successive insertion of trimer molecules into a macrocyclic unit.

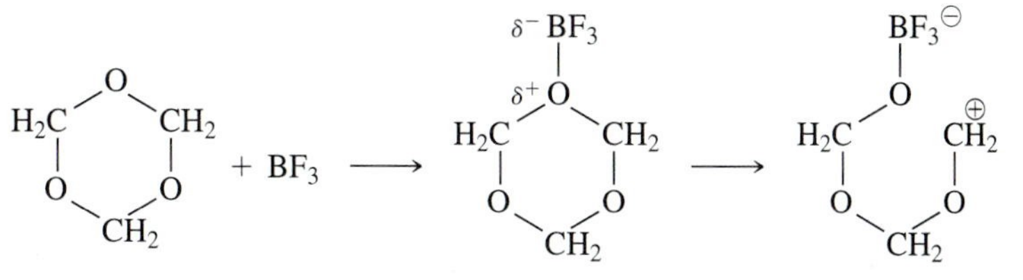

(7)

12

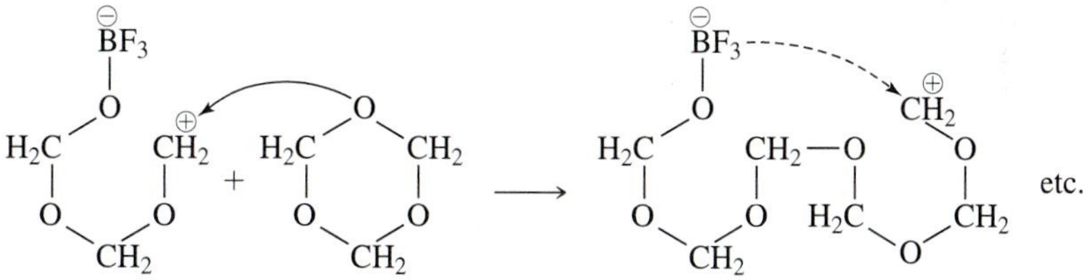

etc.   (8)

The mechanism of the radiation-induced *solid-state* polymerization of trioxane is still a subject for debate. An ionic mechanism has been proposed, but radical or cation–radical mechanisms must also be considered (see Chapter 5).

Trithiane and Tetrathiane

Trithiane (13) and tetrathiane (14) are, respectively, the cyclic trimer and tetramer of thioformaldehyde. Both cyclic compounds can be converted to polythioformaldehyde (15). The cyclic trimer (13) is a stable, crystalline solid, m.p. 215 to 216°C. This material polymerizes in the solid state when irradiated with γ-rays or when the molten material is treated with cationic-type catalysts, such as boron or antimony trifluorides. Tetrathiane behaves similarly, as does the cyclic pentamer. A cationic polymerization mechanism is believed to operate.

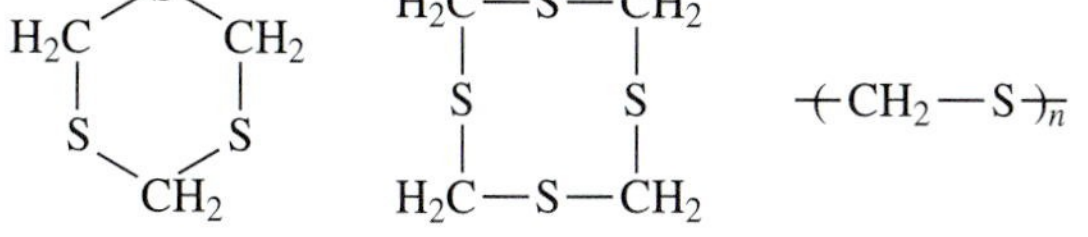

13   14   15

Tetrahydrofuran

Although tetrahydropyran (16) and 1,4-dioxane (17) are unreactive under polymerization conditions, tetrahydrofuran (18) can be induced to polymerize in the presence of phosphorus- or antimony pentafluorides or $[Ph_3C]^+[SbCl_6]^-$ as catalysts. The presence of the five-membered ring in tetrahydrofuran is apparently responsible for these differences.

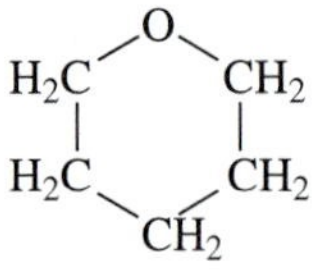

$$
\begin{array}{cc}
\text{16} & \text{17}
\end{array}
$$

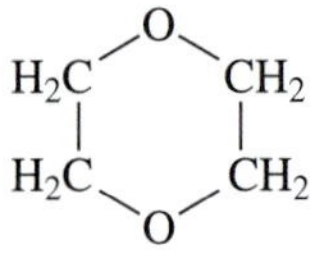

In practical terms, tetrahydrofuran must be purified rigorously before the polymerization is carried out. The compound must be boiled at reflux over sodium hydroxide pellets, distilled in a nitrogen atmosphere, refluxed over lithium aluminum hydride, and then distilled immediately before use. A suitable polymerization catalyst is a coordination complex of tetrahydrofuran with phosphorus pentafluoride. Polymerization is effected after about 6 h at 30°C to yield poly(tetramethylene oxide) (19) with a molecular weight of about 300,000. The overall process is shown in reaction (9). The use of antimony pentachloride as a polymerization catalyst yields a lower-molecular-weight polymer. Poly(tetramethylene oxide) is a tough, film-forming material, with a crystalline melting temperature of 45°C.

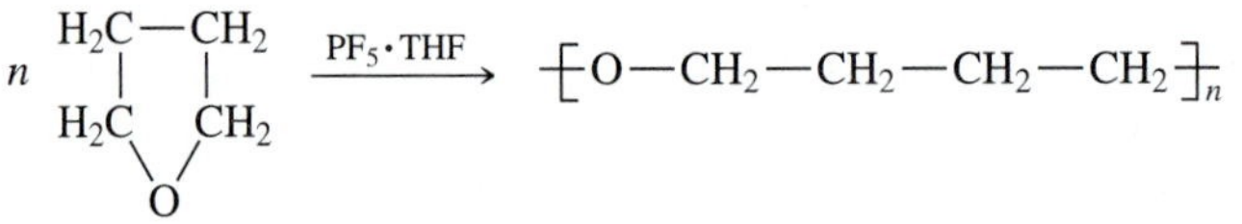

$$
n \; \underset{\text{18}}{\overset{H_2C-CH_2}{\underset{H_2C\diagdown\diagup CH_2}{\big|\quad\big|}}} \xrightarrow{\;PF_5\cdot THF\;} \underset{\text{19}}{\left[\!O-CH_2-CH_2-CH_2-CH_2\!\right]_n} \qquad (9)
$$

In the presence of trifluoromethanesulfonic acid (CF_3SO_3H) as an initiator, two types of products are formed: high-molecular-weight poly(tetramethylene oxide) polymers and macrocyclic oligomers (crown ethers). In this reaction the growing chains are believed to bear hydroxyl units and oxonium ions at opposite ends (20).

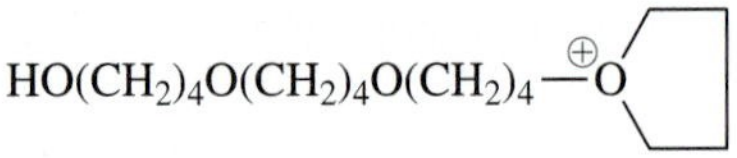

$$
HO(CH_2)_4O(CH_2)_4O(CH_2)_4\!-\!\overset{\oplus}{O}\!\diagup\!\diagdown
$$

20

Crown ethers are formed if the head and tail on the same oligomeric chain react. High polymers are formed by a pathway in which the head of one linear oligomer reacts with the tail of another.

Lewis acids in the presence of traces of water may initiate the polymerization of tetrahydrofuran by protonation of the etheric oxygen atom (reaction 10). Propagation would then occur by insertion of tetrahydrofuran molecules into the ionic bond of **21**. However, some tentative evidence exists that polymerization can occur even in the absence of water. If this is true, catalysts such as phosphorus pentafluoride or pentachloride may function as the ionic complexes $PF_4^{\oplus} PF_6^{\ominus}$ and $PCl_4^{\oplus} PCl_6^{\ominus}$. The latter formulation, in particular, is well known for PCl_5 in

$$
\begin{array}{c}
\text{H}_2\text{C}\!-\!\text{CH}_2 \\
| \qquad | \\
\text{H}_2\text{C} \qquad \text{CH}_2 \\
\diagdown \text{O} \diagup
\end{array}
+ [PF_5OH]^{\ominus}H^{\oplus} \longrightarrow
\begin{array}{c}
\text{H}_2\text{C}\!-\!\text{CH}_2 \\
| \qquad | \\
\text{H}_2\text{C} \qquad \text{CH}_2\cdots[PF_5OH]^{\ominus} \\
\diagdown \text{O} \diagup \\
| \\
\text{H}
\end{array}
\qquad (10)
$$

21

the solid state. If such ionic complexes are depicted symbolically as $X^+ Y^-$, a polymerization mechanism of the type shown in (11) can be formulated, and propagation can be visualized as an insertion of tetrahydrofuran molecules into the ionic $-\text{CH}_2^{\oplus}\cdots Y^{\ominus}$ bond of species such as **22**.

$$
\begin{array}{c}
\text{H}_2\text{C}\!-\!\text{CH}_2 \\
| \qquad | \\
\text{H}_2\text{C} \qquad \text{CH}_2 \\
\diagdown \text{O} \diagup
\end{array}
+ X^{\oplus}Y^{\ominus} \longrightarrow
\begin{array}{c}
\text{H}_2\text{C}\!-\!\text{CH}_2 \\
| \qquad | \\
\text{H}_2\text{C} \qquad \text{CH}_2^{\oplus}\cdots Y^{\ominus} \\
\diagdown \text{O} \diagup \\
| \\
\text{X}
\end{array}
\qquad (11)
$$

22

When cocatalysts are absent, it is possible that Lewis acid/Lewis base complexes, such as **23**, function as the real catalytic species.

$$
\begin{array}{c}
\text{CH}_2\!-\!\text{CH}_2 \\
| \qquad\quad \diagdown \overset{\delta^+}{}\ \overset{\delta^-}{} \\
| \qquad\quad\ \ \text{O}\cdots\text{BF}_3 \\
\text{CH}_2\!-\!\text{CH}_2 \diagup
\end{array}
$$

23

Oxetanes and Oxepanes

Substituted oxetanes, such as **24**, can be polymerized (sometimes violently) in the presence of Lewis acid catalysts such as phosphorus pentafluoride (reaction 12). Release of the ring strain in **24** almost certainly provides the driving force for polymerization. Polymer **25** is a crystalline, film-forming material that melts at 177°C. Oxetane itself **(26)** polymerizes readily at temperatures of 0°C or below to give high yields of polymer **27** (reaction 13).

As shown in reaction (14), oxepane **(28)** polymerizes slowly in the presence of catalysts, such as $[(C_2H_5)_3O]^+(BF_4)^-$ or $[(C_2H_5)_3O]^+(SbCl_6)^-$, even

$$n \text{ ClCH}_2 - \underset{\underset{\text{H}_2\text{C}-\text{O}}{|}}{\overset{\overset{\text{CH}_2\text{Cl}}{|}}{\text{C}}} - \text{CH}_2 \xrightarrow{\text{PF}_5} \left[\text{O} - \text{CH}_2 - \underset{\underset{\text{CH}_2\text{Cl}}{|}}{\overset{\overset{\text{CH}_2\text{Cl}}{|}}{\text{C}}} - \text{CH}_2 \right]_n \qquad (12)$$

$$\begin{array}{c} \text{CH}_2 - \text{CH}_2 \\ | \qquad\quad | \\ \text{CH}_2 - \text{O} \end{array} \xrightarrow{\text{BF}_3} \left(\text{O} - \text{CH}_2 - \text{CH}_2 - \text{CH}_2 \right)_n \qquad (13)$$

$$\begin{matrix} \mathbf{26} & & \mathbf{27} \end{matrix}$$

though only a minimal amount of ring strain must be released in this process. However, this polymerization is reversible, since depolymerization of **29** back to **28** takes place to yield an equilibrium mixture containing 2 to 3% of **28** and 97 to 98% **29** at 30°C. It has been shown that the polymerization reactivity falls in the order oxetane > tetrahydrofuran > oxepane.

$$n \begin{array}{c} \text{CH}_2 - \text{CH}_2 \\ | \qquad\quad | \\ \text{CH}_2 \qquad \text{CH}_2 \\ | \qquad\quad | \\ \text{CH}_2 \quad\; \text{CH}_2 \\ \backslash \quad / \\ \text{O} \end{array} \xrightarrow{\text{Et}_3\text{O}^+\text{BF}_4^-} \left(\text{O} - (\text{CH}_2)_6 \right)_n \qquad (14)$$

$$\begin{matrix} \mathbf{28} & & \mathbf{29} \end{matrix}$$

Epoxides

Epoxide polymerization is a subject of considerable technological importance. Here again, the ease of ring-opening polymerization reflects a release of ring strain. Ethylene oxide **(30)**, in particular, polymerizes readily to poly(ethylene oxide) **(31)** in the presence of both anionic- and cationic-type catalysts (reaction 15). Anionic catalysts that are suitable include alkoxide ions, hydroxides, metal oxides, and some organometallic derivatives. Cationic polymerizations are initiated by Lewis acids and protonic reagents.

$$n \; \text{H}_2\overset{\overset{\displaystyle O}{\diagdown\diagup}}{\text{C}} - \text{CH}_2 \longrightarrow \left(\text{CH}_2 - \text{CH}_2 - \text{O} \right)_n \qquad (15)$$

$$\begin{matrix} \mathbf{30} & & \mathbf{31} \end{matrix}$$

The polymerization of ethylene oxide can be carried out at 50°C in a sealed polymerization tube in the presence of strontium carbonate as a catalyst. Initially, there is an induction period which is followed by a very rapid exothermic reaction, so rapid, in fact, that *explosions may occur*. The polymerization reaction is normally complete within 2 h. The polymer can then be cast from solution to give highly crystalline films.

Other expoxides, such as propylene oxide **(32)** can also be induced to undergo ring-opening polymerization, and alkylene sulfides behave similarly. On the other hand, the well-known epoxy *resins* are usually prepared by the base-catalyzed reaction between an epoxide, such as epichlorohydrin **(33)** and a polyhydroxy compound, such as bisphenol A. The reaction yields a prepolymer by an initial base-catalyzed ring cleavage of the epoxide ring by the hydroxyl groups **(34)**. The overall process is illustrated in reaction (16). The ultimate products contain both terminal epoxy groups and pendent hydroxyl groups. Cross-linking of the prepolymer is then effected by addition of reagents such as amines. Thus, epoxy resins are characterized more by ring cleavage and condensation than by simple ring-opening polymerization.

$$\underset{\textbf{32}}{H_2C\!-\!\overset{\displaystyle O}{\overset{\displaystyle \diagup\!\diagdown}{CH}}\!-\!CH_3} \qquad \underset{\textbf{33}}{H_2C\!-\!\overset{\displaystyle O}{\overset{\displaystyle \diagup\!\diagdown}{CH}}\!-\!CH_2Cl}$$

$$HO\!-\!R\!-\!OH + CH_2\!-\!\overset{\displaystyle O}{\overset{\displaystyle \diagup\!\diagdown}{CH}}\!-\!CH_2Cl \longrightarrow HO\!-\!R\!-\!O\!-\!CH_2\!-\!\overset{\displaystyle OH}{\overset{\displaystyle |}{CH}}\!-\!CH_2Cl$$

$$\textbf{34}$$

$$\left|\begin{array}{l} NaOH \\ -NaCl \end{array}\right.$$

$$HO\!-\!R\!-\!O\!-\!CH_2\!-\!\overset{\displaystyle O}{\overset{\displaystyle \diagup\!\diagdown}{CH}}\!-\!CH_2$$

$$(16)$$

The anionic polymerization of epoxides is initiated by alkoxides, hydroxides, metal oxides, and organometallic species such as zinc alkyls. Each of these catalysts can be depicted symbolically as X^+Y^-, for example, as M^+OR^-, M^+OH^-, and so on. The anionic initiation process then operates, as shown in (17). Since, for the catalysts used, the $-O^\ominus \cdots X^\oplus$ bond is more ionic than the $-CH_2-Y$ bond, propagation occurs by insertion of monomer molecules into the $-O^\ominus \cdots X^\oplus$ bond, as depicted in (18).

$$R\!-\!\overset{\displaystyle O}{\overset{\displaystyle \diagup\!\diagdown}{CH}}\!-\!CH_2 + X^+Y^- \longrightarrow R\!-\!\overset{\displaystyle O^\ominus \cdots X^\oplus}{\overset{\displaystyle \diagup}{CH}}\!-\!CH_2\!-\!Y \qquad (17)$$

$$Y\!-\!CH_2\!-\!\overset{\displaystyle R}{\overset{\displaystyle |}{CH}}\!-\!O^\ominus \cdots X^\oplus + \quad \overset{\displaystyle H_2C\!-\!CH_2}{\underset{\displaystyle O}{\diagdown\diagup}} \longrightarrow \qquad (18)$$

$$Y\!-\!CH_2\!-\!\overset{\displaystyle R}{\overset{\displaystyle |}{CH}}\!-\!O\!-\!CH_2\!-\!CH_2\!-\!O^\ominus \cdots X^\oplus \quad etc.$$

Termination may not occur unless protonic reagents are added. However, the polymer molecular weights are often low because of chain transfer. Chain transfer can occur by proton abstraction by the terminal anion from an alkyl group, R, with the concurrent formation of an allyl ether anion, **(35)**, as shown in reaction (19).

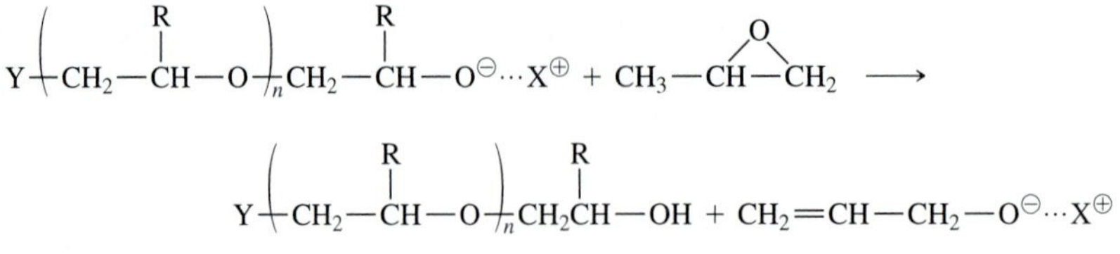

$$Y \left(CH_2-\underset{\underset{R}{|}}{CH}-O \right)_n CH_2-\underset{\underset{R}{|}}{CH}-O^{\ominus}\cdots X^{\oplus} + CH_3-\overset{\overset{O}{\diagup \diagdown}}{CH}-CH_2 \longrightarrow \tag{19}$$

$$Y \left(CH_2-\underset{\underset{R}{|}}{CH}-O \right)_n CH_2\underset{\underset{R}{|}}{CH}-OH + CH_2{=}CH-CH_2-O^{\ominus}\cdots X^{\oplus}$$

35

The cationic polymerization of epoxides probably proceeds through mechanisms that are similar to those described for trioxane and tetrahydrofuran. For example, the use of CF_3SO_3H as an initiator probably results in an initial ring cleavage followed by oxonium ion formation **(36)** (Scheme 1). Again, depending on the statistics of ring closure versus propagation, crown ethers may be formed

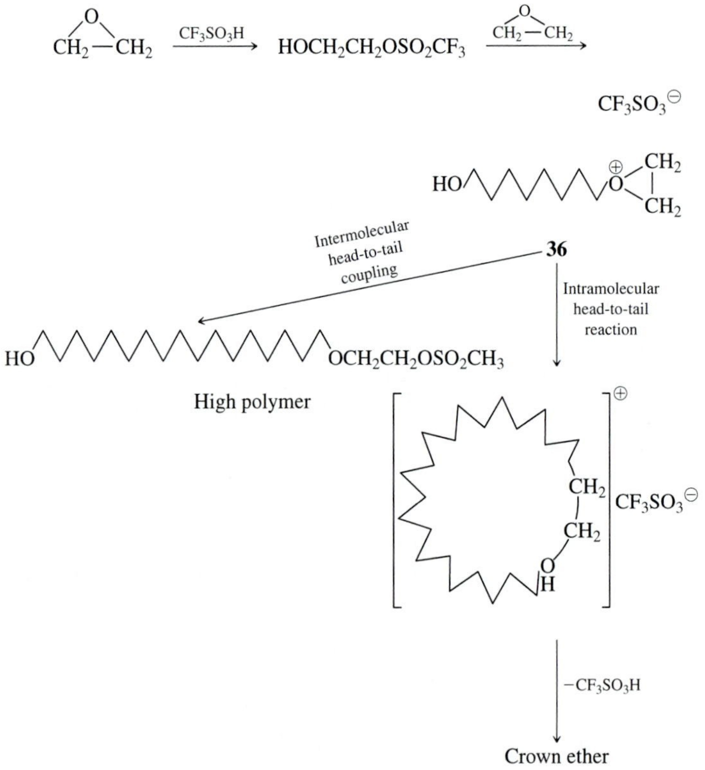

Scheme 1

by an intramolecular head-to-tail reaction, or high polymers may result from the intermolecular counterpart.

LACTONES

Lactones (37) are polymerized to polyesters with the use of either anionic or cationic catalysts. Example initiators include alcohols, amines, organometallic compounds, and alcohol–titanium alkoxide mixtures. However, it should be noted that ring size has an important and rather curious influence on the polymerizability of lactones. γ-Butyrolactone, which contains a five-membered ring, apparently does not polymerize, although δ-valerolactone, with a six-membered ring, does polymerize.

$$n \ \overline{O + CH_2 \overrightarrow{)_x} C} = O \longrightarrow \left[O - (CH_2)_x - \overset{\overset{\displaystyle O}{\|}}{C} \right]_n \tag{20}$$

37

GLYCOLIDES AND LACTIDES

Poly(glycolic acid), poly(lactic acid), and their copolymers are important materials used in biomedicine because of their susceptibility to hydrolysis and bioerosion (see Chapter 24). The obvious way to produce these polymers is by a traditional condensation polymerization from glycolic or lactic acids. However, higher molecular-weight polymers are accessible via the ring-opening polymerization of the cyclic dimers—known as glycolides and lactides. This reaction is shown in equation (21).

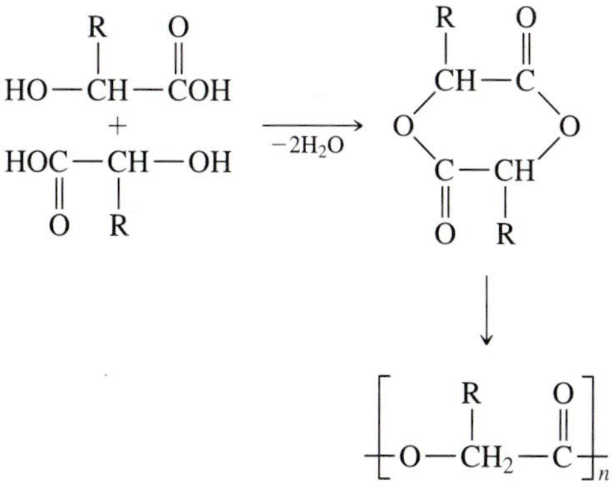

$$\begin{array}{c} \overset{R}{\underset{|}{}} \quad \overset{O}{\underset{\|}{}} \\ HO - CH - COH \\ + \\ HOC - CH - OH \\ \overset{\|}{\underset{O}{}} \quad \overset{|}{\underset{R}{}} \end{array} \quad \xrightarrow{-2H_2O} \quad \begin{array}{c} \overset{R}{\underset{|}{}} \quad \overset{O}{\underset{\|}{}} \\ CH - C \\ O \qquad O \\ C - CH \\ \overset{\|}{\underset{O}{}} \quad \overset{|}{\underset{R}{}} \end{array} \tag{21}$$

$$\downarrow$$

$$\left[O - \overset{\overset{\displaystyle R}{|}}{CH_2} - \overset{\overset{\displaystyle O}{\|}}{C} \right]_n$$

R = H (glycolic) or CH$_3$ (lactic)

The polymerizations are catalyzed by stannous chloride, stannous octoate or, less frequently, by antimony trifluoride or para-toluene sulfonic acid. Copolymers are produced by the use of mixtures of the glycolide and lactide. Bioerosion results from hydrolysis to glycolic and/or lactic acid, both of which are considered to be biologically acceptable. The rates of erosion can be controlled through the glycolide to lactide ratios.

CYCLIC ANHYDRIDES

Cyclic anhydrides (**38**) such as oxepane-2,7-dione (adipic anhydride) undergo ring-opening polymerization in solution to yield polyanhydrides. Anionic initiators such as sodium hydride or potassium acetate, cationic initiators such as aluminum chloride or boron trifluoride etherate, and coordination initiators such as stannous 2-ethylhexanoate give high polymers. As discussed elsewhere in this book, polyanhydrides are of interest as bioerodible materials.

$$n \ O{=}C{+}CH_2{\xrightarrow{}_x}C{=}O \longrightarrow \left[C{-}(CH_2)_x{-}C{-}O \right]_n \qquad (22)$$

38

CYCLIC CARBONATES

Polycarbonates, which can be prepared by classical emulsion condensation techniques (Chapter 2), can also be produced by ring-opening polymerization. The cyclic monomers are synthesized by the procedure shown in equation (23).

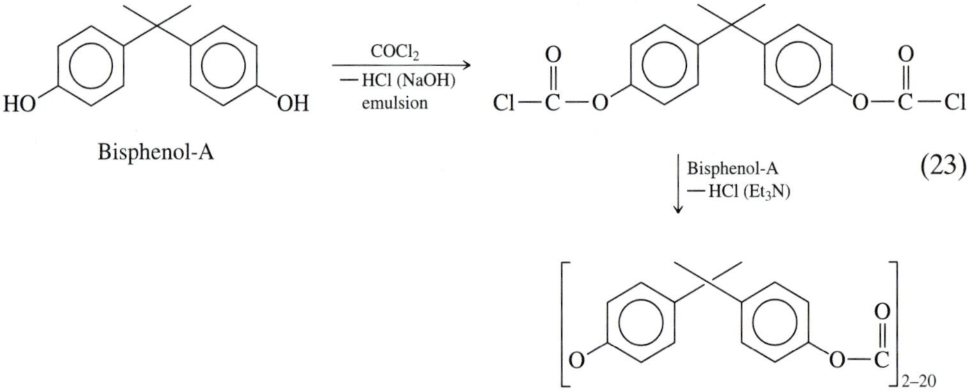

Bisphenol-A

(23)

The mixture of cyclic species formed in this reaction has a melting point below 200°C, which facilitates a solvent-free ring-opening polymerization at 250 to 275°C. Lithium stearate is one of the initiators that give high molecular-weight polymers. The advantage of the ring-opening approach over normal condensation processes is that no side-products are released during polymerization. Hence, the polymerization can be carried out in a mold or extruder so that the liquid starting material is converted directly to a high-melting solid.

LACTAMS

The polymerization of lactams **(39)**, especially caprolactam **(40)**, provides a valuable noncondensation route to the synthesis of nylons (reaction 25). The polymerization can be initiated by reagents such as strong bases (metal hydrides, alkali metals, or metal amides), protonic acids, aromatic amines, or by water. The base-catalyzed initiation is often applied to N-acylated lactams, since these species are not subject to the long induction periods that characterize the polymerization of the parent lactams.

$$n \ \text{H} \!-\! \text{N} \!\!\overbrace{+\text{CH}_2\!\!}^{}\!\!\!\overset{}{\underset{x}{)}}\!\!\text{C}\!=\!\text{O} \quad \longrightarrow \quad \left[\overset{\text{H}}{\underset{\mid}{\text{N}}} \!-\! (\text{CH}_2)_x \!-\! \overset{\text{O}}{\overset{\|}{\text{C}}} \right]_n \tag{24}$$

39

$$n \ \begin{array}{c} \text{O} \\ \| \\ \text{C} \\ \diagup \ \diagdown \\ \text{CH}_2 \quad \text{NH} \\ \mid \qquad \mid \\ \text{CH}_2 \quad \text{CH}_2 \\ \mid \qquad \mid \\ \text{CH}_2 \!-\! \text{CH}_2 \end{array} \quad \longrightarrow \quad \left[\overset{\text{H}}{\underset{\mid}{\text{N}}} \!-\! (\text{CH}_2)_5 \!-\! \overset{\text{O}}{\overset{\|}{\text{C}}} \right]_n \tag{25}$$

40   Nylon 6

The water-catalyzed polymerization of caprolactam can be carried out on a laboratory scale provided that suitable safety precautions are taken[1] A mixture of purified caprolactam and water in about a 50:1 weight ratio is sealed under nitrogen in a thick-walled polymerization tube. This tube is a potential bomb when heated, and intelligent precautions should be taken to provide shielding. The tube is heated to 250°C for about 6 h, cooled, and the end of the tube is then removed cautiously, again with the use of adequate shielding. The tube contents are now heated to 250 to 255°C as a stream of nitrogen is allowed to bathe the polymer surface. Most of the water will volatilize from the system, and the molten reaction mixture will undergo a viscosity increase. About 2 h of heating are usually sufficient to generate a polymer (nylon 6) of suitable molecular weight for melt spinning into fibers.

A variety of mechanisms have been proposed for lactam polymerization, depending on the type of initiator. Here we will consider only the commercially important processes—catalysis by bases and by water. Strong bases probably initiate polymerization by the replacement of hydrogen in the N—H residue of **41** by a cation to give **42**, as illustrated in reaction (26). However, the subsequent mech-

---

[1]Sorenson, W. R., and Campbell, T. W., *Preparative Methods of Polymer Chemistry*, 2nd ed. (New York: Wiley-Interscience, **1968**), p. 344.

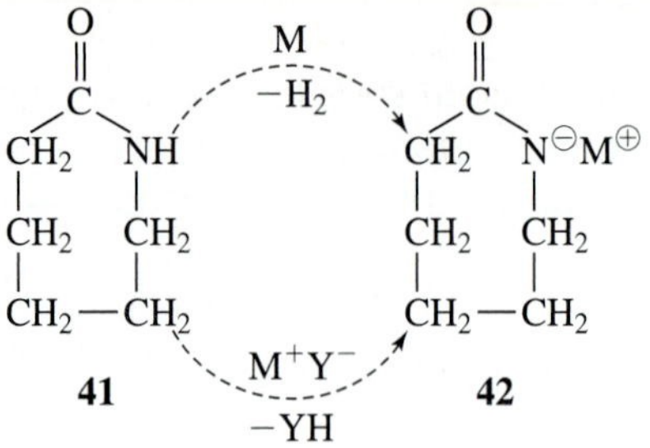

$$\begin{array}{c}\text{(structures } \mathbf{41} \text{ and } \mathbf{42}\text{, with } \overset{M}{-H_2},\ M^+Y^-,\ -YH)\end{array} \tag{26}$$

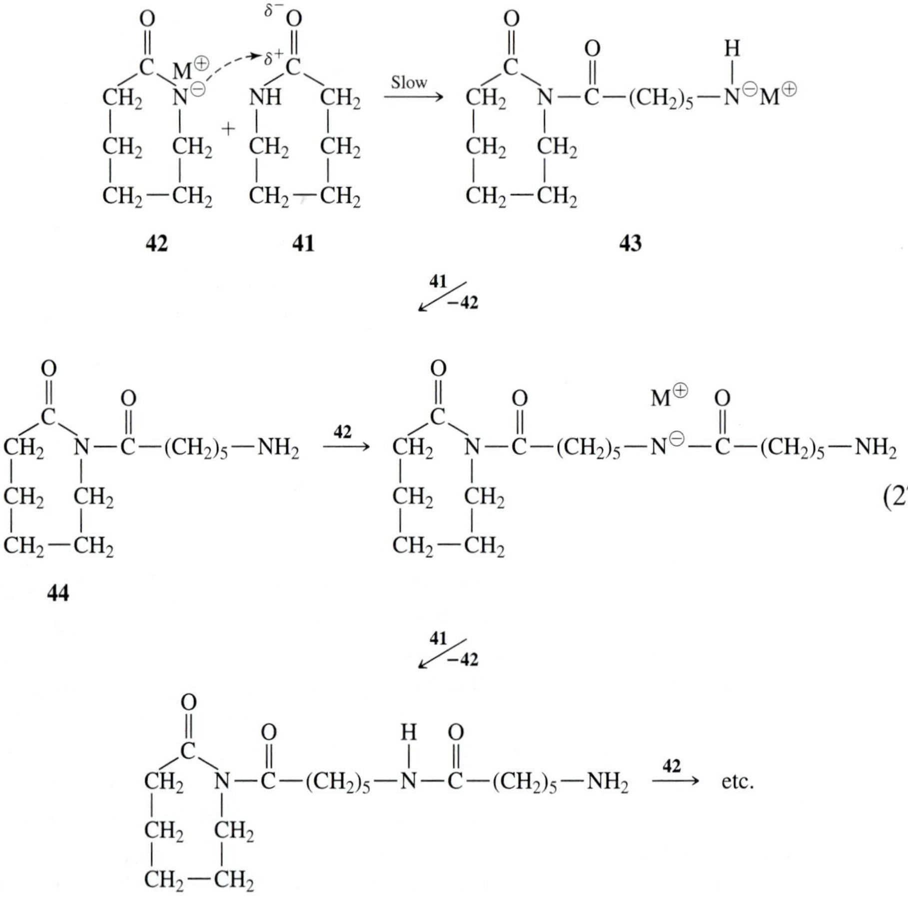

Structure **41**:
O=C–CH₂–CH₂–CH₂–CH₂–NH (ring with NH)

Structure **42**:
O=C–CH₂–CH₂–CH₂–CH₂–N⁻M⁺ (ring)

$$\mathbf{42} + \mathbf{41} \xrightarrow{\text{Slow}} \mathbf{43}$$

Structure **43**:
(ring)C(=O)–CH₂–CH₂–CH₂–CH₂–N–C(=O)–(CH₂)₅–N̄⁻ ... H ... M⁺

$$\mathbf{44} \xrightarrow{\ \mathbf{42}\ } \text{(product)} \tag{27}$$

Structure **44**:
(ring)–N–C(=O)–(CH₂)₅–NH₂

Product of (27):
(ring)–N–C(=O)–(CH₂)₅–N⁻(M⁺)–C(=O)–(CH₂)₅–NH₂

$$\xleftarrow[\ -\mathbf{42}\ ]{\ \mathbf{41}\ }$$

Structure **45**:
(ring)–N–C(=O)–(CH₂)₅–N(H)–C(=O)–(CH₂)₅–NH₂ $\xrightarrow{\ \mathbf{42}\ }$ etc.

anism is complicated. Ring opening of the initiated monomer is presumed not to occur. Instead, the anionic center can attack the carbonyl carbon of another ring by a *slow* process, as shown in the formation of **43**.

The long induction periods observed for this polymerization are probably a consequence of the slowness of this step. However, compound **43** is assumed not to be the real initiating species. Instead, compound **44** is believed to fulfill that function by reaction with **42**. Propagation appears to take place by the unusual process of *insertion* of molecules of **42** into an —NH—CO— bond, followed by a remetallation of another monomer molecule by the polymer to yield, for example, **45**. The base-catalyzed polymerization is more conveniently applied to *N*-acylcaprolactams, such as **46**, since the long induction periods are not encountered in these systems.

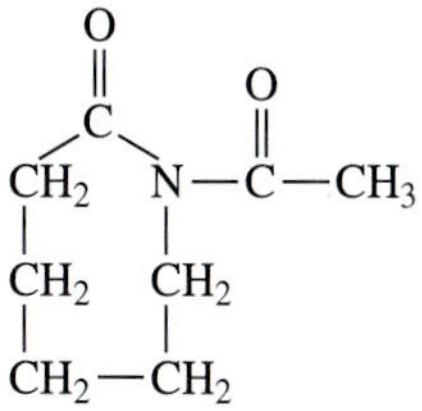

$$
\begin{array}{c}
O \\
\parallel \\
C \\
CH_2 \quad N-C-CH_3 \\
| \qquad | \qquad \parallel \\
CH_2 \quad CH_2 \quad O \\
| \qquad | \\
CH_2-CH_2
\end{array}
$$

46

The water-catalyzed reaction of caprolactam has a more straightforward mechanism. The primary step involves the hydrolysis of caprolactam to the amino acid, **(47)** (reaction 28). Propagation then involves either the direct, ring-opening attack of **47** on caprolactam, or a process in which the amino acid zwitterion, $H_3N^\oplus—(CH_2)_5C(O)O^\ominus$, undergoes a ring-opening attack on the cyclic monomer.

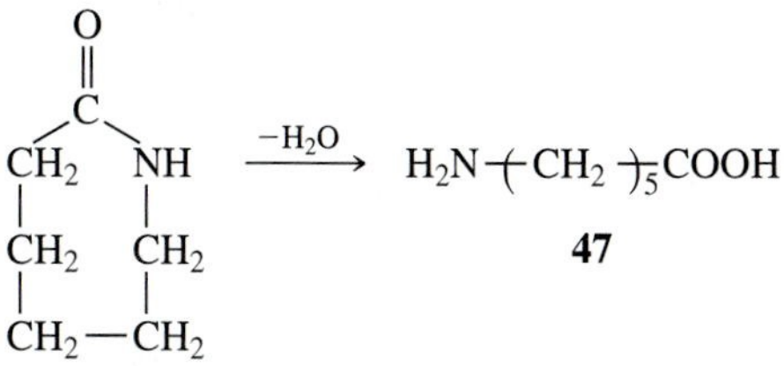

$$
\begin{array}{c}
O \\
\parallel \\
C \\
CH_2 \quad NH \\
| \qquad | \\
CH_2 \quad CH_2 \\
| \qquad | \\
CH_2-CH_2
\end{array}
\xrightarrow{-H_2O} H_2N \!\!-\!\!(CH_2)_5COOH
\qquad (28)
$$

47

ETHYLENIMINE

Ethylenimine **(48)** polymerizes very rapidly in the presence of cationic initiators, a result that probably reflects the release of ring strain. Restrictions have been placed on the use of this monomer because of its carcinogenicity.

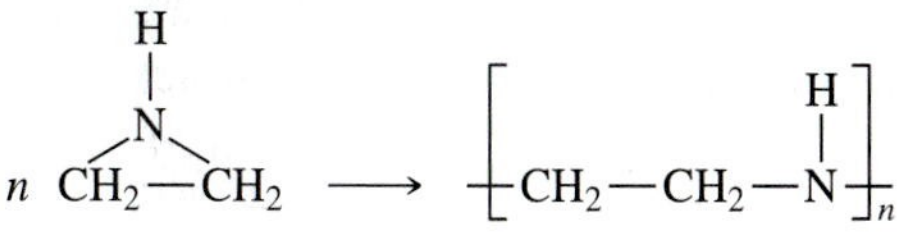

$$
n \;
\begin{array}{c}
H \\
| \\
N \\
CH_2-CH_2
\end{array}
\longrightarrow
\left[CH_2-CH_2-\underset{|}{\overset{H}{N}} \right]_n
\qquad (29)
$$

48

CYCLOALKENES

Cycloalkenes such as cyclopentene readily undergo ring-opening polymerizations under the influence of metathesis catalysts. These reactions are discussed in Chapter 4.

THE SPECIAL CASE OF OXAZOLINE POLYMERIZATIONS

Oxazolines (cyclic imino ethers) are unusual monomers since they undergo a number of different types of polymerization reactions and yield different polymer structures, depending on the reaction conditions. Vinyloxazolines are particularly interesting monomers, as illustrated in Scheme 2.

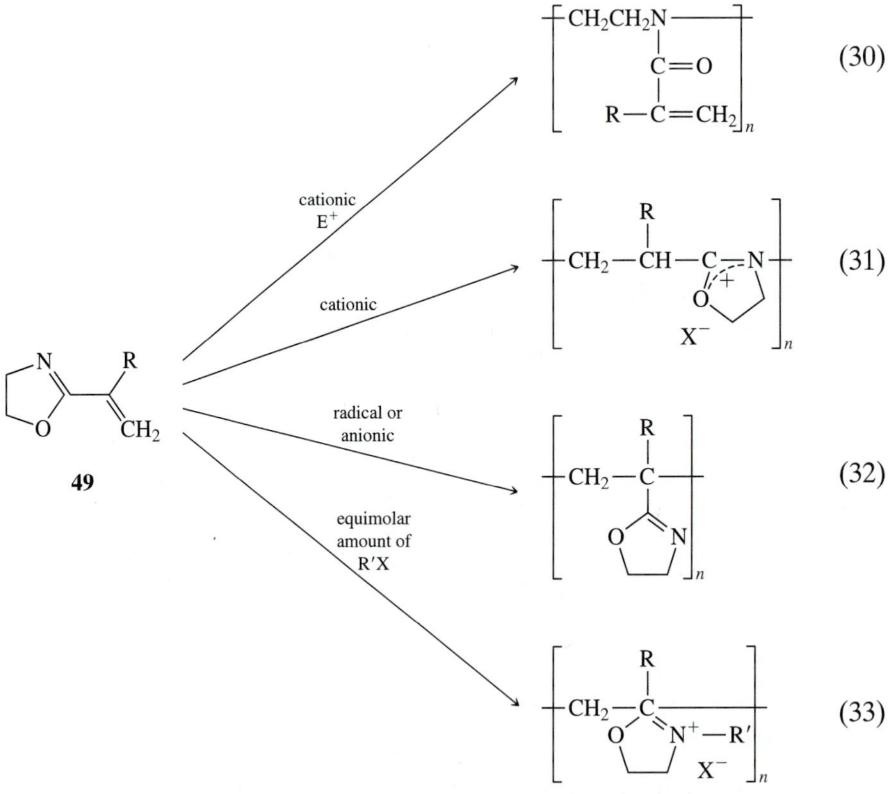

$$\left[\begin{matrix} CH_2CH_2N \\ | \\ C=O \\ | \\ R-C=CH_2 \end{matrix}\right]_n \qquad (30)$$

$$\left[\begin{matrix} & R \\ & | \\ CH_2-CH-C=N \\ & O \\ & X^- \end{matrix}\right]_n \qquad (31)$$

$$\left[\begin{matrix} R \\ | \\ CH_2-C \\ O \quad N \end{matrix}\right]_n \qquad (32)$$

$$\left[\begin{matrix} R \\ | \\ CH_2-C \\ O \quad N^+-R' \\ X^- \end{matrix}\right]_n \qquad (33)$$

Scheme 2

The processes shown in Scheme 2 include cationic ring-opening polymerization with incorporation of ring nitrogen into the chain reaction (30); cationic polymerization without ring opening but with incorporation of both a ring C—N and the vinyl group into the chain reaction (31); free-radical or anionic polymerization through the vinyl group reaction (32) to give a polymer with a pendent oxazoline unit; and a "spontaneous" polymerization through the vinyl group in the presence of equimolar amounts of an alkyl halide reaction (33).

The cationic ring-opening polymerization reaction takes place with a wide range of nonvinyl oxazolines as well. For example, species of type **50** (where X = H, CH_3, OCH_3, Cl, NO_2) yield poly(*N*-acylalkyleneimines)**(51)**. These are precursors for hydrolysis to poly(ethylenimine), a route that bypasses the need to handle ethylenimine monomer. The electrophiles (E^+) $CH_3OSO_2CH_3$ or $CH_3OSO_2C_6H_4CH_3$ are used as initiators. The polymerization mechanism is believed to follow the pathway shown in reaction (34).

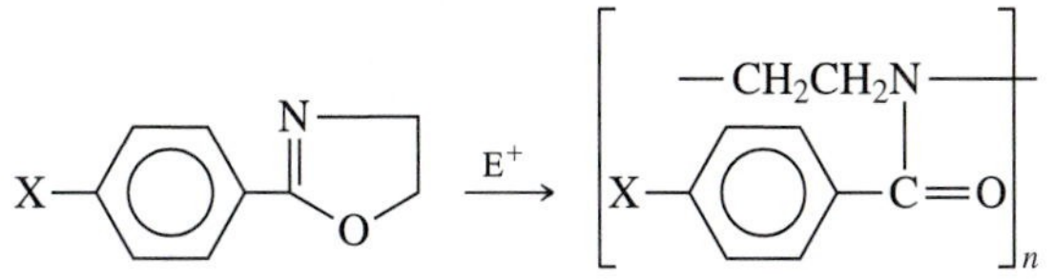

50   **51**

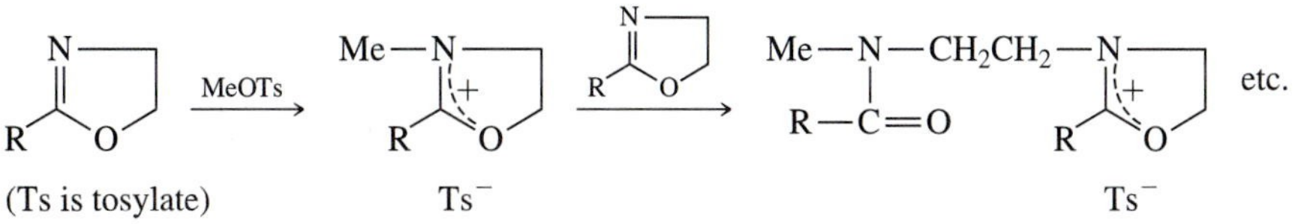

(34)

(Ts is tosylate) Ts^- Ts^-

Thus, the methyl group from the initiator remains at the terminus, and the positive charge is transferred to the incoming oxazaline ring. We comment further about oxazoline polymerization in the next section.

NO-CATALYST COPOLYMERIZATIONS

If two cyclic monomers are brought together, one a nucleophile and the other an electrophile, the possibility exists that each will initiate ring-opening polymerization of the other to generate a 1:1 alternating copolymer. No additional initiator would be needed. Such processes are known. They proceed through zwitterion intermediates.

An example is provided by the copolymerization between oxazoline **(52)** and β-propiolactone **(53)**. The copolymerization proceeds spontaneously at room temperature, following the reaction pattern shown in Scheme 3.

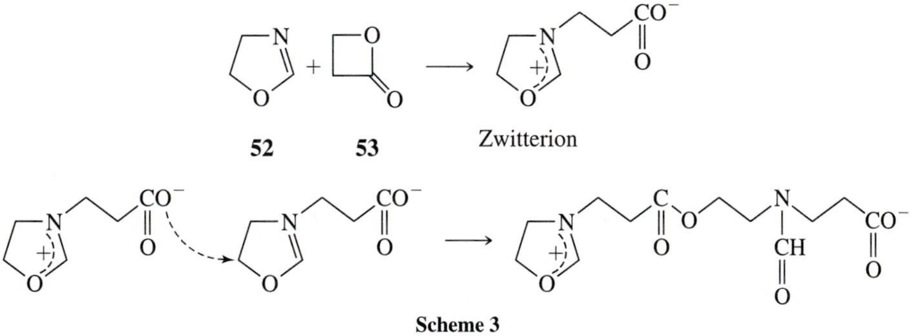

52   **53**   Zwitterion

Scheme 3

The number of nucleophiles and electrophiles that might participate in this type of reaction is quite large. One additional example will be given here, a process that involves the interaction of a cyclic phosphonite (54) with either β-propiolactone or acrylic acid (Scheme 4). Polymerization of zwitterion 55 then proceeds as in the previous example to generate an alternating copolymer. This polymerization requires the use of elevated temperatures (over 120°C).

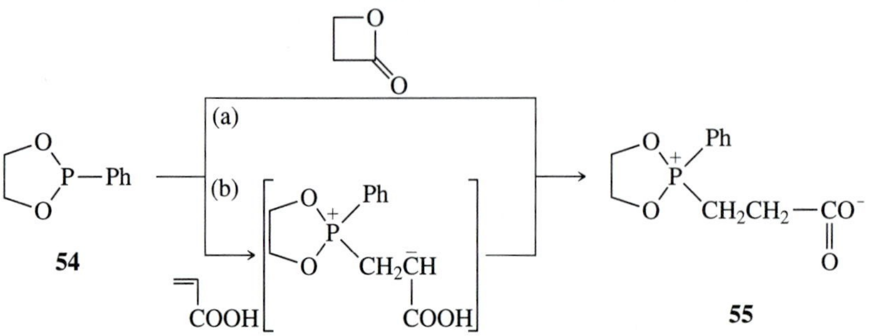

Scheme 4

FREE-RADICAL RING-OPENING POLYMERIZATION

As discussed, most ring-opening polymerizations involve ionic processes, and until recently only a few examples were known of ring-opening polymerizations that take place via radical pathways. These examples usually involve the opening of strained rings such as those of vinylcyclopropane or bicyclobutane derivatives. However, a class of unstrained cyclic monomers has been shown to undergo a free-radical ring-opening polymerization. These species are 1,3-dioxepanes of the type shown in 56 (Scheme 5).

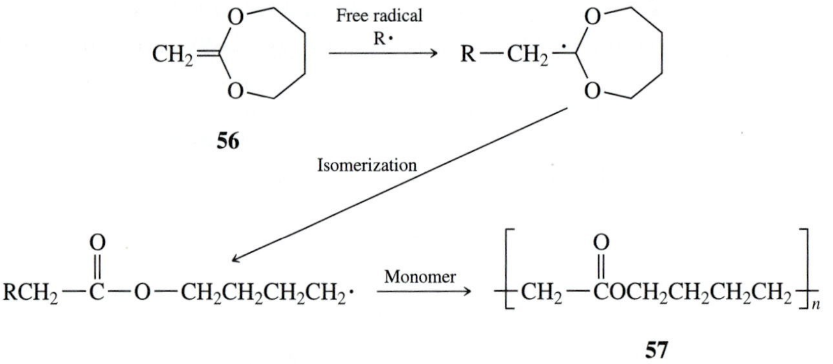

Scheme 5

The mechanism shown in Scheme 5 illustrates how radical addition is followed by a rearrangement to a linear species with a terminal free-radical site, which then begins the chain propagation process. The product (57) is poly(ε-caprolactone). By contrast, the five-membered ring analogue of 56 is exceedingly sensitive to cationic initiators, and under free-radical conditions, undergoes a polymerization in which only 50% of the monomer units have undergone ring opening.

A related free-radical ring-opening polymerization is of special interest because unlike most polymerizations it leads to a volume *expansion*. This involves the unsaturated spiro ortho carbonate **(58)**, which undergoes double ring-opening polymerization when treated with di-*tert*-butyl peroxide at 130°C. The process is believed to proceed by the mechanism shown in Scheme 6. At less than 30% conversion to polymer, the product is a polycarbonate formed by an extension of the steps shown in Scheme 6. At higher conversions, the side-group unsaturated units generate cross-links.

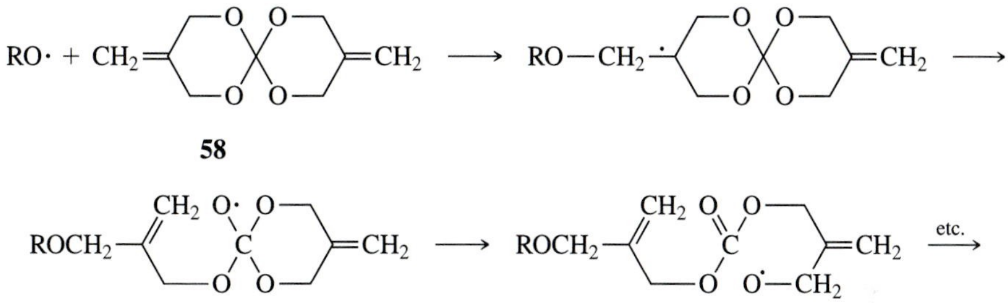

Scheme 6

CYCLOPOLYMERIZATIONS

The term *cyclopolymerization* is used to describe the addition reactions of non-conjugated dienes that lead to the *generation* of rings during polymerization. Such reactions are generally believed to occur through an alternating intra-intermolecular propagation process. These are ring-closing rather than ring-opening polymerizations, but it is instructive to compare them with ring-opening processes.

A general example of a cyclopolymerization is shown in Scheme 7. Here, a 1,6-diene **(59)** is initiated by a radical R· to generate a linear olefin **(60)** that undergoes intramolecular cyclization to form radical **61**. This then adds to another molecule of 1,6-diene in the same manner as did R·, and a propagation reaction then occurs to give **62**. Propagation may produce five-membered as well as six-membered rings. This is a complicated polymerization and much effort

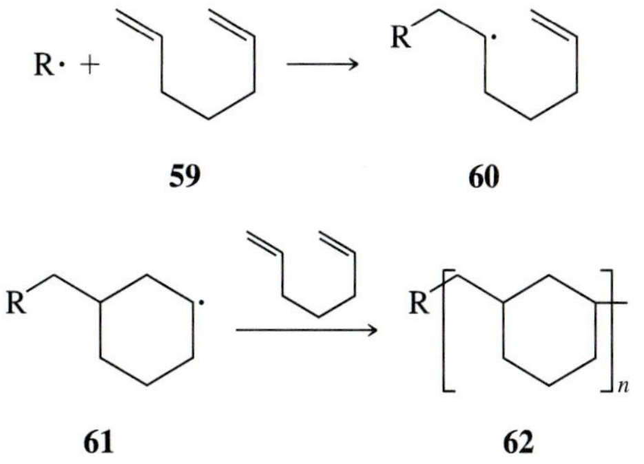

Scheme 7

has been devoted to elucidating the mechanism. It appears that the course of the reaction depends on the statistics and energetics of ring closure, interactions of radical intermediates with nonconjugated double-bonded systems, and so on.

Cyclopolymerizations can also be induced by γ-irradiation of a crystalline monomer. An example is shown in reaction (35), in which both six- and five-membered rings are formed. Normal free-radical initiation yields mainly a six-membered aliphatic ring system. In this system steric effects that involve the methyl groups would be expected to influence the pattern of cyclization.

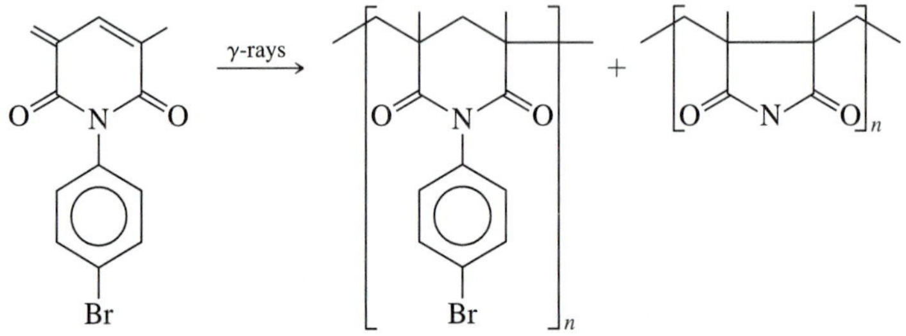

$$(35)$$

Finally, the reaction between divinyl ether and maleic anhydride leads to a radical-induced cyclopolymerization, as shown in Scheme 8. The polymeric product from this reaction **(63)** is known as DIVEMA. It and polyanions derived from it are interferon-inducing agents and possess antitumor activity.

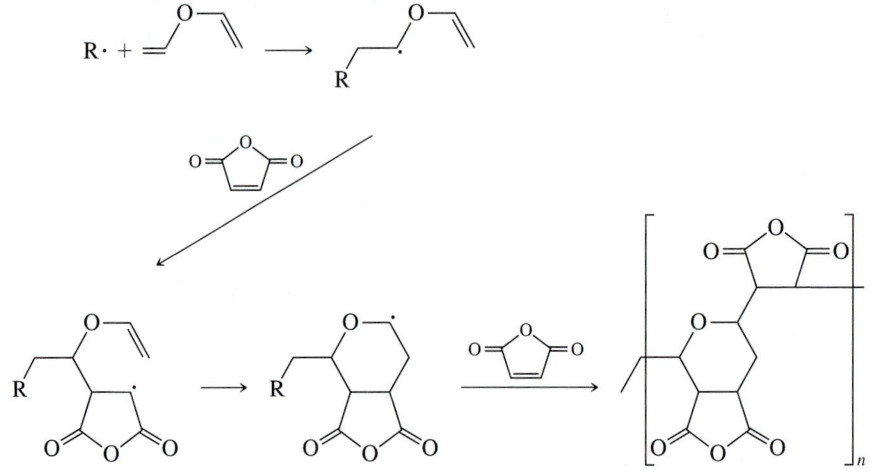

63

Scheme 8

POLY (*PARA*-XYLYLENE)

We end this chapter with an unusual polymerization process that starts with a cyclization reaction, proceeds through a ring cleavage to generate an unsaturated monomer, and ends with the free-radical polymerization of that monomer. The overall reaction sequence is shown in Scheme 9.

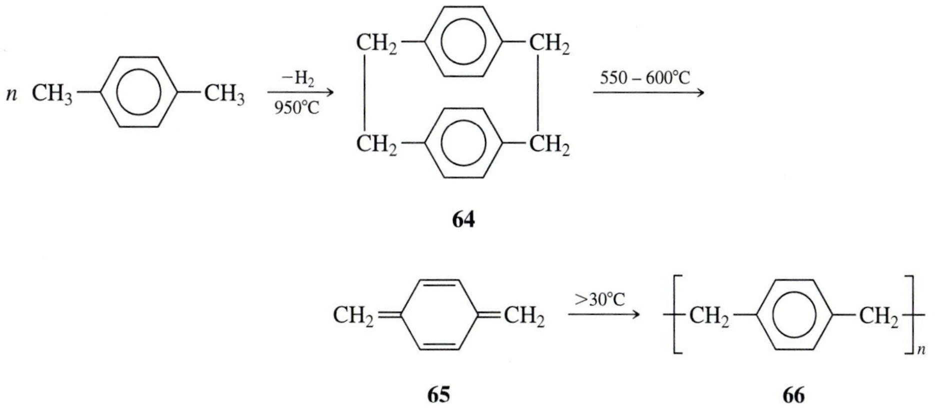

64

65   **66**

Scheme 9

p-Xylene can be oxidatively pyrolyzed at temperatures up to 950°C to give a cyclic dimer known as di-*p*-xylylene **(64)**. At 550 to 650°C this compound dissociates in vacuum to yield *p*-xylylene **(65)** and this monomer polymerizes spontaneously on a surface in vacuum at temperatures near 30°C to yield poly(*p*-xylylene) **(66)**. Thus the final step is an addition polymerization, presumably operating by a radical mechanism. The rate of polymerization is exceedingly fast, and the process can be used to coat objects (especially biomedical devices) essentially by vapor deposition.

STUDY QUESTIONS

1. Speculate on the possibility that some of the ring-opening polymerizations discussed in this chapter might fall into the category of "living" polymerizations. Which structural factors would favor or prevent such a possibility?

2. Most of the reactions discussed in this chapter follow ionic mechanisms, although a few use radical pathways. Which monomer structures favor one mechanism or the other, and why? How would you prove the existence of an ionic or a free-radical pathway?

3. Outline a cationic polymerization mechanism that might be applicable to epoxides.

4. Without referring to Chapter 10, speculate in detail on the reasons why polyethylene is not manufactured by the ring-opening polymerization of cyclohexane.

5. Some polymers (e.g., polyoxymethylene) can be prepared either from the monomer (formaldehyde) or from a cyclic species (trioxane). What are the main practical advantages or disadvantages to these alternative routes?

6. Glance through an organic chemistry textbook and compile a list of classes of organic cyclic compounds that are not mentioned in this chapter. Then suggest possible methods that might be used to induce their polymerization.

7. Suppose that you suspected that the polymerization of trioxane proceeded only by prior dissociation to formaldehyde, followed by polymerization of this monomer. Suggest ways in which you might distinguish between this mechanism and one that involved a prior trimerization of formaldehyde to trioxane, followed by a polymerization of trioxane.

8. Tetrahydrofuran is a common organic solvent that is often used in large quantities in the laboratory or in manufacturing. What reagents, other than PF_5, should you *not* bring into contact with tetrahydrofuran if you wish to avoid a (possibly dangerous) polymerization process? How could you ensure that polymerization of tetrahydrofuran would be unlikely during normal laboratory use?

9. Speculate on the prospect that an epoxide could be copolymerized with caprolactam. What reaction conditions might you choose for this process? What complications do you foresee?

SUGGESTIONS FOR FURTHER READING

AOI, K., and OKADA, M., "Polymerization of Oxazolines," *Progr. in Polymer Sci.*, **1996**, *21*(1), 151.

BAILEY, W. J., "Free Radical Ring-Opening Polymerization," *ACS Symp. Ser.*, **1985,** *286*, 47.

BAILEY, W. J., NI, Z., and WU, S.-R., "Free Radical Ring-Opening Polymerization of 4,7-Dimethyl-2-methylene-dioxepane and 5,6-Benzo-2-methylene-1,3-dioxepane," *Macromolecules*, **1982**, *15*, 711.

BRUNELLE, D. J. "Preparation and Polymerization of Cyclic Carbonates," *Ring-Opening Polymerization* (D. J. Brunelle, Ed.). New York: Hanser, **1993**.

BRUNELLE, D. J. (Ed.), *Ring-Opening Polymerization*. New York: Hanser, **1993**, Ch. 11.

BUTLER, G. B., XING, Y., GIFFORD, G. E., and FLICK, D. A., "Physical and Biological Properties of Cyclopolymers Related to DIVEMA," in *Macromolecules as Drugs and as Carriers for Biologically-Active Materials* (D. A. Tirrell, L. G. Donaruma, and A. B. Turek, Eds.), *Ann. N.Y. Acad. Sci.*, **1985**, *446*, 149.

CHUJO, Y., and SAEGUSA, T., "Polymerization of Oxazoline Family," in *Ring-Opening Polymerization* (D. J. Brunelle, Ed.) New York: Hanser, **1993**, Ch. 8.

CRIVELLO, J. V., LAI, Y. L., and MALIK, R., "The Synthesis and Photoinitiated Cationic Polymerization of Cyclic Ketene Acetals," *Cationic Polymerization*, **1997**, *665*, 83.

GOETHALS, E. J., "Telechelic Polymers by Ring-Opening Polymerization," in *Ring-Opening Polymerization* (D. J. Brunelle, Ed.) New York: Hanser, **1993**, Ch. 10.

HALL, H. K., "Bond-Forming Initiation in Spontaneous Addition and Polymerization Reactions of Olefins," in *Synthetic Polymers* (Proc. Robert A. Welch Foundation Conf.), **1983**.

HASHIMOTO, K., "Ring-Opening Polymerization of Lactams. Living Anionic Polymerization and its Applications," *Progress in Polymer Sci.*, **2000** *25*(10), 1411.

INOUE, S., and AIDA, T., "Living Polymer Systems: Epoxides and Lactones," in *Encyclopedia of Polym. Sci. & Eng., Suppl. Vol.* New York: Wiley, **1989**, p. 412.

IVIN, K. J., and SAEGUSA T., (Eds.), *Ring-Opening Polymerization.* New York: Elsevier, **1984**.

KOBAYASHI, S., and SAEGUSA, T., "Alternating Copolymerization involving Zwitterions," in *Alternating Copolymerization* (J. M. G. Cowie, Ed.). New York: Plenum Press, **1985**, pp. 189–238.

KOBAYASHI, S., TOKUZAWA, T., and SAEGUSA, T., "Cationic Ring-Opening Isomerization Polymerization of 2-[*p*-(Substituted)phenyl]—2-oxazolines: Effects of the Substituent on the Reactivities," *Macromolecules,* **1982**, *15*, 707.

MCGRATH, J. E. (Ed.), *Ring Opening Polymerization, Kinetics, Mechanisms, and Synthesis, ACS Symp. Ser.,* **1985**, *286*.

MIYAMOTO, M., SANO, Y., KIMURA, Y., and SAEGUSA, T., "'Spontaneous' Vinyl Polymerization of 2-Vinyl-2-oxazolines," *Macromolecules,* **1985**, *18*, 1641.

PENCZEK, S., and KUBISA, P., "Living Polymer Systems: Cationic," in *Encyclopedia of Polym. Sci. & Eng., Suppl. Vol.* New York: Wiley, **1989**, p. 380.

PENCZEK, S., and KUBISA, P., "Cationic Ring-Opening Polymerization," in *Ring-Opening Polymerization,* D. J. Brunelle (ed). New York: Hanser, **1993**, Ch. 2.

ROKICKI, G., "Aliphatic Cyclic Carbonates in Spiro-orthocarbonates as Monomers," *Progr. in Polymer Sci.,* **2000**, *25*(2), 259.

SAEGUSA, T., "Spontaneous Copolymerization of Phosphorus(III) Compounds via Zwitterion Intermediates: Redox Copolymerization and Deoxy Polymerization," *Pure Appl. Chem.,* **1981**, *53*(3), 691.

SAEGUSA, T., and KOBAYASHI, S., "Cyclic Imino Polymerization Chemistry and Polymer Characteristics," *Makromol. Chem., Macromol., Chem. Symp.* **1986**, *1*(1), 23.

SANDA, F., and ENDO, T., "Radical Ring-Opening Polymerization," *J. Polym. Sci. (A), Polym. Chem.,* **2001**, *39*(2), 265.

SEMLYEN, J. A., "Cyclic Siloxane Polymers," in *Siloxane Polymers,* (S. J. Clarson and J. A. Semlyen, Eds.), Englewood Cliffs, NJ: Prentice Hall, **1993**, p. 135.

SEMLYEN, J. A. (Ed.), *Large Ring Molecules,* New York: John Wiley & Sons, **1996**.

SLOMOWSKI, S., and DUDA, A., "Anionic Ring-Opening Polymerization," in *Ring Opening Polymerization* (D. J. Brunelle, Ed.). New York: Hanser, **1993**, Ch. 3.

STEVELS, W. M., DIJKSTRA, P. J., and FEIJEN, J., "New Initiators for the Ring-Opening Polymerization of Cyclic Esters," *Trends in Polymer Sci.,* **1997**, *5*(9), 300.

STONE-ELANDER, S. A., BUTLER, G. B., DAVIS, J. H., and PALENIK, G. J., "Conformational Effects on the Cyclopolymerization of N-(*p*-Bromophenyl)-dimethacrylamide," *Macromolecules,* **1982**, *15*, 45.

SURENDRAN, G., GAZICKI, M., JAMES, W. J., and YASUDA, H., "Polymerization of *Para*-xylylene Derivatives (Parylene Polymerization). VI. Effects of the Sublimation Rate of Di-*p*-xylylene on the Morphology and Crystallinity of Parylene Deposited at Different Temperatures," *J. Polymer Sci. (A), Polymer Chem. Ed.,* **1987**, *25*, 1481.

TIRRELL, D. A., "Progress in Ring-Opening Polymerization of Cyclic Ethers and Cyclic Sulfides," in *Polymer Science Technology.* New York: Plenum Press, **1984**, pp. 431–441.

VOGEL, O. and FURUKAWA, J., *Polymerization of Heterocycl.es.* New York: Dekker, **1973**.

7

Reactions
of Synthetic Polymers

INTRODUCTION

Synthetic high polymers undergo a variety of reactions either when they are brought into contact with chemical reagents or when heated to high temperatures. For convenience, these reactions will be divided into two classes: those that involve reactions of the main chain and those that involve the side groups.

REACTIONS INVOLVING THE MAIN CHAIN

Reasons for Interest in Main-Chain Reactions

There are three principal reasons why the reactions of a polymer skeleton are important. First, reactions of the skeleton may allow one type of polymer chain to be converted into another. For example, an unsaturated carbon skeleton may be converted to a saturated one, or vice versa. Second, an understanding of skeletal cleavage reactions may provide clues about why one polymer is thermally or chemically unstable but others are not. This information can help in the design of more stable polymers. Finally, the ability of a polymer chain to break down efficiently to small molecules, for example when irradiated with ultraviolet light or an electron beam, is important for the choice of materials for "resists" in integrated circuit manufacture.

Addition Reactions

Chlorine can be added across the double bonds of unsaturated polymers such as natural rubber or polybutadiene. Natural rubber is dissolved in carbon tetrachloride, and chlorine is bubbled into the system until the polymer contains 66% or more of chlorine. The chlorinated rubber is nonflammable. Mechanistically, the reaction is very complex, with halogen addition, substitution, and skeletal cyclization reactions taking place. By contrast, polybutadiene apparently reacts with chlorine almost exclusively by an addition mechanism, as shown in the conversion of **1** to **2**.

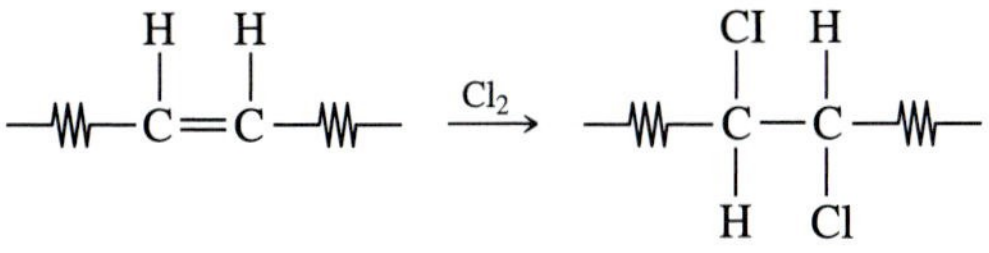

Hydrogen chloride adds across the double bonds of polyisoprene in chloroform solution to yield a material known as "rubber hydrochloride." The hydrogenation of unsaturated polymers, such as polybutadiene, has been accomplished with the use of nickel or noble metal catalysts. Saturation of all the double bonds in the polymer does not normally occur. Increasing the degree of hydrogenation leads principally to an increase in the polymer crystallinity. However, drastic hydrogenation, especially at elevated temperatures causes skeletal cleavage. This has been suggested as a method for the conversion of old rubber tires to gasoline.

Reactions that Generate Skeletal Unsaturation

Poly(vinyl acetate) **(3)** decomposes thermally to liberate acetic acid. The colored residue is believed to have a polyacetylene type of structure **(4)**. Colored polymers are also formed when poly(vinyl chloride) is heated at temperatures near 130°C. Hydrogen chloride is liberated and, again, polyacetylene-like residues **(5)** are formed.[1] Exposure of the polyunsaturated polymer to the air causes a fading of the color. This has been attributed to air oxidation of the unsaturated structures.

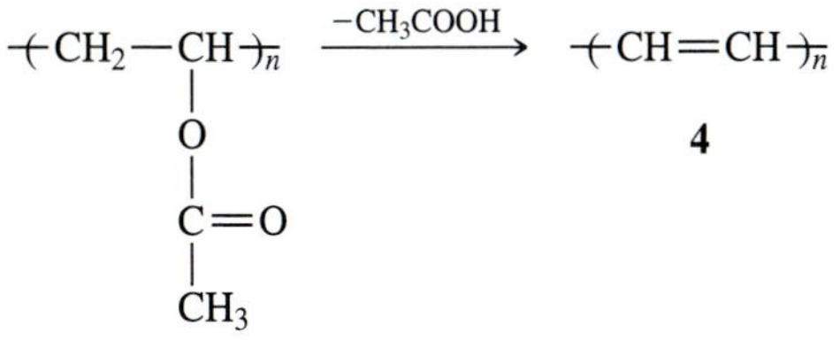

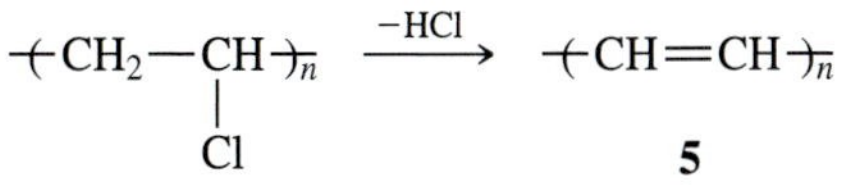

[1]Starnes, W. H., and Girois, S., *Polymer Yearbook*, **1998,** *12,* 105–131.

Another method for the preparation of polyacetylene **(5)** via an elimination reaction is found in the facile loss of 1,2-disubstituted benzene derivatives from polymer **6**. This method is one of the sources of polyacetylene for studies of electroactivity in polymers (see Chapter 23).

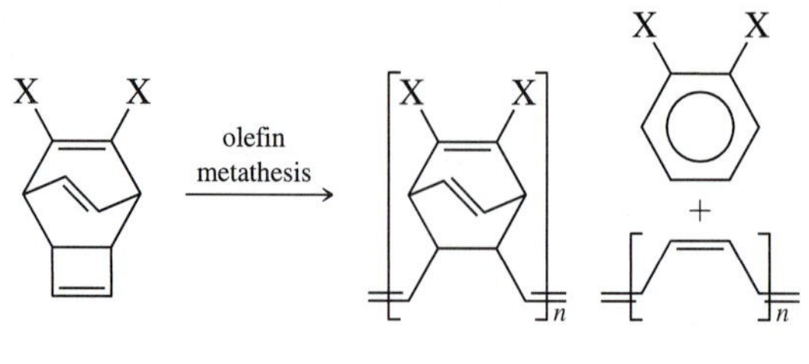

6   **5**

Hydrolytic Chain Cleavage

In Chapter 8, we will see that polysaccharides and proteins undergo hydrolytic skeletal breakdown. Some synthetic polymers behave similarly.

Polyesters can be chain-cleaved by hydrolysis. Poly(ethylene terephthalate), for example, can be hydrolyzed in acidic, neutral, or basic media. Because the reaction is acid-catalyzed, the prospect exists that the rate of neutral hydrolysis may be speeded up by the formation of carboxylic acid end groups **(7)**. The basic hydrolysis of this polymer takes place even in heterogeneous systems, and this can generate problems when thin fibers of the polymer are passed through alkaline dye baths. Synthetic polyamides are also susceptible to hydrolytic cleavage, especially in acidic media.

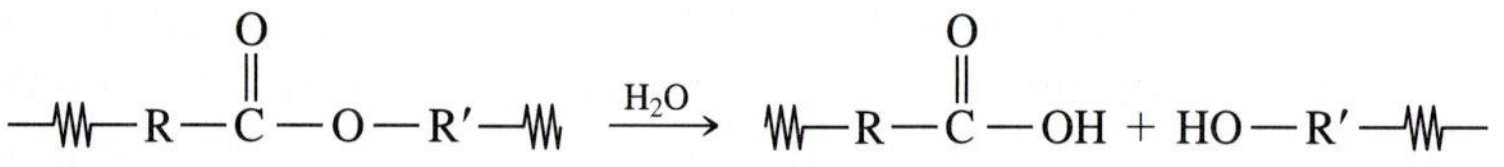

7

Enzymatic Degradation of Synthetic Polymers

One of the main advantages of synthetic polymers over naturally occurring polymeric materials such as cellulose or leather is their resistance to bacterial or fungal attack. Hence, the synthetic materials are, in general, more permanent. However, a few synthetic polymers are susceptible to biological breakdown and it is clearly important, from an applications point of view, to know which polymers are the most labile in a biological environment. Furthermore, ecological considerations have focused attention on the need for polymers that are deliberately designed to degrade when discarded.

Polyurethanes in particular (see Chapter 2) appear to be susceptible to microbial attack. Polyether polyurethanes are more resistant to biological degradation than are polyester polyurethanes. The precise mechanisms of these degradations are not fully understood. However, susceptibility to biological degradation is the exception rather than the rule. Polyamides, fluorocarbons, polyethylene, polypropylene, polyfluorocarbons, polycarbonates, and many other polymer systems appear to be resistant to biological attack. The possibility always exists that mutant bacteria or fungi may arise or be developed that could attack most synthetic polymer systems, but that phenomenon has not yet been observed.

Oxidation Reactions

Many synthetic organic polymers are oxidized in contact with the atmosphere. At room temperature in the absence of light the reaction may be very slow. But at elevated temperatures or during exposure to ultraviolet light the rate of oxidation is often quite rapid. Appreciable decomposition of polyethylene occurs when the material is exposed to outdoor daylight for less than 2 years, and the preliminary effects of photooxidation are evident after only a few months. Polypropylene is even more susceptible to photo-oxidative breakdown. However, polyisobutylene is more stable under these conditions than is polyethylene, and the same is true for polystyrene. Mechanical stress or contact of the polymer with radical-producing reagents may accelerate the oxidation process. Oxidation of a polymer usually leads to increasing brittleness and a deterioration in strength as well as a yellowing in color. Clearly, the utility of a polymer for a particular application may depend on its resistance to oxidation.

The oxidative degradation of an organic polymer generally proceeds through free-radical reactions. Free radicals are formed by the thermal or photolytic cleavage of bonds. The radicals then react with oxygen to yield peroxides and hydroperoxides by processes such as those shown in reactions (1) to (7) (here $R\cdot$ represents a polymer radical). Such reactions lead to both chain cleavage and to cross-linking. Cross-linking can be visualized as resulting from the combination of radical sites on adjacent chains. Chain cleavage can occur either by primary homolytic skeletal cleavage or by a backbiting attack by a terminal radical unit on its own chain.[1]

$$RH \longrightarrow R\cdot + H\cdot \tag{1}$$

$$RR \longrightarrow 2R\cdot \tag{2}$$

$$R\cdot + O_2 \longrightarrow ROO\cdot \tag{3}$$

$$ROO\cdot + RH \longrightarrow ROOH + R\cdot \tag{4}$$

$$ROOH \longrightarrow RO\cdot + HO\cdot \tag{5}$$

$$RO\cdot + RH \longrightarrow ROH + R\cdot \tag{6}$$

$$HO\cdot + RH \longrightarrow H_2O + R\cdot \quad \text{etc.} \tag{7}$$

Polystyrene is especially susceptible to photooxidative degradation. The phenyl groups absorb ultraviolet radiation from sunlight and transfer the energy

---

[1]Step (1) in this sequence is an oversimplification since it can involve more than one molecule of RH.

to nearby units on the polymer to generate a cascade of reactions. Oxygen radicals, hydroperoxide units, carbonyl group formation, chain cleavage, and even phenyl group ring cleavage reactions occur. The overall effect is for the polymer to become yellow and brittle.

Polymers such as polyisoprene or polybutadiene, which contain unsaturated linkages, can be attacked by atmospheric ozone as well as by oxygen. Again, free-radical cleavage and cross-linkage processes are responsible for the loss of advantageous polymer properties following oxidation. Fluorine-containing organic polymers are surprisingly stable to oxidation.

Various compounds are added to polymers to retard free-radical-induced decompositions. These additives include ultraviolet absorbers such as substituted benzophenones, which reduce the rate of photolytic oxidation. Phenolic compounds are added as radical chain terminators. Carbon black functions both as an ultraviolet screening agent and as a chain terminator. Sulfur compounds are added as peroxide deactivation reagents. Still other additives are employed to inactivate traces of metals which can participate in radical formation.

Polymers that contain unsaturated organic groups can be deliberately oxidized to form epoxy polymers. For example, polybutadiene or polyisoprene yield polymeric epoxides **(8)** when treated with hydrogen peroxide or aliphatic peracids.

8

High-Temperature Degradation Reactions

Most organic polymers decompose when heated to moderate or high temperatures. It is for this reason that few synthetic polymers can be used for long periods of time at temperatures above 150 to 200°C. This fact is largely responsible for the persistent use of metals and ceramics for many applications, even though synthetic polymers may be cheaper and, in some cases, stronger on a weight-for-weight basis.

The thermal instability of most polymers has perplexed many investigators, especially those who have attempted to predict thermal stabilities on the basis of bond strengths or by comparisons of polymers with low-molecular-weight model compounds. Polyethylene, poly(vinyl chloride), and many polacrylates decompose at least 200°C below the corresponding decomposition temperatures of short-chain paraffins, chloroparaffins, or simple esters.

The thermal instability of organic high polymers can be traced to one or more of the following reasons: (1) degradation of a polymer to a low-molecular-weight compound is favored at high temperatures by entropy effects (see Chapter 10); (2) carbon-carbon single bonds are relatively weak; (3) carbon-carbon bonds are oxidatively unstable; (4) structural abnormalities, such as branch points, exist along the chains; (5) terminal catalytic sites may initiate depolymerization; (6) a long *chain* of atoms may facilitate decomposition chain reactions, such as monomer "unzipping" processes.

The search for thermally stable high polymers has generally followed two lines of attack. The first involves the synthesis of polymers that contain inorganic elements in the backbone. This approach is discussed in Chapter 9. The second strategy has been to study the thermal decomposition mechanisms of organic polymers in the hope that the data may suggest ways in which the decomposition mechanism might be inhibited. Here we deal with the second approach.

Three types of thermal decomposition mechanisms can be recognized: (1) depolymerizations—reactions that yield monomer from a vinyl-type polymer; (2) chain cleavage reactions, which yield random chain fragments; and (3) degradation reactions that are initiated by decomposition of the side-group structures. Mechanism (3) often follows oxidation of the side-group structures, and this was discussed briefly in the preceding section. In practice, mechanisms 1 and 2 blend into one another, with many polymers showing evidence of both processes.

A number of vinyl-type polymers decompose thermally to yield the vinyl monomer **(9)**. These depolymerization reactions constitute the reverse of the original polymerization process. In fact, such reactions can occasionally be used to "recycle" polymers by offering a method for the clean regeneration of the monomer. In practice, only a few polymers degrade in such a way that they are 100% converted to the monomer. Most polymers yield some monomer and some higher fragments, and a variety of situations are known, which range all the way from pure depolymerization to pure random fragmentation. The following examples will make this clear.

$$
\left(\begin{array}{c} R \\ | \\ CH_2-CH \end{array} \right)_n \longrightarrow n\,CH_2{=}\!\!\begin{array}{c} R \\ | \\ CH \end{array}
$$

9

Poly(methyl methacrylate), poly(α-methylstyrene), and poly(tetrafluoroethylene) are three polymers which undergo 100% conversion to the monomer at elevated temperatures. However, the precise circumstances which lead to this effect are different in the three cases. Poly(methyl methacrylate) is a classical case which represents the extreme of a "pure" depolymerization. The depolymerization is a free-radical process that is initiated from the chain ends. Each initiated chain "unzips" rapidly to yield monomer. Thus, at any instant the system contains only unreacted polymer and monomer. Since whole chains apparently depolymerize in one rapid chain reaction, it is said that the "zip length" is large. The nature of the active chain ends is a question for debate. At moderate temperatures (220°C), only half the polymer chains unzip, and higher temperatures (350°C) may be needed to decompose the remaining polymer. Apparently chains which unzip at 220°C are terminated by unsaturated groups, whereas those which depolymerize only at higher temperatures have saturated end groups.

Poly(α-methylstyrene) also yields the monomer by an unzipping process. However, the chain reaction starts not from the chain ends, but from random fragmentation sites. Because of this, the rate of production of monomer depends on the molecular weight of the polymer. Presumably, the longer polymer chains are more likely to incur a random cleavage than are the shorter chains. Poly(tetrafluoroethylene) depolymerizes totally to monomer only at low pressures and high temperatures. At atmospheric pressure, the monomer molecules recombine to form dimer and other species. This polymer is one of the most thermally stable polyolefins known, but even so, it cannot withstand prolonged exposure to temperatures above about 350 to 400°C.

Polystyrene represents a case in which monomer is only one of several species formed by the thermal degradation process at 350°C. In fact, monomer, dimer, trimer, and tetramer are formed in the relative proportions of 40:10:8:1. The thermal breakdown process is believed to be initiated at weak links along the chain. Unsaturated linkages probably constitute the weak points. After the initial chain cleavage occurs at these sites, a free-radical mechanism leads to liberation of the monomer and to an intramolecular back-biting process. The latter process liberates dimer, trimer, and so on, by a mechanism such as the one shown in equation (8).

$$\text{(8)}$$

Polyethylene yields virtually no monomer. Above about 300°C, the decomposition products form a continuous spectrum of unsaturated hydrocarbons which contain from 1 to at least 70 carbon atoms. Clearly, this suggests a random chain cleavage process. It is believed that the products represent the combined results of chain cleavage initiated at weak links (possibly oxygenated sites) followed by both inter- and intramolecular chain transfer. The existence of chain branch points may facilitate the transfer process. Polypropylene behaves in a very similar manner to polyethylene.

The foregoing comments have applied specifically to vinyl addition polymers, but some observations on the thermal behavior of condensation polymers are also appropriate. Polyamides can decompose during melt spinning or molding procedures. Such decomposition, although slight, can affect the physical properties of the polymer. Apparently, the degradation process is initiated by free radicals formed by the homolytic cleavage of —NH—CH$_2$— skeletal bonds. Water and carbon dioxide are also liberated. The water serves to hydrolyze amide linkages [—NH—C(O)—] to further shorten the chains. Branches are also formed by reaction of terminal $\text{\Large W\hspace{-1mm}—}$ NH$_2$ groups with carbonyl units (reaction 9). Ultimately, the branches cause gelation of the molten polymer.

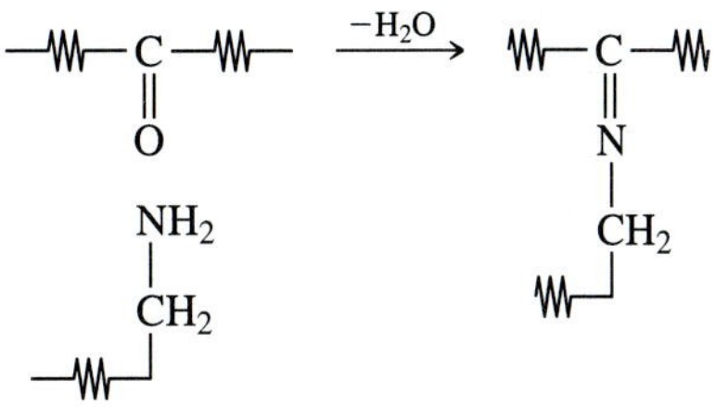

$$\text{(9)}$$

Polyesters, such as poly(ethylene terephthalate), are fairly stable at temperatures just above the melting point. However, at temperatures between 300 and 550°C, decomposition of this polymer occurs to yield carbon dioxide, acetaldehyde, and terephthalic acid, together with smaller amounts of other decomposition species, such as water, methane, acetylene, and so on.

Electron Beam Depolymerization

Several of the polymers that depolymerize to monomer at high temperatures undergo similar reactions when irradiated with a beam of X-rays, γ-rays, or electrons. This phenomenon is used for the micro-etching of resist coatings in the preparation of integrated circuits. The process works on the principle that a silicon chip surface will be exposed if an electron beam depolymerizes the protective resist that covers the surface. The exposed silicon can then be doped. Because a beam of high-energy radiation has a shorter wavelength than ultraviolet light, it is possible to etch finer details in the resist than via the alternative photocross-linking process (Chapter 5). Poly(methyl methacrylate) had been studied extensively as an etch-type resist. Polysilanes (see Chapter 9) also appear to be promising materials for this application.

Conclusion on Main-Chain Reactions

A few reactions of synthetic high polymers are beneficial in the sense that they provide a means for the modification or improvement of the polymer properties. However, some of the most facile reactions, particularly those involving oxidation and chain cleavage are, in nearly all cases, detrimental to the polymer. For this reason, a considerable technology has developed around the techniques that retard the oxidation or thermal decomposition of polymers. Some of these techniques are based on fundamental thermochemical and mechanistic principles of the types discussed in Chapter 10.

REACTIONS INVOLVING THE SIDE GROUPS

General Considerations

Although reactions of the main chains can have a profound effect on the properties of polymers, it is the side-group reactions that have the greatest potential for diversification of macromolecular structure. Thus, side-group modification following polymerization provides the fourth main method of structural modification after condensation, addition, and ring-opening polymerization.

The reasons for attempting to develop side-group reactions are many and varied. An obvious reason is to change the bulk or solution properties of a polymer by changing the side groups. This may be necessary if the desired side groups would not survive the monomer polymerization conditions. A second reason is to develop grafting or cross-linking reactions. Third, side-group reactions are usually necessary to permit the attachment of "active" units to a polymer, groups that contain catalytic transition metals, for example, or biologically active units. Significant advances are occurring in the binding of enzymes to synthetic substrates. The goal is to carry out biochemical reactions on a large manufacturing scale by allowing biological "feedstocks" to pass through columns of immobilized enzymes. This aspect is discussed in Chapter 8. Finally, reactions carried out at the *surface* of a solid polymer may change the surface character without affecting the bulk properties. Fluorination of the surface of a hydrocarbon polymer, and the linkage of enzymes or other bioactive agents to a polymer surface, are examples.

In theory, the vast arsenal of conventional organic and inorganic reaction chemistry could be used to modify the side groups in a polymer. In practice, some limitations exist. Reactions that proceed rapidly and efficiently at the small-molecule level may not take place effectively with a high polymer. This is despite the fact that, in theory, there should be little or no reactivity differences between large and small molecules in dilute solution.

There are two reasons for this discrepancy. First, high polymer molecules are usually randomly coiled in solution. This coiling may sterically inhibit the approach of reagent molecules to the most shielded monomer residues in the chain. Thus, reactions between a reagent and a polymer may be incomplete or unacceptably slow. Second, the introduction of one new substituent group may retard the introduction of a second group at adjacent monomer residue sites because of neighboring group polar or steric effects. The generally low reactivity of organic side groups to reagents presents a further problem.

On the other hand, many polymer reactions are known that proceed at similar or even faster rates than the analogous nonpolymeric reactions. Enzymatic processes provide striking examples of this effect. In such cases it appears that the role of the polymer is to generate a more favorable collision efficiency between the reacting species so that the reaction becomes one between a highly mobile reagent and a relatively immobile substrate rather than one between two highly mobile reagents. As an analogy, it is easier to fire a bullet accurately at a moving elephant than at a moving bullet. Moreover, the initial introduction of a new substituent group may serve to catalyze the introduction of more new units at adjacent sites along the chain.

There exists one polymer system—the polyphosphazenes—in which macromolecular substitution reactions are so efficient that such reactions are the main method of polymer synthesis and structural diversity (see Chapter 9). This is because the inorganic nature of the backbone allows the existence of a set of highly reactive macromolecular intermediates. Unfortunately, such intermediates are all too rare among organic polymer systems.

However, an enormous number of different reactions have been attempted with polymer molecules, and in this book we can review only a few examples. Several important substitution reactions on polymers are discussed elsewhere in this volume (especially in Chapters 8 and 9). The following examples have been chosen to illustrate briefly some of the possibilities.

Hydrolysis of Side-Group Structures

The hydrolyses of poly(vinyl esters) **(10)** or poly(vinyl amides) to a poly(vinyl carboxylic acid **(11)** (reaction 10) are processes which differ in one important respect from the conventional hydrolyses of small molecule esters or amides to carboxylic acids. Hydrolysis of a few ester or amide linkages yields a polymer in which the remaining uncharged groups are flanked by charged carboxylate groups **(12)**. The presence of these negatively charged groups would be expected to retard the approach of the reagent (OH^-) to the adjacent ester or amide groups, with a corresponding decrease in the reaction rate.

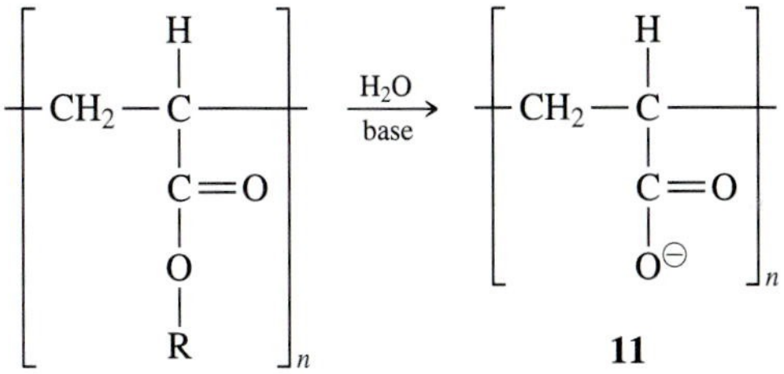

10   (10)

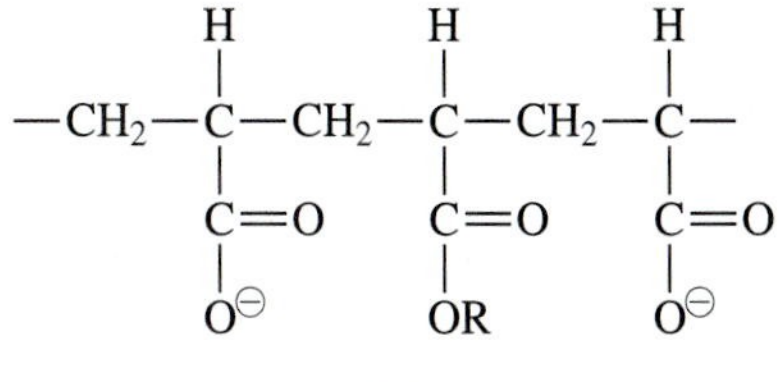

12

The hydrolysis of poly(methacrylamide) **(13)** appears to follow this pattern (reaction 11). This polymer undergoes hydrolysis in basic media to form a product with carboxylate side groups. However, all the amide groups cannot be removed, and this effect has been ascribed to the inability of hydroxide ion to penetrate the polar field of the flanking carboxylate units (**14**).

$$
\left[\begin{array}{c} \text{CH}_3 \\ | \\ -\text{CH}_2-\text{C}- \\ | \\ \text{C}=\text{O} \\ | \\ \text{NH}_2 \end{array}\right]_n \xrightarrow[\text{OH}^-]{\text{H}_2\text{O}} \begin{array}{ccc} \text{CH}_3 & \text{CH}_3 & \text{CH}_3 \\ | & | & | \\ -\text{C}-\text{CH}_2-\text{C}-\text{CH}_2-\text{C}- \\ | & | & | \\ \text{C}=\text{O} & \text{C}=\text{O} & \text{C}=\text{O} \\ | & | & | \\ \text{O}^{\ominus} & \text{NH}_2 & \text{O}^{\ominus} \end{array} \qquad (11)
$$

13   **14**

However, several cases are known where the presence of a charged carboxylate ion actually *accelerates* the rate of hydrolysis of a neighboring ester function. For example, copolymers of acrylic acid and *p*-nitrophenyl methacrylate are hydrolyzed much more rapidly than are the *p*-nitrophenyl esters of small molecule carboxylic acids. This is because the carboxylate ion itself can attack the adjacent ester function (**15**). Such intramolecular effects should be dependent on geometric factors. It has, in fact, been shown that isotactic poly(methyl methacrylate) undergoes hydrolysis more rapidly than do the syndiotactic or atactic modifications.

$$
\begin{array}{cc}
\text{R} & \text{H} \\
| & | \\
-\text{CH}_2-\text{C}-\text{CH}_2-\text{C}- \\
| & | \\
\text{O}=\text{C} & \text{C}=\text{O} \\
\text{O} \quad \text{O} \\
| \\
\text{R}'
\end{array}
$$

15

One of the most important macromolecular hydrolysis reactions is the conversion of poly(vinyl acetate) (**16**) to poly(vinyl alcohol) (**17**). Poly(vinyl alcohol) cannot be obtained by polymerization of the monomer (vinyl alcohol is unstable), hence the macromolecular hydrolysis process is critical for the preparation of this commercially important polymer. However, even under these circumstances where the polymer product has uncharged side groups, it is usually difficult to hydrolyze all the ester units.

$$
\left[\begin{array}{c} \text{H} \\ | \\ -\text{CH}_2-\text{C}- \\ | \\ \text{O} \\ | \\ \text{C}=\text{O} \\ | \\ \text{CH}_3 \end{array}\right]_n \xrightarrow[-\text{CH}_3\text{COOH}]{\text{H}_3\text{O}^+} \left[\begin{array}{c} \text{H} \\ | \\ -\text{CH}_2-\text{C}- \\ | \\ \text{OH} \end{array}\right]_n
$$

16   **17**

A variety of studies, including the acylation of side-group aromatic amino groups and the aminolysis of side groups containing activated ester residues, indicate that polymer side-group reactions are dependent on the *local* polarity of the medium in the vicinity of the polymer. A coiled polymer chain in solution can dominate the behavior of solvent molecules within its domain. This in turn can create unexpected side-group reactivity effects. The influence may explain many of the reactivity differences between polymers and their small-molecule analogues.

A further development of the idea that the local macromolecular effect has a strong reaction influence is found in the use of polymers to influence the reactions of small molecules that are located within the solvation domain of the macromolecule. For examples, poly(vinylimidazole) is much more effective than imidazole itself in catalyzing the hydrolysis of *p*-nitrophenyl acetate. This is ascribed to a cooperativity effect by adjacent imidazole units, and leads to the concept of synthetic macromolecules as enzyme-like catalysts.

Polymer-Supported Organic Reactions

One of the most important and rapidly developing areas of polymer chemistry is the use of polymers as "carriers" or "supports" for chemical reagents, catalysts, or substrates. In general, the polymer is in the form of an *insoluble* particle that may be a solvent-swollen, crosslinked gel, or a surface-active solid. Numerous advantages accrue from having a reagent, catalyst, or substrate molecule attached to an insoluble support. First, the purification of products is simplified because the immobilized species, being insoluble, can simply be filtered or centrifuged from the products. Moreover, if the reaction product remains attached to the support, large excesses of (soluble) reactants can be used without incurring product separation problems. Second, a reagent or catalyst attached to an insoluble support can be easily recovered and recycled. Third, the ease of separation of reagents or catalysts from products allows automation of complex reaction chemistry. (See the description of the Merrifield protein-synthesis process in Chapter 8.) Similarly, immobilized reagents can be packed into columns and used in continuous flow reactors. Finally, reagents that are toxic or noxious can be handled safely when they are immobilized.

This field separates into two aspects: the binding of the reagent, catalyst, or substrate to a polymer, and the reactions that the active agent undergoes while attached to the polymer. Both aspects are broad subjects, detailed treatments of which are outside the scope of this book. However, a limited number of immobilization methods will be mentioned here to provide a starting point for further study. Other examples are given in Chapters 8 and 24 and in the Suggestions for Further Reading.

Chloromethylation of polystyrene The first example is one in which a readily available starting material (polystyrene—either linear or cross-linked) is modified to provide access to a number of interesting and useful derivative polymers. Polystyrene is chloromethylated in the presence of chloromethyl ether and Friedel–Crafts catalysts such as aluminum chloride or zinc chloride. The overall reaction is shown in equation (12). Lightly chloromethylated polystyrene resins are used as a substrate in the controlled laboratory synthesis of proteins (see Chapter 8). The chloromethylation reaction can be forced to high conversions, with the ultimate product corresponding closely to poly(vinylbenzyl chloride) **(18)**. Such polymers can be quaternized with tertiary amines to yield water soluble polymers, ionomers and ion-exchange resins **(19)** (reaction 13). Alternatively, as shown in reaction (14), the chloromethylated polystyrene can be treated with a phosphide to introduce phosphinic ligands **(20)** for the binding of metal coordination complexes. These have been used as polymer bound catalyst systems, as discussed in a later section. Chloromethylated polystyrene also provides access to thioether units (by reaction with $KSCH_3$) and aldehyde functions ($Me_2S{=}O/NaHCO_3$). The latter can be converted by oxidation to carboxylic acid groups.

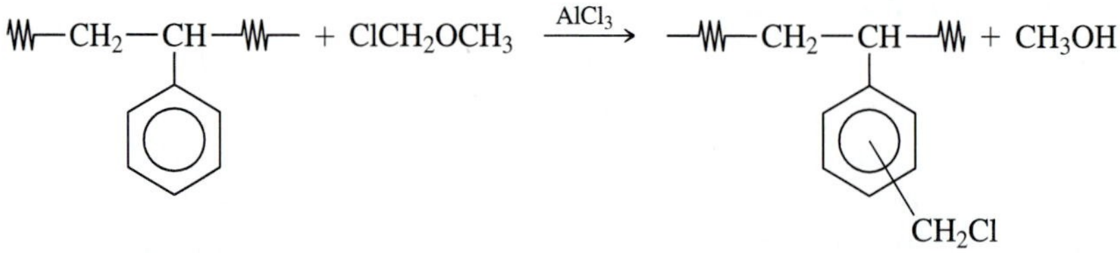

$$\text{\textasciitilde}CH_2{-}CH\text{\textasciitilde} + ClCH_2OCH_3 \xrightarrow{AlCl_3} \text{\textasciitilde}CH_2{-}CH\text{\textasciitilde} + CH_3OH \tag{12}$$

with CH_2Cl substituent on the second benzene ring.

$$\left[CH_2{-}CH\right]_n \text{ (18)} + (CH_3)_3N \longrightarrow \left[CH_2{-}CH\right]_n \text{ (19)} \tag{13}$$

with CH_2Cl on 18 and $CH_2\overset{\oplus}{N}(CH_3)_3\ Cl^{\ominus}$ on 19

18 **19**

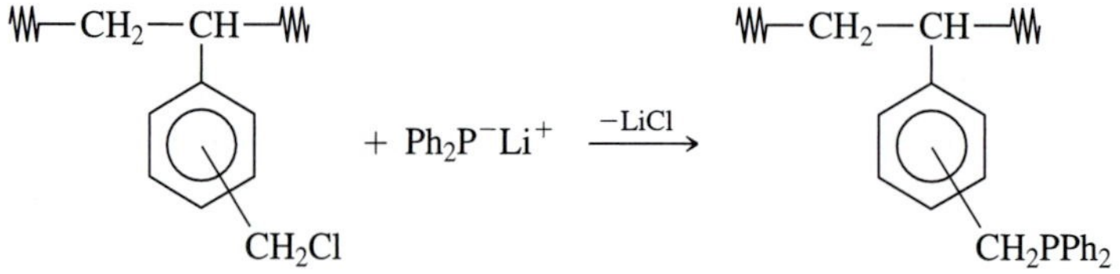

$$\text{\textasciitilde}CH_2{-}CH\text{\textasciitilde} + Ph_2P^-Li^+ \xrightarrow{-LiCl} \text{\textasciitilde}CH_2{-}CH\text{\textasciitilde} \tag{14}$$

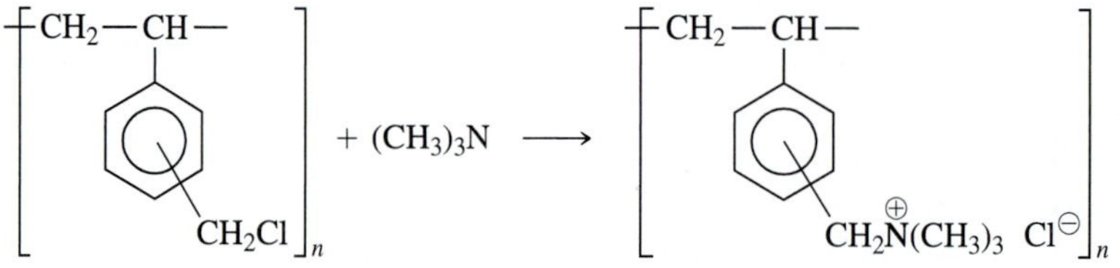

with CH_2Cl substituent converting to CH_2PPh_2

20

Reactions based on lithiopolystyrene derivatives The importance of lithioaryl derivatives in small-molecule organic chemistry is well known. Similar synthetic pathways have been developed from lithiopolystyrene **(21)**, prepared either by direct lithiation of polystyrene using *n*-butyllithium and tetramethyleth-

ylenediamine (TMEDA), or by the reactions of *n*-butyllithium with poly(*p*-brom-ostyrene).

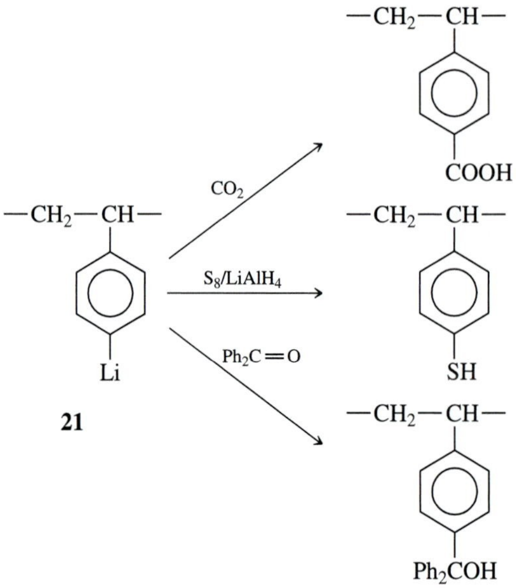

21

Polymeric protecting groups for organic synthesis   The concept of "protecting groups" in organic synthesis is widely used. Protecting groups that are insoluble macromolecules have all the advantages mentioned earlier, particularly the case of isolation of products from other species in the reaction mixture. Perhaps the best known applications of this concept are in the laboratory synthesis of proteins and nucleic acids (see Chapter 8). However, the principle has a much broader application, such as in the selective protection of hydroxyl groups in glucose derivatives by an aldehydic polystyrene derivative **(22)**.

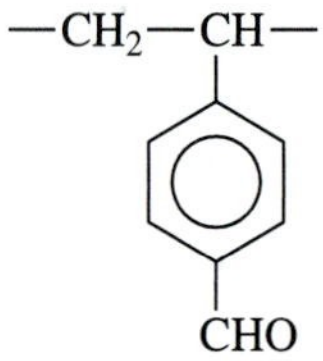

22

Polymer-Bound Transition Metals

Many important organic chemical reactions require the presence of homogeneous organometallic catalysts. However, large-scale manufacturing processes based on these reactions suffer from the extreme difficulty of recovery of the catalyst after the reaction is complete. Because such catalysts often contain expen-

sive rare metals, the loss of the catalyst, or even a small part of it, constitutes an unacceptable cost.

Cross-linked polymers are insoluble in organic or aqueous media. Hence, if the catalyst can be bound to a cross-linked polymer substrate without loss of the catalytic activity, then recovery of the catalyst may be accomplished simply by removal of the polymer particles by filtration.

One of the most common methods for binding a catalyst to a polymer is through a phosphine ligand that is itself bound to cross-linked polystyrene. The cross-linked polymer can be prepared by copolymerization of styrene with a small amount of divinylbenzene. The phosphine is attached to the polymer via chloromethylated polystyrene, as discussed in the preceding section. A typical system is shown in structure **23**, in which the Wilkinson hydrogenation catalyst, $(Ph_3P)_3RhCl$, is bound to a pendent phosphine ligand attached to the polymer substrate. A wide variety of catalysts containing iridium, cobalt, chromium, nickel, and titanium as well as rhodium have been bound to polystyrene not only through phosphine ligands, but also as π-complexes involving the aromatic rings.

$$-\!/\!\!/\!\!/\!-CH_2-CH-\!/\!\!/\!\!/-$$

$$CH_2-\overset{\displaystyle Ph}{\underset{\displaystyle Ph}{\overset{|}{\underset{|}{P}}}}\cdots RhCl(PPh_3)_2$$

23

A highly swelled but insoluble polystyrene particle will process catalytic sites *within* the interior of each polymer "bead." Hence, reactant molecules will penetrate the swelled matrix, react, and diffuse out almost as readily as in a homogeneous system. To a large extent, the catalyst molecules will be separated from each other and will function by homogeneous rather than heterogeneous mechanisms. It should be noted that phosphine-metal coordination bonds are labile. Indeed, this may be a prerequisite for catalytic activity. Unfortunately, this characteristic means that small amounts of the catalyst may be lost from the substrate during many reaction cycles or in a continuous flow reactor. Catalyst loss can also occur from phosphine-bound systems under hydroformylation reaction conditions. With cobalt carbonyl catalysts bound through—aryl—PPh_2 groups, the catalyst is lost because the aryl-phosphorus bond cleaves, allowing the free phosphine–metal complex to escape.

Transition metals have been coordinated to polymers through nitrogen-donor or cyclopentadienyl units at the terminus of side groups. Systems of this kind are of wide interest for both general chemical catalysis reactions and for deposition on electrode surfaces, where they act as electrode mediator films for electrocatalysis processes.

Graft Polymer Formation

Graft polymerization takes place when a preformed polymer provides initiation sites for the growth of branches formed from a new monomer. Thus, a graft copolymer can be represented by structure **24**.

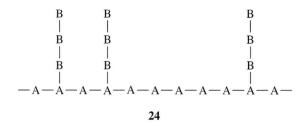

24

A number of different techniques can be used to form graft copolymers. Most of these fall into one of the following four categories.

1. A free-radical catalyst may be used to abstract a hydrogen atom from an organic polymer chain. The polymeric radical then initiates growth of a vinyl monomer.
2. A propagating vinyl polymer chain may abstract a hydrogen atom from a preformed polymer during a chain transfer step. The polymeric radical then initiates chain growth of the monomer.
3. The polymer can be irradiated by ultraviolet, gamma, or X-rays to generate free-radical sites along the chain. In the presence of a monomer, these free-radical sites initiate graft polymerization (see Chapter 3).
4. Two preformed polymers are blended together mechanically or subjected to high-speed stirring in solution. The chain backbones are mechanically cleaved by this process. Radical recombination and abstraction processes generate block and graft copolymers.

The techniques described in categories 1 to 3 generally lead to the formation of homopolymer from the monomer as well as to the synthesis of graft copolymers. It should also be noted that polymers which contain unsaturated substituent groups are particularly suitable for graft polymerization reactions. Experimentally, the likelihood of achieving a successful graft polymerization depends on the nature of the polymer, the monomer, and the catalyst. Some monomers will not polymerize from the radical sites of certain polymers. One example of a system that readily gives graft copolymers is the one formed from polystyrene and methyl methacrylate monomer in the presence of benzoyl peroxide. Radiation grafting of one polymer to another is a powerful technique, especially when one of the polymers is a biopolymer such as cellulose (see Chapter 8).

Cross-linking Reactions

The cross-linking of polymers through the side groups is an important aspect of elastomer technology, integrated circuit resist technology, immobilized reagent research, and so on. The simplest method of cross-linking, used especially on a large scale, is to heat a polymer that bears aliphatic side groups with a free-radical initiator such as benzoyl peroxide. Carbon radical sites are generated on the side groups, and these combine with others on nearby chains to form carbon-carbon cross-links. High-energy irradiation performs the same function. A variant of this method is to introduce unsaturated aliphatic units into the side groups and by means of free-radical initiation, induce dimerization or oligomerization of the active sites from different chains. The cross-linking ("curing") of rubber and many other polymers depends on this principle.

It is also possible to form such network structures using more controlled reactions. One reaction of this type is based on linking chains that are reactive only at their ends, with a multifunctional end-linking agent. The most important examples of this type involve hydroxyl-terminated polysiloxane chains end-linked with an organosilicate (room-temperature vulcanizates), and hydroxyl-terminated polyoxides end-linked with a tri-isocyanate (polyurethane elastomers).

Microlithography and Polymer Reactions

As mentioned in Chapter 5, the photocross-linking of polymers such as poly(vinyl-cinnamate) is an important process in lithography and microlithography. There are, in fact, three different approaches to the use of polymers in microlithography. The light-induced network formation yields a so-called *negative resist*. In this, the irradiated areas of the polymer remain, while the unirradiated areas are removed by dissolution in a solvent. The photodepolymerization approach mentioned earlier in this chapter gives a *positive* resist in the sense that the irradiated areas of polymer are lost (by vaporization). The third approach gives either a negative or a positive resist, depending on the details of the process. Poly(*p*-formyloxystyrene) **(25)** on irradiation with ultraviolet light is converted, by loss of carbon monoxide, to poly(*p*-hydroxystyrene) **(26)**.

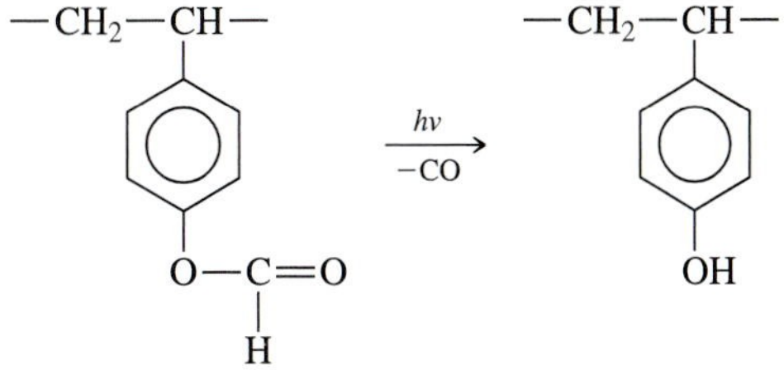

This side-group transformation changes the *solubility* of the polymer (no cross-linking is involved). The initial polymer **(25)** is insoluble in alcohols or base, but

the product polymer will dissolve in these media to give a positive resist. On the other hand, treatment with a nonpolar solvent such as chlorobenzene or anisole removes polymer **25** but does not affect polymer **26**. Excellent resolution of fine details in an integrated circuit can apparently be achieved by this method.

SURFACE REACTIONS OF POLYMERS

The surface properties of polymers are almost as important as their bulk properties. Whereas the bulk properties may determine the flexibility, elasticity, rigidity, strength, or weight, the surface characteristics often control the resistance to corrosive reagents, solvents, fuels, sunlight, and (in the short term) to oxidation. In addition, many of the uses of polymers as immobilization supports involve surface reactions. Finally, the biomedical uses of polymers (see Chapter 24) as blood vessels, heart pumps, and even catheters are severely limited by the surface character of polymers rather than by their bulk properties. For all these reasons, the subject of reactions at surfaces constitutes a topic of intense interest, one that is still in its infancy. Nearly all surface reactions are processes that involve transformations of the side groups.

Surface Fluorination

Carbon–fluoride bonds confer a special set of surface properties on polymers. They generate hydrophobicity, biomedical compatibility, oxidation resistance, and solvent or fuel resistance. Unfortunately, the bulk properties and ease of fabrication of fluoropolymers [such as poly(tetrafluoroethylene)] are less than ideal. Hence, interest exists in taking nonfluorinated polymers and, by surface fluorination reactions, converting the carbon–hydrogen bonds at the surface to carbon–fluorine bonds. This is accomplished by exposure of the polymer surface to 5 to 10% fluorine gas diluted in nitrogen at room temperature for 1 to 15 min. The surface transformation **(27)** is believed to take place by a free-radical process. Excessive exposure of the surface to fluorine must be avoided to prevent skeletal cleavage and general decomposition. Fluorine can penetrate below the surface; the fluorination may extend to a depth of 30 to 800 Å. This process can be carried out on a large scale. Surface-fluorinated polyethylene gloves are made, and polyethylene or polypropylene automobile fuel tanks are now surface fluorinated to improve their fuel resistance.

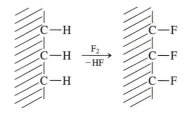

27

Surface Nitration and Sulfonation

Styrene in solution can be nitrated by mixtures of nitric and sulfuric acid. The poly(nitrostyrene) **(28)** formed in this way can then be reduced to poly(amino-styrene) **(29)** by conventional techniques (reaction sequence 15). However, reactions such as these yield complex mixtures of products, and in many cases are not necessary. Aryl groups on the surface of a solid polymer can be readily and efficiently nitrated and reduced without affecting the bulk properties. This method has been used for modification of the surface of a poly(diphenoxyphosphazene) as a prelude to the immobilization of enzymes on the surface. Other biological and biomedical examples are given in Chapters 8 and 24.

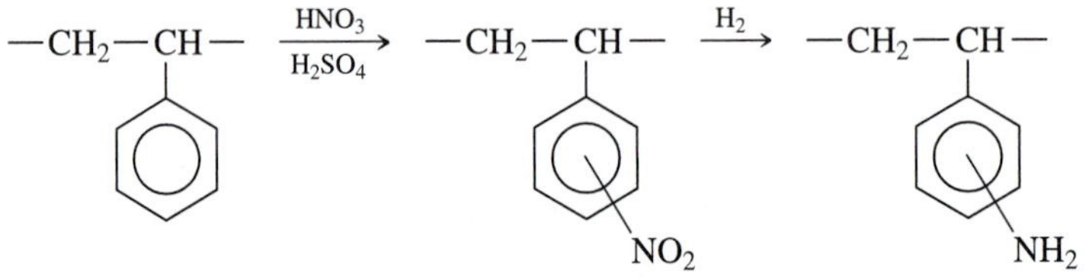

$$\begin{array}{ccccc} -CH_2-CH- & \xrightarrow[H_2SO_4]{HNO_3} & -CH_2-CH- & \xrightarrow{H_2} & -CH_2-CH- \\ & & NO_2 & & NH_2 \\ & & \mathbf{28} & & \mathbf{29} \end{array}$$

(15)

Solution sulfonation of polymers such as polystyrene usually yields complex cross-linked products. On the other hand, *surface* sulfonation generates hydrophilic SO_3H sites without affecting the bulk properties.

Surface Oxidation

The surfaces of hydrocarbon polymers, such as polyethylene, can be modified by oxidation with chromic acid. These reactions introduce carboxylic acid units on the surface, a change that alters the hydrophilicity, adhesion, metal-binding characteristics, and other properties. Strictly speaking, in the case of polyethylene, this is a skeletal oxidation reaction, but side-group oxidation can be envisaged with polypropylene and other polymers.

Other Systems

Finally, other polymers such as polyphosphazenes (see Chapter 9) can undergo a wide variety of surface reactions without altering the properties of the bulk material. Typically, these reactions are used to modify the hydrophobic or hydrophilic character of the surface, or to provide functional sites for the immobilization of biologically active molecules or metal ions. Some examples are shown in Chart 7.1.[1]

Thus, surface trifluoroethoxy groups (introduced by the method described in Chapter 9) can be replaced by ^-OH or $O^- NBu_4^+$ groups to convert a hydrophobic surface to a hydrophilic one. Exchange of trifluoroethoxy groups by

---

[1] Allcock, H. R., Fitzpatrick, R. J., *Chemistry of Materials,* **1991,** *3,* 450–454.

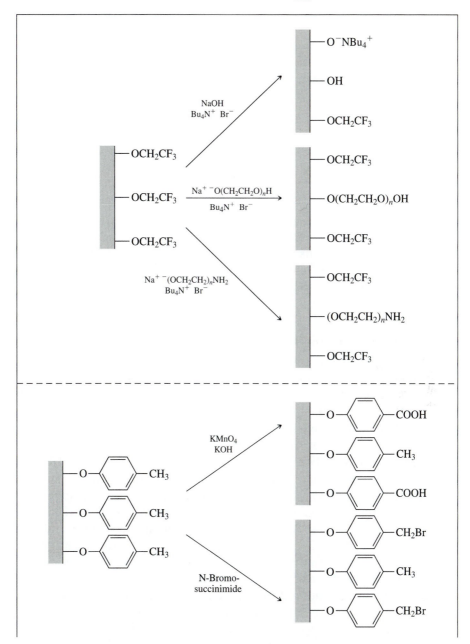

Chart 7.1 Surface reactions carried out on polyphosphazenes

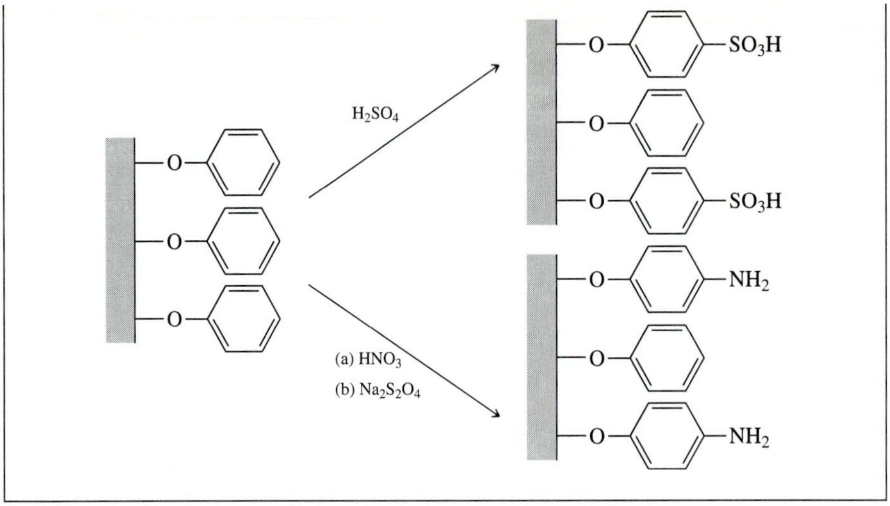

Chart 7.1 (*continued*)

other alkoxy groups that bear hydroxy or amino terminal units provides functional sites for the linkage of other molecules. Methylphenoxy groups at the surface of a polyphosphazene can be oxidized to aryl carboxylic acid groups, or the methyl units can be brominated. Phenoxy groups can be surface sulfonated or nitrated, and the resultant nitro groups reduced to amino functionalities.

Surface Characterization

The changes that occur when chemical reactions are restricted to the surface of a polymer can be studied by a range of characterization techniques designed to measure the depth of the reaction and the density of altered groups in the surface region. These methods include contact angle measurements of water droplets to determine hydrophobicity or hydrophilicity (water spreads out on a hydrophilic surface, but beads up on a hydrophobic one). Other techniques include total internal reflectance Fourier transform infrared spectroscopy (TIR-FTIR), which can identify new chemical entities in the surface region, and X-ray photoelectron spectroscopy (XPS), which indicates the elements present at the surface and the environment of those atoms. Scanning electron microscopy (SEM) is used to examine the roughness or smoothness of the modified surface, and coupled with energy dispersive X-ray spectroscopy (EDX), can also give information about the elemental composition. Optical microscopy of cross-sections of a stained polymer provide additional information. For example, a basic dye may penetrate into an acidic surface but not into the bulk material. Hence, the depth of an acid-forming reaction, such as sulfonation, can be estimated. Biological evaluation of altered surfaces may be conducted through bacterial or mammalian cell culture tests to determine if a surface favors or inhibits cell colonization and spreading.

STUDY QUESTIONS

1. Discuss in detail the validity of the "model compound" approach to the study of polymer reactions, that is, the study of the reactions of small molecules as a substitute for a direct study of the macromolecules.

2. Design a synthesis route starting from polyethylene or polystyrene that would allow the introduction of water-solubilizing side groups. Which steps in the process are likely to be the most difficult to perform, and why? Suggest possible uses for the final product of the reaction sequence.

3. Suggest applications in which the hydrolysis of a polyamide might be considered to be an advantage.

4. Discuss the possibility that the reactions of a polymer might be used to measure the stereoregularity of that polymer.

5. Comment on the fact that polymer substitution reactions are not generally used for the modification of poly(dimethylsiloxane).

6. Devise a reaction scheme for the oxidative degradation of an aromatic ladder polymer (see Chapter 2) in the atmosphere at elevated temperatures.

7. What advantages or disadvantages can you foresee for the preparation of microcircuits (**a**) by electron beam decomposition of polymers, and (**b**) by photopolymerization of a monomer or photo-crosslinking of a polymer?

SUGGESTIONS FOR FURTHER READING

Reactions Involving the Main Chain

FEAST, W. J., "Synthesis and Properties of Some Conjugated, Potentially Conductive, Polymers," *Chem. Ind. (London)*, **1985**, *8*, 263.

GUILLET, J. E., "Studies of the Mechanism of Polyolefin Photodegradation," *Pure Appl. Chem.*, **1980**, *52*, 285.

MILLER, R. D., RABOLT, J. F., SOORIYAKUMARAN, R., FLEMING, W., FICKES, G. N., FARMER, B. L., and KUZMANY, H., Soluble Polysilane Derivatives. Chemistry and Spectroscopy, in *Inorganic and Organometallic Polymers* (M. Zeldin, K. J. Wynne, and H. R. Allcock, Eds.), *ACS Symp. Ser.*, **1988**, *360*, 43.

RANBY, B., and LUCKI, J., "New Aspects of the Photodegradation and Photooxidation of Polystyrene," *Pure Appl. Chem.*, **1980**, *52*, 295.

SUGITA, K., "Application of Photodegradable Polymers to Imaging and Microfabrication Technologies: A Review of Recent Research Papers in the Last 4 Years," *Progress in Organic Coatings*, **1997**, *31*(1–2), 87–95.

TROZZOLO, A. M., and WINSLOW, F. H., "A Mechanism for the Oxidative Photodegradation of Polyethylene," *Macromolecules*, **1968**, *1*, 98.

Reactions of the Side Groups

BHATTACHARYYA, S., "Polymer-Supported Reagents and Catalysts: Recent Advances in Synthetic Applications," *Combinatorial Chemistry and High Throughput Screening*, **2000**, *3(2)*, 65.

COLLMAN, J. P., HEGEDUS, L. S., COOKE, M. P., NORTON, J. R., DOLCETTI, G., and MARQUARDT, D. N., "Resin-Bound Transition Metal Complexes," *J. Am. Chem. Soc.*, **1972**, *94*, 1789.

DUBOIS, R. A., GARROU, P. E., LAVIN, K. D., and ALLCOCK, H. R., "Cobalt Hydroformylation Catalyst Supported on a Phosphinated Polyphosphazene: Identification of Phosphorus-Carbon Bond Cleavage as a Mode of Catalyst Deactivation," *Organometallics*, **1986**, *5,* 460.

EVANS, D., and CROOK, M. A., "Irradiation of Plastics: Damage and Gas Evolution," *Mater. Res. Soc. Bull.*, **April 1997**, 36.

FRECHET, J. M. J., "Synthesis Using Polymer-Supported Protecting Groups," in *Polymer-Supported Reactions in Organic Synthesis* (P. Hodge and D. C. Sherrington, eds.). New York: Wiley, **1980**, p. 293.

HEITZ, W., "Polymeric Reagents: Polymer Design, Scope, and Limitations," *Advan. Polymer Sci.*, **1977**, *23*, 1.

HODGE, P., "Polymer-Supported Reagents," in *Polymer-Supported Reactions in Organic Synthesis* (P. Hodge and D. C. Sherrington, eds.). New York: Wiley, **1980**, p. 83.

HODGE, P., and SHERRINGTON, D. C., (eds.), *Polymer-Supported Reactions in Organic Synthesis*. New York: Wiley, **1980**.

HOLMES-FARLEY, S. R., BAIN, C. D., and WHITESIDES, G. M., "Wetting of Functionalized Polyethylene Film Having Ionizable Organic Acids and Bases at the Polymer–Water Interface: Relations between Functional Group Polarity, Extent of Ionization, and Contact Angle with Water," *Langmuir*, **1988**, *4*(4), 921.

HOLMES-FARLEY, S. R., and WHITESIDES, G. M., "The Thermal Stability of a Surface Modified Solid Organic Polymer, *Polymer Mater. Sci. Eng.*, **1985**, *53*, 127.

INOUE, K., "Functional Hyperbranched and Star Polymers," *Progress in Polymer Science*, **2000**, *25(4)*, 453.

KANG, A., and ZHANG, Y., "Surface Modification of Fluoropolymers Via Design," *Adv. Mater.*, **2000**, *12(20)*, 1481.

KUNITAKE, T., "Enzyme-like Catalysis by Synthetic Linear Polymers," in *Polymer-Supported Reactions in Organic Synthesis* (P. Hodge and D. C. Sherrington, eds.). New York: Wiley, **1980**, p. 195.

LAN, M.-J., and OVERBERGER, C. G., "Synthesis of Linear Poly(ethylenimine) Containing Nucleic Acid Pendants as Polynucleotide Analogs," *J. Polymer Sci. (A), Polymer Chem. Ed.*, **1987**, *25*, 1909.

MARK, J. E., and ERMAN, B., *Rubberlike Elasticity: A Molecular Primer*. New York: Wiley-Interscience, **1988**.

MORAWETZ, H., Comparative Studies of the Reactivity of Polymers and Their Low Molecular Weight Analogs," *J. Polymer Sci., Polymer Symp.*, **1978**, *62*, 271.

MURRAY, R. W., "Polymer Modification of Electrodes," *Ann. Rev. Mater. Sci.*, **1984**, *14*, 145.

SMITH, S. D., and ALEXANDRATOS, S. D., "Ion-Selective Polymer-Supported Reagents," *Solvent Extraction and Ion Exchange*, **2000**, *18(4)*, 779.

WARD, W. J., and MCCARTHY, T. J., "Surface Modification," in *Encyclopedia of Polymer Sci. and Eng., Suppl. Vol.* New York: Wiley, **1989**, 674.

YANG, W., and RANBY, B., "Radical Living Graft Polymerization on the Surface of Polymeric Materials," *Macromolecules*, **1996,** *29*, 3308.

8

Biological Polymers and Their Reactions

INTRODUCTION

Living things exist and reproduce because they contain macromolecules. Very early in the evolutionary process, the chemical (i.e., nonreplicative) synthesis of large molecules under natural conditions provided a mechanism for the reproductive processes on which even primitive life depends. Throughout the subsequent evolution of living organisms, polymers have been used as protective coatings, membranes, energy storage systems, skeletal systems, pathways for electrical conduction, and for countless other purposes.

There are three main classes of biological polymers:

1. Polysaccharides
2. Proteins and polypeptides
3. Nucleic acid polymers

In this chapter, we consider these three types of macromolecules. Polysaccharides are used mainly as skeletal reinforcement molecules in plants or as energy-storage molecules in plants and animals. Proteins function as protective or supportive molecules in animals but are also employed extensively as catalysts, as oxygen-transport molecules, and in many other roles that accelerate and direct the chemical reactions of the cell or organism. Nucleic acids are used almost entirely for the purposes of information storage, cell replication, and protein synthesis.

It should be noted that all these macromolecules are *condensation* polymers that, in the simplest chemical sense, are produced by the elimination of water between diols, amino acids, or inorganic acids. These condensation reactions take place rapidly at only moderate temperatures in an aqueous environment. In this respect, the efficiency of the biological syntheses is a matter of considerable interest to the synthetic chemist as well as to the biologist.

POLYSACCHARIDES

General Composition of Polysaccharides

Polysaccharides are cyclolinear polyethers formed by the condensation reactions of sugars. Cellulose, starch, or glycogen are well-known examples, although a large number of other polysaccharides have been identified. Both homopolymers and copolymers are known in linear or branched sequences. Small amounts of noncarbohydrate components may also be present, including ester residues derived from phosphate, sulfate, malonate, or pyruvate units. The degree of polymerization may range from as little as 30 to as high as about 10^5.

A wide variety of sugar monomer residues are found in polysaccharides, but two of the most frequently occurring are the cyclic forms of glucose (**1** and **2**). These cyclic molecules are known as glucopyranoses (the pyran-type form of the sugar). The glucose molecule depicted in **1** is called α-D-glucose (or α-D-glucopyranose). The isomer shown in **2** is β-D-glucose. Starch, glycogen, and cellulose are homopolymers of glucose. Chitin, the structural polysaccharide found in insect exoskeletons, is a homopolymer of *N*-acetyl-D-glucosamine (**3**).

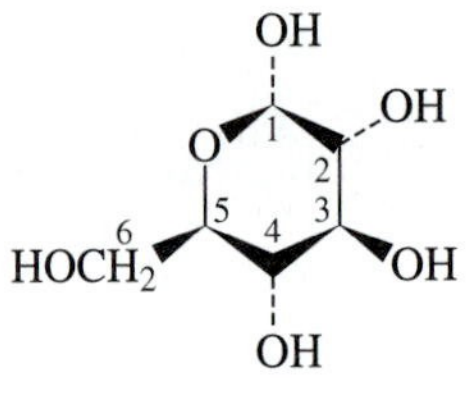

1

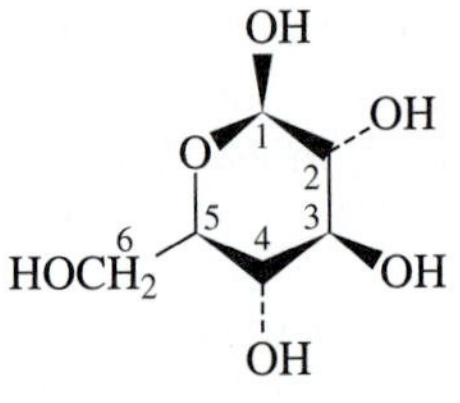

2

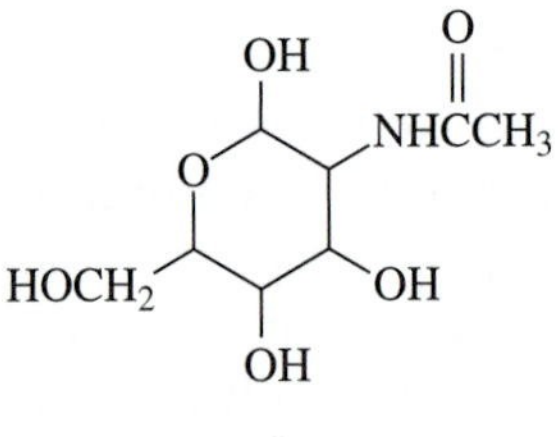

3

Specific Polysaccharides

Starch, glycogen, and cellulose are closely related homopolymers that are composed of glucopyranose residues (**1** or **2**) condensed via the hydroxyl groups at the 1 and 4 positions. Thus, the fundamental repeating structure in all three polymers is the one shown in **4**. The main differences between these three polymers are attributed to different monomer residue configurations or to different degrees of chain branching.

$$\left[\! \! \begin{array}{c} \text{OH} \quad \text{OH} \\ \text{O} \\ \text{O} \\ \text{CH}_2\text{OH} \end{array} \! \! \right]_n$$

4

Starch is the principal energy-storage polysaccharide of the photosynthetic plants. It is mainly a homopolymer of α-D-glucose (**1**). Most starches contain two structurally different components that are designated as the *amylose* and *amylopectin* fractions. The amylose fraction is a linear polymer, whereas the amylopectin fraction is highly branched. This branched component constitutes 70 to 80% of most starches.

The amylose component forms hydrated micelles in water, in which the polysaccharide chain twists into a helical conformation (Figure 8.1). The channel generated within this helix can accommodate iodine molecules to form the well-known deep blue starch–iodine complex. The molecular weights of the chains vary over a wide range, with $\overline{DP}$ values varying from 100 to 6000. However, some of this apparent molecular-weight variation may reflect hydrolytic chain cleavage that occurs during the isolation and fractionation of the polymers.

In amylopectin, a branch point exists at about every sixth to twelfth backbone residue, with each branch being roughly 12 to 15 glucose residues long. The branches involve the CH_2OH side groups—that is, they involve condensations at the 1, 4, and 6 carbon atoms of a backbone residue. Amylopectin gives a purple-brown color with iodine.

Glycogen is a storage polysaccharide found in animal tissues. It closely resembles starch in its general composition, but it is more highly branched than the amylopectin component of starch. It is estimated that an average of only three glucose residues separate each branch point in glycogen, although the molecular weights of glycogen and amylopectin are very similar (ca. 10^7). Glycogen yields a red-brown color with iodine.

Cellulose is the most abundant structural material used by plants. Cotton is nearly 100% cellulose and wood is roughly 50% cellulose. Like starch and glycogen, cellulose is a homopolymer of glucose (roughly 3500 residues per chain), but the glucose configuration is β rather than α (see structure **2**). This configu-

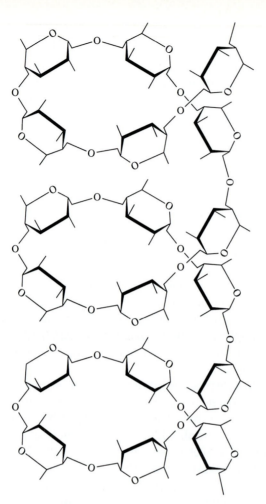

Figure 8.1 Helical conformation assumed by the amylose fraction of starch. [From A. L. Lehninger, *Biochemistry* (New York: Worth, 1970), p. 229.]

rational difference is responsible for the different properties. Cellulose consists of fully extended chains that are hydrogen bonded into sheets to form a highly crystalline matrix. Hence, cellulose is insoluble in water and is an ideal material for structural reinforcement. Cellulose cannot be used as an energy source by primates, because the hydrolytic enzymes needed to degrade it to glucose or maltose are not present in the digestive tract. However, ruminant animals (such as cows) can utilize cellulose as a food via the action of symbiotic bacteria that exist in the gut. These secrete the enzymes required to hydrolyze cellulose to glucose.

Biological Synthesis of Polysaccharides

Starch and cellulose are available in such plentiful quantities from plant life that there has been virtually no incentive to devise laboratory syntheses of polysaccharides from sugars. However, the biological pathways by which starch, glycogen, and cellulose are formed from glucose are of considerable interest. The starting material is glucose-6-phosphate **(5)**, which is first converted enzymatically to glucose-1-phosphate **(6)**. The enzyme-induced reaction of glucose 1-phosphate

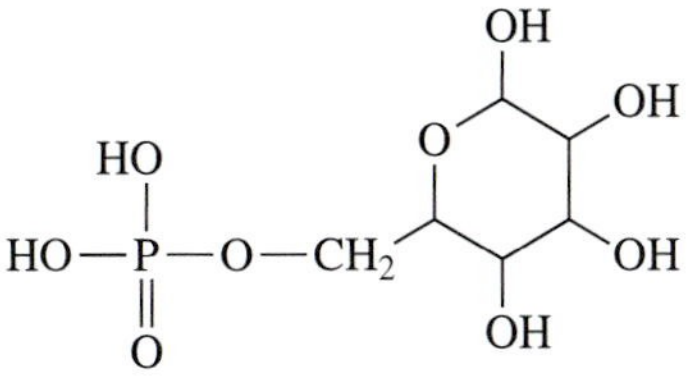

5 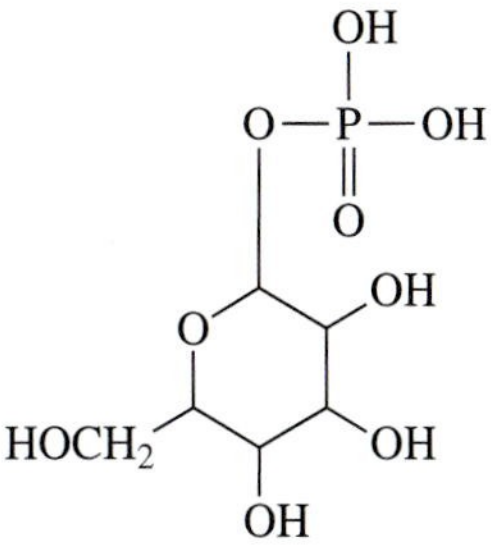   **6**

with a nucleoside-5'-triphosphate, such as uridine-5'-triphosphate or adenosine-5'-triphosphate (ATP), yields a nucleoside *di*phosphate (NDP)-1-sugar plus pyrophosphate (reaction 1). The pyrophosphate is lost from the system by hydrolysis, and this provides a

$$\text{NTP + sugar 1-phosphate} \rightleftharpoons \text{NDP-sugar} + H_2O_3P-O-PO_3H_2 \qquad (1)$$

driving force for the reaction. Thus, it is the energy-yielding break down of a nucleoside *tri*phosphate, such as ATP, to the energy-rich nucleoside *di*phosphate that enables condensation to occur even in a hydrolytic environment. Chain growth occurs by an enzyme-catalyzed loss of NDP from the 1-position of the glucose ring as that unit is condensed with the 4-position of a second sugar molecule (reaction 2). Chain branching by a condensation reaction at carbon 6 requires the presence of a special "branching enzyme."

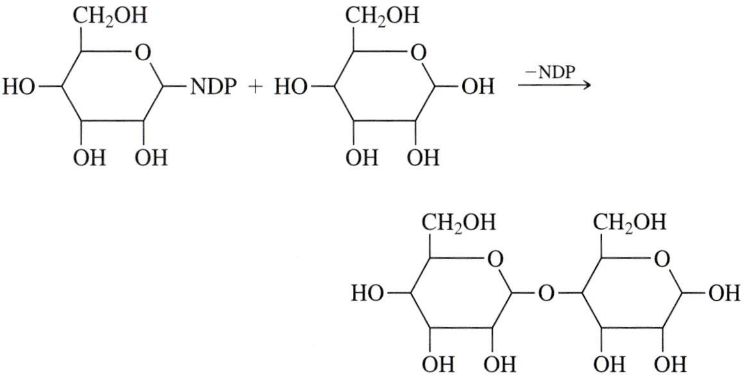

(2)

Reactions of Polysaccharides

Hydrolysis   Two enzymes, α-amylase and β-amylase, are capable of catalyzing the hydrolysis of the amylose fraction of starch. α-Amylase, produced in saliva and pancreatic juice, functions within the gastrointestinal tract to convert amylose first to medium-molecular-weight "dextrins," and then to glucose and maltose. β-Amylase, a plant enzyme found in malt, catalyzes the hydrolysis of amylose to maltose. The amylopectin fraction of starch, being highly branched, is hydrolyzed under the influence of α- and β-amylase with conversion of the outer *branches only* to glucose or maltose. However, the residual branched "core" (the limit dextrin) can be hydrolyzed by other "debranching" enzymes. Cellulose enzymes secreted by bacteria or fungi are responsible for the deterioration of cotton in warm, humid climates.

Polysaccharides can also be hydrolyzed to sugars by nonenzymatic processes. For example, cellulose undergoes skeletal hydrolysis reactions in aqueous acid which lead to a decrease in molecular weight and a loss of tensile strenth. The crystalline regions of the polymer are apparently more resistant to hydrolysis than are the amorphous domains. The initial hydrolysis yields oligosaccharides, but glucose is the end product of hydrolytic degradation. The process is summarized in reaction (3).

$$\xrightarrow[\text{H}^+]{\text{H}_2\text{O}} \quad (3)$$

Esterification   Because each macromolecule in a polysaccharide bears pendent hydroxyl groups, substitution reactions can be performed on those hydroxyl sites. These reactions include esterification, etherification, crosslinking, and grafting processes. The first two types of reaction will be discussed here.

Esterification of cellulose can be accomplished with the use of inorganic or organic acids. Nitration occurs when cellulose is treated with a mixture of nitric and sulfuric acids at room temperature for about half an hour. The product is known as *cellulose nitrate*. In practice, some unreacted hydroxyl groups and some sulfate linkages will also be present. The actual nitrating agent is the nitronium ion, which reacts with both the crystalline and amorphous regions of the polymer. Cellulose nitrates that contained 12.5 to 13.4% nitrogen are explosive and are known as "gun cotton." Related polymers which contain 11 to 12% nitrogen have been used as lacquers, films, and plastics, although nowadays they have been largely displaced by newer, less flammable, synthetic polymers.

The reaction of cellulose with acetic acid or acetic anhydride yields *cellulose acetate*. The reaction usually requires a pretreatment of cellulose with acetic acid, which is followed by a reaction with acetic anhydride and sulfuric acid. The product from this reaction approximates in composition to a triacetate. It has a high melting point and a low solubility. Partial hydrolysis and replacement of sulfate ester groups is brought about by treatment with aqueous acetic acid. This secondary process yields a polymer which has broad industrial use.

Rayon manufacture   Rayon is simply regenerated cellulose made from wood pulp. Cellophane is a film of the same material. Two processes exist for the manufacture of rayon—the xanthate method and the cuprammonium process. The xanthate method makes use of a cellulose esterification reaction. The process requires the treatment of the crude cellulosic material with aqueous sodium hydroxide and carbon disulfide to form a soluble ester by the process shown in (4).

$$
\underset{\text{H}}{\overset{\text{OH}}{\text{\char`\~\char`\~—C—\char`\~\char`\~}}} + \text{NaOH} + \text{CS}_2 \xrightarrow{-\text{H}_2\text{O}} \underset{\text{H}}{\overset{\overset{\displaystyle \text{S}}{\overset{\|}{\text{O—C—S}^{\ominus}}}\ \text{Na}^{\oplus}}{\text{\char`\~\char`\~—C—\char`\~\char`\~}}} \tag{4}
$$

(soluble in base)

Treatment of the xanthate ester with acid regenerates the cellulose. In practice, the cellulose may be precipitated as a fiber by extrusion of the xanthate solution (the "viscose") through spinnerettes into a bath of sulfuric acid that contains both sodium and zinc sulfates.

The cuprammonium process makes use of the fact that cellulose is soluble in solutions made from ammonium hydroxide and copper oxide. Spinning is then accomplished by extrusion of the solution into water. The cuprammonium process, although simpler in principle than the xanthate method, is little used at the present time because of the cost of copper lost during processing. One problem with rayon and its derivatives as textile materials is the high degree of flammability. This necessitates the addition of flame retardants for certain uses.

Etherification   Ethers of cellulose can be formed by the treatment of alkaline cellulose compositions with an alkyl halide, such as methyl chloride. The reaction is summarized by equation (5).

$$
\underset{\text{H}}{\overset{\text{OH}}{\text{\char`\~\char`\~—C—\char`\~\char`\~}}} \xrightarrow{\text{NaOH}} \underset{\text{H}}{\overset{\text{O}^-\text{Na}^+}{\text{\char`\~\char`\~—C—\char`\~\char`\~}}} \xrightarrow[-\text{NaCl}]{\text{CH}_3\text{Cl}} \underset{\text{H}}{\overset{\text{O—CH}_3}{\text{\char`\~\char`\~—C—\char`\~\char`\~}}} \tag{5}
$$

Methyl iodide or dimethyl sulfate may also be employed as methylating agents. Commercial methylcellulose, as it is called, contains some residual hydroxyl groups, and is soluble in water. Ethyl cellulose is prepared commercially by the reaction of ethyl chloride with alkaline cellulose. Carboxymethylcellulose is isolated from the reaction of alkaline cellulose with chloroacetic acid or sodium chloroacetate (6).

$$
\underset{\text{H}}{\overset{\text{O}^-\text{Na}^+}{\text{\char`\~\char`\~—C—\char`\~\char`\~}}} \xrightarrow[-\text{NaOH}]{\text{ClCH}_2\text{COOH}} \underset{\text{H}}{\overset{\text{O—CH}_2\text{COOH}}{\text{\char`\~\char`\~—C—\char`\~\char`\~}}} \tag{6}
$$

Other reactions of cellulose Attempts have been made to combine the properties of cellulose with those of synthetic polymers by grafting one polymer onto the other. The use of gamma-ray grafting for this purpose is of some interest. Finally, it should be remembered that a variety of reactions exist which permit the covalent binding of dyestuff molecules to cellulose. For example, colored aromatic vinyl sulfones undergo addition reactions to the hydroxyl groups of cotton (reaction 7). Alternatively, dyestuffs may be bound to *s*-triazines that contain active chlorine atoms. The chlorine atoms undergo reaction with the cellulosic hydroxyl groups (reaction 8).

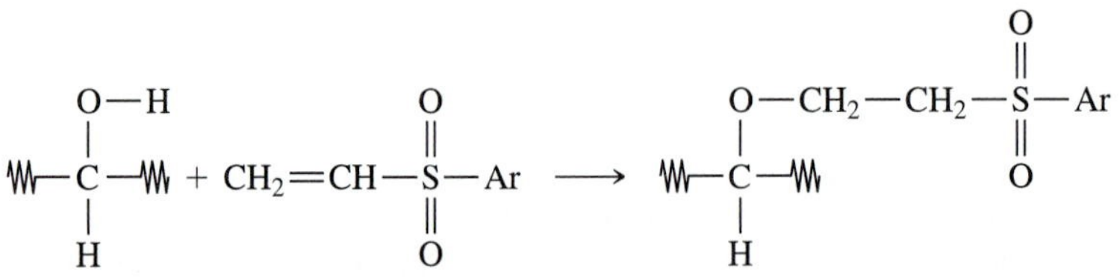

(7)

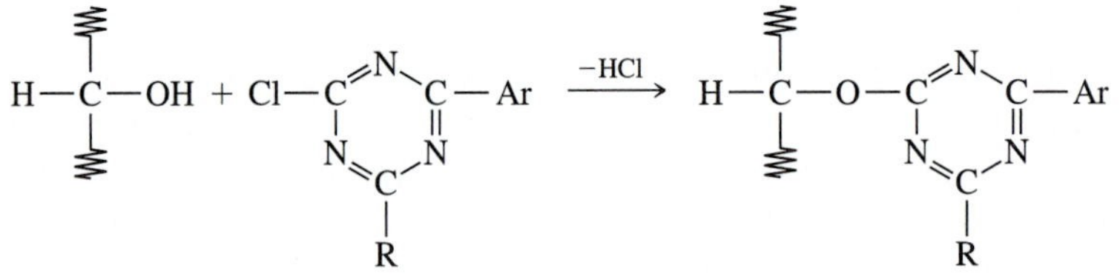

(8)

PROTEINS AND POLYPEPTIDES

General Composition of Proteins and Polypeptides

Proteins are complex polypeptide copolymers formed by the condensation reactions of amino acids (9). Synthetic polypeptide copolymers and homopolymers have also been made. Proteins exhibit an enormous variety of structures. The molecular weights of different proteins range from about 6000 to 1,000,000, with degrees of polymerization ranging from about 50 to over 8000. Some proteins consist of supramolecular agglomerates of two or more separate chains and, in such cases, the total molecular weight of the agglomerate may reach 40,000,000.

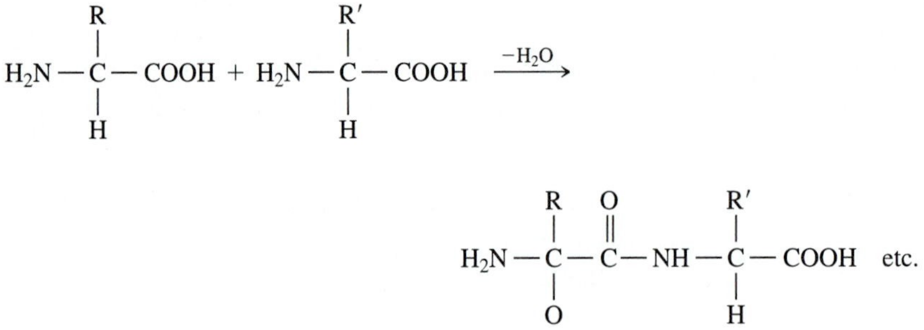

(9)

In addition to the polypeptide chains, some proteins contain nonproteinaceous components, known as *prosthetic groups*. The prosthetic groups vary in structure from the iron porphyrins found in hemoglobin, myoglobin, and cytochrome c, to the ferric hydroxide, zinc, or copper in certain other metalloproteins. Phosphate prosthetic residues are present in casein. Ribonucleic acids constitute the prosthetic component of viruses and ribosomes. Proteins that contain prosthetic groups often fall into the category known as globular proteins—the materials that perform the chemical work of the living system. For example, the iron porphyrin prosthetic group is responsible for the oxygen binding and transport function of hemoglobin and myoglobin. The metallo component of many enzymes is the actual site of the catalytic reactions.

Different proteins are different, not only because of the presence or absence of prosthetic groups, but also because of the different sequences of amino acid residues that can exist in polypeptide chains. There are only 20 commonly occurring amino acids (see Table 8.1), but the number of sequential permutations possible with 20 different residues is enormous. No less than 10^{300} different amino acid sequences could exist if only 12 different amino acids are present in equal amounts in a small protein that contains fewer than 300 total residues.

The amino acid sequence within a polypeptide chain is known as the *primary structure*. This, in turn, governs the *conformation* of the polymer chain (see Chapter 18), also known as the *secondary structure*, and determines the existence of "random coil," bending, or helical segments. The location of the random coil, bend, and α-helix regions along a given chain, and the location of internal crosslinks or mutually attractive residues, determines the overall *shape* of the molecules—called the *tertiary structure*. It follows that the secondary and tertiary structural characteristics will in turn influence the manner in which individual protein chains will agglomerate in supramolecular systems. This mode of agglomeration is called the *quaternary structure*. Hence, the precise sequence of amino acids along a polypeptide chain is one of the main keys to the properties and function of that particular protein.

Amino Acids

As mentioned above, only 20 different monomer molecules are found in all the commonly occurring proteins, and it is the sequential "coding" of these residues along a chain that determines the shape and, to a large extent, the function of a protein. The common amino acids are listed in Table 8.1.

The two most important differences between the different amino acids are the ability of a particular residue to generate an internal S—S crosslink (cystein units) and, more important, the degree to which the side groups attached to different residues on the same chain have the ability to attract or repel each other and thereby determine the conformation of the chain. The amino acids listed in Table 8.1 are grouped according to the hydrophobic, hydrophilic, basic, or acidic character of their side groups. In an aqueous environment, hydrophobic groups will tend to attract each other in the same way that oil droplets tend to coalesce

TABLE 8.1   NATURALLY OCCURRING AMINO ACIDS

Name	Structure	Symbol	Intramolecular Interaction Character
Alanine	CH_3 — H_2N—CH—COOH	Ala (or A)	Nonpolar, hydrophobic side groups
Phenylalanine	CH_2Ph — H_2N—CH—COOH	Phe (F)	
Valine	$CH(CH_3)_2$ — H_2N—CH—COOH	Val (V)	
Leucine	$CH_2CH(CH_3)_2$ — H_2N—CH—COOH	Leu (L)	
Isoleucine	$CH(CH_3)CH_2CH_3$ — H_2N—CH—COOH	Ile (I)	
Methionine	$CH_2CH_2SCH_3$ — H_2N—CH—COOH	Met (M)	
Proline	CH_2 ring (CH_2, CH_2) — HN—CH—COOH	Pro (P)	
Tryptophane	indole ring: HC=N(H), CH_2—C, H_2N—CH—COOH	Trp (W)	
Glycine	H — H_2N—CH—COOH	Gly (G)	Polar, hydrophilic side groups
Serine	CH_2OH — H_2N—CH—COOH	Ser (S)	
Cysteine	CH_2SH — H_2N—CH—COOH	Cys (C)	
Threonine	$CH(CH_3)OH$ — H_2N—CH—COOH	Thr (T)	
Tyrosine	CH_2—(C$_6$H$_4$)—OH, H_2N—CH—COOH	Tyr (Y)	
Asparagine	$CH_2C(O)NH_2$ — H_2N—CH—COOH	Asn (N)	

TABLE 8.1 *CONTINUED*

Name	Structure	Symbol	Intramolecular Interaction Character
Glutamine	$CH_2CH_2C(O)NH_2$ | $H_2N-CH-COOH$	Gln (Q)	Polar, hydrophilic side groups
Histidine	(imidazole ring structure) $CH_2-C=CH$ | $H_2N-CH-COOH$	His (H)*	
Lysine	$(CH_2)_4NH_2$ | $H_2N-CH-COOH$	Lys (K)	Basic side groups
Arginine	$(CH_2)_3N(H)C(=NH)NH_2$ | $H_2N-CH-COOH$	Arg (R)	
Aspartic acid	CH_2COOH | $H_2N-CH-COOH$	Asp (D)	Acidic side groups
Glutamic acid	CH_2CH_2COOH | $H_2N-CH-COOH$	Glu (E)	

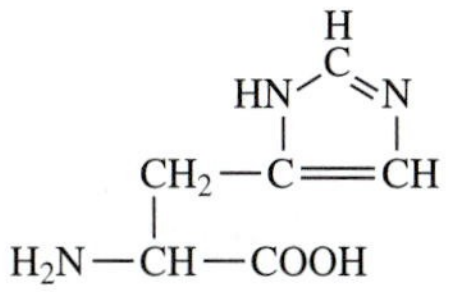

* A borderline amino acid which, at pH 7.0, functions more as a polar, hydrophilic unit than a basic residue.

in water. Moreover, highly dipolar or charged side groups may be attracted to other dipolar or oppositely charged residues. Specific amino acids provide the binding sites for prosthetic groups. For example, the imidazole component of histidine is responsible for the coordinative binding between the protein and the Fe(II)-porphyrin unit in hemoglobin and myoglobin.

Different Proteins and Their Functions

Proteins can be divided into two large classes: fibrous and globular proteins. This classification is closely related to the conformational characteristics of polymers in the two groups, and also to the function served by the protein in a living system.

Fibrous proteins Fibrous proteins are tough, insoluble materials found in the protective and connective tissues of animals, birds, and reptiles. They were among the first high polymers to be studied by X-ray diffraction techniques (see Chapter 19), and their secondary structure is much simpler than those of globular proteins. Fibrous proteins can be divided into three main classes: the α-keratins, the collagens, and the β-keratins.

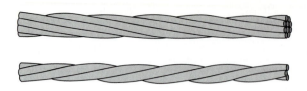

Figure 8.2 Supercoiling of α-helical polypeptide coils to form "ropes." [From A. L. Lehninger, *Biochemistry* (New York: Worth, 1970), p. 58.]

α-Keratins are found in hair, wool, horn, nails, feathers, and leather. In general, these materials are nonelastic. They are tough or flexible substances that have been used by man since antiquity as technological raw materials. The molecular conformational structure found in α-keratins is the hydrogen-bonded α-helix. In many cases, a number of α-helical chains are wound together to form "ropes." This is illustrated in Figure 8.2. The α-helical arrays in the fibrous α-keratins are comprised of L-amino acid residues arrayed in right handed helices. The strength and toughness of α-keratin materials can be traced to the multiple hydrogen bonding which holds the helix in its conformation and to the presence of —S—S— crosslinks between cystein residues. Thus, reducing agents will soften α-keratin structures by cleavage of the crosslinks and the formation of S—H bonds.

Collagen is a tough, fibrous protein found in animal connective tissue and tendons. The normal α-helical arrangements are not found in this protein. Instead, each polypeptide chain forms a loose helix that is hydrogen-bonded *inter*molecularly to two other chains to form a triple-strand rope. This arrangement prevents slippage of the chains past each other.

The third fibrous protein structure that has been studied in detail is the β-keratin system or the silk fibroin arrangement. β-Keratins, when subjected to heat and moisture by "steaming," can be stretched to a considerable degree. This process involves a cleavage of intramolecular hydrogen bonds and the conversion of the α-helix coils to an extended zigzag conformation. The extended chains are now linked to their neighbors by *inter*molecular hydrogen bonds to form a pleated sheet arrangement.

Globular proteins   Globular proteins are soluble in aqueous media. They are responsible for organizing the chemical reactions that take place in a living system. Enzymes, antibodies, hemoglobin, myoglobin, serum albumin, and some hormones are all globular proteins. Globular proteins differ from the fibrous type because they are intricately folded molecules that contain both α-helix and non-helical segments. Water molecules may occupy sites within the protein, and prosthetic groups are often present.

A detailed analysis of globular protein structures is beyond the scope of this chapter. Three lines of attack are necessary before the structure and function of a globular protein can be partly understood. First, the sequence of amino acid residues along each chain must be established, usually by chemical means. Second, the conformation of the chain and the location of prosthetic groups must be determined, usually by X-ray diffraction techniques, although nmr, esr, and other spectroscopic methods may also be used. Third, attempts can be made to *rationalize* the secondary and tertiary structure in terms of the attractive or repulsive forces between different amino acid residues on the same chain. The degree of difficulty increases as one passes through these three successive steps. It is probably true to say that no globular protein is yet fully understood in all three ways.

Three examples only of globular proteins will be mentioned—myoglobin, cytochrome c, and lysozyme. Myoglobin is an iron-containing protein present in all animal tissues, but found in substantial quantities in the muscles of aquatic diving vertebrates, such as whales or seals. The function of the protein is to store oxygen. Myoglobin is a relatively simple protein, with a molecular weight of only 16,700, made up of 153 amino acid residues. It consists of one polypeptide chain to which is attached one Fe(II)-porphyrin or heme unit. One face of the heme unit is bound coordinatively through the iron to the imidazole component of a histidine residue, while a second histidine residue loosely occupies a similar position on the opposite face. Figure 8.3 illustrates this arrangement. The polypeptide

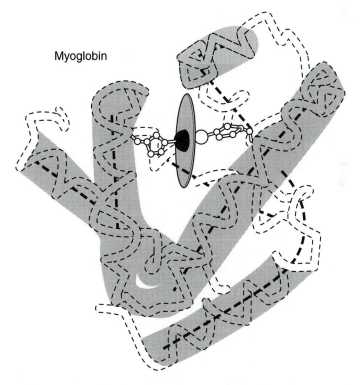

Myoglobin

Figure 8.3   Location of the Fe(II)-porphyrin unit within the coiled protein chain of myoglobin. [From R. E. Dickerson and I. Geis, *The Structure of Proteins* (New York: Harper & Row, **1969**), p. 47.]

chain is divided into eight helical regions separated by nonhelical bends. The whole tertiary structure forms a flexible box surrounding the heme component, and slight changes in the shape of the box and its central cavity, as the pH is changed, may explain why the molecule binds oxygen at pH 7.5 but releases it when carbon dioxide or lactic acid lowers the pH of the surrounding tissue. Hemoglobin contains four myoglobinlike molecules agglomerated into a supramolecular structure. Synthetic models and possible temporary substitutes for myoglobin and hemoglobin have been investigated (see Chapter 24).

Cytochrome c is a widely distributed protein in animals and plants. Its function is to act as an oxidation–reduction intermediary in the terminal oxidation chain. Cytochrome c resembles myoglobin in general composition. It possesses a single polypeptide chain of 104 amino acid residues and a heme unit. However, it contains no α-helical sections, but possesses instead many residues that occupy the extended β-conformation described earlier for β-keratins. Moreover, the heme unit of cytochrome c is not bound coordinatively to histidine residues as it is in myoglobin and hemoglobin. Instead, it is bound *covalently* through two side groups attached to the prophyrin ring that themselves are connected to two cystein residues on the main chain.

Lysozyme is an enzyme present in mucous or egg white. It kills bacteria by attacking the bacterial cell wall. Lysozyme contains no metal atom or other prosthetic group. It consists of only one polypeptide chain made up of 129 amino acid residues. The chain possesses four intramolecular cystein crosslinks (disulfide bridges), and only three α-helical segments. The tertiary structure is such that the hydrophobic amino acid residues are on the inside of the molecule, and the hydrophilic residues are on the outside. The structure is characterized by the presence of an obvious crevice or cleft in the overall globular shape. The active polysaccharide cleavage site is believed to be located in this cleft.

Anti-freeze proteins   A very impressive type of protein is used by fish to keep the water in their aqueous fluids from freezing when they swim into very cold water. Part of the required decrease in melting point T_m parallels that described in Chapter 14, where measurements of T_m depressions for determining molecular weights are mentioned. Applying these equations to the known amounts and molecular weights of the anti-freeze proteins in the body fluids, however, gives only a small fraction of the observed decrease in T_m. Most of the suppression of the freezing is kinetic, with the anti-freeze proteins binding to the faces of the water crystals to keep them from growing larger!

Biological Synthesis of Proteins

In the living cell, proteins are synthesized from amino acids at ribosome sites under the influence of enzymes and in sequences determined by the nucleic acid arrangement in messenger RNA. The energy required for the polymerization is supplied by ATP.

Protein synthesis takes place in four stages. In the first step, amino acids become enzymatically bound to a *transfer* RNA molecule by an esterification reaction. Second, an *initiation* complex is formed between the amino acid-transfer RNA ester and messenger RNA. Next, polypeptide synthesis begins by a sequential buildup of amino acid residues according to the code specified by the messenger RNA. After each amino acid residue has been introduced, both messenger RNA and the transfer RNA-peptide chain are moved along the ribosome to position the next coded site. Finally, when the last coded instruction from the messenger RNA has been followed, the completed protein separates from the ribosome. Construction of the protein begins with a reaction between the *carboxylic acid end* of the first amino acid being condensed with the amino residue of the second amino acid, and so on.

The controlled biological synthesis of proteins is now a major component of the field of biotechnology. Gene-splicing techniques allow modifications to be made to the DNA of bacteria or higher organisms. The altered DNA within the bacterial cell then induces the formation of a modified protein, with an amino acid sequence different from that of the native material. This can be used as a research tool for the study of the effect of different amino acid sequences on protein folding and other features of the tertiary structure. Alternatively, if a known foreign gene from another organism is introduced into the bacterial DNA, the bacterial cell will produce a protein characteristic of the foreign gene. In this way proteins such as human insulin can be produced on a large scale.

Laboratory Synthesis of Polypeptides

Homopolymer synthesis Homopolypeptides can be synthesized readily by conventional condensation reactions (Chapter 2) carried out with simple amino acids, such as glycine, alanine, or phenylalanine. Such polymers are not proteins—they are often only poorly soluble in water, may be highly crystalline, have polydisperse molecular weights, and do not adopt discrete folded conformations in solution or the solid state. However, homopolypeptides are of technological interest as fibers or absorbable surgical sutures, and they provide valuable models for estimating the role played by specific amino acid residues in determining the conformations of naturally occurring polypeptides, As such, they provide the raw data for conformational energy calculations on proteins, as discussed in Chapter 18.

Copolymer synthesis Simple, naturally occurring polypeptides have been synthesized by careful, sequential, copolymerization reactions. The main problem in protein synthesis is the prevention of condensation at *both* ends of the growing molecule. The polypeptide chain must be built up step by step, with the condensation normally taking place at one terminal group only. Such a process requires the use of "blocking groups" or "protective groups" to prevent unwanted, nonsequential reactions from occurring. The technique is illustrated in (10), where

X and Y are protecting groups. A commonly used dehydrating agent for the condensation step is dicyclohexylcarbodiimide (see page 40).

$$
\begin{array}{ccccc}
\begin{matrix}
X \\
| \\
NH \\
| \\
H-C-R \\
| \\
COOH \\
+ \\
NH_2 \\
| \\
H-C-R' \\
| \\
C=O \\
| \\
Y
\end{matrix}
&
\xrightarrow{-H_2O}
&
\begin{matrix}
X \\
| \\
NH \\
| \\
H-C-R \\
| \\
C=O \\
| \\
NH \\
| \\
H-C-R' \\
| \\
C=O \\
| \\
Y
\end{matrix}
&
\xrightarrow[\text{both protecting groups}]{\text{remove one or}}
&
\begin{matrix}
X \\
| \\
NH \\
| \\
H-C-R \\
| \\
C=O \\
| \\
NH \\
| \\
H-C-R' \\
| \\
C=O \\
| \\
OH
\end{matrix}
\quad \text{etc.}
\end{array}
\qquad (10)
$$

Normal laboratory procedures have been used to apply this method to the synthesis of simple peptide hormones containing 9 to 39 residues, as well as to the synthesis of insulin chains. However, the extension of such processes to the synthesis of complex proteins is a formidable task. One of the most significant breakthroughs in protein synthesis was the development between 1959 and 1964[1] of a machine that was capable of automatically performing the various sequential steps in high yield to generate complex polypeptide copolymers.

The fundamental idea underlying the whole procedure is this: the polypeptide chain is synthesized with an end of the molecule attached to an insoluble but solvent-swollen resin particle. This allows all side products, solvents, and so on, to be removed at each step in the process by the simple expedient of filtering off the solid particles. This procedure avoids the time-consuming and laborious techniques that were required earlier to purify each peptide at each step by crystallization. The machine consists of a series of amino acid, solvent, and reagent reservoirs connected via selector valves and metering pumps to a single reaction chamber. The insoluble resin is crosslinked polystyrene that has been chloromethylated by treatment with chloromethyl ether and stannic chloride (reaction 11). The core of the machine is a computer that controls a series of mi-

$$
\text{Polystyrene bead} \; \text{—} \; \bigcirc \quad \xrightarrow[\text{SnCl}_4]{\text{CH}_3\text{OCH}_2\text{Cl}} \quad \text{●} \; \text{—} \; \bigcirc \text{—CH}_2\text{Cl} \qquad (11)
$$

croswitches that allow specific amounts of reagents and solvents to enter the reaction chamber (Figure 8.4). The computer also controls the reaction time and the filtration process that removes unused reagents and solvent at the conclusion of

[1]R. B. Merrifield, *Science*, **1965,** *150,* 178.

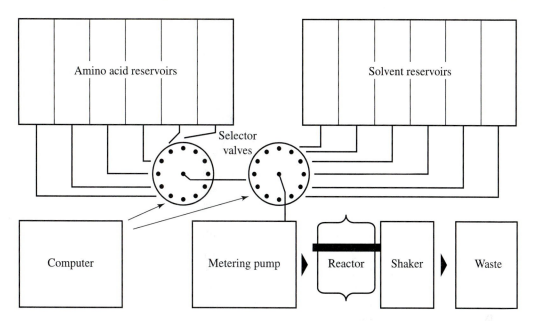

Figure 8.4 Schematic diagram of a Merrifield polypeptide synthesizer. (Modified from *Chem. Eng. News*, Aug. 2, 1971, p. 26; courtesy of the American Chemical Society.)

each reaction step. A schematic diagram of the apparatus is shown in Figure 8.4. The operator programs the sequence of chemical operations into the computer, and the synthesis then proceeds under computer control. The synthesis steps proceed through the following cycle.

The amino acid starting materials are first blocked at the amino residue, as shown in **7**, and the first amino acid is coupled to the resin through the carboxylic acid group **(8)**. Treatment of **8** with acid removes the blocking group as carbon dioxide and isobutylene. The second amino-blocked amino acid can then be coupled to the first with the use of the powerful condensing agent—dicyclohexyl-carbodiimide. These steps are repeated many times and the polypeptide is finally released from the resin by treatment with a reagent, such as HBr in trifluoroacetic acid.

The automatic nature of the sequence and the high yields at each step allowed the synthesis of the nonapeptide, bradykinin, in 27 h (3 h per peptide bond). The A-chain of insulin, with 21 residues, was synthesized in 8 days, and the B-chain (30 residues) in 11 days. Ribonuclease, with 124 residues, was synthesized in 6 weeks. The main limitation of the process is the failure of every single reaction site to give a 100% conversion to the product. Immobilized chains that have failed to undergo a chain growth reaction at any step in the cycle will accumulate as defect-containing polypeptides. Thus, a key requirement is that the maximum effort must be made to ensure as close to 100% conversion as possible at every step in the synthesis. Use of a large excess of reagent at each step encourages high yields, and the excess can easily be removed from the immobilized product by filtration.

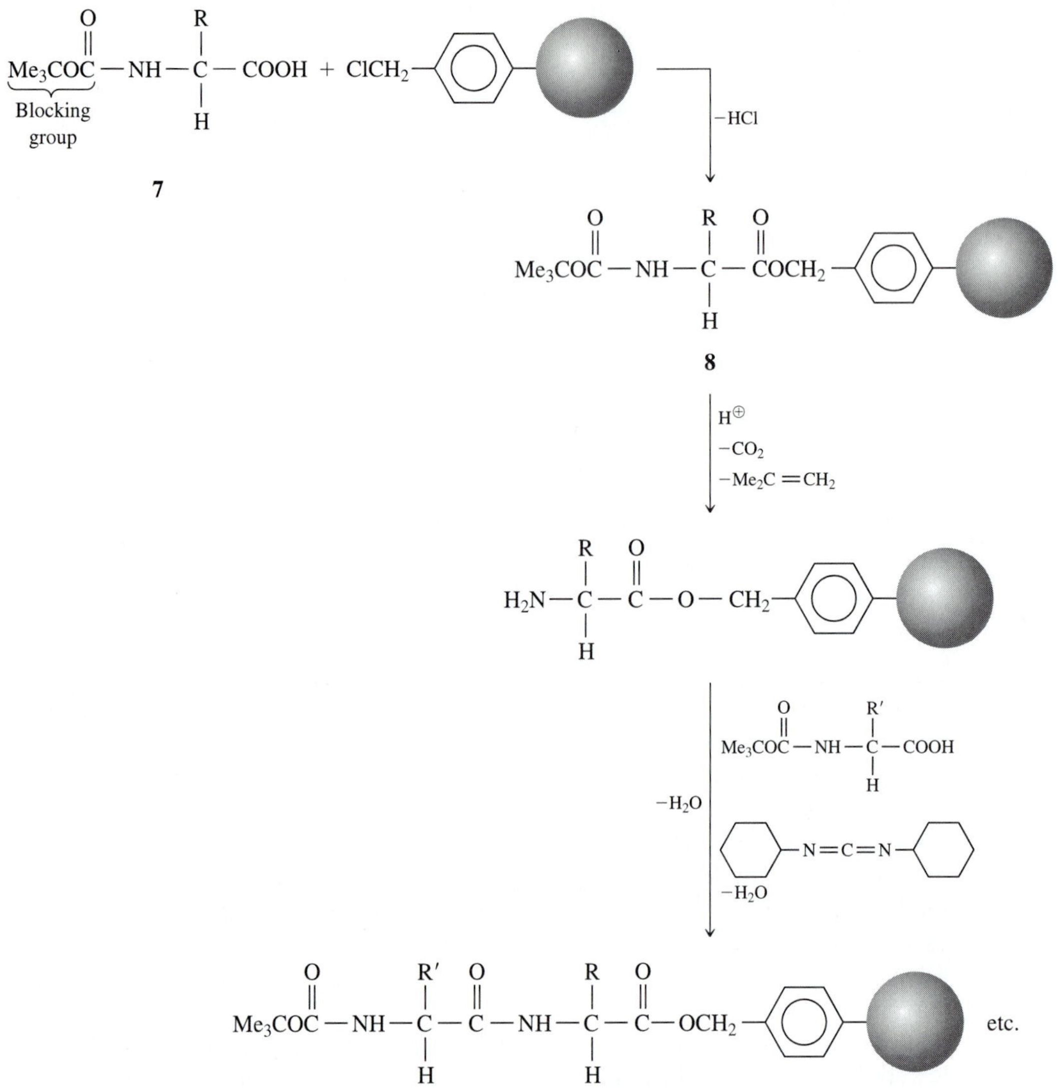

$$\underbrace{Me_3COC}_{\substack{\text{Blocking} \\ \text{group}}} \overset{O}{\underset{}{\parallel}} - NH - \overset{R}{\underset{H}{\overset{|}{\underset{|}{C}}}} - COOH + ClCH_2 - \bigcirc\!\!-\!\bullet \longrightarrow$$

7

$-HCl$

$$Me_3COC \overset{O}{\underset{}{\parallel}} - NH - \overset{R}{\underset{H}{\overset{|}{\underset{|}{C}}}} - \overset{O}{\underset{}{\overset{\parallel}{C}}} OCH_2 - \bigcirc\!\!-\!\bullet$$

8

$H^{\oplus}$
$-CO_2$
$-Me_2C=CH_2$

$$H_2N - \overset{R}{\underset{H}{\overset{|}{\underset{|}{C}}}} - \overset{O}{\underset{}{\overset{\parallel}{C}}} - O - CH_2 - \bigcirc\!\!-\!\bullet$$

$$Me_3COC \overset{O}{\underset{}{\parallel}} - NH - \overset{R'}{\underset{H}{\overset{|}{\underset{|}{C}}}} - COOH$$

$-H_2O$

$$\bigcirc\!\!-\!N=C=N\!-\!\bigcirc$$

$-H_2O$

$$Me_3COC \overset{O}{\underset{}{\parallel}} - NH - \overset{R'}{\underset{H}{\overset{|}{\underset{|}{C}}}} - \overset{O}{\underset{}{\overset{\parallel}{C}}} - NH - \overset{R}{\underset{H}{\overset{|}{\underset{|}{C}}}} - \overset{O}{\underset{}{\overset{\parallel}{C}}} - OCH_2 - \bigcirc\!\!-\!\bullet \quad \text{etc.}$$

Use of a Merrifield synthesizer is now routine in molecular biology and biochemistry laboratories. Its use for the controlled synthesis of nonbiological condensation polymers is just beginning.

One of the most recent novel efforts in the area of protein synthesis is to expand the genetic code beyond the choices provided by existing organisms. This is done by inserting "unnatural" amino acids to give customized proteins, which could even lead to new life forms. The two major approaches to date are based on stop codon suppression, and avoiding the editing scheme transfer RNA synthetase uses to delete incorrect amino acids.

Reactions of Proteins

Denaturation and recoiling   The biological activity of proteins is a direct consequence of their structure and molecular conformation. This is especially true of globular proteins. Thus, any influence that breaks up the secondary or tertiary structure will destroy the biological activity. Heat, excessive cooling, pH changes, or the presence of many chemical reagents will have this effect. This process is called *denaturation*, and it may be reversible or irreversible. The coagulation of the white of an egg when heated is an irreversible denaturation.

It should be noted that denaturation often does not change the chemical composition of the protein. The amino acid residues are present in the same sequence as before. Only the conformation of the chain has been changed from a discrete pattern of helix and bends to a more random coil arrangement. As we have seen, the tertiary structure depends mainly on hydrophobic and hydrophilic intramolecular interactions (plus the presence of disulfide linkages in some proteins). Hence, it might be expected that the biologically active conformation could reform once the perturbing conditions are removed. This can, in fact, be accomplished in the laboratory, outside a living cell. Refolding of the chains occurs spontaneously (if slowly) as the various attractive portions of the flexing, twisting chain find each other and pull the molecule back into the biologically active shape. The same phenomenon can be observed when the individual chains of supramolecular proteins are first separated and then brought together. Hemoglobin rapidly reassembles itself from mixtures of the α and β chains. The heme residues may also be removed and then reintegrated into the structure. Moreover, the metal atoms of enzymes can be removed to cause a total loss of activity, but the reintroduction of the metal restores the catalytic powers. In some enzymes, *different* metal atoms may be introduced at the renaturation stage, sometimes to yield metalloproteins that have a biological activity comparable to that of the original. Such metal substitutions are valuable for probing the mechanism of enzyme action.

A major area of modern polymer research is to understand in detail why and how protein folding occurs. It is assumed that, once a correlation is made between the amino acid sequence and the shape of a folded protein, the protein function can be altered by changing one or a small number of amino acid residues along the chain. Because protein folding depends on the hydrophobicity and hydrophilicity of individual amino acid residues, and on their steric, hydrogen bonding, and internal cross-linking abilities, it should be possible to predict the shape and function of any globular protein from first principles. A hint of how this can be accomplished for simpler macromolecules is given in Chapter 18. Computer programs are now available that help predict protein conformations for different amino acid sequences, and these are being improved continuously through comparisons of the calculated conformations with those determined experimentally.

Ultimately it will be possible to design a new and desirable polypeptide sequence through computer simulation, and then modify the DNA of a living cell to generate that protein.

Hydrolysis reactions   The hydrolytic conversion of proteins to oligomeric peptides and amino acids is a reaction that has been used by man since prehistory for the conversion of animal tissues to glues, or readily digestible foodstuffs such as gelatin. On a more subtle level, the selective hydrolysis of proteins can be used as a method of structural identification. The fundamental hydrolysis reaction is shown in reaction (12). Acidic hydrolysis (e.g., with 6 N HCl at 100 to 120°C for 10 to 24 h in a sealed system) will generate the hydrochloride salts of the free amino acids. Tryptophane is destroyed by this process and glutamine and asparagine are converted to glutamic and aspartic acids, but in general, the amino acid composition of the hydrolysis mixture will reflect that of the protein. Alkaline hydrolysis does not cause decomposition of tryptophane, but destroys other amino acids.

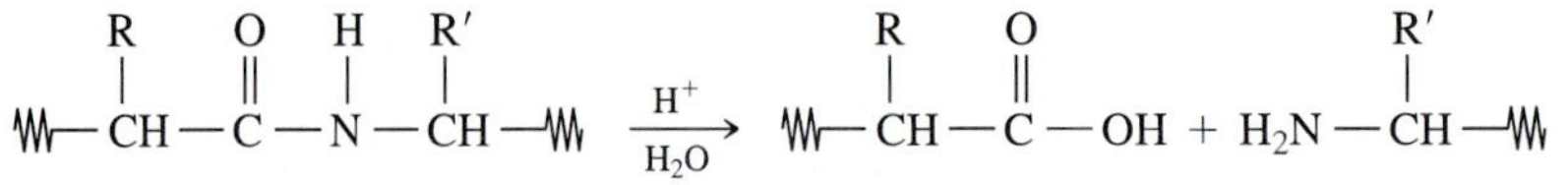

$$
\underset{\displaystyle \text{\scriptsize W}}{\text{\small—CH—C—N—CH—}}\;\;\xrightarrow[\text{H}_2\text{O}]{\text{H}^+}\;\; \text{—CH—C—OH} + \text{H}_2\text{N—CH—} \tag{12}
$$

The analysis of a protein is accomplished at two levels of detail. First, the gross amino acid composition of the hydrolysis mixture may be determined by paper chromatography, electrophoresis, or ion-exchange chromatography. Second, the *sequence* of amino acid residues along the chain must be determined. This sequencing operation involves (1) an identification of the NH_2-terminal or COOH-terminal residues at the ends of the chain, and (2) cleavage of the protein into oligopeptides by the use of trypsin, followed by identification of the terminal residues in these fragments. This process is then continued until the sequence is known.

Identification of the NH_2-terminal amino acid can be accomplished by reaction of the protein with 2,4-dinitrofluorobenzene. This reagent reacts with terminal or pendent NH_2 groups, as shown in (13), to yield a linkage that will survive hydrolysis.

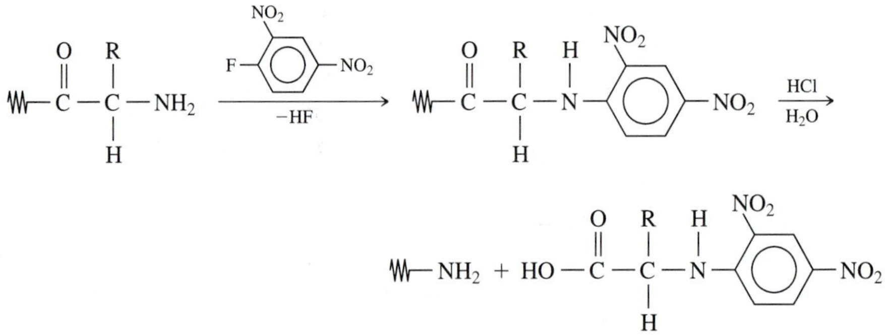

$$\tag{13}$$

Subsequent hydrolysis will yield a mixture of amino acids, only one of which bears the 2,4-dinitrofluorophenyl (DNP) groups. Phenyl isothiocyanate is also used as an —NH$_2$ end-group reagent, as illustrated in (14). The advantage of this reagent is that the hydrolysis to yield the phenylthiohydantoin (**9**) removes only the terminal amino acid. Hence, the process can be repeated to determine the actual amino acid sequence. The —COOH terminal residue can be identified by reduction to the alcohol, followed by hydrolysis of the protein and identification of the amino alcohol, or by enzymatic cleavage of the terminal residue with carboxypeptidase.

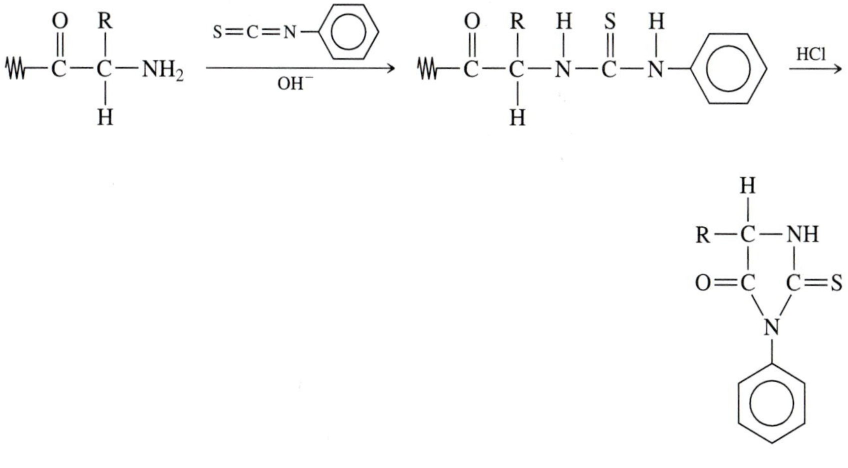

$$
\text{(14)}
$$

9

Other reagents are known which selectively cleave the skeletal peptide bonds adjacent to specific amino acid residues. Bromine or N-bromosuccinimide selectively cleaves the tyrosine peptide bonds. Methionine peptide bonds are cleaved by iodoacetamide or cyanogen bromide. Many other selective cleavage processes are known.

Reactions of protein side groups   First, and perhaps most obvious, it is possible to exchange the hydrogen atoms of peptide linkages by treatment of the polymer with deuterium oxide or tritium oxide (15). However, the rate of exchange depends on the position and role of a particular peptide link in the chain. Hydrophobic side groups (R) or hydrogen bonding involving the N—H group will retard exchange. Thus, deuterium-exchange studies can be used to examine the environment of a particular amino acid residue in an undenatured protein.

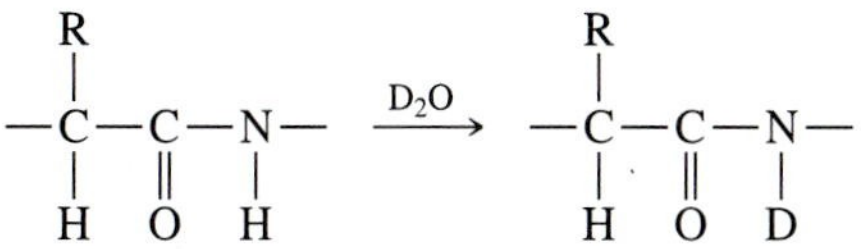

$$
\text{(15)}
$$

Proteins and Polypeptides

The different substituent groups (R) in proteins include those that contain —OH, —COOH, —NH$_2$, —NH—, —CONH$_2$, —SH, or —SCH$_3$ functional groups, and these can be induced to undergo reactions with appropriate reagents. Such reactions can be used for structure determination, or they can be carried out deliberately to modify the protein at specific sites in order to observe the resultant changes in physiological behavior. On an industrial level, side-group reactions can be employed to improve the overall physical properties of a protein-containing product.

Halogen-containing reagents react with —SH, —OH, —NH$_2$ or —COOH groups. For example, iodoacetic acid, ICH$_2$COOH, reacts with —SH or —NH$_2$ groups. Cysteine residues may be identified and modified by this procedure (16).

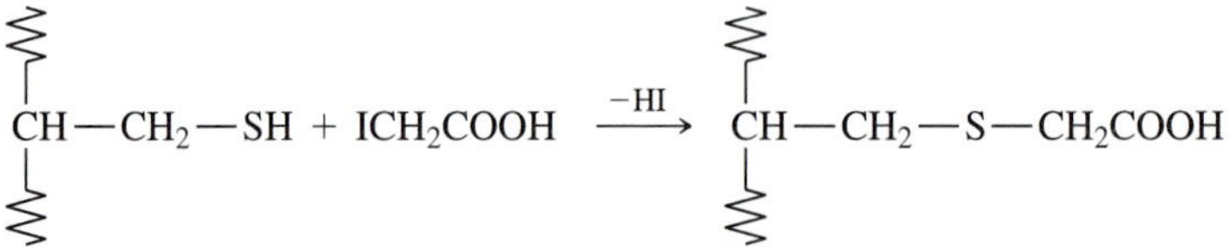

$$\text{CH—CH}_2\text{—SH} + \text{ICH}_2\text{COOH} \xrightarrow{-\text{HI}} \text{CH—CH}_2\text{—S—CH}_2\text{COOH} \qquad (16)$$

Acyl halides or anhydrides react with —OH, —NH$_2$—COOH groups (17). Acetic anhydride acetylates at least 90% of the pendent amino groups of wool.

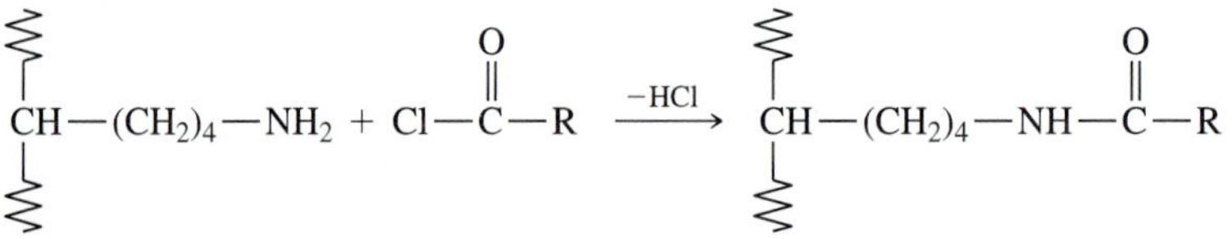

$$\text{CH—(CH}_2)_4\text{—NH}_2 + \text{Cl—}\overset{\overset{\text{O}}{\|}}{\text{C}}\text{—R} \xrightarrow{-\text{HCl}} \text{CH—(CH}_2)_4\text{—NH—}\overset{\overset{\text{O}}{\|}}{\text{C}}\text{—R} \qquad (17)$$

Carboxylic acid groups may be esterified with the use of alcohols in acidic media. Sulfuric acid reacts with —OH or —SH groups to generate sulfate esters or thioesters (18).

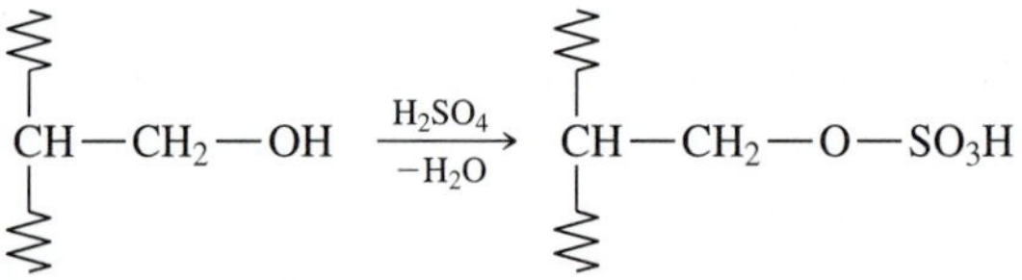

$$\text{CH—CH}_2\text{—OH} \xrightarrow[-\text{H}_2\text{O}]{\text{H}_2\text{SO}_4} \text{CH—CH}_2\text{—O—SO}_3\text{H} \qquad (18)$$

Treatment with nitric acid under mild conditions may be used to nitrate aromatic residues as, for example, in tyrosine units. Drastic nitration conditions cause decomposition of the protein. Iodination of a number of proteins in neutral or basic media leads to the conversion of tyrosine residues to 3,5-diiodotyrosine units.

Many proteins react with diazonium salts to yield diazo proteins (19). Alternatively, the protein itself may be diazotized with nitrous acid and then coupled, for example, to an amine (20).

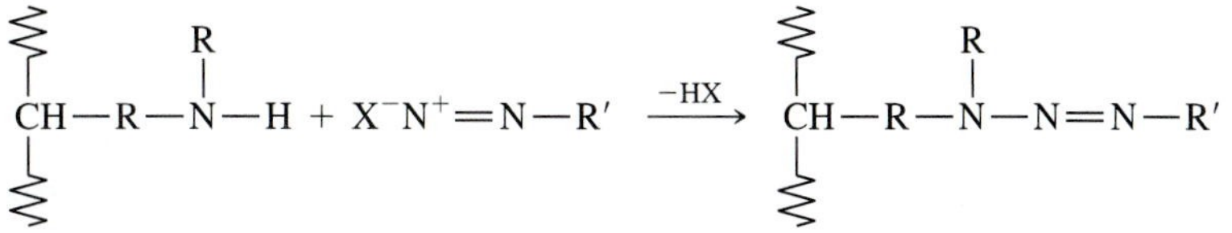

$$\text{CH}-\text{R}-\overset{\overset{\text{R}}{|}}{\text{N}}-\text{H} + \text{X}^-\text{N}^+\!\!=\!\!\text{N}-\text{R}' \xrightarrow{-\text{HX}} \text{CH}-\text{R}-\overset{\overset{\text{R}}{|}}{\text{N}}-\text{N}\!=\!\text{N}-\text{R}' \qquad (19)$$

$$\text{CH}-\text{R}-\text{NH}_2 \xrightarrow[-\text{HCl}]{\text{HNO}_2} \text{CH}-\text{R}-\text{N}\!=\!\text{N}^+\text{Cl}^- \xrightarrow[-\text{HCl}]{\text{RNH}_2}$$

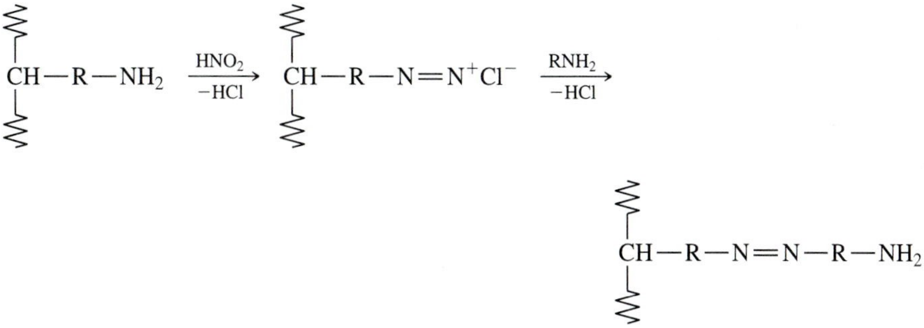

$$(20)$$

$$\text{CH}-\text{R}-\text{N}\!=\!\text{N}-\text{R}-\text{NH}_2$$

A number of unsaturated reagents undergo addition reactions with proteins. For example, as shown in reaction (21), formaldehyde adds to amino or amido groups to form a methylol derivative. As discussed in Chapter 2, methylol groups show a strong tendency to condense to generate crosslinks between the polymer molecules. Isocyanates can add to —OH, —SH, or —NH$_2$ groups. An example is shown in reaction (22).

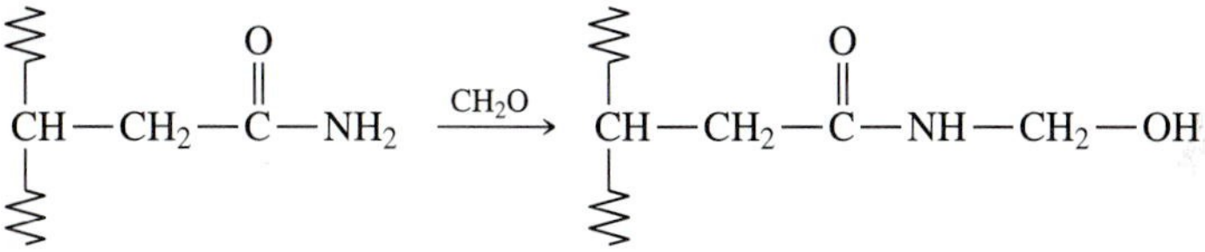

$$\text{CH}-\text{CH}_2-\overset{\overset{\text{O}}{\|}}{\text{C}}-\text{NH}_2 \xrightarrow{\text{CH}_2\text{O}} \text{CH}-\text{CH}_2-\overset{\overset{\text{O}}{\|}}{\text{C}}-\text{NH}-\text{CH}_2-\text{OH} \qquad (21)$$

$$\text{CH}-\text{CH}_2-\text{SH} + \text{O}\!=\!\text{C}\!=\!\text{N}-\text{Ph} \longrightarrow \text{CH}-\text{CH}_2-\text{S}-\overset{\overset{\text{O}}{\|}}{\text{C}}-\text{NH}-\text{Ph} \qquad (22)$$

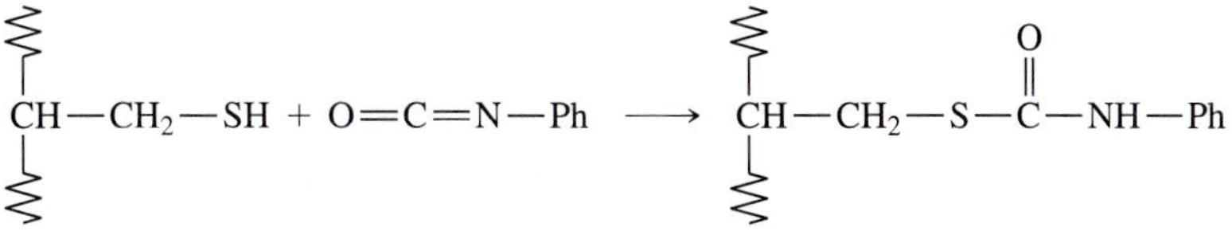

Reactions that involve cross-links The principal cross-link unit found in proteins is the disulfide bridge **(10)**. Disulfide bridges are derived from cysteine residues by the oxidation process illustrated in equation (23). Interest in the cross-linkage of proteins stems from two sources: (1) the use of reagents to cleave disulfide bridges, and (2) the study of reactions that can introduce new cross-links into a protein system.

$$\text{CH—CH}_2\text{—SH} + \text{HS—CH}_2\text{—CH} \xrightarrow{\text{O}} \text{CH—CH}_2\text{—S—S—CH}_2\text{—CH} \qquad (23)$$

10

Disulfide bridges can be cleaved by a variety of reagents, such as peracids (performic or peracetic acids), peroxides, or chlorine, which oxidize the bond, or reducing agent, such as borohydrides, bisulfites, or thiols. Water or hydroxide ion may cleave the disulfide bridges by hydrolysis. Even heat or ultraviolet light may break the bond. Of course, the destruction of native cross-links increases the ability of the protein to swell in water or even to dissolve.

The introduction of additional cross-links into proteins may increase their technological usefulness. For example, the stability of leather, wool, or silk to acids, alkalis, oxidizing agents, or to attack by insects, is improved if additional cross-linking sites can be introduced. A wide variety of reagents have been used in attempts to achieve this end. Many of the reactions mentioned in the preceding section can be modified to generate cross-links. The tanning of animal skins to form leather by treatment with tannins (phenolic tree bark extracts), chromium salts, or formaldehyde, takes place by cross-linking of collagen chains through hydrogen bonds, covalent linkages, or metal coordination binding.

Immobilization of Enzymes

The ability to carry out enzyme-catalyzed chemistry outside a biological cell system offers many advantages in both biochemistry and general organic synthesis. However, most enzymes have only limited stability outside the cell and are difficult to separate from reactants and products. The immobilization of enzymes on solid or gel polymeric supports has been developed widely, both in the laboratory and on a manufacturing scale. Surprisingly, the activity of the enzyme is often retained after immobilization. In some cases, the resistance of the enzyme to denaturation is actually improved.

Various strategies have been developed for the covalent attachment of enzymes to polymer surfaces or to crosslinked polymer gels. Only selected examples will be given here. First, it is necessary to begin with a support polymer that has functional groups either at the surface or throughout a crosslinked gel matrix. Although intuition might suggest that hydrophilic surfaces would provide the most compatible environment for an enzyme, in fact hydrophobic surfaces often work quite well. A wide variety of functional groups on the support polymer can be employed (OH, NH_2, COOH groups, and so on). Ideally, the coupling reaction should be specific for an exposed end group of the protein, for example the NH_2 terminal unit. Example coupling reactions are shown in Scheme 1.

Enzymes immobilized in this way include trypsin and glucose-6-phosphate dehydrogenase[1] Coupling of enzymes to solvent-swellable gels can be accom-

---

[1] Allcock, H. R., and Kwon, S., *Macromolecules*, **1986**, *19*, 1502.

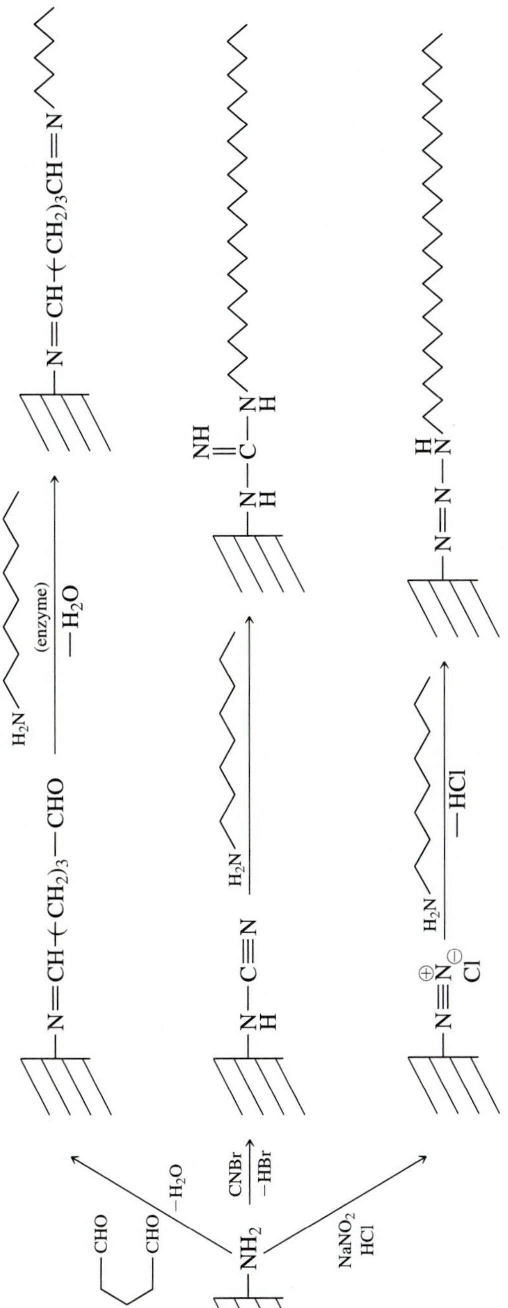

Scheme 1

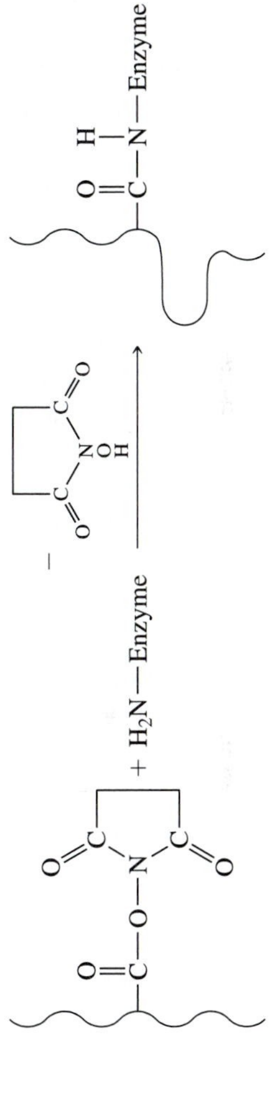

Scheme 2

plished by the use of polyacrylamide copolymers that contain *N*-acryloxysuccin-imide repeating units. The pendent groups react with the amino terminus of an enzyme, as shown in Scheme 2. This coupling method is useful for the immobilization of a wide variety of relatively delicate enzymes. Alternatively, radiation cross-linking of a solid, water-soluble polymer that contains dispersed enzyme molecules can yield hydrogels in which the enzyme is physically trapped in the water-filled regions between the cross-link sites.[1] Some radiation-induced coupling of the protein to the gel polymer may also occur.

A major advantage of enzyme immobilization is that the enzyme-support conjugate particles can be packed into a column and used as a continuous flow reactor. This lends itself to the continuous, large-scale production of biochemical or chemical products. A related development is the immobilization of bacterial, plant, fungal, or animal cells on polymeric supports, again with a view to the utilization of enzymatic processes in large-scale production.

POLYNUCLEOTIDES

General Composition of Polynucleotides

Polynucleotides are the macromolecules, found within living cells, that are responsible for the storage of genetic information, for its replication, and for providing templates that direct the synthesis of proteins. They are complex copolymers formed by the condensation of only a small range of closely related monomer molecules.

Two main types of polynucleotides are known: deoxyribonucleic acid (DNA) and ribonucleic acids (RNA). DNA molecules are the storage sites for genetic information within the chromosomal genes of the nucleus, and are responsible for the replication of that information during cell division. RNA molecules transmit information coded in their structure from the DNA molecules to the ribosomes and there direct the pattern of protein synthesis.

The monomer units from which polynucleotides are built are called mononucleotides. Each "monomer" is made up of three parts: a phosphoric acid residue plus a sugar residue in the main chain, and a heterocyclic organic base which forms a side group that is attached to the sugar (11). Different mononucleotide monomers contain one of two sugars—the cyclic, five-carbon ribose (12) or deoxyribose (13)—and one of the five heterocyclic bases, as shown in 14 to 18. A few other minor bases are found in some polynucleotides. The sites of attachment of the sugar to phosphate in the main chain are denoted by **, and the linkage sites between sugar and heterocyclic base are indicated by †. Thus, a typical nucleotide monomer residue would appear as shown in 19.

RNA contains only the sugar ribose and four bases, adenine, guanine, cytosine, and uracil, plus small amounts of methylated forms of these bases. The various forms of DNA contain only the sugar deoxyribose and four bases, adenine, guanine, cytosine, and thymine, together with minor amounts of less common

---

[1]Allcock, H. R., Pucher, S. R., and Visscher, K. B., *Biomaterials*, **1994,** *15,* 502.

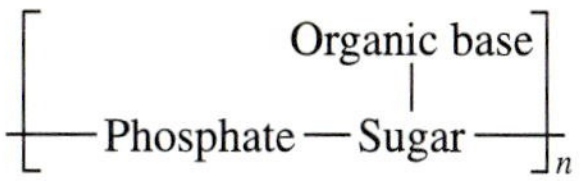

11

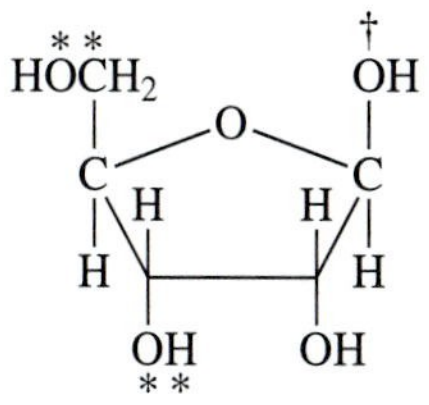

$$\left[-\text{Phosphate}-\underset{\displaystyle |}{\underset{\displaystyle \text{Organic base}}{\text{Sugar}}}-\right]_n$$

12
Ribose

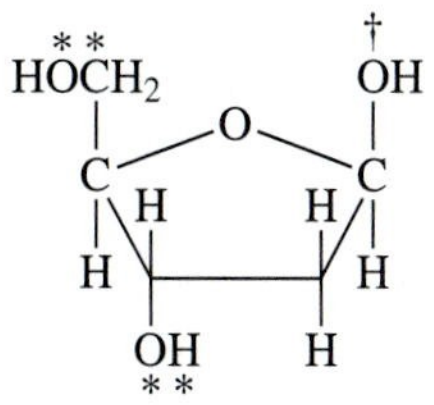

13
Deoxyribose

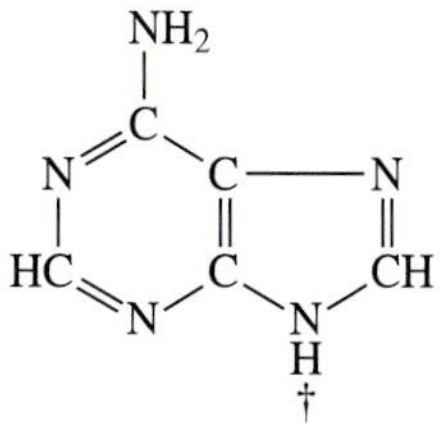

14
Adenine (A)

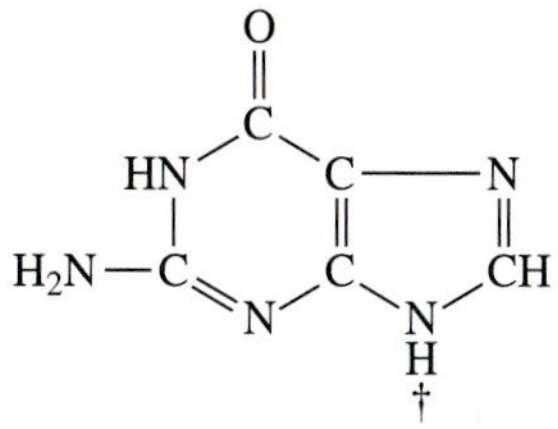

15
Guanine (G)

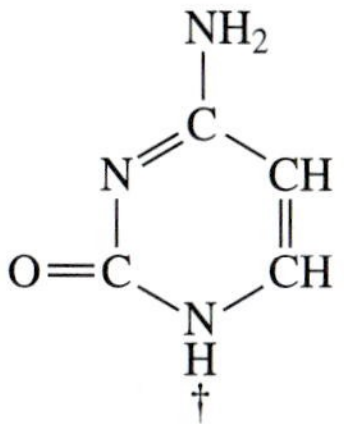

16
Cytosine (C)

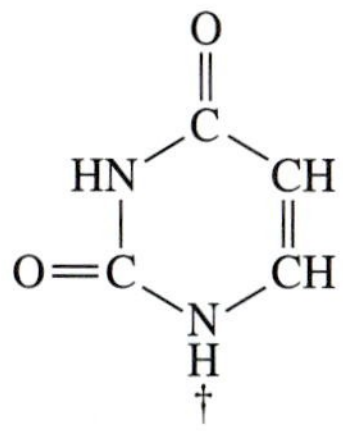

17
Uracil (U)

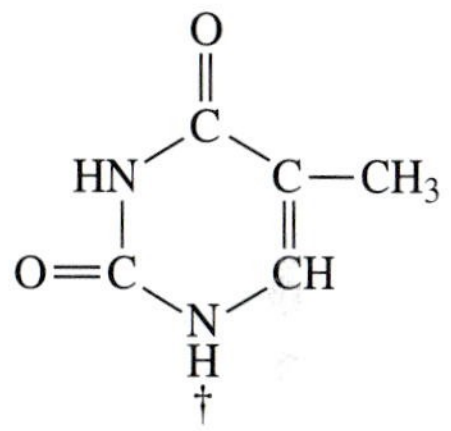

18
Thymine (T)

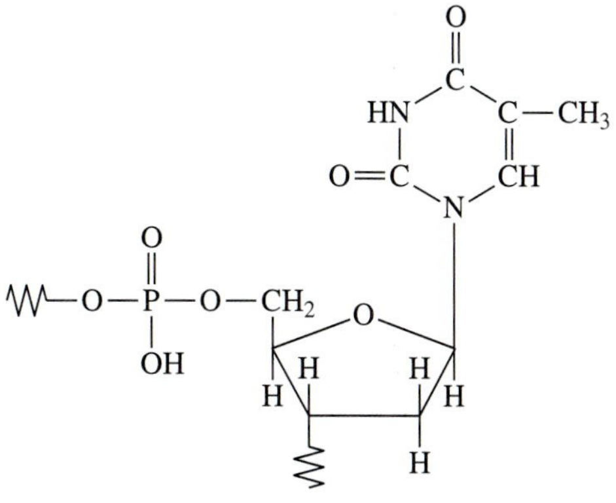

19

bases that are often methylated derivatives of the major bases. In most cells, the amount of RNA is 5 to 10 times more than the amount of DNA.

The arrangement of DNA molecules in the cell is complex. Human cells contain 23 pairs of chromosomes. About 30,000 genes are distributed unevenly over this chromosomal material. Each gene is believed to code for one protein (although this has recently been questioned), with a single gene containing from 1000 to more than 100,000 nucleotide monomer units. However, this accounts for only 1 to 1.5% of the base pairs present, since human chromosomes contain a total of about 3 billion base pairs. The purpose of most of the remaining 98.5 to 99% is not yet certain.

A multilaboratory project to map the entire human genome yielded its first conclusions in the year 2000 and revealed the approximate number of genes in 2001. The technique employed was to cleave the DNA into fragments and then, by computer correlations of overlapping end-regions in different fragments, assemble the sequences into a representation of each of the 23 human chromosomes. A major objective of this work is to correlate the nucleotide sequence in each gene with the protein generated from it in order to access methods to activate or deactivate the production of specific proteins for the benefit of human health.

The modification of plant and animal DNA by the introduction of genes from other organisms is now well-established but is currently a controversial practice in agriculture and animal biology.

Roles Played by Different Polynucleotides

DNA Deoxyribonucleic acid (DNA) provides a sequential polymeric transcription code that ultimately is responsible for the pattern of protein synthesis. Each DNA supramolecular unit consists of two polymeric strands that are wound together in the form of the well-known "double helix." The double helix is held together by the coplanar stacking of base pairs within the interior of the helix (Figure 8.5). The only allowed base-pairing combinations are adenine-thymine and guanine-cytosine, each base pair being held together by hydrogen bonding linkages of the types shown in **20** and **21**. This stacking arrangement places the bases on the inside of the helix and the sugar-phosphate main chains on the outside. Replication is accomplished by separation of the two strands of the double helix, each of

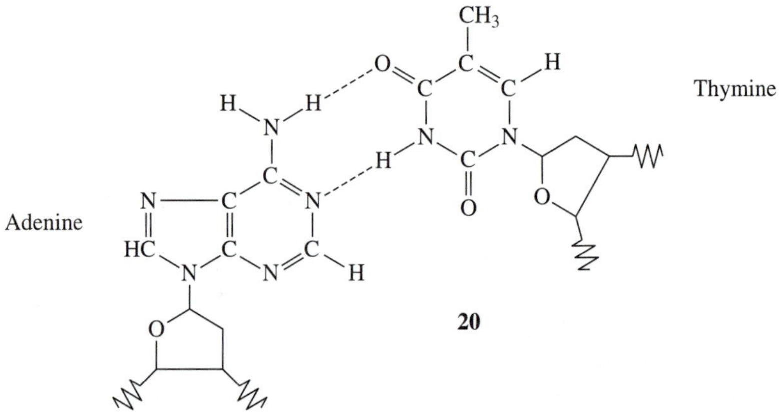

Adenine

Thymine

20

H
|
N
|
H ·····N
|
O   H   C   H   Cytosine
\\   ‖ /   ‖
C   C   C
/   / ‖   ‖ \
N   C   N · · · H   N   H
‖   \   /   \   /
C   N · · · · H   C
/ \   /   / \
H   C   C   O · · · H   N
\   /   \\   \
N   N   C   H
| \   21
             N · · · · H   N
              | |
              O   H
                          |
                          H

Guanine

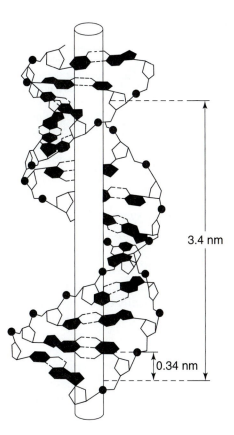

3.4 nm

0.34 nm

Figure 8.5 DNA double helix. The pentagons represent sugar residues, the black dots show the location of the phosphate units, and the short dashed lines depict the hydrogen bonds. [From J. N. Davidson, *Living Molecules*, The Royal Institute of Chemistry (Lecture Series), No. 1, 1963.]

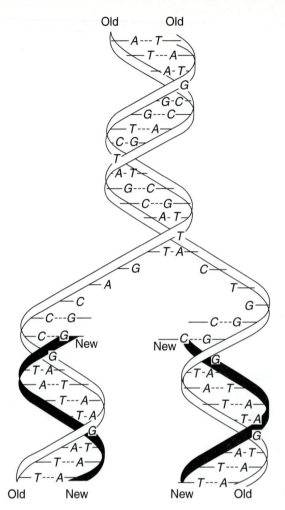

Figure 8.6 Schematic representation of the replication mechanism of DNA. Both of the old strands of DNA can serve as a template to orient monomer units from a pool of triphosphates. Polymerization is effected under the influence of the enzyme DNA synthetase. [From R. J. Light, *A Brief Introduction to Biochemistry* (New York: Benjamin, 1968), Chap. VI.]

which then functions as a template for the synthesis of a new complementary partner (Figure 8.6). Some DNA molecules are cyclic rather than linear.

Messenger RNA This form of RNA is a single-strand polynucleotide that originates in the cell nucleus. It functions as a template for protein synthesis at the ribosome sites. It is synthesized enzymatically on a single strand of DNA used as a template. Thus, messenger RNA contains bases that are complementary to those in the corresponding strand of DNA.

Transfer RNA These molecules are short-chain polynucleotides (75 to 90 monomer units) that transport amino acids to the messenger RNA at the ribosome sites. It appears that either each type of transfer RNA molecule is responsible for the transport of one type of amino acid, or several different transfer RNAs may be involved in the transfer of each type of amino acid. In other words, some amino acids are coded by two or three different transfer RNA molecules, and other amino acids by only one transfer RNA. The amino acids appear to be bound to the polynucleotide through ester linkages formed between the carboxylic acid residue and the 2'- or 3'-hydroxyl group of a *terminal* adenylic acid residue.

Ribosomal RNA   As the name suggests, this type of polynucleotide is found in the ribosomes, probably as a single-strand polymer. Its function is not understood. Ribosomes consist of RNA associated with protein.

Viruses   Viruses are supramolecular structures that contain a polynucleotide core surrounded by a protein coat. Depending on the type of polynucleotide present, they may be classified into RNA or DNA viruses. They function by invasion of a living cell and displacement of the host messenger RNA molecules from the ribosomes. The virus polynucleotide then forms the template for the synthesis by the host cell of the protein coat for the virus and also induces the synthesis of more virus polynucleotide. Many viruses can be crystallized and studied by X-ray diffraction or electron microscopy (Figure 8.7).

Polymeric Coding in Polynucleotides

It is well known that the linear coding sequence of the bases along a DNA molecule is ultimately responsible for the linear sequence of amino acid residues in a particular protein. As we have seen, the base sequence in DNA is "transcripted"

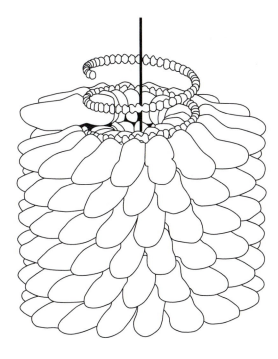

Figure 8.7   Drawing depicting part of the structure of the tobacco mosaic virus as determined by X-ray structure analysis. The RNA chain is depicted as the internal helix surrounded by protein subunits. [From Klug, A. and Casper, D. L. D., *Advan. Virus Res.*, **1960,** *7,* 225.]

to a complementary base sequence along a chain of messenger RNA, which in turn acts as the real template on which the protein is constructed. The exact nature of coding along a polynucleotide chain required to specify a particular amino acid has been worked out. Only three base molecules in sequence are needed to specify an amino acid. Thus, the GCU sequence along a messenger RNA chain is the code for alanine, CUG for leucine, and so on. However, the system is "degenerate" in the sense that *several* similar nucleotide "words" are codes for the *same* amino acid. This is believed to provide a safety factor in case a mutation at one point along a polynucleotide chain interrupts the message. Triplet code messages exist for the termination of a protein chain and the start of the next chain, but no "punctuation" codes exist *between* the triplet codes for individual amino acids along a chain. Hence, an advantage exists if a mutation in one codon (a C, A, G, or U) of the triplet yields a sequence that is the code for the same amino acid or at least a similar one. Thus, following a mutation, the degeneracy of the system may prevent the synthesis of the wrong protein or, at the worst, it may provide a very minor change in the amino acid composition that could have an evolutionary survival value. It will be clear that investigations of the effects of deliberate mutational changes to the nucleotide sequence has had, and will continue to have, a profound influence on research in this area.

Synthesis and Modification of Polynucleotides

Synthesis within the living cell   DNA is synthesized within the living cell from nucleotide monomers under the influence of the remarkable enzyme, DNA polymerase. The process occurs only in the presence of preformed, single-strand DNA,[1] which acts both as a template and as a reaction "primer." DNA polymerase as isolated from *Escherichia coli* bacteria is a single-chain polypeptide containing about 1000 peptide residues. The enzyme can not only build polynucleotide chains at a rate of about 1000 residues per minute, but it can also catalyze the hydrolytic breakdown of polynucleotide chains when excess pyrophosphate is present. A second enzyme, DNA ligase, is not capable of catalyzing the synthesis of complete DNA molecules, but serves to repair breaks in the polynucleotide chains or to cyclize linear bacterial polynucleotides to form circular DNA. Other enzymes appear to be capable of breaking the cross-links that are formed when DNA chains are exposed to ultraviolet radiation.

The RNA-type polynucleotide chains are synthesized from mononucleotide monomers by the enzyme RNA polymerase. In this case, DNA serves as the template and chain primer in the presence of magnesium ion. Once again, the same enzyme can cleave the polynucleotide chains when high concentrations of pyrophospate are present.

"Gene splicing"   Enzymes are available that will cleave the DNA molecules at specific sites (restriction enzymes). Other enzymes are capable of join-

[1]Double-helical DNA is apparently completely inactive as a polymerization template.

ing polynucleotide end units. A combination of these processes allows nucleic acid components to be inserted into the DNA of a host cell. The inserted material may be derived from living material or it may be a synthesized polymer. The modified DNA then directs the production of a "foreign" protein. The impact of this procedure on biotechnology is well known.

Laboratory synthesis of polynucleotides The enzyme polynucleotide phosphorylase has been employed in laboratory experiments to synthesize both homopolymers and copolymers from mononucleotides. These synthetic nucleotides were used in the critical experiments that led to the unraveling of the triplet code employed for protein synthesis.

The nonenzymatic construction of oligonucleotide chains has been accomplished by the use of condensation reactions carried out in automated synthesizers. The basic principles employed are similar to those used in polypeptide synthesizers, although the chemistry is very different and a more complex protection-deprotection protocol is required. The solid phase for immobilization of the growing chain can be silica particles or a synthetic polymer. The protected base–sugar–phosphorus compound building blocks are assembled first, and these are then coupled together in sequence. A number of different chemical approaches are now in use, but the one shown in Scheme 3 is reported to be particularly useful for automated synthesis. This, the so-called phosphoramidite route, requires that the trivalent phosphorus atoms be oxidized to phosphate linkages after the oligonucleotide has been constructed. Short blocks of oligonucleotide units synthesized in this way can then be linked together to form longer chains.

Reactions of Polynucleotides

Coiling and uncoiling reactions The double-helix structure of DNA is held together largely by hydrogen bonds between the base pairs. Hydrogen bonds are relatively weak—they can often be broken by raising the temperature in an aqueous system, by the addition of reagents that can more readily coordinate to oxygen or nitrogen than can hydrogen, or by the addition of alcohols, ketones, or urea. For these reasons, the DNA double helix can unwind and separate into individual single strands in a denaturation process that is known as "melting." The melting phenomenon can be monitored by the sudden decrease in viscosity or the increase in ultraviolet absorption at 260 nm that occurs on denaturation. DNA samples from different cell sources have different melting temperatures.

Perhaps the most extraordinary feature of the coiling-uncoiling process is the ease with which *re*coiling occurs to form the double helix after the denatured DNA has been returned to its original environmental conditions. If the separation of the strands has proceeded to the point at which about 12 or more residues are still helically coiled, then recoiling of the remaining segments is extremely rapid. If, on the other hand, *total* separation of the strands has occurred, the reformation of the double helix is much slower. Presumably, random collisions be-

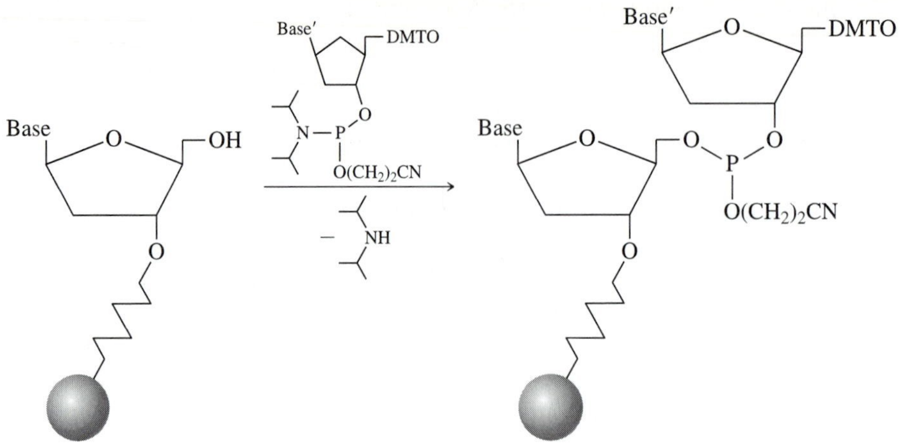

(Base′ = protected base, DMT = dimethyoxytrityl)

Scheme 3

tween the chain ends of complementary strands are needed before the end structure can align itself to generate the duplex helical structure.

Metal ions have a profound effect on the coiling-uncoiling process. Metal ions that preferentially bind to the P—OH components of the backbone stabilize the helical structure and raise the melting point. This effect is a direct result of a neutralization of P—O$^{\ominus}$ charges. On the other hand, divalent metal ions, such as Cu(II) or Zn(II), that bind to the bases, generate spurious coordination crosslinks that assist unwinding at elevated temperatures, lower the melting point, but at the same time assist in the rewinding process when, for example, the temperature is again lowered. Effects such as these have been used to explain the toxicity, mutagenicity, or anticancer properties of some metals.

Mutations   Any change in the nucleotide sequence or in the chemical composition of a particular mononucleotide will upset the replication pattern and alter the structure of the protein derived from that template. Occasional errors in the normal replication process can generate defective DNA or RNA. Moreover, living cells are constantly exposed to environmental influences that can damage the polynucleotides. Cosmic-ray bombardment may cause chain cleavage. Peroxides formed within the cell, following cosmic ray, X-ray, or ultraviolet irradiation, may modify individual nucleotide residues or crosslink the chains. Many chemical reagents can react with DNA or RNA.

Simple cleavage or crosslinking of the polynucleotide chains may not constitute a permanent mutation since enzymes exist that can rejoin the ends of cleaved chains and can break crosslinks. Furthermore, the replacement in the DNA double helix of one nucleotide complementary pair by another may be a relatively benign change. Ultimately, it could result in the synthesis of a protein molecule that has one incorrect amino acid residue in the chain—possibly a harmless modification or one that could be beneficial in a changing environment.

However, if an *extra* nucleotide residue becomes incorporated into the DNA chain or if a nucleotide is missing, the DNA code will be severely misread beyond this point. Hence, proteins synthesized from such mutants will interfere with the biochemical mechanisms of the cell, causing death of the cell or malignant changes.

Hydrolysis of polynucleotides   DNA and RNA are degraded by specific enzymes, and such degradations are used to analyze the base sequence in polynucleotides. However, nonenzymatic hydrolysis also yields valuable structural information. For example, the hydrolysis of DNA with dilute acid causes removal of the side group bases without cleavage of the backbone bonds. DNA is stable to basic hydrolysis, but RNA can be hydrolyzed by base. The use of enzymatic and nonenzymatic hydrolysis as a sequencing technique for polynucleotides is a more complex problem than, for instance, the sequencing of a protein, because the amount of *detail* that can be derived from the hydrosylate is much less. Only five principal bases are involved in polynucleotide structures, compared to the 20 or so amino acids found in proteins.

BIOLOGICAL SYNTHESIS OF POLYESTERS

Although polyesters are normally produced by conventional polycondensation techniques (Chapter 2), it is known that certain microorganisms can synthesize polyesters from a variety of aliphatic hydrocarbons and aliphatic carboxylic acids. The polyesters are deposited within the cell as an intracellular energy and carbon source storage material.[1-3] As an example, *Pseudomonas oleovorans* can produce poly(3-hydroxyalkanoates from octane, nonanoic, or 10-undecanoic acid (equation 24).[4]

$$
\begin{array}{c}
CH_3 \\
| \\
(CH_2)_5 \\
| \\
CH_2CH_2COOH
\end{array}
\xrightarrow{\textit{P. oleovorans}}
\left[
\begin{array}{c}
CH_3 \\
| \\
(CH_2)_5 \quad\quad\; O \\
| \qquad\quad\; \| \\
O\!-\!CH\!-\!CH_2\!-\!C
\end{array}
\right]_n
\tag{24}
$$

The biosynthesis also occurs when a functional group such as cyano, halogeno, olefinic, aryl, or aryloxy is present at the terminal carbon atom. For example, 1-chloro-octane can be used as a carbon source, together with octane, to yield copolymers with C_5H_{11}, $C_5H_{10}Cl$, C_3H_7, and C_3H_6Cl side chains. The mi-

[1]Anderson, J. A., and Dawes, E. A., *Microbiol. Rev.*, **1990**, *54*, 450.
[2]Doi, Y., and Abe, C., *Macromolecules*, **1990**, *23*, 3705.
[3]Kim, Y.-B., Lenz, R. W., and Fuller, R. C., *Polymer J.*, **1992**, *24*, 1852.
[4]Kim, Y.-B., Rhee, Y.-H., Lenz, R. W., and Fuller, R. C., *Polymer J.*, **1997**, *29*, 894.

croorganism has a similar response to nonanoic acid and undecanoic acid, and gives polymers with M_n values between 50,000 and 70,000, with polydispersity indices between 2 and 3. The same number of repeating units appears to be assembled irrespective of the side chain. The polymers generally are elastomers with T_g's below $-20°C$ and melting temperatures below $60°C$. However, aryl groups in the side chains raise the T_g value into the $0°$ to $10°C$ range. The properties can be varied by preparing copolymers, for example hydroxybutynol-δs-co-valerates.

An advantage of polyesters produced in this way is that they can be degraded by microorganisms in the environment. Hence, they are of interest as environmentally friendly polymers for use as degradable containers, wrapping materials, matrices for the controlled release of herbicides, and as bioerodible medical materials.

STUDY QUESTIONS

1. Cellulose is one of the few polymers that form the starting point for a "substitutive" route to other derivatives. Suggest other reactions, not mentioned in this chapter, that might be used to modify cellulose and broaden its usefulness.

2. In your opinion, could starch be used as a starting point for the synthesis of useful, nonfood polymeric derivatives by the use of substitutive techniques?

3. Taking into account the helical structure of amylose (Figure 8.1) and its capacity to accommodate iodine molecules, what other small molecules or polymers might be induced to occupy the helical channels? What experimental conditions can you devise in order to make such adducts?

4. What advantages or disadvantages can you foresee in the introduction into a synthetic polyamide system of small amounts of histidine, lysine, aspartic acid, cysteine, or methionine units as copolymer residues?

5. Why does a conventional synthetic polymer, such as Nylon 66, occupy an extended zigzag conformation in the solid state, when α-keratins form a helical array?

6. Explain why the steam pressing of cotton or wool textiles is used to remove or introduce creases. Speculate on the reactions that take place when human hair is "permanently waved."

7. In your opinion, could synthetic macromolecules be constructed that would possess the intricately folded structures seen in globular proteins? If so, how would such polymers be synthesized? What uses can you anticipate for such materials?

8. What functions are served by the protein chains in myoglobin and hemoglobin? To what extent might these chains be modified without loss of the oxygen-carrying ability of the system? Could synthetic macromolecules be used in place of the protein chains?

9. What other polymers might be used as the stationary phase in the Merrifield synthesis?

10. A number of reactions of proteins are mentioned in this chapter. Discuss the possible uses of these reactions to modify synthetic polyamides.

11. Discuss the prospect that synthetic polymers might be prepared that could function as templates (possibly stereoregular templates) for the synthesis of new polymer molecules in a manner reminiscent of polynucleotide reactions.

12. Proteins, with repeat unit [—CHR—CO—NH—], can be considered members of the aliphatic Nylon series [—(CH$_2$)$_n$—CO—NH—], specifically substituted types of Nylon-2. Why are the properties of proteins generally very different from those of more typical Nylons, such as Nylon-6?

SUGGESTIONS FOR FURTHER READING

ALLCOCK, H. R., and KWON, S., "Covalent Linkage of Proteins to Surface-Modified Poly(organophosphazenes): Immobilization of Glucose-6-Phosphate Dehydrogenase and Trypsin," *Macromolecules*, **1986,** *19,* 1502.

ALLCOCK, H. R., PUCHER, S. R., and VISSCHER, K. B., "The Activity of Urea Amidohydrolase Immobilized Within Poly[dl(methoxyethoxyethoxy)-phosphazene] Hydrogels," *Biomaterials*, **1994,** *15,* 502.

BOHR, H., and BRUNAK, S., (eds.), *Protein Folds—A Distance-Based Approach,* Boca Raton, Fla.: CRC Press, **1966.**

BORMAN, S., "Doing What Comes Unnaturally," *Chem. & Eng. News*, **2001,** *May 7,* 57.

CHOI, M. H., YOON, S. C., and LENZ, R. W., "Production of Poly(3-hydroxybutyric acid) and Poly(4-hydroxybutyric acid) without Subsequent Degradation by Hydrogenophaga Pseudoflava," *Applied and Environmental Microbiology*, **1999,** *65*(4), 1570.

CLELAND, J. F. (ed.), *Protein Folding: In Vivo and In Vitro*, Washington, D. C.: *American Chemical Society, Symp. Ser.,* **1993,** *526* .

DANIEL, J. R., "Cellulose, Structure, and Properties," *Encyclopedia of Polymer Science and Engineering*, Vol. 3, 2nd ed. (H. F. Mark, N. M. Bikales, C. G. Overberger, G. Menges, and J. I. Kroschwitz, Eds.). New York: Wiley, **1985,** p. *90.*

DORING, V., MOOTZ, H. D., NANGLE, L. A., HENDRICKSON, T. L., CRECY-LAGARD, V. D., SCHIMMEL, P., and MARLIERE, P., "Enlarging the Amino Acid Set of Escherichia Coli by Infiltration of the Valine Coding Pathway," *Science*, **2001,** *292*, 501

DUTTON, G. G. S., "Polysaccharides," *Encyclopedia of Polymer Science and Engineering*, Vol. 3, 2nd ed. (H. F. Mark, N. M. Bikales, C. G. Overberger, G. Menges, and J. I. Kroschwitz, Eds.). New York: Wiley, **1985,** p. 87.

FINDLAY, J. B. C., and GEISOW, M. J., (Eds.), *Protein Sequencing: A Practical Approach.* Oxford; New York: IRL Press, **1988.**

FOSTER, L. J. R., FULLER, R. C., and LENZ, R. W., "Activities of Extracellular and Intracellular Depolymerases of Polyhydroxyalkanoates," *Hydrogels and Biodegradable Polymers for Bioapplications*, **1996,** *627,* 68.

GRUBB, D. T., and JELINSKI, L. W., "Fiber Morphology of Spider Silk: The Effects of Tensile Deformation," *Macromolecules*, **1997,** *30*, 2860.

HAASE, W. C., and SEEBERGER, P. H., "Recent Progress in Polymer-Supported Synthesis of Oligosaccharides and Carbohydrate Libraries," *Current Organic Chemistry*, **2000,** *4*(5), 481.

HARTMEIER, W., *Immobilized Biocatalysts—An Introduction*, Heidelberg; New York: Springer-Verlag, **1986.**

KAWAGOE, Y., and DELMER, D. P., "Recent Progress in the Field of Cellulose Synthesis," *Trends in Glycoscience and Glycotechnology*, **1988,** *10*(54), 291.

KINTER, M., and SHERMAN, N. E., *Protein Sequencing and Identification Using Tandem Mass Spectrometry*. New York: Wiley-Interscience, **2000.**

LEHNINGER, A. L., NELSON, D. L., and COX, M. M., *Principles of Biochemistry*, 3rd. ed. New York: Worth Publishing, **2000.**

MANECKE, G., and SCHLIISEN, J., "Immobilized Enzymes," in *Polymeric Drugs* (L. G. Donaruma and O. Vogl, eds.). New York: Academic Press, **1978**, p. 39.

Materials Research Society Bulletin, "Reprocessing Paper and Wood-Based Materials," (multiple articles and authors), **February 1994,** *19*(2).

MERZ, K. M., Jr., and LeGRAND, S. M., (Eds.), *The Protein Folding and Tertiary Structure Prediction*. Boston: Birkhauser, **1994.**

MERRIFIELD, B., "Solid Phase Synthesis," *Science*, **1986,** *232,* 241.

O'BRIEN, J. P., FAHNSTOCK, S. R., TERMONIA, Y., GARDNER, K. H., "Nylons from Nature: Synthetic Analogs to Spider Silk," *Advanced Materials*, **1998,** *10*(15), 1185.

POLLACK, A., BLUMENFELD, H., WAX, H. M., BAUGHN, R. L., and WHITESIDES, G. M., "Enzyme Immobilization by Condensation into Crosslinked Polyacrylamide Gels," *J. Am. Chem. Soc.,* **1980,** *102,* 6324.

SMITH, J. B. (ed.), *Protein Sequencing Protocols*. Totowa, N. J.: Humana Press, **1997.**

STEWART, J. M., "Polymer-Supported Synthesis and Degradation of Peptides," in *Polymer-Supported Reactions in Organic Synthesis* (P. Hodge and D. C. Sherrington, eds.). New York: Wiley, **1980,** p. 343.

TIRRELL, J. G., and TIRRELL, D. A., "Synthesis of Biopolymers, Polyesters, and Polynucleotides," *Current Opinion in Solid State & Mater. Science*, **1996,** *1*(3), 407.

TISCHER, W., and WEDEKIND, F., "Immobilized Enzymes: Methods and Applications," *Top. Curr. Chem.*, **1999,** *200,* 95.

VLASSOV, V. V., VLASSOVA, V. V., and PAUTOVA, L. V., "Oligonucelotides and Polynucleotides as Biologically-Active Compounds," *Prog. in Nucleic Acid Res. and Molec. Bio.*, **1997,** *57,* 95.

WANG, L., BROCK, A., HERBERICH, B., and SCHULTZ, P. G., "Expanding the Genetic Code of Escherichia Coli," *Science*, **2001,** *292,* 498.

WHITESIDES, G. M., and WONG, C.-H., "Enzymes in Organic Synthesis," *Angew. Chem.*, **1985,** *97,* 617.

9

Inorganic Elements
in Polymers

REASONS FOR THE INCORPORATION OF INORGANIC ELEMENTS INTO POLYMERS

Most of the polymers that have been synthesized or studied during the past 50 years are organic polymers derived from petroleum or from living things. There are several reasons for the predominance of carbon-backbone polymers. Low-cost starting materials are available from the petrochemicals industry, and the full scope of organic chemistry has been used to convert these precursors to a broad range of polymers with a wide variety of different properties. Organic polymers are light in weight, resistant to corrosion, and are easily fabricated into useful objects at moderate temperatures. Most of them are also excellent electrical insulators.

Nevertheless, carbon-based polymers do have limitations. Except for some fluorocarbon polymers and a number of aromatic polymers (see Chapter 2), few organic polymers can be heated for prolonged periods above 150°C without either melting or decomposing. Most organic polymers will burn. When decomposition occurs, it usually results from the reaction of the carbon atoms with atmospheric oxygen. Many organic polymers dissolve or swell in hot organic liquids or lubricating oils, or in hydraulic fluids, and for this reason cannot be used in advanced engineering applications. Moreover, few organic polymers remain flexible or rubbery over a wide enough temperature range for them to be useful at both low and high temperatures. In the arctic or in high-flying aircraft this low-temperature hardness and brittleness of organic polymers can create serious hazards. A glaring example occurred when the space shuttle Challenger exploded shortly after liftoff because an O-ring failed, probably because of low temperature embrittlement. There are other applications for which no suitable polymers have yet been found—for example, where prolonged resistance to ultraviolet radiation is essential, in the textile industry where flame-retardant fabrics are needed, and in medicine for the fabrication of artificial organs that will not lead to blood clotting or other undesirable effects. These are some of the reasons why increasing numbers of polymer chemists have been exploring the possibility that the replacement of carbon atoms in polymers by inorganic elements may expand the range of properties and overcome some of the disadvantages.

The logic behind this thinking is illustrated in Figure 9.1. Organic polymers are one of the four main types of "materials." The other three are ceramics, metals, and a wide range of inorganic semiconductors and optical materials.

Ceramics are complicated inorganic systems in which three-dimensional covalently cross-linked "ultrastructures" coexist with linear inorganic chains and crystalline domains. Most ceramics are made by the high-temperature processing of mineralogical silicates, silicon carbide, silicon nitride, or related inorganic

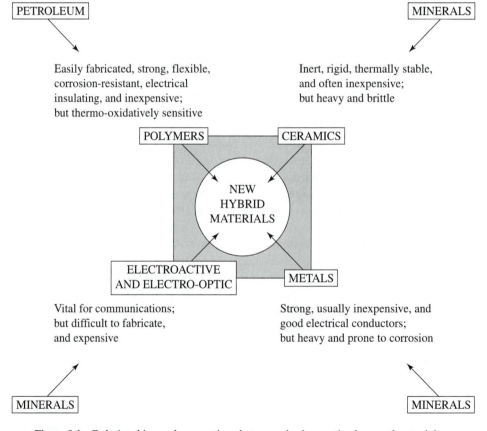

Figure 9.1 Relationships and connections between the four main classes of materials.

species. Ceramics are useful because they are heat and oxidation resistant. Some are exceedingly hard and chemical-resistant. However, they are also heavy and brittle and require large amounts of energy for their preparation and fabrication.

The third group of materials are the metals and their alloys. Metals are tough and ductile and generally have satisfactory high-temperature behavior (but not as good as ceramics). Most metals are also good conductors of electricity. On the other hand, they are heavy, prone to corrosion, and require large amounts of energy for their isolation from minerals and for their fabrication.

The fourth class of materials includes the classical semiconductors like silicon or gallium arsenide, superconductors, and a variety of optical and electro-optical glasses, all of which are based on the inorganic elements. However, most of these materials are expensive to produce, difficult to fabricate, and are relatively fragile.

The important point is that each of these four types of materials have advantages and disadvantages not shared by the other three. Thus, it seems likely that hybrid systems should be accessible that combine the advantages of all four while, at the same time, minimizing the disadvantages. An approach to this objective is to design and synthesize macromolecules that have all the attributes of organic polymers, but which also have the advantages of ceramics, metals, electronic conductors, or optical materials because they contain some of the inorganic elements that give rise to the beneficial properties of these materials (heat stability, electrical conductivity, etc.). The area shown in the middle of the square in Figure 9.1 illustrates the opportunities that exist for molecular design and the synthesis of new materials. Moreover, the synthetic chemist may be able to move the properties anywhere within the square or along its boundaries by intelligent molecular design.

SCOPE OF THE FIELD

In the rest of this chapter, we review some of the options that exist for molecular design and synthesis within the format of Figure 9.1. In general, the approach followed here will be to show how the new territory is being developed, starting from the organic polymer viewpoint. (Texts on ceramic science, metallurgy, or semiconductor science would begin at the other points of the square.)

Here, we will develop the subject at three levels. First, inorganic elements can be incorporated into the *side groups* of an organic polymer. This is perhaps the simplest approach in the sense that the synthesis methods of organic polymer chemistry can still be used. The organic-type properties are usually not changed drastically by this approach, but the presence of, for example, a transition metal organometallic unit in the side group could give rise to catalytic properties not found in the parent polymer. Similarly, a flame-retardant element in the side group might impart flame resistance to an otherwise flammable polymer. On the other hand, dramatic breakthroughs and the discovery of entirely new phenomena would not be expected by this approach.

Second, methods are now available for the incorporation of inorganic elements into the *backbone* of polymer molecules either together with skeletal carbon or without any carbon being present in the skeleton at all. Macromolecules of this type are known as *inorganic polymers* or *inorganic-organic polymers* since most of the known examples have organic side units attached to the inorganic chain. Polymers of this type can be visualized as occupying an area somewhere near the middle of the square in Figure 9.1. Changes in side group or skeletal structure can move the properties toward those of organic polymers, ceramics, or metals. This provides options not available if the synthetic work starts from a system at any of the four corners.

Third, we will consider inorganic-organic polymers that are specifically designed to serve as precursors to ceramics. Polymers of this type have inorganic elements in the skeleton and either inorganic or organic side groups that react at high temperature to form a covalently crosslinked ultrastructure. These are the so-called "pre-ceramic" polymers that are used as precursors to boron nitride, silicon nitride, silicon carbide, or even to silicates or aluminosilicates. The discussion in this section also covers the connection to mineralogical polymers. Thus, the third section concentrates on the area in Figure 9.1 near the upper-right-hand region of the square. The region near the lower-right-hand area of the square, that is, polymers as "metals," is discussed in a separate chapter (Chapter 23), dealing with electrically conducting polymeric materials.

ORGANIC POLYMERS WITH INORGANIC ELEMENTS IN THE SIDE GROUPS

Two different strategies exist for the synthesis of organic polymer chains with inorganic or organometallic units in the side groups. The first approach involves the polymerization of unsaturated monomers that bear an inorganic element in the side unit. An advantage of this method is that the "loading" of the inorganic side units can often be controlled by copolymerization with a conventional organic monomer. In this way, properties such as solubility, glass transition temperature, electroactivity, and so on, can be fine-tuned to generate an optimum combination of characteristics.

The second method requires a preformed organic polymer that provides coordination or substitution functionality for the attachment of the inorganic unit.

Polymerization of Unsaturated Monomers that Bear Inorganic Substituent Groups

The presence of a heteroelement linked to a vinyl or allyl group as in compounds **1** to **7** generally decreases the reactivity of the unsaturated group to free-radical addition polymerization. Various reasons have been proposed to explain this influence, including steric effects and delocalization of a radical into the side group.

However, some of these compounds will undergo low reactivity free-radical copolymerization with monomers such as styrene, methyl methacrylate, or acrylonitrile. Ionic initiation is somewhat speculative since many inorganic units are sensitive to anions or cations. However, compound **2** undergoes cationic polymerization, and **4** polymerizes by a "living" mechanism, following Grignard initiation. An improvement in the polymerizability generally results when the heteroelement is separated from the vinyl unit by an aliphatic or aromatic spacer group.

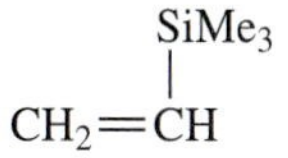

$$
\begin{array}{ccc}
\text{SiMe}_3 & \text{SiMe}_3 & \text{SiMe}_3 \\
| & | & | \\
 & \text{O} & \\
 & | & \\
\text{CH}_2\!=\!\text{CH} & \text{CH}_2\!=\!\text{CH} & \text{CH}_2\!=\!\text{CR} \\
\mathbf{1} & \mathbf{2} & \mathbf{3}
\end{array}
$$

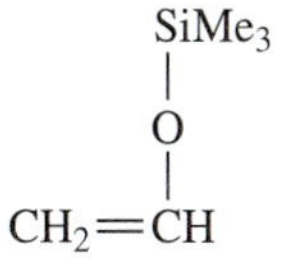

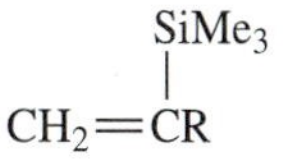

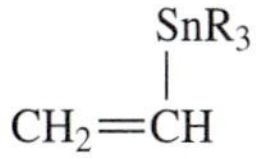

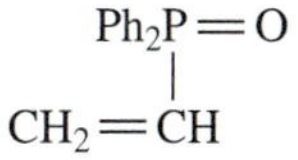

$$
\begin{array}{cc}
\text{SnR}_3 & \text{Ph}_2\text{P}\!=\!\text{O} \\
| & | \\
\text{CH}_2\!=\!\text{CH} & \text{CH}_2\!=\!\text{CH} \\
\mathbf{4} & \mathbf{5}
\end{array}
$$

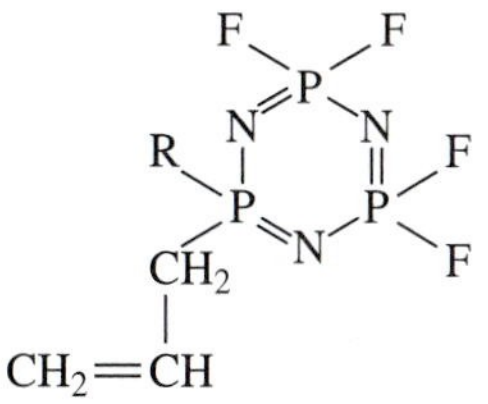

6 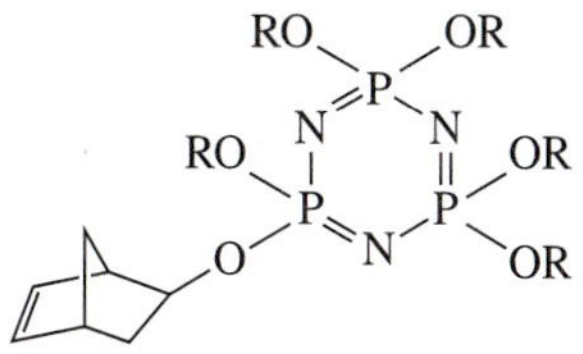   **7**

Compound **6** copolymerizes with vinyl monomers using free-radical initiation. Monomer **7** can be polymerized by ROMP techniques (see Chapter 4) to give polynorbornenes with a wide range of side groups linked to the cyclophosphazene rings.

Vinylferrocene (**8**) polymerizes under free-radical (AIBN), cationic, or Ziegler–Natta conditions. However, it is unreactive to anionic initiation. Peroxides cannot be used as free-radical initiators because they oxidize the ferrocenyl unit. The co-polymerization reactions of **8** with organic vinyl monomers have been studied extensively, and it seems clear that the polymerization reactivity results from strong electron supply from the organometallic unit into the vinyl group. Vinylferrocene polymers are of interest as oxidation-reduction mediators, especially at electrode surfaces (see Chapter 23). Species **9** undergoes radical polymerization. Compound **10** polymerizes on the reducing electrode of an electrochemical cell.

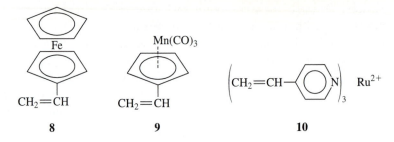

8   **9**   **10**

Reactions of Organic Polymers with Inorganic or Organometallic Reagents

Organic polymers can be synthesized that bear coordination sites for metals (e.g., **11** or **12**), or sites for salt formation (**13**). Inorganic or organometallic compounds may then form strong coordination linkages to these sites. This is an effective strategy for the attachment of inorganic elements to polymers. The method is discussed in more detail in Chapter 23. An alternative approach—via the *substitution* reactions of inorganic or organometallic reagents with organic polymers—is less effective, mainly because of the low reactivity of most organic polymers.

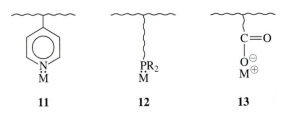

11   **12**   **13**

Finally, as mentioned in Chapter 7, cyclophosphazene rings can be linked to styryldiphenylphosphine to give monomer **14** which homopolymerizes following free-radical initiation or can be copolymerized with styrene or methyl methacrylate. In this case, the phosphorus and nitrogen atoms serve as fire-retardants for the organic components.

14

POLYMERS WITH INORGANIC ELEMENTS IN THE MAIN CHAIN

These comprise by far the largest class of polymers based on the inorganic elements. They cover the range of properties summarized in Figure 9.1, and they represent one of the most promising areas for the future development of polymer chemistry and technology. Polymers in this category are of two fundamentally different types—those based on a skeletal structure of *covalent* bonds between nonmetallic inorganic elements (Si, P, S, O, N, and so on), and those in which the main chain contains *coordinated* metals. The covalent chain structures have received the most attention so far, and these will be the main focus of our attention. However, a few examples of the coordination type will be mentioned to illustrate the possibilities.

POLY(ORGANOSILOXANES) (SILICONE POLYMERS)

Synthesis

Polysiloxane chains are found in glass and in a wide variety of mineralogical silicates. These materials are brittle, ceramic-type substances because individual chains are heavily cross-linked by ionic units or covalent bonds (see a later section of this chapter). Thus, it might be anticipated that the replacement of these charged side groups or covalent cross-links by uncharged organic groups should allow the full molecular flexibility of the polysiloxane chains to become manifest.

This concept underlies the design of organosiloxane polymers. The most widely used of these, poly(dimethylsiloxane), contains chains of alternating silicon and oxygen atoms, with two methyl groups attached to each silicon **(16)**.

Oligo- and poly-organosiloxanes were first produced inadvertently by Kipping in the 1930s during his unsuccessful search for organosilicon analogues of ketones. His idea was to hydrolyze compounds such as R_2SiCl_2 to $R_2Si(OH)_2$, which would then eliminate water to give "silicones," $R_2Si{=}O$. This did not happen. Instead, condensation occurred to give oligomers and polymers of structure $(R_2Si{-}O)_n$. This chemistry led eventually to the start of an industry based on poly(organosiloxanes) by the Corning Glass Company, the Mellon Institute in Pittsburgh, General Electric, Dow Corning, and Union Carbide. The inaccurate name "silicone polymers" is a holdover from the early work. The highest polymers are not produced directly from the hydrolysis of R_2SiCl_2, but rather by ring-opening polymerization of cyclic oligomers such as $(R_2Si{-}O)_4$.

The principle cyclic oligomer employed for ring-opening polymerization is octamethylcyclotetrasiloxane **(15)**, which is itself obtained by the hydrolysis of dimethyldichlorosilane. This hydrolysis yields a mixture of small rings and short chains.

The cyclic tetramer **(15)** is a colorless, oily material. When heated above 100°C with a trace of acid or base, it polymerizes to form a highly viscous liquid or a gum **(18)**. The molecular weight of the polymer may be as high as 2×10^6, which corresponds to over 25,000 silicon–oxygen repeating units per chain. "Silicone rubber" is made from the gum by cross-linking the chains by free-radical-type processes. Cross-linking can also take place via hydrolysis reactions (silicone sealants) or by the addition of Si—H bonds to vinyl side groups. Silicone stopcock grease contains lower-molecular-weight dimethylsiloxane polymers or oligomers plus silica as a filler.

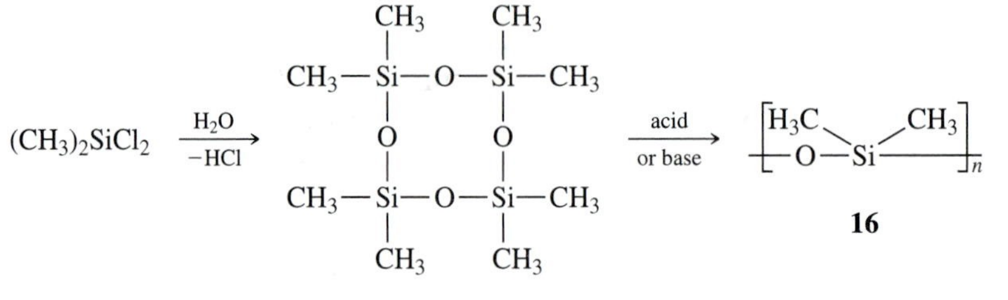

15 → (acid or base) → **16**

The mechanism of siloxane polymerization follows an ionic process. Basic catalysts, such as alkali metal hydroxides or alkoxides, cleave silicon–oxygen skeletal bonds to yield linear species which can function as chain propagation sites (**17**). Insertion of cyclic molecules takes place into the —O⁻K⁺ ionic "bond." In other words, the —O⁻ of the initiated species attacks a ring silicon atom in $[OSi(CH_3)_2]_4$ and the K^+ ion coordinates to the adjacent ring oxygen atom. The mechanism is illustrated by the sequence shown in the conversion of **15** to **18**.

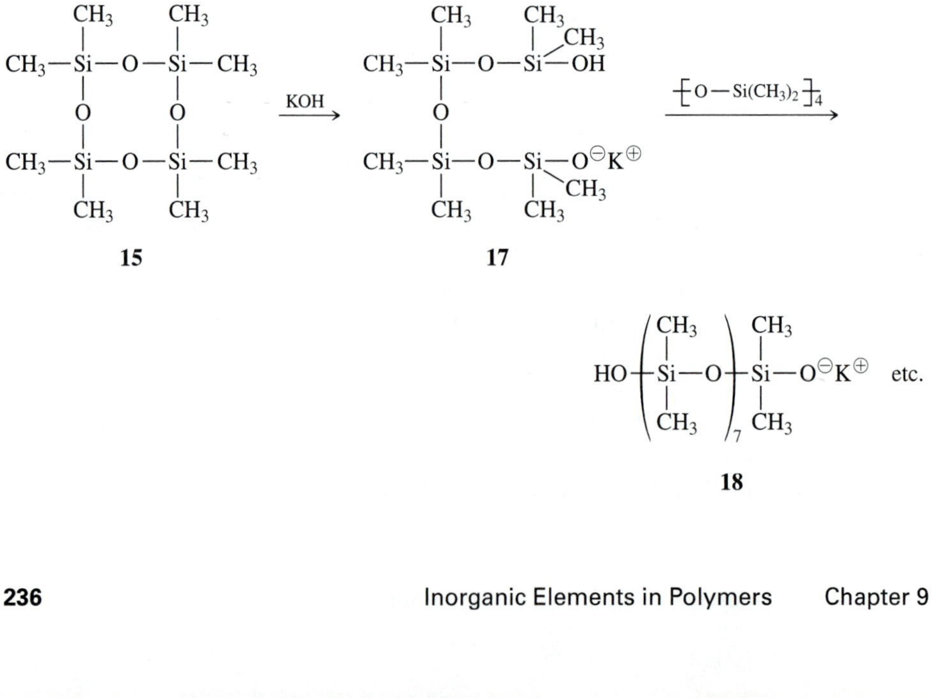

15   **17**

18

The higher the catalyst concentration, the lower will be the average chain length. In practice, it is frequently necessary both to reduce the chain length and provide chain stabilization by the addition of an end-capping reagent to the polymerization system. The most convenient terminator in common use is hexamethyldisiloxane (20). A propagating siloxane chain (19) can attack this reagent to generate an end-capped polymer and a new catalyst molecule. Thus, this reaction functions both as an end-capping procedure and as a chain transfer step, with the polymer chain length being inversely proportional to the amount of hexamethyldisiloxane added.

Catalysis by acidic reagents, such as hydrogen chloride, is less well understood. However, the acid is presumed to initiate polymerization by cleavage of a silicon–oxygen bond. A mixture of cyclic and polymeric homologues is ultimately formed irrespective of whether a basic or an acidic catalyst is employed.

$$\underset{\textbf{19}}{-\!\!\!\wedge\!\!\!\wedge\!\!\!\wedge\!\!-\!\underset{\underset{CH_3}{|}}{\overset{\overset{CH_3}{|}}{Si}}\!-\!O^-K^+} + \underset{\textbf{20}}{(CH_3)_3Si\!-\!O\!-\!Si(CH_3)_3} \longrightarrow$$

$$-\!\!\!\wedge\!\!\!\wedge\!\!\!\wedge\!\!-\!\underset{\underset{CH_3}{|}}{\overset{\overset{CH_3}{|}}{Si}}\!-\!O\!-\!\underset{\underset{CH_3}{|}}{\overset{\overset{CH_3}{|}}{Si}}\!-\!CH_3 + (CH_3)_3Si\!-\!O^{\ominus}K^{\oplus}$$

Although the dimethylsiloxane structure forms the basis of most silicone polymers, other substituent groups have also been introduced. These include vinyl, ethyl, trifluoropropyl, p-cyanoethyl, phenyl, and biphenyl groups. The introduction of specific groups improves the oil resistance, strength, and toughness, flame resistance, or compatibility of the polymer. Cosubstituent groups can be introduced by one of two methods. First, cyclotetrasiloxanes that contain two or more different substituent groups can be synthesized by the cohydrolysis of two different chlorosilanes, such as $PhMeSiCl_2$ and Me_2SiCl_2. Second, copolymerization of two or more different cyclosiloxanes can be carried out. For example, $(O\!-\!SiPh_2)_3$, can be copolymerized with $(O\!-\!SiMePh)_3$ or $(O\!-\!SiMePh)_4$. Although such trimers or tetramers react rapidly to form polymer, the polymers frequently depolymerize to cyclic tetramers as the reaction proceeds. This is a consequence of the thermodynamic problems mentioned in Chapter 10.

Thus, certain limitations exist to the variety of side groups that can be incorporated into a poly(organosiloxanes) via the polymerization process. However, some additional structural diversity is possible by post-polymerization reactions. For example, polysiloxanes with Si-H side-group units are accessible via the polymerization or copolymerization of cyclosiloxanes that bear such side units. The Si-H bonds provide functional sites that can be used to introduce new side groups.

Properties of poly(organosiloxanes)   Perhaps the most surprising feature of dimethylsiloxane high polymers is their flexibility and elasticity over a very broad temperature range. The glass transition temperature is $-130°C$. The temperature range of elasticity for silicone rubber is from -30 or $-40°C$ to $250°C$; the lower temperature marks the onset of crystalization. The flexibility of the bulk polymer is evidence of the ease with which the backbone bonds can undergo torsion. Indeed, organosiloxane polymers are among the most flexible macromolecules known.

Reasons for the unusual properties of poly(organosiloxanes) have been debated for years, and similarities have been noted with some poly(organophosphazenes), $(NPR_2)_n$ (see later), even though the backbone bonds in the two systems are quite different. One similarity is the shortness and high strength of the $Si—O$ and $P=N$ skeletal bonds and the wide angles at oxygen and nitrogen, all of which may reflect donation of lone-pair electrons from oxygen or nitrogen to the adjacent silicon or phosphorus atoms. Another peculiarity of these two classes of polymers is the ability of some examples to form liquid-crystalline phases, in spite of their flexibility.

The high torsional mobility of silicones can be attributed to the lack of charge on the side groups, to the fact that the side groups are attached to *every other* skeletal atom instead of to every skeletal atom, and to the wide-bond angle at skeletal oxygen ($\sim 144°$). In this way they differ structurally from many organic polymers. Thus, there are fewer opportunities for the side groups to "collide" with each other or even to attract or repel each other as the backbone bonds go through their torsional motions. It appears that the extreme flexibility of the siloxane backbone is responsible for the high permeability of silicone rubber to oxygen. Thus, silicone rubber films have been tested in "artificial gill" devices that would extract dissolved oxygen from water for diving purposes.

Other examples include the use of silicone rubber in "soft" lithography, silicone contact lenses, devices for the slow release of drugs, medical catheters, and water-repellent coatings for objects as diverse as semiconductor chips and pacemakers. The water-repellency is also utilized in fluids for protecting the masonry of buildings that are exposed to severe air pollution. More applications have been developed commercially for silicones than for any other inorganic polymer systems.

Because poly(organosiloxanes) repel water strongly they are used in car polishes, in antistick formulations for cooking purposes, and in biomedical devices. For example, artificial heart valves and experimental heart bypass pumps are often fabricated from silicone rubber because the polymer has a lower tendency than most organic polymers to trigger the clotting of blood or to irritate tissues (see Chapter 24).

Organosiloxane ladder polymers It has been indicated that one of the motivations for the development of polymers with inorganic backbones was the belief that these materials would be more stable at high temperatures than organic polymers. Although poly(organosiloxanes) are certainly resistant to oxidation at temperatures up to 200°C, they suffer from one drawback—at temperatures above 250°C the siloxane chains break down to form rings, and the advantageous properties of the polymer are eventually lost. The depolymerization is similar to that observed with polymeric sulfur (see later).

One solution to this problem is to design polymers that resemble the amphiboles, or double-chain silicates. Nonionic analogues of such polymers were first made with phenyl groups in place of the charged side oxygen atoms by the hydrolysis of phenyltrichlorosilane, $C_6H_5SiCl_3$. The resultant materials are called *silicone ladder polymers* (**21**) or poly(phenylsesquisiloxanes). As might be expected, the double-chain structure restricts the mobility of the silicon–oxygen bonds in the backbone and the polymers are high-melting, nonelastomeric materials. However, when dissolved in organic solvents they yield viscous solutions. Moreover, silicone ladder polymers swelled by the addition of small amounts of solvents can be stretched and oriented. Phenylsilicone ladder polymers remain stable up to a temperature of 300°C.

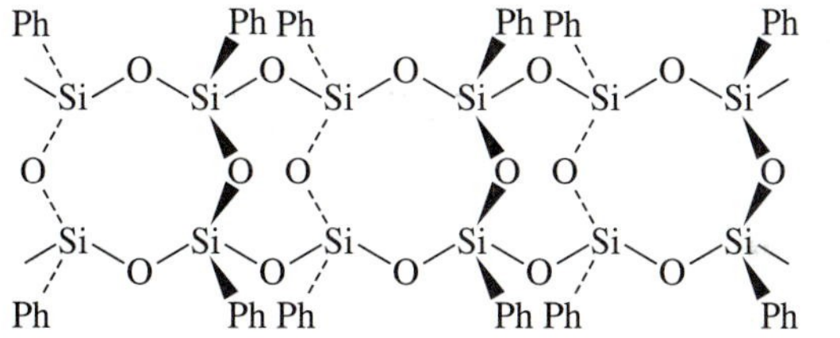

21

SILOXANE–ARYLENE AND SILOXANE–CARBORANE POLYMERS

The thermal instability of linear organosiloxane polymers has prevented their use in a number of severe-environment applications. For this reason, attempts have been made to modify the poly(organosiloxane) structure in ways that will retain the advantageous chain flexibility and yet will raise the thermal stability. One approach has been to introduce arylene rings or carborane cages into the skeleton to retard cyclization and depolymerization. An example is shown in the formation of **22**. This polymer is stable at temperatures up to 450°C and, although it has a higher glass transition temperature (-25°C) than poly(dimethylsiloxane) (-126°C), it is still an elastomer.

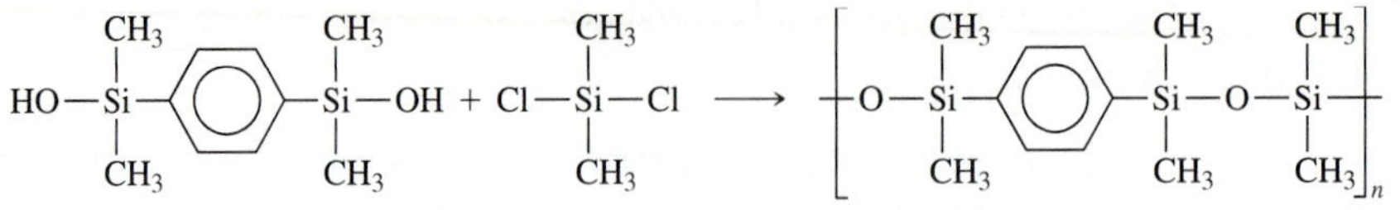

22

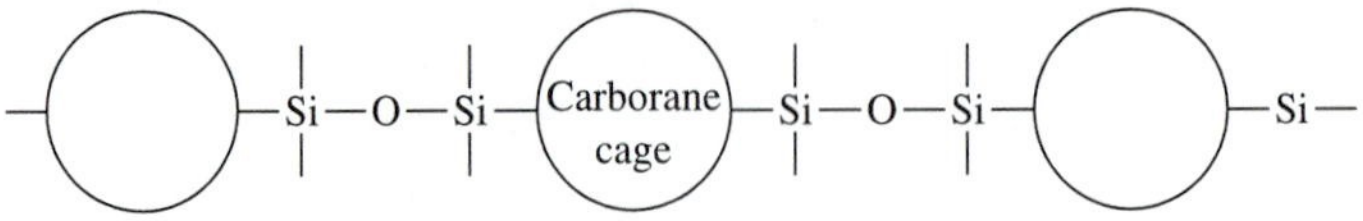

23

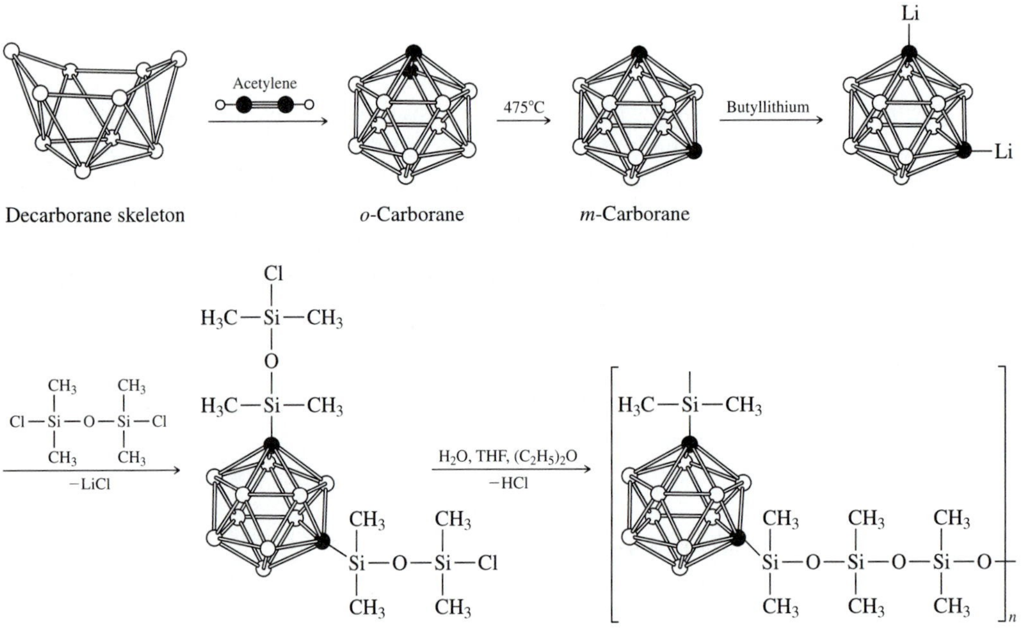

Decarborane skeleton   *o*-Carborane   *m*-Carborane

Scheme 1

Carboranes are cage-type molecules with a framework of boron and carbon atoms. *meta-Carborane* (shown in Scheme 1) is an example of this class of compounds. Each boron (open circle) and carbon (black circle) atom bears a hydrogen atom.

Carboranes have a unique stability that results in part from the fact that they are "electron sinks." In other words, the bonding in carboranes is such that the electrons are free to move widely over the whole cage. As a consequence, a carborane cage can stabilize adjacent units bonded to it, such as, for example, a siloxane chain. For this reason, polymers have been developed that contain carborane cages linked together through one or more siloxane bridging units (**23**). They can be synthesized by the route shown in Scheme 1.

PHTHALOCYANINE "SHISHKEBAB" POLYMERS

Phthalocyanines are macrocyclic rings that form exceedingly stable complexes with metals or metalloids, as shown in **24**. Silicon, germanium, or tin diol units can be coordinated into the ring, and the diol functionality may then be condensed to a stacked phthalocyanine polysiloxane etc. chain, as shown in **25**. Polymers of this structure dissolve in strong acids and can be spun into fibers. When doped with iodine, they conduct electricity (see Chapter 23).

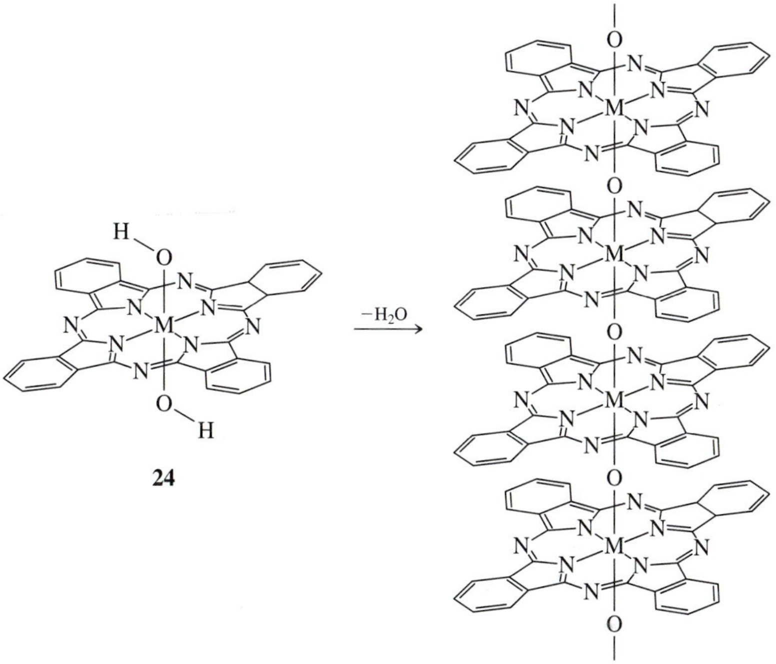

24

$\xrightarrow{-H_2O}$

25

where M = Si, Ge, Sn

POLYORGANOSILANES AND POLYCARBOSILANES

A relatively new class of silicon-containing polymers are polyorganosilanes—macromolecules that contain a backbone of Si—Si bonds, with two organic groups attached to each silicon. A typical example of a polysilane is shown in **26**. Such polymers are synthesized by the action of an alkali metal on a diorganodichlorosilane. Typically, cyclic oligomers, such as $(SiR_2)_6$, are formed most readily, but heating of these leads to the formation of higher oligomers and polymers. Poly(dimethylsilane) (**26**) is an insoluble, infusible, and intractible material. However, it is of interest as a preceramic polymer for conversion to silicon carbide. This aspect is considered in more detail in a later section.

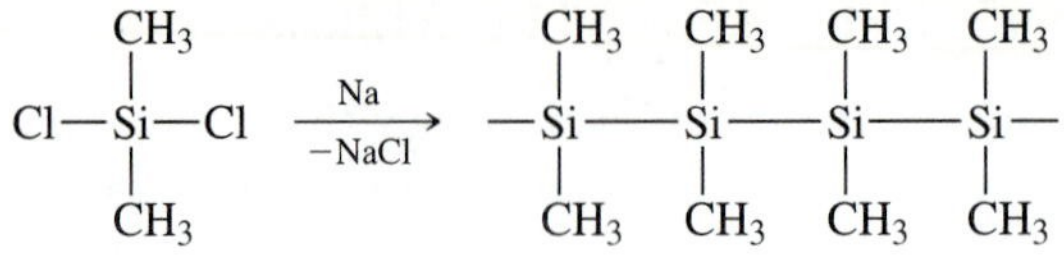

CH₃ structures shown as image.

$$\underset{\overset{|}{CH_3}}{\overset{\overset{CH_3}{|}}{Cl-Si-Cl}} \xrightarrow[-NaCl]{Na} \underset{\overset{|}{CH_3}}{\overset{\overset{CH_3}{|}}{-Si}}\underset{\overset{|}{CH_3}}{\overset{\overset{CH_3}{|}}{-Si}}\underset{\overset{|}{CH_3}}{\overset{\overset{CH_3}{|}}{-Si}}\underset{\overset{|}{CH_3}}{\overset{\overset{CH_3}{|}}{-Si-}}$$

26

Soluble polysilanes can be prepared if phenyl side groups as well as methyl groups are introduced into the molecular structure. Thus the co-condensation of the two monomers **27** and **28** gives the soluble and meltable polymer shown as **29**. The same copolymerization process has allowed access to a wide variety of polymers with various alkyl groups, as well as methyl, attached to the chain.

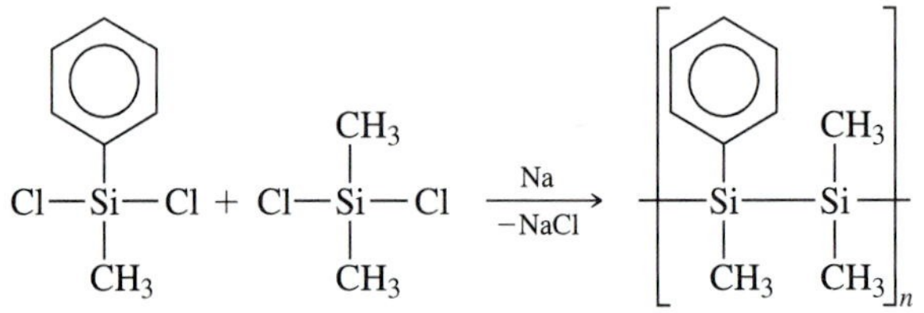

$$Cl-Si-Cl + Cl-Si-Cl \xrightarrow[-NaCl]{Na} \left[\begin{array}{c} Si-Si \end{array}\right]_n$$

27   **28**   **29**

Numerous polysilanes have been prepared with different organic side groups attached to the skeletal silicon atoms. In addition to methyl groups, a range of alkyl groups, from ethyl to octyl, have been incorporated, with the properties changing with different side units. For example, n-hexyl groups give a polymer with a T_g of $-75°C$, while n-propyl side groups increase the T_g to $+25°C$, and cyclohexyl groups increase it to $+120°C$. The mixed-substituent polymer, $(SiPhMe)_n$ has a T_g of $+70°C$. Polysilanes are interesting because of their semiconductivity when oxidized and their ease of photolytic Si—Si bond cleavage when irradiated with ultraviolet light. The spectral absorption of polysilanes varies with the length of the organic side groups in ways that suggest that the conformation of the main chain is altered by the steric characteristics of the side units and this, in turn, affects the electron delocalization along the backbone.

Although the main interest in polysilanes stems from their use as ceramic precursors, they have also been found to have unusual electronic spectra and to undergo photolytic cross-linking as well as chain scission. They have been investigated as resist materials for integrated circuits, as photoinitiators, and as photoconductors.

Heating of the polysilanes induces rearrangement reactions in which a methyl carbon atom migrates into the skeleton as shown in the conversion of **30** to **31**. This is not nearly so clean a reaction as is implied by this scheme. Nevertheless, an approximation of this rearrangement appears to be the first step in the conversion of polysilanes to silicon carbide (see later).

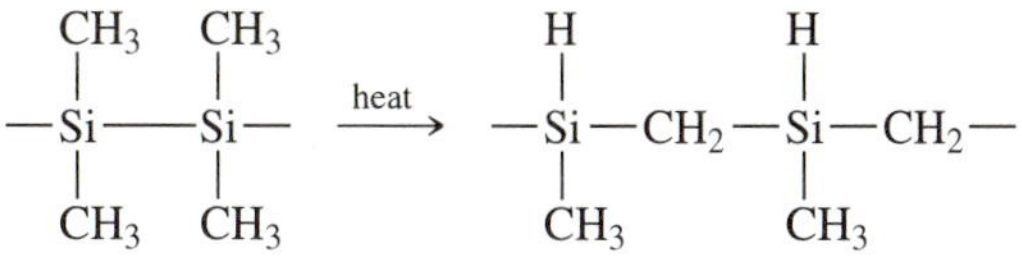

$$\underset{\mathbf{30}}{\overset{\displaystyle \mathrm{CH_3 \quad CH_3}}{\underset{\displaystyle \mathrm{CH_3 \quad CH_3}}{-\mathrm{Si}-\mathrm{Si}-}}} \xrightarrow{\text{heat}} \underset{\mathbf{31}}{\overset{\displaystyle \mathrm{H \qquad\quad H}}{\underset{\displaystyle \mathrm{CH_3 \qquad CH_3}}{-\mathrm{Si}-\mathrm{CH_2}-\mathrm{Si}-\mathrm{CH_2}-}}}$$

However, polycarbosilanes can be prepared by an alternative route in which a starting material such as 32 reacts with magnesium in tetrahydrofuran to give a silacyclobutane (**33**). Pyrolysis of the silacyclobutane at 300°C yields a polycarbosilane. Polymers (or oligomers) of this type are apparently less useful as preceramic precursors because they lack the Si—H bonds.

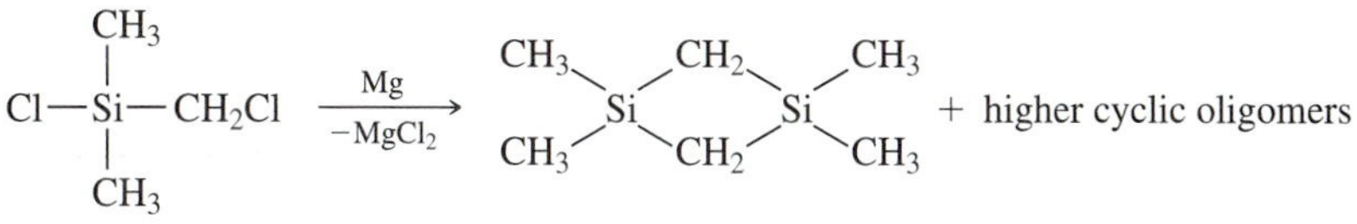

$$\underset{\mathbf{32}}{\overset{\displaystyle \mathrm{CH_3}}{\underset{\displaystyle \mathrm{CH_3}}{\mathrm{Cl}-\mathrm{Si}-\mathrm{CH_2Cl}}}} \xrightarrow[-\mathrm{MgCl_2}]{\mathrm{Mg}} \underset{\mathbf{33}}{\left[\text{silacyclobutane ring}\right]} + \text{higher cyclic oligomers}$$

$$+ \ \mathrm{Cl}\!\left(\!\!\begin{array}{c} \mathrm{CH_3} \\ | \\ \mathrm{Si}-\mathrm{CH_2} \\ | \\ \mathrm{CH_3} \end{array}\!\!\right)_{\!x}\!\!\mathrm{Cl}$$

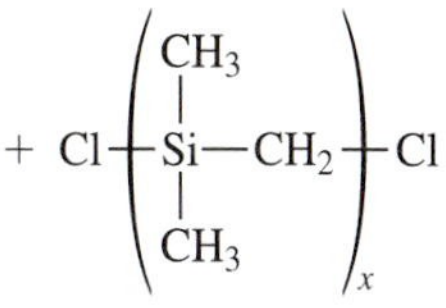

In these properties, polysilanes have more in common with polyacetylene and its derivatives than with polyalkanes such as polyethylene, polypropylene, or polyisobutylene. In some ways, these differences also mimic the contrasting characteristics of semiconductor silicon and insulating diamond.

Dendrimeric silicon-carbon polymers are also shown. These are produced by treatment of $RSiCl_3$ with sodium-potassium alloy under sonication conditions. The products are called polysilynes. These show electrical semiconduction behavior, and can also be pyrolyzed to ceramics.

Germanium analogues of polysilanes have also been synthesized. Specifically, $(n\text{-}Bu_2Ge)_n$ is accessible via the reaction of $n\text{-}Bu_2GeCl_2$ with sodium. The polymer is soluble in organic media. Copolymers that contain germanium and silicon in the chain have also been prepared.

In addition, polystannanes, $(R_2Sn)_n$, are known.[1] Like polysilanes, these too have a delocalized electronic structure in the backbone, and become semiconductors when doped. They are also sensitive to visible light. They are decomposed by oxygen and atmospheric moisture.

[1] Inori, T., Lu, V., Cai, H., and Tilley, T. D., *J. Am. Chem. Soc.*, **1995**, *117*, 9931.

A related class of polymers are the polysilazanes, species with alternating silicon and nitrogen atoms in the skeleton. Because the interest in these oligomers and polymers is almost completely focused on their preceramic character, they are discussed in that context later in this chapter.

PHOSPHAZENE POLYMERS

Some of the most rapid advances in inorganic polymer research and technology are occurring in the field of phosphazene polymers.[1] Polyphosphazene molecules consist of an alternating sequence of phosphorus and nitrogen atoms, with two substituent groups attached to each phosphorus (**34**). The molecular weights of these polymers are generally very high, with 15,000 or more repeating units being typical for most examples.

$$\left[\begin{array}{c} R \\ N=P \end{array} \begin{array}{c} R \\ \end{array} \right]_n$$

34

More than 700 different polyphosphazenes have been synthesized and studied, which makes this area by far the largest and most diverse in the field of inorganic–organic polymers. This diversity is caused by the broad range of different side groups that can be linked to the polyphosphazene skeleton.

The range of properties generated by the introduction of different side groups (R) is perhaps broader than in any other polymer system. Thus, different polyphosphazenes are low-temperature elastomers, biomaterials, polymeric drugs, hydrogels, liquid crystalline materials, nonburning fibers, or electrical semiconductors. In many respects, the polyphosphazenes epitomize the opportunities to be found through research work within the central area shown in Figure 9.1. Different side groups attached to the polymer chain move the properties toward those of organic polymers, ceramics, metals, or semiconductors. This is accomplished by an exceedingly fertile series of synthesis reactions, based on (1) ring-opening polymerizations, (2) macromolecular substitution reactions, and (3) direct synthesis by condensation reactions.

Synthesis via Ring-Opening Polymerization

The main method for poly(organophosphazene) synthesis is via a macromolecular substitution process carried out on reactive high polymeric polyphosphazene intermediates. The macromolecular intermediates are made by ring-opening polymerization. It has been known for more than a century that phosphorus pentachloride and ammonium chloride react to form a series of cyclic inorganic compounds of formula $(NPCl_2)_n$, where n is 3, 4, 5, 6, ... The principal product is the cyclic trimer (**35**), which is called hexachlorocyclotriphosphazene or "phosphonitrilic chloride trimer." This compound is a white, crystalline solid which

[1]Allcock, H. R., *Chemistry and Application of Polyphosphazenes,* New York, John Wiley, **2003**.

melts at 114°C and is soluble in organic solvents. When heated at 230 to 300°C in an evacuated glass tube or reactor, it polymerizes to a transparent rubbery high polymer (**36**). The polymer is called poly(dichlorophosphazene). It is also known as "inorganic rubber." The polymer may contain 15,000 or more repeating units, with a molecular weight of over 2 million. It is the principal macromolecular intermediate used in polyphosphazene synthesis.

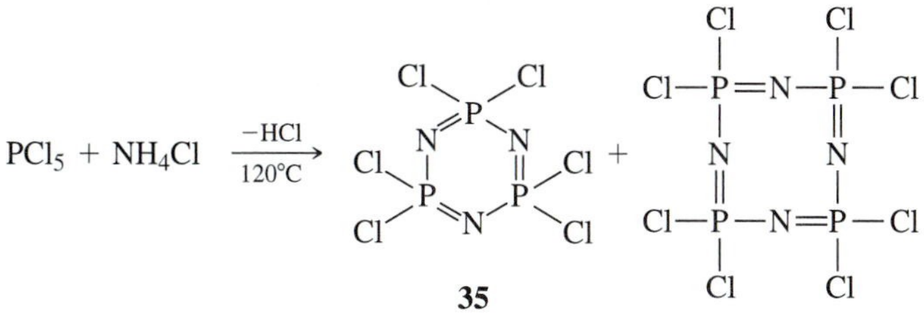

$$PCl_5 + NH_4Cl \xrightarrow[120°C]{-HCl}$$

35

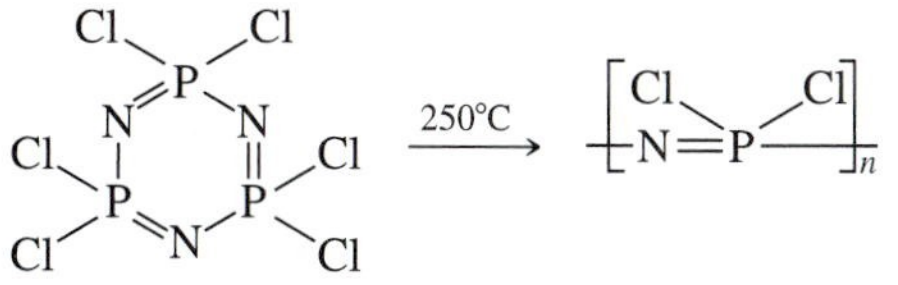

The mechanism of this polymerization is still not fully understood. Strong evidence exists that the mechanism is ionic and that the ionization of a phosphorus–chlorine bond takes place during initiation, as shown in **37**. An ionized species, such as **37**, could, in effect, function as an ionic initiator by attacking the nitrogen atom of another ring to initiate chain polymerization. However, it is also known that traces of water accelerate the polymerization rate, and it is now believed that the mechanism may be catalyzed by traces of moisture or other protonic reagents.

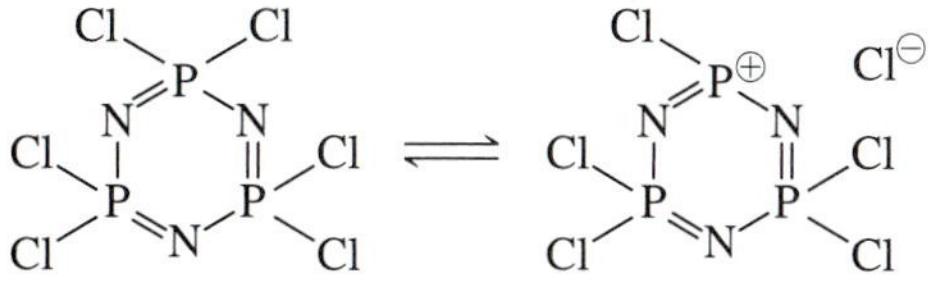

37

Considering that poly(dichlorophosphazene) is made from purely inorganic materials, it is a remarkable compound. In its stress-relaxation behavior, it is a more ideal elastomer than natural rubber. Perhaps more interesting is the observation that it remains rubbery at low temperatures and hardens only when the temperature falls to near the glass transition at −63°C. This behavior is indicative of a high degree of chain mobility.

The polymer formed under nonrigorous polymerization conditions is actually a cross-linked modification of poly(dichlorophosphazene). It swells in organic solvents, but it does not dissolve. It also shows a marked resistance to thermal degradation at temperatures up to 350°C. Poly(dichlorophosphazene) would itself be a valuable technological material were it not for its tendency to react slowly with atmospheric moisture to yield phosphoric acid, ammonia, and hydrochloric acid. During this reaction, the elastomer crumbles to a powder. However, its principal defect in this respect is also its principal attribute. The reactivity to moisture is an indication of the sensitivity of P—Cl bonds to nucleophilic agents. Thus, organic nucleophiles are used to replace the chlorine atoms and generate water-stable polymers. In the presence of organic side groups, the phosphorus–nitrogen skeleton is highly stable to water. Thus, the main route to stable poly(organophosphazenes) actually requires the prior synthesis of a polymer that has a higher side-group substitution reactivity than almost any other known macromolecule.

Poly(dichlorophosphazene) is only one of a number of macromolecular intermediates that have been used in this way. Poly(difluorophosphazene), $(NPF_2)_n$, prepared by the ring-opening polymerization of the cyclic trimer, $(NPF_2)_3$, is also used, especially when organometallic nucleophiles are to be employed in the subsequent polymer substitution reactions. In addition, a range of organo-substituted phosphazene cyclic trimers or tetramers, such as the examples shown in **38** to **40**, have been polymerized thermally, and the polymers are then used as intermediates for replacement of the halogen atoms.

38	**39**	**40**

The Macromolecular Substitution Step

Attempts to perform substitution reactions on organic high polymers are often disappointing because of the low reactivity of the side groups. Moreover, the coiling of a macromolecular chain in solution and adjacent effects along the chain may retard substitution reactions relative to their counterparts at the small molecule level.

These limitations do not apply to the substitution reactions of poly(dichlorophosphazene) and related polymers. The reactivity of the P—Cl or P—F bond is so high that macromolecular substitution reactions are usually rapid and complete. Thus, uncross-linked poly(dichlorophosphazene) readily undergoes halogen replacement when treated in solution with a wide variety of alkoxides or aryloxides, such as sodium ethoxide, sodium trifluoroethoxide, or sodium phenoxide, or with amines, such as aniline or butylamine (Scheme 2). Organometallic reagents, such as organolithium reagents, can be employed to re-

place fluorine in poly(difluorophosphazene) by alkyl or aryl groups, but this is accompanied by chain cleavage.

The remarkable feature of these macromolecular substitution reactions is that, typically, 30,000 chlorine atoms are replaced on each macromolecule with, in most cases, little or no evidence of side reactions such as cleavage of the main chain. This factor places polyphosphazenes apart from nearly all other polymer systems, and is responsible for the wide range of different derivatives that are accessible.

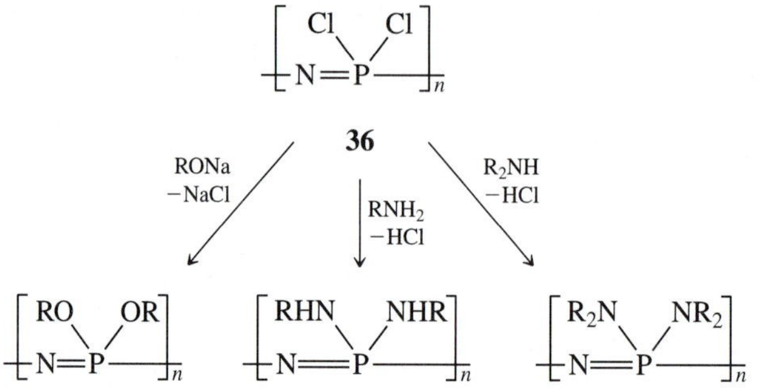

$$\left[\begin{array}{c} Cl \quad Cl \\ N=P \end{array} \right]_n$$

36

RONa / −NaCl

RNH$_2$ / −HCl

R$_2$NH / −HCl

$$\left[\begin{array}{c} RO \quad OR \\ N=P \end{array} \right]_n \qquad \left[\begin{array}{c} RHN \quad NHR \\ N=P \end{array} \right]_n \qquad \left[\begin{array}{c} R_2N \quad NR_2 \\ N=P \end{array} \right]_n$$

Scheme 2

In addition, mixed-substituent polymers can be prepared by sequential or simultaneous reactions of the substrate polymer with two or more different nucleophiles as shown in Scheme 3.

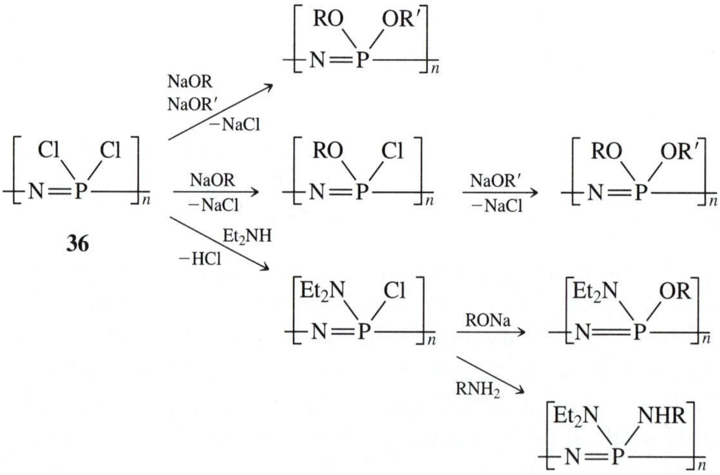

$$\left[\begin{array}{c} RO \quad OR' \\ N=P \end{array} \right]_n$$

NaOR / NaOR' / −NaCl

$$\left[\begin{array}{c} Cl \quad Cl \\ N=P \end{array} \right]_n$$

36

NaOR / −NaCl

$$\left[\begin{array}{c} RO \quad Cl \\ N=P \end{array} \right]_n$$

NaOR' / −NaCl

$$\left[\begin{array}{c} RO \quad OR' \\ N=P \end{array} \right]_n$$

Et$_2$NH / −HCl

$$\left[\begin{array}{c} Et_2N \quad Cl \\ N=P \end{array} \right]_n$$

RONa →

$$\left[\begin{array}{c} Et_2N \quad OR \\ N=P \end{array} \right]_n$$

RNH$_2$ →

$$\left[\begin{array}{c} Et_2N \quad NHR \\ N=P \end{array} \right]_n$$

Scheme 3

More than 250 different reagents have been used to introduce a wide variety of side groups into polyphosphazenes. The side groups introduced range from linear alkoxides, fluoroalkoxides, aryloxides including those with a wide range of

substituents on the phenyl ring, polyaromatic aryloxides, aromatic azo compounds, steroids, aromatic and aliphatic amines, amino acid esters, oligopeptides, glucose, adamantyl units, carboranes, borazine residues, and numerous organometallic groups, such as metal carbonyls and metallocenes. Moreover, many of these side groups have been subjected to secondary reactions to deprotect or introduce functional units that might not survive the primary macromolecular substitution step.

These same types of substitution reactions can be employed if the polymeric intermediate is one that was obtained from a cyclic trimer that has organic or organometallic as well as halogeno side groups, as shown in the conversion of **39** to **41** and **42**.

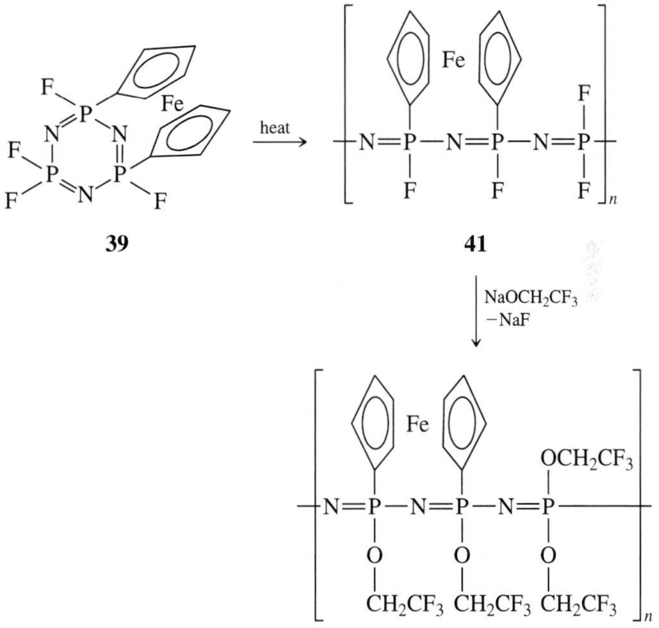

The important point is that the macromolecular substitutive mode of synthesis, coupled with the ability to polymerize trimers such as **38** to **40**, allows access to an enormous range of derivative polymers, in which the properties can be varied by subtle or gross changes in side group type, or by changes in the ratio of different side groups. Coupled with this are the opportunities for further structural modification by tertiary reactions carried out on the organic or organometallic side groups after they are attached to the polymer chain.

A few of the possibilities for macromolecular design and property variations are illustrated by the following examples. When poly(dichlorophosphazene) reacts with sodium trifluoroethoxide in a tetrahydrofuran–benzene medium, the chlorine atoms are replaced by trifluoroethoxy groups to yield poly[bis(trifluoroethoxy)phosphazene] (**43**) (Figure 9.2). This polymer can be solution-cast from

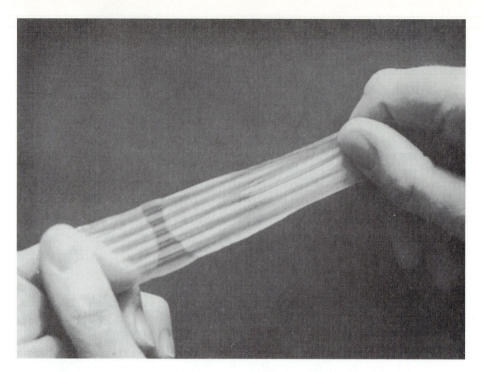

Figure 9.2 Oriented film of poly[bis(trifluoroethoxy)phosphazene], [NP(OCH$_2$CF$_3$)$_2$]$_n$.

$$\left[\begin{matrix} & \text{OCH}_2\text{CF}_3 \\ & | \\ \text{N}{=}\text{P} & \\ & | \\ & \text{OCH}_2\text{CF}_3 \end{matrix}\right]_n$$

43

acetone or methyl ethyl ketone solutions to give colorless, opalescent, flexible films, which superficially resemble polyethylene films in appearance. Solutions of the polymer can be extruded into a nonsolvent to yield flexible, slightly elastic fibers. The polymer has a low glass transition temperature ($-66°$C), and it remains flexible from this temperature up to its melting point at $242°$C. The presence of the fluorinated side groups makes the polymer highly water-repellent, more so in fact than Teflon or silicones. The high crystallinity of this polymer is responsible for the opalescent appearance. The crystallinity can be enhanced by stretching and orientation to yield strong fibers or films. In this form, the polymer yields excellent X-ray diffraction patterns (see Chapter 19). Polymer crystallinity results from the regular arrangement of substituent groups along a chain. The absence of crystallinity is often associated with rubbery or elastomeric properties.

Thus, the random introduction of two or more different substituent groups should favor the appearance of elastomeric properties. This does indeed happen. The reaction of poly(dichlorophosphazene) with a mixture of sodium trifluoroethoxide and a longer-chain fluoroalkoxide yields a new class of low-temperature, solvent-resistant elastomers (Figure 9.3).

Other examples of property variations are found with the water solubility imparted by methylamino, glucosyl, alkoxy ether, and glyceryl side groups, and by the biological activity of polyphosphazenes that bear steroidal, antibacterial, or peptide side groups. Some of these structure–property relationships are shown in Table 9.1.

Condensation Synthesis of Polyphosphazenes

Alternative methods for the synthesis or poly(organophosphazenes) have been developed based on the condensation polymerization of phosphoranimines. These are compounds of general formula, $Me_3SiN = PR_3$, where R is a halogen atom or an organic group. The overall reaction is shown in the following equation.

$$Me_3Si-N{=}\underset{\underset{R^3}{|}}{\overset{\overset{R^1}{|}}{P}}-R^2 \xrightarrow{-Me_3SiR^2} \left[N{=}\underset{\underset{R^3}{|}}{\overset{\overset{R^1}{|}}{P}} \right]_n$$

Three variations of this process have been developed—a room-temperature living cationic process, a high temperature uncatalyzed reaction, and a moderate temperature anionic polymerization. A related reaction involves the thermal polymerization of $Cl_2P(O)N = PCl_3$, with loss of $P(O)Cl_3$, to give $(NPCl_2)_n$.

Living cationic polymerization   Treatment of a phosphoranimine with a trace of PCl_5 at room temperature in solution or in the undiluted state brings about polymerization by a living process. Although this technique can be used to produce *organo*phosphazenes directly (R^1 and R^3 = organic groups), its main use so far has been to produce poly(dichlorophosphazene) for subsequent macromolecular substitution reactions (Scheme 4). It differs from the ring-opening polymerization of $(NPCl_2)_3$ because the polymer chain length can be controlled through the monomer to initiator ratio, the molecular weight distribution is narrow, and because the living nature of the process allows phosphazene–phosphazene block copolymers to be prepared with two or more different side groups. Perhaps more important is the access it provides to telechelic polyphosphazenes that can be coupled to functional end groups on organic telechelic polymers to yield phosphazene-organic block copolymers. Block copolymers with polystyrene, poly(ethylene oxide), and even with with poly(dimethylsiloxane) have been synthesized.

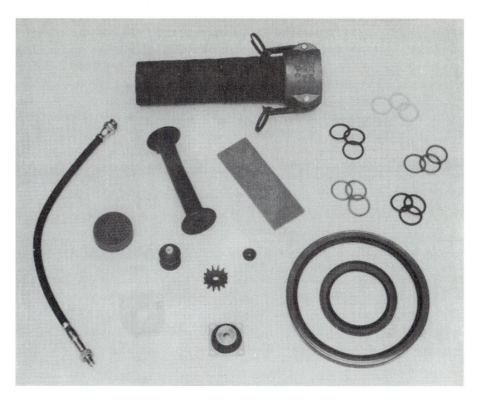

Figure 9.3 Gaskets, pipe, and sheet elastomer made from mixed substituent fluoroalkoxyphosphazene elastomers. (Courtesy of the Firestone Tire and Rubber Company, Akron, Ohio, and Ethyl Corporation, Baton Rouge, Louisiana.)

TABLE 9.1 CHANGES IN PROPERTIES THAT ACCOMPANY SIDE-GROUP VARIATIONS IN POLYPHOSPHAZENES

OC_2H_5	Elastomer
OCH_2CF_3	Hydrophobic, microcrystalline thermoplastic
OCH_2CF_3 plus $OCH_2(CF_2)_xCF_2H$	Elastomer (low T_g)
OC_6H_5	Hydrophobic, microcrystalline thermoplastic
OC_6H_5 plus OC_6H_4R	Elastomer
$NHCH_3$	Water-soluble
$OCH_2CH_2OCH_2CH_2OCH_3$	Water-soluble
$OCH_2CH(OH)CH_2OH$	Water-soluble, bioerodable
Glucosyl	Water-soluble
$NHCH_2COOC_2H_5$	Bioerodable
$OC_6H_4P(C_6H_5)_2$	Coordination ligand for transition metals
Ferrocenyl groups	Electrode mediator polymer

Phosphazene Polymers

Initiation

$$Cl_3P{=}NSiMe_3 \xrightarrow[-ClSiMe_3]{2\ PCl_5} Cl_3P{=}N{-}PCl_3^+\ PCl_6^-$$

Propagation

$$Cl_3P{=}N{-}PCl_3^+\ PCl_6^- \xrightarrow[-ClSiMe_3]{Cl_3P{=}NSiMe_3} Cl_3P{=}(PCl_2{=}N)_n{-}PCl_3^+\ PCl_6^-$$

"Living" polymer end unit
initiates more of same monomer
or a different monomer

The ClSiMe$_3$ can be fed back into the monomer synthesis procedure

Scheme 4

Thermal polymerization The second approach involves the heat-induced elimination of Me$_3$SiOCH$_2$CF$_3$ from an organophosphoranimine, as shown in the following equation.

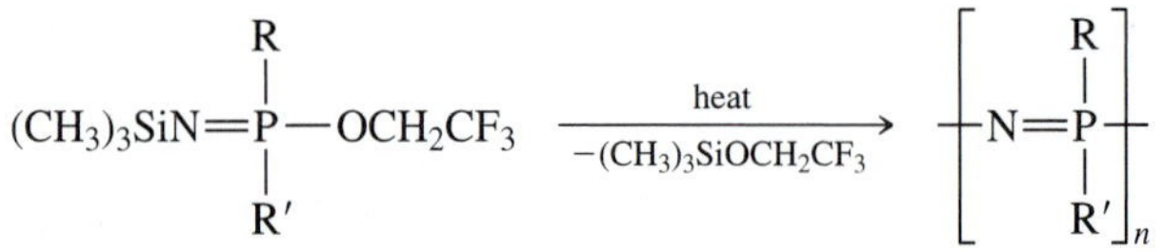

$$(CH_3)_3SiN{=}\underset{\underset{R'}{|}}{\overset{\overset{R}{|}}{P}}{-}OCH_2CF_3 \xrightarrow[-(CH_3)_3SiOCH_2CF_3]{heat} \left[{-}N{=}\underset{\underset{R'}{|}}{\overset{\overset{R}{|}}{P}}{-}\right]_n$$

This method yields poly(organophosphazenes) directly, without the need for a macromolecular substitution step. The resultant polymers, with carbon–phosphorus bonds linking the side groups to the skeleton, are precisely those polymers that are the most difficult to prepare from poly(dichlorophosphazene).

Examples are known in which R and R′ are CH$_3$ or C$_2$H$_5$, where R is CH$_3$ or C$_2$H$_5$, and R′ is Ph. These polymers can be used as macromolecular intermediates since the methyl groups can be lithiated and the —CH$_2$Li side units allowed to react with electrophiles such as organosilicon halides, chlorophosphines, and ferrocene derivatives. The polymer, [NP(CH$_3$)$_2$]$_n$, prepared by this route is a structural analogue of poly(dimethylsiloxane). However, it is not an elastomer, but is more stable to side-group modification.

Anionic polymerization The third approach builds on the second. It makes use of an anionic initiator such as Bu$_4$NF to induce polymerization at lower temperatures, and by a living process. This approach has been used to prepare alkoxy-substituted phosphazene polymers.

Azide decomposition route Finally, an entirely different condensation method has been employed which involves the elimination of nitrogen from a phosphinous azide. These monomers should never be isolated in the pure state because they are explosive but, if prepared and used with great care in solution, they yield poly(arylphosphazenes), which are difficult to prepare by any of the other methods.

It is the synthetic versatility of polyphosphazenes that sets them apart from poly(organosiloxanes). So far no reactive macromolecular intermediates comparable to poly(dichlorophosphazenes) have been found for the polysiloxane system; hence, the substitutive route cannot be used for polysiloxanes. Moreover, it appears that the phosphorus–nitrogen skeletal bond is more stable to reagent attack than is the silicon–oxygen bond. Hence, tertiary reactions on the side groups are possible with poly(organophosphazenes) that would lead to skeletal cleavage with polysiloxanes.

Structural Features of Polyphosphazenes

Two of the many questions that remain to be answered about polyphosphazenes are: (1) What is the electronic structure of the backbone, and (2) Why do polyphosphazenes show such unusual physical properties—low T_g values, elasticity, unusual crystalline transitions, and so on?

The electronic structure of the backbone is not well understood. If a normal covalent framework is assumed for these polymers, three electrons on each skeletal nitrogen atom and one electron on each phosphorus still remain to be accounted for. One explanation (Figure 9.4) is that two of the electrons on each skeletal nitrogen form part of a "lone pair" that is responsible for the basicity and coordination ability of these molecules. The remaining electrons on nitrogen and phosphorus then generate a π-system by overlap of a nitrogen p-orbital with a phosphorus d-orbital (Figure 9.4). This strengthens the backbone bonds while permitting free torsion of these bonds as the p-orbital on each nitrogen "switches" from one phosphorus d-orbital to another. However, repeated attempts to find support for this model by various theoretical techniques have been almost equally divided between investigators who conclude that the phosphorus d-orbitals are involved and those who do not.

The unusual torsional freedom of the backbone bonds (and hence the flexibility and elasticity of many polyphosphazenes) appears to be connected with the

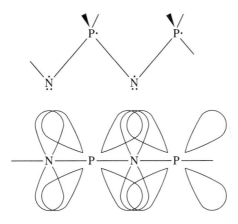

Figure 9.4 Possible bonding arrangements in polyphosphazenes. In the upper diagram the electrons that must be accounted for after the σ-bond framework has been generated from electron pairs are shown. The lower diagram shows how a $d_x - p_x$ system can be visualized as a stabilized excited state by utilization of one electron from each phosphorus and one from each nitrogen atom. The orbital overlap depicted is that of a phosphorus $3d$ orbital with a nitrogen $2p$ orbital.

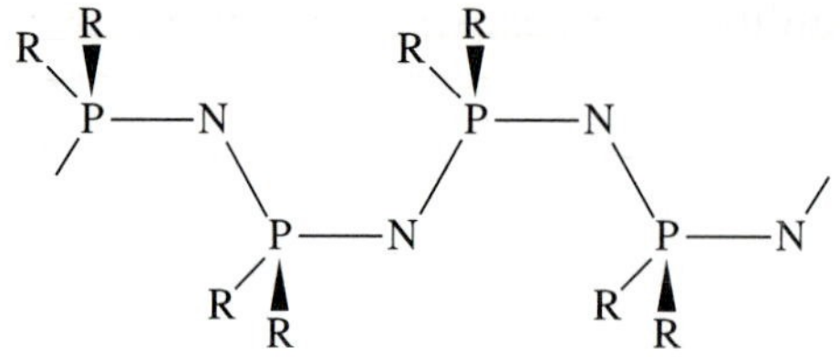

Figure 9.5 The *cis-trans*-planar skeletal arrangement that is characteristic of a number of polyphosphazenes in the solid state. This conformation places the side groups at the greatest distance apart.

wide bond angles at nitrogen, and with the absence of substituent groups on every other skeletal atom. Polyphosphazenes also show a tendency to assume a *cis-trans*-planar (0°, 180°) conformation (Figure 9.5), which represents the arrangement that places the side groups as far away from each other as possible.

Uses of Poly(organophosphazenes)

Organophosphazene elastomers have been developed for use as fuel lines, hoses, gaskets, and O-rings and as nonburning foam rubber articles (Figure 9.3). Most of the current technological applications depend on the oil resistance, nonflammability, and low glass transition temperatures of many phosphazene polymers. Other developments are taking place in the use of polyphosphazene as biomedical polymers that degrade in the body to phosphate, ammonia, and an amino acid (released from the side group).

Other polyphosphazenes are used as solid ionic conductors in experimental rechargeable lithium batteries, as proton conduction membranes in fuel cells, and as high-refractive index optical materials. The fire-resistance of many polyphosphazenes has stimulated interest in their use either alone or as additives to classical organic polymers for use in aerospace and automotive applications.

POLY(CARBO-, THIO-, AND THIONYL-PHOSPHAZENES)

Closely related to polyphosphazenes are polymers that contain phosphorus, nitrogen, and another element in the main chain. Three examples are the poly(carbophosphazenes) (**44**), poly(thiophosphazenes) (**45**), and poly(thionylphosphazenes) (**46**).[1–3]

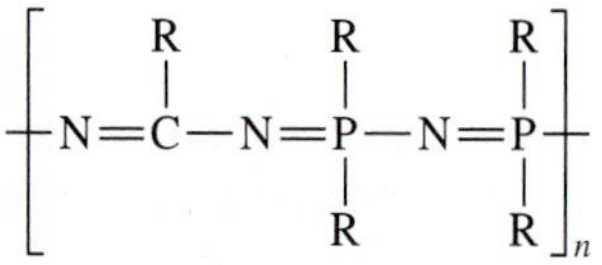

44 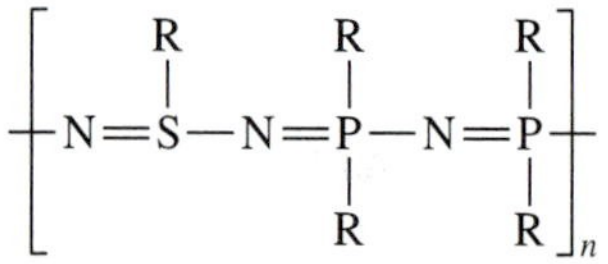   45 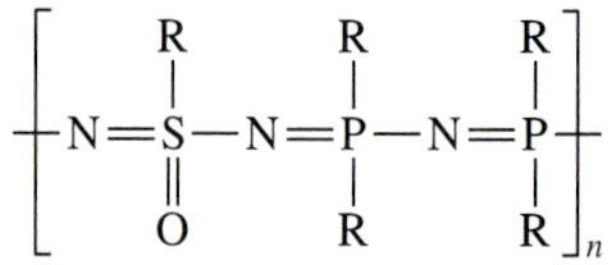   46

[1]Allcock, H. R., Coley, S. M., Manners, I., Renner, G. and Nuyken, O., *Macromolecules.*, **1991**, *24*, 2024.

[2]Allcock, H. R., Dodge, J. A., and Manners, I., *Macromolecules*, **1993**, *26*, 11.

[3]Gates, D. P., Mc Williams, A., and Manners, I., *Macromolecules*, **1998**, *31*, 3494.

All three are synthesized by the ring-opening polymerization of the appropriate chlorocyclic trimer, $N_3P_2CCl_5$, $N_3P_2SCl_5$, or $N_3P_3S(O)Cl_5$, followed by replacement of the halogen atoms by alkoxy, aryloxy, or amino side groups. Changes in the organic nucleophile generate polymers with different properties, although the range of different side groups is not as large as for the classical polyphosphazenes. The carbophosphazenes have higher glass transition temperatures than their classical phosphazene counterparts with the same side groups because the barrier to torsion of a $C=N$ bond is higher than that of a $P=N$ bond. Poly(thiophosphazenes) appear to be as flexible as classical polyphosphazenes. The thionyl (sulfur VI) derivatives (**46**) are more stable than the thio (sulfur IV) species (**45**). All three systems allow regiospecific substitution because the reactivities of C-halogen, S(IV)-halogen, or S(VI)-halogen bonds differ from that of the P-halogen bond.

POLYMERIC SULFUR

One of the earliest covalent inorganic polymers to be made, and still one of the most fascinating, is polymeric sulfur. The stable form of sulfur at room temperature is rhombic sulfur, which contains cyclic molecules with eight sulfur atoms in a ring (Figure 9.6).

Rhombic sulfur is a brittle, crystalline material that melts at 113°C to form a yellow-red liquid. Liquid sulfur has a curious property that above 159°C its viscosity increases as the temperature is raised, contrary to the behavior of nearly

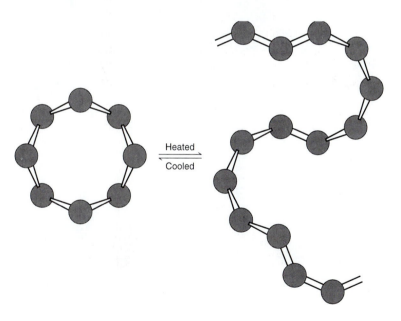

Figure 9.6 Rhombic sulfur contains cyclic octameric rings of S_8. When heated above 140°C, the ring opens and polymerizes to polymeric sulfur. The process is reversed on cooling.

all other liquids. The viscosity increase results from the opening of the eight-membered rings and their conversion to long chains by a free-radical polymerization process. Because long chains can become entangled more effectively than rings, the viscosity rises. As the temperature is raised above about 175°C, however, the viscosity begins to decrease, an indication that depolymerization is now occurring to convert the chains back to rings (see Chapter 10).

The high-polymeric form of sulfur can be "quenched" or isolated in one of two ways. First, if the molten high polymer is "quick-quenched" by pouring into a dry ice-acetone cooling bath (−78°C), the polymer is isolated as a yellow, translucent glass. This material remains noncrystalline at temperatures below −30°C. When heated above the apparent glass transition temperature at −30°C, it becomes highly elastic. This form of the polymer is believed to consist of polymeric sulfur plasticized by S_8 rings. Removal of the S_8 molecules by extraction with carbon disulfide yields an unplasticized polymer that has a glass transition temperature of +75°C. At temperatures above −10°C, the crude, quenched polymer hardens as crystallization of the S_8 rings occurs.

Second, if the molten polymer is quenched to room temperature by pouring it into water (Figure 9.7), it forms a mixture of semicrystalline polymer and S_8 crystals, known as S_w. This material undergoes very rapid reversion to rhom-

Figure 9.7 Polymeric sulfur can be prepared by heating rhombic sulfur to a temperature between 140 and 170°C. If the polymer is quenched by pouring it into cold water (a), it will retain its flexible and elastomeric properties for a short time at room temperature. However, eventually the polymerized material will revert to its original cyclic oligomeric form. [Photos by Ben Rose. From H. R. Allcock, "Inorganic Polymers," *Sci. Am.*, **Mar. 1974,** *230,* 16. Copyright © 1974 by Scientific American, Inc. All rights reserved.]

bic sulfur when heated to temperatures above 90°C. However, at room temperature the rate of cyclization-depolymerization is apparently low. Because of this behavior, polymeric sulfur has only limited practical uses. Nevertheless, its thermal behavior illustrates a characteristic feature of many inorganic polymer systems; rings can be converted to high-polymeric chains, and at elevated temperatures this process tends to be reversed.

Polyselenium can be obtained from cyclic Se_8 by techniques that are similar to those described for the preparation of polysulfur. Because the sulfur polymerization is a free-radical process, it might be assumed that copolymerization might take place with vinyl monomers. Apparently, this does not occur, perhaps because of the low reactivity of a terminal sulfur radical toward organic unsaturated species. (Tetrafluoroethylene may be an exception.) However, sulfur does copolymerize with organic cyclic sulfides such as propylene sulfide under anionic conditions.

POLY(SULFUR NITRIDE) (POLYTHIAZYL)

Synthesis and Appearance

One of the most remarkable inorganic polymer systems is poly(sulfur nitride), $(SN)_n$.[1] This polymer is prepared by a sequence of reactions that start from cyclic "tetrasulfur tetranitride" (47). Tetrasulfur tetranitride itself is prepared from elemental sulfur and liquid ammonia, from SF_4 or S_2F_{10} and ammonia or, more commonly, by the reaction of S_2Cl_2 with ammonia. The cyclic tetramer is an orange-yellow, crystalline solid, m.p. 178°C. It can explode if stored as a solid.

When heated to the sublimation temperature (85°C) in vacuum and when the vapor is passed through heated silver wool at 200 to 300°C, the cyclic tetramer is converted to the potentially explosive cyclic dimer (48), which can be condensed as a white solid on a cold finger cooled in liquid nitrogen. The cyclic dimer is then purified by vacuum sublimation from the cold trap at 25°C to another trap at 0°C to form colorless crystals. A solid-state polymerization of the cyclic dimer then occurs at 25°C during 3 days, followed by heating in vacuum at 75°C for 2 h. During polymerization, the crystals change from colorless (diamagnetic), to blue-black (paramagnetic), and then to lustrous gold (diamagnetic). The polymerization to (49) appears to begin at the surface of the crystals and proceed inward. The overall space-group symmetry of the crystals does not change during polymerization.

47   48   49

[1]Banister, A. J., Gonell, I. B., Adv. Mater. 1998, 10, 1415–1429.

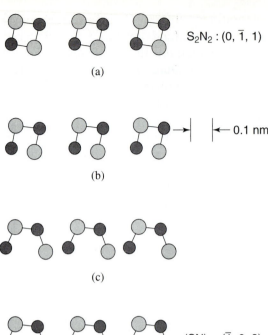

$S_2N_2 : (0, \bar{1}, 1)$

(a)

$\leftarrow$ 0.1 nm

(b)

(c)

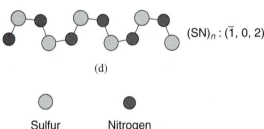

$(SN)_n : (\bar{1}, 0, 2)$

(d)

Sulfur   Nitrogen

Figure 9.8 Diagrammatic representation of the polymerization of $(SN)_2$ to $(SN)_n$ by lateral coupling of the open-chain diradical generated from the cyclic dimer. (From MacDiarmid, A. G., Heeger, A. J., and Garito, A. F., *McGraw-Hill Yearbook of Science and Technology: Polymers,* **1977**.)

The solid-state polymerization process appears to take place by the cleavage of one bond in the cyclic dimer to form a diradical and a linkage of adjacent molecules to form the polymer (Figure 9.8). The polymer can be fabricated as thin films on glass, oriented poly(tetrafluoroethylene), or poly(ethylene terephthalate) by an epitaxial polymerization. The technique involves a depolymerization of $(SN)_n$ at 145°C, possibly to linear S_4N_4, which then repolymerizes on the glass or organic polymer surface. The $(SN)_n$ fibers are aligned parallel to each other along the orientation direction of the polymer support. The films are dark blue by transmitted light.

Crystals of $(SN)_n$ have a parallel fibrous structure, and long fibrous strands may be mechanically peeled from an individual crystal. The ends of the crystal always appear black rather than gold. The gold color constitutes a metallic-type reflectance which is indicative of a metallic-type arrangement of electrons. The crystals are soft and malleable and can be flattened at right angles to the fiber axis to form gold-colored sheets. Compression can cause detonation.

At 25°C the polymer is relatively inert to air and water and may be heated in air for short periods without tarnishing. However, the polymer decomposes (to a white-gray powder) during long exposure to air or water, and it breaks down to sulfur, nitrogen, and other species on long heating at elevated temperatures. It is insoluble in all solvents with which it does not react. Great interest has been shown in $(SN)_n$ because it conducts electricity like a metal at room temperature and becomes a superconductor when cooled to 0.3 K. This behavior is discussed in Chapter 23.

POLYMERS WITH METAL COORDINATION IN THE MAIN CHAIN

Skeletal Bonding Through Classical Donor Ligands

A considerable amount of research has been carried out on the synthesis of polymers in which metal atoms form part of the chain. Some of these structures contain skeletal metal–oxygen or metal–carbon covalent bonds and are perhaps closer to those discussed in the preceding section. However, many of the known metallo polymers possess skeletal structures held together by coordination bonding, in which a donor atom, such as oxygen or nitrogen, supplies electrons to the metal. Large numbers of these structures have been described. The schematic structure shown in **50** is typical of this class of polymer.

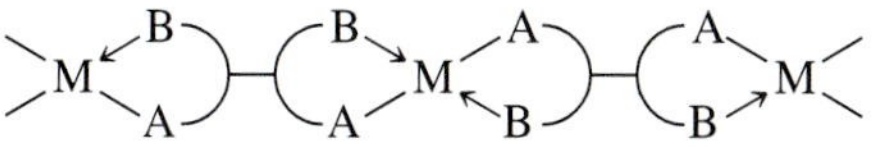

50

   Organic ligands, such as Schiff bases, δ-quinolinol, and β-diketones or phosphinate units have been used in this way with metals such as copper, nickel, cobalt, zinc, manganese, palladium, and so on. However, most examples become insoluble once the degree of polymerization reaches 10 or 20 repeating units. Intermolecular coordination and ligand exchange reactions also contribute to the difficulties experienced with characterizing such materials. The polymeric structure may be disrupted in contact with coordination liquids. In a few cases coordination polymers behave like other macromolecules if the ligands are highly "organic" in nature. For example, the polymers formed by coordination of Ni^{2+} or Fe^{2+} ions with 5-phenyltetrazolate give viscous solutions and can be fabricated into fibers or flexible films.

Polymers with Metallocene Units in the Main Chain

Attempts to introduce ferrocene units into the main chain of a polymer have been made since the 1970s, initially by condensation reactions and, more recently, by ring-opening polymerization techniques. Interest in these polymers stems from two characteristics of the ferrocene unit—its stability to heat and its facile electrochemical oxidation behavior. The latter property can be the basis of electronic semiconduction.

Typical methods for condensation polymerization are based on the reactions of species such as **51** with di-acid chlorides, as shown in the following equation. Similar processes have been developed for the reactions of titanocene derivatives with diols.

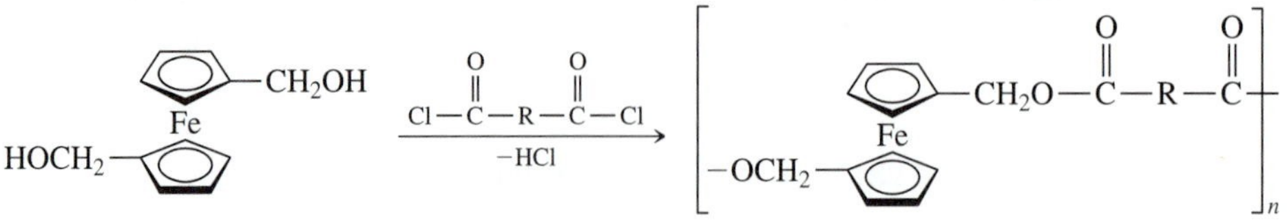

Two ring-opening approaches have been described. The example shown in the following equation is an unusual desulfurization-polymerization process which (when R is butyl) yields polymers with molecular weights in the 12,000 to 260,000 range.[1]

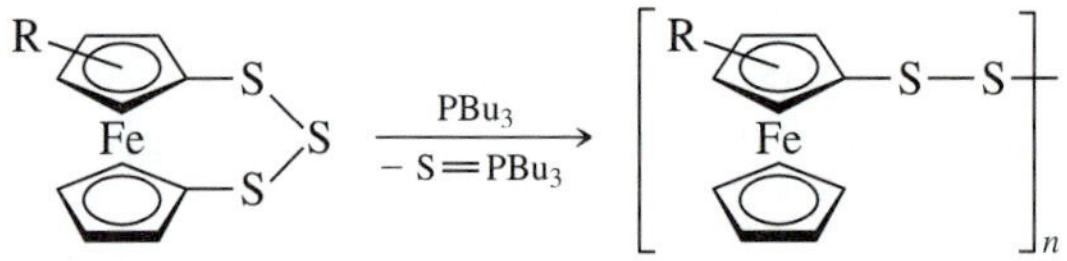

Reversible electrochemical oxidation occurs in two separate stages, thought to be caused by the successive oxidation of alternating ferrocene units along the chain. These polymers photodegrade during exposure to ultraviolet light in air. Similar species with selenium in place of sulfur have also been described.

However, some of the most stable ferrocene-based polymers are obtained by the ring-opening polymerization of ring-strained ferrocenophanes such as **52** by the process summarized in the following equation.[2]

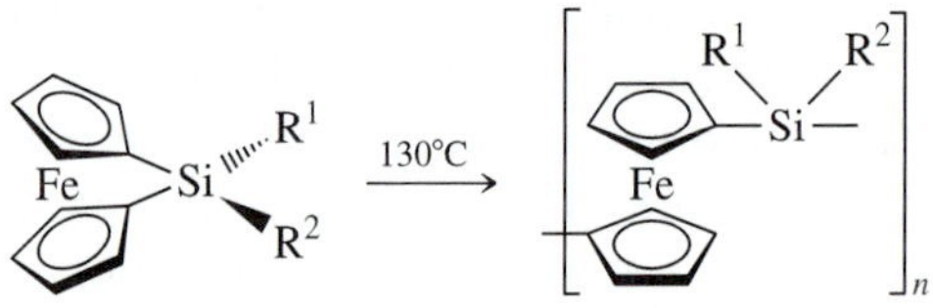

52

Red-orange
crystalline material

53

Poly(ferrocenophanes)

The ring-strain in **52** is evident from the tilting of the cyclopentadienyl rings from coplanarity. Polymerization allows release of this strain and formation of amber-colored polymers of type **53**, with M_w values in the range of 10^5 to 10^6 and $M_n > 10^5$ (~400 repeating units). The side groups, R^1 and R^2

[1]Brandt, P. F., and Rauchfuss, T. B., *J. Am. Chem. Soc.*, **1992,** *114,* 1926.
[2]Foucher, D. A., Tang, B.-Z., and Manners, I., *J. Am. Chem. Soc.*, **1992,** *114,* 6246.

can be varied from methyl to hexyl. Polymers are also accessible when R^1 is methyl and R^2 is chlorine, with the chlorine atoms being available for macromolecular substitution. The T_g values of these polymers depend on the side groups, R, with values varying from +99°C (when R^1 = methyl and R^2 = ferrocenyl) to −51°C (when R^1 = R^2 = hexoxy). The main interest in these polymers so far is as electronic (hole) conductors and as preceramic compounds for conversion to magnetic ceramics. Block copolymers with polystyrene or polysiloxanes have also been prepared.[1] Those with polysiloxane blocks undergo self-assembly in hexane to yield worm-like micelles with the ferrocenyl units forming an inner core. Related polymers with germanium or $-CH_2-CH_2-$ units in place of silicon have also been made.

Other Systems

Finally, an interesting class of metallo-backbone polymers is formed when platinum or palladium alkynes (**54**) undergo oxidative coupling reactions of the type shown in **55**. The yellow polymers are stable in air, soluble in a variety of organic solvents, and have molecular weights in the region of 150,000. The polymers are rodlike. The main interest in these materials is as electrically conducting polymers (Chapter 23).

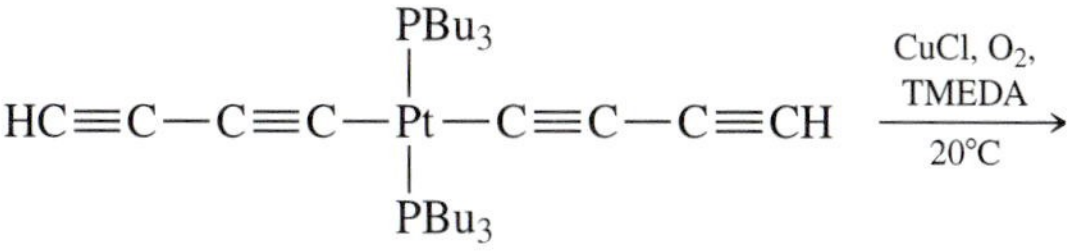

$$HC\equiv C-C\equiv C-\underset{\underset{PBu_3}{|}}{\overset{\overset{PBu_3}{|}}{Pt}}-C\equiv C-C\equiv CH \xrightarrow[20°C]{\substack{CuCl,\ O_2, \\ TMEDA}}$$

54

$$\left[\underset{\underset{PBu_3}{|}}{\overset{\overset{PBu_3}{|}}{Pt}}-C\equiv C-C\equiv C-C\equiv C-C\equiv C \right]_n$$

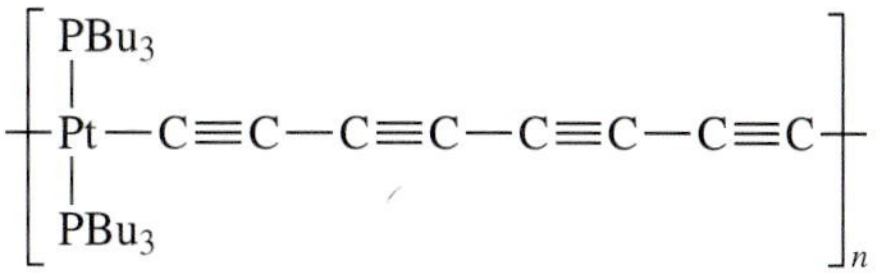

55

MINERALOGICAL AND PRECERAMIC POLYMERS

Mineralogical Polymers and Ceramics

A wide variety of inorganic polymers, in the form of silicates, are found throughout the earth's crust. Some are linear polymers, others are ladder structures and sheets, while many are three-dimensional cross-linked ultrastructures. All three types of silicates have been used by human beings since antiquity—as building materials, for heat insulation, and even as tools and weapons. For most of their histories, polymer chemistry and geochemistry developed in almost complete

[1]Massey, J., Power, K. N., Ninnik, M. A., Manners, I., *J. Am. Chem. Soc.,* **1998**, *120*, 9533–9540.

isolation from each other, separated by the conceptual barrier that "organic" chemistry and mineralogy were somehow alien disciplines that had little in common. In recent years this misconception has begun to disappear as the similarities between the two fields have become more obvious, and indeed as research programs have begun to cross the border from both sides.

Why do mineralogical polymers such as glass, linear silicates, or asbestos differ from their organic counterparts? The presence of noncarbon elements as such has little direct effect, for as we have seen, poly(organosiloxanes) and poly(organophosphazenes) have much in common with classical organic polymers. The rigidity and high-temperature stability of polysilicates is more a result of the presence of *ionic* side-group structures and of the absence of oxidizable elements, than of anything else.

Polysilicates can be visualized as the products formed by the condensation of hydroxysilicon monomers. Because silicic acid (**56**) (the primary monomer) can form mono-, di-, tri-, or tetra-salt structures, the polymerization functionality may be 0, 1, 2, or 3 (Scheme 5). The situation is usually more complex than is implied by Scheme 5. At elevated temperatures redistribution reactions may yield rings, chains, and ultrastructures of immense complexity. Nevertheless, the chemistry outlined in the scheme provides a good working picture.

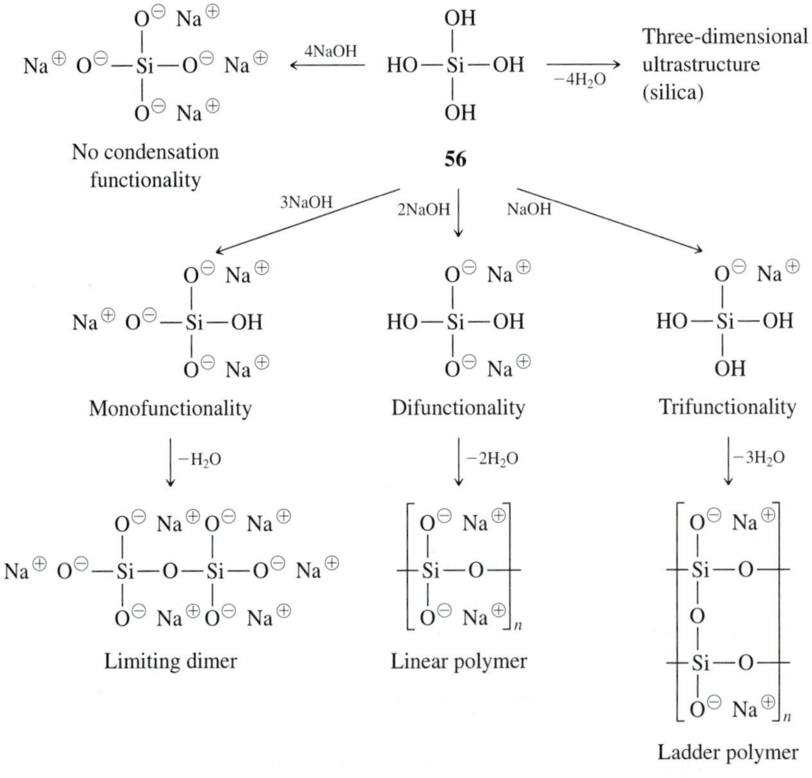

Scheme 5

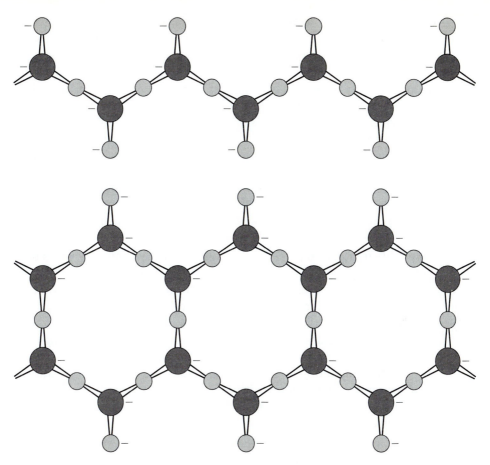

Figure 9.9 Many mineralogical polymers contain silicate chains either as double-strand structures [as, for example, in the amphiboles (bottom)] or single-chain systems [as in pyroxenes (top)]. Connected to the skeletal silicon atoms are charged side-group oxygen atoms. These side groups are bound together in ionic cross-links by di- or higher-valent metallic cations to generate a rigid, high-melting structure. [From H. R. Allcock, "Inorganic Polymers," *Sci. Am.*, **230**, 16 (Mar. 1974). Copyright © 1974 by Scientific American, Inc. All rights reserved.]

Consider the two structures shown in Figure 9.9. Based on our knowledge of organic polymers, we might expect the single-strand silicate to form flexible or elastomeric materials [as in poly(dimethylsiloxane)]. Instead, it forms rigid materials of the type associated with minerals. This is a consequence of ionic interactions between the side groups, especially when the negative charges are neutralized by di- or trivalent cations such as Ca^{2+} or Al^{3+}. The ionic linkages function like covalent cross-links, binding the molecules into a rigid matrix. The ladder silicates, like those found in amphibole minerals such as one form of asbestos, also are bound to their neighbors by ionic forces. Hence, flexibility at the molecular level is very limited. Of course, three-dimensional covalent cross-linking (Scheme 5) would yield an intractible material, irrespective of the elements involved. The advantages of these structures is their heat stability and structural rigidity, and these are the attributes that underlie all their uses.

The ionic cross-links that hold individual chains or silicate sheets together can sometimes be broken by chemical reagents. For example, layer silicates such as vermiculite or montmorillonite will delaminate when treated with alkylammonium cations. Other silicates have been converted to silicone-like ladder polymers by treatment with trimethylchlorosilane. Even chrysotile asbestos reacts with trimethylchlorosilane to give an organic-solvent swellable polymer in which the ionic groups have been removed and only the covalent cross-links remain. These reactions illustrate the close connections between mineral chemistry and polymer science.

Many mineralogical materials are used without further chemical modification. However, since the beginning of civilization, human beings have found ways to modify minerals to fabricate cookingware, ornaments, vases, and building materials. These modified silicates are known under the general category of *ceramics*.

Ceramics are made by melting and re-forming silicates or other inorganic materials, by sintering (heating below the melting point) or by chemical modification (the addition of sodium carbonate to silica, for example, to cleave cross-links and form glass). Ceramics have one significant advantage over other materials—their heat stability. But large amounts of energy are needed for their preparation and fabrication. This is especially true for the preparation of nonsilicate ceramics, such as boron nitride, silicon carbide, or aluminum nitride. Moreover, such materials are exceedingly difficult to shape after they are formed. These two problems are the reasons for the development of the sol-gel synthesis and preceramic polymer routes discussed in the following sections.

The Sol-Gel Process

Ceramic silicates are not only found in nature; they can also be synthesized. Synthesis allows shaped objects to be formed at lower temperatures than are needed during conventional ceramic processing. Hence, significant energy savings are possible. Moreover, the synthesized materials may be porous, thus allowing access to light-weight ceramics.

The sol-gel process involves the hydrolysis of a reactive, multifunctional inorganic monomer. An example is the hydrolysis of tetraethylsilicate (**57**). This compound can be hydrolyzed in solution to silicic acid [$Si(OH)_4$], a monomer that will condense rapidly to a cross-linked matrix (an ultrastructure) and ultimately to silica (Scheme 6).

In practice, the hydrolysis is carried out in ethanol solvent under controlled conditions that allow the solution ("sol") to form a formable, loosely cross-linked matrix ("gel"). This, after molding into a prescribed shape, is heated to remove the remaining water and alcohol and convert the matrix to a heavily cross-linked ceramic with a composition similar to that of silica. The method is also applicable to the alkoxides of titanium, aluminum, and many other elements. The main drawback of this process is that a considerable contraction of volume occurs during the final condensation steps.

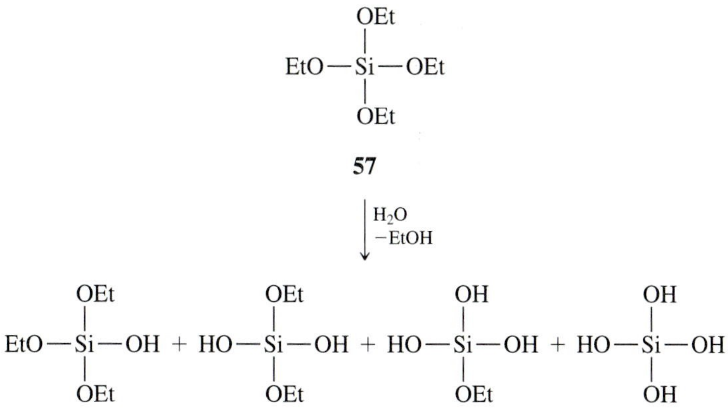

$$\begin{array}{c} OEt \\ | \\ EtO - Si - OEt \\ | \\ OEt \end{array}$$

57

$$\Big\downarrow \begin{array}{l} H_2O \\ -EtOH \end{array}$$

$$\begin{array}{ccccc} OEt & & OEt & & OH & & OH \\ | & & | & & | & & | \\ EtO-Si-OH & + & HO-Si-OH & + & HO-Si-OH & + & HO-Si-OH \\ | & & | & & | & & | \\ OEt & & OEt & & OEt & & OH \end{array}$$

Scheme 6

The brittleness and the volume contraction of sol-gel ceramics can be reduced somewhat by the incorporation of water-soluble or hydrophilic organic polymers into the silicate network. Thus, the final product is an interpenetrating composite material in which the ceramic component provides strength and the polymer provides impact resistance. Such materials are known as either *ceramers* or ORMOSILS (ORganically MOdified SILicas).

Hybrid organic/inorganic composites are also discussed in Chapter 21.

Preceramic Polymers—Purpose

Linear polymers can readily be fabricated into fibers and films. Small molecules and cross-linked ceramics cannot. However, if a polymer could be fabricated into the desired shape (see Chapter 20) and then cross-linked and pyrolyzed, the final product may be a ceramic with the same shape as the original polymer. Of course, pyrolysis results in a loss of side-group components, and this will cause a volume contraction. Thus, the objective in this process is to use a polymer that can be converted to a ceramic with the minimum loss of weight and volume. The first example shows that totally organic polymers can be used as ceramic precursors. However, considerable effort is being made to prepare ceramics from polymers that contain silicon or boron. A further objective is to prepare ceramics that contain the minimum amount of oxygen so as to approximate the compositions of nonoxide ceramics, such as silicon carbide, silicon nitride, or boron nitride.

Carbon Fibers

Although carbon atoms form the principal building blocks of organic polymers, inorganic-type carbon polymers are also known. Diamond, charcoal, and graphite fall into this category. In recent years, new developments have led to the preparation of fibers made from "inorganic" carbon. The starting point for the preparation of a carbon fiber is a conventional organic fiber, such as rayon or polyacrylonitrile, either in the form of a monofilament or a woven textile. Pyrolysis of the organic fiber results in the removal of hydrogen and the formation of polyaromatic structures.

For example, when polyacrylonitrile (**58**) is pyrolyzed, internal addition takes place to yield a condensed polycyclic known as "black Orlon" (**59**). Further pyrolysis results in removal of the hydrogen atoms and the presumed generation of a polyaromatic structure (**60**). In general, pyrolysis reactions are carried out at about 1000°C, but the ultimate "graphitization" may require temperatures as high as 2800°C.

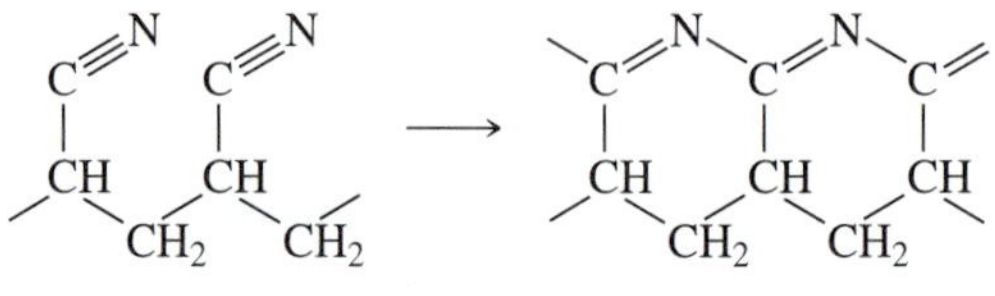

Fibers made in this way are light in weight, chemically inert, and form electrical semiconductors or conductors. They have exceedingly high thermal stabilities. Carbon fiber fabrics can be heated to red heat with the flame from a propane torch without burning and without noticeable degradation taking place. Carbon fibers can be stiffer than glass fibers and they have found increasing use a reinforcement materials for plastics.

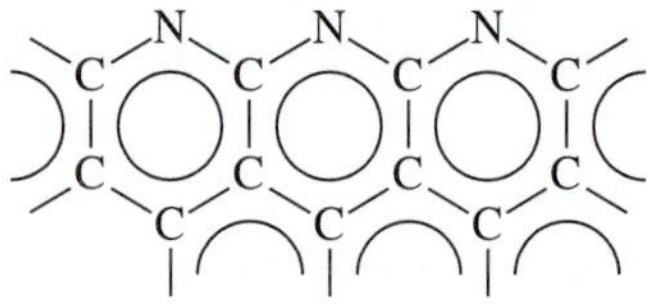

Polysilanes and Carbosilanes as Precursors to Silicon Carbide

In principle, any inorganic backbone polymer should be pyrolyzable to an ultrastructure ceramic. In practice, certain limitations exist. For example, if the polymer depolymerizes to cyclic oligomers before cross-linking takes place, the material will liquefy or volatilize. Thus, the rapidity of cross-linking in the initial stages of the pyrolysis is critical for a preceramic polymer. A slow raising of the temperature, rather than an immediate exposure to extreme temperatures, also aids this process. Finally, as mentioned earlier, the most efficient preceramic processes are those that involve the minimum loss of side-group material, while at the same time yielding the required elemental ratio in the pyrolyzed material.

The conversion of polysilanes via polycarbosilanes to silicon carbide has been the subject of intense activity in recent years, and this system illustrates many of the features that apply to preceramic polymers in general.

As mentioned in an earlier section, polydimethylsilane (**26**) undergoes a rearrangement at 450°C to a poly(methylcarbosilane) (**31**) which, on pyrolysis, first in air at 350°C and then in nitrogen at 1300°C, is converted to β-silicon carbide. The process is a complicated one. The carbosilane is soluble in hexane. It has a molecular weight of about 8000. Melting spinning (see Chapter 20) is used to prepare fibers that are surface-oxidized at 350°C to provide enough rigidity to survive the subsequent processing. The pyrolysis step at 800°C causes loss of methane and hydrogen, and yields an amorphous ceramic. Heating at 1300°C induces crystallization to β-SiC. The fibers have very high tensile strengths (350 kg/mm^2) and are among the strongest substances known. Some evidence exists that the final ceramic contains carbon in excess of that needed to form SiC.

Soluble polysilanes $[(\text{PhMeSi})(\text{Me}_2\text{Si})]_n$ can be spun into fibers directly and pyrolyzed. Even small-molecule cyclosilanes can be used, provided that the initial thermolysis yields a polymeric material that can be fabricated into fibers or shaped objects. Indeed, the polysilanes and polycarbosilanes used in the commercial process may contain linked or fused-ring cyclic structures. Advanced work in this area involves the synthesis of polysilanes or polycarbosilanes that bear more sensitive cross-linking units (vinyl groups, hydrogen atoms, and so on) to speed the initial strengthening of the matrix. However, side groups with many carbon atoms are unsuitable because of the unacceptably high loss of weight and volume during pyrolysis.

Polysilazanes as Precursors to Silicon Nitride

Silazanes are molecules that contain the silicon–nitrogen linkage. They are generally prepared by the reactions or organochlorosilanes with ammonia or amines. Depending on the nonhalogen groups attached to silicon and the nature of the amine, the products can be cyclic silazanes or more complex cyclic, cyclolinear, or cyclomatrix materials. These are precursors for the formation of silicon nitride (Si_3N_4) ceramics.

The reaction of $(\text{CH}_3)_2\text{SiCl}_2$ with ammonia yields mainly cyclic oligomers of formula $[(\text{CH}_3)_2\text{Si}-\text{NH}]_x$. These are not ideal preceramic materials because of the high carbon content. However, H_2SiCl_2 reacts with ammonia to yield complex cyclolinear and cyclomatrix oligomers that contain structures of the type shown in **61**. Heating of these species to 1200°C results in polymerization, loss of ammonia and (at the higher temperatures) of $\text{H}_3\text{SiNHSiH}_3$, and finally formation of Si_3N_4 in 69% yield. A similar sequence initiated by the reaction of H_2SiCl_2 with methylamine yields a composite of Si_3N_4 and elemental silicon. The reactivity of the Si—H bonds apparently plays a major role in facilitating ultrastructure formation.

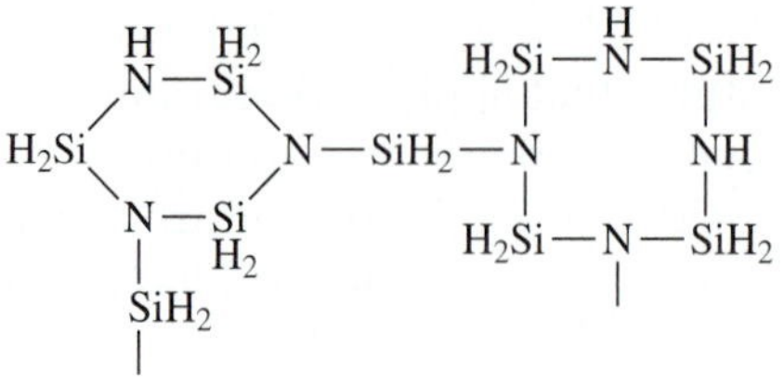

61

Boron Nitride Precursors

Boron nitride (BN) is a remarkable material. It is thermally stable in excess of 2000°C, is an electrical insulator up to 1800°C, and has exceptional resistance to chemical attack. It is also one of the most difficult materials to fabricate into fibers, films, or shaped objects.

One of the earliest preceramic polymer routes to boron nitride fibers made use of an ingenious reaction and pyrolysis process. Fibers of boron oxide (B_2O_3) polymer, made by the dehydration of boric acid [$B(OH)_3$] (**62**) are heated in an atmosphere of ammonia. Water is evolved as boron–oxygen rings are converted to boron–nitrogen (probably borazine) rings. At temperature above 600°C, these rings coalesce to form a layer (graphite) structure, and at 1800°C the structure rearranges to give polycrystalline (diamond-type) boron nitride (**63**), still in the form of a fiber.

More recently, borazine rings have been linked together, first to form cyclolinear species (**64**), and then, on pyrolysis to graphitic (**65**), and later diamond-type boron nitride.

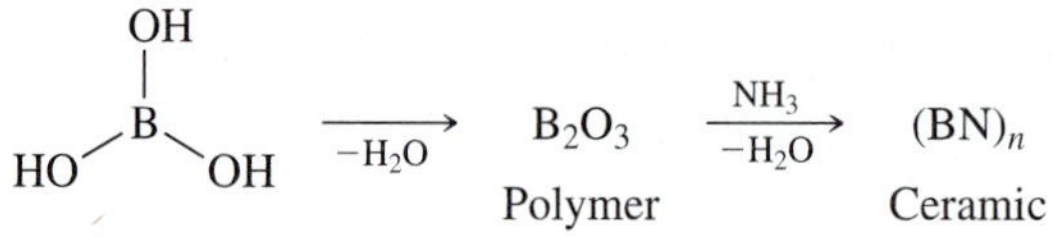

62 **63**

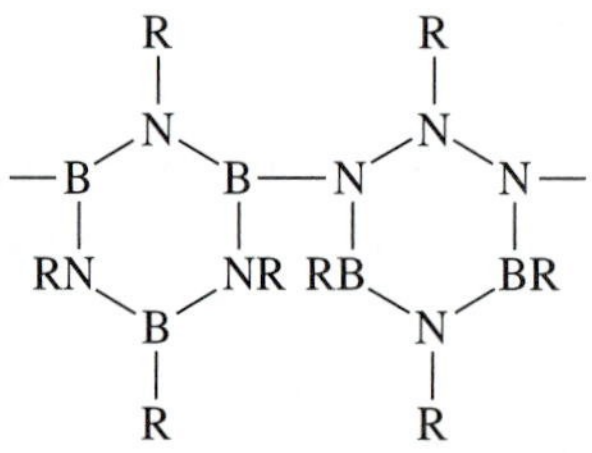

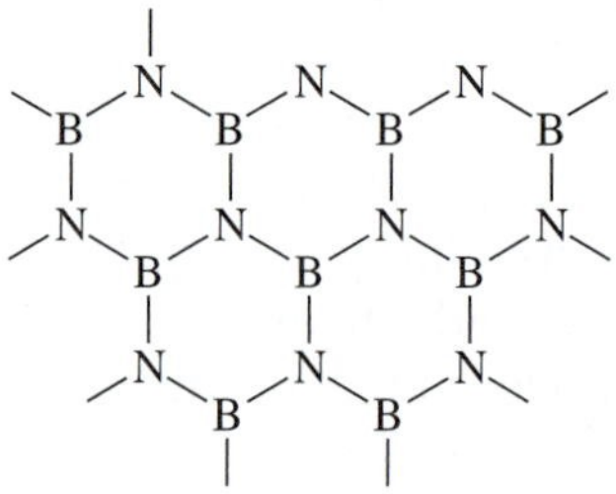

64 **65**

CONCLUSIONS

One of the main directions of effort in polymer chemistry is to synthesize new materials that are easily fabricated, but once fabricated, are rigid, tough, and heat resistant. We see a sequence of developments in this direction starting with the use of noncrystalline glasses [polystyrene, poly(methyl methacrylate)], moving to microcrystalline materials (polyamides, polyesters), then to rigid-rod polymers (aromatic polyamides and polyesters, polyimides, and aromatic ladder polymers), and thence to ceramics. The ceramic-type approach has been present in polymer science almost from the beginning. Phenol–formaldehyde and melamine–formaldehyde resins are organic ceramics. Cyclomatrix phosphazenes have been studied since the 1960s. Nevertheless, the hybridization of polymer chemistry, inorganic synthesis, and ceramic science represents one of the major opportunities for the future.

STUDY QUESTIONS

1. What options are available to the synthetic chemist who wishes to increase the thermal stability of those inorganic polymers that show a tendency to depolymerize to cyclic oligomers at elevated temperatures?

2. Relatively few polymer systems have been developed that contain backbone structures derived from the main-group elements other than carbon, nitrogen, oxygen, phosphorus, sulfur, or silicon. In an essay-type answer, suggest some of the reasons for this restriction and describe ways in which such polymers might be synthesized. What practical advantages might result from the incorporation of, say, tin or antimony into a polymer skeleton?

3. What are the main problems that might be encountered in the synthesis of macromolecules that contain transition metals coordinated to organic ligands in a backbone structure?

4. Poly(sulfur nitride) is an electrical conductor. Speculate on the underlying chemical and physical reasons for this phenomenon and suggest other high-polymeric systems that might behave in the same way.

5. Design synthesis routes to polyphosphazenes that would have the properties needed for use as (a) elastomers for use in artificial heart valves; (b) electrical wire insulators; (c) compounds to reduce turbulence in a recirculating water-cooling system; (d) polymers that are stable to ultraviolet light; (e) carrier molecules for chemotherapeutic drugs.

6. Large amounts of elemental sulfur are available at very low prices. What is the source of this material? How could it be converted to high-polymeric sulfur on a large scale? How might the polymer be stabilized against (a) reversion (depolymerization) and (b) crystallization? What potential uses can you foresee for polymeric sulfur if these two problems can be solved?

7. Discuss the prospect that polymers other than polyacrylonitrile or cellulose might be converted into carbon fibers. What advantages or disadvantages can you think of for the use of other polymers?

8. The three fields of polymer chemistry, ceramic science, and metallurgy are slowly co-alescing to provide new materials for use in technology. In an essay-type answer dis-cuss why these fields remained separate for the past 150 years and what conceptual hurdles need to be overcome before scientists in these areas can truly pool their resources.

9. The macromolecular substitution route is a key component of the synthesis of polyphosphazenes. What other inorganic-backbone polymers might be amenable to this approach? What are the prospects for using this method for organic backbone polymers (*Hint*: See Chapter 7.)

10. Carboranes and metal phthalocyanines are two stable units that have been incorporated into polymers in an attempt to generate special properties. What are those properties? What advantages might be expected if ferrocene units form part of a polymer chain? Speculate about the effect on the macromolecular properties if borazine units could be incorporated into a siloxane chain or an organic chain.

11. Does a counterpart of "sol-gel" polymer synthesis exist for organic polymers? Could hybrid organic-inorganic polymers be synthesized by combining these two processes? If so, what complications might need to be overcome?

12. As part of a group project, debate the reasons for the low temperature flexibility of poly(organosiloxanes) and poly(organophosphazenes), taking into account both the chemical bonding and molecular mechanics aspects.

13. Discuss the following observation: Property diversity in poly(organosiloxanes) is achieved using very few alternative side groups through changes in polymer-chain length and degree of cross-linking, whereas diversity in poly(organophosphazenes) is accomplished mainly through the use of a very broad range of different side groups.

SUGGESTIONS FOR FURTHER READING

(a) General Overviews

ARCHER, R. D., *Inorganic and Organometallic Polymers*. New York: Wiley-VCH, **2001**.

ALLCOCK, H. R., "Inorganic Macromolecules," *Chem. & Eng. News*, **1985**, *63*, 22.

CARRAHER, C. E., SHEATS, J. E., and PITTMAN, C. U., (Eds.), *Organometallic Polymers*. New York. Academic Press, **1978**.

COCKE, D. L., and CLEARFIELD, A., (Eds.), *Design of New Materials*. New York: Plenum Press, **1987**.

HARROD, J. E., and LAINE, R. M., (Eds.) *Inorganic and Organometallic Oligomers and Polymers*, Dordrecht: Kluwer, **1990**.

LAINE, R. M. (Ed.), *Inorganic and Organometallic Polymers with Special Properties*, Dordrecht: Kluwer, **1992**.

MANNERS, I., "Polymers and the Periodic Table: Recent Developments in Inorganic Poly-mer Science," *Angew. Chem. Int. Ed. Eng.*, **1996**, *35*, 1602.

MARK, J. E., ALLCOCK, H. R., and WEST, R., *Inorganic Polymers*, Englewood Cliffs, NJ: Prentice Hall, **1992**.

PITTMAN, C. U., CARRAHER, C. E., ZELDIN, M., SHEATS, J. E., and CULBERTSON, B. M., (Eds.), *Metal-Containing Polymeric Materials*, New York, Plenum, **1996**.

WISIAN-NEILSON, P., ALLCOCK, H. R., and WYNNE, K. J., *Inorganic and Organometallic Polymers II. Advanced Materials and Intermediates, ACS Symp. Ser.*, **1994**, *572.*

ZELDIN, M. (Ed.), *Journal of Inorganic and Organometallic Polymers*, (Plenum, New York), **1991**, *1,* and subsequent issues.

ZELDIN, M., WYNNE, K. J., and ALLCOCK, H. R., (Eds.), *Inorganic and Organometallic Polymers, ACS Symp. Ser.*, **1988**, *360.*

(b) Organic Polymers with Inorganic Elements in the Side Groups

ALLCOCK, H. R., HARTLE, T. J., TAYLOR, J. P., and SUNDERLAND, N. J., "Organic Polymers with Cyclophosphazene Side Groups," *Macromolecules,* **2001**, *34,* 3896.

ALLCOCK, H. R., LAREDO, W. R., DE DENUS, C. R., and TAYLOR, J. P., "Ring-Opening Metathesis Polymerization of Phosphazene-Functionalized Norbornenes," *Macromolecules*, **1999**, *32*, 1719.

ALLEN, C. W., BROWN, D. E., HAYES, R. F., TOOZE, R., and POYSER, G. L., "Copolymerization Reactions of Inorganic Rings Containing Olefinic Substituents. Quantitative Reactivity Studies," *ACS Symp. Ser.*, **1994**, *572*, 389

ALLEN, C. W., "Hybrid Inorganic-Organic Polymers Derived from Organo-functional Phosphazenes," in *Inorganic and Organometallic Polymers* (M. Zeldin, K. J. Wynne, and H. R. Allcock, Eds.), *ACS Symp. Ser.* **1988**, *360.*

ASAMI, R., OKU, J., TAKEUCHI, M., NAKAMURA, K., and TAKAKI, M., "Anionic Polymerization of Vinylsilanes, I. Novel Isomerization in the Anionic Polymerization of Trimethylvinylsilane," *Polymer J. (Tokyo)*, **1988**, *20*(8), 699.

BRYANTSEVA, I. S., KHOTIMSKII, V. S., DURGAR'YAN, S. G., and PETROVSKII, P. V., "Branching during Anionic Polymerization of Vinyltrimethylsilane." *Vysokomol. Soedin., Ser. B*. **1985**, *27*(2), 149.

DENISEVICH, P., ABRUNA, H. D., LEIDNER, C. R., MEYER, T. J., and MURRAY, R. W., "Electropolymerization of Vinylpyridine and Vinylbipyridine Complexes of Iron and Ruthenium," *Inorg. Chem.* **1982**, *21*, 2153.

HARTLE, T. J., SUNDERLAND, N. J., MCINTOSH, M. B., and ALLCOCK, H. R., "Phosphinimine Modification of Organic Polymers," *Macromolecules*, **2000**, *33*, 4307.

PITTMAN, C. U., "Vinyl Polymerization of Organometallic Monomers containing Transition Metals," in *Organometallic Polymers* (C. E. Carraher, J. E. Sheats, and C. U. Pittman, Eds.). New York: Academic Press, **1978**, p.1.

PITTMAN, C. U., CARRAHER, C. E., and REYNOLDS, J. R., "Organometallic Polymers," in *Encyclopedia of Polymer Science and Engineering*, 2nd ed. (H. F. Mark, N. Bikales, C. G. Overberger, G. Menges, and J. I. Kroschwitz, Eds.). New York: Wiley, **1987**.

RICKLE, G. K., "The Anionic Polymerization of Trimethylvinylsilane," *J. Macromol. Sci. Chem.* **1986**, *A23*(11), 1287.

(c) Polymers with Silicon in the Main Chain

BARRY, A. J., and BECK, H. N., in *Inorganic Polymers* (F. G. A. Stone, and W. A. G. Graham, eds.) New York: Academic Press, **1962**, p. 189.

BOUQUEY, M., BROCHON, C., BRUZAND, S., MINGOSTAND, A. F., SCHAPPACHER, M., and SOUM, A., "Ring-Opening Polymerization of Nitrogen-Containing Cyclic Organosilicon Monomers," *J. Organomet. Chem.*, **1996**, *521*, 21.

BROOK, M. A., *Silicon in Organic, Organometallic, and Polymer Chemistry*. New York: John Wiley & Sons, **2000**.

CLARSON, S. J., FITZGERALD, J. J., OWEN, M. J., and SMITH, S. D., (Eds.), *Silicones and Silicone-Modified Materials*. Washington, D. C.: American Chemical Society, **2000**, Vol. *729*.

DAVISON, J. B., and WYNNE, K. J., "Silicon Phthalocyanine-Siloxane Polymers: Synthesis and ^{1}H Nuclear Magnetic Resonance Study," *Macromolecules*, **1978**, *11*, 186.

KOIDE, N., and LENZ, R. W., "Preparation and Properties of Poly(silylarylene) siloxanes," *J. Polymer Sci.*, **1983**, *70*, 91.

LI, H., and WEST, R., "Structures and Photophysical Properties of Silicon-Containing Phenyleneethynylene Polymers," *Macromolecules*, **1998**, *31*, 2866.

LIU, Q., SHI, W., BABONNEAU, F., and INTERRANTE, L. V., "Synthesis of Polycarbosilane/Siloxane Hybrid Polymers," *Chem. Mater.*, **1997**, *9*, 2434.

MARK, J. E., ALLCOCK, H. R., and WEST, R., *Inorganic Polymers.*, Ch. 4 and 5, Englewood Cliffs, NJ: Prentice Hall, **1992**.

MARKS, T. J., "Electrically Conductive Metallomacrocyclic Assemblies," *Science*, **1985**, *227*, 881.

McGREGOR, R. R., *Silicones and their Uses*. New York: McGraw-Hill, 1954. (Although this book is long out-of-print, the reader who can locate a copy will be rewarded with a valuable insight into the applications aspects of poly(organosiloxanes)).

MILLER, R. D., HOFER, D., FICKES, G. N., WILLSON, C. G., MANINERO, E., TREFONAS, P., and WEST, R., "Soluble Polysilanes: An Interesting New Class of Radiation Sensitive Materials," *Polym. Eng. Sci.*, **1986**, *26*, 1129.

PETERS, E. N., "The Development of Carborane-Siloxane Polymers," *Ind. Eng. Chem. Prod. Res. Develop.* **1984**, *23*, 28.

PRANGE, R., and ALLCOCK, H. R., "Telechelic Synthesis of the First Phosphazene-Siloxane Block Copolymers," *Macromolecules*, **1999**, *32*, 6390.

ROCHOW, E. G., *Silicon and Silicones*. New York: Springer-Verlag, **1987**.

RUSHKIN, H. L., and INTERRANTE, L. V., "Synthesis of Poly(silylenemethylenes) through Reactions Carried Out on Preformed Polymers," *Macromolecules*, **1996**, *29*, 3123.

Silicon-Containing Polymers: The Science and Technology of Their Synthesis and Applications (R. G. Jones, W. Ando, and J. Chojnowski, Eds.), Dordrecht: Kluwer Academic Publishers, **2000**.

WARRICK, E. L., *Forty Years of Firsts. The Recollections of a Dow Corning Pioneer*, New York: McGraw-Hill, **1990**.

WEST, R., and MAXKA, J., "Polysilane High Polymers: An Overview," in *Inorganic and Organometallic Polymers* (M. Zeldin, K. J. Wynne, and H. R. Allcock, eds.), *ACS Symp. Ser.*, **1988**, *360*, 6.

YILGOR, I., and McGRATH, J. E., "Polysiloxane-Containing Copolymers," *Adv. Polymer Sci.*, **1988**, *86*, 1.

ZEIGLER, J. M., "One-Dimensional Conjugated Polysilylenes—Science and Technology," *Mol. Cryst. Liq. Cryst.* **1990**, *190*, 265.

(d) Polystannanes

INORI, T., LU, V., CAI, H., and TILLEY, T. D., *J. Am. Chem. Soc.*, **1995**, *117*, 9931.

WOLFE, P. S., GOMEZ, F. J., and WAGENER, K. B., "Metal-Containing Polymers Synthesized via Acyclic Diene Metathesis: Polycarbostannanes," *Macromolecules*, **1997**, *30*, 714.

(e) Polyphosphazenes

ALLCOCK, H. R., *Chemistry and Applications of Polyphosphazenes*. New York: John Wiley, **2003.**

ALLCOCK, H. R., *Phosphorus-Nitrogen Compounds*. New York: Academic Press, **1972.**

ALLCOCK, H. R., and KUGEL, R. L., "Synthesis of High Polymeric Alkoxy- and Aryloxyphosphonitriles," *J. Am. Chem. Soc.*, **1965**, *87*, 4216.

ALLCOCK, H. R., SINGLER, R. E., PENTON, H. R., NEILSON, R. H., WISIAN-NEILSON, P., and HADDON, R. C., and their co-workers, a series of papers in *Inorganic and Organometallic Polymers, ACS Symp. Ser.,* **1980**, *360*, 250–302.

ALLCOCK, H. R., "Functionalized Polyphosphazenes," Ch. 18 in *Functional Polymers* (A. O. Patil, D. N. Schulz, and B. M. Novak, eds.), *ACS Symp. Ser.*, **1998**, *704.*

ALLCOCK, H. R., and KUHARCIK, S. E., "Hybrid Phosphazene-Organo-silicon Polymers, II," *J. Inorg. Organomet. Polymers*, **1996**, *6*, 1.

ALLCOCK, H. R. "Polyphosphazenes and their Diversity," Ch. 28 in *Macromolecular Design of Polymeric Materials* (K. Hatada, T. Kitayama, and O. Vogl, eds.), New York: Dekker, **1996.**

ALLCOCK, H. R., CRANE, C. A., MORRISSEY, C. T., NELSON, J. M., REEVES, S. D., HONEYMAN, C. H., and MANNERS, I., " "Living" Cationic Polymerization of Phosphoranimines as an Ambient Temperature Route to Polyphosphazenes with Controlled Molecular Weights," *Macromolecules*, **1996**, *29*, 7740.

ALLEN, C. W., "Linear, Cyclic, and Polymeric Phosphazenes," *Coord. Chem. Rev.*, **1994**, *130*, 137.

DE JAEGER, R., and GLERIA, M., "Poly(organophosphazenes) and Related Compounds: Synthesis, Properties, and Applications." *Prog. Polym. Sci.*, **1988**, *23*, 179.

GLERIA, M., and DE JEAGER, R., (Eds.), *Phosphazenes: A Worldwide Insight.* Nova Science Publishers, **2003.**

GUGLIELMI, G., BRUSATIN, G., FACCHIN, G., and GLERIA, M., "Poly-(organophosphazenes) and the Sol-Gel Technique," *Organomet. Chem.*, **1999**, *13*, 339.

MARK, J. E., ALLCOCK, H. R., and WEST, R., *Inorganic Polymers*, Englewood Cliffs, NJ: Prentice Hall, **1992**, Ch. 3.

MATYJASZEWSKI, K., FRANZ, U., MONTAGUE, R. A., and WHITE, M. L., "Synthesis of Polyphosphazenes form Phosphoranimines and Phosphine Azides," *Polymer*, **1994**, *35*, 5005.

NEILSON, R. H., and WISIAN-NEILSON, P., "Polyalkyl/arylphosphazenes) and their Precursors," *Chem. Rev.*, **1988**, *88*, 541.

SINGLER, R. E., SCHNEIDER, N. S., and HAGNAUER, G. L., "Polyphosphazenes-Synthesis-Properties-Applications," *Polymer. Eng. Sci.*, **1975** *15*, 5, 321.

TATE, D. P., "Polyphosphazene Elastomers," *J. Polymer Sci., Polymer Symp.* **1974**, *48*, 33.

WISIAN-NEILSON, and NEILSON, R. H., *J. Am. Chem. Soc.*, **1980**, *102*, 2848.

WISIAN-NEILSON, P., ZHANG, C. P., KOCH, K. A., and GRUNEICH, J. A., "Poly(alkyl/arylphosphazenes and their Derivatives," *Phosphorus, Sulfur and Silicon and the Related Elements*, **1999**, *146*, 69.

WISIAN-NEILSON, P., ZHANG, C., and KOCH, K. A., "Deprotonation-Substitution Reactions of Poly(methylalkylphosphazenes) and their N-Silylphosphoranimine Precursors," *Macromolecules*, **1998**, *31*, 1808.

(f) Polymers with Metals in the Main Chain

ANDRIANOV, K. A., *Metalorganic Polymers*. New York: Wiley-Interscience, **1965**.

ARCHER, R. S., WANG, B., TRAMONTANO, V. J., LEE, A. Y., and OCHAYA, V. O., "Soluble Metal Chelate Polymers of Coordination Numbers 6, 7, and 8," in *Inorganic and Organometallic Polymers* (M. Zeldin, K. J. Wynne, and H. R. Allcock, Eds.), *ACS Symp. Ser.*, **1988**, *360*, 463.

COMPTON, D. L., BRANDT, P. F., RAUCHFUSS, T. B., ROSENBAUM, D. F., and ZUKOSKI, C. F., "Organometallic Polymers Based on S-S and Se-Se Linked Butylferrocenes," *Chem. Mater.*, **1995**, *7*, 2342.

GONSALVES, K. E., and RAUSCH, M. D., "Cationic and Condensation Polymerization of Organometallic Monomers," in *Inorganic and Organometallic Polymers* (M. Zeldin, K. J. Wynne, and H. R. Allcock, Eds.), *ACS Symp. Ser.*, **1988**, *360*, 437.

HANACK, M., and LANG, M., "Conducting Stacked Metallophthalocyanines and Related Compounds," *Adv. Mater.*, **1994**, *6*, 819.

MANNERS, I., "New Polymers Based on Inorganic Elements," *Polymer News*, **1993**, *18*, 133.

MARCAT, L., MALDIVI, P., and MARCHON, J.-C., "Metallopolymers: Preparation of Polymer Films with a High Content of Metal Centers via Photopolymerization of Metal-Containing Liquid Crystalline Monomers," *Chem. Mater.*, **1997**, *9*(10), 2051.

NEUSE, E. W., "Polymetallocenylenes: Recent Developments," in *Advances in Organometallic and Inorganic Polymer Science* (C. E. Carraher, J. E. Sheats, and C. U. Pittman, Eds.). New York: Dekker, **1982**, p. 3

NGUYEN, P., GOMEZ-ELIPE, and MANNERS, I., "Organometallic Polymers with Transition Metals in the Main Chain," *Chem. Rev.,* **1999**, *99*, 1515.

PECKHAM, T. J., MASSEY, J. A., EDWARDS, M., MANNERS, I., and FOUCHER, D. A., "Synthesis, Characterization, and Properties of High Molecular Weight Poly(ferrocenylgermanes) and Poly(ferrocenylsilane)-Poly(ferrocenylgermane) Random Copolymers, *Macromolecules*, **1996**, *29*, 2396.

TAKAHASHI, S., MURATA, E., SONOGASHIRA, K., and HAGIHARA, N., "Studies on Polyyne Polymers Containing Transition Metals in the Main Chain," *J. Polymer Sci.*, **1980**, *18*, 661.

(g) Mineralogical and Preceramic Polymers

BIROT, M., PILLOT, J.-P., and BUNOGUES, J., "Comprehensive Chemistry of Polycarbosilanes, Polysilazanes, and Polycarbosilanes as Precursors of Ceramics," *Chem. Rev.,* **1995**, *95*, 1443.

BRINKER, C. J., and SCHERER, G. W., *Sol-Gel Science*. San Diego: Academic Press, **1990**.

CURRELL, B. R., and PARSONAGE, J. R., "Trimethylsilylation of Mineral Silicates," in *Advances in Organometallic and Inorganic Polymer Science* (C. E. Carraher, J. E. Sheats, and C. U. Pittman, Eds.). New York: Dekker, **1982**, p. 141.

ECONOMY, J., "Now, That's an Interesting Way to Make a Fiber," *Chem. Tech.*, **1980**, *10*, 240.

FRAZIER, S. E., BEDFORD, J. A., HOWER, J., and KENNEY, M. E., "An Inherently Fibrous Polymer," *Inorg. Chem.*, **1967**, *6*, 1693.

HENCH, L. L., and ULRICH, D. R., (Eds.), *Science of Ceramic Processing*, New York: Wiley-Interscience, **1986**.

HJELM, R. J., NAKATANI, A. I., GERSPACHER, M., and KRISHNAMOORTI, R., (Eds.), *Filled and Nanocomposite Polymers Materials*. Warrendale, PA: Materials Research Society, **2001**, Vol. *661*.

INTERRANTE, L. V., LIU, Q., RUSHKIN, I., and SHEN, Q., "Poly(silylene-methylenes)—A Novel Class of Organosilicon Polymers," *J. Organomet. Chem.*, **1996** *521*, 1.

LAINE, R. M., BLUM, Y. D., TSE, D., and GLASER, R., "Synthetic Routes to Organosilazanes and Polysilazanes: Polysilazane Precursors to Silicon Nitride," in *Inorganic and Organometallic Polymers* (M. Zeldin, K. J. Wynne, and H. R. Allcock, Eds.), *ACS Symp. Ser.*, **1988**, *360*, 124.

LAINE, R. M., SANCHEZ, C., GIANNELIS, E., and BRINKER, C. J., (eds), *Organic/Inorganic Hybrid Materials—2000*. Warrendale, PA: Materials Research Society, **2001**, Vol. *628*.

MACKENZIE, J. D., and ULRICH, D. R., (Eds.), *Ultrastructure Processing of Advanced Ceramics*. New York: Wiley-Interscience, **1988**.

NARULA, C. K., SCHAEFFER, R., PAINE, R. T., DATYE, A., and HAMMETER, W. F., "Synthesis of Boron Nitride Ceramics from Poly(borazinylamine) Precursors," *J. Am. Chem. Soc.*, **1987**, *109*, 5556.

PACIOREK, K. L., KRONE-SCHMIDT, J. W., HARRIS, D. H., KRATZER, R. H., and WYNNE, K. J., "Boron Nitride and Its Precursors," in *Inorganic and Organometallic Polymers* (M. Zeldin, K. J. Wynne, and H. R. Allcock, Eds.), *ACS Symp. Ser.*, **1988**, *360*, 392.

SANCHEZ, C., and RIBOT, F., "Design of Hybrid Organic-Inorganic Materials Synthesized via Sol-Gel Chemistry," *New J. Chem.*, **1994**, *18*, 1007.

SCHILLING, C. L., WESSON, J. P., and WILLIAMS, T. C., "Polycarbosilane Precursors for Silicon Carbide," *Am. Ceram. Soc. Bull.*, **1983**, *62*, 912.

SEYFERTH, D., WISEMAN, G. H., SCHWARK, J. M., YU, Y.-F., and POUTASSE, C. A., "Organosilicon Polymers as Precursors for Silicon-Containing Ceramics," in *Inorganic and Organometallic Polymers* (M. Zeldin, K. J. Wynne, and H. R. Allcock, Eds.), *ACS Symp. Ser.*, **1988**, *360*, 143.

SEYFERTH, D., "Preceramic Polymers: Past, Present, and Future," in *Materials Chemistry: An Emerging Discipline*, L. V. Interrante, and M. Hampden-Smith (Eds.). Wiley-VCH, **1995**, 132.

WEN, J., and WILKES, G., "Organic/Inorganic Hybrid Network Materials by the Sol-Gel Approach," *Chem. Mater.*, **1996**, *8*, 1667.

WRIGHT, J. D., and SOMMERDIJK, N. A. J. M., *Sol-Gel Materials. Chemistry and Applications*. Amsterdam: Gordon and Breach Science Publishers, **2001**.

YAJIMA, S., "Special Heat-Resisting Materials from Organometallic Polymers," *Am. Ceram. Bull.,* **1983**, *62*, 893.

(h) Sulfur-containing Polymers

GOETHALS, E. J., "Sulfur-Containing Polymers," *J. Macromol. Sci., Rev. Macromol. Chem.*, **1968**, *C2*, 74.

LABES, M. M., LOVE, P., and NICHOLS, L. F., "Polysulfur Nitride: A Metallic, Superconducting Polymer," *Chem. Rev.*, **1979**, *79*, 1.

MIKULSKI, C. M., RUSSO, P. J., SAVAN, M. S., MACDIARMID, A. G., GARITO, A. F., and HEEGER, A. J., "Synthesis and Structure of Metallic Polymeric Sulfur Nitride, $(SN)_x$ and Its Precursor, Disulfur Dinitride, S_2N_2," *J. Am. Chem. Soc.*, **1975**, *97*, 6360.

SCHMIDT, M. "Sulfur Polymers," in *Inorganic Polymers* (F. G. A. Stone and W. A. G. Graham, eds.). New York: Academic Press, **1962**, p. 98.

TOBOLSKY, A. V., and MACKNIGHT, W. J., *Polymeric Sulfur and Related Polymers* (*Polymer Reviews*, Vol. 13). New York: Wiley-Interscience, **1965**.

(i) Other Systems

COLQUHOUN, H. M., HERBERTSON, P. L., WADE, K., BAXTER, I., and WILLIAMS, D. J., "A Carborane-Based Analogue of Poly(*p*-phenylene)," *Macromolecules*, **1998**, *31*, 1694.

10

Polymerization and Depolymerization Equilibria

INTRODUCTION

The synthesis of organic and inorganic polymers by the ring-opening polymerization of cyclic compounds was discussed in Chapters 6 and 9. Such polymers frequently break down at high temperatures to yield cyclic oligomers, often the same cyclic compounds from which the polymers were formed in the first place. Similarly, a number of addition polymers degrade at high temperatures to regenerate the original monomer. In many cases it can be shown that the polymerization and depolymerization steps are simply different aspects of an equilibration process. Whether polymerization or depolymerization occurs at a given temperature depends on the position of the equilibrium at that temperature.

An understanding of these processes is vitally important for those who are concerned about the practical problems of thermal stability in high polymers. It is also important for an understanding of why a substantial number of cyclic systems and unsaturated compounds have not yet been converted to high polymers.

It is worthwhile to approach this subject by first considering a number of questions. For example, why do certain unsaturated monomers or cyclic compounds polymerize to linear high polymers, whereas others do not? Why are some polymers more stable at high temperatures than others? Which molecular features in different systems might stabilize chains more than rings or monomers, or vice versa? What role does the *mechanism* of polymerization or depolymerization play in these processes? Do monomeric compounds participate in ring-chain interconversion, or do cyclic species play a role in monomer–polymer equilibrations?

In order to attempt to answer these questions, let us first consider some examples of systems which undergo monomer–polymer or ring–chain equilibration and some which do not. We can then formulate a number of general observations.

MONOMER–POLYMER EQUILIBRIA

At high temperatures, poly(methyl methacrylate) depolymerizes to methyl methacrylate, and poly(tetrafluoroethylene) depolymerizes to tetrafluoroethylene monomer. This general phenomenon can be summarized by the process shown in **1**. Other polymers that behave in the same way include poly(α-methylstyrene), poly(methacrylonitrile), and poly(vinylidine cyanide). The monomers formed from these polymers cannot undergo α-hydrogen abstraction. Hence, hydrogen abstractive side reactions do not compete with simple depolymerization to the monomers.

$$\left[\begin{array}{c} H \quad R \\ | \quad | \\ C-C \\ | \quad | \\ H \quad R' \end{array}\right]_n \longrightarrow n \begin{array}{c} H \quad R \\ | \quad | \\ C=C \\ | \quad | \\ H \quad R' \end{array}$$

1

Furthermore, it is known that although monomers such as acetaldehyde, $MeCH{=}O$, propionaldehyde, $EtCH{=}O$, butyraldehyde, $PrCH{=}O$, or acetone, $Me_2C{=}O$ can be polymerized, the polymers depolymerize back to the monomer at only moderate temperatures. High-pressure, low-temperature reaction conditions are needed for polymerization, but release of the pressure and warming to room temperature results in depolymerization.

On the other hand, polystyrene undergoes thermal breakdown to yield not only styrene, but also products derived from random chain scission, α-hydrogen abstraction, and chain transfer processes. Thus, it is important to recognize that depolymerization (i.e., the reverse of polymerization) is only one process of several that are possible when a polymer is heated to high temperatures (see also Chapter 9).

EXAMPLES OF RING–POLYMER INTERCONVERSIONS

As discussed in Chapter 9, rhombic sulfur, S_8, can be polymerized to S_n at temperatures above 160°C, but depolymerization back to S_8 occurs at higher temperature. Selenium behaves similarly. Trioxane, $(OCH_2)_3$, and tetroxane, $(O{-}CH_2)_4$, polymerize to polyoxymethylene during γ- or X-ray irradiation or under the influence of cationic initiators. The polymer depolymerizes to the monomer (formaldehyde) or to cyclic oligomers above 100°C. Trithiane, $(S{-}CH_2)_3$, tetrathiane, $(S{-}CH_2)_4$, and higher cyclic homologues behave similarly.

Particularly interesting equilibria exist in the isocyanate series, where monomers (2), cyclic dimers (3), cyclic trimers (4), and high polymers (5) can all participate in an equilibration process. In the dimethylsiloxane system, the cyclic trimer, $(O—SiMe_2)_3$, and tetramer, $(O—SiMe_2)_4$, polymerize readily with anionic or cationic initiators to yield poly(dimethylsiloxane), $(O—SiMe_2)_n$. The polymer depolymerizes at temperatures above 300 to 400°C to yield mainly the cyclic tetramer.

Halogen-substituted cyclophosphazenes, such as $(NPCl_2)_{3 \, or \, 4}$, $(NPF_2)_3$, or $(NPBr_2)_3$, polymerize thermally, but organic-substituted cyclic derivatives, such

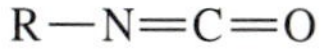

| 2 | 3 | 4 | 5 |

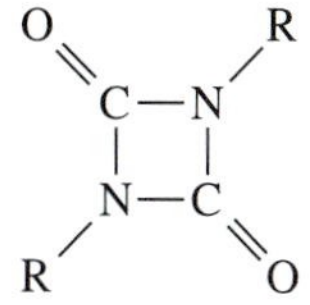

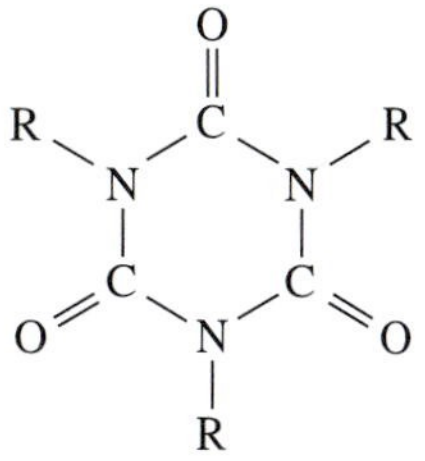

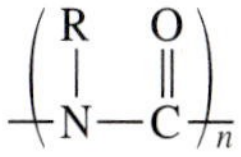

as $(NPPh_2)_3$ or $[NP(OCH_2CF_3)_2]_3$, do not. However, organic-substituted high polymers, such as $[NP(OCH_2CF_3)_2]_n$ or $[NP(OPh)_2]_n$, prepared by an alternative route (see Chapter 7) depolymerize at elevated temperatures to the appropriate cyclic trimers or higher oligomers. A few other inorganic systems appear to generate polymerization-depolymerization systems. For example, the cyclic dimer (6), cyclic tetramer (7), and high polymer (8) of sulfur nitride may form an equilibrating system.

| 6 | 7 | 8 |

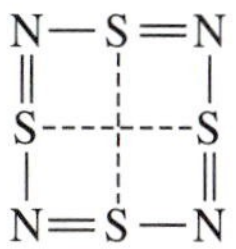

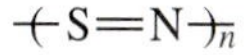

"UNPOLYMERIZABLE" COMPOUNDS

So many cyclic compounds and olefin derivatives polymerize that the fact is often overlooked that a number of compounds which should polymerize have so far proved highly resistant to polymerization. In Chapter 6, it was pointed out that cyclic compounds, such as benzene, cyclohexane, s-triazines, 1,4-dioxane, tetrahydropyran, and borazines resist polymerization. Yet acetylene can be trimerized to benzene or polymerized to polyacetylene. Benzene is an energy trap in the polymeric series (scheme 9). Similarly, s-triazines are the principal products formed from the attempted polymerization of nitriles (scheme 10). Even some cyclic dimers, such as tetramethylcyclodisilthiane (11), equilibrate to the cyclic trimer (12) but do not yield high polymer. An analysis of the reasons for the unpolymerizability of systems such as these can provide valuable clues about the factors which influence polymerization–depolymerization equilibria in other systems. Some of these factors will be considered throughout this chapter.

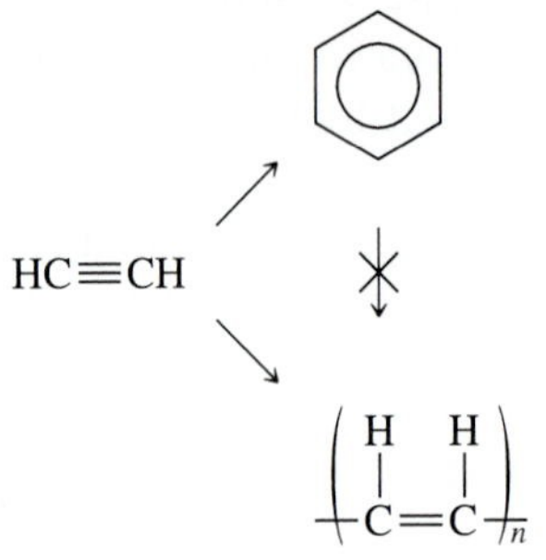

$$HC\equiv CH$$

$$\left(\begin{array}{c} H \quad H \\ | \quad | \\ C=C \end{array}\right)_n$$

9

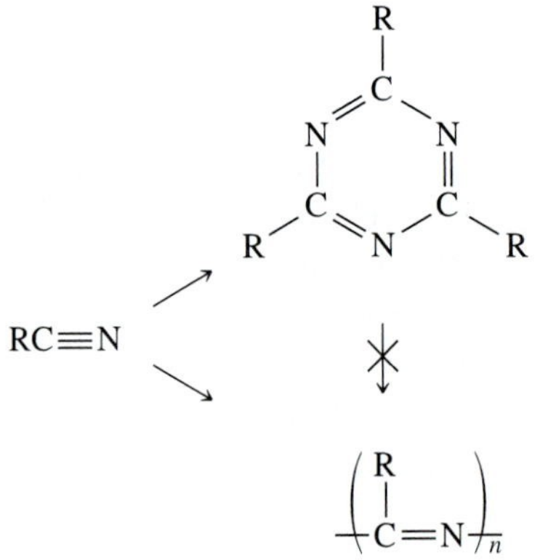

$$RC\equiv N$$

$$\left(\begin{array}{c} R \\ | \\ C=N \end{array}\right)_n$$

10

$$\begin{array}{c} Me \\ | \\ Me-Si-S \\ | \quad | \\ S-Si-Me \\ | \\ Me \end{array} \rightleftharpoons$$

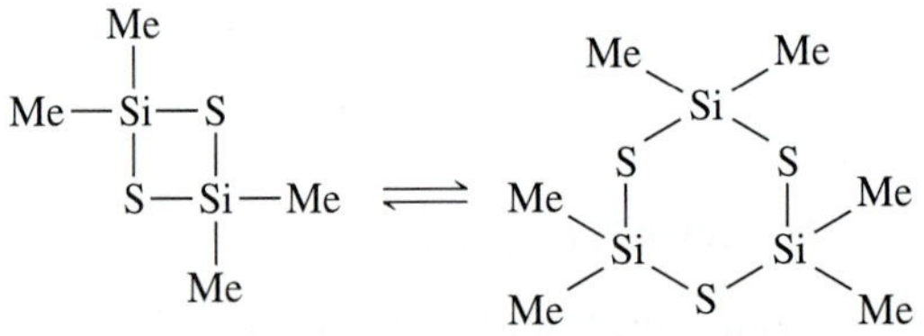

11   **12**

THE GENERAL THERMODYNAMIC PROBLEM

If we consider the polymerization of a monomer or a cyclic oligomer and the depolymerization of a polymer to be governed by thermodynamic factors,[1] some of the observations discussed in the preceding sections begin to make sense.

[1] As discussed later in this chapter, in practice, finite kinetic factors may override the thermodynamic effects.

First, let us consider a hypothetical series of compounds which extends from the monomer, A=B, to the high polymer, $(A\!-\!B)_n$. It will be assumed initially that the lower oligomers in the series are cyclic and that the higher polymers are either macrocyclic or linear. If we make these assumptions, the series can be formulated as:

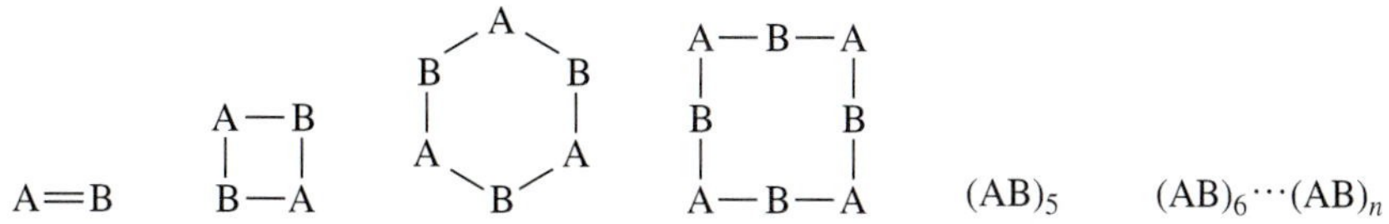

Let us also consider for a moment that equilibration can occur between all the members of this series. For this to happen, every compound in the series would need to possess a similar free energy. In other words, every member of the series can coexist in the equilibrium only if each compound has a similar stability to the others at a given temperature and pressure.

However, in practice, one or several of the compounds will usually be more stable than the others and will be present in greater amounts at equilibrium. In extreme cases, only one or two homologues will exist, with the total exclusion of all other species. Apparently, this is the situation found for benzene, s-triazines, borazines, and the other systems mentioned earlier. On the other hand, if an equilibrium does exist, it should, in principle, be possible to calculate an equilibrium constant for each interconversion, and this could be correlated with the free-energy change in the usual way.

SPECIFIC THERMODYNAMIC EFFECTS

The well-known thermodynamic expression

$$\Delta G = \Delta H - T\Delta S \tag{1}$$

can be used as a basis for understanding the polymerization–depolymerization behavior. In order for a polymerization to be thermodynamically feasible, the Gibbs free-energy change must be negative, that is, $\Delta G_p < 0$. If ΔG_p is positive, depolymerization will be favored. *Any factor that lowers the enthalpy or raises the entropy of a particular species in the system will shift the equilibrium to favor that species.* This elementary consideration allows us to understand many of the puzzling features of monomer–polymer or ring–polymer equilibria.

Polymerization is an *association* reaction in which many molecules come together to form one molecule. Regardless of the mechanism by which this occurs, this process results in a large loss in the number of translational and rotational degrees of freedom in the system. It is fairly obvious that the conversion of three monomer molecules to one cyclic trimer molecule is accompanied by a decrease in translational entropy (there are now fewer molecules in the system). Similarly, the polymerization of a thousand cyclic trimers to one linear polymer molecule reduces the translational entropy further. This major loss of entropy is not compensated by the small entropy increase associated with the torsional and vibrational degrees of freedom in a nonrigid polymer chain. The result is that the entropy change in the polymerization process is nearly always negative, that is, $\Delta S_p < 0$. This is true both for the conversion of olefinic monomers to high polymers and for the polymerization of cyclic compounds. Hence, on entropic grounds alone, depolymerization should always be favored over polymerization. Clearly, if polymerization is to predominate over depolymerization, it must do so under conditions where ΔH_p is sufficiently negative that it compensates for the entropy loss and yields a negative ΔG_p term. In practice, polymerization is often favored at low temperatures where the $T\Delta S_p$ term is small, but depolymerization often occurs at high temperatures where $T\Delta S_p$ is large. This is one of the main reasons for the thermal instability of many polymers at high temperatures.

On a slightly more detailed level, it is possible in many polymer systems to detect a *ceiling temperature* (T_c) above which no polymer can exist. This situation is found when the $T\Delta S_p$ term increases rapidly as the temperature is raised and sharply overtakes the ΔH_p term at the ceiling temperature. The ceiling temperature can be understood in a slightly different way by a consideration of Le Châtelier's principle, which states that a system will respond to a stress in such a way as to relieve that stress. It is clear from the discussion above that most known polymerizations are exothermic processes. Thus, in such cases, depolymerization will be endothermic. Therefore, as the temperature of the system is increased by supplying heat, depolymerization (which absorbs heat) becomes more probable than polymerization. At the ceiling temperature the thermodynamic tendencies for polymerization and depolymerization become equal. Above this temperature polymerization is thermodynamically unfavorable.

However, in other systems, a *floor temperature* (T_f) exists *below* which polymer cannot be detected. The exact behavior of any monomer–polymer or ring–polymer system as the temperature is raised depends on the relationship between ΔH_p and ΔS_p for that particular system. The real problem arises when one needs to know how different magnitudes of ΔH_p and ΔS_p will influence the system, or when it is necessary to make an intelligent guess about the way in which molecular structural changes will affect ΔH_p and ΔS_p. The first problem is discussed in the next two sections, and the latter problem will be considered later.

STANDARD ENTHALPIES, ENTROPIES, AND FREE ENERGIES OF POLYMERIZATION

The general process of polymerization may be described by the stoichiometric equation

$$M(l) \rightarrow \frac{1}{n} P_n(a) \qquad (2)$$

This states that 1 mol of monomer in the liquid state is converted to $1/n$ mol of polymer with an average degree of polymerization, n, in the amorphous or slightly crystalline state. These states of aggregation of monomer and polymer are chosen because they are representative of many polymerizations. An actual polymerization of interest may in fact be carried out with the reactants and products in physical states other than the above (i.e., both in solution and in an inert solvent) and over a wide range of temperatures. However, for the purpose of tabulation and comparison, it is necessary to choose standard states such as those described above, together with a single temperature, which is conveniently taken to be 25°C. Standard thermodynamic methods generally permit straightforward conversions to be made between these standard states and other conditions of interest in actual polymerizations. It should also be noted that equation (2) assumes that the equilibrium lies so far to the side of the polymer that the system can be considered as an irreversible process.

We will now consider how experimental values are obtained for standard enthalpy, entropy, and free-energy changes for polymerization. Following that, we will discuss some specific examples of systems for which equilibrium considerations are important and some for which they are not.

Standard enthalpies of polymerization, ΔH_p°, are most generally obtained by direct calorimetric measurement of the amount of heat evolved when a known amount of monomer is converted to a known amount of polymer. Alternatively, enthalpies of polymerization may be determined by direct calorimetric measurement of heats of combustion of monomer and polymer to yield standard enthalpies of formation ΔH_f° for the monomer and polymer. The enthalpy of polymerization is then obtained, according to (2), by the well-known relationship

$$\Delta H_p^\circ = \frac{1}{n} \Delta H_f^\circ(P_n) - \Delta H_f^\circ(M) \qquad (3)$$

Standard entropies of polymerization ΔS_p° are generally calculated from the absolute entropies of the monomer and polymer:

$$\Delta S_f^\circ = \frac{1}{n} S^\circ(P_n) - S^\circ(M) \qquad (4)$$

The absolute entropies are determined from calorimetric measurements of the heat capacities of the monomer and polymer over a wide range of temperature using the expression

$$S^\circ(T) = \int_0^T \frac{C_p}{T} \, dT \qquad (5)$$

where C_p is the molar heat capacity at constant pressure and T the absolute temperature. The integral is usually evaluated graphically from the experimental data. Much less information is available on ΔS_p than on ΔH_p.

The standard free energies of polymerization, ΔG_p°, are easily obtained by calculation, using (1), once ΔH_p° and ΔS_p° are known.

SPECIFIC MONOMER–POLYMER EQUILIBRIA

In practical terms, information is usually needed about the temperature range in which the monomer can be polymerized and about the ceiling temperature. Below the ceiling temperature (T_c), conversion of the monomer to a polymer involves a free-energy decrease, hence polymerization should occur. Above T_c, depolymerization involves a free-energy decrease and, therefore, depolymerization occurs. At T_c, the tendencies of polymerization and depolymerization should be equal. Thus, the question really revolves around the actual values of ΔH° and ΔS° for polymerization, since $T_c = \Delta H_p^\circ / \Delta S_p^\circ$.[1] Table 10.1 lists ΔH_p°, and calculated values of ΔG_p° for several olefin and aldehyde polymerization systems, together with calculated T_c values. The data in Table 10.1 indicate that the ΔH_p° values for a variety of monomer–polymer systems vary over a considerable range, from -5.1 kcal/mol for *n*-butyraldehyde through about -13 kcal/mol for the

TABLE 10.1 STANDARD ENTHALPIES, ENTROPIES, FREE ENERGIES, AND CEILING TEMPERATURES FOR POLYMERIZATION OF VARIOUS MONOMER-POLYMER SYSTEMS AT 25°C[*]

Monomer	$-\Delta H_p^\circ$ (kcal/mol)	$-\Delta S_p^\circ$ (cal/deg-mol)	$-\Delta G_p^\circ$ (kcal/mol)	Ceiling Temperature (°C) $(T_c = \Delta H_p^\circ / \Delta S_p^\circ)$
Acetaldehyde	—	—	—	-35 to -40[†]
Butadiene	17.6	20.5	11.5	585
n-Butyraldehyde	5.1[†]	22.3[‡]	—	-45[‡]
Chloral	8.0[‡]	28.0[‡]	—	13[‡]
Ethylene	21.2	24	14.0	610
Formaldehyde	7.4	19	1.7	116
Isobutylene	12.9	28.8	4.3	175
Isoprene	17.9	24.2	10.7	466
Methyl methacrylate	13.2	28	4.9	198
α-Methylstyrene	8.4	24.8	1.0	66
Styrene	16.7	25.0	9.2	395
Tetrafluoroethylene	37	26.8	29	1100
2,4,6-Trimethylstyrene	16.7	—	—	—

Source: Mortimer, C. T., *Reaction Heats and Bond Strengths* (New York: Pergamon Press, 1962), Chap. 5; Joshi, R. M. and Zwolinski, B. J., in *Vinyl Polymerization*. Part I, G. H. Ham, ed. (New York: Dekker, 1967), Chap. 8; Dainton, F. S. and Ivin, K. J., *Rev. Chem. Soc. (London)*, 1958, 22, 61; T. Ohtsuka and C. Walling, *J. Am. Chem. Soc.*, 1966, 88, 4167.

[*] Unless otherwise specified, data refer to standard states of pure liquid for the monomer and amorphous or slightly crystalline polymer for the polymer.

[†] Estimated by experimental observations.

[‡] Data refer to standard state of 1 mol/liter in solution.

[1] Or $T_c = \Delta H_0 / [\Delta S_0 + R \ln [M]/[M]_0]$.

methacrylic esters, -17 kcal/mol for styrene, to the very high exothermicity of -37 kcal/mol for tetrafluoroethylene.

The decreasing exothermicity of polymerization through the series ethylene, propylene, and isobutylene can be correlated with the increasing steric hindrance toward polymerization that results from the successive replacement of hydrogen atoms on the same carbon atom by methyl groups. Very little effect is noted for replacement of the first hydrogen by a methyl, but a drastic reduction in exothermicity occurs for replacement of the second hydrogen to yield isobutylene. The same phenomena can be observed in the series ethylene, styrene, α-methylstyrene. However, substitution on the aromatic ring of styrene has very little or no effect on ΔH_p°, as may be seen by a comparison of styrene with 2,4,6-trimethylstyrene. Obviously, steric effects play an important role in the thermochemistry of polymerization.

Other significant influences on the magnitudes of ΔH_p° can be attributed to energy differences in the monomer and polymer that arise from the resonance stabilization that is lost on polymerization or from changes in bond hydridization that occur on polymerization. Differences in the extent of hydrogen bonding in the monomer and polymer can also have an influence on ΔH_p°.

On the other hand, the values for ΔS_p° show much less variation than do the values for ΔH_p°. The range for ΔS_p° values (Table 10.1) is from -19 to -28 cal/deg mol. The reason for this narrow range is that the dominant factor in ΔS_p° is the loss of translational entropy brought about by the large reduction in the number of molecules present, and this factor is relatively constant from system to system.

For most of the monomer–polymer systems listed in Table 10.1, the equilibria at 25°C lie on the side of the polymer rather than the monomer. This fact can be verified by substitution of the ΔG_p° values from Table 10.1 into the well-known equation (6) that connects the equilibrium constant with the standard free-energy change:

$$\Delta G^\circ = -RT \ln K_{eq} \tag{6}$$

But as the temperature is raised, the equilibrium shifts to favor the monomer until, at sufficiently high temperatures, depolymerization predominates. The ceiling temperature (T_c) values in Table 10.1 reflect this change. An enormous spread exists in the T_c values. Thus, polyformaldehyde and poly(methyl methacrylate) depolymerize between 100 and 200°C. Poly(n-butyraldehyde) is unstable above -45°C, while poly(tetrafluoroethylene) should have an equilibrium ceiling temperature about 1100°C. In practice, of course, poly(tetrafluoroethylene) would undergo *fragmentation* reactions in a closed system well below this temperature. It is important to recognize that the ceiling-temperature concept applies only to *closed* systems at equilibrium. If the system is not closed, monomer can be lost by volatilization, and total depolymerization can occur well below the ceiling temperature. Thus, in actual practical use, few polymers demonstrate the thermal stability predicted from the ceiling temperature.

Specific Monomer–Polymer Equilibria

INFLUENCE OF ΔH AND ΔS ON RING-CHAIN EQUILIBRIA (NONRIGOROUS APPROACH)

An approach described by Gee[1] allows a qualitative prediction to be made of the effect on a ring-polymer equilibrium of changes in the magnitudes of ΔH and ΔS. This approach is perhaps the most useful one for systems where only minimal thermodynamic data are available.

Consider the polymerization of a ring to a long polymer chain:

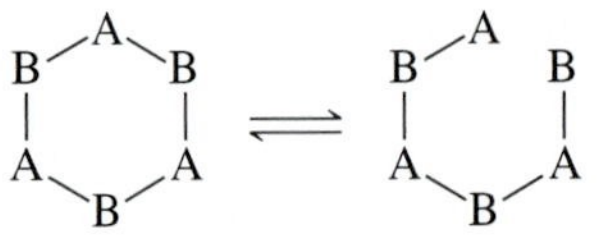

$$\mathrm{R}_x \rightleftharpoons \mathrm{C}_x$$

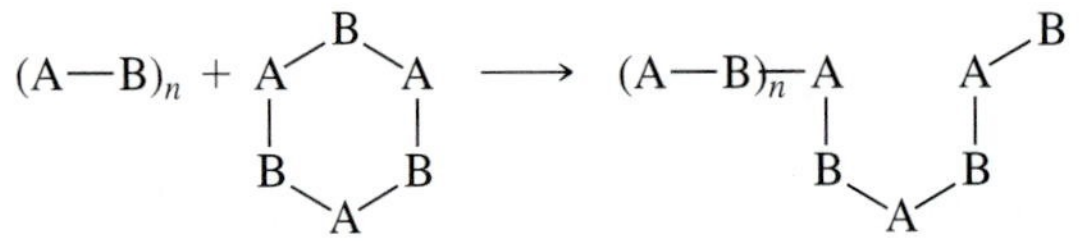

$$(\mathrm{A{-}B})_n + \mathrm{C}_y \longrightarrow (\mathrm{A{-}B})_{\overline{n}}\mathrm{A} \qquad \mathrm{C}_{x+y}$$

The mechanism of this process is not important since we are only concerned with the final position of an equilibrium.

The polymerization process can be viewed as a two-step sequence; (a) the opening of a ring (R_x) to yield short chain (C_x), and (b) the ring-opening reaction of a ring with the end of an existing chain (C_y). The equilibrium state of the system is then given by

$$\mathrm{R}_x \rightleftharpoons \mathrm{C}_x \tag{7}$$

$$\mathrm{R}_x + \mathrm{C}_y \rightleftharpoons \mathrm{C}_{x+y} \tag{8}$$

Reaction (7) involves scission of a bond. Hence, ΔH_7 will be much greater than zero, since bond cleavage requires the input of energy. The entropy change (ΔS_7) for this step will also be significantly greater than zero because the linear fragment C_x will have greater torsional and vibrational entropy than the cyclic species, R_x.

However, the enthalpy and entropy changes inherent in reaction (8) are more difficult to predict. One bond is broken and one bond is formed so that ΔH_8 could be close to zero. Furthermore, although a ring is opened in step (8) with a consequent increase in entropy, two "molecules" are combined into one so that there is a loss of three translational degrees of freedom. This means that ΔS_8 is

[1]Gee, G., *Chem. Soc. (London), Spec. Publ.*, **1961,** *15,* 67.

usually less than zero. We can say with certainty, that, if reaction (8) takes place spontaneously, then the free energy, G_8, must decrease in the process and therefore, ΔH_8 must be less than $T\Delta S_8$.

Let us consider an equilibration in which rings have been partly converted to chains. For reaction (8) we can then define an equilibrium constant K_8 given by

$$K_8 = \frac{[C_{x+y}]_{eq}}{[C_y]_{eq}[R_x]_{eq}}$$

where the subscript "eq" refers to *equilibrium* concentrations. If equilibration has progressed to the point where rings are being added to the ends of *long* chains, then $[C_y]$ will be approximately the same as $[C_{x+y}]$. Therefore, K_8 will become

$$K_8 \approx \frac{1}{[R_x]_{eq}}$$

The relationship between the *standard* free-energy change, ΔG_8°, and the equilibrium constant is given by

$$\Delta G_8^\circ = -RT \ln K_8$$

and by substitution for K_8 we obtain

$$\Delta G_8^\circ = RT \ln [R_x]_{eq}$$

Now, to define the state of polymerization or depolymerization of this system we need to know (a) the weight fraction of molecules that are in the form of chains (denoted by ϕ), and (b) the degree of polymerization of the chains. If $[R_x]_0$ is the concentration of rings in the pure ring compound (i.e., in the starting material), it is possible to relate ΔG_8° to ΔG_8 by the standard methods of thermodynamics. Thus, for reaction (8),

$$\Delta G_8 = \Delta G_8^\circ + RT \ln \frac{[C_{x+y}]}{[C_y][R_x]}$$

If now the approximation that $[C_{x+y}] \approx [C_y]$ can be made at all stages of the equilibration, including the initial stages, then

$$\Delta G_8 = \Delta G_8^\circ - RT \ln [R_x]_0$$

and by substitution of the expression for ΔG_8°,

$$\Delta G_8 = RT \ln \frac{[R_x]_{eq}}{[R_x]_0}$$

If no solvent is present, $[R_x]_{eq}/[R_x]_0 = 1 - \phi$, and substituting for ΔG_8 in $\Delta G_8 = \Delta H_8 - T\Delta S_8$ gives

$$-R \ln (1 - \phi) = \Delta S_8 - \frac{\Delta H_8}{T} \qquad (9)$$

This approximate equation should yield reasonably accurate values for the weight fraction of chains (ϕ) provided that the calculated values for ϕ lie between 0 and 1. If ϕ falls outside this range, the approximate treatment obviously fails and must be replaced by a more rigorous approach. The results of substituting various values for ΔH_8 and ΔS_8 into this equation and assuming that they are independent of temperature are shown in Figure 10.1.

In general terms, there are three main equilibrium situations that can be anticipated.

1. If step (8) is exothermic (ΔH_8 is less than zero), then substitution in the preceding equation indicates that when ΔS_8 is zero or positive, the polymer concentration (as defined by ϕ) will fall with increasing temperature. This

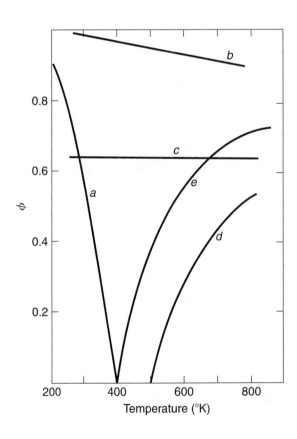

Figure 10.1 Equilibrium fraction of chain polymer (theoretical). [From G. Gee, *Chem. Soc. (London), Spec. Publ.*, **15**, 67 (1961).]

Curve	a	b	c	d	e
ΔH_2 (kcal/mol)	−2	−2	0	2	2
ΔS_2 (cal/deg-mol)	−5	2	2	4	5

situation is summarized by case (b) in Figure 10.1. However, if ΔS_8 is also negative, no polymer can exist above a *ceiling temperature* defined by the value of $\Delta H_8/\Delta S_8$. This possibility is depicted by curve (a) in Figure 10.1.

2. If ΔH_8 is zero, the polymer concentration will be independent of temperature, and given by the term $-R \ln (1 - \phi) = \Delta S_8$. This situation is depicted by curve (c) in Figure 10.1. Significant values of ϕ are obtained only if ΔS_8 is positive. If ΔS_8 is zero or negative, the polymer will always be unstable relative to rings.

3. If ΔH_8 is positive and if ΔS_8 is also positive, the polymer concentration *increases* with increasing temperature and no polymer can exist below a *floor temperature*, T_f (curves d and e, Figure 10.1). The value of the floor temperature is given by $\Delta H_8/\Delta S_8$, and equation (9) can be written as

$$ - \ln (1 - \phi) = \frac{\Delta H_8}{R} \left(\frac{1}{T_f} - \frac{1}{T} \right) \tag{10} $$

However, if ΔS_8 is equal to or less than zero, no polymer can exist at any temperature.

The presence of a solvent will favor the formation of rings at the expense of polymer. This is because an additional term, $R \ln c$ (where c is the weight fraction of total solute in the solvent) must be added to the right-hand side of equation (9).

If two or more sizes of ring are in simultaneous equilibrium with the same polymer chain, equation (9) must be modified to replace $(1 - \phi)$ by the weight fraction ϕ_x of the ring under consideration. Thus, if two rings (R_{x1} and R_{x2}) are present, two simultaneous equations are generated:

$$ -\ln \phi_{x1} = \frac{\Delta H_{81}}{R} \left(\frac{1}{T_{\phi 1}} - \frac{1}{T} \right) \tag{11} $$

$$ -\ln \phi_{x2} = \frac{\Delta H_{82}}{R} \left(\frac{1}{T_{\phi 2}} - \frac{1}{T} \right) \tag{12} $$

Since $\phi_{x1} + \phi_{x2}$ must remain less than 1, this treatment is restricted to the situation where the two rings and the polymer are simultaneously present in significant amounts.

This approach gives particularly favorable results for the $S_8 \rightleftharpoons S_n$ equilibrium, where a floor temperature is known to exist at 159°C. It is also a valuable approach for use in systems where accurate thermodynamic data are not available. If an *estimate* of ΔH_8 and ΔS_8 can be made for a new ring–chain equilibration system, then the existence or absence of ceiling or floor temperatures can perhaps be predicted.

THE SULFUR EQUILIBRIUM (RIGOROUS APPROACH)

A more rigorous approach to ring–chain equilibration calculations was developed by Tobolsky,[1-3] and this method also has been very successfully applied to the sulfur system. The derivation is as follows. Because the polymerization of S_8 is a diradical reaction, the process can be considered in terms of three steps:

(a) $S_8 \overset{k}{\rightleftharpoons} \cdot SSSSSSSS \cdot$ (or S_8^*)   (initiation)

(b) $S_8^* + S_8 \overset{k_3}{\rightleftharpoons} S_{16}^*$ (initial propagation)

(c) $S_{8n}^* + S_8 \overset{k_3}{\rightleftharpoons} S_{8(n+1)}^*$ (general propagation)

It is assumed that the equilibrium constants for the initial and general propagation steps are identical for this reaction, but that these are different from K for the initiation step. Thus if S_8 (ring) is designated as M, S_8^* as M_1^*, and S_{8n}^* as M_n^*, the equilibrium constants for steps (a), (b), and (c) are:

(a) $K = \dfrac{[M_1^*]}{[M]}$ or $[M_1^*] = K[M]$

(b) $K_3 = \dfrac{[M_2^*]}{[M_1^*][M]}$ or $[M_2^*] = K_3[M_1^*][M]$

$$= K[M](K_3[M])$$

(b) $K_3 = \dfrac{[M_3^*]}{[M_2^*][M]}$ or $[M_3^*] = K_3[M_2^*][M]$

$$= K[M](K_3[M])^2$$

Therefore,

$$[M_n^*] = K[M](K_3[M])^{n-1}$$

This expression allows the calculation of K and K_3 at any temperature if [M] and the number-average degree of polymerization, P, are known. Since $P = [W]/[N]$, where [W] is the total concentration of S_8 units in the polymer mixture, and [N] is the total concentration of polymer molecules, [W] and [N] can be obtained from

[1]Tobolsky, A. V., *J. Polymer Sci.*, **1957**, *25*, 220; **1958**, *31*, 126.
[2]Tobolsky, A. V., and Eisenberg, A., *J. Am. Chem. Soc.*, **1959**, *81*, 780, 2303 ; **1960**, *82*, 289.
[3]Tobolsky, A. V., and MacKnight, W. J., in *Polymeric Sulfur and Related Polymers*, H. F. Mark, and E. H. Immergut, Eds. (New York: Wiley-Interscience, **1965**).

$$[N] = [M_1^*] + [M_2^*] + [M_3^*] + \cdots$$

$$[N] = K[M]\{1 + K_3[M] + (K_3[M])^2 + \cdots\}$$

$$[N] = \frac{K[M]}{1 - K_3[M]}$$

$$[W] = [M_1^*] + 2[M_2^*] + 3[M_3^*] + 4[M_4^*] + \cdots$$

$$[W] = K[M]\{1 + 2(K_3[M]) + 3(K_3[M])^2 + \cdots\}$$

$$[W] = K[M]/(1 - K_3[M])^2$$

whence

$$P = \frac{1}{1 - K_3[M]}$$

The total concentration of S_8 units, $[M_0]$, in monomer and polymer is given by

$$[M_0] = [M] + [W] = [M] + \frac{K[M]}{(1 - K_3[M])^2}$$

$$= [M](1 + KP^2)$$

[M] can be eliminated, and the final equation becomes

$$[M_0] = \frac{P - 1}{PK_3} + \frac{KP(P - 1)}{K_3}$$

where $[M_0]$ is the number of moles of S_8 units per kilogram of sulfur and has the value of 3.90 mol/kg at all temperatures.

Thus, determination of K and K_3 at two temperatures allows the calculation of $\Delta H°$, $\Delta S°$, $\Delta H_3°$ and $\Delta S_3°$ from the assumed van't Hoff equation for each equilibrium constant:

$$\ln K = -\left(\frac{\Delta H°}{RT}\right) + \left(\frac{\Delta S°}{R}\right)$$

$$\ln K_3 = -\left(\frac{\Delta H_3°}{RT}\right) + \left(\frac{\Delta S_3°}{R}\right)$$

and the linearity of plots of $\ln K$ against $1/T$ and $\ln K_3$ against $1/T$ allows K and K_3 and, hence, [M] and [P] to be evaluated at all temperatures. As shown in Figures 10.2 and 10.3, there is a striking correspondence between the experimental and calculated values, both above and below the floor temperature.

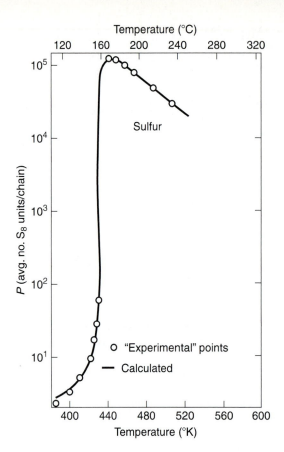

Figure 10.2 Plot of P versus T for liquid sulfur. [From A. V. Tobolsky and A. Eisenberg, *J. Am. Chem. Soc.*, **81,** 780, 2303 (1959); **82,** 289 (1960); © the American Chemical Society; and A. V. Tobolsky and W. J. MacKnight, *Polymeric Sulfur and Related Polymers*, H. F. Mark and E. H. Immergut, eds. (New York: Wiley-Interscience, 1965); © the American Chemical Society.]

Similar expressions have been derived for ring-chain equilibria which proceed through other mechanisms,[1,2] but insufficient experimental data are available for most systems to allow this approach to be more widely applied.

It will be clear from the preceding discussion that techniques are available for the interpretation of ring-chain equilibria in terms of thermodynamic data only for systems which have been subjected to considerable experimental investigation. Unfortunately, these approaches generally do not permit detailed predictions to be made about the expected thermal behavior of new or superficially studied polymers.

THE STATISTICAL INFLUENCE

A ring-chain equilibrium can conveniently be viewed from the statistical point of view. Two approaches to this aspect have been developed. In one, the concentration of any cyclic species in equilibrium with its open-chain homolog is considered

[1]Tobolsky, op. cit.
[2]Tobolsky and Eisenberg, op. cit.

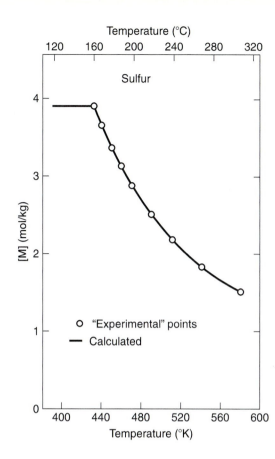

Figure 10.3 Plot of [M] versus T for liquid sulfur. [From Tobolsky, A. V. and Eisenberg, A., *J. Am. Chem. Soc.*, **1959**, *81*, 780, 2303; **1960**, *82*, 289; © the American Chemical Society; and A. V. Tobolsky and W. J. MacKnight, *Polymeric Sulfur and Related Polymers*, H. F. Mark and E. H. Immergut, eds. (New York: Wiley-Interscience, **1965**); © the American Chemical Society.]

in terms of the *probability* of ring closure. In the other, equilibrations are viewed as scrambling reactions between monomer and end-capping units. Both of these viewpoints will now be discussed.

Consider a growing polymer chain

$$AB \xrightarrow{AB} ABAB \xrightarrow{AB} ABABAB \xrightarrow{AB} ABABABAB \xrightarrow{AB} etc.$$

At each step in the formation of open-chain species, there is a possibility that cyclization can take place. Although not indicated in the diagram, cyclization often involves the loss of the end-capping catalyst components. The probability that cyclization will occur from an open-chain homologue depends on the ease with which the chain ends can come together to form a ring, and this, in turn, is a function of the bond angles, bond lengths, and torsional flexibility of the skeletal bonds. Because many skeletal bond angles are in the region of 109 to 120°, six- and eight-membered rings are often heavily represented among the cyclic oligomers but, as the degree of polymerization increases, linear chains appear in greater concentrations at the expense of cyclic species.

The polymerization of rings and the depolymerization of chains are simply different aspects of the same backbone scrambling process. This is one reason why the phenomenon is frequently found in inorganic backbone systems, because the polar skeletal bonds can scramble easily at elevated temperatures.

Several investigators have examined the probability of cyclization in terms of chain length and molecular parameters. In one approach,[1] a model is used which assumes that the distribution of polymer end-to-end distances, r, in a randomly coiled polymer is given by a Gaussian function. For polymers of the type discussed here, the following equation has been derived:

$$R_n = BV x^n n^{-5/2}$$

where R_n is the number of rings with a degree of polymerization, n; $B = (3/2\pi v)^{3/2}/2b^3$, V is the volume of the system; x is the fraction of reacted end groups in the chains; v is the number of skeletal atoms per monomer unit; and b is the "effective link length" of the chain, which equals the individual bond lengths times $(1 + \cos \alpha)/(1 - \cos \alpha)$, α being the skeletal bond angle. A plot of $\log R_n$ against $\log n$ should be a straight line with a slope of $-\frac{5}{2}$ if this model is valid.

The poly(dimethylsiloxane) system has been studied sufficiently that an appreciable amount of experimental ring-chain equilibration data has been accumulated. Thus, this system is particularly suitable for an analysis in terms of the theory discussed above. Experimental data are available[2] which provide the ring size of the species present at equilibrium. These data are shown in Figure 10.4, in which $K_n = x^{-n} V^{-1} R_n$. The cyclic trimer, $(Me_2Si{-}O)_3$, is always present in smaller amounts than the tetramer. The expected distribution is found from $n = 35$ at least up to $n = 200$, but deviations from the theory occur when n is 3 to 35. A small trough in the concentration versus n curve is found between $n = 12$ to $n = 16$, which indicates a smaller proportion of these species present than is predicted by the theory.

A major weakness of the theory is the presumed free torsion of the skeletal bonds. An alternative approach has been used in which a threefold rotational isomeric model for the chain is postulated.[3] This model gives a satisfactory explanation of the relative concentrations of trimer and tetramer at equilibrium, but

[1]Jacobson, H., and Stockmayer, W. H., *J. Chem. Phys.*, **1950,** *18,* 1600.
[2]Brown, J. F., and Slusarczuk, G. M. J., *J. Am. Chem. Soc.*, **1965,** *87,* 931.
[3]Carmichael, J. B., and Kinsinger, J. B., *J. Polymer Sci.*, **1963,** *A1,* 2459.

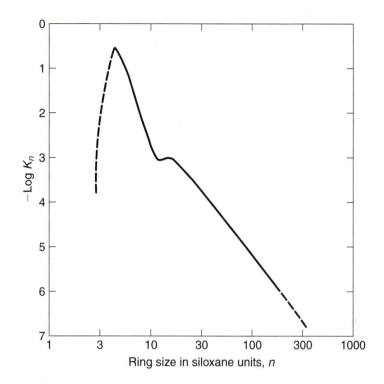

Figure 10.4 Dependence of the molar cyclization constant, K_n, on the number of siloxane units per ring, n, for poly(dimethylsiloxanes) in toluene at 110°C. The dashed sections of the curve are considered less reliable than the remainder. [From Brown, J. F. and Slusarczuk, G. M. J., *J. Am. Chem. Soc.*, **1965**, *87*, 931, © the American Chemical Society.]

not of the higher homologues. However, equilibrium ratios of the $n = 15$ to 200 homologues have been predicted correctly by statistical weighting of the calculation in favor of a *trans*-planar conformation.[1]

It seems clear that the deviations from theory result mainly from specific molecular effects which predominantly influence cyclization of the lower-molecular-weight cyclic homologues. Ring strain, intramolecular repulsions and attractions, and electrostatic effects associated with the polar chain ends must all affect the ease of cyclization, particularly in the range $n = 3$ to $n \approx 20$. Unfortunately, comparable data are not yet available for other skeletal systems, and valid extrapolations of the theory cannot be made.

The second statistical approach is concerned with the random scrambling of chain end units and middle units to create rings or chains.[2] An equilibrating system can be viewed as a scrambling mixture of chain units, end units, and ligands. For example, in terms of the terminology used previously, a simple equilibration

[1]Flory, P. J., and Semlyen, J. A., *J. Am. Chem. Soc.*, **1966**, *88*, 3209.
[2]Van Wazer, J. R., *J. Macromol. Sci.*, **1967**, *A1(1)*, 29.

between rings of formula $(A-B)_n$ and chains of formula $X(A-B)_n Y$ might be resolved into statistical scrambling reactions between $A-B$, XY, $X(A-B)_n Y$, and so on, units.

An analysis of the equilibrium constants, assuming random scrambling, suggests that pure chain-chain equilibria should be unaffected by dilution, but that ring-chain equilibria should be shifted toward small rings and the XY species at high dilutions.[1] Inorganic polymer systems are especially suited to this kind of analysis, and experimental correlations have been made for systems that contain $Si-O$, $Si-S$, and $S-O$ skeletal bonds.

MOLECULAR STRUCTURAL EFFECTS (QUALITATIVE APPROACH)

So far in this chapter, we have considered the general thermodynamic factors which influence polymerization–depolymerization equilibria. Unfortunately, for many organic systems and for nearly all inorganic systems, insufficient thermodynamic data are available to draw accurate conclusions or, indeed, to make even crude predictions about the thermal equilibration behavior. However, it is fortunate that certain qualitative trends can be recognized that enable molecular structural features to be correlated with polymerization or depolymerization behavior. These trends provide a nonrigorous, but useful, method of interpretation. Specific molecular factors that can be interpreted in this way include (1) an analysis of skeletal bond energies, (2) the influence of skeletal bond angles, (3) the presence or absence of pseudoaromaticity, and (4) the effect of side-group interactions. Each of these factors will now be considered.

SKELETAL BOND ENERGIES

In any polymerization system, it is possible to visualize a complete series of compounds which would contain all species from the monomer through the cyclic or linear oligomers to high polymers. As we have already seen, the amount of each species present at equilibrium will depend on the free energy of that species. Although the statistical influences are important, a major factor which influences the amount of each species present will be the enthalpy. This, in turn, will depend largely on the skeletal bond energy.

Many systems are known in which the skeletal bond energy varies very little from cyclic oligomers to high polymers, and it is in these systems that statistical factors tip the balance in favor of one species or another. However, significant bond energy differences usually exist between monomers or cyclic dimers and the higher homologues in the series. Cyclic dimers are often destabilized by ring strain, and this factor will be discussed in the next section.

However, an olefin-type monomer will have a higher enthalpy than the cyclic oligomers or high polymers derived from it because of the energy stored

[1]Ibid.

in the double bond. Thus, most unsaturated monomers are expected to polymerize exothermically and with a negative entropy change (see Table 8.1). The degree to which the polymerization is exothermic depends on the particular energy of the π-bond in the monomer, and this, in turn, is a function of the types of substituent group present.

The type of multiple bonding is also important. No stable unsaturated monomers are known of structure, $R_2Si{=}O$, $R_2Si{=}NR'$, $R_2P{\equiv}N$, $S{\equiv}N$, and so on, and species such as these either do not participate in the equilibration process or at best exist only as transient high-energy intermediates. Silicon, in particular, shows a strong tendency to avoid the formation of $p_\pi - p_\pi$ bonds. Thus, the depolymerization products from poly(organosiloxanes) or poly(organophosphazenes) are cyclic oligomers rather than monomers.

SKELETAL BOND ANGLES

The phenomenon of ring strain is well known. It occurs when the formation of cyclic compounds requires the distortion of bond angles from their preferred values. Thus, the destabilization of ring-strained species can be ascribed to an increased enthalpy term, although the probability of cyclization occurring to form such species would also be reduced. We may describe the "preferred" bond angle as the bottom of an energy well, with angular distortions from this value represented by higher enthalpy values.

It is obvious that if, for example, all the skeletal bond angles have a minimum energy at a 120° angle, then the six-membered ring will probably be a favored homologue during equilibration, whereas a four-membered ring will be a higher-energy species and will either be present at equilibrium in very small amounts or may not be found at all. Some ring strain may also be present in cyclic tetramers, pentamers, or hexamers, but the conformational mobility of higher homologues should provide for a release of the strain.

Nevertheless, some bonds have a much greater angular flexibility than others, and the bond angle may vary over as much as a 30° range before molecular destabilization becomes evident. In general, angular flexibility is often associated with skeletal atoms which can rehybridize easily, which have a large atomic radius, or which form bonds with a high ionic character. Thus, bonds to tetrahedral carbon (109.5° angle) or tetrahedral silicon often have little angular flexibility, whereas bonds to dicoordinate skeletal oxygen, nitrogen, or sulfur can vary over a wide range before appreciable strain enthalpy becomes manifest.

To illustrate the powerful destabilizing influence of ring strain, it is only necessary to note the nonexistence of a formaldehyde dimer, $(H_2C{-}O)_2$ in the oxymethylene equilibrate, and the absence of cyclic dimers in the siloxane, phosphazene, or vinyl-type equilibrations. Those cyclic dimers which do exist, such as $(S{=}N)_2$ or $[(CF_3)_2C{-}S{-}]_2$, polymerize readily (and sometimes violently). Ring strain can also be a factor in reducing the concentration of some cyclic trimers in an equilibrium mixture. For example, it has been suggested that the low equilibrium concentration of $(Me_2Si-O)_3$ in the dimethylsiloxane equilibrate may reflect 3 to 9 kcal/mol of ring strain. Higher cyclic oligomers can usually relieve bond-angle strain by puckering of the ring.

AROMATICITY AND DELOCALIZATION

The concepts of aromaticity and delocalization are well known. Delocalization of π-bonding electrons over several centers generally results in a lowering of the enthalpy of the molecule. If a polymeric series of compounds contains certain species in which the delocalization per repeating unit is particularly favored, then that molecule may be stabilized sufficiently to be a predominant species at equilibrium.

The most obvious example is benzene. The delocalization in this molecule is overwhelmingly favored by the 120° skeletal bond angles and by the planarity of the ring. By contrast, cyclobutadiene is unstable and is considered to be "antiaromatic," and cyclooctatetraene is puckered and nonaromatic. Furthermore, the well-known $4n + 2$ π-electron rule (Huckel's rule) provides an additional rationalization for the particular stability of benzene. A similar situation is found with s-triazines and borazines.

However, the same arguments do not apply if the skeletal multiple bonding is of the $d_\pi - p_\pi$ type. In phosphazenes (**13**), for example, the "unsaturation" results from the use of phosphorus $3d$ orbitals. Because $d_\pi - p_\pi$ bonds have different symmetry requirements from $p_\pi - p_\pi$ bonds, the usual aromatic restrictions do not apply to phosphazenes, and no one member of the series appears to be stabilized by this effect. Similarly, although siloxanes are also believed to be stabilized by $d_\pi - p_\pi$ skeletal bonding, the trimer is not especially stable when compared to the other homologues.

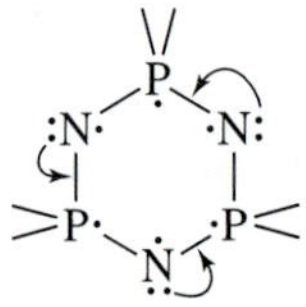

13

Electron delocalization phenomena may also stabilize certain vinyl monomers (e.g., styrene) more than others, and this would be expected to become manifest as a lower than expected heat of polymerization.

SIDE-GROUP INTERACTIONS

Bulky side groups attached to a polymer chain can destabilize a polymer relative to the monomer or cyclic oligomers. This effect is so far-reaching that it can overpower many of the other factors that have already been discussed.

When an equilibrium exists between a cyclic oligomer and a high polymer, the position of the equilibrium is usually determined by the relative enthalpies of the two species. If the side groups attached to the skeleton are small, then the intramolecular repulsions in the oligomer and the polymer will be comparable. However, if bulky side groups are present, polymerization of a cyclic oligomer (a trimer or tetramer, for example) to the polymer will be accompanied by an enthalpy increase—a result of intramolecular steric repulsions within the polymer. This can be illustrated by the polymerization schemes shown in Figure 10.5 and by the energy profile shown in Figure 10.6.

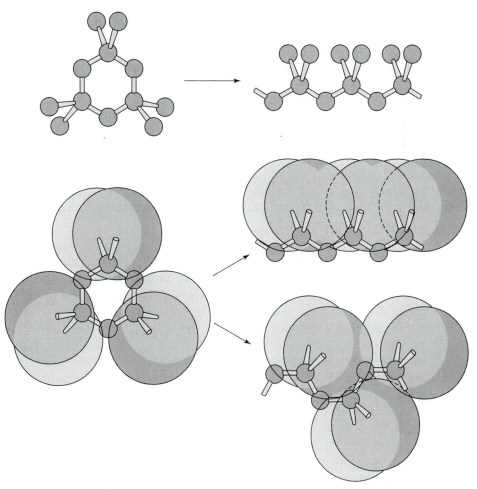

Figure 10.5 Chain stability is favored by small side groups (top), which do not interfere with each other when a ring is polymerized to a chain. If a ring contains bulky side groups (bottom), polymerization becomes difficult, if not impossible, because the side groups repel each other. Moreover, chains with bulky side groups tend to depolymerize at lower temperatures than chains with small ones.

Side-Group Interactions

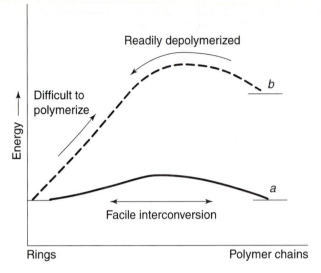

Figure 10.6 Those rings that can be polymerized yield polymer chains that are comparable in energy (per repeating unit) to the original cyclic molecules (a). However, if bulky substituent groups are present (b), the energy of the polymer may be higher than that of the small rings. In such a case, at high temperatures the polymer will be converted easily over a small activation energy barrier to rings, but the rings cannot be polymerized to chains.

The side group–side group and side group–chain distances are always shorter in polymers than in cyclic dimers, trimers or tetramers, irrespective of the preferred conformation of the polymer chain. The shorter intramolecular distances in the polymer nearly always result in more serious van der Waals repulsions, and the ultimate outcome is the destabilization of the polymer relative to the cyclic dimer, trimer, or tetramer. When such a situation exists, polymerization of the cyclic oligomers may be thermodynamically impossible. Such reasons are believed to underlie both the failure of many cyclic compounds to polymerize, and the facile depolymerization of hindered polymers prepared by alternative routes (see Chapter 7). It is, in fact, possible to predict roughly which side-group units cause destabilization for a particular skeletal system with the use of molecular models or hard-sphere van der Waals radii approaches or by a rough calculation of the energies involved within a short polymer segment.

Much difficulty has been reported with the preparation of high-molecular-weight polysiloxanes which contain side groups larger than methyl groups. For example, although equilibration of dimethylsiloxanes yields a mixture of 87% linear polymers and approximately 13% cyclic tetramer plus pentamer, the presence of one trifluoropropyl group per silicon atom in $(F_3CC_2H_4Si(CH_3)-O)_n$ markedly shifts the equilibrium to favor the cyclic oligomers.[1] Thus, the $n = 3$ to 6 cyclic content is 86.5% at equilibrium. The same type of behavior has been reported when ethyl, aryl, and trialkylsilyl side groups are present, although mechanistic influences may be partly responsible in these cases.

Evidence exists that there is a rough correlation between the ease of depolymerization of poly(organophosphazenes) and the dimensions of the sub-

[1]Brown, E. D., and Carmichael, J. B., *J. Polymer Sci.*, **1965,** *B3,* 473.

stituent group. For example, poly(dichlorophosphazene) appears to resist de-polymerization at temperatures below 350°C, whereas $[NP(OCH_2CF_3)_2]_n$ and $[NP(OPh)_2]_n$ depolymerize at lower temperatures. Linear organophosphazene polymers, such as $[NP(OCH_2CF_3)_2]_n$ or $[NP(OPh)_2]_n$, suffer appreciable in-tramolecular steric hindrance between the side groups themselves and between the side groups and the chain atoms. Van der Waals radii-type molecular models for these polymers can be constructed only with difficulty. The destabilization inherent in this crowding must be especially serious at elevated temperatures. On the other hand, *cyclic* trimers or tetramers with the same substituent groups suffer negligible intramolecular steric repulsions because the side groups are oriented away from each other and away from the nearby skeletal atoms. Thus, a strong argument can be made that many organosubstituted linear polyphosphazenes are thermodynamically less stable than the related cyclic trimers or tetramers. At moderate temperatures the polymers are kinetically stabilized against depolymerization, but at high temperatures a trimer-favoring equilibrium will be established. The failure of many organo*cyclo*phosphazenes to polymerize at high temperatures can be readily understood in these terms.

If an unsaturated *monomer* can be formed by depolymerization, the side-group influence may be even more serious. Conversion of a cyclic oligomer or polymer to an unsaturated monomer not only relieves the repulsions illustrated in Figure 10.5, but it also relieves the repulsions between two side groups attached to the same skeletal atom **(14)**. The relatively low enthalpy of polymerization of isobutylene (-12.6 kcal/mol) may reflect the influence of methyl-group hindrance in the conversion of the monomer to the polymer. For example, conversion of a polyaldehyde to an aldehyde changes the hybridization at carbon from sp^3 to sp^2. This means that the R—C—R bond angle changes from 109.5° to 120°. Hence, the repulsions from the presence of bulky R groups can be relieved especially easily by depolymerization.

14

Although formaldehyde can be polymerized to trioxane, tetroxane, and polyoxymethylene, aldehydes or ketones with side groups larger than hydrogen polymerize only with difficulty. For example, equilibration of acetaldehyde at 15°C yields 94.3 wt % of the cyclic trimer, $(MeHC—O)_3$, and at 0°C the equili-brate contains 5.6% monomer, 91% cyclic trimer, and 3.2% cyclic tetramer. High polymers are formed only below the ceiling temperature of -40°C, and they de-polymerize at room temperature. The thermal instability of this polymer con-trasts strikingly with that of poly(oxymethylene), and the differences must be ascribed to steric repulsions by the methyl side groups. When the side group is larger than methyl, the facile depolymerization of the polymer becomes even more noticeable. Aldehydes with ethyl, *n*-propyl, isopropyl, *n*-butyl, and cyclohexyl groups can be polymerized only at high pressures and often at low temperatures. Once formed, the polymers depolymerize to monomer under ambient conditions. Polyacetone depolymerizes readily at room temperature, as does poly(hexafluorothioacetone), $[(F_3C)_2C{-}S]_n$.[1]

THE MECHANISTIC ASPECT

Polymerization of an unsaturated monomer or a cyclic oligomer, or depolymerization of a high polymer, depends not only on the free-energy change, but also on the activation energy for the reaction. It is, in fact, quite possible for a polymerization or depolymerization to be energetically feasible, yet kinetically inhibited. If the activation energy is high, the rate of equilibration will be infinitely slow at moderate temperatures. This raises the interesting point that, if the activation energy for polymerization is high, and if depolymerization is thermodynamically preferred at high temperatures, then it may be impossible to find conditions for the uncatalyzed polymerization of a monomer or cyclic oligomer. Thus, catalysts play an important role in facilitating most monomer-chain and ring-chain equilibration processes. It must also be emphasized that serious experimental difficulties are involved in the separation of mechanistic influences from thermodynamic factors. For example, in many systems the observed change in concentration of one homologue as the temperature is raised results not from changes in the equilibrium constant but from side reactions. Thus, extreme care must be exercised in the interpretation of equilibration data.

Most ring-chain interconversions are strongly influenced by the presence of a catalyst or end-capping species. Thus, the initiation for a heterolytic ring cleavage generally involves a reaction with the initiator. A general mechanism can be depicted as

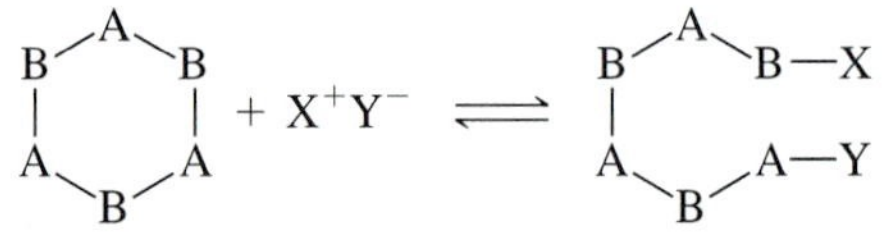

where X^+Y^- is the initiator and B is the most electronegative element in the skeleton. Propagation then involves attack by the BX or AY bonds on another oligomer molecule.

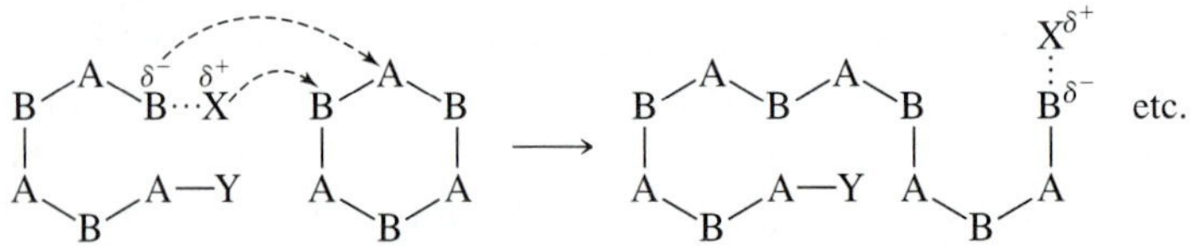

[1]Middleton, W. J., Jacobson, H. W., Putnam, R. E., Walter, H. C., Pye, D. G., and Sharkey, W. H., *J. Polymer Sci.*, **1965**, *A3*, 4115.

If the chain is short, cyclization can occur by the process

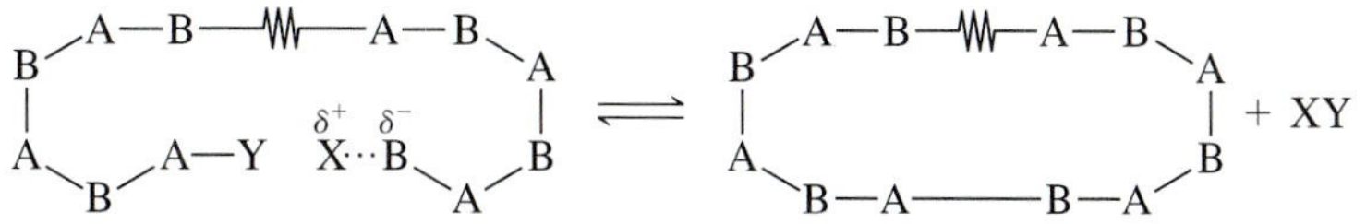

But, if the chain is long, cyclization may occur only infrequently, and end-capped high-molecular-weight chains may be present in appreciable quantities. Of course, if the chain ends are held together by electrostatic forces, cyclization will always be preferred. However, the presence of high-molecular-weight chains with active end groups is a prime reason for the thermal instability of many polymers, since facile cyclization–depolymerization reactions occur at moderate or high temperature, often by a "back-biting" mechanism such as

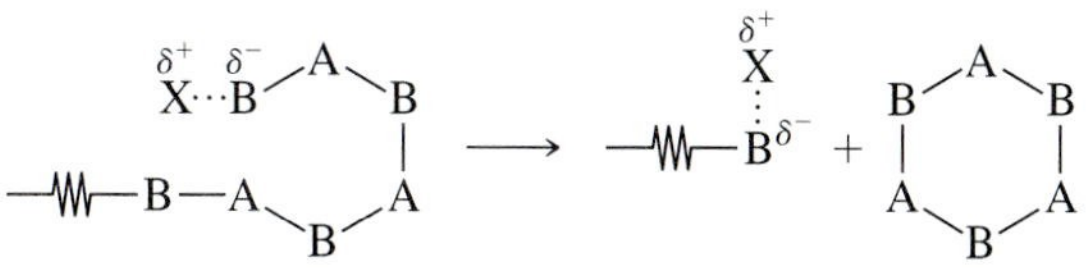

It will be recognized that this general mechanism operates in the polymerization and depolymerization of trioxane (Chapter 6), siloxanes, and phosphates (Chapter 7). Since the mechanism provides a low-energy pathway for depolymerization, polymers susceptible to this type of mechanism often depolymerize at moderate temperatures. For this reason, the thermal stability of a high polymer can frequently be improved by total removal of the initiator or by replacement of the end groups by nonionic substituents.

Free-radical cleavage processes of carbon–carbon bonds occur readily at 200 to 300°C. However, it is important to distinguish between the *random* fragmentation of polymer chains (which does not represent an equilibration process) and the "unzipping" of monomer molecules from the chain ends that takes place with depolymerization. The latter process can be understood readily in terms of the discussion in this chapter.

STUDY QUESTIONS

1. Discuss possible reasons why a cyclic compound such as trioxane can be induced to polymerize, but benzene cannot.
2. A cyclic trimer, $(A—B)_3$, polymerizes to form a linear macromolecule, $(A—B)_n$, in the melt at 100°C. At equilibrium, 40 wt % of the total species present is high polymer. Calculate ΔG and ΔH for the addition of a trimer unit to the end of a growing chain. Assume that $\Delta S = 28$ cal/deg-mol.
3. In what way does an initiator influence a polymerization–depolymerization equilibrium, and why?

4. Define the terms "ceiling temperature" and "floor temperature." Why do these definitions generally apply only to closed systems?

5. Assume that the polymerization of S_8 proceeds through the formation of a zwitterion $^+S-(S_6)-S^-$, instead of through the diradical, $\cdot S-S_6-S\cdot$. How would this be expected to affect the overall character of the equilibrium, T_c, T_f, ΔS_p, and ΔH_p?

6. What effects on the equilibration behavior of organosiloxanes would be expected from the replacement of methyl side groups by ethyl, phenyl, chloro, fluoro, cyano, or trimethylsilyl groups?

7. A certain ring-opening polymerization reaction (cf. equation 8) of a cyclic compound is exothermic by 7.0 kcal/mol and occurs with a negative entropy change of -21 cal/deg-mol.

    (a) Using the Gee approach, construct a plot of the weight fraction of molecules that exist as chains as a function of temperature.

    (b) Evaluate the ceiling temperature for the polymerization.

    (c) At what temperature is the weight fraction of chains equal to 0.95?

    (d) Specify any implicit or explicit assumptions that you have made in your calculations.

8. In a polymerization of rhombic sulfur from the monomeric form of S_8 rings, the following data were found at equilibrium [Tobolsky, A. V. and Eisenberg, A., *J. Am. Chem. Soc.*, *81*, 780, 2303; **1960**, *82*, 289]:

	267°C	167°C
Average degree of polymerization	1.0×10^4	9.5×10^4
Monomer concentration (mol/kg)	1.82	3.63

    (a) Using the approach of Tobolsky, evaluate $\Delta H°$, $\Delta S°$, $\Delta H_3°$, and $\Delta S_3°$, as these quantities are defined on page 245, for the sulfur equilibrium.

    (b) Construct van't Hoff plots (i.e., $\ln K$ versus $1/T$) for both K and K_3.

    (c) Calculate the average degree of polymerization and the monomer concentration in this system at equilibrium at 200°C.

9. Discuss how the synthesis of cyclic polynucleotides such as DNA by living organisms might differ from the cyclization methods described in this chapter.

SUGGESTIONS FOR FURTHER READING

BILLMEYER, F. W., Jr. *Textbook of Polymer Science*, 3rd ed. New York: John Wiley, **1984**, Chaps. 2, 3, 6.

CLARSON, S. J., FITZGERALD, J. J., OWEN, M. J., and SMITH, S. D., (Eds.), *Silicones and Silicone-Modified Materials*. Washington: American Chemical Society, **2000**.

DAINTON, F. S., and IVIN, K. J., "Some Thermodynamic and Kinetic Aspects of Addition Polymerization," *Quart. Rev. Chem. Soc. (London)*, **1958**, *22*, 61.

ELIAS, H. F., *Macromolecules*, Vol. 2, 2nd ed. New York: Plenum, **1984**, Chap. 16.

GRASSIE, N., "Degradation," *Encycl. Polymer Sci. Technol.* (H. F. Mark, N. G. Gaylord, and N. M. Bikales, Eds.), **1966**, *4*, 647.

JELLINEK, H. H. G., "Depolymerization," *Encyl. Polymer Sci. Technol.* (H. F. Mark, N. G. Gaylord, and N. M. Bikales, Eds.), **1966**, *4,* 740.

LENZ, R. W., *Organic Chemistry of Synthetic High Polymers,*. New York: Interscience Publishers, **1967**.

ODIAN, G., *Principles of Polymerization*, 3rd ed. New York: Wiley-Interscience, **1991**.

PLATE, N. A., LITMANOVICH, A. D., and NOAH, O. V., *Macromolecular Reactions. Peculiarities, Theory, and Experimental Approaches*. Chichester: John Wiley & Sons, **1995**.

SAUVAGE, J.-P., and DIETRICH-BUCHECKER, C. (Eds.), *Molecular Catenanes, Rotaxanes, and Knots*. Weinheim: Wiley-VCH, **1999**.

SAWADA, H., "Thermodynamics of Polymerization. V. Thermodynamics of Copolymerization, Part 1," *J. Macromol. Sci., Rev. Macromol. Chem.,* **1974**, *C10*, 293.

SAWADA, H., "Thermodynamics of Polymerization. VI. Thermodynamics of Copolymerization, Part 2, *J. Macromol. Sci., Rev. Macromol. Chem.*, **1974**, *C11*, 257.

SEMYLEN, J. A., "Ring-Chain Equilibria and the Conformations of Polymer Chains." *Advan. Polymer Sci.*, **1976**, *21*, 41.

SEMYLEN, J. A., "Cyclic Siloxane Polymers," in *Siloxane Polymers* (S. J. Clarson and J. A. Semylen, Eds.) Englewood Cliffs: Prentice Hall, **1993**, p. 135.

SEMYLEN, J. A. (Ed.), *Large Ring Molecules.* New York: John & Wiley & Sons, **1996**.

TOBOLSKY, A. V., *Properties and Stucture of Polymers.* New York: Wiley, **1960**.

11

Kinetics of Condensation (Step-Growth) Polymerization

INTRODUCTION

As discussed in Chapter 2, condensation (or step-growth) polymerization occurs by consecutive reactions in which the degree of polymerization and average molecular weight of the product increase as the reaction proceeds. Usually, although not always, the reactions involve the elimination of a small molecule such as water, and hence the name *condensation polymerization* is used as a general term. We may represent the reactions occurring in a water-elimination polymerization by processes (1) to (6).

$$\text{Monomer} + \text{monomer} \quad \rightarrow \quad \text{dimer} + H_2O \qquad (1)$$

$$\text{Monomer} + \text{dimer} \quad \rightarrow \quad \text{trimer} + H_2O \qquad (2)$$

$$\text{Monomer} + \text{trimer} \quad \rightarrow \quad \text{tetramer} + H_2O \qquad (3)$$

$$\text{Dimer} + \text{dimer} \quad \rightarrow \quad \text{tetramer} + H_2O \qquad (4)$$

$$\text{Dimer} + \text{trimer} \quad \rightarrow \quad \text{pentamer} + H_2O \qquad (5)$$

$$\text{Trimer} + \text{trimer} \quad \rightarrow \quad \text{hexamer} + H_2O \qquad (6)$$

etc.

Generally, the reactions are reversible, so that the eliminated water must be removed if a high-polymeric product is to be formed. To describe the course of these reactions (which produce, consecutively, stable dimers, trimers, tetramers, pentamers, ..., high polymers) in terms of reaction kinetics would seem at first

306

sight to be a very complicated task. However, fortunately, it is possible to introduce a simplifying approximation that makes the kinetic problem tractable.

For example, consider a polyesterification in which a dibasic acid condenses with a glycol. No matter what the degree of conversion or the molecular weight of the reactants happens to be, the chemical interaction in each step is the same and may be written as in (7). The simplification that renders the kinetic problem soluble is the assumption that the rate constant of (7) is independent of the size of the molecules to which the functional groups are attached. The effect of molecular size on the rate constants of several small molecule reactions that are quite analogous to stepwise polymerization is shown in Figure 11.1. It may be seen in this figure that, except for the very small reactants ($n \leq 2$), the rate constants or reactivities are independent of molecular size. Thus, as a reasonable approximation, each and every condensation reaction in a polyesterification may be assumed to proceed with the same rate constant. Only in the very early stages, that comprise only a minor part of the polymerization, is this assumption of questionable validity.

$$\text{\Large$\wedge$\small$\wedge$}-\text{COOH} + \text{HO}-\text{\Large$\wedge$\small$\wedge$} \longrightarrow \text{\Large$\wedge$\small$\wedge$}-\overset{\overset{\textstyle O}{\|}}{C}-O-\text{\Large$\wedge$\small$\wedge$} + H_2O \qquad (7)$$

THE REACTIVITY OF LARGE MOLECULES

The assumption that the reaction rate may be considered to be constant in spite of changes of molecular size of the reactants is, at first sight, startling. Therefore, it is necessary to inquire into a physical rationalization for the validity of this assumption. There appear to be two major reasons:

1. Consider a molecule undergoing a polyesterification reaction to be a long chain with many possible conformations and having a COOH functional group on the end. The center of mass of the molecule is far removed from the COOH group. It is true that the larger (and heavier) the molecule, the slower will be the diffusion of the center of mass through the solution. However, the rate of movement of the COOH and OH functional groups through the solution may be very different from the rate of movement of the center of mass, and in the final analysis it is the encounter of functional groups that results in chemical reactions. Thus, through changes in the polymer chain conformations, the functional groups COOH and OH may encounter each other at much higher rates than is suggested by the masses of the molecules of which they are a part. This enhanced encounter rate, which is caused by changes in the polymer conformation, would be expected to be approximately independent of molecular size.

2. According to the liquid cage theory of Rabinowitch and Wood,[1] a colliding pair of particles (or functional groups) will be surrounded by a confining cage of solvent molecules. Before the colliding pair can escape from this

[1]E. Rabinowitch and W. C. Wood, *Trans. Faraday Soc.*, **1936,** *32,* 1381.

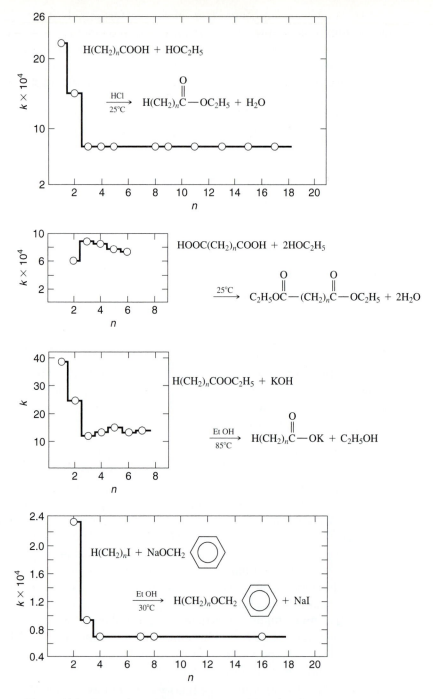

Figure 11.1 Dependence of rate constants of functional group reactions on molecular size. [After P. J. Flory, *Principles of Polymer Chemistry* (Ithaca, N.Y.: Cornell University Press, **1953**), pp. 70–71.]

solvent cage by diffusion away from each other, they will collide frequently. Slower diffusion rates are found in liquids, as compared with gases, and this means that two potential reactants will not come together rapidly, but once brought together will not separate rapidly either. Thus, each encounter leads to multiple collisions. Both the average number of collisions per encounter and the time between encounters increase as the molecular size increases (i.e., as the diffusion rate decreases).

We may represent this in a kinetic sense for two reactants, A and B, by

$$A + B \underset{k_{-8}}{\overset{k_8}{\rightleftharpoons}} (A + B) \tag{8}$$

$$(A + B) \xrightarrow{k_9} P \tag{9}$$

where $(A + B)$ represents the pair of reactants trapped in the liquid cage and P is a product molecule (i.e., polymer). k_8 and k_{-8} represent diffusion rate constants of the reactants into and out of the liquid cage, respectively, while k_9 is the rate constant for chemical reaction. If we assume a steady state for the concentration of trapped pairs, the observed rate of product formation, $d[P]/dt$, is easily shown to be given by

$$\frac{d[P]}{dt} = \frac{k_8 k_9}{k_{-8} + k_9} [A][B] \tag{10}$$

There are two cases to consider:

(a) Diffusion is much more rapid than chemical reaction, $k_{-8} \gg k_9$, and the observed rate constant is given by

$$\text{Case (a): } \frac{d[P]}{dt} = \frac{k_8}{k_{-8}} k_9 [A][B] \tag{11}$$

Since k_8 and k_{-8} represent diffusion into and out of the liquid cage, they are affected in the same way by increases in the size of the reactants. Therefore, the effect of the size of the reactants on the reaction rate will be determined by the effect of size on k_9. The rate constant for a reaction of two functional groups *contained in the same solvent cage* should be independent of the size of the molecule to which they are attached. Hence, for case (a), namely when diffusion is fast compared to chemical reaction, the observed rate constant should be independent of molecular size.

(b) If the chemical reaction is very fast relative to diffusion, $k_9 \gg k_{-8}$, and (10) reduces to

$$\text{Case (b): } \frac{d[P]}{dt} = k_8 [A][B] \tag{12}$$

The Reactivity of Large Molecules

In this case the observed rate constant will be the same as the diffusion rate constant, and this itself will depend on molecular size.

It is apparent from (10) to (12) that, on the basis of liquid cage theory, the lack of dependence of the reaction rate constant on molecular size requires that the chemical reaction be slow compared to diffusion. Those reactions that are characteristic of condensation polymerization are typically the slow reactions of organic chemistry, with appreciable activation energies of at least 12 kcal/mol. It may therefore be concluded that the rate constants of these chemical reactions will, in general, be significantly lower than the diffusion rate constants. Hence, according to (11), the rate constants observed in condensation polymerization will be independent of molecular size. In other words, it is a valid approximation to look upon polyesterification, polyamidation, or polyurethane-forming reactions, and so on, as general reactions of the forms shown by (13) to (15), respectively, in which the rate constants k, k', and k'' do not depend on the sizes of the molecules to which the functional groups are attached.

$$\text{\small{$\mathcal{W}$}}\text{—COOH} + \text{HO—}\text{\small{$\mathcal{W}$}} \xrightarrow{k} \text{\small{$\mathcal{W}$}}\text{—}\overset{\overset{\text{O}}{\|}}{\text{C}}\text{O—}\text{\small{$\mathcal{W}$}} + \text{H}_2\text{O} \tag{13}$$

$$\text{\small{$\mathcal{W}$}}\text{—}\overset{\overset{\text{O}}{\|}}{\text{C}}\text{—Cl} + \text{H}_2\text{N—}\text{\small{$\mathcal{W}$}} \xrightarrow{k'} \text{\small{$\mathcal{W}$}}\text{—}\overset{\overset{\text{O}}{\|}}{\text{C}}\text{—NH—}\text{\small{$\mathcal{W}$}} + \text{HCl} \tag{14}$$

$$\text{\small{$\mathcal{W}$}}\text{—N}=\text{C}=\text{O} + \text{HO—}\text{\small{$\mathcal{W}$}} \xrightarrow{k''} \text{\small{$\mathcal{W}$}}\text{—NH}\overset{\overset{\text{O}}{\|}}{\text{C}}\text{O—}\text{\small{$\mathcal{W}$}} \tag{15}$$

RATES OF POLYCONDENSATION REACTIONS

Since the rate constant of a condensation polymerization reaction (or reactivity of two functional groups) is independent of molecular size, it is possible to measure the rate of reaction simply by determining the concentration of functional groups as a function of time. For example, this may be done easily by titration of the unreacted carboxylic acid groups during a polyesterification reaction. Thus, in a polyesterification, the general reaction at any time t is as shown by (13) and, as the reaction proceeds, the functional groups—COOH and —OH disappear at the same rate. Therefore, samples can be removed from the reaction mixture at various intervals and the concentration of carboxylic acid groups can be determined. The rate of the reaction is then defined as

$$\text{reaction rate} = -\frac{d[\text{COOH}]}{dt} \tag{16}$$

KINETICS OF POLYESTERIFICATION

It is well known in organic chemistry that esterification reactions are catalyzed by acids. Polyesterification is no exception. The rate law may then be written

$$-\frac{d[\text{COOH}]}{dt} = k[\text{COOH}][\text{OH}][\text{acid}] \qquad (17)$$

To proceed further it is necessary to distinguish between those systems in which no acidic catalyst is added and those to which a catalyst has been added and in which its concentration remains constant throughout the polymerization.

Case 1: No Acidic Catalyst Added

In this situation the carboxylic acid groups themselves must function as the acid catalyst and (17) becomes

$$-\frac{d[\text{COOH}]}{dt} = k_3[\text{COOH}]^2[\text{OH}] \qquad (18)$$

where the subscript to k denotes a third-order rate constant. Assume now that stoichiometric quantities of the reactants were present initially. In other words, at $t = 0$,

$$[\text{COOH}]_0 = [\text{OH}]_0 = 2[\text{HOOC}-\text{R}-\text{COOH}]_0 = 2[\text{HOR}'\text{OH}]_0 \quad (19)$$

As shown by the general reaction, (13), COOH and OH groups disappear at the same rate. Therefore, at all times $[\text{COOH}] = [\text{OH}]$ and the rate equation (18) for the uncatalyzed reaction becomes

$$-\frac{d[\text{COOH}]}{dt} = k_3[\text{COOH}]^3 \qquad (20)$$

Integration of (20) leads immediately to

$$\frac{1}{[\text{COOH}]^2} = \frac{1}{[\text{COOH}]_0^2} + 2k_3t \qquad (21)$$

It is convenient to express the conversion, or extent of reaction, in terms of the fraction of COOH groups (or OH groups) that have reacted. Thus, if P is the fraction of COOH groups reacted, then

$$P = 1 - \frac{[\text{COOH}]}{[\text{COOH}]_0} \qquad (22)$$

or

$$[\text{COOH}] = [\text{COOH}]_0(1 - P) \qquad (23)$$

Substitution of (23) into (21) gives the result

$$\frac{1}{(1 - P)^2} = 1 + 2[\text{COOH}]_0^2 k_3 t \qquad (24)$$

When plots are made of experimental values of $1/(1 - P)^2$ versus time, it is found that (24) is not obeyed from $P = 0$ up to about $P = 0.80$ (i.e., for the first

80% of the esterification of —COOH and —OH groups). After 80% conversion the integrated rate expression (24) is obeyed very well. The deviations below 80% conversion are not unique to *poly*esterifications, however, because they are also observed for the simple esterifications that result when the dicarboxylic acid is replaced by a monocarboxylic acid.

Apparently, the major reason for the nonadherence to (24) below about 80% conversion is that the reaction medium is changing from one of pure reactants initially to one in which the ester product is the solvent. The prevalent[1] (although not universal) view among polymer chemists is that the kinetics of condensation polymerization have meaning only for the last 20% of the reaction when the reaction medium has become essentially invariant. Hence, the *true* reaction rate constants are to be obtained from the linear portion of plots of $1/(1 - P)^2$ versus time. For example, typical plots of approximately the last 20% of the uncatalyzed polyesterification of adipic acid and 1,10-decanediol, namely,

$$n\text{HOOC}-(\text{CH}_2)_4-\text{COOH} + n\text{HO}(\text{CH}_2)_{10}\text{OH} \longrightarrow$$

$$\text{HO}\left[\overset{O}{\underset{\|}{\text{C}}}-(\text{CH}_2)_4\overset{O}{\underset{\|}{\text{C}}}-\text{O}(\text{CH}_2)_{10}\text{O}\right]_n\text{H} + (2n - 1)\text{H}_2\text{O} \tag{25}$$

are shown in Figure 11.2. The rate constants obtained from the slopes and initial concentrations of [COOH] are shown in Table 11.1, along with those from some other esterifications of adipic acid. Note that the concentration units of the rate constants are in terms of "equivalents per kilogram"; this is a more convenient measure of concentration than the usual "moles per liter" because the volume of the system decreases significantly. The Arrhenius parameters A and E of the equation $k = Ae^{-E/RT}$ are also tabulated in Table 11.1 for those reactions that have been studied kinetically at more than one temperature.

Case 2: Acid-Catalyzed Polyesterification

If an acid catalyst is added to a polyesterification system (which contains equal quantities of COOH and OH), the general equation, (17), becomes

$$-\frac{d[\text{COOH}]}{dt} = [\text{COOH}]^2(k_3[\text{COOH}] + k_{\text{cat}}[H^+]) \tag{26}$$

where k_3 is the rate constant for the uncatalyzed reaction and k_{cat} is the rate constant for the catalyzed process. By the definition of a catalyst, the [H$^+$] does not change throughout the course of the reaction and generally $k_{\text{cat}}[H^+] \gg k_3[\text{COOH}]$. As a result, (26) usually can be approximated by (27), in

[1]Solomon, O. H., in *Step-Growth Polymerizations*, D. H. Solomon, Ed. (New York: Dekker, **1972**), Chap. 1.

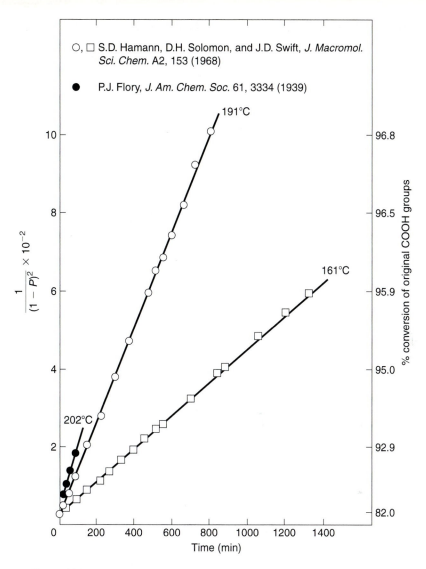

Figure 11.2 Later stages of the uncatalyzed polyesterification of adipic acid and 1,10-decamethylene glycol.

TABLE 11.1 RATE CONSTANTS FOR UNCATALYZED POLYESTERIFICATION OF ADIPIC ACID

Glycol	$A(\text{kg/equiv})^2$ min^{-1}	$E(\text{kcal/mol})$	k at 202°C $(\text{kg/equiv})^2$ min^{-1}
HO—(CH$_2$)$_2$—OH	—	—	~0.005
HO—(CH$_2$)$_{10}$—OH	4.8×10^4	14	0.0175
HO—(CH$_2$)$_{12}$—OH	—	—	0.0157
HO—(CH$_2$)$_2$—O—(CH$_2$)$_2$—OH	4.7×10^2	11	0.0041

Kinetics of Polyesterification

which the second-order rate constant k_2 is related to k_{cat} by the expression $k_2 = k_{cat}[H^+]$.

$$-\frac{d[COOH]}{dt} = k_2[COOH]^2 \qquad (27)$$

Integration of (27) and substitution of (23) leads to

$$\frac{1}{1-P} = 1 + k_2[COOH]_0 t \qquad (28)$$

which is a description of the dependence of the conversion on reaction time for a catalyzed polyesterification. Second-order plots according to (28) are shown in Figure 11.3 for the last 20% of the polyesterifications of adipic acid by 1,10-decanediol and diethyleneglycol catalyzed by p-toluene sulfonic acid. The increase in reaction rate resulting from the presence of the acid catalyst can be seen by comparing the respective conversions as a function of time in Figures 11.2 and 11.3. The second-order rate constants for catalyzed polyesterifications are obtained from the slopes of such plots and the initial concentration of carboxyl groups in accordance with (28).

Second-order rate constants for some acid-catalyzed polyesterifications and polyamidations obtained in this way are shown in Table 11.2.

TIME DEPENDENCE OF THE AVERAGE DEGREE OF POLYMERIZATION AND THE AVERAGE MOLECULAR WEIGHT

Consider a polyesterification of bifunctional monomers in which initially equal amounts of dibasic acid and glycol are present. Under such conditions, the initial number of COOH groups is equal to the total number of molecules present initially in the system. As a consequence of each esterification reaction, namely (7),

TABLE 11.2 RATE CONSTANTS FOR SOME ACID-CATALYZED POLYESTERIFICATIONS AND POLYAMIDATIONS

Monomer System	Catalyst	$T(°C)$	k_2 (kg/equiv-min)	A (kg/equiv-min)	E (kcal/mol)
$HO(CH_2)_2O(CH_2)_2OH$ + $HOOC(CH_2)_4COOH$	0.4% p-toluene sulfonic acid	109	0.013	—	—
$HO(CH_2)_{10}OH$ + $HOOC(CH_2)_4COOH$	0.4% p-toluene sulfonic acid	161	0.097	—	—
$H_2N\!-\!(CH_2)_{\overline{16}}COOH$	m-Cresol (solvent)	175	0.012	1.7×10^{12}	29
$H_2N\!-\!(CH_2)_{\overline{10}}COOH$	m-Cresol (solvent)	176	0.011	1.4×10^{13}	31

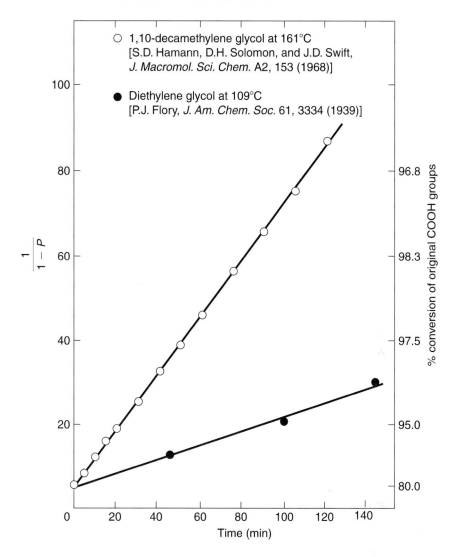

Figure 11.3 Later stages of the polyesterification of adiptic acid catalyzed by *p*-toluene sulfonic acid. (The reaction time of zero corresponds to about 82% esterification of the original COOH groups present.)

$$\mathsf{W\!\!-\!COOH} + \mathsf{HO\!\!-\!\!W} \longrightarrow \mathsf{W\!\!-\!\!\overset{\displaystyle O}{\overset{\|}{C}}\!\!-\!O\!\!-\!\!W} + H_2O \qquad (7)$$

one COOH group disappears, but the total number of molecules present is unchanged. However, if the water formed in the reaction is removed (and this must be done to obtain high polymer), then, for each COOH group lost, one molecule is removed from the system. Thus, with an efficient removal of water, the number of COOH groups present is equal to the number of molecules present, not only initially, but throughout the reaction. If N is the total number of molecules in the system and V is the volume, it is possible to write

$$\frac{N}{V} = [COOH] = [COOH]_0(1 - P) \tag{29}$$

The *repeating unit* of each polyester molecule formed from the monomers $HOOC{-}R{-}COOH$ and $HOR'OH$, namely

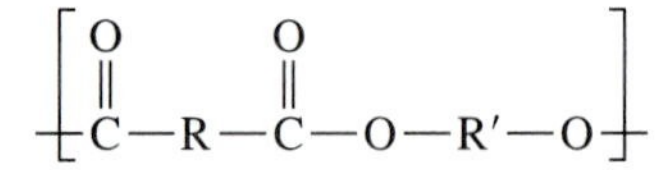

$$\left[\begin{matrix} O & O \\ \parallel & \parallel \\ {-}C{-}R{-}C{-}O{-}R'{-}O{-} \end{matrix} \right]$$

contains one *structural unit* from the glycol, $-O-R'-O-$, and one *structural*

$$unit \text{ from the dibasic acid, namely } \begin{matrix} O & O \\ \parallel & \parallel \\ -C-R-C- \end{matrix}.$$ Structural units are never removed from the system. Therefore, the total number of structural units present at all times is a constant and is equal to the initial number of molecules. Hence, in view of (29) it is possible to write

$$\frac{N_{\text{structural units}}}{V} = [COOH]_0 \tag{30}$$

The average degree of polymerization of the system, $\overline{DP}$, is defined as the average number of structural units per molecules. Therefore, in view of (23), (29), and (30), the $\overline{DP}$ can be defined by

$$\overline{DP} = \frac{[COOH]_0}{[COOH]} = \frac{1}{1 - P} \tag{31}$$

Note that, for condensation polymers prepared from two reactants, the average number of *repeating units* per molecule is one-half of the average degree of polymerization.

If $\overline{M}_0$ is the average molecular weight of the structural units that make up a repeating unit, then the number-average molecular weight of the polyester is given by

$$\overline{M}_n = \frac{\overline{M}_0}{1 - P} + 18 \tag{32}$$

where 18 is added to account for unreacted groups at the ends of each polyester chain.

As an example of the dependence of the average molecular weight on the conversion, consider the polyesterification of adipic acid, $HOOC{-}(CH_2)_4{-}COOH$, and 1,10-decanediol, $HO{-}(CH_2)_{10}{-}OH$. The average molecular weight of a structural unit is: $\overline{M}_0 = (112 + 172)/2 = 142$. Hence, the dependence of the number-average molecular weight on conversion P is, from (32),

$$\overline{M}_n = \frac{142}{1 - P} + 18$$

A plot of $\overline{M}_n$ versus conversion, P, for this system is shown in Figure 11.4. This illustrates the fact that very high conversions are required to produce useful polymers having molecular weights above 10,000.

By combining (32) with (24) and (28), the dependence of the molecular weight on reaction time for uncatalyzed and catalyzed polyesterifications, respectively, is obtained as shown by

$$\overline{M}_n = \overline{M}_0(1 + 2[\text{COOH}]_0^2 k_3 t)^{1/2} + 18 \tag{33}$$

$$\overline{M}_n = \overline{M}_0(1 + [\text{COOH}]_0 k_2 t) + 18 \tag{34}$$

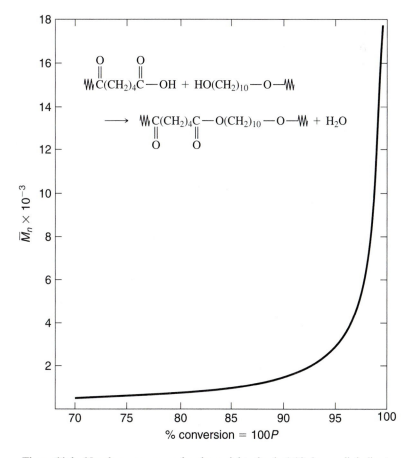

Figure 11.4 Number-average molecular weight of poly-1,10-decanedioladipate as a function of conversion.

The kinetic expressions (24) and (28) are obeyed for conversions above ~80%. For conversions of this magnitude it is generally true that the values of t are sufficiently large that unity in the parentheses of (33) and (34) may be neglected. The approximate equations

$$\overline{M}_n(\text{uncatalyzed}) \approx \overline{M}_0[\text{COOH}]_0(2k_3)^{1/2}t^{1/2} \qquad (35)$$

$$\overline{M}_n(\text{catalyzed}) \sim \overline{M}_0[\text{COOH}]_0 k_2 t \qquad (36)$$

are then obtained. These equations may be used with the rate-constant data of Table 11.1 and 11.2 to construct curves of number-average molecular weights as a function of time.

While most of the kinetic relationships derived in this section have referred to polyesterification reactions between a dicarboxylic acid and a glycol, an extension to other step-growth polymerizations of bifunctional monomers is straightforward.

MOLECULAR-WEIGHT DISTRIBUTIONS OF LINEAR CONDENSATION POLYMERS

The preceding section has covered the average molecular weight and its dependence on reaction time. However, it is also of interest to determine the *distribution* of molecular weights and the dependence of this distribution on the reaction time. An evaluation of the molecular-weight distribution is readily accomplished with the general assumption that the reactivity of functional groups is independent of the size of the molecule to which they are attached.

Consider, for example, the polyamidation of an amino acid of structure $H_2N{-}R{-}COOH$. With a single-component system, the equality of reacting functional groups can be guaranteed at all times (i.e., $[NH_2] = [COOH]$) and only one structural unit is present (i.e., ${-}NH{-}R{-}C{-}$). Of course, the analysis will also be valid for polycondensations or step-growth polymerizations that involve the interaction of two bifunctional monomers, and an extension of the argument to such cases is straightforward.

What is the probability that a molecule selected randomly from the polymerizing mixture will be found to contain *exactly* x structural units? In other words, if the terminal NH_2 group of this randomly selected polymer molecule is considered, what is the probability that it is connected to exactly x structural units? To answer this question, recall that P is the fraction of COOH groups that have reacted in time t. Then, $1 - P$ is the fraction of COOH groups remaining at time t. Expressed in another way, P is the probability that at time t a given COOH group will have reacted; $1 - P$ is then the probability that at time t a given group will *not* have reacted. It is instructive at this state to consider Table 11.3, describing the progress of the polyamidation.

Examination of Table 11.3 shows that in the x-mer (i.e., the randomly selected polymer molecule containing exactly x structural units), $(x - 1)$ reacted

TABLE 11.3 PROGRESS OF POLYAMIDATION

Molecule	Number of Structural Units Present	Number of Reacted COOH Groups
$\overset{\displaystyle O}{\overset{\displaystyle \|}{H-NHRC}}-OH$	1	0
$\overset{\displaystyle O}{\overset{\displaystyle \|}{H-NHRC}}-\overset{\displaystyle O}{\overset{\displaystyle \|}{NHRC}}-OH$	2	1
$\overset{\displaystyle O}{\overset{\displaystyle \|}{H-NHRC}}-\overset{\displaystyle O}{\overset{\displaystyle \|}{NHRC}}-\overset{\displaystyle O}{\overset{\displaystyle \|}{NHRC}}-OH$	3	2
$H-NHRC-NHRC-NHRC-NHRC-OH$ (each with $\overset{O}{\|}$)	4	3
$\vdots$		
$H-\left[NHRC\right]_{x-1}-NHRC-OH$ (each with $\overset{O}{\|}$)	x	$x-1$

COOH groups and one unreacted COOH group will form part of the residue that is connected to each NH_2 group. Beginning at the terminal NH_2 group of a randomly selected polymer molecule, the probability that $(x-1)$ reacted COOH groups and one unreacted COOH group will be found is the probability that the molecule will contain exactly x structural units, and this is given by

$$\text{Prob}(x) = P^{x-1}(1-P) \tag{37}$$

The chance that a randomly selected polymer molecule contains exactly x structural units is given by (37) and is equal to the fraction of molecules that is composed of x-mers. Hence, the number of x-mers, N_x, in a system of N molecules is given by

$$N_x = NP^{x-1}(1-P) \tag{38}$$

The total number of structural units in the system, as discussed earlier, is equal to the initial number of molecules present. Furthermore, if the water produced by the reaction is removed, the number of COOH groups is at all times equal to the number of molecules present. Therefore, we may write

$$N_{COOH} = N = N_0(1-P) \tag{39}$$

which on substitution into (38) gives the number of x-mers present in terms of the initial number of molecules, N_0, as

$$N_x = N_0(1-P)^2 P^{x-1} \tag{40}$$

Molecular-Weight Distributions of Linear Condensation Polymers   **319**

The distribution in (40) shows that, for any given conversion, P (or reaction time t), the highest probability is that low-molecular-weight polymers will be present. However, the distribution becomes broader and the average molecular weight increases as the conversion increases. This is shown clearly in Figure 11.5, in which the fraction N_x/N_0 is plotted as a function of x for conversions of 90, 95, and 99%. According to the definition of N_x/N_0, the area under each curve is $1 - P$.

The number-average molecular weight may be obtained from the probability function in (37) and the usual definition of an arithmetic average. Neglecting the weight of water contained in the terminal groups, the molecular weight of an x-mer is given by $M_{x-\text{mer}} = xM_0$, where M_0 is the molecular weight of the structural unit. Therefore, we may write

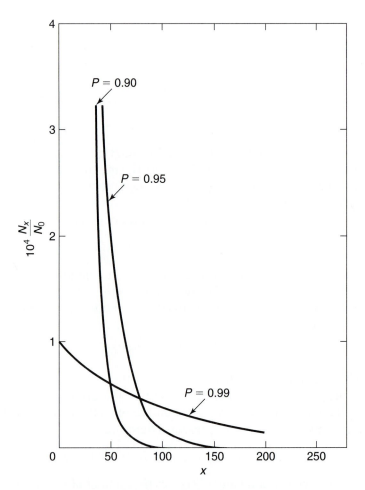

Figure 11.5 Numerical distribution of the number of structural units in a condensation polymer for conversions of 90, 95, and 99% (see equation 40).

$$\overline{M}_n = \sum_{x=1}^{N_0} x M_0 \, \mathrm{Prob}(x) \tag{41}$$

Substitution of (37) into (41) yields

$$\overline{M}_n = M_0 \sum_{x=1}^{N_0} x(1 - P)P^{x-1} = M_0(1 - P) \sum_{x=1}^{N_0} xP^{x-1} \tag{42}$$

The weight fraction of x-mers, W_x, is the weight of molecules containing exactly x structural units divided by the total weight of the polymer, as shown by

$$W_x = \frac{N_x M_x}{\sum_{x=1}^{\infty} N_x M_x} = \frac{M_0 x N_x}{M_0 \sum_{x=1}^{\infty} x N_x} = \frac{x(1 - P)^2 P^{x-1}}{\sum_{x=1}^{\infty} x(1 - P)^2 P^{x-1}} \tag{43}$$

As will be shown later, $\sum_{x=1}^{\infty} xP^{x-1} = (1 - P)^{-2}$, so that from (43) we obtain the weight fraction distribution of x-mers as a function of conversion P. This distribution is given by (44) and is plotted for several values of the conversion in Figure 11.6.

$$W_x = x(1 - P)^2 P^{x-1} \tag{44}$$

The *weight-average molecular weight*, $\overline{M}_w$, is defined by (45), which is, again, simply the concept of an average.

$$\overline{M}_w = \sum_x W_x M_x = M_0 \sum_x x W_x \tag{45}$$

Substitution of (44) into (45) then gives

$$\overline{M}_w = (1 - P)^2 M_0 \sum_{x=1}^{N_0} x^2 P^{x-1} \tag{46}$$

The summations appearing in the expressions for number-average molecular weight (42) and weight-average molecular weight (46) are well known for values of $P < 1$. Because P is less than unity, by definition, the summations may be written as

$$\sum_{x=1}^{N_0} xP^{x-1} = 1 + 2P + 3P^2 + 4P^3 + \cdots + nP^{n-1} + \cdots = \left(\frac{1}{1 - P} \right)^2 \tag{47}$$

$$\sum_{x=1}^{N_0} x^2 P^{x-1} = 1 + 4P + 9P^2 + 16P^3 + \cdots + n^2 P^{n-1} + \cdots = \frac{1 + P}{(1 - P)^3} \tag{48}$$

Substitution of (47) and (48) into (42) and (46), respectively, then gives (49) and (50) for the average molecular weights:

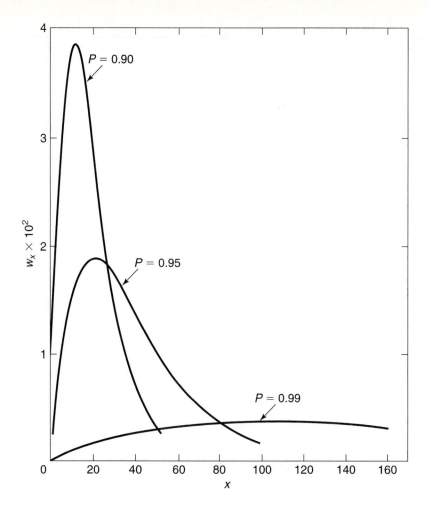

Figure 11.6 Distribution by weight of the number of structural units in a condensation polymer for conversions of 90, 95, and 99% (see equation 44).

$$\overline{M}_n = \frac{M_0}{1 - P} \tag{49}$$

$$\overline{M}_w = M_0\left(\frac{1 + P}{1 - P}\right) \tag{50}$$

The weight-average molecular weight is seen to be always greater than the number-average molecular weight. In fact, for polymers produced by step-growth (condensation) mechanisms, (49) and (50) combine to show this relationship is

$$\overline{M}_w = \overline{M}_n(1 + P) \tag{51}$$

EFFECT OF NONSTOICHIOMETRIC REACTANT RATIOS ON LINEAR CONDENSATION POLYMERS

To produce a high polymer (i.e., $\overline{M} > 10^4$) in condensation polymerization, it is necessary to allow the reaction to proceed to a very high degree of conversion or, in other words, to a product that contains a very small number of *chain ends*. In general, this will be possible only when equal concentrations of reactive functional groups are maintained throughout the course of the reaction. This means that not only must the reaction be initiated with stoichiometric reactant ratios, but the system must also be free of impurities that contain the same functional groups. The system must also be free from side reactions that might selectively consume functional groups and thereby destroy the equality of the functional-group concentrations.

Consider a condensation polymerization of a monomer that has two functional groups of type A (i.e., COOH) with a monomer having two functional groups of type B (i.e., OH). The general reaction is an esterification reaction, and it may be written as

$$A \text{---} A + B \text{---} B \longrightarrow A \text{---} ab \text{---} B + a'b' \qquad (52)$$

Let N_A^o and N_B^o be the respective number of functional groups that are present initially, and assume that their ratio, r, is less than unity (i.e., $r = N_A^o/N_B^o < 1$). Because one structural unit is present for every two functional groups, the *total* number of structural units, N_{su}, is given by

$$N_{su} = \tfrac{1}{2}(N_A^o + N_B^o) = \tfrac{1}{2}N_A^o\left(\frac{1 + r}{r}\right) \qquad (53)$$

Let P be the fraction of A groups that have reacted at time t. Because the functional groups A and B destroy each other on a $1:1$ basis (i.e., $\Delta N_A = \Delta N_B$), the fraction of B groups that have reacted at time t is rP as shown by

$$\frac{\Delta N_B}{N_B^o} = \frac{\Delta N_A}{N_B^o} = \frac{r\Delta N_A}{N_A^o} = rP \qquad (54)$$

At any time t, the number of chain ends, N_{ce}, must be equal to the sum of the numbers of *unreacted* A and B groups, or

$$N_{ce} = (1 - P)N_A^o + (1 - rP)N_B^o \qquad (55)$$

which is easily converted to (56), since $N_B^o = N_A^o/r$.

$$N_{ce} = N_A^o\left[2(1 - P) + \frac{1 - r}{r}\right] \qquad (56)$$

For bifunctional monomers and linear polymers, the number of chain ends, N_{ce}, is at all times equal to twice the total number of molecules present (i.e.,

$N = \frac{1}{2}N_{ce}$). Following substitution of this result into (56), the total number of molecules is given by

$$N = N_A^o \left(1 - P + \frac{1 - r}{2r} \right) \tag{57}$$

As discussed previously, we express the average degree of polymerization $\overline{DP}$ in terms of the numbers of structural units and of molecules as in (58):

$$\overline{DP} = \frac{N_{su}}{N} \tag{58}$$

Substitution of (53) and (57) into (58) leads to (59), which describes the average degree of polymerization in terms of the conversion, P, and the ratio, r.

$$\overline{DP} = \frac{1 + r}{2r(1 - P) + 1 - r} \tag{59}$$

The reader may verify by inspection that (59) reduces to (31) in the limit of $r = 1$ (i.e., at the stoichiometric reactant ratio).

For $r < 1$, an excess of B groups will be present and the reaction is, therefore, complete when all the A groups have reacted. Hence, the maximum average degree of polymerization possible corresponds to a complete conversion of the A groups (i.e., to $P = 1$). According to (59), the maximum $\overline{DP}$ is then given by

$$(\overline{DP})_{max} = \overline{DP}(P = 1) = \frac{1 + r}{1 - r} \tag{60}$$

Equation (60) illustrates clearly the necessity for the use of very pure reagents, and for ensuring exact stoichiometric functional-group concentrations. Similarly, an absence of side reactions that destroy the functional groups selectively must be guaranteed if high-molecular-weight polymers are to be formed in good yield. As an example, consider the effect of a 1% monocarboxylic acid impurity in the adipic acid on the polyesterification of adipic acid with ethylene glycol. The general reaction is

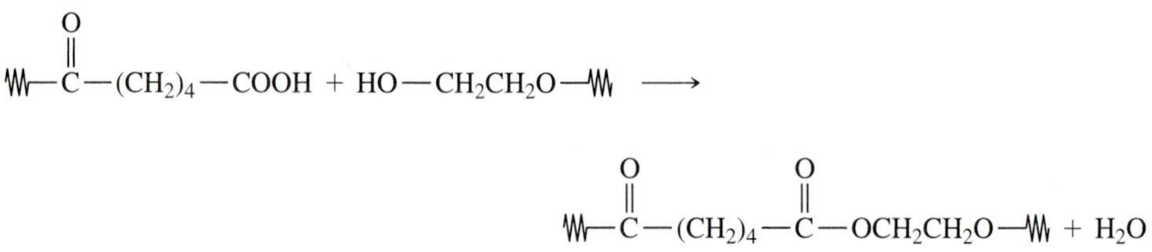

$$\begin{array}{c} O \\ \parallel \\ \text{\Wchar—C—(CH}_2)_4\text{—COOH} + \text{HO—CH}_2\text{CH}_2\text{O—\Wchar} \longrightarrow \end{array}$$

$$\begin{array}{c} O \qquad\qquad\qquad O \\ \parallel \qquad\qquad\qquad \parallel \\ \text{\Wchar—C—(CH}_2)_4\text{—C—OCH}_2\text{CH}_2\text{O—\Wchar} + \text{H}_2\text{O} \end{array}$$

For this polymer $\overline{M}_0 = 86$ g/mol. A 1% impurity in the acid corresponds to a value of r of

$$r = \frac{[\text{Acid}]_0}{[\text{Glycol}]_0} = 0.99$$

Then, according to (60), the maximum average degree of polymerization and average molecular weight are

$$(\overline{DP})_{\text{max}} = \frac{1.99}{0.01} = 199$$

$$(\overline{M}_n)_{\text{max}} = (199)(86) = 17{,}114 \text{ g/mol}$$

BRANCHED AND CROSSLINKED CONDENSATION POLYMERS

In the preceding sections the reactions of *bifunctional* monomers to yield *linear* condensation polymers have been described. However, if at least one of the reactants is a tri- or multifunctional species, the polymerization will generate a branched polymer.

For example, if glycerol, $HOCH_2CH(OH)CH_2OH$, is allowed to react with a dicarboxylic acid or its anhydride, each glycerol molecule that has reacted completely will generate a branch point. Let B be an unreacted OH group on glycerol and b a reacted OH group on glycerol. Similarly, let A and a be unreacted and reacted COOH groups on the acid. Then a reaction in which 9 mol of acid has reacted with 4 mol of glycerol may be written as in (61). The product molecule (**1**) has six unreacted acid groups which may condense with more glycerol. Hence, such branched polymer molecules can grow to very high molecular weights and may ultimately form an infinite network. If internal coupling occurs (reaction of a hydroxyl group and an acid function from branches of the same or different molecule), then the polymer will become cross-linked. In practice, extensive branching and cross-linking cause "gelation" of the polymer. In this state the polymer is swellable by solvents, but it does not dissolve. Highly cross-linked polymers are totally unaffected by solvent.

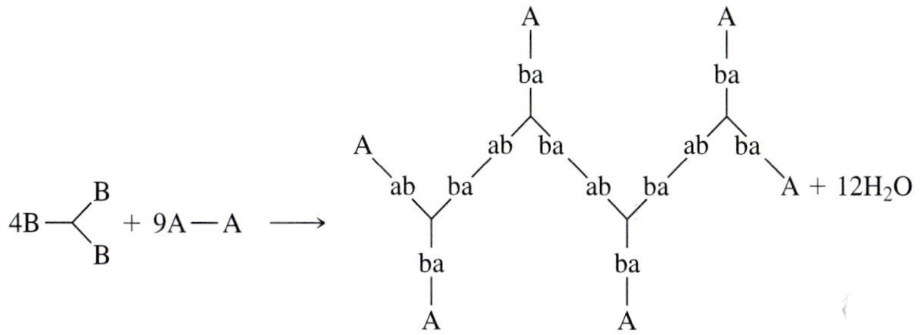

1

If a diol B—B were present, condensation of A—A with B—B would produce linear chains. Clearly, the degree of branching or crosslinking in (61) can be controlled by the relative amounts of triol and diol added to the system. It is possible to calculate the conditions needed to avoid or ensure the reaching of the gel point by use of the Carothers equations,

$$P = \frac{2}{F_{av}} - \frac{2}{\overline{DP}F_{av}} \tag{62}$$

where P is the extent of reaction, F_{av} is the average functionality of the system, and $\overline{DP}$ is the average degree of polymerization. Gelation is presumed to occur when average degree of polymerization becomes infinitely large, in which case, (62) reduces to

$$P_c = \frac{2}{F_{av}} \tag{63}$$

For example, suppose that we wish to compare the behavior of two polyester systems, one of which contains 2 mol of glycerol and 3 mol of a diacid (system X) and another which contains 2 mol of glycerol, 1 mol of a diol, and 4 mol of diacid (system Y). In system X, 12 functional groups are present for every 5 monomer molecules. Hence, F_{av} is $\frac{12}{5}$, or 2.4. Thus, for system X, gelation should occur at the critical point (P_c) of 83.3% reaction. On the other hand, in system Y, there are 16 functional groups per 7 monomer molecules and F_{av} is 2.29. Gelation should occur in this system at 87.3% reaction. In practice, this approach overestimates the reaction point at which gelation occurs, because polymer molecules exist that have molecular weights higher than the average value. These will reach the gelation point before those which have the average value of the molecular weight.

An alternative approach to the prediction of gelation points is based on a statistical treatment. The fundamental argument behind this approach is that the presence of three or more branch points in one molecule is sufficient to cause gelation. Thus, the problem revolves around a calculation of the probability that one of the four terminal branches in the segment in **2** will eventually yield another branch point.

2

If the functionality of the branch-point monomer is f, then the residual functionality of this unit at the chain end will be $f - 1$. Thus, the probability that one of the four branches will generate another branch point is given by $1/(f - 1)$. Because this also represents the probability that gelation will occur, we can write

$$\alpha_c = \frac{1}{f - 1}$$

where α_c is the critical branching coefficient required for gel formation.

Several assumptions are needed to simplify the treatment to a manageable degree. First, it is assumed that all functional groups are equally reactive. This is, in fact, an erroneous assumption even for glycerol, where the secondary OH group is known to be less reactive than the primary ones. Second, it is assumed that all condensation steps take place between different molecules and that internal condensations to give species such as **2** do not occur. The deviations from these assumptions probably account for the numerical discrepancies between experiment and theory.

Again an important practical problem is to calculate the point at which gelation should occur for different ratios of tri- (or multi-) and difunctional reagents. When a trifunctional monomer is present, gelation will occur if α (the branching coefficient) is greater than $\frac{1}{2}$. The value $\alpha = \frac{1}{2}$ therefore constitutes the critical condition for the formation of an infinite network. In a more general sense, the critical value of α is defined, as mentioned previously, by the term: $\alpha(f - 1) = 1$ or $\alpha = 1/(f - 1)$.

Thus, an estimate of α is the key requirement. This estimate may be made using a similar statistical approach to that discussed earlier. The result is

$$\alpha = \frac{P_B^2 p}{r - P_B^2(1 - p)} = \frac{r P_A^2 p}{1 - r P_A^2(1 - p)} \tag{64}$$

As discussed previously, P_A and P_B represent the fractions of the original A and B functional groups that have reacted. Similarly, r is the ratio of the initial number of A groups to the initial number of B groups. The term p is the fraction of original A groups that are contained in the tri- (or higher) functional monomer. It can be shown[1] that α_c can be attained only when $1/(1 + p) < r < 1 + p$.

There are three interesting special cases of (64).

1. No bifunctional monomer that contains A groups is present. In this case $p = 1$ and (64) simplifies to

$$\alpha = r P_A^2 = \frac{1}{r} P_B^2 \tag{65}$$

2. The initial numbers of A and B groups are equal. In this case, $r = 1$ and $P_A = P_B = P$ and (64) simplifies to

$$\alpha = \frac{P^2 p}{1 - P^2(1 - p)} \tag{66}$$

3. If the initial number of A and B groups are equal *and* no bifunctional monomer that contains A groups is present, (64) simplifies to

$$\alpha = P^2 \tag{67}$$

[1]Flory, P. J., *J. Am. Chem. Soc.*, **1941**, *63*, 3083.

As an example, let us apply the expressions above to predict the gel points of the hypothetical systems mentioned earlier:

For system X, with 2 mol of glycerol and 3 mol of bifunctional acid, $r = 1$ and $p = 1$. The critical branching coefficient for this trifunctional system is $\alpha_c = \frac{1}{2}$. Thus, from (67), the gel point will occur at a degree of reaction of

$$P = \sqrt{\tfrac{1}{2}} = 0.707$$

or a conversion of 70.7%.

For system Y with 2 mol of glycerol, 1 mol of diol, and 4 mol of acid, $r = 1$ and $p = \frac{3}{4}$. Again $\alpha_c = \frac{1}{2}$, so that from (66) the gel point will occur at

$$P = \sqrt{\tfrac{8}{14}} = 0.756$$

or at a conversion of 75.6%.

STUDY QUESTIONS

1. Three samples of monodisperse poly(1,10-decanediol adipate) are mixed together as follows: 10 g of A, having a molecular weight of 40,000; 5 g of B, having a molecular weight of 100,000; and 3 g of C, having a molecular weight of 200,000. Calculate:

   (a) The number-average molecular weight of the polymer mixture.

   (b) The weight-average molecular weight of the polymer mixture.

2. A sample of a condensation polymer has the following experimental distribution of molecular weights:

Range of $M \times 10^{-3}$	Weight (mg)	Range of $M \times 10^{-3}$	Weight (mg)
0–20	0.35	120–140	14.43
20–40	2.52	140–160	14.70
40–60	5.70	160–180	13.77
60–80	8.75	180–200	12.16
80–100	11.52	200–220	8.40
100–120	13.31	220–240	5.75

Determine the weight-average and number-average molecular weights of the polymer.

3. The hydroxy acid $HO-CH_2CH_2CH_2CH_2COOH$ undergoes a condensation polymerization to form the polymer

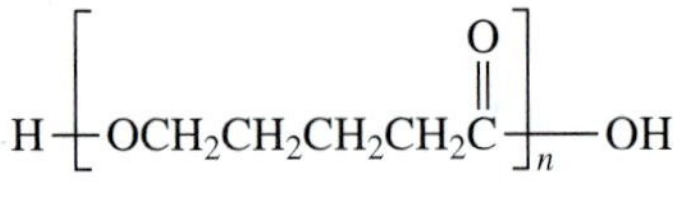

$$H-\left[OCH_2CH_2CH_2CH_2\overset{\overset{\displaystyle O}{\|}}{C}\right]_n-OH$$

It is found in a certain polymerization the product has a weight-average molecular weight of 18,400 g/mol. Calculate:

(a) The percentage of carboxyl groups that have esterified. [*Ans.:* 98.9%.]

(b) The number-average molecular weight of the polymer. [*Ans.:* 9250 g/mol.]

(c) The average number of structural units in the polymer molecules. [*Ans.:* 92.5.]

(d) The probability that a polymer molecule chosen at random will contain twice the average number of structural units.

4. It has been reported [Zhubanov et al., *Izv. Akad. Nauk Kaz. SSR Ser. Khim.*, **1967**, *17*, 69] that aminoheptanoic acid, $H_2N-(CH_2)_5CH_2COOH$, in *m*-cresol solution undergoes condensation polymerization to a polyamide. The reaction was found to be second-order in the amino acid concentration, with the following rate constants:

$T(°C)$	150	187
k (kg/mol-min)	1.0×10^{-3}	2.74×10^{-2}

(a) Write a balanced chemical reaction that describes the conversion of monomer to a polyamide of molecular weight equal to 12,718 g/mol.

(b) Calculate the activation energy of the reaction and the preexponential factor of the Arrhenius expression for the rate constant.

(c) What percent conversion of monomer is necessary to produce a polyamide with a number-average molecular weight of 4.24×10^3 g/mol?

(d) What percent conversion is necessary to produce a polyamide with a weight-average molecular weight of 2.22×10^4 g/mol?

5. A solution of aminoheptanoic acid (see Problem 4) in *m*-cresol, having a concentration of 3.3 mol of amino acid per kilogram of solution, is prepared and quickly brought to a temperature of 187°C.

(a) Derive an expression for the degree of polymerization (i.e., the average number of structural units per polymer molecule) as a function of reaction time. Modify your expression to apply to a system in which caproic acid is present at a level of 0.65% of the amino acid.

(b) Calculate the time required to form polyamide with a number-average molecular weight of 6340 g/mol.

(c) If the polymerization is carried out in a solution weighing 1 kg and for the reaction time calculated in (b), what would be the weight of polyamide formed that has a molecular weight of 12,718 g/mol?

6. Derive expressions that relate the mole fraction of *x*-mer to the reaction rate constant, initial concentration of carboxyl groups, and reaction time for (a) an uncatalyzed polyesterification, and (b) an acid-catalyzed polyesterification. Plot the mole fractions of polymer with $x = 10$ and $x = 100$ as a function of reaction time.

7. The following data were obtained by Hamann, Solomon, and Swift [*J. Macromol. Sci., Chem.*, **1968**, *A2*, 153; reproduced by permission of Marcel Dekker, Inc.] for the polyesterification of an equal molar mixture of

$$HO-(CH_2)_{10}-OH + HOOC-(CH_2)_4-COOH$$

The kinetics were studied by taking time zero to correspond to 82% esterification of the original COOH groups (e.g., see Figure 11.2). $[COOH]_0$ may be taken as 1.25 equiv/kg of mixture.

FURTHER POLYMERIZATION
OF POLY(1,10-DECANEDIOL ADIPATE)

(a)		(b)		(c)	
				Temp. 161°C, Catalyzed by *p*-Toluene Sulfonic Acid (0.004 mol Per Mole of Polymer)	
Temp. 190°C		Temp. 161°C			
Time (min)	% Reaction	Time (min)	% Reaction	Time (min)	% Reaction
0	0	0	0	0	0
30	20.6	20	9.1	5	34.6
60	39.0	40	16.0	10	54.7
90	50.2	100	31.6	15	65.5
150	61.2	150	41.1	20	70.8
225	66.8	210	47.9	30	77.9
300	71.5	270	52.5	40	82.9
370	74.4	330	57.0	50	85.7
465	77.2	390	60.0	60	87.9
510	78.2	450	62.6	75	90.1
550	78.8	510	64.6	90	91.5
600	79.6	550	65.5	105	92.6
660	80.6	700	69.2	120	93.6
730	81.7	840	71.9		
800	82.5	880	72.4		
		1060	74.8		
		1200	76.2		
		1320	77.2		

(a) Determine the rate constants and the activation energy for the uncatalyzed reaction.

(b) Determine the rate constant for the catalyzed polymerization.

8. From the rate constant obtained in Problem 7 for the catalyzed polyesterification of 1,10-decanediol and adipic acid, calculate and plot the mole fraction and weight fraction distribution of *x*-mer [N_x/N versus x, and W_x versus x] at 120 min and at 240 min.

9. Suppose that in the polyesterfication reaction in Problem 7, the adipic acid has an impurity of 0.85 mol%, of which you were unaware. What are the maximum values of the number-average and weight-average molecular weights that you could obtain in the reaction?

10. The following data were obtained [P. J. Flory, *J. Am. Chem. Soc.*, **1939**, *61*, 3334] for the polyesterification of diethylene glycol with adipic acid at 166°C:

Time (min)	% Conversion	Time (min)	% Conversion
6	13.79	321	86.72
12	24.70	398	88.37
23	36.75	488	89.74
37	49.75	596	90.84
59	60.80	690	91.63
88	68.65	793	92.20
170	78.94	900	92.73
203	81.61	1008	93.03
235	83.49	1147	93.54
270	85.00	1370	94.05

(a) Determine the order of the reaction with respect to the concentration of carboxyl groups.
(b) Determine the rate constant of the reaction in appropriate units. (*Hint:* Take the concentration units to be moles of COOH per kilogram of reactant mixture and require that $[OH]_0 = [COOH]_0$.)
(c) Calculate and plot the weight-fraction distribution of molecular weights obtained at reaction times of 270 and 1370 min.

11. Calculate the percentage conversion of acid at which gelation will occur in a mixture of 50 mol % adipic acid, 40 mol % ethylene glycol, and 10 mol % glycerol.

12. Write a condensation reaction that could be used to prepare the polypeptide poly-alanine.

13. Consider the condensation reaction between $HOOC(CH_2)_2COOH$ and either:
(a) $HOCH_2C(CH_3)_2CH_2OH$, or
(b) $HOCH_2C(CH_3)_2CH_2CH_2OH$.
Why would one expect use of (b) to yield a polyester of low crystallinity?

SUGGESTIONS FOR FURTHER READING

ALLEN, P. E. M., and PATRICK, C. R., *Kinetics and Mechanisms of Polymerization Reactions.* New York: Wiley, **1974**, Chap. 5.

BILLMEYER, F. W., JR., *Textbook of Polymer Science*, 3rd ed. New York: Wiley, **1984**, Chaps. 1, 2.

CARRAHER, C. E., JR., *Seymour/Carraher's Polymer Chemistry*, 5th ed., New York: Marcel Dekker, Inc., **2000**.

CRAVER, C., and CARRAHER, C. E., JR. (Eds.), *Applied Polymer Science—21st Century*, Washington: American Chemical Society, **2000**.

FLORY, P. J., *Principles of Polymer Chemistry*. Ithaca, N.Y.: Cornell University Press, **1953**, Chap. 3.

HIEMENZ, P. C., *Polymer Chemistry*. New York: Dekker, **1984**, Chaps. 1, 5.

LENZ, R. W., *Organic Chemistry of Synthetic High Polymers*. New York: Wiley-Interscience, **1967**, Chap. 3.

ODIAN, G., *Principles of Polymerization*, 3rd ed., New York: Wiley-Interscience, **1991**.

REMPP, P., and MERRILL, E. W., *Polymer Synthesis*, Basel: Huthig & Wepf, **1986**.

12

Kinetics
of Free-Radical
Polymerization

INTRODUCTION

The speed of a polymerization reaction, the molecular weight of the product, the composition of a copolymer, and the formation of unwanted side products are factors that have an enormous significance both in laboratory experiments and in the manufacture of polymers. Before the course of a chemical reaction can be predicted, it is necessary to understand at least some of the underlying principles that determine the influence of different reaction conditions on the yields and products.

The study of polymerization mechanisms and reaction kinetics aims to uncover these underlying principles in order to permit predictions to be made. This chapter deals with free-radical polymerizations, about which a great deal is known. Chapters 11 and 13 cover condensation and ionic-type processes.

The elementary reactions involved in a free-radical polymerization were discussed in Chapter 3. In that discussion the following mechanism was used to describe a polymerization under conditions where the conversion of monomer to polymer is sufficiently low that the polymeric product molecules do not undergo reactions with the free radicals:

$$\text{Initiator} \xrightarrow{k_i} 2R' \qquad \text{Formation of free radicals}$$

$$R' + M \xrightarrow{k'} R_1 \qquad \text{Initiation of chains}$$

$$R_1 + M \xrightarrow{k_{1p}} R_2 \qquad \text{Propagation of chains}$$

$$R_2 + M \xrightarrow{k_{2p}} R_3 \qquad \text{Propagation of chains}$$

$$R_n + M \xrightarrow{k_{np}} R_{n+1} \qquad \text{Propagation of chains}$$

$$R_n + R_m \xrightarrow{k_{tnmc}} P_{n+m} \qquad \text{Termination of chains}$$

$$R_n + R_m \xrightarrow{k_{tnmd}} P_n + P_m \qquad \text{Termination of chains}$$

In this mechanism, M represents a monomer molecule, R' is an initiating radical, R_n is a propagating radical of degree of polymerization n, P_n is a polymer molecule of degree of polymerization n, and so on. The rate constants for each elementary reaction are shown above the arrows. For simplicity at this state we omit chain transfer reactions from the mechanism. An extension of this simple mechanism to include chain transfer to the monomer, to the solvent, and to added chain transfer agents will be made in a later section of this chapter.

APPROXIMATIONS

A rigorous kinetic treatment of this mechanism leads to immense complexities and, in order to derive useful and tractable results, it is necessary to introduce some simplifying assumptions and approximations.

Kinetic Chain Length

It is assumed that the kinetic chain lengths are very large. This means that the number of monomer molecules consumed in the chain initiation process (i.e., that react with R') is negligible compared with the number consumed in the chain propagation reactions. Since we are, by definition, dealing with products that are high polymers containing many monomer units, this approximation is a very good one.

The Direction of Radical Addition to the Monomer

Only one type of radical is assumed to be present as a chain carrier. This will be true if each radical addition to the monomer occurs in the same way. As discussed in Chapter 3, the head-to-tail addition greatly predominates over head-to-head addition. It is assumed therefore, that head-to-tail addition is the sole type occurring. This means that, in the polymerization of $CH_2\!=\!CHX$, the structure of all the propagating radicals is of the type shown in **1**, where the zigzag line denotes a polymeric chain of $-\!(CH_2CHX)\!-$ units.

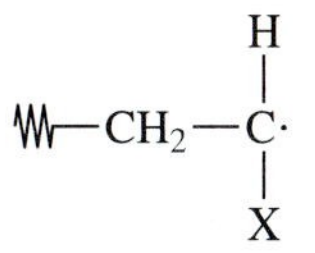

$$\text{W}-CH_2-\overset{\displaystyle H}{\underset{\displaystyle X}{\overset{|}{\underset{|}{C}}}}\cdot$$

1

Radical Reactivity and Size

It is assumed that the reactivity of the propagating radicals is independent of the size or degree of polymerization of the radical. The result of this assumption is that the rate constants for propagation, that is, $k_{1p}, k_{2p}, k_{3p}, \ldots, k_{np}, \ldots$ are taken to be equal and are written simply as k_p. Similarly, for termination reactions, the rate constants k_{nmc} and k_{nmd} are assumed to be independent of n and m and are written simply as k_{tc} and k_{td}.

The application of this assumption to the propagation reactions may be rationalized in terms of simple collision theory. According to this theory, the rate constant for propagation, k_p, may be written as

$$k_p = \xi_p \sigma_p \left(\frac{RT}{\pi \mu} \right)^{1/2} e^{-E_p/RT} \tag{1}$$

where ξ_p is the steric factor, σ_p is the cross section of the collision, μ is the reduced mass of the colliding pair, E_p is the activation energy for propagation, R is the gas constant, and T is the absolute temperature. Since the only effective collisions will be those of monomer M with the growing end of the propagating radical R_n, the product $\xi_p \sigma_p$ should be roughly independent of the degree of polymerization of the radical. Because the chemical nature of the reactive end of the radical is independent of the degree of polymerization, E_p should also be independent of the degree of polymerization of the radical. The reduced mass is defined by

$$\mu = \frac{M_R M_M}{M_R + M_M} \tag{2}$$

where M_R and M_M are the masses of the propagating radical and monomer, respectively. For most of the propagation reactions, M_R is much larger than M_M. Therefore, to a good approximation, equation (2) may be written

$$\mu \approx M_M \tag{3}$$

Approximations

and the conclusion may be drawn that k_p should be a constant that is independent of radical size. This will generally be a valid approximation except in the initial stages of the propagation.

However, the assumption that k_t^1 is independent of size is usually less valid. This is because, for most termination reactions, the rate constant is determined by the collision frequency of radicals (diffusion-controlled reactions), and the reduced mass does not become independent of radical size as in (3). Nevertheless, this assumption must be made in order to obtain tractable results. The reader should keep in mind the limitation of this assumption and recognize that in most cases an *average* termination rate constant is being used.

The Steady-State Approximation

All free radicals present in the system are assumed to be at steady-state concentrations. This means that the total concentration of propagating radicals (which is the sum of the concentrations of propagating radicals of all degrees of polymerization) is also a steady-state value. Expressed algebraically this means that

$$\frac{d[\mathrm{R}']}{dt} = \frac{d[\mathrm{R}_n]}{dt} = \frac{d[\mathrm{R}]}{dt} = 0 \tag{4}$$

where $[\mathrm{R}] = \sum_n [\mathrm{R}_n]$ and the other symbols are as described earlier.

Justification of this assumption may be seen as follows. Let $[\mathrm{M}]_0$ be the initial concentration of the monomer. Monomer molecules that have reacted must be contained either in the propagating radicals or in the polymer (i.e., product molecules). Therefore, the stoichiometry requires that, at all times, equation (5) must hold.

$$[\mathrm{M}]_0 = [\mathrm{M}] + \sum_n n[P_n] + \sum_n n[R_n] \tag{5}$$

If (5) is differentiated with respect to time and then rearranged, the expression shown in (6) is obtained.

$$-\frac{d[\mathrm{M}]}{dt} = \sum_n n\left(\frac{d[P_n]}{dt}\right) + \sum_n n\left(\frac{d[R_n]}{dt}\right) \tag{6}$$

Experimentally, it is found that, except in the very earliest (and generally negligible) stages of the reaction, the loss of monomer is, in fact, accounted for quantitatively by the appearance of the polymeric product. Thus, we can write equation (6) in the terms shown in (7).

$$-\frac{d[\mathrm{M}]}{dt} = \sum_n n\left(\frac{d[P_n]}{dt}\right) \tag{7}$$

Equation (7) is generally valid only if (4) is also valid.

$^1 k_t = k_{tc} + k_{td}.$

STEADY-STATE CONCENTRATIONS OF THE PROPAGATING RADICALS

The steady-state approximation (4) is probably the most important single assumption required to find a tractable solution to the rate equations needed for the mechanism described on page 334. This approximation enables us to express free-radical concentrations in terms of the concentrations of *stable* substances by the solution of simple *algebraic* equations. Without this approximation, a set of nonlinear simultaneous differential equations would have to be solved.

In each propagation step of the mechanism, one radical is destroyed but another is created. Therefore, the entire sequence of chain propagation reactions has no effect on the *total* concentration of propagating radicals. If this total concentration is to be in a steady state, as assumed by (4), the rate of initiation of propagating radicals must be equal to their rate of termination, or

$$\frac{d[R]}{dt} = r_i - r_t = 0 \tag{8}$$

where r_i and r_t are the rates of initiation and termination, respectively, of the propagating radicals.

According to the mechanism on page 334, and the assumption that the initiating radicals are at a steady state, the rate of initiation is given by

$$r_i = k'[R'][M] \propto 2k_i[I] \tag{9}$$

Actually, as discussed in Chapter 3, some loss of initiating radicals always occurs through side reactions that are not shown in the mechanism. To recognize this fact, the rate of initiation is usually written as

$$r_i = 2fk_i[I] \tag{10}$$

where f represents the fraction of initiating radicals produced that actually add to monomer to form R_1.

In considering the rate of termination, r_t, it is necessary to distinguish between the reactions of *like* radicals and those of *unlike* radicals. Thus, for both combination and disproportionation, we must distinguish between the two termination processes

$$R_n + R_n \xrightarrow{\ k_{tnn}\ } \text{polymer} \tag{11}$$

$$R_n + R_m \xrightarrow{\ k_{tnm}\ } \text{polymer} \tag{12}$$

In a chemical sense, all the propagating radicals are the same: they differ only in size. Therefore, because we have assumed that reactivity is independent of size, the rate constants k_{tnn} and k_{tnm} can differ only by a factor of 2. This is a consequence of the relative collision frequencies of like and unlike species (i.e., $k_{tnm} = 2k_{tnn}$).

This well-known result may be seen as follows. Consider the case in which two types of radicals, R_n and R_m, are present. The total rate of collisions may be written

$$\text{collision rate} = \alpha[R]^2 \tag{13}$$

where α is the rate constant. Substitution of the total radical concentration $[R] = [R_n] + [R_m]$ into (13) gives the collision rate as

$$\text{collision rate} = \alpha[R_n]^2 + 2\alpha[R_n][R_m] + \alpha[R_m]^2 \tag{14}$$

The first and last terms in (14) are the rates of collisions of like radicals, while the second term gives the collision rate of unlike species. The rate constant for unlike collisions, namely 2α, is twice that for like collisions. In terms of chain termination rate constants, then, $k_{tnm} = 2k_{tnn}$.

Application of the mass action law to (11) and (12) gives the termination rate for radicals of degree of polymerization n:

$$r_t(n) = 2k_{tnn}[R_n]^2 + [R_n]\sum_{m \neq n} k_{tmn}[R_m] \tag{15}$$

The factor 2 in (15) arises because two radicals of degree of polymerization n are consumed by the process shown in (11), while only one is consumed in (12). Since $k_{tnm} = 2k_{tnn}$, (15) becomes

$$r_t(n) = 2k_{tnn}[R_n]\left\{[R_n] + \sum_{m \neq n}[R_m]\right\} \tag{16}$$

or

$$r_t(n) = 2k_{tnn}[R_n]\sum_m[R_m] = 2k_t[R_n][R] \tag{17}$$

where k_t is the rate constant for termination of like radicals and $[R]$ is the total radical concentration. The subscript nn has been dropped because of the assumption on pages 334–336. The total rate of termination of all radicals is obtained by summation over n as

$$r_t = \sum_n r_t(n) = 2k_t[R]\sum_n[R_n] = 2k_t[R]^2 \tag{18}$$

Substitution of (10) and (18) into (8) gives, for the steady-state concentration of propagating radicals,

$$[R] = \left(\frac{r_i}{2k_t}\right)^{1/2} = \left(\frac{fk_i[I]}{k_t}\right)^{1/2} \tag{19}$$

In the case of photolytic or radiation-induced polymerization, $2k_i[I]$, which is the rate of formation of initiating radicals, may simply be replaced by the appropriate initiating radical formation rates given in Chapter 5.

RATE OF POLYMERIZATION

The rate of polymerization, r_p, is defined as the instantaneous decrease of monomer concentration, [M], with respect to time, t, as shown by (20). However, polymerization rates,

$$r_p = -\frac{d[M]}{dt} \qquad (20)$$

generally depend on the monomer concentrations and, since the monomer concentration is known most accurately at zero time, it is useful to work with *initial* rates of polymerization, as given by (21). Unless otherwise specified all rates mentioned

$$r_p^0 = -\left(\frac{d[M]}{dt}\right)_0 = \lim_{\Delta t \to 0} \frac{[M]_0 - [M]}{\Delta t} \qquad (21)$$

in the following sections will be initial rates, and the superscript and subscript of r_p and $(d[M]/dt)$, respectively, in (21) will be dropped.

It is easy to determine the rate of polymerization since, according to (21), it is merely necessary to measure [M] as a function of t and then determine the *initial* slope of a plot of [M] versus t. Moreover, such plots are usually sufficiently linear up to about 10 to 15% monomer depletion that only one or two experimental points are necessary. Despite the simplicity of the measurements, rates of polymerization measured under a variety of reaction conditions provide a basis for the determination of a large amount of practical and theoretical kinetic information.

Experimental Measurement of Rates of Polymerization

In order to determine the rate of polymerization it is necessary to measure the monomer concentration [M] as a function of the reaction time. The most obvious and straightforward way to do this is to stop the reaction at some predetermined time (e.g., by a sudden chilling of the reaction mixture) and then separate the monomer reactant from the polymeric product. The amount of monomer remaining or the amount of polymer formed at this reaction time may then be determined simply by weighing. This process can then be repeated for other reaction times until sufficient information has been obtained to permit a plot of [M] as a function of t. The initial slope, $\lim_{\Delta t \to 0} \Delta[M]/\Delta t$, is the initial rate, r_p.

However, the measurement of [M] by this method is time-consuming and requires the preparation of a new reaction mixture for each experimental point on the concentration–time plot. The method is useful as a standard method since *absolute* concentrations are obtained from it. But it is much faster and more convenient to measure some *physical property* of the reaction mixture that changes as the polymerization proceeds and which may be related to the concentration of the monomer. Although a number of suitable physical methods of analysis are

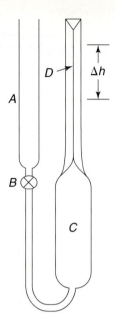

Figure 12.1 Simple dilatometer.

available, *dilatometry*[1] is most often used in the measurement of polymerization rates. In this method a dilatometer is employed to measure the volume contraction that occurs as the monomer is converted to high polymer.

The principle of operation of a dilatomer may be seen from the simple apparatus shown in Figure 12.1. The monomer or a solution of the monomer in a solvent are introduced, along with an initiator, into the apparatus through filling tube A until the liquid is drawn well up into the capillary tube D. The stopcock B is then closed and the apparatus is placed in a thermostatic bath maintained at the desired reaction temperature and regulated to $\pm 0.002°C$. The diameter and total height of the capillary depend on whether the polymerization is carried out in pure monomer or in a solution of the monomer. During the reaction the change in the height of the liquid in the capillary, Δh, *is* measured with a cathetometer (a rigidly mounted, vertically sliding telescope).

The total volume of the reaction system is given by

$$V = V_{BC} + \pi r^2 h \tag{22}$$

where V_{BC} is the volume contained in the space from the stopcock B through the bulb C to the entrance to the capillary tube D, r is the radius of the capillary, and h is the height of the liquid in the capillary above some arbitrary reference point. The change in volume is proportional to the change in h, as in

$$\Delta V = V_0 - V = \pi r^2 \Delta h \tag{23}$$

[1]Rubens, L. C., and Skochdopole, R. E., *J. Appl. Polymer Sci.*, **1965**, *9*, 1487.

where V_0 is the initial volume of the system and V is the volume at time t.

The change in volume of the system may be related to the monomer concentration, reaction yield, and rate of reaction as follows. The total volume of the system, V, at any time t is given by

$$V = w_m \bar{v}_m + w_p \bar{v}_p + w_s \bar{v}_s \tag{24}$$

where w_m, w_p, and w_s are the weights and $\bar{v}_m$, $\bar{v}_p$, and $\bar{v}_s$ are the partial specific volumes[1] of monomer, polymer, and solvent, respectively. To a very good approximation, $w_p = w_m^o - w_m$, where w_m^o is the initial weight of monomer. Solving (24) gives the weight of monomer at time t, as in

$$w_m = \frac{V - w_m^o \bar{v}_p - w_s \bar{v}_s}{\bar{v}_m - \bar{v}_p} \tag{25}$$

Because no polymer is present initially, the initial volume is given by

$$V_0 = w_m^o \bar{v}_m + w_s \bar{v}_s \tag{26}$$

and, provided that all the monomer is converted to polymer at the completion of the reaction, the final volume is

$$V_\infty = w_m^o \bar{v}_p + w_s \bar{v}_s \tag{27}$$

Equations (26) and (27) may be used to eliminate $\bar{v}_m - \bar{v}_p$ and $w_s \bar{v}_s$ from (25) and, when this is done, equation (28) is obtained.

$$w_m = \frac{V - V_\infty}{V_0 - V_\infty} w_m^o \tag{28}$$

The yield of the reaction on a wt % basis, Y, may then be written as

$$Y = 100 \frac{w_m^o - w_m}{w_m^o} = 100 \frac{V_0 - V}{V_0 - V_\infty} \tag{29}$$

or

$$Y = \frac{100 \Delta V}{V_0 - V_\infty} = \frac{100 \, \Delta h(t)}{\Delta h(t = \infty)} \tag{30}$$

Similarly, it is possible to proceed from (28) and utilize the definition of the initial rate of polymerization (21) to obtain the expression

$$r_p = \frac{Y}{100} \frac{[M]_0}{t} = \frac{\Delta h(t)}{\Delta h(t = \infty)} \frac{[M]_0}{t} \tag{31}$$

Some polymer yields from the polymerization of methyl methacrylate at 50°C [measured dilatometrically and calculated by equation (30)] and the cor-

---

[1]Lewis, G. N., and Randall, M., *Thermodynamics*, revised by K. S. Pitzer and L. Brewer (New York: McGraw-Hill, **1961**), p. 208.

TABLE 12.1 POLYMERIZATION OF METHYL METHACRYLATE AT 50°C; INITIATOR: 1% BENZOYL PEROXIDE

Reaction Time (s)	Yield (wt %)	$r_p \times 10^4$ (mol·liter^{-1} s^{-1})
0	0	—
480	0.77	1.5
1200	2.17	1.64
1920	3.50	1.65
4080	7.27	1.62
5880	10.38	1.60
8280	14.60	1.60
9600	17.16	1.62

responding polymerization rates calculated from (31) are shown in Table 12.1. The constancy of r_p in Table 12.1 shows that, up to the point of 17% conversion, depletion of the reactant does not significantly affect the measurement of the initial rate.

The validity of this simple method may be seen in Figure 12.2, in which the yields are plotted versus time for values determined both dilatometrically and absolutely (by the separation of polymer and monomer).

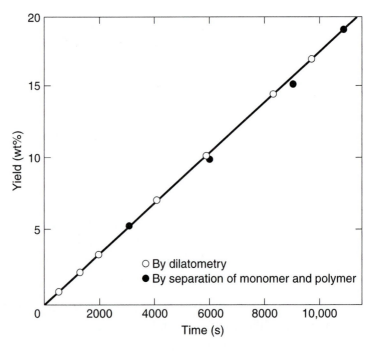

Figure 12.2 Reaction yields in polymerization of methyl methacrylate at 50°C. [From G. V. Schulz and G. Harborth, *Angew. Chem.*, **1947**, *59A*, 90.]

Despite the success of the dilatometric method with many polymerization systems, it is not an *absolute* method and complications can arise. The applicability of the method to an untried system cannot be assumed but must first be tested against an absolute method. Once the dilatometric method has been shown to be appropriate for the particular system under study as, for example, by data of the type shown in Figure 12.2, it provides probably the simplest and fastest method for the determination of polymerization rates. It is largely on the basis of such experimental rates, determined under a variety of conditions, that the free-radical mechanisms of polymerization have been developed.

Theoretical Rates of Polymerization

Applying the mass action law to the mechanism given on page 334 for the rate of disappearance of monomer leads to the expressions shown in (32) and (33) for the *theoretical* rate of polymerization.

$$r_p = k'[R'][M] + k_{1p}[R_1][M] + k_{2p}[R_2][M] + \cdots \tag{32}$$

or

$$r_p = [M]\left\{ k'[R'] + \sum_n k_{np}[R_n] \right\} \tag{33}$$

According to the "assumption of long chains" made earlier, $k'[R']$ is very much less than $\sum_n k_{np}[R_n]$ and may be neglected in the bracketed term of (33). Similarly, according to the assumption made regarding radical size and reactivity, $k_{1p} = k_{2p} = \cdots k_{np} = k_p$. Incorporation of these two approximations into (33) yields

$$r_p = k_p[M]\sum_n [R_n] = k_p[M][R] \tag{34}$$

where [R] is the total concentration of propagating radicals. We have already evaluated the steady-state radical concentration, (19), in terms of the rate of initiation, and substitution of (19) into (34) gives the steady-state rate of polymerization:

$$r_p = \left(\frac{k_p^2}{2k_t}\right)^{1/2} r_i^{1/2}[M] = \left(\frac{k_p^2}{2k_t}\right)^{1/2} (2fk_i)^{1/2}[I]^{1/2}[M] \tag{35}$$

where the last term in (35) refers specifically to initiation by the thermal dissociation of an initiator.

Equation (35) predicts that the *rate of polymerization at a given temperature should vary with the square root of the initiation rate or initiator concentration and with the first power of the monomer concentration.* The validity of this equation has been verified many times. An example of the dependence of r_p on the initiator concentration is shown in Figure 12.3, while an example of the dependence of r_p on monomer concentration is given in Figure 12.4.

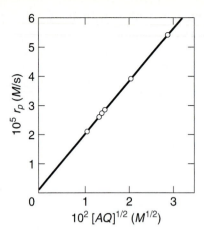

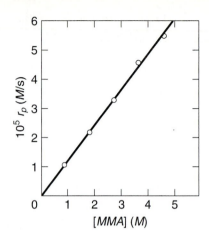

Figure 12.3 Variation in initial rates of polymerization with concentration of anthraquinone in THF at 30°C ([MMA] = 4.68). [Reproduced from Ledwith, A., Ndaalio, G., and Taylor, A. R., *Macromolecules*, **1975**, *8*, 1; with permission of the American Chemical Society, Washington, D.C.]

Figure 12.4 Variation in initial rates of polymerization with concentration of MMA in THF at 30°C ([AQ] = 2.91 10^{-4} M). [Reproduced from Ledwith, A., Ndaalio, G., and Taylor, A. R., *Macromolecules*, **1975**, *8*, 1; with permission of the American Chemical Society, Washington, D.C.]

According to (35), the slope of the straight line resulting from a plot of r_p versus [M], such as in Figure 12.4, is $(k_p^2/2k_t)^{1/2}r_i^{1/2}$. If r_i is known, which is usually the case, then plots such as the one shown in Figure 12.4 lead to numerical values of the ratio $(k_p^2/2k_t)^{1/2}$. Values of this important ratio for the polymerization of a number of monomers at 60°C are shown in Table 12.2.

The value of $(k_p^2/2k_t)^{1/2}$, determined from a plot such as that in Figure 12.4, refers to a single temperature. If such values are obtained at several temperatures, a combination of these data with the Arrhenius formulation of a rate constant yields more information. Thus writing the individual rate constants as

$$k_j = \frac{A_j e^{-E_j}}{RT} \tag{36}$$

TABLE 12.2 RATE CONSTANT RATIOS AND TEMPERATURE DEPENDENCE FOR FREE-RADICAL POLYMERIZATION

Monomer	$\left(\dfrac{k_p^2}{2k_t}\right)^{1/2}$ [liters/mol-s]$^{1/2}$ at 60°C]	$\left(\dfrac{A_p^2}{2A_t}\right)^{1/2}$ [liters/mol-s]$^{1/2}$	$E_p - \frac{1}{2}E_t$ (kcal/mol)
Methacrylonitrile	0.053	4.3×10^4	9.0
Methyl acrylate	0.99	1.03×10^3	4.6
Methyl methacrylate	0.12	2.0×10^2	4.9
Styrene	0.029	6.2×10^2	6.6
Vinyl acetate	0.13	1.6×10^2	4.7

where A_j is the preexponential factor and E_j is the activation energy of the jth reaction, the rate-constant ratio may be written as

$$\left(\frac{k_p^2}{2k_t}\right)^{1/2} = \left(\frac{A_p^2}{2A_t}\right)^{1/2} - \exp\left(\frac{E_p - E_t/2}{RT}\right) \qquad (37)$$

or

$$\ln\left(\frac{k_p^2}{2k_t}\right)^{1/2} = \ln\left(\frac{A_p^2}{2A_t}\right)^{1/2} - \frac{E_p - E_t/2}{R}\frac{1}{T} \qquad (38)$$

According to (38), the Arrhenius parameters $(A_p^2/2A_t)^{1/2}$ and $(E_p - E_t/2)$ may be obtained, respectively, from the intercept and slope of a plot of $\ln (k_p^2/2k_t)^{1/2}$ versus $1/T$. Values of these parameters for several monomers are also shown in Table 12.2.

From the Arrhenius parameters in Table 12.2, it is possible to calculate values for the ratio $(k_p^2/2k_t)^{1/2}$ at any temperature. Such information is of great utility because values of $(k_p^2/2k_t)^{1/2}$ and the rate of initiation r_i permit the prediction not only of rates of polymerization but also, as will be seen, of average kinetic chain lengths, average degrees of polymerization, and the distribution of the degree of polymerization.

Average Kinetic Chain Length

The average kinetic chain length, v, is defined as the average number of monomer molecules polymerized per chain initiated, or equivalently, as the rate of polymerization per unit rate of initiation. Thus from (18) and (34), equation (39) is easily obtained.

$$v = \frac{r_p}{r_i} = \frac{k_p[\text{M}]}{2k_t[\text{R}]} = \left(\frac{k_p^2}{2k_t}\right)^{1/2}\frac{[\text{M}]}{r_i^{1/2}} \qquad (39)$$

This expression shows that given a rate of initiation, kinetic data of the type shown in Table 12.2 can be used to calculate average kinetic chain lengths of polymerization. Combination of (35) and (39) leads to the useful expression

$$(r_p v)^{1/2} = \left(\frac{k_p^2}{2k_t}\right)^{1/2}[\text{M}] \qquad (40)$$

which in the absence of information on the initiation rate can be used to determine values of $(k_p^2/2k_t)^{1/2}$.

AVERAGE DEGREE OF POLYMERIZATION AND AVERAGE MOLECULAR WEIGHT

According to the mechanism shown on page 334, termination by *combination* produces a polymer molecule larger than the terminating free radicals. On the other hand, termination by *disproportionation* yields two polymer molecules of the same size as the terminating free radicals. Because the average degree of polymerization, $\overline{DP}$, is defined as the average number of monomer molecules per polymer molecule, a distinction must be made between these two types of termination. Thus the total rate of formation of polymer is defined by (41) (the factor of 2 appears because the two polymer molecules are formed in termination by disproportionation,

$$\frac{d[P]}{dt} = (k_{tc} + 2k_{td})[R]^2 \qquad (41)$$

whereas only one is formed by combination). The instantaneous average degree of polymerization is then given by

$$\overline{DP} = \frac{-d[M]/dt}{d[P]/dt} = \frac{k_p[M]}{(k_{tc} + 2k_{td})[R]} \qquad (42)$$

From (40), (42), and the relationship $k_t = k_{tc} + k_{td}$, we may express $\overline{DP}$ in terms of the average kinetic chain length:

$$\overline{DP} = 2v\left(\frac{k_{tc} + k_{td}}{k_{tc} + 2k_{td}}\right) \qquad (43)$$

Often, one type of termination will predominate. If so, the following simplifications of (43) can be assumed.

1. If termination by combination predominates, $k_{tc} \gg k_{td}$, and

$$\overline{DP} = 2v \qquad (44)$$

2. If termination by disproportionation predominates, $k_{td} \gg k_{tc}$, and

$$\overline{DP} = v \qquad (45)$$

Thus, for any free-radical chain polymerization in which chain transfer does not occur, $\overline{DP}$ must be between v and $2v$.

Because the average degree of polymerization represents the average number of monomer molecules contained in a polymer molecule, the average molecular weight may at once be written as

$$\overline{M} = M_0\overline{DP} \qquad (46)$$

where M_0 is the molecular weight of the monomer. The average molecular weight obtained in this way is the *number*-average molecular weight, as will be discussed further in Chapters 14 and 15. A measurement of the number average molecular weight is the most straightforward way to determine $\overline{DP}$.

DISTRIBUTION OF THE DEGREE OF POLYMERIZATION AND OF MOLECULAR WEIGHT[1]

According to the mechanism given on page 334 and the steady-state approximation, the rate equation for propagating radicals of *degree of polymerization, n,* may be written

$$\frac{d[R_n]}{dt} = k_p[M][R_{n-1}] - k_p[M][R_n] - 2k_t[R_n][R] = 0 \qquad (47)$$

which after division by $[R_{n-1}]$, followed by arrangement, yields

$$\frac{[R_n]}{[R_{n-1}]} = \frac{k_p[M]}{k_p[M] + 2k_t[R]} = \left(1 + \frac{2k_t[R]}{k_p[M]}\right)^{-1} \qquad (48)$$

In view of the definition of v, (48) may be written conveniently as

$$\frac{[R_n]}{[R_{n-1}]} = \left(1 + \frac{1}{v}\right)^{-1} \qquad (49)$$

We may obtain an expression for the ratio $[R_n]/[R_1]$ by successive multiplication of the ratios in (49). Thus,

$$\frac{[R_n]}{[R_1]} = \left(\frac{[R_n]}{[R_{n-1}]}\right)\left(\frac{[R_{n-1}]}{[R_{n-2}]}\right)\left(\frac{[R_{n-2}]}{[R_{n-3}]}\right) \cdots \frac{[R_2]}{[R_1]} \qquad (50)$$

or

$$[R_n] = [R_1]\left(1 + \frac{1}{v}\right)^{-(n-1)} \qquad (51)$$

If f_n is defined as the fraction of propagating radicals that have a degree of polymerization n, then, in view of (51), we have the relationship

$$f_n = \frac{[R_n]}{[R]} = \frac{[R_1]}{[R]}\left(1 + \frac{1}{v}\right)^{1-n} \qquad (52)$$

Consider now the steady-state expressions for the total radical concentration, $[R]$, and for the concentration of the smallest propagating radical, $[R_1]$ shown in (53) and (54).

$$\frac{d[R]}{dt} = r_i - 2k_t[R]^2 = 0 \qquad (53)$$

$$\frac{d[R_1]}{dt} = r_i - k_p[M][R_1] - 2k_t[R_1][R] = 0 \qquad (54)$$

[1]North, A. M., *The Kinetics of Free Radical Polymerization* (New York: Pergamon Press, **1966**), pp. 14–16.

Equating (53) and (54) and rearranging gives

$$\frac{[R_1]}{[R]} = \frac{1}{\nu}\left(1 + \frac{1}{\nu}\right)^{-1} \tag{55}$$

Substitution of (55) into (52) then yields the distribution function shown in (56) for the degree of polymerization of the propagating radicals.

$$f_n = \frac{1}{\nu}\left(1 + \frac{1}{\nu}\right)^{-n} \tag{56}$$

The distribution of the degree of polymerization (or of the molecular weight) in the polymer will be determined by f_n and by the mechanism of formation of the polymer. In the simple mechanism described on page 334, polymer molecules are formed only by combination and disproportionation of propagating radicals.

The mole fraction of polymer having a $\overline{DP}$ of n (denoted by X_n) will be given by the rate ratio

$$X_n = \frac{d[P_n]/dt}{\sum_n d[P_n]/dt} = \frac{d[P_n]/dt}{d[P]/dt} \tag{57}$$

where the denominator has already been given in (41). To evaluate the numerator we need only sum the rates of all the reactions in which P_n is formed. These reactions are:

	Combination	Disproportionation	
	$R_{n-1} + R_1 \rightarrow P_n$	$R_n + R_1 \rightarrow P_n + P_1$	(58)
	$R_{n-2} + R_2 \rightarrow P_n$	$R_n + R_2 \rightarrow P_n + P_2$	(59)
	$\vdots$	$\vdots$	
	$R_{n-m} + R_m \rightarrow P_n$	$R_n + R_m \rightarrow P_n + P_m$	(60)
	$\vdots$		
	$R_1 + R_{n-1} \rightarrow P_n$		(61)

Note that in summing the rates of the combination reactions in (58) to (61) [i.e., in allowing the summation index, m, to go from the value 1 in (58) to $(n-1)$ in (61)], every reaction is counted twice, with one exception. The one exception occurs for that reaction in which $m = n - m = n/2$ and such a reaction between like radicals is possible only if the integer n is even. If n is even, then

$$\frac{d[P_n]}{dt} = k_{tnnc}[R_{n/2}][R_{n/2}] + \frac{1}{2}\sum_{\substack{m=1 \\ m \neq n/2}}^{n-1} k_{tnmc}[R_{n-m}][R_m] + 2k_{td}[R_n][R] \tag{62}$$

where the factor 1/2 corrects for the addition of each reaction twice. Recall now that the rate constant for termination of unlike radicals is twice that for like radicals (i.e., $k_{tnm} = 2k_{tnn} = 2k_t$). Therefore, (62) may be written

$$\frac{d[P_n]}{dt} = k_{tc} \sum_{m=1}^{n-1} [R_{n-m}][R_m](1)^m + 2k_{td}[R_n][R]$$ (63)

There can be no combination of like radicals to produce a polymer having an odd number for $\overline{DP}$ and (63) follows directly from the summation of the rates. Substitution of (56) leads to

$$\frac{d[P_n]}{dt} = \left[\frac{k_{tc}}{v^2} \left(1 + \frac{1}{v} \right)^{-n} \sum_{m=1}^{n-1} (1)^m + \frac{2k_{td}}{v} \left(1 + \frac{1}{v} \right)^{-n} \right] [R]^2$$ (64)

The summation is, of course, equal to $(n - 1)$, so we have

$$\frac{d[P_n]}{dt} = \frac{1}{v} \left(1 + \frac{1}{v} \right)^{-n} \left(\frac{n-1}{v} k_{tc} + 2k_{td} \right) [R]^2$$ (65)

Dividing by (41), we finally obtain the mole fraction of polymer with $\overline{DP}$ of n, that is,

$$X_n = \frac{1}{v} \left(1 + \frac{1}{v} \right)^{-n} \frac{(n-1)/v + 2(k_{td}/k_{tc})}{1 + 2(k_{td}/k_{tc})}$$ (66)

As seen in (66), the distribution function for the degree of polymerization of polymer formed by a free-radical chain mechanism, in which chain transfer is absent depends only on the kinetic chain length and the ratio of disproportionation to combination.

The two extreme cases mentioned on page 334 are easily derived from (66):

1. If $k_{td} = 0$ and termination is solely by combination,

$$X_n = \frac{n-1}{v^2} \left(1 + \frac{1}{v} \right)^{-n}$$ (67)

2. If $k_{tc} = 0$ and termination is solely by disproportionation,

$$X_n = \frac{1}{v} \left(1 + \frac{1}{v} \right)^{-n}$$ (68)

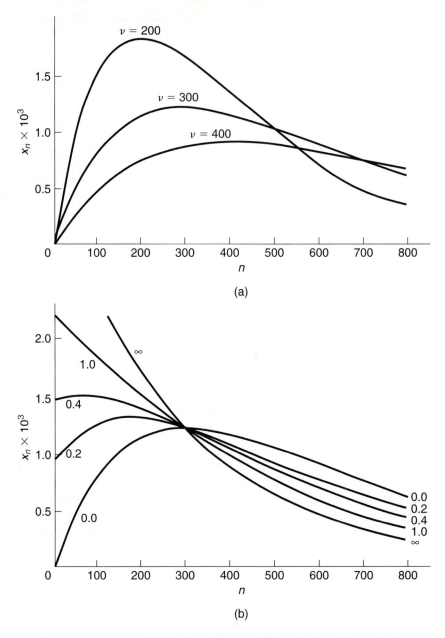

Figure 12.5 Number distribution of degree of polymerization: (a) termination by combination for several values of average kinetic chain length; (b) distribution for $\bar{\nu} = 300$ for several values of k_d/k_c, shown explicitly on the figure.

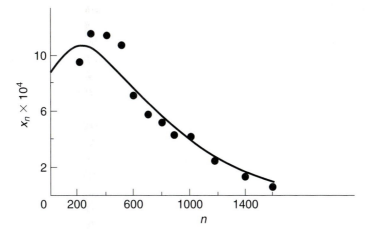

Figure 12.6 Experimental and theoretical number distribution of the degree of polymerization (calculated using equation 66).

The functional dependence of the distribution function in (66) on v and k_{td}/k_{tc} is shown graphically in Figure 12.5. Note in Figure 12.5b that all the distribution curves pass through the same point. This means there is a value of n for which the mole fraction of polymer with $\overline{DP} = n$ is independent of k_{td}/k_{tc}. It is left as an exercise for the student to show that this value is: $n = v + 1$.

An experimentally determined distribution for polystyrene formed in benzene solution at 50°C and having an average degree of polymerization of 626 is shown by the points in Figure 12.6. The solid line in Figure 12.6 is calculated from (66), using $v = 374$ and $(k_{td}/k_{tc}) = 0.24$ (see equation 42 and Problem 6, Chapter 3). The agreement between the theoretical and experimental distributions attests to the validity and utility of the derived distribution functions. The molecular-weight distribution is easily obtained from X_n since $M = M_0 X$, where M_0 is the molecular weight of the monomer.

The reader should be aware that the distributions in (66) to (68) and in Figures 12.5 and 12.6 are *number* or *mole-fraction* distributions. It is left as an exercise (Problem 8) for the student to show that the weight-fraction distribution W_n may be obtained by multiplication of the mole-fraction distribution (66) by $n/\overline{DP}$. When that is done, one has

$$W_n = \frac{n}{2v^2}\left(1 + \frac{1}{v}\right)^{-n} \frac{(n-1)/v + 2(k_{td}/k_{tc})}{1 + (k_{td}/k_{tc})} \tag{69}$$

CHAIN TRANSFER

In practice it is often observed that the average degree of polymerization is less than v, and sometimes considerably less than v, a fact that cannot be reconciled with the simple mechanism described on page 334. Moreover, it has been found that the addition of certain reagents to a free-radical polymerization reduces the average degree of polymerization from values greater than ν to values considerably less than ν without affecting the rate of polymerization.

The simplest explanation of these facts is that a reaction is occurring in which a growing free radical reacts with a stable molecule to form a polymer molecule with the simultaneous generation of a new (and generally small) free radical. Such a process, which is shown by (70), has not been included in the mechanism on page 334. In (70), $X \cdot$ is a free radical derived

$$R_n + X \xrightarrow{k_x} P_n + X \cdot \tag{70}$$

from the molecule X. If $X \cdot$ has a reactivity toward the monomer that is similar to that of R_n, so that (70) is nearly always followed by

$$X \cdot + M \longrightarrow R_1 \tag{71}$$

there will be no change in the rate of polymerization, r_p, or in the average kinetic chain length, ν, because the concentration of propagating radicals, [R], is unchanged. However, there will be a definite change in the average degree of polymerization, and therefore also in the average molecular weight and distribution of molecular weights. The reaction (70) is termed chain transfer when it is followed nearly always by (71), and X is then called a chain transfer agent.

If $X \cdot$ [formed in (70)] has a sufficiently lower reactivity toward the monomer than does R_n, so that other reactions of $X \cdot$ can compete effectively with (71), then (70) represents inhibition, and X is called an inhibitor or retarder. If the free radical X does not react with monomer (M), it is not likely to react with the other stable molecules present in the system, such as polymer (P_n), initiator (I), or added reagent (X). The ultimate fate of an unreactive $X \cdot$ in such a severe case of inhibition is generally to react with other free radicals to terminate the kinetic chain. We shall consider here only the process that we have called chain transfer, namely, the case in which $X \cdot$ is reactive toward M.

If (70) is always followed by (71), the occurrence of (70) results in no change in the total radical concentration [R]. Therefore, the rate of polymerization and the kinetic chain length will be unchanged from the expressions given in (35) and (39), respectively. However, the average degree of polymerization and the distribution of the degrees of polymerization are altered significantly.

Effect of Chain Transfer on Average Degree of Polymerization

We may incorporate the occurrence of chain transfer into the expression for the average degree of polymerization simply by recognizing that reaction (70) produces a stable polymer molecule. Because chain transfer can occur for radicals that have any degree of polymerization, the total rate of polymer formation becomes

$$\frac{d[P]}{dt} = \sum_n \frac{d[P_n]}{dt} = (k_x[X] + k_{tc}[R] + 2k_{td}[R])[R] \tag{72}$$

By analogy to (42), which refers to $k_x = 0$, the average degree of polymerization becomes

$$\overline{DP} = \frac{k_p[M]}{k_x[X] + (k_{tc} + 2k_{td})[R]} \tag{73}$$

It is more convenient to consider the *reciprocal* of the average degree of polymerization. Utilizing (34) or (39) to eliminate [R], we obtain the expressions

$$(\overline{DP})^{-1} = \frac{k_x[X]}{k_p[M]} + \frac{(k_{tc} + 2k_{td})r_p}{k_p^2[M]^2} = \frac{k_x[X]}{k_p[M]} + \frac{k_{tc} + 2k_{td}}{2v(k_{tc} + k_{td})} \tag{74}$$

It must be recognized that the chain transfer reaction is not restricted to substances that are added specifically for that purpose. Chain transfer may also occur with monomer (M), polymer (P), solvent (S), or initiator (I), as well as with impurities that may be present. Thus, the term $k_x[X]$ is a composite that should actually be written as

$$k_X[X] = k_M[M] + k_S[S] + k_I[I] + k_{\text{polymer}}[P] + k_Y[Y] \tag{75}$$

where Y represents a chain transfer agent added specifically for this purpose and k_i represents chain transfer to the species i. Substitution of (75) into (74) and definition of the *chain transfer constant*, C_i, as $C_i \equiv k_i/k_p$, leads to

$$(\overline{DP})^{-1} = (\overline{DP})_0^{-1} + C_1 \frac{[I]}{[M]} + C_S \frac{[S]}{[M]} + C_Y \frac{[Y]}{[M]} + C_{\text{polymer}} \frac{[P]}{[M]} \tag{76}$$

where

$$(\overline{DP})_0^{-1} = C_M + \frac{(k_{tc} + 2k_{td})r_p}{k_p^2[M]^2} \tag{77}$$

Very often, conditions can be chosen that eliminate or at least minimize chain transfer to the initiator or to the polymer, and further considerations here assume that this is the case. Therefore, the corresponding terms in (75) may be neglected.

According to (76), the chain transfer constants C_S and C_Y may be evaluated from experimental measurements of the average degree of polymerization as a function of the concentrations of the solvent and chain transfer agent at fixed values of the ratio $r_p/[M]^2$. Thus, a plot of $(\overline{DP})^{-1}$ versus [S]/[M] or [Y]/[M] should yield a straight line with an intercept of $(\overline{DP})_0^{-1}$ and a slope of C_S or C_Y, respectively. Typical plots for chain transfer to various solvents in vinyl acetate polymerization at 60°C are shown in Figure 12.7.

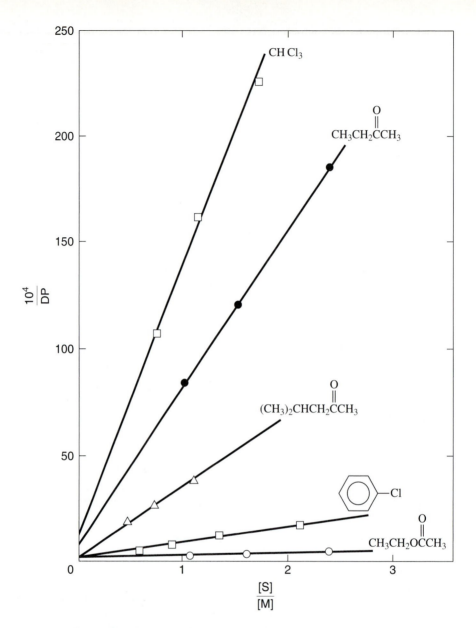

Figure 12.7 Chain transfer to various solvents in vinyl acetate polymerization at 60°C. [From S. R. Palit and S. K. Das, *Proc. Roy. Soc. (London)*, **1954,** *A226, 82.*]

Because the chain transfer constants are actually *ratios of rate constants* of elementary reactions, they will be temperature dependent, except in the improbable case that the activation energies of propagation and of chain transfer are the same. Thus, the temperature dependence of the chain transfer constant to the solvent is given by

TABLE 12.3 CHAIN TRANSFER CONSTANTS IN FREE-RADICAL POLYMERIZATION

Polymerizing Monomer	Chain Transfer Agent Y	$C \times 10^4$ (60°C)	$E_y - E_p$ (kcal/mol)
Styrene[*]	Styrene	0.6	7.9
	Benzene	0.018	15
	Cyclohexane	0.024	13
	Triphenylmethane	35	5.1
	Carbon tetrachloride	92	5.0
	Carbon tetrabromide	13,600	3.0
Vinyl acetate[†]	Vinyl acetate	2.5	3.1
	Benzene	2.2	15
	Cyclohexane	6.6	
	Triphenylmethane	850	
	Carbon tetrachloride	960	1.4
	Carbon tetrabromide	390,000	−7.6
Methyl methacrylate[*]	Methyl methacrylate	0.07	
	Benzene	0.075	
	Cyclohexane	0.1	
	Carbon tetrachloride	2.4	
	Carbon tetrabromide	2700	

[*] Ham, G. E., in *Vinyl Polymerization*, G. E. Ham, Ed. (New York: Dekker, **1967**), Chap. 1.

[†] Lindemann, M. K., in *Vinyl Polymerization*, G. E. Ham, Ed. (New York: Dekker, **1967**), Chap. 4.

$$C_s = \frac{A_s e^{-E_s/RT}}{A_p e^{-E_p/RT}} = \left(\frac{A_s}{A_p}\right) e^{(E_p - E_s)/RT} \tag{78}$$

where A_s and A_p are the frequency factors of the transfer and propagation reactions, respectively, and E_s and E_p are the respective activation energies. According to (78) the kinetic parameters A_s/A_p and $E_s - E_p$ may be determined from the intercept and slope, respectively, of plots of the logarithm of C_s as a function of $1/T$. The C_s values are determined at each temperature as described above.

Chain transfer constants have been determined for the reactions of polymerizing monomers with a large number of reagents. A small collection of such constants at 60°C, and the corresponding activation energy differences, $E_s - E_p$ (or $E_y - E_p$), are shown in Table 12.3.

Effect of Chain Transfer on the Distribution of the Degree of Polymerization

Chain transfer does not affect the rate of polymerization, r_p, or the average kinetic chain length, v. However, as shown in the preceding section, it can have a marked effect on the average degree of polymerization and hence on the average molecular weight. Obviously, the distribution of the degree of polymerization and the molecular weight will also be affected. Chain transfer may be introduced into the distribution function for the degrees of polymerization with the use of the same procedure employed in the development of (66) to (69). Although these derivations are left as exercises for the student, the distribution functions are presented in (79a) and (79b).

$$f_n = \frac{[R_n]}{[R]} = \frac{1 + \gamma v}{v}\left(1 + \frac{1}{v} + \gamma\right)^{-n} \tag{79a}$$

$$X_n = \frac{1 + \gamma v}{v}\left[\frac{(n - 1)\left(\dfrac{1 + \gamma v}{v}\right) + 2\gamma v\left(1 + \alpha + \dfrac{\alpha}{\gamma v}\right)}{1 + 2\gamma v\left(1 + \alpha + \dfrac{\alpha}{\gamma v}\right)}\right]\left(1 + \gamma + \frac{1}{v}\right)^{-n} \tag{79b}$$

where γ is the chain transfer term defined in (80a),

$$\gamma = C_M + C_S\frac{[S]}{[M]} + C_Y\frac{[Y]}{[M]} \tag{80a}$$

and

$$\alpha = \frac{k_{td}}{k_{tc}} \tag{80b}$$

In (80a), S represents solvent and Y represents a reagent added for the purpose of chain transfer.

Example of the Effect of Chain Transfer on $\overline{DP}$ and Distribution of $\overline{DP}$

Chain transfer can have a drastic effect on the average degree of polymerization and on the molecular-weight distribution. As an example, consider the polymerization of 1 M styrene at 25°C in benzene and carbon tetrachloride solutions. For an initiation rate of $r = 10^{-8}$ mol/L-s (appropriate for 0.4 M azobisisobutyronitrile at 25°C), the data in Problem 6, Chapter 3, Tables 12.2 and 12.3, and equations (39) and (76) yield the following values:

$$v = 90$$
$$\overline{DP}(\text{benzene}) = 162 = 1.80v$$
$$\overline{DP}(\text{CCl}_4) = 24 = 0.27v$$

The theoretical weight distributions of the degrees of polymerization, as calculated from (79b), and the relationship between W_n and X_n given earlier (see Problem 8) are shown in Figure 12.8. On the scale of this figure, the distribution for the total absence of chain transfer ($\gamma = 0$) is indistinguishable from that with benzene as solvent.

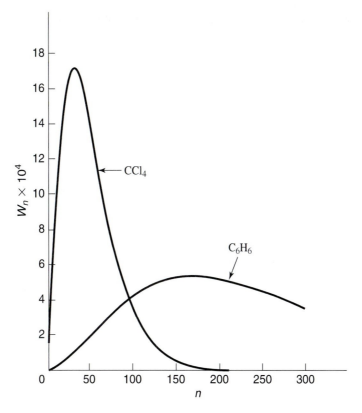

Figure 12.8 Effect of chain transfer to solvent on the distribution of *DP* for polymerization of 1 *M*. styrene at 25°C ($r_1 = 10^{-8}$ mol/L-s).

DEPENDENCE OF THE DEGREE OF POLYMERIZATION ON TEMPERATURE

The manner in which the average degree of polymerization varies with temperature depends on the degree to which the initiation is temperature dependent (e.g., thermal dissociation of an initiator) or temperature *in*dependent (e.g., photolytic or radiolytic initiation) and on the extent of chain transfer.

Consider first the situation where chain transfer may be neglected. Then according to (44) and (45), the average degree of polymerization will lie between v and $2v$, depending on the relative contributions of combination and disproportionation to the chain termination step. Suppose, for simplicity, that $k_{td} = 0$, so that (44) applies. Then expression (81) is valid.

$$\overline{DP} = 2v = \frac{2(k_p^2/2k_t)^{1/2}[M]}{(2fk_i[I])^{1/2}} \tag{81}$$

If the rate constants are written according to the Arrhenius formulation and f is taken to be independent of temperature,[1] equation (82) or (83) follows:

$$\overline{DP} = \frac{A_p[M]}{(A_t A_i f[I])^{1/2}} \exp\left[\frac{-(E_p - \frac{1}{2}E_t - \frac{1}{2}E_i)}{RT}\right] \tag{82}$$

or

$$\ln \overline{DP} = \ln K - \frac{E_p - \frac{1}{2}E_t - \frac{1}{2}E_i}{RT} \tag{83}$$

where the temperature-independent term of (82) has been written as K. Typical values for $(E_p - \frac{1}{2}E_t)$ and E_i for several monomers and thermal initiators can be obtained from Tables 12.2 and 3.3, respectively. From the magnitudes of these activiation energies it is apparent that when thermal initiation occurs the $\overline{DP}$ *decreases* as the temperature is increased. On the other hand, for photolytic, radiolytic (or any temperature-independent method of initiation), $E_i = 0$. Then, according to (83), the $\overline{DP}$ will *increase* with increasing temperature.

As a specific example, consider a styrene polymerization, in which $(E_p - \frac{1}{2}E_t) = 6600$ cal/mol, initiated as follows: (1) by the thermal dissociation of azobisisobutyronitrile for which $(E_i = 30,800$ cal/mol); and (2) by the photolytic dissociation of azopropane, for which $E_i = 0$. For case (1), expression (84) applies,

$$\frac{d \ln (\overline{DP})_1}{dT} = -\frac{8800}{RT^2} < 0 \tag{84}$$

while for case (2), expression (85) is appropriate.

$$\frac{d \ln (\overline{DP})_2}{dT} = +\frac{6600}{RT^2} > 0 \tag{85}$$

When chain transfer cannot be neglected, equation (74) applies. If the two terms in this equation are comparable, the temperature dependence is complicated, and an expression that is linear in $1/T$ cannot be written. In the very extreme case, where chain transfer to an added substance Y is the dominant mode of polymer formation, equation (86) is applicable:

$$\ln \overline{DP} = \ln \frac{[M]}{[Y]} + \ln \frac{A_P}{A_Y} + \frac{E_Y - E_P}{RT} \tag{86}$$

where the notation is the same in Table 12.3. Since in most cases $E_y - E_p$ for chain transfer agents or solvents is positive, $\overline{DP}$ would *decrease* with increasing temperature.

[1]Berger, K. C., *Makromol. Chem.*, **1975**, *176*, 3575.

ABSOLUTE PROPAGATION AND TERMINATION RATE CONSTANTS

All the kinetic quantities derived in the preceding treatment depend on the rate of initiation, the monomer concentration, and the rate constant ratios $(k_p^2/k_t)^{1/2}$, C_M, and C_Y. As we have seen, if the initiation rate is known, the rate constant *ratios* given in Tables 12.2 and 12.3 can be determined quite easily from simple experimental measurements of the rates of polymerization and the average degrees of polymerization for varying concentrations of monomer and chain transfer substances. However, it is much more difficult to obtain *absolute* values for k_p, k_t, and k_y, since to do so requires a departure from the steady-state situation that led to the simple kinetic expressions for r_p, ν, $\overline{DP}$, and so on.

A method that has been used frequently to obtain the individual rate constants involves the intermittent or pulsed illumination of a photolytically initiated polymerization. The intermittent illumination produces a periodic departure of the propagating radical concentration from its steady-state value in such a way that a study of the polymerization rate (and therefore the mean propagating radical concentration) as a function of the time of interruption of the illumination yields the steady-state *lifetime* of the radical chain, $\overline{\tau}_s$. Because it can be shown that the steady-state lifetime is given by (87).

$$\overline{\tau}_s = \frac{[R]}{2k_t[R]^2} = \frac{k_p}{2k_t}\left(\frac{[M]}{r_p}\right) \tag{87}$$

a measurement of $\overline{\tau}_s$, $[M]$, and r_p permits computation of the ratio k_p/k_t. A combination of this latter ratio with the corresponding $(k_p^2/k_t)^{1/2}$ of Table 12.2 yields k_p and k_t separately. A combination of k_p with C_Y of Table 12.3 then yields k_y.

For further details of this method, the reader is referred to more specialized treatments.[1] Individual rate constants at 60°C for propagation and termination are shown in Table 12.4. As mentioned, absolute rate constants for chain transfer at 60°C may be obtained by combining k_p from Table 12.4 with the appropriate chain transfer constant from Table 12.3.

TABLE 12.4   RATE CONSTANTS FOR PROPAGATION AND TERMINATION AT 60°C

Monomer	$k_p[(\text{liters-mol}^{-1}\text{-s}^{-1}) \times 10^{-3}$	$k_t[(\text{liters-mol}^{-1}\text{-s}^{-1}) \times 10^{-7}$
Methacrylonitrile	0.36	2.3
Methyl acrylate	1.6	0.13
Methyl methacrylate	0.71	1.8
Styrene	0.074	0.33
Vinyl acetate	1.0	3.2

[1]Calvert, J. G., and Pitts, J. N., Jr., *Photochemistry* (New York: Wiley, **1966**), pp. 651 ff.

COPOLYMERIZATION

The Copolymer Composition Equation: Reactivity Ratios

The kinetic considerations of the preceding sections have been restricted to the polymerization of a single monomer. Suppose now that two monomers are present in the reaction mixture and that both are susceptible to free-radical polymerization. Four types of propagation reactions will now exist, as shown in (88) to (91),

$$\text{\Large\wedge}\!\!\!\!-\, M_1\cdot + M_1 \xrightarrow{k_{p11}} \text{\Large\wedge}\!\!\!\!-\, M_1\cdot \tag{88}$$

$$\text{\Large\wedge}\!\!\!\!-\, M_1\cdot + M_2 \xrightarrow{k_{p12}} \text{\Large\wedge}\!\!\!\!-\, M_2\cdot \tag{89}$$

$$\text{\Large\wedge}\!\!\!\!-\, M_2\cdot + M_2 \xrightarrow{k_{p22}} \text{\Large\wedge}\!\!\!\!-\, M_2\cdot \tag{90}$$

$$\text{\Large\wedge}\!\!\!\!-\, M_2\cdot + M_1 \xrightarrow{k_{p21}} \text{\Large\wedge}\!\!\!\!-\, M_1\cdot \tag{91}$$

where it is assumed that the reactivity of a particular radical is independent of its size and also independent of the nature of the polymeric chain bound to the radical sites $M_1\cdot$ and $M_2\cdot$. Generally, the polymer obtained will be one in which both monomer units are incorporated together in the polymer molecules, namely into a *copolymer*. We can recognize from (88) to (91) two extreme types of kinetic behavior for a copolymerization. In the first, $k_{p12}/k_{p11} = k_{p21}/k_{p22} = 0$, and no copolymerization will occur. Instead, parallel polymerizations of the two monomers M_1 and M_2 will lead to the formation of two homopolymers $(M_1)_n$ and $(M_2)_m$. In the second extreme case, $k_{p11}/k_{p12} = k_{p22}/k_{p21} = 0$ and a copolymerization will take place to produce an alternating copolymer $(M_1M_2)_n$. Most copolymerizations fall between these two extremes.

Using the "long-chain assumption," that monomer molecules are consumed solely by the propagation reactions, the rates of monomer depletion in (88) to (91) are

$$-\frac{d[M_1]}{dt} = k_{p11}[M_1\cdot][M_1] + k_{p21}[M_2\cdot][M_1] \tag{92}$$

$$-\frac{d[M_2]}{dt} = k_{p12}[M_1\cdot][M_2] + k_{p22}[M_2\cdot][M_2] \tag{93}$$

Let us now define $\gamma = n_{M_1}/n_{M_2}$ as the ratio of the number of molecules of M_1 to the number of molecules of M_2 in the copolymer. Then from (92) and (93) we have

$$\gamma = \frac{-d[M_1]/dt}{-d[M_2]/dt} = \frac{[M_1]}{[M_2]}\left(\frac{k_{p11}[M_1\cdot] + k_{p21}[M_2\cdot]}{k_{p12}[M_1\cdot] + k_{p22}[M_2\cdot]}\right) \tag{94}$$

where $[M_1]/[M_2]$ is the mole ratio of monomers in the reactant mixture.

The steady-state approximation applied to the radicals $M_1\cdot$ and $M_2\cdot$ leads to the expression

$$-\frac{d[M_1\cdot]}{dt} = \frac{d[M_2\cdot]}{dt} = k_{p12}[M_1\cdot][M_2] - k_{p21}[M_2\cdot][M_1] = 0 \qquad (95)$$

from which rearrangement yields

$$\frac{[M_2\cdot]}{[M_1\cdot]} = \frac{k_{p12}[M_2]}{k_{p21}[M_1]} \qquad (96)$$

Substitution of (96) into (94) leads finally to the expression

$$\gamma = \frac{1 + r_1([M_1]/[M_2])}{1 + r_2([M_2]/[M_1])} \qquad (97)$$

in which the *reactivity ratios*, r_1 and r_2, are defined by $r_1 = k_{p11}/k_{p12}$ and $r_2 = k_{p22}/k_{p21}$. Equation (97) is known as the *copolymer composition equation*, since it gives the composition of the copolymer being formed, at a given instant, in terms of the composition of the monomer mixture and the reactivity ratios of the two monomers.

The copolymer composition equation (97) has been well tested, and the following predicted properties have been verified.

1. The equation is independent of dilution because r_1 and r_2 and the concentration ratios are dimensionless.
2. It is independent of the initiation rate because no rate constants for initiation of termination appear.
3. It is independent of the medium in which the reaction is carried out because no changes in r_1 and r_2 are found when reagents are added that change significantly the nature of the medium.

In order to show the dependence of the polymer composition on the composition of the reaction mixture, it is convenient to write the copolymer composition equation in terms of the mole fractions in the polymer, f_i, and in the reactant mixture, F_i. Thus,

$$\frac{f_1}{f_2} = \frac{1 + r_1(F_1/F_2)}{1 + r_2(F_2/F_1)} = \frac{F_1(F_2 + r_1F_1)}{F_2(F_1 + r_2F_2)} \qquad (98)$$

is obtained which, following the use of the relationships $f_2 = 1 - f_1$ and $F_2 = 1 - F_1$, yields

$$f_1 = \frac{F_1(1 + [r_1 - 1]F_1)}{(r_1 + r_2 - 2)F_1^2 + 2(1 - r_2)F_1 + r_2} \qquad (99)$$

The functional dependence of f_1 on F_1 is shown graphically in Figure 12.9 for several sets of values of r_1 and r_2.

Copolymerization

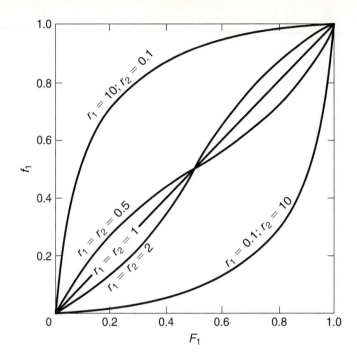

Figure 12.9 Dependence of copolymer composition on composition of reactant mixtures.

For the special case of $r_1 = r_2$, a family of curves is obtained that pass through the single point $(0.5, 0.5)$. The shape depends on whether the value of r is greater or less than unity. In such a system all reactant mixtures show azeotropic properties. By this we mean that the mole fractions of the reactants in any mixture will approach 0.5 as the polymerization proceeds. When the mole fraction of each reactant becomes 0.5, no further change will occur in monomer composition, and the monomer ratio in the polymer is the same as the concentration ratio of the monomers in the reaction mixture.

It is also possible to have azeotropic reaction mixtures for systems in which $r_1 \neq r_2$. Thus, an application of the azeotropic condition, namely, $\gamma = [M_1]/[M_2]$, leads to the result

$$\left(\frac{[M_1]}{[M_2]}\right)_{\text{azeotropic}} = \frac{1 - r_2}{1 - r_1} \tag{100}$$

Because the ratio of monomer concentrations must be a positive number, equation (100) shows that azeotropic reaction mixtures are possible when r_1 and r_2 are both less than unity or both greater than unity. In practice, many azeotropic systems are known for which both r_1 and r_2 are less than unity, but cases in which both reactivity ratios are greater than unity are very rare.

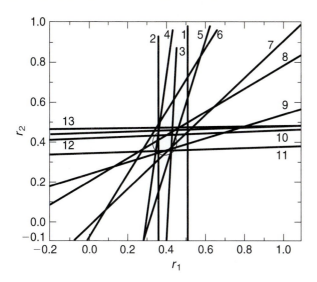

Figure 12.10 Determination of reactivity ratios by the slope-intersection method. The circle represents the most probable area for r_1, r_2 values. [Reproduced from Alfrey, T., Jr., Bohrer, J. J., and Mark, H., *Copolymerization* (New York: John Wiley & Sons, Inc. **1952**), p. 20.]

Experimental Determination of Reactivity Ratios

Actual numerical values of the reactivity ratios are determined from experimental measurements of the molar ratios of the monomers in the copolymer that is formed from reactant mixtures of known initial monomer concentration ratios. Thus, if (97) is solved explicitly for r_2, equation (101) is obtained.

$$r_2 = \frac{1}{\gamma}\left(\frac{[M_1]}{[M_2]}\right)^2 r_1 + \frac{[M_1]}{[M_2]}\left(\frac{1}{\gamma} - 1\right) \tag{101}$$

For a chosen value of $[M_1]/[M_2]$, it is possible to determine *by chemical analysis* a corresponding value of γ. Substitution of these values into (101) yields a linear equation of positive slope for r_2 as a function of r_1. Each separate experiment at different $[M_1]/[M_2]$ values yields a straight-line equation of positive slope. The coordinates of the intersection of these straight lines on a graph of r_2 versus r_1 is the reactivity ratio for the particular monomer pair under investigation. Of course, because r_1 and r_2 are ratios of reaction rate constants, their values will depend on the temperature at which the copolymerization is carried out. Figure 12.10 shows a typical plot of this type with the "intersection" for an actual case being an area instead of a point. This is mainly caused by experimental error. Table 12.5 shows some representative values of r_1 and r_2.

Individual Monomer Reactivity in Copolymerization

Each copolymerization reactivity ratio r_j describes the relative tendency of two monomers to add to a particular growing chain. The reactive end of the growing chain is a free radical derived from one of the two monomers. Obviously, two types of reactive ends can exist and, for this reason, reactivity ratios must be de-

TABLE 12.5 FREE-RADICAL COPOLYMERIZATION REACTIVITY RATIOS

M_1	M_2	Temperature (°C)	r_1	r_2
Methyl methacrylate	Acrylonitrile	80	1.22	0.15
Methyl methacrylate	Butadiene	90	0.25	0.75
Methyl methacrylate	p-Chlorostyrene	60	0.42	0.89
Styrene	Butadiene	60	0.78	1.39
Styrene	Methyl methacrylate	60	0.52	0.46
Styrene	Methyl methacrylate	131	0.59	0.54
Styrene	Vinyl acetate	60	55	0.01
Vinyl acetate	Acrylonitrile	70	0.07	6.0
Vinyl acetate	Methyl methacrylate	60	0.015	20
Vinyl acetate	Vinyl chloride	60	0.23	1.68

Source: Mark, H., Immergut, B., Immergut, E. H., Young, L. J., and Beynon, K. I., in *Copolymerization,* G. E. Ham, Ed. (New York: Wiley-Interscience, **1964**), pp. 695 ff.

termined in pairs. Moreover, because the values obtained experimentally are *relative* values, they pertain only to *one particular pair* of monomers. Thus, in the absence of any correlation procedure, reactivity ratios must be determined experimentally for each pair of monomers. On the other hand, a correlation procedure that permits the assignment to each monomer of a reactivity parameter that is applicable to its copolymerization with all other monomers would represent a great economy in data accumulation and tabulation.[1]

Several such correlations having varying degrees of complexity and theoretical foundation have been proposed, the best known and most widely used being the Alfrey–Price *Q–e* scheme. In this correlation procedure, each propagation rate constant is written as

$$k_{ij} = P_i Q_j \exp\left(-e_i e_j\right) \tag{102}$$

where P_i describes the reactivity of radical i, Q_j describes the reactivity of monomer j, and e_i and e_j describe the polarity interactions of radical and molecule, respectively. From (102) the reactivity ratios become

$$r_1 = \frac{Q_1}{Q_2} \exp(-e_1[e_1 - e_2]) \tag{103}$$

$$r_2 = \frac{Q_2}{Q_1} \exp(-e_2[e_2 - e_1]) \tag{104}$$

and the product of the reactivity ratios is

$$r_1 r_2 = \exp(-[e_1 - e_2]^2) \tag{105}$$

[1]For a collection of n monomers, there are $n(n - 1)/2$ copolymerization pairs. Thus, taking a number as conservative as $n = 100$ indicates that with a correlation procedure employing two parameters, we need to tabulate only 200 numbers, whereas without such a procedure, we require 9900 numbers.

TABLE 12.6 ALFREY–PRICE Q-E VALUES AT 60°C

Monomer	Q	e
Acrylonitrile	0.60	1.20
Butadiene	2.39	−1.05
p-Methoxystyrene	1.36	−1.11
p-Cyanostyrene	1.61	0.30
Methyl acrylate	0.42	0.60
Methyl methacrylate	0.74	0.40
Methyl vinyl ketone	1.0	0.7
Styrene	(1.0)	(−0.8)
Vinyl acetate	0.026	−0.22
Vinyl chloride	0.044	0.20

If one monomer is chosen as a reference, and arbitrary values of Q and e are assigned to it, it is possible to construct a table of Q, e values from experimental r_1, r_2 values. This is analogous to the construction of tables of standard electrode potentials in electrochemical cells. With this procedure we have two parameters for each monomer. These may be used as input to equations (103) and (104) to calculate reactivity ratios for the copolymerization of any monomer pair for which Q, e values are available. Of course, the Q, e values determined in the manner described depend on the temperature, since r_1 and r_2 are temperature-dependent. Table 12.6 shows some representative values at 60°C, in which the reference monomer, styrene, has been assigned the values $Q_{\text{styrene}} = 1.0$ and $e_{\text{styrene}} = -0.8$. The data in this table may be used to calculate reactivity ratios and also to construct copolymer–monomer composition diagrams for 45 copolymerization pairs.

Although the theoretical significance of the $Q - e$ scheme is still not clear (despite numerous attempts to provide it with a theoretical basis), it remains, as it was originally proposed, an extremely useful correlation framework for relative reactivities in copolymerization.

Distributions of the Monomers in a Copolymer[1]

Although chemical analysis of the copolymeric product indicates the mole ratio of monomers present, it does not reveal the manner in which the monomer units are distributed in the copolymer. Thus, for two monomers M_1 and M_2, the mole ratio $\gamma = (N_{M_1}/N_{M_2})_{\text{polymer}}$ yields no information concerning the average lengths of the $-(M_1)_n$ and $-(M_2)_n$ sequences in a typical copolymer as illustrated by

$$\underline{-M_1-M_1-M_1-M_1}-\underline{M_2-M_2}-\underline{M_1}-\underline{M_2-M_2-M_2}-\underline{M_1-M_1-M_1}$$

where the sequences are underlined. It is now necessary to consider how kinetic reasoning may be used to calculate the mean sequence lengths of M_1 and M_2 units from reactivity ratios.

[1]North, op. cit., pp. 90–92.

Let P_{11} be the probability that a growing radical chain $M_1 \cdot$ will add to monomer M_1. To a good approximation the only two possible fates of the growing chain $M \cdot_1$ are addition of M_1 or addition of M_2. Hence, it is possible to write this probability as

$$P_{11} = \frac{k_{p11}[M_1 \cdot][M_1]}{k_{p11}[M_1 \cdot][M_1] + k_{p12}[M_1 \cdot][M_2]} \tag{106}$$

or

$$P_{11} = \frac{r_1[M_1]}{r_1[M_1] + [M_2]} = 1 - P_{12} \tag{107}$$

where P_{12} is the probability that $M_1 \cdot$ will react with M_2. Given a radical site of type $M_1 \cdot$ in the copolymer, consider now the probability of forming a sequence of *exactly* m units of monomer M_1 and denote this probability by $P_{M_1}(m)$. It is instructive for this purpose to construct Table 12.7. To form a sequence of *exactly* m units the sequence must end with the last entry in the table. This means that the last growing radical in the table must react with M_2. Thus,

$$P_{M_1}(m) = P_{11}^{m-1} P_{12} \tag{108}$$

Similarly, the probability that a sequence of m units of M_2 will be formed, given a radical site derived from M_2, is given by

$$P_{M_2}(m) = P_{22}^{m-1} P_{21} \tag{109}$$

The average sequence lengths, $\overline{m}_{M_1}$ and $\overline{m}_{M_2}$, may now be determined from (108) and (109) using the definition of an arithmetic mean. Thus,

$$\overline{m}_{M_1} = \frac{\sum_{m=1}^{\infty} m P_{M_1}(m)}{\sum_{m=1}^{\infty} P_{M_1}(m)} = \frac{\sum_{m=1}^{\infty} m P_{11}^{m-1}}{\sum_{m=1}^{\infty} P_{11}^{m-1}} \tag{110}$$

The reader may easily verify (by writing out the first few terms of the sums in (110) and comparing with known algebraic series) that the numerator is given by $(1/1 - P_{11})^2$ and the denominator by $(1/1 - P_{11})$. Therefore, expression (111) holds:

$$\overline{m}_{M_1} = \frac{1}{1 - P_{11}} = \frac{1}{P_{12}} \tag{111}$$

TABLE 12.7 BUILDUP OF A SEQUENCE OF M_1 UNITS

Reaction	Sequence Length	Probability
$M \cdot_1 + M_1 \longrightarrow M_1 M \cdot_1$	2	P_{11}
$M_1 M \cdot_1 + M_1 \longrightarrow (M_1)_2 M \cdot_1$	3	P_{11}^2
$(M_1)_{m-2} \!\!-\!\! M \cdot_1 + M_1 \longrightarrow (M_1)_{m-1} \!\!-\!\! M \cdot_1$	m	P_{11}^{m-1}

In view of the relationship between P_{11} and r_1, (107), we finally obtain (112).

$$\overline{m}_{M_1} = 1 + r_1 \frac{[M_1]}{[M_2]} \tag{112}$$

Similarly, the average sequence length of M_2 units is given by

$$\overline{m}_{M_2} = 1 + r_2 \frac{[M_2]}{[M_1]} \tag{113}$$

The *run number, R,* of the copolymer is defined as the average number of sequences of either type per 100 monomer units. To illustrate the meaning of this term, consider the segment of a hypothetical copolymer shown below in which the sequences are underlined. The number of sequences here are nine and twenty monomer units are present. Hence, $R = 9(100/20) = 45$.

$$\underline{M_1-M_1}-\underline{M_2-M_2-M_2}-\underline{M_1-M_1-M_1-M_1}-\underline{M_2}-\underline{M_1-M_1}$$
$$-\underline{M_2-M_2-M_2-M_2}-\underline{M_1-M_1}-\underline{M_2}-\underline{M_1}$$

The rate of sequence formation, *dS/dt,* regardless of length, is simply the rate at which sequences are ended. Neglecting chain termination, this is given by

$$\frac{dS}{dt} = k_{12}[M_1 \cdot][M_2] + k_{21}[M_2 \cdot][M_1] \tag{114}$$

where the subscript p on the rate constants has been dropped for convenience. The total rate of polymerization is given by

$$-\frac{d([M_1] + [M_2])}{dt} = k_{11}[M_1 \cdot][M_1] + k_{12}[M_1 \cdot][M_2] + k_{21}[M_2 \cdot][M_1] + k_{22}[M_2 \cdot][M_2] \tag{115}$$

Elimination of the time by combination of (114) and (115) and application of the steady-state approximation (95) yields

$$-\frac{d([M_1] + [M_2])}{dS} = \frac{k_{11}[M_1] + k_{12}[M_2]}{2k_{12}[M_2]} + \frac{k_{22}[M_2] + k_{21}[M_1]}{2k_{21}[M_1]} \tag{116}$$

or

$$-\frac{d([M_1] + [M_2])}{dS} = 1 + \frac{r_1}{2}\frac{[M_1]}{[M_2]} + \frac{r_2}{2}\frac{[M_2]}{[M_1]} \tag{117}$$

As mentioned above in the definition, the *run number* is the average number of sequences per 100 monomer units. This may be written as

$$R = 100 \left(-\frac{dS}{d([M_1] + [M_2])} \right) \quad (118)$$

which, after substitution of (117), yields, finally,

$$R = \frac{200}{2 + r_1([M_1]/[M_2]) + r_2([M_2]/[M_1])} = \frac{200}{\overline{m}_{M_1} + \overline{m}_{M_2}} \quad (119)$$

Equations (112), (113), and (119) indicate that a knowledge of $[M_1]$, $[M_2]$, r_1, and r_2 enables a prediction to be made, not only of the average mole ratio γ, but also of the average number of sequences of monomer units per unit length of polymer and the average sequence length of each monomer. As many significant properties of the copolymers depend on the distribution of monomer units, the ability to make such predictions from a relatively small amount of experimental data can be very useful indeed.

STUDY QUESTIONS

1. For the free-radical chain polymerization of styrene at 60°C, $k_p = 74$ and $k_i = 3.3 \times 10^6$ liters-mol^{-1}-s^{-1}. A typical rate of initiation of propagating radicals may be taken as 10^{-6} mole-liter^{-1}-s^{-1}. Assuming that the rate of initiation is constant:
   (a) Calculate the steady-state concentration of free radicals.
   (b) Derive an expression for the time dependence of the free-radical concentration and calculate the time required to reach 95% of the steady-state value.
   (c) If the initial monomer concentration is 5 mol/liter, calculate the time required to polymerize 20% of the monomer.
   (d) Use your results to discuss the validity of the steady-state hypothesis. For example, can you say how much error is involved in using the steady-state hypothesis in answering (c).

2. Initial rates of polymerization of methyl methacrylate (MMA) initiated by the decomposition of azobisisobutyronitrile in benzene at 77°C have been reported [Arnett, L. M., *J. Am. Chem. Soc.*, **1952**, *74*, 2027] to be as shown in the accompanying table.

[MMA] (mol/liter)	[ABIN] × 10⁴ (mol/liter)	Rate × 10³ (mol/liter-min)	[MMA] (mol/liter)	[ABIN] × 10⁴ (mol/liter)	Rate × 10³ (mol/liter-min)
9.04	2.35	11.61	4.75	1.92	5.62
8.63	2.06	10.20	4.22	2.30	5.20
7.19	2.55	9.92	4.17	5.81	7.81
6.13	2.28	7.75	3.26	2.45	4.29
4.96	3.13	7.13	2.07	2.11	2.49

   (a) Is the rate law expressed by (35) in accord with these data? Why or why not?

(b) Given that the rate constant for dissociation of azobisisobutyronitrile is described by $k_i \sim 10^{15.2}e^{-30,800cal/RT}\,\mathrm{s}^{-1}$ and that the efficiency of initiation is 0.7, derive from these data a value of the rate constant ratio $k_p/k_t^{1/2}$.

3. Schulz, G. V., and Harborth, G. [*Makromol. Chem.*, **1947**, *1*, 106, published by Hüthig and Wepf Verlag, Basel] reported the following data for polymerization of methyl methacrylate in benzene at 50°C and 70°C using the thermal decomposition of 0.0413 M benzoyl peroxide for initiation.

[MMA]	$r_p \times 10^5$ at 50°C		$r_p \times 10^5$ at 70°C	
(M)	M/s	$\overline{DP}$ (50°C)	M/s	$\overline{DP}$ (70°C)
0.944	1.53	630	8.2	210
1.89	3.34	1200	18.6	450
3.78	6.74	2120	38.4	840
5.66	9.72	2900	56.6	1190

(a) Show that these data are in accord with (35).

(b) Assuming that the efficiency of initiation is independent of temperature, estimate $E_p - \frac{1}{2}E_t + \frac{1}{2}E_i$ from these data.

(c) Using data given in this chapter, evaluate the activation energy for the decomposition of benzoyl peroxide.

(d) Explain the effects of monomer concentration and temperature on $\overline{DP}$.

4. Derive equation (69).

5. Calculate and plot the weight-fraction distributions of DP for **(a)** $k_{tc} = 0$ and **(b)** $k_{td} = 0$ when chain transfer is absent.

6. Derive the number distribution function for DP when chain transfer is operating and $k_{td} = 0$, namely, equation (79).

7. Derive the number distribution function for DP analogous to that in Problem 4 when chain transfer is operating and $k_{tc} = 0$.

8. Show that the weight-fraction distribution of DP, W_n, is equal to the number-fraction distribution, X_n, multiplied by $n/\overline{DP}$.

9. Vinyl acetate at a concentration of 4 M in benzene is polymerized at 60°C using benzoyl peroxide (0.05 M) as an initiator. The rate constant for benzoyl peroxide decomposition is given by $k_d = 3.0 \times 10^{13}e^{-29.600cal/RT}\,\mathrm{s}^{-1}$, and you may assume an initiation efficiency of 0.75. You may also assume that $k_{tc} = 0$ (see Table 3.5). Using data from this chapter, calculate **(a)** the rate of polymerization; **(b)** the kinetic chain length; **(c)** the average degree of polymerization; **(d)** the distribution function for DP, including an appropriate plot; **(e)** the lifetime of the kinetic chain; **(f)** the lifetime of the propagating radical of $DP = 15$. The densities of vinyl acetate and benzene are 0.93 and 0.87 g/cm^3, respectively.

10. A reaction mixture containing 8.6 mol/liter of styrene and 0.1 mol/liter carbon tetrachloride is polymerized at 0°C using photosensitization. The average kinetic chain length is observed to be 1000. Calculate, using data given in this chapter: **(a)** the rate of initiation; **(b)** the average degree of polymerization; **(c)** the average degree of polymerization at 100°C if the initiation rate remains the same.

11. A sample of polystyrene prepared by a free-radical polymerization was separated into 21 fractions of different molecular weight [Schulz, G. V., Scholz, A., Figini, R. V., *Makro mol. Chem.*, **1962**, *57,* 220, published by Hüthig and Wepf Verlag, Basel]. The data obtained were as follows:

Weight of Fraction (mg)	DP of Fraction	Weight of Fraction (mg)	DP of Fraction
25.60	138	23.05	885
51.65	274	47.55	960
55.95	365	51.15	1050
37.40	428	52.65	1160
45.05	480	50.10	1260
52.85	535	47.70	1420
31.30	605	44.75	1600
33.75	673	36.25	1890
40.10	740	27.00	2100
38.75	795	28.90	2030
42.60	835		

(a) From these data construct the weight distribution and number distribution curves of the molecular weight. (*Hint:* First construct the integral distribution curves by plotting the weight fraction of molecules with M less than a given M versus the given M. Then differentiate the curve obtained. A similar procedure is used to determine the number distribution curve.)

(b) Compare the experimental distribution points obtained in (a) with the theoretical curve calculated by equation (67).

12. A benzene solution contains vinyl acetate at 3.5 M and vinyl chloride at 1.5 M. A free-radical polymerization is initiated by adding 0.1 M azobisisobutyronitrile and heating the solution to 60°C. Calculate:

(a) Composition of the copolymer first formed.

(b) Average sequence lengths of vinyl acetate and vinyl chloride in the copolymer first formed.

(c) The run number of the copolymer first formed.

(d) The probability of forming a vinyl acetate sequence that is 8 units long.

13. Using the Alfrey–Price Q-e values, calculate the reactivity ratios for the following copolymerization systems: (a) acrylonitrile–butadiene; (b) methyl acrylate–vinyl chloride; (c) methyl vinyl ketone–*p*-methoxystyrene; (d) styrene–*p*-methoxystyrene.

14. In the thermolysis of the peroxide bond $-O-O-$, why is the product much more likely to be the two desired radicals $-O\cdot$, rather than $-O^+$ and $-O^-$?

15. Why is chain transfer to solvent relatively important, even when the solvent's chain transfer constant is small?

16. Why is chain transfer to initiator relatively unimportant, even when the initiator's chain transfer constant is large?

17. Will increasing the monomer concentration [M] make chain transfer to monomer more important?

SUGGESTIONS FOR FURTHER READING

ALLEN P. E. M., and PATRICK, C. R., *Kinetics and Mechanisms of Polymerization Reactions*. New York: Wiley, **1974**, Chaps. 2, 3, 7.

BEVINGTON, J. C., *Radical Polymerization*. New York: Academic Press, **1961**.

BILLMEYER, F. W., Jr., *Textbook of Polymer Science*. New York: Wiley, **1984**, Chaps. 3, 5.

CRAVER, C., and CARRAHER, C. E., Jr. (Eds.), *Applied Polymer Science—21st Century*, Washington: American Chemical Society, **2000**.

DAINTON, F. S., *Chain Reactions*. London: Methuen, **1956**, Chap. 7.

FISCHER, H., "The Persistent Radical Effect: A Principle for Selective Radical Reactions and Living Radical Polymerizations," *Chem. Rev.*, **2001**, *101*, 3581.

FLORY, P. J., *Principles of Polymer Chemistry*. Ithaca, N.Y.: Cornell University Press, **1953**, Chap. 4.

GRIDNEV, A. A., and ITTEL, S. D., "Catalytic Chain Transfer in Free-Radical Polymerizations," *Chem. Rev.*, **2001**, *101*, 3611.

HAWKER, C. J., BOSMAN, A. W., and HARTH, E., "New Polymer Synthesis by Nitroxide Mediated Living Radical Polymerizations," *Chem. Rev.*, **2001**, *101*, 3661.

HIEMENZ, P. C., *Polymer Chemistry*. New York: Dekker, **1984**, Chap. 6.

KAMIGAITO, M., ANDO, T., and SAWAMOTO, M., "Metal Catalyzed Living Radical Polymerization," *Chem. Rev.*, **2001**, *101*, 3689.

LOVELL, P. A., and EL-AASSER, M. S. (Eds.), *Emulsion Polymerization and Emulsion Polymers*, Chichester: John Wiley and Sons, **1997**.

MATYJASZEWSKI, K. (Ed.), *Controlled Radical Polymerization*, Washington, DC: American Chemical Society, **1998**.

NORTH, A. M., *The Kinetics of Free Radical Polymerization*. New York: Pergamon Press, **1966**.

SCOTT, G. E., and SENOGLES, E., "Kinetic Relationships in Radical Polymerization," *J. Macromol. Sci., Rev. Macromol. Chem.*, **1973**, *C9*, 49.

SEYMOUR, R. B., and CARRAHER, C. E., *Polymer Chemistry*, 2nd ed. New York: Dekker, **1988**, Chaps. 9, 10.

13

Kinetics
of Ionic Polymerization

DIFFERENCES BETWEEN IONIC AND FREE-RADICAL KINETICS

As discussed earlier in Chapter 4, ionic polymerizations take place by chain mechanisms in which many monomer molecules add to a single chain center. Thus, ionic polymerization resembles free-radical polymerization in terms of the initiation, propagation, transfer, and termination reactions. The only difference between the two modes is the nature of the active chain end, that is, whether it is a free radical, a positive ion, or a negative ion.

However, the kinetics of ionic polymerizations are significantly different from free-radical polymerizations. The initiation reactions of ionic polymerization have only very low activation energies and, thus, they more closely resemble photoinitiated free-radical polymerizations. In free-radical polymerizations, chain termination occurs by the mutual destruction of two polymeric radicals, but in ionic polymerization such a process is impossible because the charge is not neutralized in the reaction between two positive or two negative ions. Solvent effects are much more pronounced in ionic polymerizations because of the role that the solvent plays in assisting the separation of electric charge. Thus, in a solvent of high dielectric constant, a polymerization may proceed via free ions. In a solvent of low dielectric constant the chain centers may be ion pairs, or both ion pairs and free ions. No such solvent role is encountered in free-radical polymerization.

The overall result of the foregoing features is to make the kinetics of ionic polymerization much more complex than the kinetics of free-radical polymerization. Most of the complications in ionic polymerization arise from the initiation reactions. The variety of initiation possibilities often gives the appearance that each ionic polymerization is unique and that no general kinetic treatment is possible. However, some cases do exist, particularly in anionic polymerization, in which the initiator dissociates *completely* into the active ionic form and does so *before* any significant amount of polymerization has occurred. In such cases, the polymerization kinetics are so simple that it is useful to classify ionic polymerization initiators according to whether they are quantitatively and instantaneously dissociated or not.

ANIONIC POLYMERIZATION

Quantitative and Instantaneous Dissociation of Initiator: Living Polymers

As mentioned above, particularly simple kinetics are found when an ionic polymerization initiator is completely dissociated before the polymerization begins. Examples of such initiators (see Table 4.1) are (1) alkali metal suspensions in liquids that are Lewis bases; (2) organolithium compounds (also used in solvents that act as Lewis bases toward the lithium ion); and (3) sodium naphthalenide, which is prepared by the electron exchange reaction between sodium and naphthalene.

If we represent the undissociated initiator by GA, then we can assume, for the present, that in the polymerization medium the dissociation reaction (1a) is instantaneous and complete

$$GA \longrightarrow G^+ + A^- \tag{1a}$$

A monomer molecule must add to A^-, to complete the initiation of a polymerization, namely,

$$G^+ + A^- + M \longrightarrow G^+ + AM^- \tag{1b}$$

Depending on the solvent, the propagating anion may behave as a free ion, AM^-, or as an ion pair, AM^-G^+, or as both. For simplicity we will consider only one type (i.e., free ions) for the present kinetic treatment. The propagation reactions may then be written

$$A-M^- + M \longrightarrow A-M-M^- \tag{2}$$

$$A-M-M^- + M \longrightarrow A-M_2-M^- \tag{3}$$

$$A-M_{n-1}-M^- + M \longrightarrow A-M_n-M^- \tag{4}$$

It is assumed that the positive ion G^+ is always in the vicinity of the negative chain center at each step. The extent of charge separation between G^+ and the negative ions depends on the dielectric constant of the solvent. In the absence of impurities or of substances added deliberately for the purpose of chain termination, no termination reactions exist in this special case. The polymerization ceases only when the monomer is consumed. It begins again if more monomer is added. For this reason, the polymers produced in this special case are the so-called "living polymers."

In such a situation, the kinetics are especially simple because no initiation reaction takes place *during* the polymerization. The number of chain centers to which the monomer molecules may add reaches its maximum value before polymerization begins. Moreover, the number of chain centers does not change during the polymerization because there is no termination step.

Rate of polymerization   The rate of polymerization is defined in a similar manner to that described earlier for free-radical polymerization,

$$r_p = -\frac{d[M]}{dt} = k_p[A^-][M] \tag{5}$$

where $[A^-]$ is the total concentration of anions of all degrees of polymerization. The total concentration $[A^-]$ is constant and is given by the concentration of the initiator before dissociation, namely $[GA]_0$. Hence,

$$r_p = -\frac{d[M]}{dt} = k_p[GA]_0[M] \tag{6}$$

Integration of this first-order rate equation gives the time dependence of the monomer concentration as

$$[M] = [M]_0 e^{-k_p[GA]_0 t} \tag{7}$$

Rates of anionic polymerization are measured experimentally in the same way as described for free-radical polymerizations in Chapter 12, namely, by the measurement of $[M]$ at various times after the polymerization is started. The value of $[M]$ may be measured directly or by a secondary method once the relationship between the two has been established. From the experimental measurements and a knowledge of $[GA]_0$, k_p may be determined either by (6) or (7). The significance of the values obtained will be discussed later in this chapter.

Average kinetic chain length   Because no termination step exists in a true "living" polymerization, the kinetic chain growth is ended only when the monomer is completely consumed. The average kinetic chain length by definition is

$$\nu = \frac{\text{monomer consumed}}{\text{number of chain centers}} = \frac{[M]_0 - [M]}{[GA]_0} \tag{8a}$$

which on substitution of [M] from (7) becomes a function of time as given by (8b)

$$v = \frac{[M]_0}{[GA]_0}(1 - e^{-k_p[GA]_0 t})$$ (8b)

According to (8b), in the limit of $t \to \infty$, or, in other words, at the completion of reaction,

$$v_\infty = \frac{[M]_0}{[GA]_0}$$ (8c)

Average degree of polymerization The average degree of polymerization is given by the number of monomer molecules polymerized per polymer molecule formed. The number of polymer molecules formed is equal to the number of chain centers, or initiators, so that (9) is obtained.

$$\overline{DP} = \frac{[M]_0 - [M]}{[GA]_0} = \frac{[M]_0}{[GA]_0}(1 - e^{-k_p[GA]_0 t}) = v$$ (9)

In some cases, it is necessary to make a simple modification of (9) to account for the fact that dianions are involved (see Chapter 4). For example, in the sodium naphthalenide initiated polymerization of styrene, the mechanism shown in (10) and (11) operates.[1]

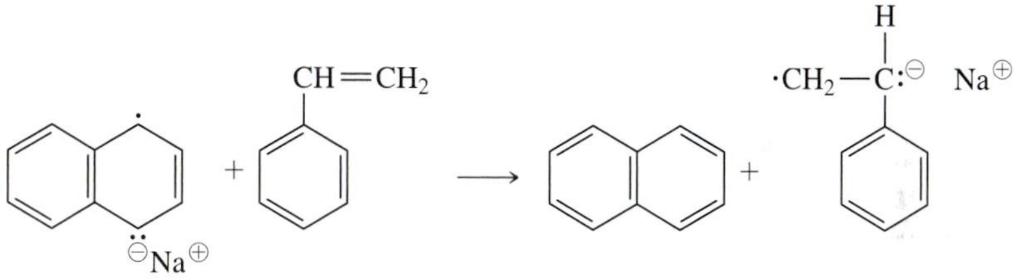

(10)

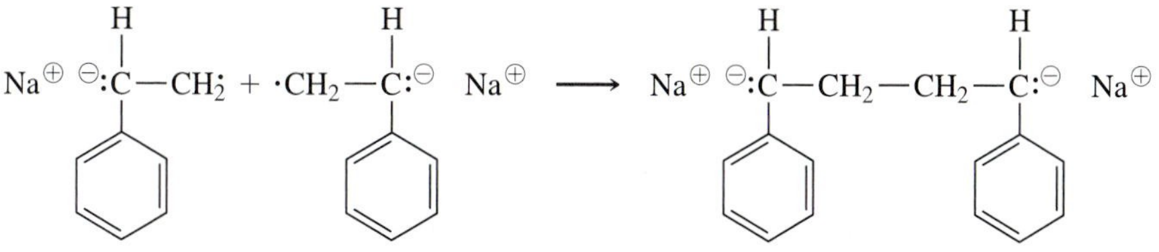

(11)

The product of (11) is the true initiating species so that, for this particular case, the number of polymer molecules formed is .50 of the number of chain centers or initiators. The degrees of polymerization is then given by

[1]Szwarc, M., Levy, M., and Milkovich, R., *J. Am. Chem. Soc.*, **1956,** *78,* 2656.

$$\overline{DP} = \frac{[M]_0 - [M]}{\frac{1}{2}[GA]_0} = 2\frac{[M]_0}{[GA]_0}(1 - e^{-k_p[GA]_0 t}) = 2v \qquad (12)$$

Distribution of the degree of polymerization   The most important difference between an ionic polymerization which has no termination or transfer mechanism and free-radical or ionic processes that do have termination or chain transfer steps is that the distributions of the degrees of polymerization are quite different. The distribution function may be derived by a kinetic treatment due to Flory,[1] which is analogous to that discussed earlier for free-radical reactions. However, the steady-state approximation with its many simplifications cannot be used in this present case.

Consider the mechanisms, (1) to (4), for this polymerization. If the usual assumption is made that k_p is independent of size and also that the initiation steps (1a and 1b) are instantaneous, application of the mass action law gives the set of rate equations shown in (13) to (16).

$$\frac{d[AM^-]}{dt} = -k_p[AM^-][M] \qquad (13)$$

$$\frac{d[AMM^-]}{dt} = k_p[M]\{[AM^-] - [AMM^-]\} \qquad (14)$$

$$\frac{d[AMMM^-]}{dt} = k_p[M]\{[AMM^-] - [AMMM^-]\} \qquad (15)$$

$$\vdots$$

$$\frac{d[AM_nM^-]}{dt} = k_p[M]\{[AM_{n-1}M^-] - [AM_nM^-]\} \qquad (16)$$

Substitution of the expression for [M], namely (7), into the first of these rate equations yields

$$\int \frac{d[AM^-]}{[AM^-]} = -k_p[M]_0 \int e^{-k_p[GA]_0 t} dt \qquad (17)$$

On integration, under the condition that at $t = 0$, $[AM^-] = [GA]_0$,

$$[AM^-] = [GA]_0 \exp\left[-\frac{[M]_0}{[GA]_0}(1 - e^{-k_p[GA]_0 t})\right] \qquad (18)$$

In view of the expression for average kinetic chain length, namely (8), this last result may be written in the compact form

[1]Flory, P. J., *J. Am. Chem. Soc.*, **1940**, *62*, 1561.

$$[AM^-] = [GA]_0 e^{-\nu} \tag{19}$$

It is useful at this point to eliminate the time, t, from the remainder of the rate equations (14) to (16). This may be done conveniently by the use of the expression for the average kinetic chain length, (8b). Differentiation of (8b) with respect to time gives

$$d\nu = k_p[M]_0 e^{-k_p[GA]_0 t} \, dt \tag{20}$$

Substitution of (7), (19), and (20) into (14), with elimination of t, transforms this rate equation for $[AMM^-]$ into the differential equation

$$\frac{d[AMM^-]}{d\nu} + [AMM^-] = [GA]_0 e^{-\nu} \tag{21}$$

The rate equation (21) is in the standard form of a linear, first-order differential equation whose solution is given by

$$e^{\nu}[AMM^-] = \int e^{\nu}[GA]_0 e^{-\nu} \, d\nu + C \tag{22}$$

where e^{ν} is the integrating factor and C is a constant of integration. After integration of (22) and evaluation of C by the condition that at $\nu = 0$ (i.e., $t = 0$), $[AMM^-] = 0$, we obtain

$$[AMM^-] = [GA]_0 \nu e^{-\nu} \tag{23}$$

The process by which the rate equation (14) was solved to yield (23) may now be repeated for the rate equation (15). Thus, elimination of t by substitution of (7), (20), and (23) transforms (15) into the differential equation (24),

$$\frac{d[AMMM^-]}{d\nu} + [AMMM^-] = [GA]_0 \nu e^{-\nu} \tag{24}$$

which may be solved in the same manner as was (21) using the integrating factor e^{ν}. The result is

$$[AMMM^-] = \tfrac{1}{2}[GA]_0 \nu^2 e^{-\nu} \tag{25}$$

Repetition of this process soon makes it evident that the concentration of the anion containing n monomer molecules is given by

$$[AM_{n-1}M^-] = [GA]_0 \frac{\nu^{n-1} e^{-\nu}}{(n-1)!} \tag{26}$$

When polymerization is complete, the kinetic chain length as given by (8c) is $[M]_0/[GA]_0$. Furthermore, each initial anion has produced one polymer species, since there is no termination. Therefore, the fraction of polymer of degree of polymerization n at the end of the reaction is

$$X_n = \frac{\text{number of anions containing } n \text{ monomers}}{\text{number of anions}} = \frac{[AM_{n-1}M^-]}{[GA]_0} \quad (27)$$

or

$$X_n = \frac{v_\infty^{n-1} e^{-v_\infty}}{(n-1)!} = \frac{1}{(n-1)!}\left(\frac{[M]_0}{[GA]_0}\right)^{n-1} e^{-[M]_0/[GA]_0} \quad (28)$$

The distribution of v (or DP) given by (28) is shown graphically for $v_\infty = 50$ in Figure 13.1. Also shown for comparison is the distribution of DP for a polymer produced by free-radical polymerization (with termination by combination) with the same value of v. The narrowness of the distribution of DP for the living polymer is very striking. Actually, the observed distributions of DP are usually some-

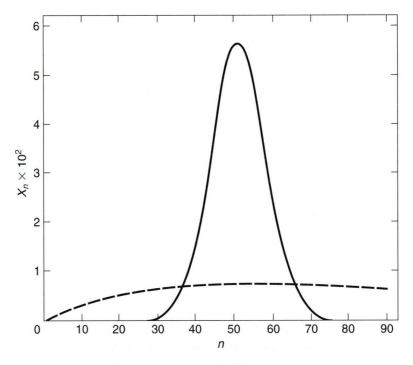

Figure 13.1   Distribution of the degree of polymerization with $v = 50$ for "living" polymers (solid line) (equation 28) and for free-radical polymerization (dashed line) (Chapter 12, equation 67).

what broader than predicted by (28). This fact is attributed to the existence of a propagation–depropagation equilibrium,

$$A\!-\!M_n\!-\!M^- + M \rightleftharpoons A\!-\!M_{n+1}\!-\!M^- \tag{29}$$

which was not considered in the derivation of (28). Nonetheless, the observed distributions are often very close to those predicted.

It should be kept in mind that the simple kinetics and the narrow distributions of DP are not a general characteristic of anionic polymerization, nor are they unique to it. They arise because the entire initiation process occurs before propagation and because no termination process is present to destroy the chain centers.

Rate Constants for Propagation

The use of initiators that dissociate quantitatively before propagation occurs and the absence of termination reactions permits the straightforward determination of k_p from measurements of the extent or rate of polymerization and equations (6) and (7). Some values obtained in this way at 25°C in tetrahydrofuran solvent with the counterion (G^+) being Na^+ are shown in Table 13.1.

It was mentioned earlier that the propagation reaction could involve free anions, ion pairs, or both. Experimentally, it is found that k_p depends on the nature of the counterion and on the solvent. Neither of these effects would exist if the *sole* chain centers in the propagation steps were free ions. It must be concluded that ion pairs are important chain carriers, and may perhaps be the sole chain centers in anionic polymerizations. Therefore, the k_p values of Table 13.1 may represent composite values for propagation by ion pairs and free ions.

In principle, the degree of dissociation in any equilibrium of the type

$$A^-G^+ \rightleftharpoons A^- + G^+ \tag{30}$$

will be shifted to the right as the system is diluted with solvent, or, in other words, as the concentration $[GA]_0$ is reduced. Extrapolations of experimental values of k_p to infinite dilution should yield the k_p for propagation that occurs solely by free

TABLE 13.1 PROPAGATION RATE CONSTANTS FOR ANIONIC POLYMERIZATION AT 25°C; SOLVENT = TETRAHYDROFURAN, COUNTERION = Na^+

Monomer	k_p(liters-mol^{-1}-s^{-1})	Monomer	k_p(liters-mol^{-1}-s^{-1})
α-Methylstyrene	2.5	Styrene	950
p-Methoxystyrene	52	1-Vinylnaphthalene	850
o-Methylstyrene	170	2-Vinylpridine	7300
p-t-Butylstyrene	220	4-Vinylpyridine	3500

Source: Szwarc, M., and Smid, J., in *Progress in Reaction Kinetics*, Vol. 2, G. Porter, Ed. (New York: Pergamon Press, **1964**), p. 249.

Anionic Polymerization

ions. Where such studies have been possible they have indicated that the value of k_p for a free-ion propagation is much greater than that for propagation by ion pairs. Therefore, the contribution of free ions to the overall propagation may be significant even in solvents of low dielectric constant. Hence, it must be concluded that the experimental values of k_p determined from polymerization rates represent a composite value for mechanisms that involve ion pairs *and* free ions. These complicating effects are typical of ionic reactions. Unfortunately, they reduce the utility of the k_p values because such rate constants can be used to make predictions only for polymerizations carried out under exactly the same set of conditions (i.e., exactly the same solvent, initiator, concentration of initiator, and temperature). The k_p values measured in free-radical polymerizations are independent of such conditions, except for the influence of temperature.

Incomplete Dissociation of Initiator

The simple kinetics of the anionic polymerization discussed earlier exist because the initiator is converted *completely* from the inactive form, GA, to the active form, G^+A^-, (or $G^+ + A^-$) before any propagation reactions take place. However, some initiators (i.e., lithium alkyls and aryls) maintain an equilibrium between the active form and the inactive form. Moreover, this equilibrium may extend to the growing anionic chains also. In such a situation we must write the initiation steps as

$$ GA \rightleftharpoons G^+A^- \rightleftharpoons G^+ + A^- \qquad (1a') $$

$$ G^+A^- + M \longrightarrow AM^-G^+ \rightleftharpoons AMG \qquad (1b') $$

and the propagation steps as

$$ AM^-G^+ + M \longrightarrow AMM^-G^+ \rightleftharpoons AM_2G \qquad (2') $$

$$ AM_{n-1}-M^-G^+ + M \longrightarrow AM_n-M^-G^+ \rightleftharpoons AM_{n+1}G \qquad (4') $$

We have included here propagation by ion pairs only. Propagation by free anions in mechanisms that correspond to $(2')$ and $(4')$ must also be considered in the complete scheme.

The mechanism described here is much more complex than that of "living" polymerization because of existence of the equilibrium

$$ AM_nG \rightleftharpoons AM_n{}^-G^+ \qquad (31) $$

As discussed, such equilibria exist not only in the initiation but in the propagation reactions as well. Furthermore, the equilibria may involve solvation contributions by the solvent, although this is not shown explicitly in the mechanism. Finally, as a further complicating feature, the monomer may also affect the equilibria. In such a case, the initiation rate would depend on the nature of the monomer, even though all other factors, such as solvent, temperature, concentrations, initiator, and so on, were fixed.

The complexities described above make it virtually impossible to write explicit general equations for the rate of polymerization, kinetic chain length, average degree of polymerization, and distribution of degree of polymerization as was done in Chapter 12 for free-radical polymerization, and earlier in this chapter for a completely dissociated ionic initiator. Hence, with the exception of those cases discussed above, in anionic polymerization (and in cationic reactions as well) each system represents a kinetically unique problem that must be solved separately.

Anionic Copolymerization

For the polymerization of two monomers by an anionic mechanism, we may write a set of elementary propagation reactions analogous to those described in Chapter 12:

$$\text{\text{\wasysym{W}}}-M_1^- + M_1 \xrightarrow{k_{11}} M_1^- \tag{32}$$

$$\text{\text{\wasysym{W}}}-M_1^- + M_2 \xrightarrow{k_{12}} M_2^- \tag{33}$$

$$\text{\text{\wasysym{W}}}-M_2^- + M_2 \xrightarrow{k_{22}} M_2^- \tag{34}$$

$$\text{\text{\wasysym{W}}}-M_2^- + M_1 \xrightarrow{k_{21}} M_1^- \tag{35}$$

Similarly, we may define reactivity ratios: $r_1 = k_{11}/k_{12}$ and $r_2 = k_{22}/k_{21}$, and determine such ratios from the composition of the copolymer product. However, a serious complication exists. The propagation rate constants, k_{ij}, are *composite* rate constants, being composed of free-ion contributions and ion-pair contributions. Therefore, the reactivity ratios will also be composite quantities, having contributions from both ion pairs and free ions. Because the relative abundances of free ions and ion pairs are strongly dependent on the reaction conditions, the reactivity ratios will also depend on these conditions. Therefore, the utility of such ratios is much more limited in anionic than in free-radical polymerization, because they can be applied only to systems identical to those for which they were determined.

Typical reactivity ratios for the anionic copolymerization of styrene with several monomers are shown in Table 13.2. Most of the values shown in this table were determined from measurements of the copolymer compositions as a function of monomer concentrations, as discussed in Chapter 12. The use of "living copolymerizations" (in which one monomer is first polymerized and the second monomer then added) permits reactions (33) and (35) to be studied independently and allows the respective rate constants k_{12} and k_{21} to be measured directly. The rate constants k_{11} and k_{22} are known or can be determined, in principle, from the polymerization of the pure monomers. Thus, in some anionic systems, the simplicity afforded by complete initiator dissociation (i.e., living polymers) permits not only a determination of the rate-constant ratios but also the mea-

TABLE 13.2   REACTIVITY RATIOS IN THE ANIONIC COPOLYMERIZATION OF STYRENE (M_1)

M_2	Initiator	Solvent	Temperature (°C)	r_1	r_2
Acrylonitrile[*]	C_6H_5MgBr in toluene	Cyclohexane	−45	0.05	15.0
Methyl methacrylate[*]	C_6H_5MgBr in ether	Toluene	−30	0.01	25.0
Methyl methacrylate[*]	C_6H_5MgBr in ether	Ether	−30	0.05	14.0
Methyl methacrylate[*]	C_6H_5MgBr in ether	Ether	−78	0.02	20.0
Methyl methacrylate[*]	C_6H_5MgBr in ether	Ether	+20	0.30	2.0
p-Methoxystyrene[†]	C_4H_9Li	Toluene	0	10.9	0.05
p-Methoxystyrene[†]	Li	Tetrahydrofuran	0	2.9	0.23
α-Methylstyrene[‡]	Na-K alloy	Tetrahydrofuran	+25	35	0.003
p-Methylstyrene[§]	Na-K alloy	Tetrahydrofuran	+25	5.3	0.18
p-Methylstyrene[†]	Na	Tetrahydrofuran	0	1.97	0.38

[*] Dawans, F., and Smets, G., *Makromol. Chem.*, **1963**, *59*, 163.

[†] Tobolsky, A. V., and Boudreau, R. J., *J. Polymer Sci.*, **1961**, *51*, S53.

[‡] Lee, C. L., Smid, J., and Szwarc, M., *J. Am. Chem. Soc.*, **1961**, *83*, 2961.

[§] Shima, M., Bhattacharyya, D. N., Smid, J., and Szwarc, M., *J. Am. Chem. Soc.*, **1963**, *85*, 1306.

surement of individual rate constants. The reactivity ratios for the copolymerizations of styrene with α-methylstyrene and p-methylstyrene in Table 13.2 were determined in this manner.

The drastic effect that the reaction medium and the reaction conditions have on the anionic copolymerization reactivity ratios may be appreciated by inspection of the data for styrene–methyl methacrylate in Table 13.2.

CATIONIC POLYMERIZATION

Rate of Polymerization

The elementary reactions involved in addition polymerization via positive ions are formally identical to those found in anionic polymerization. The only difference is that the charges of the propagating chain center and the counterion are reversed. As in anionic polymerization, propagation may proceed by both ion pairs and free ions. Despite this similarity, the kinetics of cationic polymerizations are less well understood than those of anionic polymerizations. This is probably because of the relatively small number of cationic polymerization systems that are known to show the phenomena of complete and prior dissociation of the initiator (i.e., "living" polymerization). Essentially all the known cationic polymerizations involve initiation reactions that occur simultaneously with the propagation steps. As we have seen in anionic polymerization, the variety of initiation mechanisms and the accompanying complex association–dissociation equilibria preclude the writing of a general kinetic scheme. Moreover, the initiation of a cationic polymerization is often complicated by the need for cocatalysts. Cocatalysis occurs when traces of certain substances are required along with the initiator before polymerization can proceed (see pages 106 to 109).

For the purpose of discussion, we may depict a cationic polymerization by the following set of elementary reactions, in which C is the catalyst or initiator, RX is the cocatalyst, and M is the monomer:

$$\left. \begin{aligned} C + RX \; &\underset{k_{-1}}{\overset{k_1}{\rightleftharpoons}} \; R^+CX^- \; \underset{k_{-2}}{\overset{k_2}{\rightleftharpoons}} \; R^+ + CX^- \\ R^+CX^- + M \; &\overset{k_i}{\longrightarrow} \; RM^+CX^- \end{aligned} \right\} \; \text{Initiation} \qquad \begin{aligned} &(36)\\[1.2em] &(37) \end{aligned}$$

$$\left. \begin{aligned} RM^+CX^- + M \; &\overset{k_p}{\longrightarrow} \; RMM^+CX^- \\ &\cdots\cdots\cdots\cdots\cdots\cdots\cdots\cdots\cdots\cdots \\ RM_{n-1}{-}M^+CX^- + M \; &\overset{k_p}{\longrightarrow} \; RM_n{-}M^+CX^- \end{aligned} \right\} \; \text{Propagation} \qquad \begin{aligned} &(38)\\[1.2em] &(39) \end{aligned}$$

$$R{-}M_n{-}M^+CX^- + M \;\rightarrow\; M^+CX^- + R{-}M_nM \quad \text{Chain transfer} \qquad (40)$$

$$RM_n{-}M^+CX^- \;\rightarrow\; RM_{n+1}{-}X + C \quad \text{Chain termination} \qquad (41)$$

In (37) to (41), only the polymerization that involves ion pairs is shown explicitly. It must be kept in mind that a similar polymerization chain carried by the free ions, $RM_n{-}M^+$, may be occurring simultaneously and that an equilibrium involving ion pairs and free ions, such as shown in (36), exists for ion pairs of all sizes. An example of such a mechanism is the one in which isobutylene is polymerized in methyl chloride solution in the presence of aluminum chloride, in which case

$$M = \begin{matrix} & CH_3 \\ & | \\ CH_2{=}C \\ & | \\ & CH_3 \end{matrix} \qquad RM_n{-}M^+ = CH_3\!\left(\!\begin{matrix} CH_3 \\ | \\ CH_2C \\ | \\ CH_3 \end{matrix}\!\right)_{\!n}\!\!\begin{matrix} CH_3 \\ | \\ CH_2C^+ \\ | \\ CH_3 \end{matrix}$$

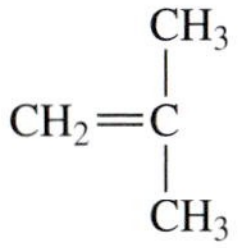

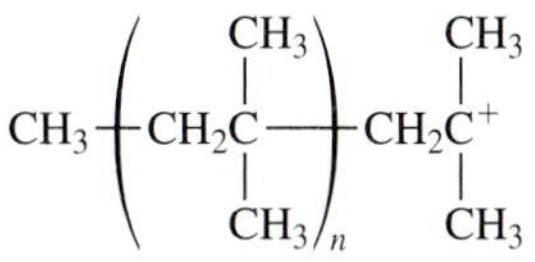

$$C = AlCl_3$$

$$R'M^+ = \begin{matrix} & CH_3 \\ & | \\ CH_3C^+ \\ & | \\ & CH_3 \end{matrix}$$

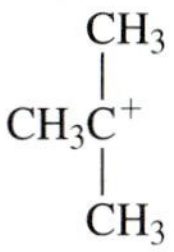

$$RX = CH_3Cl$$
$$R^+ = CH_3{}^+$$

$$RM_{n+1}X = CH_3\!\left(\!\begin{matrix} CH_3 \\ | \\ CH_2C \\ | \\ CH_3 \end{matrix}\!\right)_{\!n}\!\!\begin{matrix} CH_3 \\ | \\ CH_2C{-}Cl \\ | \\ \end{matrix}$$

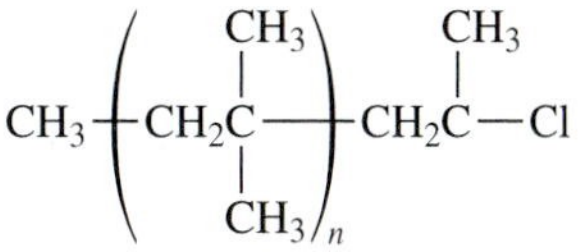

$$CX^- = AlCl_4^-$$

The rate of polymerization by ion pairs may be written

$$\frac{-d[\mathrm{M}]}{dt} = r_p = k_p[\mathrm{M}] \sum_{n=0}^{\infty} [\mathrm{RM}_n\!-\!\mathrm{M}^+\mathrm{CX}^-] \qquad (42)$$

The rate of termination in a catalyzed ionic polymerization will be first-order with respect to the growing chain concentration. This is understandable on the basis of electrostatic considerations. In the media usually employed, the counterion must remain so close to the positive chain center that the ion pair (or "free ions") behaves in a kinetic sense as a single entity. Thus,

$$r_t = k_t \sum_{n} [\mathrm{RM}_n\!-\!\mathrm{M}^+\mathrm{CX}^-] \qquad (43)$$

If we assume a steady state for the concentration of growing chains (the validity of which is not nearly as certain as for free-radical polymerizations), then $r_t = r_i$ and we have, from (43),

$$\sum_{n=0}^{\infty} [\mathrm{RM}_n\!-\!\mathrm{M}^+\mathrm{CX}^-] = \frac{r_i}{k_t} \qquad (44)$$

Therefore, the rate of polymerization by ion pairs is given by

$$r_p = \left(\frac{k_p}{k_t}\right)[\mathrm{M}]r_i \qquad (45)$$

with an analogous expression holding for the rate of polymerization by free ions.

It is the complexity and variety of the kinetic expressions for r_i, the rate of initiation, that defeat attempts to write general rate laws for cationic polymerizations. The rate of initiation at a fixed temperature and in a given solvent will be determined by the equilibria or steady-state existing among ion pairs, free ions, catalyst, cocatalyst, and monomer. Thus, it is possible to write

$$r_i = f([\mathrm{C}], [\mathrm{RX}], [\mathrm{M}]) \qquad (46)$$

but it must be recognized that the function f can take on such a variety of forms that it can be determined with confidence only by actual experiment for any given system.

Degree of Polymerization

In the discussion of free-radical kinetics in Chapter 12, the average degree of polymerization was given simply by the ratio of the rate of depletion of the monomer to the rate of formation of the polymer. The same relationship holds in cationic polymerization. Thus, because polymer is formed in both the termination and the transfer reactions, namely (40) and (41), we have

$$\overline{DP} = \frac{r_p}{r_t + r_{\mathrm{Tr}}} \tag{47}$$

In terms of the cationic mechanism shown in (36) to (41), in which transfer occurs only to monomer, equation (47) becomes

$$\overline{DP} = \frac{k_p[\mathrm{M}]}{k_t + k_{\mathrm{Tr}}[\mathrm{M}]} \tag{48}$$

or, in the usual inverted form, known as the Mayo equation.

$$\frac{1}{\overline{DP}} = \frac{k_{\mathrm{Tr,M}}}{k_p} + \frac{k_t}{k_p}\frac{1}{[\mathrm{M}]} \tag{49}$$

Figure 13.2 illustrates the use of the Mayo equation (49) to obtain the rate-constant ratios $k_{\mathrm{Tr,M}}/k_p$ and k_t/k_p in the cationic polymerizaton of

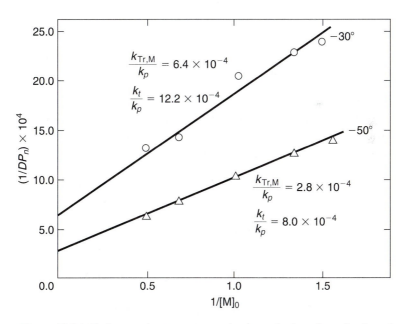

Figure 13.2 Chain transfer to monomer in the cationic polymerization of methylstyrene in dichloromethane solution. [From Kennedy, J. P., and Chou, R. T., *J. Macromol. Sci., Chem.*, **1982**, *A18*, 17.]

α-methylstyrene at -30 and $-50°C$. From the intercepts and slopes of the lines in Figure 13.2, one obtains the values

$$\frac{k_{\text{Tr,M}}}{k_p} = 6.4 \times 10^{-4} \text{ at } -30°C$$

$$\frac{k_{\text{Tr,M}}}{k_p} = 2.8 \times 10^{-4} \text{ at } -50°C$$

$$\frac{k_t}{k_p} = 1.22 \times 10^{-3} \text{ at } -30°C$$

$$\frac{k_t}{k_p} = 8.0 \times 10^{-4} \text{ at } -50°C$$

Although only approximate, because the kinetics were studied at only two temperatures, the results above suggest the following Arrhenius parameters for this reaction:

$$E_{\text{Tr,M}} - E_p = 4.5 \text{ kcal/mol}$$

$$\frac{A_{\text{Tr,M}}}{A_p} = 6.4$$

$$E_t - E_p = 2.3 \text{ kcal/mol}$$

$$\frac{A_t}{A_p} = 0.13 \text{ mol/liter}$$

Transfer can occur also to additional substances, often introduced inadvertently into the system as impurities. In such cases, (49) must be transformed to (50) to take into account the additional opportunities for chain transfer.

$$\frac{1}{\overline{DP}} = \frac{k_{\text{Tr,M}}}{k_p} + \frac{k_t}{k_p} \frac{1}{[\text{M}]} + \frac{\sum_x k_{\text{Tr,X}}}{k_p} \frac{[\text{X}]}{[\text{M}]} \tag{50}$$

In contrast to free-radical polymerization, the formation of the polymeric product in cationic polymerization occurs mainly by transfer rather than by termination. Therefore, if cationic systems are rigorously freed from all impurities (i.e., $[\text{X}] = 0$), (50) predicts that when $k_t \ll k_{\text{Tr,M}} [\text{M}]$, $\overline{DP}$ will be independent of $[\text{M}]$ and should be determined only by the value of the ratio $k_p/k_{\text{Tr,M}}$. In practice, this is often difficult to achieve because impurities need be present only at trace levels to decrease $\overline{DP}$. For example, in the stannic chloride-catalyzed polymerization of isobutylene in ethyl chloride solution at $-78°C$, it has been shown[1] that trace amounts of water can prevent $\overline{DP}$ from attaining the level $k_p/k_{Tr,M}$ because the reaction shown in (51) is highly probable.

$$\underset{\underset{CH_3}{|}}{\overset{\overset{CH_3}{|}}{W\!\!-\!\!CH_2C^+}} + H_2O \longrightarrow W\!\!-\!\!CH\!\!=\!\!\overset{\overset{CH_3}{|}}{C}\!\!-\!\!CH_3 + H_3O^+ \cdot \qquad (51)$$

Effect of Temperature

The rate of the initiation of a cationic polymerization via the formation of ion pairs and free ions is only very slightly dependent on the temperature. This means that the effect of temperature on the rate of polymerization will be determined by the temperature dependence of k_p/k_t or k'_p/k'_t, where the primes denote rate constants for free ions and the unprimed units signify rate constants for ion pairs. If the rate constants are written in the Arrhenius formulation, then when one form of propagation dominates (i.e., free ions or ion pairs),

$$r_p \propto \frac{A_p}{A_t} e^{(E_t - E_p)/RT} \qquad (52)$$

In cationic polymerization, the activation energy for termination, (41), is usually greater than that of propagation (39). As a result, the rate of cationic polymerization will generally *increase* as the temperature is *lowered*. This is in direct contrast to free-radical polymerization. The effect is caused primarily by the much lower activation energies for cationic initiation and propagation than is found for free-radical initiation and propagation.

The effect of temperature on the degree of polymerization (and therefore on the molecular weight of the polymer) in impurity-free systems can be seen from a consideration of (48). If termination is more important than chain transfer to monomer, then $k_t \gg k_{Tr}[M]$. The degree of polymerization will then show the same dependence on temperature as the rate of polymerization, namely that given by (52). Because E_t is usually greater than E_p, as already mentioned, $\overline{DP}$ will increase as the temperature is lowered.

On the other hand, if chain transfer is more important than chain termination, $k_{Tr,M}[M] \gg k_t$ and (48) reduces to

$$\overline{DP} = \frac{k_p}{k_{Tr,M}} = \frac{A_p}{A_{Tr,M}} e^{(E_{Tr} - E_p)/RT} \qquad (53)$$

[1]Norish, R. G. W., and Russell, K. E., *Trans. Faraday Soc.*, **1952**, *48*, 91.

Usually, the activation energy for chain transfer to the monomer is greater than that for propagation, so $\overline{DP}$ will again increase as the temperature is lowered.

Rate Constants for Propagation

In catalyzed cationic polymerizations, the observed rate constants are composite quantities that contain contributions from free ions and from ion pairs. The relative importance of the two types of contribution depends on the experimental conditions. It has been found possible to obtain rate constants for cationic propagation only for systems in which it is thought that the propagation is solely by free ions. Such systems are (1) polymerizations initiated by ionizing radiation, where no counterions exist, and (2) polymerizations initiated by stable carbonium ion salts, such as $(C_6H_5)_3C^+SbCl_6^-$ and $C_7H_7^+SbCl_6^-$, in which dissociation to the free ions is instantaneous and may be complete. Typical rate constants for free ions, k_p', determined from such systems are shown in Table 13.3. It will be noted that these rate constants are very large. In fact, even at the low temperatures used, all the rate constants are comparable to or greater than the rate constants for free-radical polymerizations at 60°C.

While it has been stated that the rate constants in Table 13.3 refer to free-ion propagation, it should be noted that much larger values are obtained by radiation initiation than by initiation using stable carbonium salts. This suggests that ion-pair propagation may be playing a significant role in the initiation of those cationic polymerizations by the salt $C_7H_7^+SbCl_6^-$ that are shown in Table 13.3.

Cationic Copolymerizations

Cationic copolymerization can be treated in an identical manner to anionic copolymerization. The mechanistic scheme for propagation is obtained simply by replacing the negative signs in (32) to (35) by positive signs. The limitations discussed previously for reactivity ratios in anionic copolymerization apply also to cationic reactivity ratios. However, in cationic systems, the living-polymer tech-

TABLE 13.3 RATE CONSTANTS FOR PROPAGATION BY FREE CATIONS

Monomer	Solvent	Temperature (°C)	Initiator	k'_p(liters/mol-s)
Styrene	None	15	Radiation	3.5×10^6
α-Methylstyrene	None	0	Radiation	4×10^6
Isobutyl vinyl ether	None	30	Radiation	3×10^5
Isobutyl vinyl ether	CH_2Cl_2	0	$C_7H_7^+SbCl_6^-$	5×10^3
t-Butyl vinyl ether	CH_2Cl_2	0	$C_7H_7^+SbCl_6^-$	3.5×10^3
Methyl vinyl ether	CH_2Cl_2	0	$C_7H_7^+SbCl_6^-$	1.4×10^2

Source: Ledwith, A. and Sherrington, D. C., in *Reactivity Mechanism and Structure in Polymer Chemistry*, A. D. Jenkins and A. Ledwith, Eds. (New York: © John Wiley and Sons, Ltd., **1974**), p. 278.

TABLE 13.4 REACTIVITY RATIOS IN THE CATIONIC COPOLYMERIZATION OF STYRENE (M_1)

M_2	Initiator	Solvent	Temperature (°C)	r_1	r_2
α-Methylstyrene	$BF_3O(C_2H_5)_2$	SO_2	−40	<0.1	>20
	$BF_3O(C_2H_5)_2$	CH_2Cl_2	−20	0.2–0.5	12 ± 2
	$TiCl_4 — CCl_3COOH$	CH_2Cl_2	−78	0.24 ± 0.05	1.12 ± 0.09
Isobutylene	$TiCl_4$	Toluene	−78	1.20 ± 0.10	1.78 ± 0.10
	$TiCl_4$	n-Hexane	−20	1.20 ± 0.11	0.54 ± 0.24
	$SnCl_4$	SO_2	−78	1.1	3.1
Isoprene*	$SnCl_4$	C_2H_5Cl	−30 to 0	0.8	0.1

Source: Tsukamoto, A. and Vogl, O., *Progr. Polymer Sci.* (A. D. Jenkins, Ed.), **1971**, *3*, 199.
*Lipatova, T. E., Gantmakher, A. R., and Medvedev, S. S., *Dokl. Akad. Nauk SSSR*, **1955**, *100*, 925.

nique of measuring k_{11}, k_{12}, k_{21}, and k_{22} independently has not yet been exploited. Thus, in cationic systems even less is known, in an absolute sense, than in anionic systems.

Some typical reactivity ratios for the cationic copolymerization of styrene are shown in Table 13.4.

STUDY QUESTIONS

1. Styrene is added to a solution of sodium naphthalenide in tetrahydrofuran so that the initial concentrations of styrene and sodium napthalenide in the reaction mixture are 0.2 M and 1×10^{-3} M, respectively. After 5 s of reaction at 25°C, the styrene concentration is determined to be 1.73×10^{-3} M. Calculate:
   (a) The rate constant for propagation of the polymerization. [*Ans.*: 950 liters/mol-s.]
   (b) The initial rate of polymerization.
   (c) The rate of polymerization after 10 s.
   (d) The number-average molecular weight of the polymer after 10 s.
   (e) The width at half-height of the maximum in the mole fraction distribution of the molecular weight at the completion of reaction.

2. After completion of the reaction in Problem 1, p-methoxystyrene is added to the mixture so that the initial concentration is 0.15 M. Using data from this chapter, calculate:
   (a) The initial rate of polymerization.
   (b) The rate of polymerization after 10 s.
   (c) The average degree of polymerization of the copolymer after 100 s.

3. A copolymerization of p-methoxystyrene (1 M) with styrene (0.5 M) in tetrahydrofuran solution initiated by C_4H_9Li is carried out at 25°C. Calculate:
   (a) The composition of the copolymer first formed.
   (b) The mean sequence lengths of p-methoxystyrene and styrene in the copolymer.
   (c) The probability of forming a styrene sequence that has a length of 5 units.

4. For the copolymerization of Problem 3, calculate and plot the monomer concentrations versus time, making use of the following assumptions: (1) the concentrations of

the growing anionic centers derived from the two monomers does not change; and (2) the relative probabilities of addition of $C_4H_9^-Li^+$ to styrene and p-methoxystyrene are equal.

5. 1-Vinylnaphthalene is polymerized anionically at 25°C in a tetrahydrofuran solution containing initially 5×10^{-3} M C_4H_9Li and 0.75 M 1-vinylnaphthalene. Calculate:
   (a) The average degree of polymerization.
   (b) The number-fraction and weight-fraction distributions of the degree of polymerization.

6. In studies of the low-temperature polymerization of isobutylene using $TiCl_4$ as catalyst and H_2O as cocatalyst [R. H. Biddulph, P. H. Plesch, and P. P. Rutherford, *J. Chem. Soc.*, **275** (1965)], the following results have been obtained at $-35°C$ for the effect of monomer concentration on the average degree of polymerization:

$[C_4H_8]$ (mol/liter)	0.667	0.333	0.278	0.145	0.059
$\overline{DP}$	6940	4130	2860	2350	1030

From these data, evaluate the rate constant ratios k_{Tr}/k_p and k_t/k_p.

7. In similar studies over a range of temperatures, Biddulph, Plesch, and Rutherford (see Problem 6) found the following values for the intercepts of plots of $(\overline{DP})^{-1}$ versus $[C_4H_8]^{-1}$:

$T(°C)$	+18	-14	-35	-48
$10^3/\overline{DP}$	4.37	0.50	0.098	0.027

   (a) Evaluate from these data the difference in activation energy between chain propagation and chain transfer to monomer.
   (b) Evaluate the ratio of the preexponential factor for transfer to that for propagation.
   (c) Assuming chain transfer to be much more important in producing polymer than termination, calculate and plot a curve showing the dependence of $\overline{DP}$ on temperature.

8. What is the simple physical picture for the narrowness of the molecular-weight distributions frequently obtained in anionic polymerizations?

SUGGESTIONS FOR FURTHER READING

Anionic Polymerization

ALLEN, P. E. M., and PATRICK, C. R., "The Kinetics of Addition Polymerization," in *Kinetics and Mechanisms of Polymerization Reactions*. New York: Wiley, **1974**, Chap. 7.

BILLMEYER, F. W., JR., *Textbook of Polymer Science*, 3rd ed. New York: Wiley, **1984**, Chap. 4.

BYWATER, S., "Anionic Polymerization," in *Progress in Polymer Science*, Vol. 4 (A. D. Jenkins, ed.). New York: Pergamon Press, **1975**, Chap. 2.

CARRAHER, C. E., JR., *Seymour/Carraher's Polymer Chemistry*, 5th ed., New York: Marcel Dekker, Inc., **2000**.

CUBBON, R. C. P., and MARGERISON, D., "The Kinetics of Polymerization of Vinyl Monomers by Lithium Alkyls," in *Progress in Reaction Kinetics*, Vol. 3 (G. Porter, ed.). New York: Pergamon Press, **1965**, Chap. 9.

HIEMENZ, P. C., *Polymer Chemistry*, New York: Dekker, **1984**, Chap. 6.

MORTON, M., "The Mechanism of Stereospecific Polymerization of Propylene," in *Vinyl Polymerization*, Vol. 1, Part 2 (G. E. Ham, ed.). New York: Dekker, **1969**, Chap. 5.

MULVANEY, J. E., OVERBERGER, C. G., and SCHILLER, A. M., "Anionic Polymerization," *Fortschr. Hochpolymer-Forsch.*, **1961**, *3*, 106.

PARRY, A., "Anionic Polymerization," in *Reactivity, Mechanism and Structure in Polymer Chemistry* (A. D. Jenkins and A. Ledwith, eds.). New York: Wiley, **1974**, Chap. 11.

SZWARC, M., and SMID, J., "The Kinetics of Propagation of Anionic Polymerization and Copolymerization," in *Progress in Reaction Kinetics*, Vol. 2 (G. Porter, ed.). New York: Macmillan, **1964**, Chap. 5.

VAN BEYLEN, M., BYWATER, S., SMETS, G., SZWARC, M., and WORSFOLD, D. J., "Developments in Anionic Polymerization: A Critical Review," *Advan. Polymer Sci.*, **1988**, *86*, 87–143.

WILSON, A. D., and PROSSER, H. J. (Eds.), *Developments in Ionic Polymers—2*, London: Elsevier Applied Science Publishers, **1986**.

Cationic Polymerization

ALLEN, P. E. M., and PATRICK, C. R., "The Kinetics of Addition Polymerization," in *Kinetics and Mechanisms of Polymerization Reactions*. New York: Wiley, **1974**, Chap. 7.

ALLEN, P. E. M., and PLESCH, P. H., "A Comparison of the Radical, Cationic and Anionic Mechanisms of Addition Polymerization," in *The Chemistry of Cationic Polymerization* (P. H. Plesch, ed.). New York: Macmillan, **1963**, Chap. 3.

BILLMEYER, F. W., JR., *Textbook of Polymer Science,* 3rd ed. New York: Wiley, **1984**, Chap. 4.

CARRAHER, C. E., JR., *Seymour/Carraher's Polymer Chemistry*, 5th ed., New York: Marcel Dekker, Inc., **2000**.

FAUST, R., and SHAFFER, T. D. (Eds.), *Cationic Polymerization. Fundamental and Applications,* Washington, DC: American Chemical Society, **1997**.

HIEMENZ, P. C., *Polymer Chemistry*, New York: Dekker, **1984**, Chap. 6.

LEDWITH, A., and SHERRINGTON, D. C., "Reactivity and Mechanism in Cationic Polymerization," in *Reactivity, Mechanism and Structure in Polymer Chemistry* (A. D. Jenkins and A. Ledwith, Eds.). New York: Wiley, **1974**, Chap. 9.

MATYJASZEWSKI, K. (Ed.), *Cationic Polymerizations. Mechanisms, Synthesis, and Applications*, New York: Marcel Dekker, Inc., **1996**.

PEPPER, D. C., "Ionic Polymerisation," *Quart. Rev.*, **1954**, *8*, 88.

PLESCH, P. H., "The Propagation Rate Constants in Cationic Polymerisations," *Advan. Polymer Sci.*, **1971**, *8*, 137.

WILSON, A. D., and PROSSER, H. J. (Eds.), *Developments in Ionic Polymers—2*, London: Elsevier Applied Science Publishers, **1986**.

ZLAMAL, Z., "Mechanisms of Cationic Polymerization," in *Vinyl Polymerization*, Vol. 1, Part 2 (G. E. Ham, Ed.). New York: Dekker, **1969**, Chap. 6.

14

Determination of Absolute Molecular Weights

INTRODUCTION

Two major differences exist between the molecular weights of polymers and those of small molecules. First, polymerizations are random processes, in the sense that different growing chains can terminate at very different points in their growth. This gives a *distribution* of molecular weights, rather than the single value that is characteristic of small-molecule products. Thus, for polymers it is necessary to estimate the *breadth* of this distribution, since it generally has a profound effect on polymer properties, even at constant average molecular weight. The second complication arises from the fact that the molecular weights of polymers are generally estimated in an indirect way. Thus, the polymer is dissolved in a solvent, some property of the resultant solution is measured, and finally thermodynamics are used to interpret this property in terms that give the molecular weight of the macromolecules. However, different measurable properties depend on the molecular weight in different ways, and this affects the type of average molecular weight obtained. For example, the light scattered from a polymer chain in solution depends on the *square* of the molecular weight. Thus, the larger molecules in a sample scatter out of proportion to their number. If the sample consisted of only two chains, one with a molecular weight of 1 million and the other with molecular weight 2 million, then the second chain would scatter four times as much as the first, and thus contribute disproportionately to the averaging process. This would be analogous to an election in which heavier people or richer people had more votes than others!

The types of average molecular weight of greatest importance are the:

(a) number-average, $\overline{M}_n$ (from end-group analysis and colligative properties)
(b) weight-average, $\overline{M}_w$ (primarily from light-scattering)
(c) Z-average, $\overline{M}_z$ (primarily from ultracentrifugation)
(d) viscosity-average, $\overline{M}_\eta$ (from solution viscometry)
(e) GPC-average, $\overline{M}_{GPC}$ (from gel-permeation chromatography)

The fact that these averages are different *for the same polymer sample* is exploited to provide the simplest characterization of the molecular-weight polydispersity, as described in Chapter 15. The specific quantity used is the ratio of the weight-average to number-average molecular weight, called the "polydispersity index," PI. The extent to which a value of $\overline{M}_w/\overline{M}_n$ exceeds unity (the monodisperse limit) is a measure of the breadth of the distribution. Values for typical polymerizations are around 2.0 but can be much higher in unusual cases, such as the high-temperature, high-pressure polymerization of ethylene. In this example, branching can broaden the distribution by further increasing the molecular weight of the longest chains in the sample (since they have the largest number of potential sites for the branching).

Figure 14.1 provides one of the reasons why a newly prepared polymer should be investigated immediately with regard to its molecular weight. It shows how many polymer properties increase with increasing molecular weight before leveling off at some asymptotic value. In the low molecular-weight part of the sketch, the materials are not really polymeric, and they have low values of the property of interest. An example would be low molecular-weight (oligomeric) polyethylene, which would be more like a grease or wax than the material we appreciate in commercial applications. It would be possible to produce a polymer

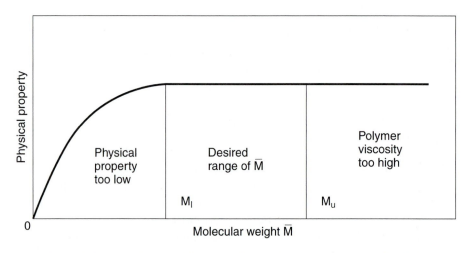

Figure 14.1 Typical effect of the molecular weight of a polymer on its physical properties, for example its modulus.

with physical properties well above the lower threshold value M_l shown in the sketch, by permitting the molecular weight to increase without limit. However, practical problems would result from exceeding the upper limit M_u because the bulk viscosity of the polymer would increase to the point where processing would be exceedingly difficult.

Thus, a knowledge of the molecular weight of a polymer is vital for even a preliminary understanding of the relationship between structure and properties. Moreover, as discussed in the preceding chapters, the molecular-weight *distribution* provides valuable clues to the polymerization reaction mechanism. Two fundamentally different approaches are used for the measurement of polymer molecular weights—absolute and secondary methods. Absolute methods give values that provide a *direct* estimate of the molecular weight. Secondary methods yield *comparisons* between the molecular weights of different polymers, and must be *calibrated* by reference to a system that has been studied by one of the absolute approaches. Absolute methods are considered in this chapter. Secondary methods are discussed in Chapter 15.

The measurement of absolute molecular weights of polymers is not an easy task. The experimenter is faced with the problem of studying materials which are nonvolatile, of very high molecular weight, and sometimes poorly soluble in organic media. Moreover, the samples are not homogeneous in molecular weight, but contains molecules whose molecular weights may span a broad range. The physical methods commonly used to determine the average molecular weights of high-polymer samples on an absolute scale require that the polymer sample should first be dissolved in a solvent. It is appropriate, therefore, to consider briefly some characteristics of polymer solubility.

SOLUBILITY OF HIGH POLYMERS

Some Qualitative Aspects

For any process to take place, whether chemical or physical, the Gibbs free-energy change under the usual conditions of constant temperature and pressure has to be negative.[1]

$$\Delta G_{T,p} = \Delta H - T \Delta S < 0 \tag{1}$$

In the present context, the process of interest is the dissolution of a polymer into a solvent to yield a molecular dispersion called a solution. As already mentioned, the preparation of such a solution is important for a number of reasons, in particular for studying the properties it generates and then interpreting this information to establish the molecular weight of the solute.

Unfortunately, ΔH for almost all solute-solvent combinations is positive, which reflects the well-known rule-of-thumb that "like dissolves like." In fact,

---

[1]Atkins, P. W., *Physical Chemistry*, 4th ed. Oxford: Oxford University Press, **1990**.

TABLE 14.1 INTERPRETATIONS OF SOME COMMON SOLUBILITY OBSERVATIONS

Observation	Interpretation
Polymers are generally difficult to dissolve	Connectivity of repeat units in the polymer reduces ΔS_{mix}
Two polymers are seldom miscible	Connectivity in both components further reduces ΔS_{mix}
Rigid-rod polymers are particularly hard to dissolve	Highly regular conformations greatly reduce ΔS_{mix}
Increasing temperature generally increases chances for solubility	Favorable (negative) $-T\Delta S_{mix}$ term increases in magnitude
Crystalline polymers are very hard to dissolve	Total ΔH for dissolution contains a large positive heat of fusion

even small differences in the structures of solute and solvent increase the energy, and this is an unfavorable contribution to the inequality given above. The relatively rare occurrences where ΔH is negative are apparently caused by unusually strong favorable interactions between the solute and solvent, as might occur for hydrogen-bond donors and acceptors.[1] This means that when dissolution does occur, it is an entropically driven process. Specifically, the positive entropy of mixing ΔS_{mix} multiplied by some positive absolute temperature makes the negative $-T\Delta S$ term dominate the positive ΔH term.

This simple argument can be used to provide a molecular interpretation of a number of commonplace observations, as described in Table 14.1.

The first observation is that it is harder to find a solvent for a polymer than it is for its low molecular-weight analogue. An example is the dissolution of polystyrene compared with the dissolution of styrene monomer. This results from the fact that monomer molecules can sample any parts of the solution volume, whereas their transformation into repeat units in a contiguous chain structure severely constrains them. These constraints reduce the entropy of mixing ΔS_{mix} in the case of the polymer, and diminish the favorable $-T\Delta S_{mix}$ term being relied on to give a negative ΔG.

Another observation is that two polymers are only rarely miscible in one another. In this case, the "solvent" is now also polymeric instead of being low molecular weight. Because this chain connectivity now exists in both components, ΔS_{mix} is reduced further.

It has also been observed that rigid-rod polymers are particularly hard to dissolve. This is because such chains have highly regular conformations, and this can also greatly reduce ΔS_{mix}.

A third illustration is related to the well-known observation that increases in the temperature generally increase the solubility. Because $-T\Delta S_{mix}$ is directly proportional to the absolute temperature, the magnitude of this favorable term increases with increase in temperature.

[1]Kamide, K. and Dobashi, T., *Physical Chemistry of Polymer Solutions. Theoretical Background*. Amsterdam: Elsevier, **2000**.

A final example is the observation that it is more difficult to find a solvent for a partially crystalline polymer than for its amorphous counterpart. An illustration is the poor dissolution of crystalline (isotactic) polypropylene compared with dissolution of amorphous (atactic) polypropylene. Here the difficulty results from the thermodynamic requirement that ΔH in equation (1) is for the total process, which now includes melting of the crystallites in the partially crystalline polymer. This corresponds to the addition of a large positive heat of fusion to the already positive ΔH_{mix}. No matter how the process is visualized, melting has to occur somewhere in the process of converting polymer chains located in crystallites to polymer chains dispersed as free molecules in a solvent.

In order to understand dissolution thermodynamically, it is useful to compare it with polymer crystallization. They are quite opposite, because dissolution involves a *disordering* of the chains into a solvent, while crystallization involves *ordering* of parts of chains into crystallites. Thus, as is shown in the first row in Table 14.2, the signs of ΔS_{mix} are reversed in the two cases. The sign of $-T\Delta S$ is also reversed, and this changes the situation from favorable to unfavorable when comparing dissolution with crystallization. The heat of the process also changes sign; the repulsions between unlike solvent and polymer segments are replaced by attractions between the chains as they pack more efficiently in the crystalline state. (This is analogous to the decrease in energy when a chemical bond is formed between two atoms; the unfavorable effects of this ordering are more than offset by the energy decrease from the covalent bonding.) Thus, dissolution is an entropically driven process, while crystallization is energetically driven. As a result, dissolution is facilitated by increases in temperature, while crystallization requires lowering it.

The dissolution of a polymer sample in a solvent takes place in two distinct stages. In the first stage, the polymer sample "takes up" or imbibes solvent and expands to a swollen gel. This first stage is exhibited by all amorphous linear, branched, or lightly cross-linked polymer samples, regardless of whether or not they will ultimately form a true solution. The second stage of dissolution consists of a breakdown of the swollen gel to give an actual solution of polymer molecules in the given solvent. This second stage will not be shown by polymers that are cross-linked and may not be shown by polymers that contain microcrystalline domains. Cross-linked polymers do not dissolve in any solvent (unless the cross-

TABLE 14.2   SOME COMPARISONS BETWEEN DISSOLUTION AND CRYSTALLIZATION

	Dissolution	Crystallization
ΔS	Positive	Negative
$-T\Delta S$	Negative	Positive
ΔH	Positive	Negative
Nature of process	Entropically driven	Energetically driven
Temperature change required to facilitate	Increase	Decrease

links are broken). Hence, cross-linked network polymers will be excluded from our consideration of molecular-weight determination.

A liquid will be a "good" solvent for a polymer if the molecules of the liquid chemically and physically resemble the structural units of the polymer. If this situation exists, the adhesive forces between the solvent and the polymer are similar to the cohesive forces that exist between solvent molecules or between polymer molecules. An exchange of a solvent molecule by a polymer structural unit can then occur with little or no change in the interaction forces that exist between solvent or polymer molecules. In a thermodynamic sense, we can say that the dissolution of a polymer in a "good" solvent occurs with a negligibly small heat of mixing. This is a more elegant way in which to state the well-known chemist's rule that "like dissolves like." For example, this rule suggests that cumene and ethylbenzene (**1** and **2**) should be good solvents for polystyrene (**3**) because of the similarity between the solvent molecules and the structural units of the polymer.

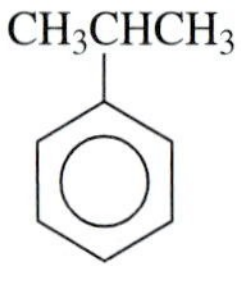

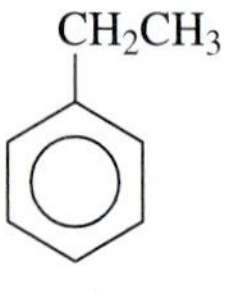

Cohesive Energy Densities and Solubility Parameters

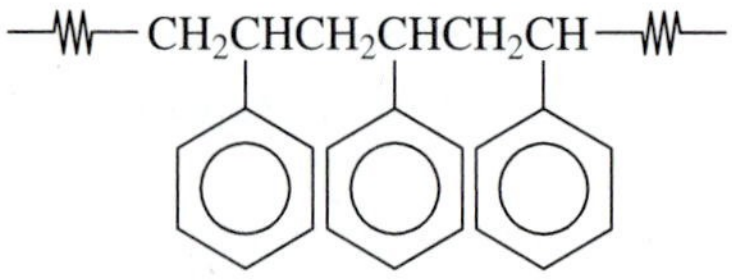

The criterion of "like dissolves like" can be interpreted more quantitatively. When a substance vaporizes, energy is needed to overcome the cohesive forces between molecules. The magnitude of these cohesive forces may be described simply by the amount of energy needed to vaporize a certain volume of the substance. Such a quantity is called the *cohesive energy density*[1] and is defined by (2), in which E_0

$$\delta^2 = \frac{E_0}{V_0} \qquad (2)$$

is the latent energy of vaporization for a volume V_0 of the substance in question. The quantity δ itself is called the "solubility parameter." If the adhesive forces between the solvent and the polymer are to be similar to the solvent–solvent and

[1]Hildebrand, J. H., and Scott, R. L., *Solubility of Non-electrolytes,* 3rd ed. (New York: Reinhold, **1950**), p. 124.

polymer–polymer cohesive forces, then it is obvious that the cohesive energy densities of the solvent and the polymer should be nearly the same. In general, for endothermic dissolution (the usual case with polymers), the enthalpy change per unit volume is given by (3), in which ϕ_1 is the volume fraction of solvent

$$\Delta H_{\text{dissolution}} = \phi_1\phi_2(\delta_1 - \delta_2)^2 \tag{3}$$

and ϕ_2 is the volume fraction of polymer in the solution. A good solvent for a given polymer is one in which ΔH approaches as closely as possible to zero. In other words, δ_1^2 and δ_2^2—the *cohesive energy densities*—should be nearly equal.

It is a fairly straightforward matter to obtain an experimental value of the cohesive energy density or solubility parameter δ_1 of a solvent. If values of the specific volume or density have not been reported in the literature, they can be measured easily by pycnometry or dilatometry. Molar heats of vaporization ΔH_{vap} have also been reported for most common solvents or if not available, they can be measured by calorimetric measurements at constant temperature and atmospheric pressure. However, they do require minor modifications because the quantity required in the definition of δ is the molar energy of vaporization ΔE_{vap}. Note that $\Delta H_{\text{vap}} = \Delta E_{\text{vap}} + \Delta(pV)$, and $\Delta(pV) = p\Delta V$ at constant pressure. The term $p\Delta V$ is $p(V_{\text{vapor}} - V_{\text{liq}})$ and, since the volume of the vapor is approximately 10^3 that of the liquid, it is simply equal to pV_{liq}. If the vapor is assumed to be an ideal gas, then ΔH_{vap} can be simply converted to ΔE_{vap} by subtraction of RT.

However, obtaining an experimental value of the solubility parameter δ_2 for a polymer is not straight-forward, since polymers cannot be vaporized to give the required value of ΔH_{vap}. The methods for determining δ_2 are therefore indirect, or computational. The simplest involves a choice of a series of solvents that cover a wide range of values of δ_1. Each solvent is placed in a test tube with a thin piece of the polymer to be characterized, and visual observation used to determine if dissolution has occurred. The results are then plotted as a simple "Yes or No" bar-type plot as a function of δ_1. Solvents with values of δ_1 that are significantly lower than the unknown value δ_2 of the polymer will not dissolve it. (In general, the values of $\delta(\text{cal/cc})^{1/2}$ should not differ by more than 1.5 units.) Similarly, solvents having values of δ_1 significantly greater than δ_2 will also fail. By indirect argument, the solvents that did dissolve the polymer must have values of δ_1 very close to δ_2 of the polymer. Correspondingly, the average value of δ_1 for successful solvents is taken to be δ_2 for the polymer.

A somewhat more precise method involves use of the same solvents, but heating them until each does successfully dissolve the polymer. The temperature at which this occurs is then plotted against δ_1. The value of δ_2 is taken to be the value of δ_1 that corresponds to the smallest increase in temperature required for dissolution, which occurs at the minimum on the curve. As an alternative, it would be possible to plot the solubility itself at some constant temperature, and it would go through a maximum. In those unusual cases where a great deal of solution viscosity data are available for various solvents, plots can be made of the viscosity against δ_1. The value of δ_2 is taken to be the value of δ_1 that gave the largest value of the viscosity (maximum in the curve). This is based on the best solvent giving the largest value of the viscosity, because of chain expansion effects discussed in Chapter 16.

If the polymer to be characterized is cross-linked, it will not dissolve in any solvent, but the extent to which it swells in a solvent is significant, as described in Chapter 21. Of interest here is its use to estimate δ_2. The solvents with values of δ_1 closest to δ_2 will swell the polymer the most, and thus an estimate of δ_2 can be derived from the maximum degree of swelling. An actual experimental curve of this type is shown in Figure 14.2.

An interesting extrapolation technique has also been used to estimate δ_2. It is based on the availability of low molecular weight analogues, "oligomers," of the polymer of interest. These molecules can be characterized by vaporization measurements of ΔH_{vap}, and thus give values of δ_1 directly. These values of δ_1 can then be plotted against values of the reciprocal of their molecular weights M, and the resultant curve extrapolated to infinite M. This should give the value of δ_2 of the corresponding high polymer. An example would be an estimation of δ_2

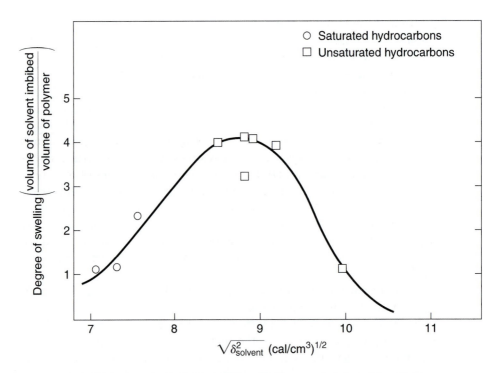

Figure 14.2 Degree of swelling of natural rubber as a function of the cohesive energy density of the solvent. [Data from Scott, R. L. and Magat, M., *J. Polymer Sci.*, **1949**, *4*, 555.]

for polyethylene by extrapolation of results obtained for n-hexane, n-octane, n-decane, n-hexadecane, etc.

The group additivity scheme is an example of a computational approach. The goal is to estimate the contributions that various atoms and groups make to the δ_2 of a polymer. An example is the estimate of the contribution of a methylene group, by noting that it must be the difference between the solubility parameters of two consecutive members of the n-alkane series, for example n-nonane and n-octane: $\delta_1(CH_2) = \delta_1(C_9H_{18}) - \delta_1(C_8H_{16})$. Another example would be from estimates of the value contributed by a chlorine atom from values for symmetrical dichloroethylene and ethylene: $2\delta_1(Cl) = \delta_1(C_2H_4Cl_2) - \delta_1(C_2H_4)$. The contributions for all of the atoms and groups in the repeat unit are then summed and divided by the molar volume of the unit, giving the estimated value of δ_2. Software for directly calculating solubility parameters is now available from companies such as Accelrys (San Diego). It involves calculations of interaction energies between molecules (and thus ΔE_{vap}), and the packing of the molecules in the liquid state (and thus molar volumes). The values of δ_1 and δ_2 calculated in this way seem to be good agreement with experiment.

Some typical values of cohesive energy densities for solvents and polymers are given in Table 14.3. The values are seen to increase in the change from solvents that are nonpolar, to those that are polar, and to those that are hydrogen bonding as well as polar. These trends are caused by increases in ΔE_{vap} that arise from increased favorable attractions from dipolar and hydrogen-bonding interactions. (As a digression, these arguments explain why steam burns are so serious. The breaking of hydrogen bonds when liquid water is taken into the vapor phase requires a great deal of energy. This makes ΔH_{vap} unusually large. When steam condenses onto an object, the heat given off as $\Delta H_{cond} = -\Delta H_{vap}$ must then be absorbed by the object!)

These various methods for estimation of δ_2 give results in reasonably good agreement, with some typical values shown in the first column of Table 14.3. They are certainly sufficient for the main use of this parameter which is to provide guidance on which subset of all available solvents is most likely to dissolve a specific polymer near room temperature. The objective is to attempt to match solubility paramers as closely as possible, that is by minimizing $(\delta_1 - \delta_2)$, and this corresponds to making horizontal pairings between entries in the first and third columns in the table. For example, carbon tetrachloride and benzene should be

TABLE 14.3   COHESIVE ENERGY DENSITIES OF SOLVENTS AND POLYMERS

Polymer	$\delta^2(cal/cm^3)$	Solvent	$\delta^2(cal/cm^3)$
Polyethylene	62.4	n-C_6H_{14}	52.4
Polystyrene	74.0	CCl_4	73.6
Poly(methyl methacrylate)	82.8	C_6H_6	83.7
Poly(vinyl chloride)	90.3	$CHCl_3$	85.4
Nylon 66	185	$(CH_3)_2C{=}O$	94.3
Polyacrylonitrile	237	CH_3OH	210

Solubility of High Polymers

good solvents for polystyrene and poly(methyl methacrylate), respectively. However, care must be exercised because the predictions are strictly for both polymer and solvent in the liquid state. Therefore, crystalline polymers are not covered in this scheme. For example, it would not be valid to predict from such a table that polyethylene would dissolve in a solvent with a δ_1^2 of approximately 62 cal/cm^3, except when the temperature is sufficiently high to take polyethylene into the amorphous, liquid state.

More refined versions of this scheme exist in which the energies are resolved into their van der Waals, dipolar, and hydrogen-bonding components.[1] This would essentially give three tables of solubility parameters, with the goal of matching three pairs of values of $(\delta_1 - \delta_2)$. Comparisons can also be made graphically.

Another application of these results is for choosing a polymer that is least likely to be swollen when placed in contact with a specified liquid. An example is the choice of a polymer to be used in a hose for pumping this liquid or as a seal or gasket used to contain it. In this case the maximum *mis*match of solubility parameters would be needed. In other words $(\delta_1 - \delta_2)$ should be maximized. This would correspond to making diagonal pairings between entries in the first and third columns in the table.

END-GROUP ANALYSIS, COLLIGATIVE PROPERTIES, AND NUMBER-AVERAGE MOLECULAR WEIGHTS[2]

End-Group Analysis

One of the most direct ways to measure the molecular weight of a linear polymer is to determine the ratio of end groups to middle units. Examples of suitable end groups would be the hydroxyl groups on both ends ($\gamma = 2$) of a polymer synthesized via a ring-opening polymerization, or the radical fragments from the initiator attached during a free-radical polymerization (either $\gamma = 2$ for termination by coupling or $\gamma = 1$ for termination by disproportionation). These end groups would be particularly easy to detect and analyze if they were chosen to be chemically very different from the rest of the chain, had radioactivity or luminescent properties, etc. Similarly, in a condensation polymerization, the polymer from an A-B monomer could be titrated for either A or B ($\gamma = 1$), for example end groups of either carboxylic acid or amine.

In any case, analysis for the chain ends then gives their number Q, and the number of chains is $N = Q/\gamma$. The number of moles, n, of chains is N divided by Avogadro's number, and the molecular weight is simply the weight of the sam-

---

[1]van Krevelen, W., *Properties of Polymers.* (Amsterdam: Elsevier, **1997**).

[2]The number-average molecular weight is the total mass of the solute divided by the number of solute particles present.

ple analyzed divided by n. This is obviously a number-average value because direct counting was involved. The same type of average is obtained from any of the colligative properties. Colligative means "tying together," and this is a grouping of all properties that depend only on their number densities or molar concentration in the solution. They are described in detail in the following section.

COLLIGATIVE PROPERTIES AND NUMBER-AVERAGE MOLECULAR WEIGHTS

In principle, any of the colligative properties of solutions such as freezing-point depression, osmotic pressure, and so on, may be used to determine the average molecular weight of a dissolved polymer sample. A colligative property of a solution is any property that depends on the lowering of the chemical potential of a solvent by the introduction of a solute.

According to classical thermodynamics, equation (4) gives the chemical potential of a liquid solvent in a solution if the vapor of the solvent behaves as an ideal gas.

$$\mu_s = \mu_s^o + RT \ln \frac{P_s}{P_s^o} \tag{4}$$

In this equation μ_s^o is the chemical potential of the pure solvent at temperature T, P_s^o is the vapor pressure of the pure solvent at temperature T, and P_s is the pressure of the solvent vapor above the solution at temperature T.

According to Raoult's law, which is valid for all solutions provided they are sufficiently dilute, equation (5) holds:

$$\frac{P_s}{P_s^o} = X_s \tag{5}$$

Here X_s is the mole fraction of solvent in the solution. In view of equations (4) and (5), we may write the expression shown in (6) for the chemical potential of the solvent in solution

$$\mu_s = \mu_s^o + RT \ln X_s \tag{6}$$

For a solution, X_s must be less than unity and, according to equation (6), μ_s must therefore be less than μ_s^o. Hence, the presence of the solute lowers the chemical potential of the solvent. The observable quantities that depend on this effect are called *colligative properties*. These are:

1. Vapor-pressure lowering (see equation 5)
2. Boiling-point elevation
3. Freezing-point depression
4. Osmotic pressure

All four of these properties can be transposed into the form

$$\Delta y_i = K_i \frac{c}{\overline{M}_n} \tag{7}$$

in which c is the concentration of the solute (in weight per unit volume); $\overline{M}_n$ is the number average molecular weight of solute; K_i is a constant which depends on the solvent and on the effect being investigated; and Δy_i is the experimental factor actually being observed [e.g., the depression of the freezing point $(T_0 - T)$].

This general relationship can be illustrated by the melting point depression equation, in the form usually encountered:

$$\Delta T_f = (T_{f,\mathrm{o}} - T_f) = K_f m \tag{8}$$

where K_f is the freezing-point depression constant (a function of only the freezing point and enthalpy of freezing of the pure solvent) and $m = c/\overline{M}$ is the molality. Measurement of ΔT_f, obtaining K_f from the literature, and insertion of c in grams of solute per kilogram of solution then gives $\overline{M}$ directly. From equation (7) it is clear that Δy_i becomes smaller as $\overline{M}$, the average molecular weight, becomes larger. This leads to serious inadequacies in the use of the first three colligative properties mentioned above for the determination of the molecular weights of high polymers.

For example, consider the freezing-point depression of benzene brought about by addition of a high polymer. For a concentration of $10^{-3}\mathrm{g/cm}^3$ (a concentration at which the dilute solution laws begin to be obeyed approximately) a freezing-point depression of only about $0.0002°C$ would be produced if the average molecular weight of the polymer was 25,000. This temperature change is below the limits of accurate measurement. The boiling-point elevation or vapor-pressure lowering by the same polymer would be even more difficult to measure.

OSMOTIC-PRESSURE MEASUREMENT OF ABSOLUTE MOLECULAR WEIGHTS

Theory

Osmotic pressure is the only one of the four colligative properties that provides a practical method for the measurement of the average molecular weights of polymers. In fact, osmotic pressure measurements can be used to determine molecular weights in the range 3×10^4 to 1×10^6.

The principle of osmotic-pressure measurements may be illustrated[1] by consideration of the schematic apparatus shown in Figure 14.3. The dashed line represents a semipermeable membrane through which the solvent, but not the

---

[1]Klotz, I. M., *Chemical Thermodynamics* (Englewood Cliffs, N.J.: Prentice-Hall, **1950**), pp. 261–263.

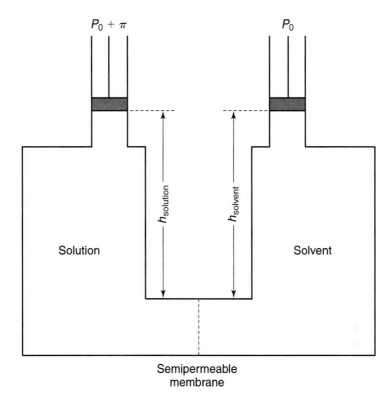

$P_0 + \pi$ P_0

h_{solution} h_{solvent}

Solution   Solvent

Semipermeable
membrane

Figure 14.3 Schematic diagram of an osmometer.

polymer, may pass. If both sides of the apparatus contained pure solvent, the liquid levels at equilibrium would be at the same height, and the external pressure on both pistons would be P_0. In terms of the chemical potential of the solvent, expression (5) for this trivial case would simplify to

$$\mu_s(\text{left}) = \mu_s(\text{right}) = \mu_s^{\circ} \tag{9}$$

If the polymer solute is now added to the left-hand side of the apparatus shown in Figure 14.3, it must remain on the left-hand side, since it cannot pass through the membrane. A system such as this, which contains a solution on the left side and pure solvent on the right side, can be described in terms of the chemical potential of the solvent, which, from equation (6), may be stated $\mu_s(\text{left}) < \mu_s(\text{right})$. If the external pressure on the solution side is maintained equal to that on the solvent side, namely P_0, the level of the liquid on the left-hand side will rise. This rise is due to the flow of solvent through the membrane into the solution, a region in which its chemical potential is lower. This flow of solvent could be prevented (in order to keep the heights of the two liquid levels fixed) by increasing the external pressure on the solution. The amount of this increase in external pressure is the *osmotic pressure*, π, of the solution.

The driving force for the flow of solvent through the membrane into the solution is the reduction of the chemical potential of a solvent that occurs when a polymer dissolves in it (equation 6). On the other hand, it is possible to prevent solvent flow and, therefore, to restore μ_s on the solution side to μ_s^o by increasing the external pressure on the solution. Obviously, then, the chemical potential of the solvent is a function of both the concentration and the external pressure (at constant temperature). Thus, the general expression shown in (10) can be employed,

$$\mu_s = f(P, X_p) \tag{10}$$

where X_p is the mole fraction of polymer in the solution. If equation (10) is differentiated (assuming constant temperature), expression (11) is obtained:

$$d\mu_s = \left(\frac{\partial \mu_s}{\partial P}\right)_{T,X_p} dP + \left(\frac{\partial \mu_s}{\partial X_p}\right)_{T,P} dX_p \tag{11}$$

By increasing the pressure on the solution to prevent flow of solvent, the value of μ_s in solution is maintained constant at μ_s^o. Hence, under this condition $d\mu_s = 0$ and equation (10) is obtained:

$$\left(\frac{\partial \mu_s}{\partial P}\right)_{T,X_p} dP = -\left(\frac{\partial \mu_s}{\partial X_p}\right)_{T,P} dX_p \tag{12}$$

The definition of the chemical potential[1] of the solvent is given by

$$\mu_s = \left(\frac{\partial G}{\partial n_s}\right)_{T,P,n_p} \tag{13}$$

where G is the Gibbs free energy, n_s is the number of moles of solvent, and n_p is the number of moles of the polymer. From elementary thermodynamics the derivative of the Gibbs free energy with respect to pressure at constant temperature and composition is given by the volume of the solution or by

$$\left(\frac{\partial G}{\partial P}\right)_{T,X_p} = V \tag{14}$$

By differentiation of (13) with respect to P at constant composition, and differentiation of (14) with respect to n_s, all other variables being kept constant, expressions (15) and (16), respectively, are obtained. In (16), $\overline{V}_s$ is the *partial molar volume* of the

[1]Moore, W. J., *Physical Chemistry*, 4th ed. (Englewood Cliffs, N.J.: Prentice-Hall, 1972), p. 205.

$$\left(\frac{\partial \mu_s}{\partial P}\right)_{T,X_p} = \frac{\partial^2 G}{\partial P \, \partial n_s} \tag{15}$$

$$\frac{\partial^2 G}{\partial n_s \, \partial P} = \left(\frac{\partial V}{\partial n_s}\right)_{T,P,n_p} = \overline{V}_s \tag{16}$$

solvent. Since the order of differentiation is of no consequence, (15) and (16) may be equated so that (17) results:

$$\left(\frac{\partial \mu_s}{\partial P}\right)_{T,X_p} = \overline{V}_s \tag{17}$$

Differentiation of (6) with respect to X_p at constant T and P leads directly to

$$\left(\frac{\partial \mu_s}{\partial X_p}\right)_{T,P} = \left(\frac{dX_s}{dX_p}\right)\left(\frac{\partial \mu_s}{\partial X_s}\right)_{T,P} = -\frac{RT}{1 - X_p} \tag{18}$$

Substitution of (17) and (18) into (12) then yields

$$\int_{P_0}^{P_0+\pi} \overline{V}_s \, dP = RT \int_0^{X_p} \frac{dX_p}{1 - X_p} \tag{15}$$

The partial molar volume of solvent may be assumed to be independent of the external pressure over the pressure range P_0 to $P_0 + \pi$, and integration of (19) then yields

$$\pi = -\frac{RT}{\overline{V}_s} \ln(1 - X_p) \tag{20}$$

for the osmotic pressure.

In the very dilute solutions to which (5) is applicable, X_p is so small that the following approximations are generally assumed to be valid:

$$\ln(1 - X_p) \approx -X_p \approx -\frac{n_p}{n_s}$$
$$n_s \overline{V}_s = V_s \approx V_{\text{solution}}$$

Use of these approximations with equation (20) then leads to the well-known approximate expression of van't Hoff:

$$\frac{\pi}{c} \approx \frac{RT}{M_n} \tag{21}$$

where c is the concentration in weight of polymer per unit volume of solution, and π and R are in the corresponding units. This equation is seen to be analogous to the rearranged ideal gas law, $p/(w/V) = RT/M$. A measurement of the colligative properties effectively is a count of the *number of particles* in solution. Hence, the average molecular weight in (21) is the *number-average molecular weight*.

As mentioned, expression (19) applies only to ideal solutions for which Raoult's law is obeyed. Generally, ideality can only be approached in the limit of infinite dilution, and this is particularly true for solutions of high polymers. Therefore, the correct expression relating osmotic pressure to the molecular weight must be written

$$\lim_{c \to 0} \left(\frac{\pi}{c} \right) = \frac{RT}{\overline{M}_n} \tag{22}$$

The form of (22) indicates that, in order to determine molecular weights by the osmotic pressure method, it is necessary to extrapolate a plot of π/c as a function of c to the value at $c = 0$. Note that the ratio π/c will not diverge, since both π and c are going to zero.

The dependence of π/c on the concentration is often written in the form of a virial equation:

$$\left(\frac{\pi}{c} \right) = \frac{RT}{\overline{M}_n}(1 + \Gamma c + g\Gamma^2 c^2 + \cdots) \tag{23}$$

This equation, and an alternative involving the coefficients A_2, A_3, ..., parallel the usual virial expansions for real gases involving either the pressure or reciprocal volume.

As will be discussed further in Chapter 16, the second virial coefficient Γ is given by the Flory–Huggins theory of polymer solutions as

$$\Gamma = \frac{\overline{M}_n \rho_s}{M_s \rho_p^2} \left(\frac{1}{2} - \chi_1 \right) \tag{24}$$

where ρ_s and ρ_p are the densities of the solvent and the polymer, respectively; M_s is the molecular weight of the solvent; and χ_1 is a polymer-solvent interaction constant. The value of the constant g in (23) is often assumed to be $\sim\frac{1}{4}$ in good solvents. With $g = \frac{1}{4}$, equation (23) reduces to (25). The latter form is sometimes convenient to use for the extrapolation of π/c data to zero concentration because, for good solvents, plots given by the expression (23) are not linear.

$$\left(\frac{\pi}{c} \right)^{1/2} = \left(\frac{RT}{\overline{M}_n} \right)^{1/2} \left(1 + \frac{1}{2}\Gamma c \right) \tag{25}$$

In poor solvents $g \approx 0$, and hence plots given by (23) are found to be linear over the usually measured ranges of concentration.

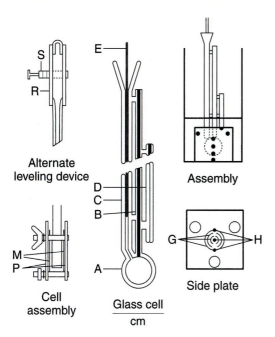

S

R

Alternate leveling device

E

D
C
B

A

Assembly

G ◄─────► H

Side plate

M
P

Cell assembly

Glass cell

cm

Figure 14.4 Simple osmometer. [Reproduced from B. H. Zimm and I. Myerson, *J. Am. Chem. Soc.* **1946,** *68,* 911; © the American Chemical Society.]

Practical Osmometry

A simple, much-used osmometer is shown in Figure 14.4.[1] In this apparatus two semipermeable membranes, M, are held against the finely ground glass walls of an open-ended cylindrical cell, A, by perforated metal plates, P. Two capillary tubes, B and C, are sealed onto the side of the cylindrical cell. C is the filler tube. This has an inside diameter of 2 mm. B is the measuring capillary tube, which has an inside diameter of 0.5 mm. The reference capillary, D, also has an inside diameter of 0.5 mm.

The cell is filled with the polymer solution through the tube C, and a snugly fitting metal rod, E, is then inserted into C to close the filling tube. The apparatus is then immersed in the solvent and osmotic pressure of the solution is measured as the difference in the levels of the liquid in capillaries B and D. Generally, the difference in the levels of liquid in the two capillaries is measured by a cathetometer.

The difference in heights of the liquid levels, Δh, in an osmometer such as this is related to the osmotic pressure of the solution by the expression

$$\pi = \rho_{\text{solution}} g' \Delta h \qquad (26)$$

in which ρ_{solution} is the density of the solution and g' is the acceleration due to gravity. If the weight is measured in grams, volume in cm^3, and height in cen-

---

[1]More elaborate commercial instruments are available.

timeters, π will be given in units of dyn/cm^2. It is customary to report osmotic pressures as π/g', which has the units grams of solution/cm^2. These units must be kept in mind clearly when converting extrapolated values of π/c to number-average molecular weights.

Some typical results of osmotic pressure measurements for a polymer solution are shown in Figures 14.5 and 14.6 for the specific case of polyisobutylene in chlorobenzene. The plot of π/c versus c in Figure 14.5 (see equation 23) shows some curvature. This indicates that chlorobenzene is a good solvent for poly-

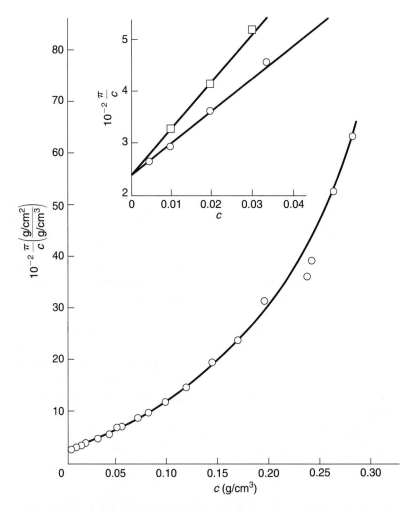

Figure 14.5 Concentration dependence of the osmotic pressure of polyisobutylene–chlorobenzene solutions. [Reproduced from Leonard, J. and Daoust, H., *J. Polymer Sci.*, **1962**, *57*, 53; with permission of John Wiley & Sons, Inc. New York.] ○, 25°C; □ 40°C.

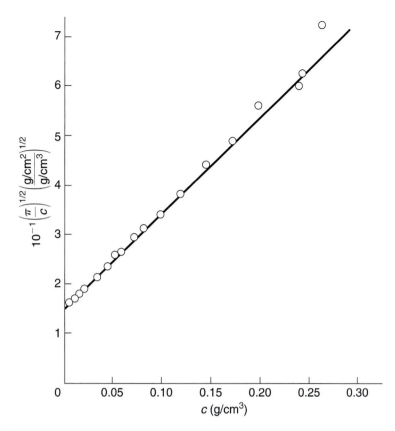

Figure 14.6 Concentration dependence of the square root of the reduced osmotic pressure of polyisobutylene–chlorobenzene solutions. [Reproduced from Leonard, J. and Daoust, H., *J. Polymer Sci.*, **1962**, *57*, 53; with permission of John Wiley & Sons, Inc., New York.]

isobutylene. However, as shown by the inset section of Figure 14.5, an extrapolation of the data for the lower concentrations to $c = 0$ can be made easily to obtain $\lim_{c\to 0}(\pi/c)$. Figure 14.6 shows that a plot of $(\pi/c)^{1/2}$ versus c for the same data (according to equation 25) yields a straight line over the entire range of concentrations measured. This indicates that g in equation (23) is $\sim\frac{1}{4}$.

It is instructive to use the data obtained from Figure 14.5 (or 14.6) to illustrate the calculation of a number-average molecular weight and the second virial coefficient. From Figure 14.5, or the inset to this figure, it is possible to obtain the following result for the intercept at 25°C:

$$\lim_{c\to 0}\left(\frac{\pi}{c}\right) = 235(\text{g/cm}^2)\text{-concn.}^{-1}$$

According to equation (22), the number-average molecular weight is

$$\overline{M}_n = \frac{RT}{\lim_{c \to 0} (\pi/c)} = \frac{(82.06 \text{ cm}^3\text{-atm/mol-deg})(298 \text{ deg})(1.013 \times 10^6 \text{ g/cm-s}^2\text{-atm})}{(235 \text{ g/cm}^2\text{-concn.})(980 \text{ cm/s}^2)}$$

or

$$\overline{M}_n = 1.07 \times 10^5 \text{ cm}^3\text{-concn./mol} = 1.07 \times 10^5 \text{ g/mol}$$

From the inset of Figure 14.5, the initial slope at 25°C is found to be 6350 g/cm²-concn.². From equation (23), the following result is then obtained for the second virial coefficient:

$$\Gamma = (6350 \text{ g/cm}^2\text{-concn.}^2)\left(\frac{\overline{M}_n}{RT}\right) \frac{(6350 \text{ g/cm}^2\text{-concn.}^2)(1.075 \times 10^5 \text{ g/mol})(980 \text{ cm/s}^2)}{(1.013 \times 10^6 \text{ g/cm-s}^2\text{-atm})(82.06 \text{ cm}^3\text{-atm/mol-deg})(298 \text{ deg})}$$

or

$$\Gamma = 27.0 \text{ cm}^3/\text{g}$$

Accuracy of molecular weights determined from osmotic pressure
Despite the excellent *precision* of the data shown in Figures 14.5 and 14.6, the absolute *accuracy* of the molecular weights of polymers obtained from osmotic pressure measurements is not nearly so good. This may be seen from inspection of Figure 14.7. This figure shows plots of π/c versus c obtained independently by eight laboratories in the United States, Canada, and Europe for samples of a single polystyrene fraction. Measurements were made in two solvents, toluene and methyl ethyl ketone (which are representative of good and poor solvents for polystyrene, respectively). Obviously, a great deal of scatter in the data exists. This can be attributed mainly to the different semipermeable membranes used by the various laboratories. The average molecular weights and second virial coefficients obtained from Figure 14.7 are:

(a) Toluene: $\overline{M}_n = 4.50 \pm 1.08 \times 10^5$
$\qquad\qquad \Gamma = 214 \pm 27 \text{ cm}^3/\text{g}$
(b) Methyl ethyl ketone: $\overline{M}_n = 4.60 \pm 0.82 \times 10^5 \text{ g/mol}$
$\qquad\qquad\qquad \Gamma = 60 \pm 32 \text{ cm}^3/\text{g}$

It is somewhat reassuring that essentially the same molecular weight was obtained in the two solvents. However, the accuracy is clearly not as good as might be assumed from an inspection of the data obtained from a single osmometer and membrane (see Figure 14.5). Also noteworthy (as illustrated by the foregoing results) is that the more accurate extrapolation to $c = 0$ that is possible in a poor solvent (i.e., a shallower slope) leads to considerably less uncertainty in the average molecular weight. Of course, this greater accuracy is obtained at the ex-

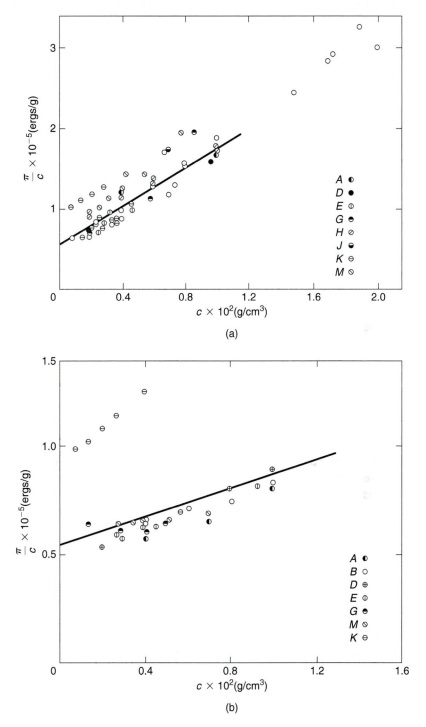

Figure 14.7 Concentration dependence of the osmotic pressure of polystyrene solutions in two solvents at 25°C: (a) toluene; (b) methyl ethyl ketone. [Reproduced from Frank, H. P. and Mark, H., *J. Polymer Sci.*, **1955,** *17,* 1; with permission of John Wiley & Sons, Inc., New York.] The symbols indicate results from different laboratories.

pense of a much greater percentage error in the second virial coefficient. This is also demonstrated by the results given above.

Semipermeable Membranes

The semipermeable membrane is the most important experimental component in the determination of molecular weights by osmometry. In fact, it is generally accepted that variations in the degree of semipermeability of the various membranes employed is the major factor for the scatter of π/c data shown in Figure 14.7.

Pretreated gel cellophane is probably the most widely used organic membrane material. Generally, such membranes are treated first with aqueous sodium hydroxide, then washed with water, and immersed successively for an hour or two first in ethanol, next in a mixture of ethanol and the solvent to be used, and finally in pure solvent. Such membranes have been found to be satisfactory for the measurement of molecular weights above 10^4 g/mol.

Other membranes commonly used with organic solvents include rubber, poly(vinyl alcohol), polyurethane, and poly(trifluorochloroethylene). Nitrocellulose membranes are often used for aqueous solutions. They are of little use with organic solvents because they are soluble in or are swelled by most organic media.

LIGHT SCATTERING FOR MEASUREMENT OF ABSOLUTE MOLECULAR WEIGHT AND SIZE

Intensity of Scattered Light: Rayleigh Ratio

When a beam of monochromatic light traverses a system and no absorption of light occurs, the transmission is still not complete. A fraction of the incident light is scattered and a resultant attenuation of the intensity of the incident light occurs. The light is scattered in all directions relative to the incident beam and the major portion of the scattered light has the same wavelength as the incident beam.

Over 100 years ago Lord Rayleigh[1] considered the scattering of light in terms of the optical properties of individual molecules. He showed that, for a dilute gas, the intensity of scattered light is given by

$$\frac{i(\theta)r^2}{I_0 (1 + \cos^2 \theta)} = R(\theta) = \frac{2\pi^2}{\lambda^4 N_0} \frac{(n - 1)^2 M}{c} \tag{27}$$

In this expression I_0 is the intensity of light of wavelength λ incident on the system; $i(\theta)$ is the intensity of scattered light per unit volume of the system that is detected at angle θ to the incident beam direction and at distance r from the center of the system; M is the molecular weight of the gas; n is the refractive index; c is the density or concentration of the gas in mass per unit volume; and N_0 is Avogadro's number. The term $R(\theta)$ is called the *Rayleigh ratio*. Note that the

---

[1]Strutt, J. W. (Lord Rayleigh), *Phil. Mag.*, **1871**, *41,* 107, 274, 447.

units of Rayleigh ratio are length^{-1}. It therefore represents the fraction of light scattered at angle θ per unit path length through the system.

Serious difficulties are encountered in attempts to extend the Rayleigh treatment of light scattering in dilute gases to liquids. The main reason for these difficulties lies in the existence of strong intermolecular forces in liquids that are absent in dilute gases. A completely different approach to light scattering in liquids, worked out by Einstein, showed how these difficulties could be avoided.[1] He considered the scattering to arise from local fluctuations in the density due to the thermal motions of the molecules. The density fluctuations lead directly to local fluctuations in the refractive index and hence to scattering of the incident light. Einstein's expression for the Rayleigh ratio of a pure liquid is given by (28). In this equation p is the hydrostatic pressure on the liquid

$$R(\theta) = \frac{i(\theta)r^2}{I_0(1 + \cos^2\theta)} = \frac{2\pi^2}{\lambda^4 N_0}\frac{RT}{\beta}\left(n\frac{dn}{dp}\right)^2 \tag{28}$$

and β is the compressibility, that is, $\beta = -(1/V)(\partial V/\partial p)_T$. The other terms are as described earlier.

High polymers do not usually exist as gases or pure liquids but they can often be dissolved in liquid solvents. However, the use of light scattering measurements to determine the molecular weights and sizes of high polymers was held up by the lack of a theoretical treatment of light scattering in liquid solutions. Such a treatment was made by Peter Debye in 1944.[2] Debye pictured the additional scattering of light by a solution (over and above that of the pure solvent) to result from local fluctuations in the concentration of the solute. In a treatment analogous to that of Einstein for the density fluctuations in pure liquids, Debye considered the local fluctuations in solute concentration, because of random thermal motion, to be opposed by the osmotic pressure of the solution. With this model he derived the expression shown in (29) for the Rayleigh ratio of the scattering caused by the solute.

$$R'(\theta) = \frac{i'(\theta)r^2}{I_0(1 + \cos^2\theta)} = \frac{2\pi^2}{\lambda^4 N_0}n_0^2(n - n_0)^2\frac{RT}{c(\partial\pi/\partial c)_T} \tag{29}$$

Here the prime denotes the excess scattering from the liquid caused by the solute, n_0 is the refractive index of the solvent, c is the concentration of solute in mass per unit volume, π is the osmotic pressure of the solution, and the other terms are as described previously.

Turbidity

The decrease in the intensity of a beam of light because of scattering is used to define the *turbidity* of a solution. The decrease or attenuation depends on the

---

[1]Einstein, A., *Ann. Physik*, **1910**, *33*, 1275.
[2]Debye, P., *J. Appl. Phys.*, **1944**, *15*, 338.

length of the light path through the system and, by analogy to the Lambert law, it is possible to write

$$\frac{I}{I_0} = e^{-\tau l} \tag{30}$$

where I_0 is the incident light intensity, I the transmitted light intensity, l the length of the light path in the solution, and τ the turbidity. We may write (30) in the form

$$e^{-\tau l} = \frac{I_0 - I_s}{I_0} = \frac{I_0 - I'_s\, l}{I_0} = 1 - \frac{I'_s\, l}{I_0} \tag{31}$$

where I_s is the total intensity of light that is scattered by the solution and I'_s the total intensity scattered per unit path length. The fraction of light scattered is generally very small and it is a good approximation to express the exponential in (31) as

$$e^{-\tau l} = 1 - \tau l + \tfrac{1}{2}(\tau l)^2 - \tfrac{1}{6}(\tau l)^3 + \cdots \approx 1 - \tau l \tag{32}$$

A combination of (31) and (32), with neglect of higher powers of τl, leads to

$$\tau = \frac{I'_s}{I_0} \tag{33}$$

as the relationship between the total intensity of scattered light per unit path and the turbidity.

The total light intensity scattered per unit path length through the solution is the intensity scattered through all angles of polar coordinates, or

$$I'_s = \int_0^\pi \int_0^{2\pi} r^2 i'(\theta) \sin\theta \, d\theta \, d\phi \tag{34}$$

Since $r^2 i'(\theta) = I_0 R'(\theta)$ (cf. 29), (33) becomes

$$\frac{I'_s}{I_0} = \tau = \int_0^\pi \int_0^{2\pi} R'(\theta) \sin\theta \, d\theta \, d\phi \tag{35}$$

or

$$\tau = \frac{2\pi^2}{\lambda^4 N_0} n_0^2 (n - n_0)^2 \frac{RT}{c(\partial\pi/\partial c)_T} \int_0^\pi (1 + \cos^2\theta) \sin\theta \, d\phi \int_0^{2\pi} d\theta \tag{36}$$

The value of the product of the definite integrals is $16\pi/3$, so that we obtain the relationship between the turbidity and the total scattered light intensity as

$$\tau = \frac{I'_s}{I_0} = \left(\frac{32\pi^3}{3\lambda^4 N_0}\right) \frac{n_0^2 (n - n_0)^2 RT}{c(\partial\pi/\partial c)_T} \tag{37}$$

Note that the relationship between the turbidity and the Rayleigh ratio is

$$\tau = \frac{16\pi}{3} R_{(90°)} = \frac{16\pi i(90°)r^2}{3I_0} \tag{38}$$

or, more generally,

$$\tau = \left(\frac{16\pi}{3}\right) R(\theta) = \left(\frac{16\pi}{3}\right) \frac{i(\theta)r^2}{I_0(1 + \cos^2\theta)} \tag{39}$$

In (38) and (39) the prime on $i(\theta)$ (that referred to excess scattering by solution over that of solvent) has been dropped. These expressions now provide us with a way to determine the turbidity by measurement of the intensity of light scattered at given angles, that is, by measurement of $i(\theta)$.

Turbidity and Molecular Weight of Polymer Solutions

The connection between turbidity and the molecular weight of a single molecular-weight solute at zero scattering angle may be obtained by substitution of $(\partial\pi/\partial c)_T$ into (37). However, before making this substitution it is convenient to rearrange (37) to (40):

$$\frac{Hc}{\tau} = \frac{1}{RT}\left(\frac{\partial\pi}{\partial c}\right)_T \tag{40}$$

where the function H is given by

$$H = \frac{32\pi^3}{3\lambda^4 N_0} n_0^2 \left(\frac{n - n_0}{c}\right)^2 \tag{41}$$

Note that there is a considerable advantage to having as large a mismatch $n - n_0$ between the indices of refraction for solution and solvent (by choosing a solvent with an index of refraction as different from that of the polymer) as possible. Doubling this difference will quadruple H and the scattering intensity.

Then substitution of $(\partial\pi/\partial c)_T$ from (21) yields

$$\frac{Hc}{\tau} = \frac{1}{M}(1 + 2\Gamma c + 3g\Gamma^2 c^2 + \cdots) \tag{42}$$

According to (41) and (42), the determination of molecular weight by light scattering requires (a) a measurement of the refractive index differences between solutions of varying concentration and pure solvent, (b) a measurement of the turbidities of the solutions, and (c) an extrapolation of Hc/τ to $c = 0$, the intercept of the plot being $1/\overline{M}$. By analogy to the treatment of osmotic pressure data, the initial slope yields the second virial coefficient Γ. Typical data for solutions of polystyrene in toluene are shown in Table 14.4, and a plot according to (42) is shown in Figure 14.8.

TABLE 14.4 LIGHT-SCATTERING DETERMINATION OF MOLECULAR WEIGHT OF POLYSTYRENE IN TOLUENE ($\lambda = 4360$ Å; $n_0 = 1.4976$)

Concn.(g/cm^3)	$n - n_0$	τ(cm^{-1})	$Hc/\tau \times 10^6$ (mol/g)
0.066	0.0075	0.0061	2.79
0.128	0.0145	0.0100	3.34
0.255	0.0288	0.0159	4.17
0.510	0.0576	0.0202	6.55

Source: Frank, H. P. and Mark, H. P., *J. Polymer Sci.*, **1955**, *17*, 1.

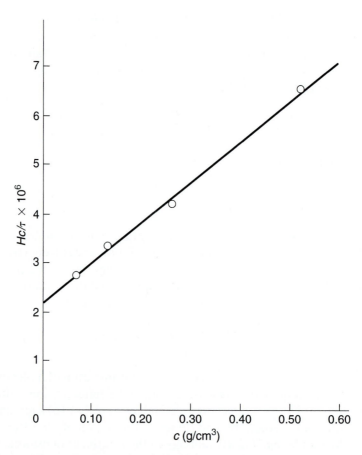

Figure 14.8 Light-scattering determination of the molecular weight of polystyrene in toluene. (Data from Table 14.4)

Weight-Average Molecular Weight

The molecular weight obtained by application of (42) to a polymer solution will be some *average* over the molecular-weight distribution characteristic of the polymer. It is instructive to inquire as to what type of average this is.

Let us define τ° as the turbidity of a polymer solution in the limit of infinite dilution. In other words, τ° represents the turbidity of a solution in which interparticle interference of the scattered light intensity has been eliminated. For a solute such as a polymer, which contains a distribution of molecular weights, we may write (in the limit of infinite dilution)

$$\tau^\circ = \sum_i \tau_i^\circ \tag{43}$$

where τ_i° is the turbidity contributed by species having molecular weight M_i. According to (42), τ approaches HMc as $c \rightarrow 0$. Therefore, (43) may be written as shown in (44), namely

$$\tau^\circ = H^\circ \sum_i M_i c_i \tag{44}$$

where H° is the value of the constant defined by (41) in the limit $c \rightarrow 0$. The concentration c_i is the weight per unit volume of solution of molecules having molecular weight M_i. Therefore, we may write

$$c_i = \frac{w_i}{V} = \frac{(N_i/N_0)M_i}{V} \tag{45}$$

in which w_i is the weight of molecules of molecular weight M_i, N_i is the number of molecules of molecular weight M_i, N_0 is Avogadro's number, and V is the volume of the solution. Substitution of (45) into (44) leads to

$$\tau^\circ = \frac{H^\circ}{N_0 V} \sum_i N_i M_i^2 \tag{46}$$

It will be obvious from (46) that the turbidity is proportional to the square of the polymer molecular weight. This means that the heavier molecules will contribute more than the lighter ones to the turbidity of the solution. *Hence, any average molecular weight defined from the turbidity will be influenced more by heavier molecules than by lighter ones, as was mentioned in Chapter 1.*

The weight-average molecular weight of a polymer is defined by

$$\overline{M}_w = \frac{\sum_i w_i M_i}{\sum_i w_i} = \frac{\sum_i N_i M_i^2}{\sum_i N_i M_i} \tag{47}$$

where the symbols are as defined previously. From (47) it is seen that

$$\sum_i N_i M_i^2 = \overline{M}_w \sum_i N_i M_i \tag{48}$$

After substitution of (48) into (46), equation (49) is obtained:

$$\tau^\circ = \frac{H^\circ}{N_0 V} \overline{M}_w \sum_i N_i M_i = H^\circ \overline{M}_w \sum_i \frac{(N_i/N_0) M_i}{V} \tag{49}$$

which, by reference to (45) may be written

$$\tau^\circ = H^\circ \overline{M}_w \sum_i c_i = H^\circ \overline{M}_w c \tag{50}$$

where c is the total concentration of polymer. Recalling now that τ° and H° refer to the limit of infinite dilution, we may write (50) as

$$\lim_{c \to 0} \frac{Hc}{\tau} = \frac{1}{\overline{M}_w} \tag{51}$$

Thus, light-scattering measurements yield the weight-average molecular weight. For any solute having a distribution of molecular weights, the weight-average molecular weight is greater than the number-average molecular weight. The ratio $\overline{M}_w/\overline{M}_n$, determined from light-scattering measurements and from osmotic pressure measurements on the same solutions, is a measure of the width of the molecular-weight distribution, as was mentioned earlier.

Polymer Dimensions and Corrections for Dissymmetry of Scattering

The relationships between light scattering and molecular weights of polymers, discussed in the previous sections, are satisfactory provided that the scattering is isotropic (i.e., symmetric about 90°). This will be true for polymer solutions if the average size of the largest dimension of the polymer is less than about 1/20 of the wavelength of the incident light. If the average largest dimension of the polymer becomes greater than $\lambda/20$, a dissymmetry of the scattered light about 90° is observed. The dissymmetry arises because of destructive interference of light scattered from different parts *of the same molecule*. This *intra*particle interference cannot be eliminated by an extrapolation to infinite dilution.

The origin of intraparticle interference is shown schematically in Figure 14.9. In this figure, two incident light rays, R_1 and R_2, of a parallel beam are scattered at an angle θ to the incident direction from different parts of the same polymer molecule. Initially, the light waves associated with the two rays are in phase but, because they travel different distances before being detected, the

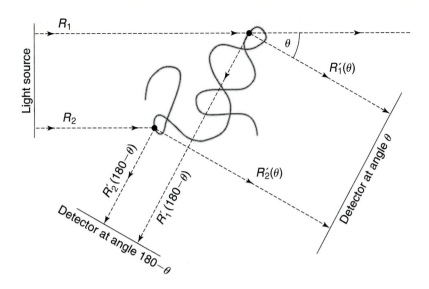

Figure 14.9 Intraparticle destructive interference for large polymer molecules.

corresponding scattered rays R_1' and R_2' are, in general, out of phase. The mutual canceling at the detector of the out-of-phase scattered rays is termed *destructive interference*. The extent of destructive interference depends on the path-length difference of the two rays reaching the detector. Geometric considerations, which the reader may easily verify, establish the following two facts:

1. For a given value of the angle θ, the path-length difference of the two scattered rays detected at θ is less than that of the two scattered rays detected at $180° - \theta$; that is, $[R_1 + R_1'(\theta) - R_2 - R_2'(\theta)] < [R_1 + R_1'(180 - \theta) - R_2 - R_2'(180° - \theta)]$. As a result, destructive interference is greater at $180 - \theta$ (backward direction) than at θ (forward direction). In turn, this results in a greater observed scattered intensity in the forward direction than in the backward direction. That is, $I_s(\theta) > I_s(180 - \theta)$.

2. As θ approaches $0°$, the path-length differences between forward scattered rays from different parts of the molecule also approach zero. Thus, in the limit $\theta = 0$, no intraparticle destructive interference of the scattered light occurs.

To take into account the reduction of scattered intensity by intraparticle interference, it is convenient to replace the turbidity, τ, by the Rayleigh ratio, $R(\theta)$ (e.g., equation 39) and further modify (42) to read

$$\frac{Kc}{R(\theta)} = \frac{1}{\overline{M}_w P(\theta)} + \frac{2\Gamma}{\overline{M}_w} c + \frac{3g\Gamma^2}{\overline{M}_w} c^2 + \cdots \tag{52}$$

where $K = (3/16\pi)H$ and $P(\theta)$ is a scattering function that has the property $P(\theta) \to 1$ as $\theta \to 0$. Thus, the determination of the molecular weights of large polymer molecules requires an extrapolation, not only to the limit $c \to 0$ to eliminate interparticle interference, but also to the limit of $\theta \to 0$ to eliminate intraparticle interference. That is,

$$\frac{1}{M_w} = \lim_{\substack{c \to 0 \\ \theta \to 0}} \frac{Kc}{R(\theta)} \tag{53}$$

Although the extrapolation of experimental scattering data to $\theta = 0$ is independent of any model of the polymer in solution, the scattering functions, $P(\theta)$, depend on the model chosen. Moreover, the determination of average polymer dimensions also depends on the model chosen. The following scattering functions have been derived for three models in terms of a polymer dimension, the wavelength of the light, and the refractive index of the solvent.

1. *Random coil polymer*

$$P(\theta) = \frac{2}{v^2}[e^{-v} - (1 - v)] \tag{54}$$

where

$$v = \tfrac{8}{3}\pi^2 n_0^2 \left(\frac{\overline{r^2}}{\lambda^2}\right) \sin^2 \frac{\theta}{2}$$

$\overline{r^2}$ = mean-square separation of polymer ends[1]

2. *Spherical polymer molecules*

$$P(\theta) = \left[\left(\frac{3}{u^3}\right)(\sin u - u \cos u)\right]^2 \tag{55}$$

where

$$u = 2\pi \left(\frac{n_0 d}{\lambda}\right) \sin \frac{\theta}{2}$$

d = diameter of sphere

3. *Rigid-rod polymer*

$$P(\theta) = \frac{1}{x} S(2x) - \left(\frac{1}{x} \sin x\right)^2 \tag{56}$$

---

[1]Most of the recent literature on chain statistics replaces the overbar symbol for the averages by the fences <...>, for example, in the mean-square radius of gyration $<S^2>$ and mean-square end-to-end distance $<r^2>$.

where

$$S(x) = \int_0^x \frac{\sin y}{y}\, dy$$

$$x = 2\pi \left(\frac{Ln_0}{\lambda} \right) \sin \frac{\theta}{2}$$

$$L = \text{length of rod}$$

Note that all the scattering functions depend on the angle θ through $\sin(\theta/2)$ or (in the important case of the random coil) on $\sin^2(\theta/2)$. This suggests that the double extrapolation to obtain molecular weights (see equation 53) can be made conveniently by plotting $Kc/R(\theta)$ versus $kc + \sin^2(\theta/2)$, where k is an arbitrary constant When this is done, a grid of points is obtained as shown in Figure 14.10 for poly(L-lactic acid) in bromobenzene. The quantity $1/\overline{M}_w$ is evaluated from the plot as the intercept on the vertical axis, for which $\theta = 0$ and $c = 0$. The molecular weights obtained by this extrapolation procedure are independent of the model chosen, because, for all models, $P(\theta) \to 1$ as $\theta \to 0$. Grid plots such as shown in Figure 14.10 are often called Zimm plots after their originator.[1]

An average size of the polymer molecule may be obtained from the dissymmetry coefficient,[2] defined as

$$Z_\theta = \frac{i(\theta)}{i(180° - \theta)} = \frac{P(\theta)}{P(180° - \theta)} \tag{57}$$

It must be realized that the polymer sizes so obtained depend on the extent to which the model chosen represents the actual polymer. For three models of a polymer in solution, the theoretical values of Z_{45}° as functions of polymer dimensions [i.e., r, d, and L of (54) to (56)] are shown in Figure 14.11. A determination of the dimensions appropriate to the particular model chosen requires an experimental measurement of the intensity ratio, i_{45}/i_{135}. The value obtained experimentally may be located on the plot in Figure 14.11 to give immediately the ratios $(\overline{r^2})^{1/2} n_0/\lambda$, $n_0 d/\lambda$, or Ln_0/λ. Since n_0, the refractive index of solvent, and λ, the wavelength of light, are known, $(\overline{r^2})$, d, or L is determined.

Experimental Apparatus and Technique

According to (41) and (42), the quantities that must be determined to calculate the molecular weight are the turbidity τ and the refractive index increment $(n - n_0)/c$. It is not practical to measure the turbidity from the attenuation of the incident light intensity (cf. equation 30.) because this attenuation is generally

[1]Zimm, B. H., *J. Chem. Phys.*, **1948**, *16*, 1099.
[2]Debye, P., *J. Phys. Colloid Chem.*, **1947**, *51*, 18.

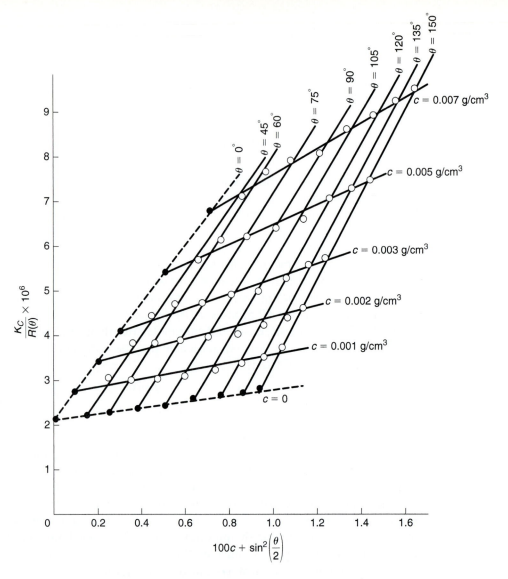

Figure 14.10 Light-scattering measurements at $\lambda = 5461$ Å for solutions of poly(L-lactic acid) in bromobenzene at 85°C. [Reproduced from Tonelli, A. E. and Flory, P. J., *Macromolecules* **1969**, *2*, 225; © the American Chemical Society.] The filled circles locate extrapolated values.

too small. The turbidity τ of a solution must be determined, then, from the Rayleigh ratio (see equations 38 and 39). This means that the intensity of light scattered per unit volume of solution at some angle θ (usually 90°) to the incident beam must be measured. Furthermore, if the polymer molecules have, on the average, a dimension that is greater than about 1/20 of the wavelength of the light

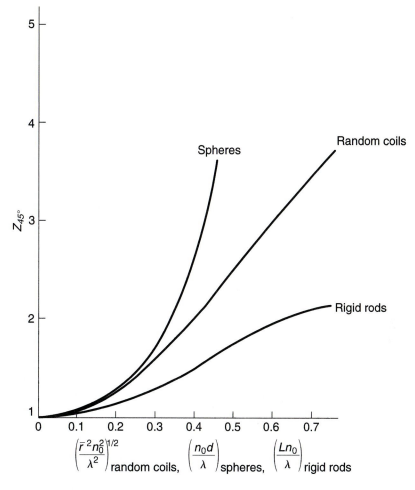

Figure 14.11 Dissymmetry coefficient at 45° as a function of polymer size.

used (which is typically, 4000 to 5000 Å), it is necessary to measure the scattered intensity at several angles.

Therefore, a light-scattering determination of the molecular weight and of the molecular size requires the measurement of (1) the intensity of light scattered at several angles by solutions of various concentrations, and (2) the difference in refractive index between solution and solvent for various concentrations.

Light-scattering photometry   A light-scattering photometer is, in principle, quite simple, and numerous designs have appeared in the literature over the past four decades. All light-scattering photometers contain a radiation source, an optical system to provide a collimated beam incident on the sample, and a second optical system to detect and measure the intensity of the scattered light as a function of scattering angle. Many photometers provide for continuous monitoring of the incident beam intensity. While the older photometers generally used only a

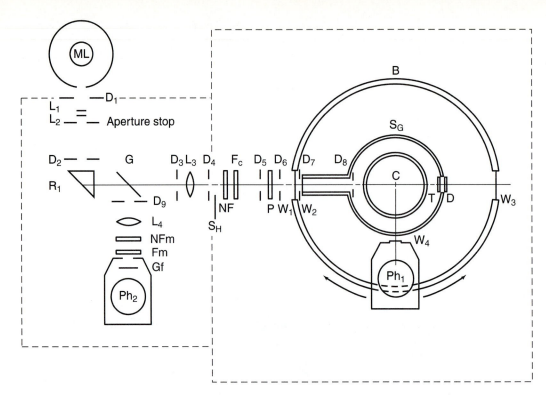

Figure 14.12 Classical light-scattering photometer. (Reproduced with permission of Applied Optics.)

mercury arc as a light source, most of the more recent instruments make provision to interchange the arc with a laser and typically Ar-ion and He–Ne lasers are used. The mercury arcs provide incident light beams of wavelengths 436, 546, and 578 nm, while the lasers above provide light of 488 and 514, and 633 nm, respectively. The precise directional properties of the laser facilitate the extrapolation of scattered intensity to zero angle (Zimm plot) by permitting measurements to be made at smaller scattering angles.

A typical classical light-scattering photometer,[1] with a mercury lamp as a light source, is shown schematically in Figure 14.12. It is a simple matter to incorporate a laser into the system. In the photometer shown, light from the source is focused by the lenses L_1 and L_2 at the aperture stop and the image of the aperture stop is focused by the lens L_3 at the center of the scattering cell, C, that contains the polymer solution. A total reflection prism, R_1, and a thin glass plate, G, are located between the aperture stop and L_3. The prism is so placed that a laser may be positioned along the direction of the incident light beam, while the glass plate effects reflection of a part of the incident beam to the monitoring photo-

[1]Utiyama, H., and Tsunashima, Y., *Appl. Opt.*, **1970,** *9,* 1330.

multiplier, Ph_2. Four neutral filters of varying transmittance, NF, are provided, as is a color diffraction filter, F_C, to isolate the various emission lines from the sources. A polarizer, P, is also incorporated into the incident beam path. The scattered light from the cell, after passing through the glass cylinder, S_G, is detected and measured by the photomultiplier, Ph_1, which can be positioned over a continuous range of angles relative to the incident beam. For further details of the instrument and its performance, the original literature should be consulted.

In addition to the use of lasers, there have been a number of other advances in light-scattering photometers. One example is the design incorporated in instruments by the Wyatt Technology Corporation (Santa Barbara). The single photocell (PC), which, in earlier experiments, had to be rotated on a platter to obtain scattering intensities as a function of angle θ, is replaced by an array of photocells located at fixed angles, as illustrated schematically in Figure 14.13. This approach permits the simultaneous gathering of the scattered intensities at the various values of θ. It also minimizes errors in the angular settings, and facilitates the use of software programs for calculations of molecular weights, second virial coefficients, and radii of gyration.

Scattering cells of a variety of cross-sectional shapes may be used. Top views of typical cells are shown in Figure 14.14. Cell (a) is used for polymer molecules that are small [i.e., $(\overline{r^2})^{1/2} < \lambda/20$], because a measurement of the scattered intensity at 90° only is necessary. For larger molecules, in which $i(\theta)$ must be mea-

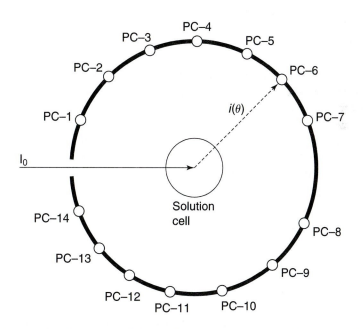

Figure 14.13 An improved photocell (PC) arrangement in a light-scattering photometer.

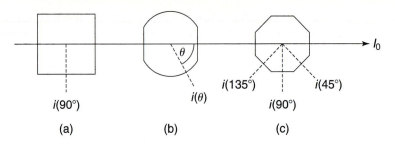

Figure 14.14 Top view of typical light-scattering cells.

sured at various values of θ, cell (b) would be preferred. Cell (c) is a dissymmetry cell which may be used to measure Z_{45} as well as $R(90°)$.

Other scattering techniques can also be used to obtain absolute values of the molecular weights of polymers, with the most prominent examples involving X-rays and neutrons. However, the equipment required is generally very expensive, the facilities can be relatively remote from one's laboratory, and the techniques can involve additional constraints such as requiring measurements under vacuum. For these reasons, X-ray and neutron scattering are more commonly used for other purposes, such as characterizing the sizes of polymer molecules or their domains in multiphase systems.

Refractive index increment A differential refractometer for measurement of the difference in refractive index between solution and solvent is shown schematically in Figure 14.15. The particular instrument shown has been reported to be capable of measurements of $n - n_0$ to an accuracy of about ±0.000003. It can also be used in gel permeation chromatography, as described in Chapter 15.

Monochromatic light produced by the mercury arc H4 and filter F illuminates the adjustable slit S. An image I of the slit S is viewed by a micrometer which is capable of measuring displacements of this image of 0.001 cm. A rectangular outer cell C, made of glass and containing the solvent, surrounds a hollow prism cell P that contains the solution. The lenses L_1 and L_2 are collimating lenses to produce a parallel beam and to focus it at the position I. Any difference in refractive index between solution and solvent, will cause the image I to be deflected. Moreover, if the refractive index difference is not too large, the deflection

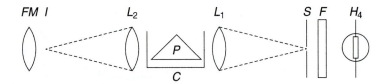

Figure 14.15 Schematic arrangement of a differential refractometer. [Reproduced from Debye, P., *J. Appl. Phys.*, **1946**, *17*, 392; with permission of the American Institute of Physics, New York.]

TABLE 14.5 SPECIFIC REFRACTIVE INDEX INCREMENTS OF TYPICAL POLYMER SOLUTIONS IN UNITS OF cm^3/g

Polymer	Solvent	$\lambda = 4360$ Å	$\lambda = 5460$ Å
Polystyrene	Methyl ethyl ketone	0.231	0.220
Poly(ethyl acrylate)	Acetone	0.109	0.106
Poly(vinyl palmitate)	Neohexane	0.122	0.120
Poly(vinyl laurate)	Neohexane	0.118	0.114
β-Lactoglobulin	0.1 M NaCl, pH 5.2	0.189	0.182
Lysozyme	0.1 M NaCl, pH 6.2	0.196	0.189

of the image is proportional to $n - n_0$. For example, if A is the apex angle of the prism cell and f is the focal length of the lenses, the deflection of the image can be shown to be given by (58). Measurement of Δd for known values of f and A thus yield $n - n_0$.

$$\Delta d = 2f(n - n_0) \tan \frac{A}{2} \tag{58}$$

For the concentration ranges generally employed, the refractive index difference, $n - n_0$, is a linear function of the concentration. In other words, $(n - n_0)/c$, which is called the *specific refractive index increment*, is a constant for a given polymer-solvent system and a given wavelength of light. This means that $n - n_0$ need be measured for only one or two different concentrations. Typical specific refractive index increments are shown in Table 14.5. As can be seen in the table, $(n - n_0)/c$ depends on the wavelength of light used, being slightly larger at the shorter wavelength.

Preparation of solutions   It is very important to remove all suspended dust particles from a solution before making measurements of the intensity of scattered light. Dust particles are generally larger than polymer molecules and may produce greater scattering. Solvents and solutions are generally clarified by filtration (usually under pressure) or by ultracentrifugation. In these procedures, care must be taken not to remove very large polymer molecules that have molecular weights at the heavy end of the distribution.

Accuracy of Light-Scattering Measurements of Molecular Weight

The accuracy of molecular weights determined by light scattering is not as good as might be expected on the basis of the precision shown in Figures 14.8 and 14.10. This may be seen from an inspection of Figures 14.16 and 14.17. These figures show plots of Hc/τ (or equivalently $Kc/R_{90°}$) at 4360 Å versus c, as determined independently by six laboratories in the United States, Canada, and Europe for samples of the same polystyrene fraction. Figure 14.16 refers to solutions of the polymer in toluene and Figure 14.17 to solutions of the polymer in methyl

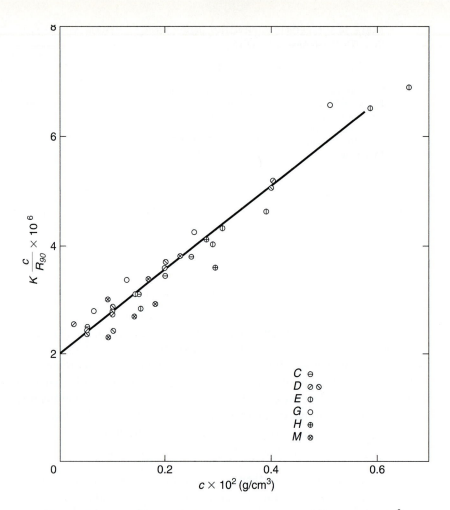

Figure 14.16 Accuracy of light-scattering measurements at λ = 4360 Å on polystyrene in toluene at 25°C. [Reproduced from Frank, H. P. and Mark, H. P., *J. Polymer Sci.*, **1955**, *17*, 1; with permission of John Wiley & Sons, Inc, New York.] The symbols indicate results from different laboratories.

ethyl ketone. Fewer discrepancies are noted than for the corresponding osmotic pressure measurements. The still-appreciable scatter of the data has been attributed to difficulties in the calibration of the light-scattering photometer.

The average molecular weights and second virial coefficients found for this polystyrene sample by light scattering are:

(a) Toluene: $\overline{M}_w = 5.90 \pm 0.58 \times 10^5$ g/mol
$\Gamma = 224 \pm 30$ cm³/g

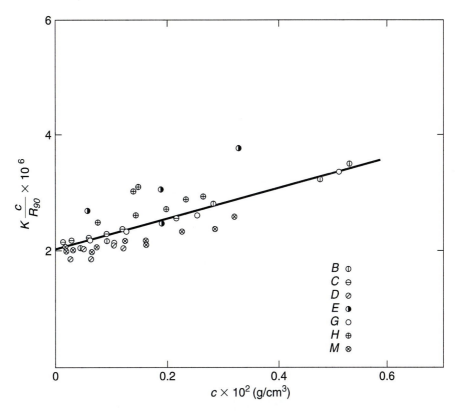

Figure 14.17 Accuracy of light-scattering measurements at $\lambda = 4360$ Å on polystyrene in methyl ethyl ketone at 25°C. [Reproduced from Frank, H. P. and Mark, H. F., *J. Polymer Sci.*, **1955**, *17*, 1; with permission of John Wiley & Sons, Inc., New York.] The symbols indicate results from different laboratories.

(b) Methyl ethyl ketone: $\overline{M}_w = 5.92 \pm 0.75 \times 10^5$ g/mol
$$\Gamma = 79.3 \pm 19 \text{ cm}^3/\text{g}$$

The agreement of the weight-average molecular weights obtained from the measurements in the two solvents is well within the experimental errors. Moreover, from a theoretical viewpoint it is reassuring that, despite the difference in weight-average and number-average molecular weights of this polymer sample, the second virial coefficients determined by the two very different methods are in agreement, within experimental error.

It should be pointed out that the polystyrene molecules in this test sample were sufficiently large to cause an asymmetry in the light scattering. The weight-average molecular weights given above were determined by the double extrapolation procedure (Zimm plot), in which values of Kc/R_θ were extrapolated to $c \rightarrow 0$ and $\theta \rightarrow 0$ (see Figure 14.10).

MASS SPECTROMETRY

Until relatively recently, mass spectrometry was limited to relatively low molecular-weight materials because of the difficulty involved in volatilization of a high molecular-weight polymer without prematurely fragmenting it. One new techique for applying this type of spectrometry to polymers involves Matrix-Assisted Laser Desorption-Ionization (MALDI), in which the polymer is first mixed into a carefully chosen solid matrix. The matrix and polymer are then vaporized into the gas phase by a powerful laser pulse, and this permits the spectrometry to be carried out on the polymer component in the usual manner.

This approach has now been used to obtain molecular weights up to around half a million. One advantage of this techniques is that only very small amounts of polymer are required for analysis; also, the accuracy is excellent, and the measurements can be carried out quite rapidly and reproducibly. Electrospray mass spectrometry in which a mist of polymer in a solvent is ionized also allows the molecular weights of polymers such as proteins to be measured.

ULTRACENTRIFUGATION AS A METHOD FOR MEASUREMENT OF ABSOLUTE MOLECULAR WEIGHTS

Principle of the Method

Sedimentation velocity A particle in the earth's gravitational field is subjected to a downward force, mg, where m is the mass of the particle and g is the acceleration due to gravity. If the particle is in a vacuum, it will fall with a velocity u that continuously increases with time. According to Newton's second law of motion, this can be expressed by

$$m \frac{du}{dt} = mg \tag{59}$$

or

$$u = gt \tag{60}$$

where t is the time. If the particle is not in a vacuum, but rather is immersed in a fluid of density ρ_s, a buoyancy force equal to the weight of the displaced fluid will oppose the gravitational force. If the volume displaced by the particle is V_p, the mass of displaced fluid will be $V_p\rho_s$ and the weight or buoyancy force will be given by $V_p\rho_s g$. In addition, a frictional force proportional to the velocity (i.e., Fu, where the proportionality constant F is called the frictional coefficient) will oppose the gravitational force. Hence, we may write for the net downward force on the particle in a solvent

$$m \frac{du}{dt} = (m - V_p\rho_s)g - Fu \tag{61}$$

Eventually, the velocity of fall will increase to such a value that the net force on the particle becomes zero and, from this time on, the particle falls with a constant velocity u_s, called the *terminal velocity* or *sedimentation velocity*. Thus, from (61) we have

$$(m - V_p\rho_s)g = Fu_s \tag{62}$$

In the special case of a spherical particle of radius r, the value of F is given by Stokes' law, which, on substitution into (62), yields

$$m(1 - \overline{v}_p\rho_s)g = \frac{M}{N_0}(1 - \overline{v}_p\rho_s)g = 6\pi\eta r u_s \tag{63}$$

in which η is the viscosity of the fluid, $\overline{v}_p$ the partial specific volume of the particle in the fluid, M the mass of the particle per mole, and N_0 is Avogadro's number. Eliminating r from (63), by the use of the relationship $\overline{v}_p = (4/3)\pi r^3 N_0/M$, and solving for M yields

$$M = 9\pi\sqrt{2\overline{v}_p}N_0\left[\frac{\eta u_s}{(1 - \overline{v}_p\rho_s)g}\right]^{3/2} \tag{64}$$

Of course, if the particles are polymer molecules, the molecular weight M in (64) is an average molecular weight. Hence, (64) provides a means to determine an average molecular weight for a spherical polymer simply by measurement of the rate of sedimentation and the partial specific volume of the polymer.

In practice the matter is not quite so simple. First, as is obvious from the fact that polymer solutions are stable, the sedimentation rate under the influence of gravity is vanishingly small. Elaborate ultracentrifugation techniques must be used to supply the much larger accelerations needed. Second, (64) applies only to spherical polymer molecules, whereas most polymer molecules exist in solution as random coils. Therefore, it is usually necessary to use (62) in which g, the acceleration caused by gravity, is replaced by the centrifugal acceleration, $\omega^2 x$, where ω is the angular velocity of rotation and x is the distance of the sample from the center of rotation. The proportionality constant, F, is not known explicitly and must be eliminated by experimental measurement of some other quantity that depends on it. Usually, this other quantity is the diffusion coefficient D, since many years ago Nernst[1] showed that the coefficient for free diffusion in an infinitely dilute solution is given by

$$D = \frac{RT}{N_0 F} \tag{65}$$

[1]Nernst, W., *Z. Physik, Chem.*, **1888**, *2*, 613.

It is expected that the proportionality constant, F, for diffusion should be the same as that for sedimentation. Combination of (62) and (65) so as to eliminate F, and replacement of g by $\omega^2 x$ leads to

$$(m - V_p \rho_s)\omega^2 x = \left(\frac{N_0 D}{RT}\right)^{-1} u_s \tag{66}$$

The *sedimentation coefficient* is defined by

$$s = \frac{u_s}{\omega^2 x} \tag{67}$$

and if we also introduce into (66) the partial specific volume of polymer in the solution, $\bar{v}_p$, the expression

$$M = N_0 m = \frac{sRT}{D(1 - \bar{v}_p \rho_s)} \tag{68}$$

is finally obtained. This is known as the Svedberg equation. According to this equation, the molecular weight of polymer particles in solution can be determined by measurement of the sedimentation coefficient, s, the diffusion coefficient, D, and the partial specific volume of the polymer in solution, $\bar{v}_p$.

Actually, it is found that the experimental values of s and D depend significantly on the concentration of polymer in the solution. Moreover, the Nernst equation (65) and therefore (68) are strictly valid only in the limit of infinite dilution. Therefore, in order to eliminate intermolecular effects and obtain reliable molecular weights, the values of s and D inserted into (68) must be values that have been extrapolated to zero concentration. Strictly speaking, the partial specific volume of the polymer $\bar{v}_p$ also depends on the concentration, but the dependence is sufficiently weak that extrapolation of measured values of $\bar{v}_p$ to zero concentration is generally not necessary. Finally, the derivation of (68) considered solution particles that have a single molecular weight M. A generalization[1] to polymer samples having a distribution of molecular weights shows that the molecular weight obtained is *approximately* equal to the weight-average molecular weight, $\overline{M}_w$.

In view of the foregoing discussion, the Svedberg equation (68) is more properly written as

$$\overline{M}_w = \frac{s_0 RT}{D_0(1 - \bar{v}_p \rho_s)} \tag{69}$$

where s_0 and D_0 represent values that have been extrapolated to zero concentration. It is obviously important to obtain as large a mismatch in densities as

[1]Meyerhoff, G., *Angew. Chem.*, **1960**, 72, 699.

possible between the solvent and polymer, in order to keep the denominator of this equation from approaching zero.

As in light-scattering and osmotic-pressure measurements on polymer solutions, it is useful to describe the concentration dependence of s and D in terms of a virial equation (see equations 23 and 42]. Thus, it can be shown[1] that the concentration dependence of (68) is well described by

$$\frac{D(1 - \bar{v}_p\rho_s)}{sRT} = \frac{1}{M_w}(1 + 2\Gamma c + 3g\Gamma^2 c^2 + \cdots) \tag{70}$$

where Γ, a fundamental parameter of the Flory–Huggins theory of polymer solutions is given by equation (24). The weight-average molecular weight $\overline{M}_w$ is obtained either from (69) or from the zero-concentration limit of the left-hand side of (70). The second virial coefficient, Γ, is determined from the initial slope of a plot of the left-hand side of (70) as a function of concentration.

The sedimentation equilibrium method   In a homogenous polymer solution the concentration gradient is zero in all directions. During centrifugation the sedimentation process produces a migration of polymer particles toward the bottom of the container and, in so doing, this creates a nonzero concentration gradient, dc/dx, in the direction from the center of rotation to the bottom of the centrifugation cell. The process of diffusion, on the other hand, produces a migration of polymer molecules from a region of high concentration to a region of lower concentration. Thus, sedimentation and diffusion operate in opposite directions during centrifugation and, after a sufficient time, they must produce a pseudo equilibrium or steady state that is characterized by a nonzero but constant value of dc/dx. The faster the rotation, the larger the gradient dc/dx. It can be shown that at equilibrium the ratio of concentrations at two positions in the cell is given by

$$\ln\frac{c_2}{c_1} = \frac{\omega^2(1 - \bar{v}_p\rho_s)(x_2^2 - x_1^2)\overline{M}}{2RT} \tag{71}$$

where c_2 and c_1 are the concentrations of polymer at positions x_2 and x_1, respectively. The concentration ratio is generally determined by optical methods that depend on the change of refractive index with concentration.

Molecular weights obtained by this method will deviate appreciably from the weight-average molecular weights. Another disadvantage in the use of this method, compared with the sedimentation velocity method, is that, because of the small diffusion coefficients of polymer molecules, very long times are often required to attain equilibrium.

Experimental measurements   Calculation of the weight-average molecular weight $\overline{M}_w$ by (69) requires the experimental measurement of the sedi-

[1]Schulz, G. V., *Z. Physik. Chem.*, **1944**, *193*, 168.

mentation constant s, the diffusion coefficient D, and the partial specific volume of the polymer $\bar{v}_p$. Moreover, s and D must be measured in solutions of varying concentrations so that the respective values at infinite dilution, s_0 and D_0, may be obtained by extrapolation.

Partial Specific Volume. The partial specific volume of a polymer in solution is generally calculated from the densities of the solutions and solvent. The respective densities are determined by a pycnometric method[1] in which an accurately known volume containing, in turn, solvent and solution is weighed. The relationship between the volume of the solution (pycnometer volume) and the partial specific volumes of polymer and solvent is given by

$$V = m_p \bar{v}_p + m_s \bar{v}_s \tag{72}$$

The partial specific volumes are defined thermodynamically by (73) and (74).

$$\bar{v}_p = \left(\frac{\partial V}{\partial m_p} \right)_{m_s, T, p} \tag{73}$$

$$\bar{v}_s = \left(\frac{\partial V}{\partial m_s} \right)_{m_p, T, p} \tag{74}$$

In actual solutions of finite concentration of polymer, $\bar{v}_s$ is not equal to the reciprocal of the solvent density (i.e., to ρ_s^{-1}). However, in dilute solutions little error is introduced by assuming that $\bar{v}_s \cong \rho_s^{-1}$. The partial specific volume of the polymer is then given by a rearrangement of (72) as

$$\bar{v}_p = \frac{V - m_s/\rho_s}{m_p} \tag{75}$$

Introduction of the density of the solution, $\rho = (m_p + m_s)/V$, and the concentration of polymer, $c = m_p/V$, into (75) transforms the expression into (76), a more useful form in terms of the solution and solvent densities.

$$\bar{v}_p = \frac{1}{\rho_s} \left(1 - \frac{\rho - \rho_s}{c} \right) \tag{76}$$

In (76) ρ_s and ρ are the densities of solvent and solution of concentration c, respectively.

Strictly speaking, the experimental values of $\bar{v}_p$ obtained from (76) should be extrapolated to zero concentration and the extrapolated value used in (69). In

---

[1]Shoemaker, D. P., Garland, C. W., and Steinfeld, J. I., *Experiments in Physical Chemistry* 3rd ed. (New York: McGraw-Hill, **1974**), p. 171.

practice, the values of $\bar{v}_p$ in the dilute solutions used do not differ appreciably from the value at infinite dilution, and extrapolation is generally not necessary.

The reader should note carefully that $\bar{v}_p$ is a property of the *polymer solution* and is not the same as the reciprocal of the density of the solid polymer. Hence, in general, $\bar{v}_p \neq \rho_p^{-1}$.

Sedimentation Constant. According to the definition in (67), the sedimentation constant may be calculated from measurements of the terminal sedimentation velocity, u_s, that is reached when the solution is subjected to a centrifugal force of $\omega^2 x$. Such measurements are carried out in an ultracentrifuge, an instrument that is capable of rotational speeds up to 80,000 revolutions per minute. The dimensions are such that these angular velocities correspond to accelerations that are in the range 7,000 to 420,000 times that of gravity.

A schematic diagram of an ultracentrifuge is shown in Figure 14.18. A dilute solution of the polymer, at a concentration c, is placed in the centrifugation cell, such as shown in Figure 14.19, and this cell, as well as a balancing cell, are placed in the rotor of the ultracentrifuge. The rotor is operated in a vacuum to reduce friction and the temperature is controlled to $\pm0.1°C$.

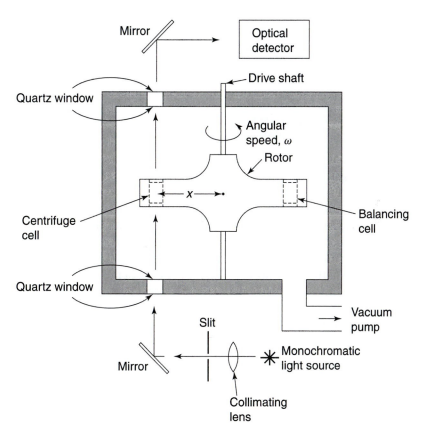

Figure 14.18 Schematic diagram of an ultracentrifuge.

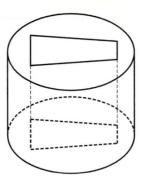

Figure 14.19 Typical centrifugation cell of the sector type.

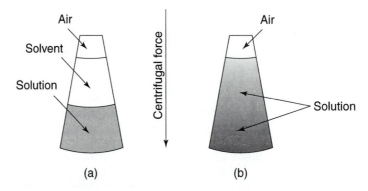

Figure 14.20 Light-beam view of a cell during centrifugation: (a) monodisperse polymer; (b) polydisperse polymer.

Initially, the solution is homogenous throughout the cell (i.e., $dc/dx = 0$), but when the centrifugal force is applied by spinning the rotor, the polymer molecules migrate, ultimately with a constant velocity, u_s, toward the bottom of the cell. This migration of the polymer molecules results in the formation of a *boundary* between solution and solvent. The sharpness of this boundary depends upon the nature of the molecular-weight distribution of the polymer and is characterized by the value of the concentration gradient, dc/dx. In Figure 14.20, the result of polymer migration in the cell due to centrifugation is shown for two extreme cases of polymer. In the left-hand picture is seen the sharp boundary that results if all polymer molecules have the same molecular weight. The right-hand diagram shows the diffuse type of boundary resulting from a sample with a broad distribution of molecular weights.

A quantitative measurement of dc/dx is generally accomplished by the use of *schlieren* or optical interference techniques. Both techniques measure the change in refractive index of the system as a function of the position in the cell.

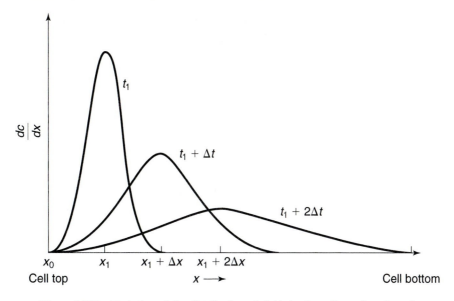

Figure 14.21 Variation of the distribution of dc/dx in the cell as a function of centrifugation time.

Because the refractive index depends on the concentration, both techniques ultimately yield the concentration gradient, dc/dx, as a function of distance x. For details of these fairly complex optical techniques, the reader should consult more specialized accounts.[1]

At each revolution the centrifugation cell is in the position at which the light beam passes through the cell (see Figure 14.18) and, therefore, a distribution curve of dc/dx as a function of x is obtained for each revolution (or equivalently for each time of measurement). For a polymer having a broad but continuous distribution of molecular weights, data such as those depicted in Figure 14.21 would be obtained. From such curves it is possible to determine the sedimentation velocity u_s from the relationship

$$u_s = \frac{x_1 + \Delta x - x_1}{t_1 + \Delta t - t_1} = \frac{\Delta x}{\Delta t} \tag{77}$$

The sedimentation coefficient, s, is then calculated by the expression

$$s = \frac{u_s}{\omega^2 \bar{x}} = \frac{1}{\omega^2}\left(\frac{\Delta x}{\Delta t}\right)\left(\frac{2}{2x_1 + \Delta x}\right) \tag{78}$$

[1]Daniels, F., Williams, J. W., Bender, P., Alberty, R. H., Cornwell, C. D., and Harriman, J. E., *Experimental Physical Chemistry*, 7th ed. (New York: McGraw-Hill, **1970**), pp. 460–465.

where $\bar{x} = (2x_1 + \Delta x)/2$ is the average distance from the center of rotation.

As has already been mentioned, the values of s calculated from the data by (78) depend on the concentration, and an extrapolation to zero concentration must be made to eliminate interparticle effects. It is found empirically that the extrapolation is most readily made according to

$$\frac{1}{s} = \frac{1}{s_0} + k_s c \qquad (79)$$

Thus, the reciprocals of the experimental s values are plotted versus the concentration, and s_0 is determined as the reciprocal of the intercept. A typical plot of $1/s$ versus c for polyisobutylene is shown in Figure 14.22.

Diffusion Coefficient. The process of diffusion acts in opposition to sedimentation. Thus, sedimentation begins with a uniform solution (i.e., $dc/dx = 0$) and produces concentration gradients. Diffusion begins with a finite value for dc/dx and the process drives the system toward the equilibrium value $dc/dx = 0$. Accordingly, the same optical techniques mentioned previously may be used to measure the distribution of dc/dx as a function of time and, from these measurements, the diffusion coefficient D may be calculated.

Consider the schematic diffusion cell in Figure 14.23a. Initially, a partition divides the polymer solution of concentration c from the solvent. The initial values of c and dc/dx are then described by

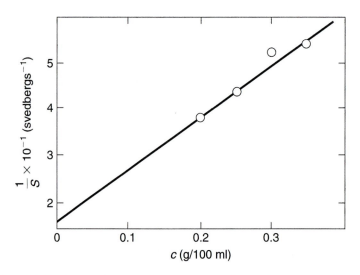

Figure 14.22 Extrapolation of the reciprocal sedimentation constant to infinite dilution. [Reproduced from Closs, W. J., Jennings, B. R., and Jerrard, H. G., "Concentration Dependence of the Sedimentation Coefficient," *Eur. Polymer J.*, **1968**, *4*, 639; with permission of Pergamon Press, New York.]

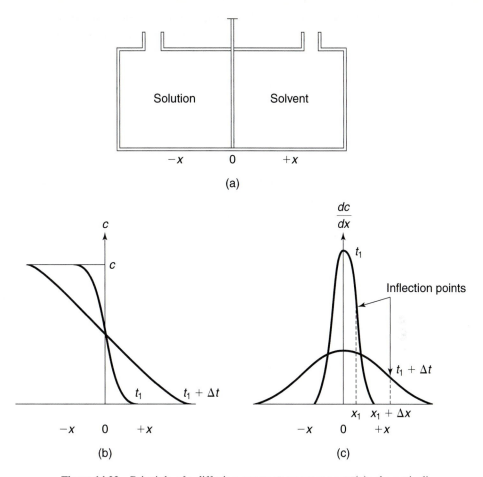

Figure 14.23 Principle of a diffusion-constant measurement: (a) schematic diagram of a diffusion cell; (b) variation of concentration with position and time; (c) variation of concentration gradient with position and time.

$$c_0 = c \qquad \text{for } x < 0$$

$$c_0 = 0 \qquad \text{for } x > 0$$

$$\left(\frac{dc}{dx}\right)_0 = \infty \qquad \text{at } x = 0$$

$$\left(\frac{dc}{dx}\right)_0 = 0 \qquad \text{anywhere else}$$

When the partition is removed, polymer diffuses from left to right, that is, from the region of high concentration to that of low concentration. The variation of concentration gradient with position x at two subsequent times after removal of the partition are shown in Figure 14.23b and c, respectively.

The position of the inflection points x_{in} of the dc/dx distribution curves (Figure 14.23c) are taken as a measure of the progress of the diffusion. The mean rate of diffusion is then given by x_{in}/t, where t is the time at which the curve with inflection point x_{in} was measured. It is found that this mean rate of diffusion is inversely proportional to the progress of the diffusion (i.e., inversely proportional to x_{in}). Therefore, it is possible to write

$$\frac{x_{in}^2}{t} = \text{constant} \tag{80}$$

It may be shown[1] that the constant in (80) can be identified with $2D$, where D is the diffusion coefficient. Thus, the diffusion coefficient may be determined by a measurement of x_{in} and t by the equation

$$D = \frac{x_{in}^2}{2t} \tag{81}$$

In cases where the inflection points are nonexistent or difficult to determine, an alternative definition of the progress can be used. One convenient method defines the average value of x_{in}^2 as in (82).

$$\overline{x_{in}^2} = \frac{\int_{-\infty}^{\infty} x^2 (dc/dx) dx}{\int_{-\infty}^{\infty} (dc/dx) dx} \tag{82}$$

The necessary extrapolation of D-values, obtained from solutions of various concentration, to infinite dilution may be made simply by plotting D versus c. Straight lines of small slope are generally obtained *at low concentrations* which may be described by

$$D = D_0 + D_0 k_D c \tag{83}$$

An example of such a plot is shown in Figure 14.24.

Molecular Weights

After the determination of $\overline{v}_p$, s_0, and D_0, as described above, these values are inserted into (69) and the weight-average molecular weight is calculated. The versatility of the method, and its comparative freedom from extraneous disturbing effects, has led to the extensive use of ultracentrifugation for the determination of molecular weight of biological polymers. Typical values of $\overline{v}_p$, s_0, D_0, and M at 20°C for a number of such biological polymers and for a synthetic polystyrene are shown in Table 14.6.

The reliability of sedimentation velocity measurements can be judged by an examination of the results obtained independently by five laboratories for methyl

[1]Moore, op. cit., pp. 159–163.

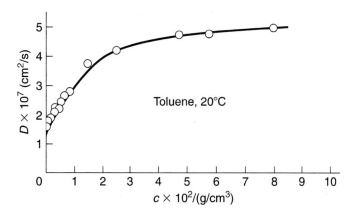

Figure 14.24   Concentration dependence of the diffusion coefficient of polystyrene ($\overline{M}_w = 670{,}000$) in toluene at 20°C. [Reproduced from Büldt, G. and Meyerhoff, G., *Makromol. Chem.*, **1975**, *176*, Suppl. 1, 359; with permission of Hüthig and Wepf Verlag, Basel, Switzerland.]

ethyl ketone solutions of the same fractionated sample of polystyrene. The results are shown in Figure 14.25, from which an extrapolated value of $s_0 = 17.9 \pm 0.4 \times 10^{-13}$ s at 20°C is obtained. The diffusion coefficient was determined by only one of the participating laboratories and a value of $3.6 \pm 0.2 \times 10^{-7}$ cm²/s was reported.

Although ultracentrifugation is the most elaborate, expensive, and time-consuming technique for the determination of polymer molecular weights, it is perhaps the most versatile of all the absolute methods. Unlike osmotic pressure, it is not limited in the range of molecular weights that can be measured. Also, unlike light scattering, it is not influenced by the presence of minute particulate matter in the solutions. In the hands of skilled experimenters, sedimentation constants are very reproducible (see Figure 14.25), and it must be concluded that the method is very reliable. However, probably because of the expense, ultracentrifugation

TABLE 14.6   PARTIAL SPECIFIC VOLUMES, SEDIMENTATION CONSTANTS, DIFFUSION COEFFICIENTS, AND MOLECULAR WEIGHTS OF SOME POLYMERS

Polymer	$\overline{v}_p(\text{cm}^3/\text{g})$	$S_0 \times 10^{13}(\text{s})$	$D_0 \times 10^7(\text{cm}^2/\text{s})$	$\overline{M}_w \times 10^{-3}(\text{g/mol})$
Hemoglobin[*]	0.749	4.48	6.9	63
Serum albumin[*]	0.736	4.67	5.9	72
Urease[*]	0.73	18.6	3.46	480
Tobacco mosaic virus[*]	0.73	185	0.53	31,400
Polystyrene[†]	0.99	17.9	3.6	592

Source: Moore, W. J., *Physical Chemistry*, 4th ed. (Englewood Cliffs, N.J.: Prentice-Hall, **1972**), p. 939.

[*]In aqueous solution.

[†]In toluene solution.

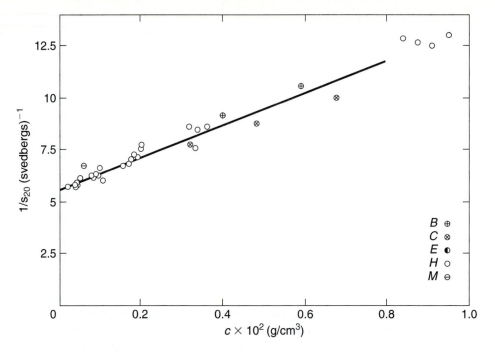

Figure 14.25 Accuracy of measurements of the sedimentation constant of polystyrene in methyl ethyl ketone at 20°C. [Reproduced from Frank, H. P. and Mark, H. F., *J. Polymer Sci.*, **1955**, *17*, 1; with permission of John Wiley & Sons, Inc., New York.] The symbols indicate results from different laboratories.

has not been used as widely for the determination of molecular weights of synthetic polymers as have the other absolute methods. On the other hand, as mentioned earlier, ultracentrifugation has been used very extensively,[1] for molecular-weight determinations of biological polymers such as proteins. Globular proteins are particularly relevant in this regard, since their very compact configurations makes their solutions less ideal than those of the usual random coil polymers, and this simplifies the extrapolations required to treat the ultracentrifugation data.

STUDY QUESTIONS

1. The densities and enthalpies of vaporization of three common solvents at 20°C are as follows:

	Toluene	Carbon Disulfide	Water
ρ (g/cm^3)	0.867	1.263	0.998
ΔH_{vap} (cal/mol)	9016	6620	10,540

[1]Freifelder, D., *Physical Biochemistry* (San Francisco: W.H. Freeman, **1976**), Chaps. 11 and 12.

Calculate the cohesive energy density of each solvent.

2. The density of a sample of poly(methyl methacrylate) is 1.18 g/cm^3. Using data available in this chapter, estimate the enthalpy change involved in dissolving 1 g of poly(methyl methacrylate) in 50 cm^3 of benzene at 20°C. What assumptions have you made in obtaining your estimate?

3. Calculate $\overline{M}_n$, $\overline{M}_w$, $\overline{M}_z$, and $\overline{M}_v(a = 0.8)$, and the ratios $\overline{M}_w/\overline{M}_n$, $\overline{M}_z/\overline{M}_n$, and $\overline{M}_v/\overline{M}_w$ for the following two distributions:

Distribution	n_i (mols)	$10^{-3}\, M_i$ (g/mol)
A	1.0	7.0
	2.0	10.0
	1.0	13.0
B	1.0	2.0
	2.0	10.0
	1.0	18.0

$M_z = \Sigma N_i M_i^3 / \Sigma N_i M_i^2$ and $M_v = (\Sigma N_i M_i^{1+a} / \Sigma N_i M_i)^{1/a}$.

Note that the second distribution covers a wider range of M, and this is reflected by a larger value of the polydispersity index $\overline{M}_w/\overline{M}_n$. Also note that $\overline{M}_v$ is closer to $\overline{M}_w$ than it is to $\overline{M}_n$.

4. A polymer having a chlorine atom at one end of each chain is analyzed and found to contain 0.250% chlorine. What is its number-average molecular weight?

5. Pure benzene at 20°C has a vapor pressure of 75.00 mmHg. If 20.0 g of polymer added to 80.0 g of benzene decreases the vapor pressure to 74.50 mmHg at this temperature, what is the number-average molecular weight of the polymer?

6. Adding 10.0 g of a polymer to 50.0 g of benzene depresses the melting point of the benzene from 5.500 to 5.400°C. If the melting point depression constant $K = 5.07$ deg kg/mol, what is the number-average molecular weight of the polymer?

7. A sample of polyisobutylene has a number-average molecular weight of 428,000 g/mol. The second virial coefficient in chlorobenzene solution at 25°C is $\Gamma = 94.5$ cm^3/g. Calculate the osmotic pressure in g/cm^2 of a 7.0×10^{-6} m solution of this polymer in chlorobenzene at 25°C. (The density of chlorobenzene at 25°C is 1.11 g/cm^3.) Compare this with the value calculated for an ideal solution.

8. Solutions of a sample of polychloroprene ($\rho = 1.25$ g/cm^3) in toluene at 30°C have been reported to have the following osmotic pressures [Sato, K., Eirich, F. R., and Mark, J. E., *J. Polymer Sci., Polymer Phys.*, **1976**, *14*, 619]:

Concn. (g/cm^3 $\times$ 10^3)	1.33	2.10	4.52	7.18	9.87
Osmotic Pressure (dyn/cm^2 $\times$ 10^{-3})	0.30	0.51	1.32	2.46	3.90

From these data determine **(a)** the number-average molecular weight of the sample; **(b)** the second virial coefficient, Γ; and **(c)** the polymer-solvent interaction constant χ_1. Give appropriate units for all calculated quantities.

9. Calculate values of the polystyrene-solvent interaction constants, χ_1, for the samples discussed in Figure 14.7 ($\rho_p = 1.050$ g/cm^3).

10. An osmotic pressure measurement of a solution of poly(vinyl chloride) ($c = 1.5 \times 10^{-3}$ g/cm^3) in toluene at 25°C($\rho = 0.867$ g/cm^3) in the apparatus shown in Figure 14.4 indicated a difference of 4.67 mm in the heights of the solution and solvent levels. (a) What is the osmotic pressure of the solution? (b) If the second virial coefficient for poly(vinyl chloride) in toluene is $\Gamma = 200$ cm^3/g, calculate the number-average molecular weight of the polymer.

11. A narrow molecular-weight fraction of poly(methyl methacrylate) gave the following osmotic pressures in acetone at 30°C:

$10^2\,c$ (g/cm^3)	π (cm of solution head)
0.275	0.457
0.338	0.592
0.334	0.609
0.486	0.867
0.896	1.756
1.006	2.098
1.119	2.710
1.536	3.725
1.604	3.978

Calculate $\overline{M}_n$ and $A_2(\text{cm}^3\text{ mol/g}^2) = \Gamma/M$ by plotting π/c vs. c. (Assume the density of the solution is the same as the density of the acetone, which is 0.780 g/cm^3 at 30°C.

12. Calculate the fraction of light of $\lambda = 546$ nm that would be scattered in a 1-cm path through a solution of poly(vinyl palmitate) in neohexane ($c = 2.0 \times 10^{-4}$ g/cm^3). The weight-average molecular weight of the polymer is 5.20×10^4 g/mol and the refractive index of the solvent is 1.3688.

13. Derive an expression for the turbidity of a gas analogous to that given in equation (37) for a solution. Use your expression to calculate the fraction of light of $\lambda = 546$ nm that is scattered by 1 cm of SO_2 at 1 atm pressure and 25°C. The refractive index of SO_2 is 1.000665. Compare your result with the fraction of light scattered by the polymer solution in Problem 12.

14. In a study of light scattering ($\lambda = 436$ nm) from toluene solutions of a polystyrene sample [D. Rahlwes and R. G. Kirste, *Makromol. Chem.*, **178**, 1793 (1977)], the following results were obtained for the Rayleigh ratio, $R(\theta)$, at various concentrations and scattering angles:

$c \times 10^3$ (g/cm^3)	$R(\theta) \times 10^4(\text{cm}^{-1})$					
	15°	45°	75°	105°	135°	150°
0.20	2.47	1.53	0.860	0.560	0.428	0.395
0.40	4.40	2.84	1.65	1.09	0.839	0.775
0.60	5.91	3.96	2.37	1.59	1.23	1.14
0.80	7.07	4.91	3.02	2.05	1.60	1.49
1.00	7.98	5.71	3.61	2.49	1.96	1.82

Construct a Zimm plot of the data and evaluate the weight-average molecular weight of the sample. Determine also the second virial coefficient Γ. (The refractive index of toluene is 1.4976 and the specific refractive index increment of polystyrene-toluene solutions is 0.1121.)

15. A sample of poly(α-methylstyrene) having a weight-average molecular weight of 4.10×10^6 g/mol behaves as a random coil in toluene solutions with a root-mean-square end-to-end separation of 2609 Å. The second virial coefficient of the solution is $\Gamma = 709$ cm^3/g, the refractive index of toluene is 1.4976, and the specific refractive index increment of the solution is 0.1370 cm^3/g. Use the appropriate scattering function and any other expressions in this chapter to calculate the Rayleigh ratio, $R(\theta)$, as a function of scattering angle in the range 15 to 150° for a solution with $c = 1.0 \times 10^{-4}$ g/cm^3. Plot the Rayleigh ratios on polar coordinates to show the dissymmetry of the scattering. What is the dissymmetry coefficient Z_{45}?

16. Using anionic polymerization techniques, D. Rahlwes and R. G. Kirste [*Makromol. Chem.*, **178**, 1793 (1977)] produced highly uniform styrene/α-methyl styrene block copolymers. The results of light-scattering studies for $\lambda = 436$ nm on one such sample in toluene ($n_0 = 1.4976$) were as follows:

$c \times 10^3$(g/cm^3)	$R(\theta) \times 10^4$(cm^{-1})					
	15°	45°	75°	105°	135°	150°
0.20	1.91	1.47	1.01	0.725	0.577	0.537
0.40	3.55	2.78	1.95	1.41	1.13	1.05
0.60	4.95	3.94	2.80	2.05	1.65	1.54
0.80	6.16	4.96	3.59	2.66	2.15	2.01
1.00	7.19	5.87	4.31	3.22	2.62	2.46

The specific refractive index increment of the block copolymer in toluene was 0.1263 cm^3/g.

(a) Construct a Zimm plot and determine the weight-average molecular weight of the polymer.

(b) Evaluate the dissymmetry coefficient Z_{45} and determine the root-mean-square end-to-end length of the polymer in toluene solution.

(c) Evaluate the second virial coefficient.

(d) Calculate the turbidity of the solution of $c = 0.2$ g/cm^3 and the fraction of light scattered by this solution.

17. Data for a fraction of cellulose acetate in acetone, given by Benoit, Holtzer, and Doty [*J. Phys. Chem.*, **1954**, *58*, 635] is tabulated below:

	VALUES OF 10^7 Kc_2/R_θ		
$10^3 c_2$, g/cm^3	$\theta = 30°$	90°	135°
0.86	19.2	49.8	74.0
0.43	16.2	46.5	70.5

For acetone, $n_0 = 1.36$, and the Hg arc used gave a filtered beam of $\lambda = 5461$ Å (mercury green line).

Make a Zimm plot using 2000 as the constant coefficient multiplying c. Determine the weight-average molecular weight, second virial coefficient A_2(cm^3 mol/g^2), and rootmean-square radius of gyration (Å).

18. A certain monodisperse polymer has a molecular weight of 10^6 g/mol. Assuming that the polymer may be considered to be a sphere of diameter 70 Å in methyl ethyl ketone, calculate the sedimentation velocity under the influence of gravity at 20°C. The viscosity of methyl ethyl ketone at this temperature is 4.284 millipoises. The density of MEK = 0.805 g/cm^3.

19. An ultracentrifuge in which the centrifugation cell is located at a distance of 5.0 cm from the center of rotation operates at a maximum speed of 65,000 rpm. **(a)** Calculate the ratio of the centrifugal force generated to the force due to gravity. **(b)** Calculate the sedimentation velocity of the spherical polymer in Problem 18 in methyl ethyl ketone at 20°C that would be observed with this ultracentrifuge.

20. A solution of polystyrene in toluene with $c = 5 \times 10^{-2}$ g/cm^3 is found by a pycnometric measurement to have a density of 0.8855 g/cm^3 at 25°C as compared with a density of 0.8788 found by the same technique with pure toluene. The sedimentation coefficient and diffusion coefficient at infinite dilution were found to be 17.9×10^{-13} s and 3.6×10^{-7} cm^2/s, respectively. **(a)** Calculate the average molecular weight of the polystyrene. **(b)** If the ultracentrifuge is that described in Problem 19 and is operating at its maximum speed, calculate the sedimentation velocity of the polymer sample.

21. The diffusion coefficient of hemoglobin in aqueous solution at 20°C is 6.9×10^{-7} cm^2/s and the partial specific volume of the polymer is 0.749 cm^3/g. If the protein molecules are assumed to be spherical, calculate the molecular weight. (The viscosity of water at 20°C is 10.05 millipoise.)

22. G. Meyerhoff [*Z. Physik. Chem.*, **4**, 336 (1955)] used ultracentrifugation techniques to measure the average molecular weight of a sample of polystyrene in toluene at 20°C. The partial specific volume of the polymer was found to be 0.91 cm^3/g and the sedimentation and diffusion coefficients measured were as follows:

$c \times 10^3$(g/cm^3)	4.3	2.2	1.1	0.6	0.4
$S \times 10^{13}$(s)	4.31	5.69	6.72	7.60	7.88

$c \times 10^3$(g/cm^3)	7.5	5.0	2.5
$D \times 10^7$(cm^2/s)	2.36	2.17	2.07

From these data determine: **(a)** the average molecular weight of the polymer, and **(b)** the second virial coefficient of the polymer in toluene solution.

23. The following results have been reported for the sedimentation coefficients of a polystyrene sample in cyclohexane at 35°C [W. J. Closs, B. R. Jennings, and H.G. Jerrard, *Eur. Polymer J.*, **4**, 639 (1968)]:

$c \times 10^3$(g/cm^3)	2.0	3.0	4.0	5.0	6.0	7.0
$s \times 10^{13}$(s)	14.8	13.9	13.1	12.4	11.8	11.2

The density of cyclohexane at 35°C is 0.765 g/cm^3 and the partial specific volume of polystyrene is 0.93 cm^3/g. The dependence of the diffusion constant of polystyrene in cyclohexane at 35°C on weight-average molecular weight has been shown to be $D_0 = 1.3 \times 10^{-4} M_w^{-0.497}$ cm^2/s. [T. A. King, A. Knox, W. I. Lee, and J. D. G. McAdam,

Polymer, **14**, 151 (1973)]. From these data determine the average molecular weight of the polystyrene sample.

SUGGESTIONS FOR FURTHER READING

BILLINGHAM, N. C., *Molar Mass Measurements in Polymer Science*. New York: Wiley, **1977**.

BILLMEYER, F. W., JR., *Textbook of Polymer Science*, 3rd ed. New York: Wiley, **1984**, Chaps. 1, 8.

CASASSA, E. F., and BERRY, G. C., in *Polymer Molecular Weights* (P. E. Slade, Jr., Ltd.). New York: Dekker, **1975**, Chap. 5.

YUAN, C. J., *Determination of Molecular Weights of High Polymers*. Jerusalem: Israel Program for Scientific Translations, **1963**.

COLLINS, E. A., BARES, J., and BILLMEYER, F. W., Jr., *Experiments in Polymer Science*. New York: Wiley-Interscience, **1973**.

FLORY, P. J., *Principles of Polymer Chemistry*. Ithaca, N.Y.: Cornell University Press, **1953**, Chap. 7.

HANTON, S. D., "Mass Spectrometry of Polymers and Polymer Surfaces," *Chem. Rev.*, **2001**, *101*, 527.

HIEMENZ, P. C., *Polymer Chemistry*. New York: Dekker, **1984**, Chaps. 1, 8, 9, 10.

HIGGINS, J. S., and BENOIT, H., *Neutron Scattering from Polymers*. Oxford: Clarendon Press, **1994**.

HUGLIN, M. B. (Ed.), *Light Scattering from Polymer Solutions*. New York: Academic Press, **1972**.

LIMBACH, P. A., "Matrix-Assisted Laser Desorption-Ionization Mass Spectrometry: An Overview," *Spectroscopy*, **1998**, *13*(10), 16.

McCAFFERY, E. L., *Laboratory Preparation for Macromolecular Chemistry*. New York: McGraw-Hill, **1970**.

McINTYRE D., and GORNICK, F. (Eds.), *Light Scattering from Dilute Polymer Solutions*. New York: Gordon and Breach, **1964**.

RAFIHOV, S. R., PAVLOVA, S. A., and TVERDOKHLEBOVA, I. I., *Determination of Molecular Weights and Polydispersity of High Polymers*. Jerusalem: Israel Program for Scientific Translations, **1964**.

ROE, R.-J., *Methods of X-Ray and Neutron Scattering in Polymer Science*, Oxford: Oxford University Press, **2000**.

SANDLER, S. R., KARO, W. BONESTEEL, J.-A., and PEARCE, E. M., *Polymer Synthesis and Characterization. A Laboratory Manual*. San Diego: Academic Press, **1998**.

SEYMOUR, R. B., and CARRAHER, JR., C. E., *Polymer Chemistry*, 2nd ed. New York: Dekker, **1988**, Chap. 4.

SPERLING, L. H., *Introduction to Physical Polymer Science*, 3rd ed. New York: Wiley Interscience, **2001**.

STACEY, K. A., *Light Scattering in Physical Chemistry*. New York: Academic Press, **1956**.

SUN, S. F., *Physical Chemistry of Macromolecules. Basic Principles and Issues.* New York: Wiley Interscience, **1994**.

SVEDBERG, T., and PEDERSEN, K. O. *The Ultracentrifuge.* London: Oxford University Press, **1940**.

TANFORD, C., *Physical Chemistry of Macromolecules.* New York: Wiley, **1961**, Chaps. 4–6.

TOMPA, H., *Polymer Solutions.* New York: Academic Press, **1956**, Chaps. 6, 10.

VOLLMERT, B., *Grundriss der makromolekularen Chemie.* Berlin: Springer-Verlag, **1962**, Chap. 3.

WILLIAMS, J. W., *Ultracentrifugation of Macromolecules.* New York: Academic Press, **1972**.

WIGNALL, G. D., "Neutron and X-Ray Scattering," in *Physical Properties of Polymers Handbook*, (J. E. Mark, Ed.), New York: Springer-Verlag New York, Inc., **1996**, p. 299.

15

Secondary Methods for Molecular-Weight Determination and Molecular-Weight Distributions

INTRODUCTION

The absolute methods for the determination of molecular weights that were discussed in Chapter 14 are well established both theoretically and experimentally. Unfortunately, the absolute measurements are difficult to carry out, are time-consuming, and often require expensive apparatus. For these reasons, most molecular-weight determinations are routinely carried out by the much faster methods of solution viscosity, vapor-phase osmometry, and gel permeation chromatography. However, these techniques are not absolute methods, and their use requires a prior determination of empirical relationships that relate the molecular weight to the viscosity of a polymer solution or to the retention volume of a polymer solution being eluted from a gel permeation column. Once the calibration has been accomplished, the secondary methods provide the polymer chemist with a fast, simple, and accurate way to obtain molecular weights. Gel permeation chromatography is also used to obtain information on molecular-weight distributions, largely replacing the earlier techniques of fractional precipitation and gradient elution.

SOLUTION VISCOSITY

Solution Viscosity and Molecular Size

In the early days of polymer chemistry, Staudinger[1] observed that even a low concentration of a dissolved polymer markedly increases the viscosity of a solution relative to that of the pure solvent. This increase in viscosity is caused principally by the unusual size and shape of the dissolved polymer and by the nature of solutions of high polymers.

Most polymer molecules are best described not as long thin rods but as random statistical coils. In dilute solution these coils are free from entanglement with other coils but are completely solvated, which means they have taken up as much solvent as they can hold. Thus, the smallest entities of solute in a polymer solution are not the actual polymer molecules but rather the large, irregularly shaped "particles" made up of polymer coils and large numbers of absorbed solvent molecules. As far as motion through the solution is concerned, the polymer coil and absorbed solvent form a single entity which is actually much heavier than the polymer molecule itself. In many respects, each polymer "particle" resembles a completely saturated sponge. On the basis of the size of these solute "particles," polymer solutions are correctly classified as *colloidal dispersions*. Each colloidal particle is a solvent-filled polymer coil; hence, they are sometimes called *molecular colloids*.

For a long time it has been known that the large particles in colloidal solutions or dispersions tend to impede the flow of adjacent layers of liquid when the liquid is subjected to a shearing force. In other words, the viscosity of the liquid is increased relative to that of the pure solvent by the presence of a colloidal or polymeric solute. As long ago as 1906, it was shown by Einstein[2] that, in the case of spherical colloid particles, the relative viscosity is given by the expression

$$\eta_r = \frac{\eta}{\eta_0} = 1 + 2.5\phi_2 \tag{1}$$

where η_r is the relative viscosity, η the viscosity of the solution, η_0 the viscosity of the pure solvent, and ϕ_2 the volume fraction of the colloidal particle. According to (1), as the overall size or volume of the colloidal particle (i.e., polymer molecule plus imbibed solvent) increases, so do the volume fraction ϕ_2 and the relative viscosity. Because the molecular weight of a polymer molecule also increases with size, it is possible to relate the increase in solution viscosity to the molecular weight.

Measurement of Viscosity

Principles According to Newton's law of viscous flow, the frictional force, *F*, that resists the flow of any two adjacent layers of liquid is given by

[1]H. Staudinger, *Kolloid-Z.*, **1930**, *51*, 71.
[2]Einstein, A., *Ann. Physik*, **1906**, *19*, 289.

$$F = \eta A \frac{dv}{dx} \qquad (2)$$

where A is the area of contact of the layers, dv/dx the velocity gradient between them, and the proportionality constant, η, is called the coefficient of viscosity or, simply, the viscosity. The unit of viscosity is the poise (i.e., 1 poise = 1 g-cm^{-1}-s^{-1}).

When an external driving force is applied to overcome the frictional resistance and cause the liquid to flow uniformly through a tube, the rate of flow is given by Poiseuille's law,[1]

$$\frac{dV}{dt} = \frac{\pi R^4 \Delta P}{8 \eta L} \qquad (3)$$

In (3), dV/dt is the volume of liquid that flows through the tube per unit time; R and L are the radius and length of the tube, respectively; and ΔP is the difference in external pressure between the ends of the tube. In practice, measurements are usually carried out in viscometer tubes in which the capillary is in a vertical position and the driving force is simply the weight of the liquid itself. Therefore, the pressure difference, ΔP, which is the driving force per unit area, is given by

$$\Delta P = h\rho g \qquad (4)$$

where h is the average height of the liquid during measurement, ρ the density, and g the acceleration due to gravity. Substitution of (4) into (3), along with the assumption of a constant flow rate, yields equation (5) for the viscosity.

$$\eta = \frac{\pi R^4 h g \rho t}{8LV} \qquad (5)$$

The applicability of (5) demands that the flow be "Newtonian" or "viscous." This will be true provided that a dimensionless quantity, called the Reynolds number, is less than 1000. In terms of the variables of (5) this condition is given by

$$\frac{2V\rho}{\pi R \eta t} < 1000 \qquad (6)$$

and is readily satisfied for the apparatus and liquids usually used for measurements of the viscosities of polymer solutions. However, in addition to the re-

[1]Moore, W. J., *Physical Chemistry*, 4th ed. (Englewood Cliffs, N.J.: Prentice-Hall, **1972**), pp. 153ff.

quirements of viscous flow, the derivation of (5) relies on the following assumptions:

1. All of the potential energy of the driving force is expended in overcoming the frictional resistance. This is not strictly true, since some energy must be expended to accelerate the liquid in the tube. When this "kinetic-energy correction" is made, equation (5) becomes

$$\eta = \frac{\pi R^4 h g \rho t}{8LV} - \frac{\rho V}{8\pi L t} \qquad (7)$$

For measurements of the *absolute* viscosity, this correction term can amount to 10 to 15% but, for measurements of the *relative* viscosities of interest in polymer chemistry (see equation 1), the error introduced by the use of (5) rather than (7) is usually less than 2%.

2. The second assumption is that the velocity of the liquid at the walls of the capillary is zero (i.e., there is no "slippage" of the liquid along the walls). This assumption is usually valid for liquids that "wet" the capillary walls. In any event, viscosities of polymer solutions are measured relative to the pure solvent and, unless the presence of small concentrations of polymer markedly affect the surface tension of the solvent, such capillary effects may be neglected.

Although expressions (5) and (7) are usually used, in practice it is not necessary to make precise measurements of the viscometer tube dimensions. Thus, if (7) is written in terms of the *kinematic viscosity*, η/ρ, we have

$$\frac{\eta}{\rho} = \alpha t - \frac{\beta}{t} \qquad (8)$$

where the constants, $\alpha = \pi R^4 hg/8LV$ and $\beta = V/8\pi L$, depend only on the geometry of the viscometer tube. Measurement of the times required for the fixed volume V of two liquids of known viscosity and density to flow through *the same* tube is sufficient to define the viscometer constants, α and β. The viscosity of a liquid depends markedly on the temperature. Hence, the calibration measurements and the measurement of the viscosities of the polymer solutions must be made at the same carefully controlled temperature ($\pm 0.1°C$).

Since only ratios of viscosities will be relevant here, it is possible to divide both sides of equation (8) by α, giving a corrected efflux time equal to $t - (\beta/\alpha)/t$. Since β/α is typically the order of 100, the correction can be avoided entirely by choosing a viscometer giving efflux times greater than 100 s, which will make the $(\beta/\alpha)/t$ corrections less than a tolerable 1%.

The viscosities and densities of some typical solvents used in polymer chemistry are given in Table 15.1. It should be noted from (8) that the viscosities ob-

TABLE 15.1 DENSITIES AND VISCOSITIES OF COMMON SOLVENTS

Solvent	$\rho(g/cm^3)$ at:		100η (poises) at:	
	20°C	30°C	20°C	30°C
Benzene	0.8737	0.8684	0.652	0.564
Toluene	0.8669	0.8577	0.590	0.526
p-Xylene	0.8610	0.8523	0.644	0.568
Cyclohexane	0.7786	0.7693	0.935	0.820
n-Hexane	0.6594	0.6505	0.326	0.293
Ethanol	0.7893	0.7808	1.200	1.003
Acetone	0.7908	0.7793	0.326	0.295
Methyl ethyl ketone	0.8047	0.7945	0.400	0.365
Carbon tetrachloride	1.5940	1.5748	0.969	0.843
Chloroform	1.4892	1.4706	0.568	0.514

tained from viscometer tube flow times are *kinematic* viscosities. The conversion of relative kinematic viscosity to relative viscosity can usually be made by assuming that the densities of solution and solvent are equal.

Experimental apparatus   A number of methods exist for the determination of the viscosity of a liquid. The most useful method from the viewpoint of simplicity, accuracy, and cost is based on a measurement of the flow rate of the liquid through a capillary tube. In practice, the capillary tube forms part of the "viscometer."

The most commonly used viscometers are of the Ostwald and Ubbelohde typer[1] shown in Figure 15.1. In the Ostwald viscometer, a given volume of liquid is introduced into B and is drawn up by suction into A until the liquid level is above the mark m_1. The suction is released and the time required for the liquid level to fall from m_1 to m_2 is measured. The average driving force during the flow of this volume of liquid through the capillary tube is proportional to the average difference in heights of the liquids in tubes B and A (i.e., proportional to h, as shown in Figure 15.1a). In order that this driving force is the same in all cases, it is clearly essential that the same amount of liquid should always be introduced into tube B.

This requirement that the same amount of liquid should always be used does not apply in the case of the Ubbelohde viscometer shown in Figure 15.1b. Here, the liquid is introduced into B. With tube 3 closed, the liquid is drawn up by suction into A so that the liquid level is above mark m_1. The suction is released and, before the liquid level in tube 2 reaches the mark m_1 tube 3 is opened to the air. Bulb C fills with air and the liquid flowing out of bulb A must do so along the walls of bulb C. In this case, the driving force for flow through the capillary is independent of the level of the liquid in B, since the average height, h, is always the same.

[1]Ubbelohde, L., *Ind. Eng. Chem., Anal. Ed.*, **1937**, *9*, 85.

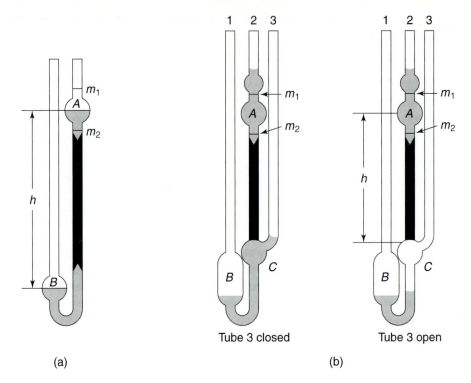

Tube 3 closed   Tube 3 open

(a)   (b)

Figure 15.1 Viscometers commonly used in polymer chemistry: (a) Ostwald viscometer; (b) Ubbelohde viscometer. [Reproduced from Ubbelohde, L., *Ind. Eng. Chem., Anal. Ed.*, **1937**, *9*, 85; with permission of the American Chemical Society, Washington, D.C.]

Typical flow times and relative viscosities for chloroform solutions of two fractions of poly(methyl methacrylate), as determined in an Ostwald viscometer, are shown in Table 15.2. The kinetic-energy correction was not made in this determination. The following sections illustrate how such data may be related to the average molecular weight of the polymer.

Definition of Solution–Viscosity Terms

The *relative viscosity,* which has already been defined in (1), may be written very simply in terms of the viscometer flow times if the kinetic energy correction is neglected:

$$\eta_r = \frac{\eta}{\eta_0} = \frac{t}{t_0} \tag{9}$$

where t and t_0 are for the flow of solution and solvent, respectively. Obviously, η and η_0 (i.e., t and t_0) must be measured under the same conditions. The relative viscosity is always greater than unity because the presence of the polymeric solute always increases the viscosity.

TABLE 15.2 VISCOSITY OF SOLUTIONS OF POLY(METHYL METHACRYLATE) IN CHLOROFORM AT 20°C

Fraction	Concentration $(g/cm^3) \times 10^2$	Flow Time* (s)	η_r (Eq. 1)	$\dfrac{\eta_{sp}}{c}$ (Eq. 10)
1	0.0000	170.1	1.000	—
	0.03535	178.1	1.047	133
	0.05152	182.0	1.070	136
	0.06484	185.2	1.089	137
	0.100	194.3	1.142	142
	0.200	219.8	1.292	146
	0.400	275.6	1.620	155
2	0.02242	180.8	1.063	281
	0.03520	187.3	1.101	287
	0.04620	192.7	1.133	288
	0.08682	214.2	1.259	298
	0.18806	273.0	1.605	322

Source: Schulz, G. V. and Blaschke, F., *J. Prakt. Chem.*, **1941**, *158*, 130.

* Ostwald viscometer, $R = 1.5 \times 10^{-2}$ cm, $L = 11$ cm.

The best range experimentally is $\eta_r = 1.2 - 1.8$, because less than a 20% increase in viscosity is too difficult to measure reliably, and more than 80% could cause curvature in some of the extrapolations to infinite dilution that will be described.

It is appropriate, then, to define the *specific viscosity*, η_{sp}, as the fractional increase in viscosity caused by the presence of the dissolved polymer in the solvent, as shown in equation (10).

$$\eta_{sp} = \frac{\eta - \eta_0}{\eta_0} = \eta_r - 1 \tag{10}$$

The specific viscosity and the relative viscosity clearly depend on the concentration of the polymer in solution; they increase in magnitude with increasing concentration. This may be seen in Table 15.2 for solutions of poly(methyl methacrylate) in chloroform. The quantity η_{sp}/c, where c is the concentration of polymer in g/cm³, is sometimes called the *reduced viscosity* or *reduced specific viscosity* and is a measure of the specific capacity of the polymer to increase the relative viscosity. Finally, the *intrinsic viscosity*, $[\eta]$, is defined as the limit of the reduced viscosity as the concentration approaches zero, and is given by

$$[\eta] = \lim_{c \to 0} \left(\frac{\eta_{sp}}{c} \right) \tag{11}$$

Note that none of the terms defined here actually has the dimensions of viscosity. The relative viscosity and the specific viscosity are dimensionless, but the reduced viscosity and the intrinsic viscosity have the dimensions of a specific volume (i.e., cm^3/g).

A linear dependence of the reduced viscosity on polymer concentration is usually found when $\eta_r < 2$. This linear dependence is described well by the expression

$$\frac{\eta_{sp}}{c} = [\eta] + k'[\eta]^2 c \tag{12}$$

where k' is a constant, usually in the range 0.35 to 0.40; it is sometimes called the Huggins constant.[1] In view of equations (11) and (12), it is evident that the intrinsic viscosity $[\eta]$ can be found by an extrapolation of the experimental values of the reduced viscosity (η_{sp}/c) to zero concentration. An alternative extrapolation replaces η_{sp}/c by in η_r/c, and k' by k''.

Examples of these viscosity terms are illustrated by the typical data for two fractions of poly(methyl methacrylate) in chloroform that are shown in Table 15.2. In addition, plots of η_{sp}/c as a function of concentration are shown in Figure 15.2. The intrinsic viscosities of the polymer samples are given by the intercepts of these plots, in accordance with (11).

Intrinsic Viscosity and Molecular Weight

Suppose that the intrinsic viscosities $[\eta]_i$ (defined by equation 11) are determined for different molecular-weight fractions of a given polymer, each fraction having a very narrow range of molecular weights. Assume that the molecular weights, M_i, of the various fractions are known from the use of an absolute method such as ultracentrifugation or light scattering. It has been found that a straight line is obtained if the logarithms of the intrinsic viscosities are plotted versus the logarithms of the molecular weights of the different fractions. Such plots for fractions of polyisobutylene in cyclohexane[2] at 30°C and for polystyrene in cyclohexane[3] at 35°C and in butanone[4] at 22°C are shown in Figure 15.3. It may be seen from this figure that a linear relationship does exist between log $[\eta]_i$ and log M_i over the useful range $10^4 < M < 10^6$. The slopes of the lines in Figure 15.3 are: polyisobutylene in cyclohexane, 0.69; polystyrene in cyclohexane, 0.50; polystyrene in butanone, 0.58. Such nonintegral values of the slopes are typical of log $[\eta]$ versus log [M] plots, with the values for all polymer-solvent combinations falling in the range 0.5 to 1.0. It is clear from the data in Figure 15.3 that, for a given polymer, the slope depends on the *solvent*. It is also found that, for a given polymer and solvent, the slope depends on the *temperature*.

[1]Huggins, M. L., *J. Am. Chem. Soc.*, **1942**, *64*, 2716.
[2]Krigbaum, W. R., and Flory, P. J., *J. Polymer Sci.*, **1953**, *11*, 37.
[3]Outer, P., Carr, C. I., and B. H., Zimm, *J. Chem. Phys.*, **1950**, *18*, 830.
[4]Cantow, H. J., *Makromol. Chem.*, **1959**, *30*, 169.

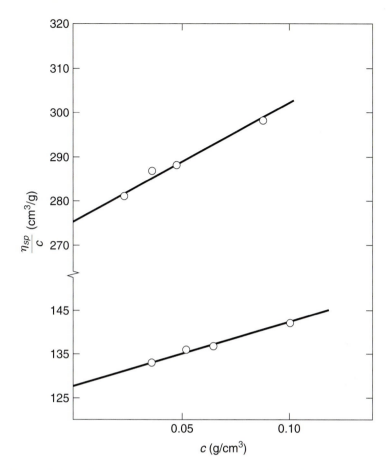

Figure 15.2 Reduced viscosities of two samples of poly(methyl methacrylate) in chloroform at 20°C as a function of concentration. [From Schulz, G. V. and Blaschke, F., *J. Prakt. Chem.*, **1941,** *158,* 130.]

Since a plot of log $[\eta]_i$ versus log M_i is linear for narrow molecular-weight fractions of a given polymer, we may write

$$\log [\eta]_i = \log K + a \log M_i \tag{13}$$

or

$$[\eta]_i = KM_i^a \tag{14}$$

where K and a are constants that are easily determined from calibration plots such as in Figure 15.3. The relationship given in (14) is usually known as the Mark–Houwink[1,2] equation.

[1]Mark, H., *Z. Elektrochem.*, **1934,** *40,* 499.
[2]Houwink, R., *J. Prakt. Chem.*, **1940,** *157,* 15.

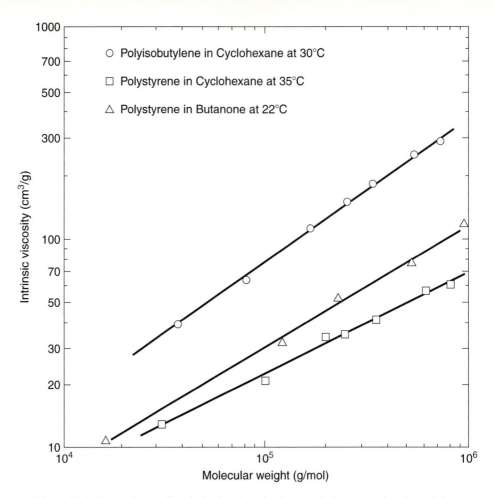

Figure 15.3 Dependence of intrinsic viscosity of polymer solutions on molecular weight.

It must be kept in mind that the data in Figure 15.3 and, hence, equations (13) and (14), refer to *fractionated* samples of a given polymer, in which molecular-weight ranges of the fractions are very small. Carefully fractionated samples must be used in order to obtain numerical values for K and a for a given polymer-solvent pair at a given temperature. However, in practice, solution viscosity measurements are used to obtain, quickly and easily, a measure of the molecular weight of an unfractionated or crudely fractionated polymer. An *average* molecular weight of the sample is obtained by this procedure and it is necessary to inquire about the type of average that is involved.

Polymer molecules of a given molecular weight, M_i, will contribute an amount $(\eta_{sp})_i$ to the total observed specific viscosity (see equation 10). The observed specific viscosity of the solution of unfractionated polymer is then obtained as a sum over all the molecular weights (i.e., all the polymer molecules) present in the solution. Thus, equation (15) applies.

$$\eta_{sp} = \sum_i (\eta_{sp})_i \tag{15}$$

According to the Huggins equation, (12), for those molecules in each narrow molecular-weight range, i, whose concentration is c_i, we may write

$$(\eta_{sp})_i = [\eta]_i c_i + k'_i [\eta]_i^2 c_i^2 \tag{16}$$

In the limit of infinite dilution, we can neglect the second term on the right-hand side of (16). Combining (14), (15), and (16) results in the expressions shown in (17) and (18).

$$(\eta_{sp})_i = [\eta]_i c_i = K M_i^a c_i \tag{17}$$

$$\eta_{sp} = K \sum_i M_i^a c_i \tag{18}$$

Equation (18) describes the specific viscosity of an infinitely dilute solution of an unfractionated sample.

Because the extrapolation to infinite dilution has, in effect, already been carried out by the neglect of the second term in (16), the observed intrinsic viscosity is obtained simply by dividing through (18) by c, the total concentration of polymer in the solution. Thus,

$$[\eta] = \lim_{c \to 0} \frac{\eta_{sp}}{c} = K \sum_i M_i^a \left(\frac{c_i}{c} \right) \tag{19}$$

The concentration ratio $c_i/c = c_i / \sum_i c_i$ is equal to the weight ratio $w_i/w = w_i / \sum_i w_i$. Hence, finally we may write expression (20) for the intrinsic viscosity of an unfractionated polymer,

$$[\eta] = K \sum_i w_i M_i^a \bigg/ \sum_i w_i = K \sum_i W_i M_i^a \tag{20}$$

where the term $W_i = w_i / \sum_i w_i$ is the *weight fraction* of polymer of molecular weight M_i. Now, if we define the *viscosity-average* molecular weight by the relationship

$$\overline{M}_v = \left(\sum_i W_i M_i^a \right)^{1/a} = \left(\sum_i w_i M_i^a \bigg/ \sum_i w_i \right)^{1/a} \tag{21}$$

then, for an unfractionated polymer, the intrinsic viscosity/molecular weight relationship (see equation 14) may be written

$$[\eta] = K(\overline{M}_v)^a \tag{22}$$

Therefore, use of the solution viscosity technique yields the *viscosity-average* molecular weight defined by (21). To see how this particular molecular weight relates to the number-average and weight-average molecular weights, consider the general expression for an average molecular weight, defined in terms of the index β, and shown in (23).

$$\overline{M}_\beta = \sum_i w_i M_i^\beta \bigg/ \sum_i w_i M_i^{\beta-1} \tag{23}$$

It is easy to show from (23) that the number-average molecular weight corresponds to $\beta = 0$, and the weight-average molecular weight corresponds to $\beta = 1$. On the other hand, in terms of the Mark-Houwink equation, (21), the weight-average molecular weight corresponds to $a = 1$, and the number-average molecular weight to $a = -1$. Thus, in terms of the definitions (21) and (23), we have for the number-average and weight-average molecular weights,

$$\overline{M}_n = \overline{M}_v(a = -1) = \overline{M}_\beta(\beta = 0) \tag{24}$$

$$\overline{M}_w = \overline{M}_v(a = 1) = \overline{M}_\beta(\beta = 1) \tag{25}$$

The range of the empirical constant, a, is found experimentally to lie in the range 0.5 to 1.0. Because this constant is never as low as -1, the viscosity-average molecular weight *cannot* correspond to the number-average molecular weight. The relationship between β and a depends on the molecular weight distribution in a complicated manner,[1] but it has been shown to be such that $0.5 < a < 1.0$ corresponds to $0.75 < \beta < 1.0$.

In view of the discussion above, we may conclude that *the viscosity-average molecular weight lies between $\overline{M}_n$ and $\overline{M}_w$ but closer to $\overline{M}_w$*. Furthermore, in the special case of $a = 1$, $\overline{M}_v = \overline{M}_w$, but because $a \neq 1$, $\overline{M}_v$ can never become identical to $\overline{M}_w$.

In principle, any of the absolute methods of molecular-weight determination may be used to establish the calibration plot and determine K and a. However, for the calibration to be strictly valid, the molecular-weight ranges of the various fractions must be very small.

[1] Meyerhoff, G., *Fortschr. Hochpolymer-Forsch.*, **1961**, *3*, 59.

In practice, incompletely fractionated polymers must normally be studied, so the molecular-weight ranges of the various fractions may not be as small as one would wish. Therefore, to obtain a clear relationship between $[\eta]$ and M, absolute methods in which a and β correspond as closely as possible should be used. The number-average molecular weight corresponds to $\beta = 0$, while the weight-average molecular weight corresponds to $\beta = 1$. Because the *physically real* values of the empirical constant a correspond to $0.75 < \beta < 1$, it is preferable to use an absolute method that yields the weight-average molecular weight for calibration. The reader should bear in mind that even this is usually only an approximation, because only in the rare case of $a = \beta = 1$ is $\overline{M}_v = \overline{M}_w$. However, the approximation is better when absolute methods that yield $\overline{M}_w$ rather than those that yield $\overline{M}_n$ are used. For this reason most determinations of K and a are carried out with the use of light-scattering and the various ultracentrifugation techniques, rather than with osmotic pressure or other colligative property measurements.

Values of K and a for a number of polymer-solvent systems are shown in Table 15.3. For more extensive tables, the reader should consult the literature.[1]

TABLE 15.3 INTRINSIC VISCOSITY/MOLECULAR-WEIGHT CONSTANTS $[\eta] = KM^a$

Polymer	Solvent	Temperature (°C)	Molecular-Weight Range	$K \times 10^2$	a
Amylose	Dimethylsulfoxide	25	$1.5 \times 10^3 - 1.2 \times 10^6$	0.850	0.76
Gelatin	Water	35	$3 \times 10^4 - 2.1 \times 10^5$	0.166	0.885
Natural rubber	Toluene	25	$4 \times 10^4 - 1.5 \times 10^6$	5.0	0.67
Polyacrylontrile	Dimethylformamide	25	$3 \times 10^4 - 3.7 \times 10^5$	2.33	0.75
Poly(p-bromostyrene)	Benzene	20	$3 \times 10^4 - 3 \times 10^5$	9.4	0.53
Polybutadiene	Cyclohexane	20	$2.3 \times 10^5 - 1.3 \times 10^6$	3.6	0.70
Poly(dimethylsiloxane)	Toluene	25	$3.6 \times 10^4 - 1.1 \times 10^6$	0.738	0.72
Polyisobutylene	Cyclohexane	30	$5 \times 10^2 - 3.2 \times 10^6$	2.88	0.69
	Diisobutylene	20	$5 \times 10^2 - 3.2 \times 10^6$	3.63	0.64
	Toluene	25	$1.4 \times 10^5 - 3.4 \times 10^5$	8.70	0.56
Poly(methyl methacrylate)	Acetone	25	$8 \times 10^4 - 1.4 \times 10^6$	0.75	0.70
	Chloroform	25	$8 \times 10^4 - 1.4 \times 10^6$	0.48	0.80
	Methyl ethyl ketone	25	$8 \times 10^4 - 1.4 \times 10^6$	0.68	0.72
Polypropylene	Benzene	25	$1 \times 10^3 - 7 \times 10^4$	9.64	0.73
	Cyclohexane	25	$1 \times 10^3 - 7 \times 10^4$	7.93	0.81
Polystyrene	Benzene	20	$1.2 \times 10^3 - 1.4 \times 10^5$	1.23	0.72
	Methyl ethyl ketone	20–40	$8 \times 10^3 - 4 \times 10^6$	3.82	0.58
	Toluene	20–30	$2 \times 10^4 - 2 \times 10^6$	1.05	0.72
Poly(vinyl acetate)	Acetone	30	$2.7 \times 10^4 - 1.3 \times 10^6$	1.02	0.72
	Methanol	30	$2.7 \times 10^4 - 1.3 \times 10^6$	3.14	0.60
Poly(vinyl alcohol)	Water	25	$8.5 \times 10^3 - 1.7 \times 10^5$	30.0	0.50
Poly(vinyl bromide)	Cyclohexanone	20	$1.9 \times 10^4 - 1.0 \times 10^5$	3.28	0.55

[1]*Physical Properties of Polymers Handbook*, Mark, J. E., Ed. (New York: Springer-Verlag, 1996); *Polymer Handbook*, 4th ed., Brandrup, J., Immergut, E. H., and Grulke, E. A., Eds. (New York: Wiley, **1999**); *Polymer Data Handbook*, Mark, J. E., Ed. (New York: Oxford University Press, **1999**).

Molecular Size from Intrinsic Viscosity

It has been mentioned previously that the intrinsic viscosity has the dimensions of a specific volume—namely, volume per unit mass. This agrees with a viscosity relationship derived by Einstein in 1906,[1] for the case of a solution of rigid spherical particles. This relationship, (1), may be expressed in terms of intrinsic viscosity as

$$[\eta] = 2.5 \frac{N_0 V_e}{M} = 2.5 \frac{N_0(\frac{4}{3}\pi R_e^3)}{M} \tag{26}$$

where N_0 is Avogadro's number, M is the molecular weight, and V_e and R_e are the volume and the radius, respectively, of the effective hydrodynamic sphere. However, polymer molecules usually are not spherical. Later theories have treated the polymer in solution more realistically as a random coil and have led to the expression

$$[\eta] = \Phi \frac{(\overline{r^2})^{3/2}}{M} \tag{27}$$

in which $\overline{r^2}$ is the mean-square end-to-end distance of the random coil solute and Φ may be regarded approximately as a universal constant.

Values for Φ may be determined by a comparison of light-scattering measurements of $\overline{r^2}$ and M with intrinsic viscosity measurements made in the same solvent and at the same temperature. If r is expressed in centimeters, $[\eta]$ in cm^3/g, and M in atomic mass units, Φ is a dimensionless quantity, the value of which (within $\pm 20\%$) may be taken as

$$\Phi = 2.0 \times 10^{23} \tag{28}$$

Using the foregoing value of Φ, values of $(\overline{r^2})^{1/2}$, the root-mean-square end-to-end distance of the polymer in solution, may be determined from a knowledge of the intrinsic viscosity and the molecular weight of the polymer. Thus, a combination of (14) with (27) yields

$$(\overline{r^2})^{1/2} = \left(\frac{KM^{1+a}}{\Phi}\right)^{1/3} \tag{29}$$

where K and a are the constants contained in Table 15.3.

In general, the end-to-end distance of a given polymer chain depends on the polymer-solvent interaction. In "poor" solvents, the polymer would be expected to coil up so as to maximize the polymer-polymer interactions. In a "good" solvent, the polymer chain would tend to stretch out in order to maximize the

---

[1]Einstein, op. cit.

polymer-solvent interactions. This is not a compelling argument since, even in a poor solvent, a polymer chain is open enough that its segments are already largely surrounded by solvent molecules. A more supportable explanation involves the fact that the osmotic pressure, π, of a solution is higher in a good solvent because of its higher virial coefficients. This increased π drives more solvent into the random coil, expanding it in the same way that solvent is driven into the solution compartment of a membrane osmometer. A solvent in which the free energies of solvent–solvent, solvent–polymer, and polymer–polymer interactions are all the same is called a *theta* solvent. In principle, any solvent will become a theta solvent for a given polymer at the theta temperature (see Chapter 16, equations 43, 51, and 52).

In the special case of a theta solvent, we may define $\overline{r_0^2}$ as the mean-square end-to-end distance of the *unperturbed* polymer, and for *any* solvent we may write

$$\overline{r^2} = \alpha^2 \overline{r_0^2} \tag{30}$$

where α is known as the expansion coefficient. The usefulness of the unperturbed mean-square end-to-end distance, $\overline{r_0^2}$, may be appreciated when the polymer is viewed as a freely jointed chain of bound monomer units or chain segments. An application[1] of classical random walk theory to such a freely jointed chain leads to the relationship

$$(\overline{r_0^2}) = L^2 \cdot x \tag{31}$$

where x is the number of chain segments, or monomer units, and L is the length of each segment. Because the molecular weight is given by

$$M = M_0 x \tag{32}$$

where M_0 is the molecular weight of a monomer unit, equation (31) yields the result

$$\frac{\overline{r_0^2}}{M} = \frac{L^2}{M_0} \tag{33}$$

Thus, according to (33), $\overline{r_0^2}/M$ is a constant that is independent of the molecular weight and the solvent. Substitution of (30) into (27) leads to

$$[\eta] = \Phi(\overline{r_0^2})^{3/2} \frac{\alpha^3}{M} = \Phi\left(\frac{\overline{r_0^2}}{M}\right)^{3/2} \alpha^3 M^{1/2} \tag{34}$$

[1]Flory, P. J., *Principles of Polymer Chemistry* (Ithaca, N.Y.: Cornell University Press, 1953), Chap. 14.

In view of the constancy of $\overline{r_0^2}/M$, (34) may be written as

$$[\eta] = K\alpha^3 M^{1/2} \tag{35}$$

where the viscosity constant K is given by

$$K = \Phi\left(\frac{\overline{r_0^2}}{M}\right)^{3/2} \tag{36}$$

As expressed by (35), the intrinsic viscosity increases with molecular weight according to the product $\alpha^3 M^{1/2}$. Because it is found experimentally that $[\eta]$ usually increases with M by a power somewhat larger than $\frac{1}{2}$, α must depend, at least weakly, on the molecular weight. Thus, by comparison of (35) and (14), this dependence must be

$$\alpha^3 = M^{a-1/2} \tag{37}$$

However, by definition, in a theta solvent $\alpha = 1$, and in such a case the intrinsic viscosity is given simply as the product $KM^{1/2}$. This square-root dependence of $[\eta]$ on M in theta solvents (i.e., $a = \frac{1}{2}$) has been verified experimentally.

Accuracy of the Determination of Intrinsic Viscosity

Figures 15.4 and 15.5 indicate the accuracy to be expected in the determination of the intrinsic viscosity of a given polymer–solvent pair. These figures show plots of η_{sp}/c versus c as determined independently by six laboratories in Europe and four laboratories in the United States and Canada on samples of the same polystyrene fraction.[1] Figure 15.4 refers to solutions of the polymer in toluene and Figure 15.5 to solutions of the polymer in methyl ethyl ketone.

The average intrinsic viscosities obtained from these plots and the viscosity-average molecular weights calculated from the data in Table 15.3 are as follows:

(a) Toluene: $[\eta] = 148.3 \pm 2.3 \text{ cm}^3/\text{g}$
$M_v = 5.60 \pm 0.08 \times 10^5 \text{ g/mol}$
(b) Methyl ethyl ketone: $[\eta] = 82.9 \pm 2.8 \text{ cm}^3/\text{g}$
$\overline{M}_v = 5.66 \pm 0.19 \times 10^5 \text{ g/mol}$

In the above values of $\overline{M}_v$, the error limits shown represent only those caused by the indicated error in $[\eta]$.

It will be seen that the average molecular weights determined from these data agree within experimental error despite a difference of nearly a factor of 2

[1]Frank, H. P., and Mark, H. F., *J. Polymer Sci.*, **1955,** *17,* 1.

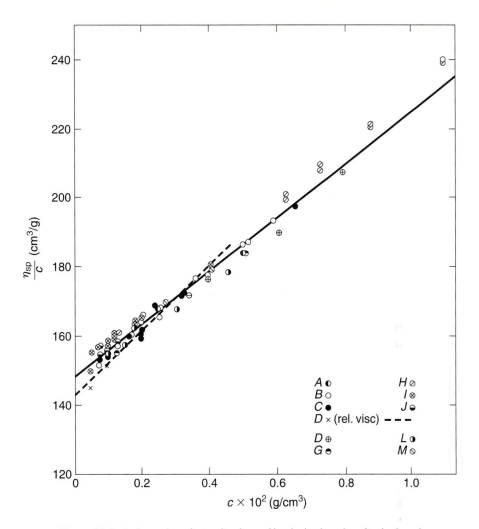

Figure 15.4   Independent determinations of intrinsic viscosity of a single polystyrene fraction in toluene at 25°C by several laboratories. [Reproduced from Frank, H. P. and Mark, H. F., *J. Polymer Sci.*, **1955,** *17,* 1; with permission of John Wiley & Sons, Inc., New York.] The symbols indicate results from different laboratories.

in the intrinsic viscosities of the polymer in the two different solvents. Moreover, in view of the discussion earlier in this chapter, it is possible to confirm by this specific example that $\overline{M}_v$ for a given polymer sample lies between $\overline{M}_n$ and $\overline{M}_w$, but much closer to $\overline{M}_w$. To confirm this the reader should compare the foregoing results for $\overline{M}_v$ with $\overline{M}_n$ and $\overline{M}_w$ determined for this same polystyrene sample by osmotic pressure and light scattering (see pages 412 and 430ff).

It is apparent that the solution viscosity method for the determination of polymer molecular weights can provide a satisfactory accuracy. Morever, pro-

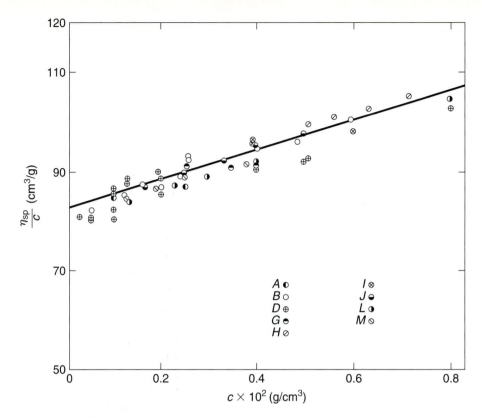

Figure 15.5 Independent determinations of intrinsic viscosity of a single polystyrene fraction in methyl ethyl ketone at 25°C by several laboratories. [Reproduced from Frank, H. P. and Mark, H. F., *J. Polymer Sci.*, **1955**, *17*, 1; with permission of John Wiley & Sons, Inc., New York.] The symbols indicate results from different laboratories.

vided that the constants *K* and *a* are known for the solvent and temperature used, the method is by far the fastest, simplest, and most inexpensive way to determine a reliable average molecular weight of a high polymer.

VAPOR-PHASE OSMOMETRY

Membrane osmometry was described in Chapter 14, along with other techniques for obtaining an absolute value for the molecular weight of a polymer. It is a rather cumbersome technique, requiring much time and patience to find a membrane with appropriate permeability, and obtaining osmotic pressures that reflect equilibrium. For this reason, a variation has been devised that removes some of these problems, but at the cost of making the molecular weights secondary or relative, with the associated need for calibration against some absolute method.

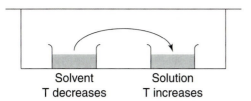

Figure 15.6 "Isothermal" distillation.

The method is based on what was once called "isothermal" distillation, which is illustrated in Figure 15.6. A closed desiccator contains a beaker of solvent along with a beaker of a solution of a non-volatile solute in the same solvent. Without the application of heat, the solvent from the pure solvent beaker distills over into the solution. This can be explained in terms of activities, vapor pressures, free energies, or chemical potentials, etc., but the simplest way is in terms of entropies. Solvent molecules that transport over into the solution can disorder themselves by mixing with the solute, while those left behind cannot, and the system moves in the direction of maximum disorder. The process does not, however, occur isothermally, since the temperature of the pure solvent decreases, while that of the solution increases. In the case of the pure solvent, the energy-rich molecules are more likely to enter the vapor state as required for the transport, leaving behind lower-energy ones (giving a lower average temperature). When these molecules condense into the solution in the other beaker, they give up their latent heat of vaporization, as demanded by the First Law of Thermodynamics, and the solution temperature increases correspondingly. This is exploited in the design of an osmometer in which the vapor phase, instead of a membrane, brings about the selection between solute and solvent molecules.

Such an osmometer is shown schematically in Figure 15.7. A drop of solvent is placed on a thermistor, which consists of a metal or semiconductor with a resistivity that is very sensitive to temperature changes. A drop of a solution of known weight concentration is then placed on another thermistor, and the difference in resistance between them is converted into a temperature difference ΔT that is displayed as a function of time. A typical curve of this type is shown in Figure 15.8. As can be seen, ΔT appears to approach an equilibrium value but,

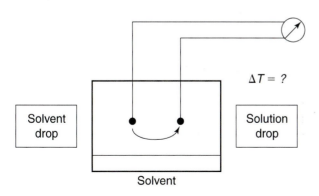

Figure 15.7 Vapor-phase osmometer.

Vapor-Phase Osmometry

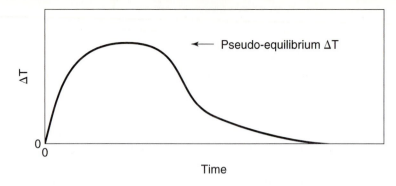

Figure 15.8 Temperature profile in vapor-phase osmometry.

after a brief plateau, begins to return to zero. This is caused by heat losses from radiation and through the wires. As a result, nonzero values of ΔT do not refer to equilibrium, which prevents the use of the relevant thermodynamics, specifically the Clausius–Clapeyron equation. Thus, rather than use this technique as an absolute method, the results are calibrated in terms of the values of ΔT at the plateau. These pseudo-equilibrium values are measured using solutions of a material of known molecular weight, and plotting them against the molarity. This gives a value for the calibration constant K in the equation $\Delta T = Km$, which is analogous to the colligative property equations discussed in Chapter 14. Such a plot is shown schematically in Figure 15.9. Interpretation then follows in the usual manner, in which the value of ΔT measured for the unknown solution is used to obtain the value of m. The molecular weight is then obtained from the fact that m times the molecular weight is simply the known weight concentration, c.

If the calibration molecule has a much lower molecular weight than the unknown solute, its solution will be less non-ideal. In this case, the reliability can be improved by extrapolating values of K to infinite dilution.

The main advantage of this technique is the speed with which measurements can be made, and the fact that only a very small amount of polymer is re-

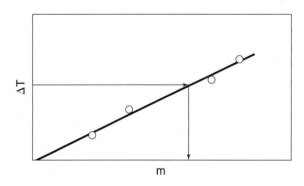

Figure 15.9 Calibration of vapor phase osmometery results.

quired. The main disadvantage is the fact that the molecular weights it gives are relative to those of some standard material.

MOLECULAR WEIGHT DISTRIBUTIONS

Importance

Many properties of a polymer show a strong dependence on molecular-weight distribution, as well as on the average molecular weight. Failure to recognize this caused considerable confusion in the early days of polymer processing. For example, a batch of polymer described as having the same molecular weight as a preceding batch would extrude completely differently, flowing either more rapidly or less rapidly than the previous batch under identical conditions. The reason was the presence of a small amount of very high molecular-weight material that escaped detection in colligative property measurements, or a small amount of very low molecular-weight material that did not contribute much in light-scattering studies. The effects are so pronounced that low molecular-weight polymers or high molecular polymers are frequently used as processing aids, as are branched polymers.

The above circumstances illustrate one reason for the study of molecular-weight distributions, specifically the goal of understanding structure–property relationships. Another reason is the evaluation of polymerization mechanisms, such as those described in several earlier chapters. A proposed mechanism might predict a certain distribution, but experiments that show a different distribution would require that mechanism to be abandoned or modified.

Distribution information generally requires the investigator to separate the parent polymer into portions or "fractions" that are more nearly monodisperse in molecular weight. If the separations are carried out on a large-enough scale (i.e., preparatively instead of simply analytically) then the resulting fractions would be large enough to yield structure–property relationships uncompromised by the problems of polydispersity.

Some llustrative Calculations

The simplest way to describe the distribution of molecular weights in a polymer sample is to specify the ratio of two different types of molecular-weight average. The ratio almost universally chosen is that of the weight-average to the number average molecular weight, $\overline{M}_w/\overline{M}_n$, which is called the "polydispersity index" or PI. The extent to which this ratio exceeds unity is a measure of the breadth of the distribution. Typical values are 1.0 for the monodisperse limit, 1.05 for the very narrow distributions obtained from some anionic polymerizations, and in the region of 2.0 for many other types of polymerization (as described in Chapter 4). Unusually high values can be found in some cases, for example in high-temperature, high-pressure polymerizations of ethylene gas. One reason for the broadening under these conditions is the relatively large amounts of branching

TABLE 15.4 ILLUSTRATIVE CALCULATIONS INVOLVING MOLECULAR WEIGHTS

$w_i(\text{g})$	$M_i(\text{g/mol})$	$n_i(\text{mol})$	$n_iM_i(\text{g})$	$n_iM_i^2 \ (\text{g}^2 \, \text{mol})$
1,000	1.0×10^3	1.000	1,000	1.0×10^6
1,000	1.0×10^6	0.001	1,000	1.0×10^9
		1.001	2,000	1.001×10^9

that can occur under these aggressive conditions. This would occur by removal of hydrogen radicals from the polyethylene chains, with the subsequent growth of branches from the corresponding radicals left on the chains at these points. Because the longer chains have more hydrogen atoms, these would tend to increase in molecular weight from branching more than the shorter chains, and this would broaden the distribution. This is another example of the large or successful becoming even larger or more successful! Ring-opening polymerizations (Chapter 6) sometimes generate very broad molecular-weight distributions.

Different types of molecular weight have very different sensitivities to the low and high molecular-weight parts of a sample. This is illustrated in Table 15.4 for the number-average and weight-average molecular weights ($\overline{M}_n$ and $\overline{M}_w$) defined in Chapter 1. It is based on a hypothetical sample made by mixing two monodisperse polymers, specifically 1,000 g of one having a molecular weight 1.0×10^3 g/mol, with 1,000 g of another having a molecular weight 1.0×10^6 g/mol. The two components are described in the first two columns of the table, with the remaining columns giving the quantities required to calculate $\overline{M}_n$, $\overline{M}_w$, and the polydispersity index. As can be seen from the third column, the low molecular-weight component predominates with regard to the number of moles n_i in the mixture: the lower the molecular weight, the more molecules in a specified weight of material. Both components contribute equally to n_iM_i, and the high molecular-weight component dominates with regard to $n_iM_i^2$. The defining equations in Chapter 1 give a number-average molecular weight of approximately 2000 g/mol, which means that the high molecular-weight part hardly contributes at all to $\overline{M}_n$. This average depends on the *number* of particles, and there are relatively few of the high molecular-weight type. The weight-average value, on the other hand, is approximately 500,000 g/mol, which illustrates the fact that the low molecular-weight component does not contribute much to the weight average. The PI is approximately 250, an extraordinarily high value that reflects the peculiar nature of this hypothetical distribution. The basic point is that it is necessary to have more than one type of molecular weight to characterize a sample.

Representations

Of course, it would be preferable to have more detailed information about molecular-weight distributions. In essence, it would be useful to know, for example, that 5% of the polymer has a molecular weight between 5,000 and 10,000

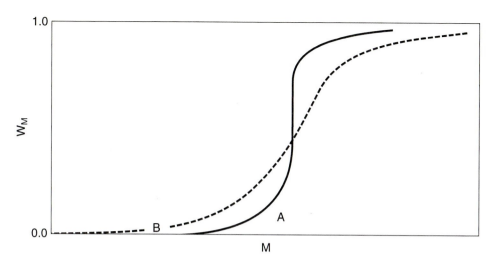

Figure 15.10   A narrow (A) and broad (B) molecular weight distribution, in the cumulative representation.

g/mol, another 10% has a molecular weight between 10,000 and 15,000 g/mol, etc. Such information is typically given in one of two ways that are interconvertible: either as a "cumulative or integral" distribution, or as a "differential" distribution. Two typical cumulative distributions are shown schematically in Figure 15.10. The ordinate is the weight fraction of polymer having a molecular weight less than or equal to the value shown on the abscissa. Thus, any point on such a curve corresponds to accumulation of the weights of all the chains having molecular weights from the minimum in the sample up to the value corresponding to the abscissa of this point. The curves are seen to be "S-shaped." The slope is very low at both small and high values of M, because of the scarcity of very short and very long chains. Between these two limits, the curves show a maximum slope (at the inflection point), which corresponds physically to the most-probable value of M in the sample. These characteristics will remind the reader of the distributions of molecular speeds in a gas. Distribution A is seen to be narrower than distribution B, as evidenced by the narrower range in M that is required for the fraction of polymer to sweep from 0.0 to 1.0. The alternative term, "integral distribution" obviously arises from the accumulation being analogous in calculus to summing the areas under a curve between two points.

The nature of a cumulative distribution is somewhat difficult to absorb at a glance and, for this reason, it is generally converted into the corresponding differential representation. This is done by taking slopes along a cumulative distribution curve, which corresponds, of course, to differentiation. These slopes are then plotted against the molecular weight, as is illustrated for the same two distributions in Figure 15.11. The resultant curves are bell-shaped, which is the more customary way to represent distributions. The ordinate now corresponds to the weight fraction of polymer in an infinitesimal interval around the specified value

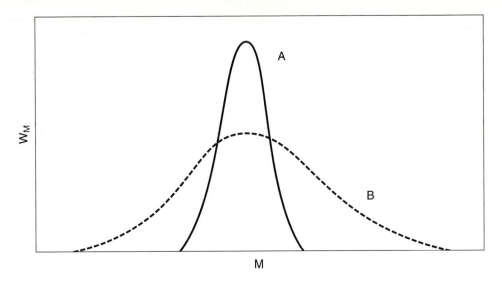

Figure 15.11 The same narrow (A) and broad (B) molecular-weight distributions, now in the differential representation.

of M. Thus, the weight fraction of polymer having a molecular weight between specified values corresponds to the relative area under this part of the curve. Accordingly, it is meaningless to inquire about the weight fraction of polymer having a specific molecular weight, since the area under a point on such a curve is zero. This is a characteristic of this continuous curve, in that there is now an infinite number of other values a chain can have. This representation shows much more directly the most important features of the distribution, specifically the broadness of the distribution as gauged by the width of the curve (for example at half height), the most probable value of M (at the maximum), and the extent to which the distribution is skewed toward larger values of M.

The differential representation can be converted into the cumulative form by summing slices of area under the differential curve. This is akin to summing the areas under nuclear magnetic resonance (NMR) peaks to give the cumulative representation, which is generally displayed as alternative output in NMR scans.

Distributions from Fractional Precipitations

The primary goal of these techniques is to separate a polymer into narrow distribution fractions, as already mentioned. This is shown schematically in Figure 15.12. The oldest technique for doing this is called "fractional precipitation," which is based on the fact that the solubility of a polymer in a given solvent at fixed temperature decreases monotonically with increase in molecular weight. This dependence is shown schematically in Figure 15.13. In this approach, the polymer is dissolved in a solvent, which is strong enough to dissolve even the

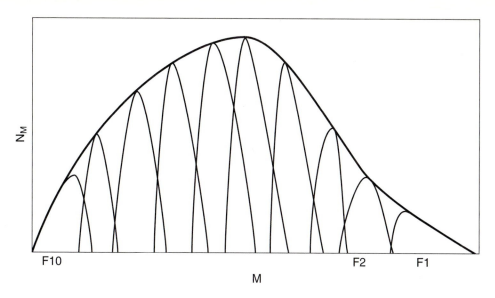

Figure 15.12 Fractionation of a polymer with a broad molecular weight distribution into a series of fractions each of which has a narrower distribution.

highest molecular-weight chains present. The solvent power is then reduced slightly, either by a lowering the temperature, or by adding a small amount of non-solvent. Both approaches are carried out in a constant temperature bath, with stirring. Because of the molecular-weight dependence of the solubility, the portion of polymer precipitated at this first stage has the highest molecular weights present in the sample, identified as fraction F1 in Figures 15.12 and 15.13.

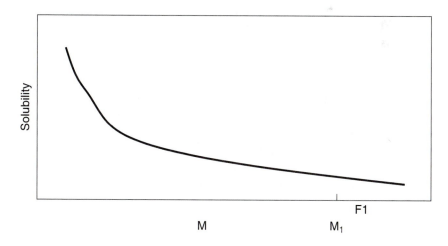

Figure 15.13 The decrease in solubility with molecular weight that is exploited in fractional precipitation.

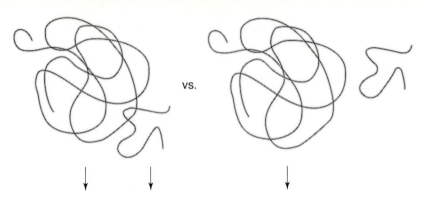

vs.

Figure 15.14 The right portion of the figure shows the use of high dilutions to prevent entangling of a small polymer chain with a larger one, which would complicate separating them in a fractional precipitation.

Decanting of the supernatant from the precipitated phase exposes the first fraction, which can easily be recovered by dissolution in the original solvent. Repetition of these steps n times would then produce n fractions, prefererably with each on average representing approximately $1/n$ of the total weight of the polymer.

The main advantage of this technique is the efficiency of the separation when the precipitations are carried out from rather dilute solutions. High dilutions separate the chains of different molecular weights, which permits them to either precipitate or stay in solution, depending on whether their molecular weights are relatively high or relatively low. High-polymer concentrations can cause problems from chain overlap or interpenetration. This could cause some low molecular-weight chains to be dragged along by being entangled with a polymer which is precipitating from solution because of its higher molecular weight. This is illustrated in the left portion of Figure 15.14. Of course, high dilutions require large amounts of solvent, and the safety problems from handing such large volumes is one of the disadvantages of the technique. Another disadvantage is the time required, since it may require an entire day to assure careful equilibration between the precipitated phase and the other, more dilute phase with which it is in contact. Finally, the technique is essentially impossible to automate. Nonetheless, all of the earliest distributions, such as those described in the classic 1953 book by Flory,[1] were obtained by this cumbersome technique.

Distributions from Gradient Elutions

Several techniques were subsequently developed to simplify the above method, frequently at the cost of considerable losses in resolution. The first of these is "gradient elution." This is illustrated by the sequence of steps shown in Fig-

---

[1]Flory, P. J., *Principles of Polymer Chemistry*. (Ithaca NY: Cornell University Press, **1953**).

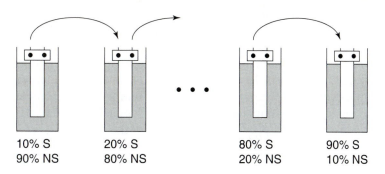

Figure 15.15 Manual version of fractionation by gradient elution.

ure 15.15. A thin strip of the initial polymer is dipped into a solvent mixture which consists largely of a non-solvent (NS). An example would be 10% toluene and 90% methanol. This would dissolve only the lowest molecular-weight material with the highest solubility, which could then be recovered as the first fraction. The strip would then be extracted in a solvent mixture of increased solvent power, for example 20% toluene and 80% methanol. This extraction would remove another fraction, with a higher molecular weight than that of the first. Repetition of these steps would yield a series of fractions, until all the material was finally dissolved, say at a composition of 90% toluene and 10% methanol. The main disadvantage of the gradient elution technique is its attempt to approach equilibrium from the wrong direction, i.e., by trying to extract short chains out of a highly viscous, extensively entangled mass of chains. The main advantage is the fact that it can be automated. The polymer to be fractionated is dissolved in a relatively poor solvent by increase in temperature. The warmed solution is then poured through a column packed with high-surface area beads, such as porous glass. The cooling of the solution precipitates the polymer as a thin coating on the particles that comprise the column packing. Relatively poor solvent is passed over the polymer-coated particles, and this extracts the lowest molecular-weight fraction. A mixing chamber is then used to replace a portion of the poor solvent with a good solvent, at a rate that can be controlled by the experimentalist. In this way, the extraction liquid is gradually changed from one in which the proportion of good solvent is relatively low to one in which it predominates to the extent that it will dissolve polymer of even the highest molecular weight. The resultant materials are collected with a fraction collector of the type long familiar to those doing chromatographic separations.

The final technique of this type is known by two names. Gel permeation chromatography (GPC) describes the apparatus (a gel being permeated by a polymer solution), while size exclusion chromatography (SEC) emphasizes the separation mechanism (molecules being excluded from pores in a gel because of their unacceptably large sizes). Its widespread use requires it to be discussed separately, in the following section.

Molecular Weight Distributions

GEL PERMEATION CHROMATOGRAPHY

The Underlying Principle

The gel permeation chromatography (GPC) method is essentially a process for *separating* macromolecules according to their *size*. The method has been used extensively in biochemistry to separate biological macromolecules from small-molecule contaminants (with the use of Sephadex columns). Its general application to synthetic polymer chemistry in the 1970s has revolutionalized the procedures for polymer characterization and molecular-weight determination.

The principle that underlies the method is as follows. Imagine that a dilute solution is available that contains a broad molecular-weight distribution of polymer chains, oligomers, and perhaps even the monomer from which the other species were derived. Assume that this solution is allowed to flow through a column that is packed with finely divided particles that contain pores. These particles can be of two types. First they may consist of particles of a solvent-swollen gel in which the space between cross-links in the gel surface generates openings of say 1000-Å diameter that correspond to pores. Alternatively they may be solid particles that are permeated by pores (tunnels) that have a diameter of the same 1000 Å. As the dissolved solute passes each particle, the smaller molecules (those with dimensions smaller than 1000 Å) will enter the openings of the pores and will "explore" the pore space under the influence of the usual thermal motions. Thus, the smaller molecules will be "delayed" in their elution through the column. On the other hand, the larger polymer molecules (those with a random coil radius of larger than 1000 Å) will be unable to penetrate the pores and will be swept along with the solvent front to be eluted before the smaller molecules (Figure 15.16).

In practice, even if the substrate particles have only one uniform pore size, the process can separate molecules that form part of a continuous molecular-weight distribution. This is because those molecules that have "diameters" below 1000 Å will be differentially delayed according to their molecular size. The very smallest molecules can presumably penetrate far into the tunnel system, whereas the medium-sized molecules may merely "sample" the openings into the tunnels. Thus, in principle, a single-pore-size column would be expected to *separate* high-molecular-weight molecules (in this example, those with diameters over 1000 Å) from the rest, and *fractionate* the smaller molecules according to their size. Clearly, the smallest molecules (the monomers, dimers, trimers, etc.) will differ very little in their retention by a 1000-Å column, and hence they will be eluted together. On the other hand, those molecules that have a coil diameter close to that of the pore diameter will be fractionated the most effectively.

The effectiveness of the fractionation process for the whole molecular-weight distribution can be improved by the use of several different columns in series, each of which contains particles with a different pore size. For example, a series of columns that contain 1×10^6, 1×10^5, 1×10^3, and 500-Å pore-diameter particles should, in principle, be capable of fractionating a molecular-weight distribution that encompasses this entire range.

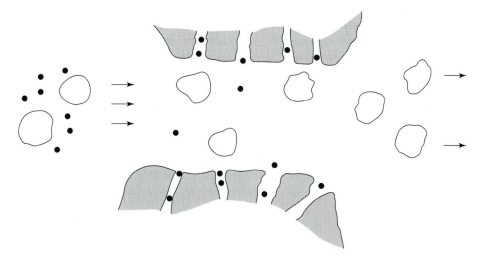

Figure 15.16 Gel permeation chromatography (GPC). The mixture of different-sized polymer and oligomer molecules is eluted in a solvent through a column of porous particles. The smaller molecules (black circles) can enter the pores and be retarded, whereas the larger molecules are swept through relatively unhindered. If a *distribution* of different molecular sizes enters the column at the same time, the molecules will emerge from the column in sequence, distributed according to molecular size.

Thus, the gel permeation method is essentially a process for the *fractionation* of polymers according to their size and, therefore, according to their molecular weight. The molecular weight as such cannot be determined directly, but only after calibration of the system in terms of the elution time (or volume of solution eluted) expected for a particular polymer molecular-weight fraction with the use of that particular piece of equipment. An alternative is to pass the solution from the GPC column directly into a light-scattering photometer, to determine absolute values of the molecular weight. In either case, the entire elution curve can be converted into the type of differential distribution curve described above.

Equipment

It will be clear that the use of this method to measure molecular weights depends critically on being able to ensure that the elution time along the column is reproducible for two different specimens of the same polymer that have the same molecular weight and molecular-weight distribution. This requirement can only be met if (1) the flow rate of the eluting solvent through the column remains the same, and (2) the size of the tunnels within the stationary particles remains the same in different experiments. Both of these factors may change if the column temperature varies or if the solvent composition changes. The equipment used for gel permeation chromatography is usually designed to avoid such problems.

A schematic layout of a typical gel permeation chromatography unit is shown in Figure 15.17. Although crude *separations* of macromolecules from small molecules can be achieved by gravity elution through a vertical column (as in conventional chromatography), the elution rates under such conditions will be slow and nonreproducible. Hence, a mechanical pump is usually employed to force the sample and the elution solvent through the columns at pressures of up to 1000 to 4000 psi and at a rate of 2 to 3 ml/min. When a reciprocating pump is employed, the individual pressure pulses must be smoothed out by some form of constrictor coil. Injection of the sample into the line is usually accomplished from a graduated hypodermic syringe (typically, 0.5 to 3 ml of a 0.05 to 0.1% solution of the polymer) by means of a mechanical inlet device.

The columns are usually $\frac{3}{8}$-in. (ca. 1-cm)-diameter stainless steel tubes with a combined length of from 3 ft to 10 ft or more, dependent on the type of packing material used. Two principle types of column packing materials have been employed—microporous glass beads and powdered, swelled, crosslinked polystyrene. The latter material is in more common use. With polystyrene particles, the pore size is determined by the amount of crosslinking, since the tunnels are formed by the solvent-swelled cavities that exist between the crosslinks (hence the name *gel* permeation). The particle size of the stationary phase has a profound effect on the *resolution* of the separation. Gel permeation is a diffusion-controlled phenomenon and, clearly, the speed and efficiency of the differential imbibition will be increased as the stationary particle size is decreased and the relative surface area is increased. Thus, small particles ($\approx$10 μm diameter, called microstyragel particles) allow faster separations with smaller samples and shorter columns than do larger particles (37 to 74μm).

After passage through the column system, the eluent passes through a detector. Two alternative detection methods are commonly employed—differen-

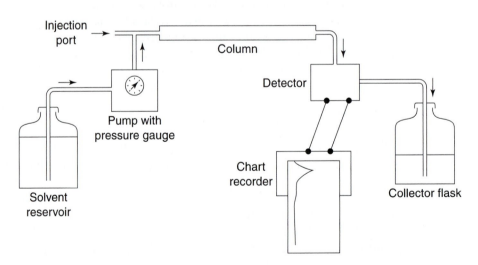

Figure 15.17 Schematic diagram of a gel permeation chromatography apparatus.

tial refractive index measurements or ultraviolet absorption. A differential refractometer measures the difference in refractive index between the eluted solution and the pure solvent. A plot of the refractive index difference as a function of time (in a chart recorder) can yield directly a plot of the molecular-weight distribution. Of course, the accuracy of this procedure will depend on the requirement that the refractive index difference between the polymer and solvent depends only on concentration and is independent of molecular weight.

When ultraviolet detection is used, the spectrometer is usually set to a particular wavelength (e.g., to the aromatic absorption region of a polymer that contains phenyl rings) and the absorbance is monitored as a function of elution time.

Problems in Gel Permeation Chromatography

Two main types of problem are often encountered with this technique—practical problems and problems of data interpretation. Perhaps the most serious practical problem is an overloading of the stationary phase by the polymer. It is fairly obvious that if all the pores in the column are occupied by polymer molecules, an effective separation will be impossible. Thus, only very small samples of very dilute polymer solutions are used—the smaller the better. A second practical problem may be encountered in conjunction with a polar stationary substrate such as porous glass. Under such circumstances, the polymer may become chemically *adsorbed* onto the surface of the substrate. In extreme cases it may be impossible to remove it by simple solvent elution. In other cases, the gel permeation elution pattern may be severely distorted by fractionation effects that depend on differential adsorption. Third, with swelled, crosslinked polystyrene used as a stationary phase, the degree of swelling of the polystyrene (and hence the pore size) will depend on the nature of the elution solvent. Fluids in which polystyrene is "insoluble" may close the pores completely. Thus, only certain specific elution solvents can be employed (e.g., tetrahydrofuran, benzene, xylene, chloroform, dimethylformamide, or fluorinated alcohols). This constitutes one of the main reasons for the continued use of porous glass as a substrate, since water, alcohols, and a wide variety of other solvents can be used with columns of this type.

The interpretation of gel permeation data can be complicated by two important factors. First, the ease with which a polymer molecule will penetrate a pore depends on whether it assumes a random coil or an extended rodlike conformation in the solvent being used. Thus, the chromatographic behavior of a polymer might be quite different in two different solvent systems. Second, one of the most valuable features of the gel permeation technique is the *speed* with which the average molecular weight and the molecular-weight distribution can be measured. Strictly speaking, an accurate assessment of the molecular weight can only be made if the average molecular weights of specific fractionated samples are first measured by another method and then used to calibrate the apparatus. This can be a time-consuming process that largely negates the speed advantage of the GPC technique. Of course, once the system is calibrated for a given polymer, other samples of the same polymer can be examined easily and rapidly.

Often, in order to obtain preliminary indications of molecular weight, the equipment is calibrated with well-characterized fractions of a different but chemically or structurally related polymer, and it is *assumed* that the calibration applies to the polymer of interest. Well-characterized "standard" samples of polystyrene and some other polymers are available commercially. A more satisfactory "universal" calibration technique has been developed, which permits the calibration of gel permeation columns for a wide range of polymers using a single set of standard samples (i.e., polystyrene fractions having narrow ranges of molecular weights). The best approach, however, may be to interface the GPC column directly with a light-scattering photometer, as already mentioned.

Universal Calibration in Gel Permeation Chromatography

The calibration of a gel permeation column for a given polymer-solvent system requires the establishment of a relationship between the volume of solution eluted (or, equivalently, the elution time for a given flow rate of solution) and the molecular weight of *monodisperse* fractions of the polymer. The main problem encountered is that monodisperse samples of most polymers are not generally available. However, such samples are available for a few specific polymers. A notable example is polystyrene for which samples having a molecular-weight range of 10^3 to 10^6 and a ratio of the weight-average molecular weight to number-average molecular weight of less than 1.15 can be obtained commercially.

If a set of monodisperse samples of a single polymer can be obtained, the remaining problem is to establish a relationship for a particular GPC column (or columns) between the volume of solution eluted and the molecular weight of some chemically different polymer. Clearly, in order to be able to do this, a calibration parameter is required which is independent of the chemical nature of the polymer, that is, a *universal calibration parameter*. Such a parameter has been found[1] experimentally to be the product of the intrinsic viscosity and the molecular weight (i.e., $[\eta]M$). This correlation succeeds because the separation occurs on the basis of the effective size of the polymer coil, and this has to be proportional to $(\overline{r^2})^{3/2}$. As shown in equation (27), this also makes it proportional to $[\eta]M$. Thus, as shown in Figure 15.18, with tetrahydrofuran used as a solvent, the logarithm of the product $[\eta]M$ plotted against the volume of solution eluted from the column provides a *single curve* from all the points determined for a wide variety of polymers. This is not possible on a plot of log M versus elution volume, which would be the simplest way to display the data.

The experimental finding that $[\eta]M$ is the same function of elution volume, V_e, for many different polymers suggests the possibility that a universal calibration procedure may be possible. First, the functional relationship between the molecular weight of the monodisperse standard samples and the elution volume of the solution must be determined for a given solvent and column under fixed conditions. Such a relationship is shown graphically in Figure 15.19 for a set of commercial polystyrene samples at 25°C using tetrahydrofuran as solvent and

[1] Grubisic, Z., Rempp, P., and Benoit, H., *Polymer Lett.*, **1967**, *5*, 753.

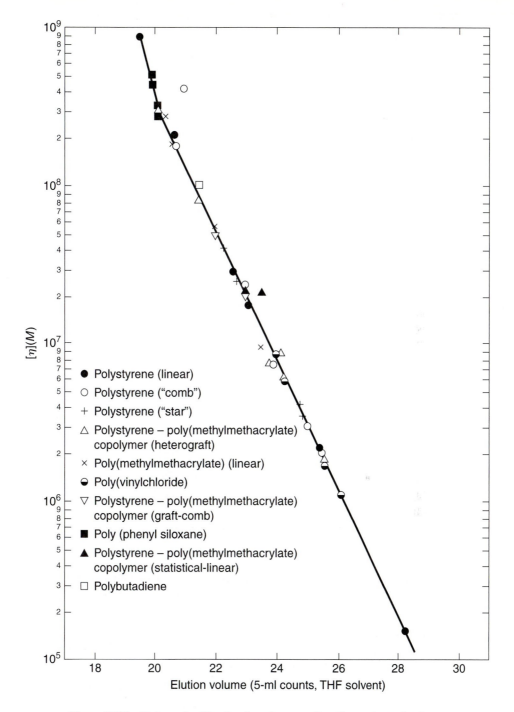

Figure 15.18 Universal calibration in gel permeation chromatography for a variety of polymers in tetrahydrofuran. [Reproduced from Grubisic, Z., Rempp, P., and Benoit, H., *Polymer Lett.*, **1967,** *5,* 753; with permission of John Wiley & Sons, Inc., New York.]

Gel Permeation Chromatography

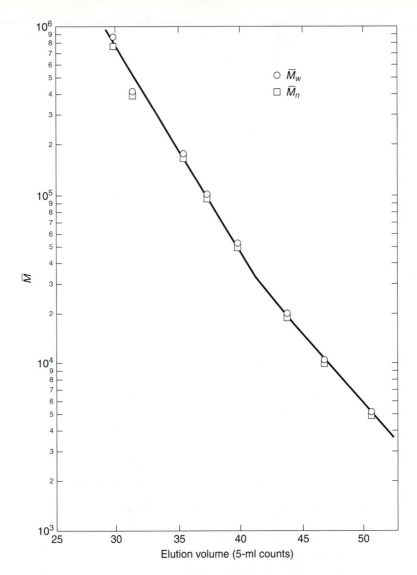

Figure 15.19 Molecular weight of monodisperse polystyrene standards as a function of elution volume in tetrahydrofuran. [From Kolinsky, M. and Janca, J., *J. Polymer Sci., Chem. Ed.*, **1974,** *12,* 1181.]

with spherical porous silica beads used as the column packing.[1] If it is assumed (as shown[2] in Figure 15.18 to be true for a wide variety of polymers) that log $[\eta]M$ is a constant for all polymers in a given solvent at a given temperature, *at the same elution volume,* it is possible to write (38), in which the

$$\log [\eta]_x M_x = \log [\eta]_s M_s \tag{38}$$

[1]Kolinsky M., and Janca, J., *J. Polymer Sci., Polymer Chem. Ed.*, **1974,** *12,* 1181.
[2]Grubisic et al., op. cit.

subscripts x and s indicate the unknown polymer and the standard polymer, respectively. If each intrinsic viscosity term in (38) is replaced by its Mark–Houwink expression,

$$[\eta]_j = K_j M_j^{a_j} \tag{14}$$

and the resulting expression is solved for $\log M_x$, (39) is obtained, which describes the elution volume calibration curve for M_x.

$$\log M_x = \left(\frac{1}{1 + a_x}\right) \log \frac{K_s}{K_x} + \frac{1 + a_s}{1 + a_x} \log M_s \tag{39}$$

The value of V_e that corresponds to a GPC peak in the unknown polymer is used to obtain a value of $\log M_s$ from Figure 15.19, and M_x is then calculated from (39). An alternative procedure is simply to choose values of V_e and construct a new calibration curve for the unknown polymer from a curve such as Figure 15.19 and equation (39). All this presupposes that the Mark–Houwink constants, K_s, a_s, K_x, and a_x, are known. K_s and a_s are available in the literature for a variety of solvents (i.e., Table 15.3), and so, in many cases are values of K_x and a_x. On the other hand, if the desired Mark–Houwink constants are not available for the polymer under study or for the standard in the solvent to be used, they can easily be determined by measurement of the intrinsic viscosity. Unfractionated or crudely fractionated samples of known average molecular weight can be used for a determination of K_x and a_x from intrinsic viscosity measurements and a reasonable accuracy can be obtained.

As an example of the universal calibration technique using monodisperse polystyrene samples as standards, suppose that it is necessary to determine the molecular weight of polyisobutylene in toluene solution by gel permeation chromatography. First, it would be necessary to construct a calibration curve, such as the one shown in Figure 15.19 from the polystyrene standards in toluene. The Mark–Houwink constants for the two polymers in toluene are given in Table 15.3 and, when these are substituted in (39), expression (40) is obtained.

$$\log M_x = -0.589 + 1.10 \log M_s \tag{40}$$

Suppose now that the elution volume of a peak in the gel permeation chromatogram of polyisobutylene was such that a value of 10^5 was deduced for M_s from the calibration curve. According to (40), this peak corresponds to polyisobutylene with $\overline{M}_x = 8.15 \times 10^4$.

Some disagreement exists in the literature about the type of molecular-weight average that is given by the positions of the peak maxima in a gel permeation chromatography measurement. Some investigators assume the peak maxima represent number average $(\overline{M}_n)$ molecular weights but others consider that the method determines the weight-average molecular weight $\overline{M}_w$. It has been

proposed[1] that the root-mean-square molecular weight, $\overline{M}_{rms} = (\overline{M}_w \cdot \overline{M}_n)^{1/2}$, correlates better than $\overline{M}_w$ with the peak maxima in a gel permeation chromatogram when the calibration technique just outlined is used. On the other hand, an elaboration[2] on the calibration technique that replaces (39) by

$$\log M_x = \left(\frac{1}{1 + a_x}\right) \log \frac{K_s}{K_x} \frac{f(\varepsilon_x)}{f(\varepsilon_s)} + \left(\frac{1 + a_s}{1 + a_x}\right) \log M_s \qquad (41)$$

where

$$f(\varepsilon_j) = 1 - 2.63\varepsilon_j + 2.86\varepsilon_j^2 \qquad (42)$$

$$\varepsilon_j = \tfrac{1}{3}(2a_j - 1) \qquad (43)$$

leads to just as good a correlation of $\overline{M}_w$ with the peak maxima.

Although it was mentioned that the universal calibration parameter in gel permeation chromatography is the product $[\eta]M$, the more fundamental universal parameter is the partial molar volume of the polymer at infinite dilution. This may be seen as follows: according to the Einstein relationship for the relative viscosity of spherical colloidal solutes,[3]

$$\eta_r - 1 = \eta_{sp} = 2.5\phi_2 \qquad (1)$$

where, as mentioned before, ϕ_2 is the volume fraction of solute (i.e., polymer) in the solution. In view of the definition of volume fraction, ϕ_2 is given by

$$\phi_2 = \frac{n_2 \overline{V}_2}{n_1 \overline{V}_1 + n_2 \overline{V}_2} \qquad (44)$$

where n_1 and n_2 are the numbers of moles of solvent and dissolved polymer, respectively, and $\overline{V}_1$ and $\overline{V}_2$ are the partial molar volumes of solvent and dissolved polymer. Since the denominator of (44) is simply the volume of solution, (44) may be written as (45)

$$\phi_2 = \frac{(w_2/M_2)\overline{V}_2}{V_{soln}} = \frac{\overline{V}_2}{M_2} c \qquad (45)$$

where c is the concentration of the solution in mass per unit volume and M_2 is the molecular weight of polymer. Substitution of (45) into (1) with rearrangement yields (46)

$$\frac{\eta_{sp}}{c} = 2.5 \frac{\overline{V}_2}{M_2} \qquad (46)$$

[1]Kolinsky and Janca, op. cit.
[2]Coll H., and Gilding, D. K., *J. Polymer Sci. (A2)*, **1970**, *8*, 89.
[3]Einstein. op. cit.

which in the limit of infinite dilution becomes

$$\lim_{c \to 0} \frac{\eta_{sp}}{c} = [\eta] = 2.5 \frac{\overline{V}_2^o}{M_2} \tag{47}$$

$\overline{V}_2^o$ is the partial molar volume of polymer at infinite dilution,

$$\overline{V}_2^o = \lim_{n_2 \to 0} \left(\frac{\partial V_{soln}}{\partial n_2} \right)_{n_1, T, p} \tag{48}$$

and is equivalent to the hydrodynamic volume of the polymer. An expression similar to (47), but with a different proportionality constant, is obtained when the polymer solute is not spherical but rather appears as a random coil (i.e., equations 26 and 27).

A simple rearrangement of (47) shows that the product $[\eta]M$ is simply proportional to $\overline{V}_2^o$. Therefore, the statement that $[\eta]M$ is the same function of elution volume for all polymers in a given solvent is equivalent to the observation that the partial molar volumes (or hydrodynamic volumes) of all polymers in a given solvent at infinite dilution are given *by the same function* of the GPC elution volume. The experimental results discussed in this section demonstrate the validity of this statement for a wide variety of polymers in a single solvent. However, the statement does not hold true for cases in which a specific chemical interaction exists between polymer and the stationary gel. Moreover, the same $[\eta]M$ versus V_e curve would not normally apply to polymers of the rigid-rod type in the same way as it does to those of the random coil variety.[1]

Finally, it should be mentioned that if the columns and sample sizes are large enough, it is possible to carry out the separation *preparatively*. This would give fractions of low polydispersity in sufficient amounts to establish structure–property relationships.

STUDY QUESTIONS

1. Show from equation (23) that $\overline{M}_{\beta=0} = \overline{M}_n$ and $\overline{M}_{\beta=1} = \overline{M}_w$.
2. The following data have been obtained for the intrinsic viscosity of polystyrene fractions in dichloroethane at 22°C using light scattering as the absolute measurement of molecular weight.

$[\eta](cm^3/g)$	260	278	142	138	12.2	4.05
$\overline{M}_w \times 10^4$	178	157	56.2	48.0	1.55	0.308

Evaluate the constants in the intrinsic viscosity/molecular weight relationship.

[1]Coll, H. and Prusinowski, L. R., *Polymer Lett.*, **1967**, *5*, 1153.

3. Suppose that you have an Ostwald viscometer with $R = 2.00 \times 10^{-2}$ cm, $L = 11.0$ cm, $V = 4.00$ cm^3, and $h = 16.0$ cm. What percentage error will be introduced by neglect of the kinetic-energy correction in equation (27) in determining:

  (a) The absolute viscosity of chloroform which has a flow time of 170 s at 20°C.

  (b) The relative viscosity of a solution of poly(methyl methacrylate) in chloroform whose flow time through the viscometer is 230 s.

4. With the viscometer described in Problem 3, a solution of poly(methyl methacrylate) in acetone, having a concentration of 0.0865 g/cm^3, is found to have a flow time of 232 s. Assuming a value of $k' = 0.40$ for the Huggins constant in equation (12), calculate the viscosity-average molecular weight of the polymer.

5. Show that extrapolating $\ln \eta_r/c$ to infinite dilution gives the same intercept $[\eta]$ as does η_{sp}/c.

6. Given the following data, calculate the intrinsic viscosity $[\eta]$ (dl/g) and Huggins constants, k' and k'':

c (g/dl)	Efflux time (sec)
0.0000	66.78
0.0780	84.03
0.1428	101.18
0.1788	110.57
0.2336	127.89

The kinetic energy correction factor β/α is 110 for this viscometer. dl is deciliter.

7. Using data given in this chapter, calculate the viscosity-average molecular weights of the poly(methyl methacrylate) fractions to which Table 15.2 and Figure 15.2 refer.

8. Calculate the Mark–Houwink parameters, K and a, for the polymer-solvent system described below:

M 10^{-6} (g/mol)	$[\eta]$ (dl/g)
0.0743	0.325
0.119	0.493
0.175	0.604
0.261	0.838
0.272	0.864
0.509	1.32

9. A vapor-phase osmometer is calibrated using benzil, $(C_6H_5CO)_2$, with the results given below (where c is in g of solute per kg of solvent, and the change in resistance ΔR is in arbitrary units). What is the number-average molecular weight of the polymer that gave the results also shown in the table?

Sample	c	ΔR (arbitrary units)
Benzil	20.0	1.00
	40.0	2.10
	60.0	3.20
Polymer	10.0	0.080
	20.0	0.170
	30.0	0.270

10. A sample (19.03 g) of syndiotactic poly(isopropyl acrylate) was fractionated by fractional precipitation from chlorobenzene at 100°C. The results are given below:

Fraction	Wt (g)	$10^{-6} M$ (g/mol)
1	0.32	2.58
2	2.03	2.31
3	1.21	2.16
4	1.50	1.69
5	0.87	1.31
6	2.32	1.15
7	1.51	0.946
8	1.93	0.798
9	1.30	0.737
10	1.50	0.557
11	0.92	0.441
12	0.55	0.325
13	0.76	0.236
14	0.63	0.146

Plot the integral (cumulative) distribution (W_M vs. M) and the differential distribution curve (w_M vs. M) for this polymer. (Assume that the distribution within each fraction is symmetrical, e.g., 0.32 g of fraction 14 has a molecular weight less than 0.146×10^6 g/mol). [*Ans.:* Inflection point in the integral distribution and maximum in the differential distribution should occur at around 0.8×10^6 g/mol (the most-probable value)].

11. The following data on intrinsic viscosities and elution volumes from a gel permeation chromatograph at 25°C for standard polystyrene samples dissolved in tetrahydrofuran have been reported [Kolinsky, M. and Janca, J., *J. Polymer Sci., Chem. Ed.*, **1974**, *12*, 1181]:

$\overline{M}_w \times 10^{-3}$ (g/mol)	867	411	173	98.2	51	19.85	10.3	5.0
$[\eta]$ (cm^3/g)	206.7	125.0	67.0	43.6	27.6	14.0	8.8	5.2
$V_{e\frac{1}{5}}$ (cm^3)	29.8	31.4	35.4	37.3	39.9	43.8	46.8	50.7

The Mark–Houwink constants for poly(vinyl bromide) in tetrahydrofuran at 25°C may be taken as $K = 1.59 \times 10^{-2}$ cm^3/g and $a = 0.64$ [Ciferri, A., Kryezewski, M.,

and Weil, G., *J. Polymer Sci.*, **1958**, *27*, 167]. Construct an appropriate GPC calibration curve (i.e., log $\overline{M}_w$ as a function of V_e) for poly(vinyl bromide) in tetrahydrofuran.

12. A sample of poly(vinyl bromide) is dissolved in tetrahydrofuran and introduced at a liquid flow rate of 2 cm^3/min into the GPC column of Problem 11. When the refractive index difference between the eluted solution and pure solvent was plotted versus the elution time, the result was a broad peak, the maximum of which occurred at an elution time of 90 min. Calculate the average molecular weight, corresponding to the peak maximum, of the poly(vinyl bromide) sample.

13. The following elution volumes were obtained in a gel permeation chromatograph at 35°C for a set of monodisperse polystyrene standards dissolved in chloroform [Dawkins, J. V. and Hemming, M., *Makromol. Chem.*, **1975**, *176*, 1777]:

$M\,(g/mol) \times 10^{-3}$	1900	867	670	411	160	98.2	51	19.8	10.3	3.7
$V_e\frac{1}{5}(cm^3)$	23.75	24.55	25.20	25.80	27.30	28.20	29.40	31.30	32.50	34.00

The Mark–Houwink constants for polystyrene in chloroform at 35°C may be taken as $K = 4.9 \times 10^{-3}$ cm^3/g and $a = 0.79$. Assuming that universal calibration is valid, construct a calibration curve for the molecular weight-elution volume of poly(dimethylsiloxane) in chloroform at 35°C; the Mark–Houwink constants for this polymer in chloroform at 35°C are $K = 5.4 \times 10^{-3}$ cm^3/g and $a = 0.77$.

14. A sample of poly(dimethylsiloxane) in chloroform is injected into the same gel permeation chromatograph used in Problem 13. Using a differential refractometer as detector, the following primary data of refractive index difference, Δn, and elution volume, V_e, were obtained:

$\Delta n \times 10^5$	0.6	3.4	12.4	15.0	11.7	4.1	1.0
$V_e(cm^3) \times \frac{1}{5}$	32.00	31.20	30.41	29.72	29.02	28.19	27.40

Using your results from Problem 13, calculate and show graphically the molecular-weight distribution of the poly(dimethylsiloxane). You may assume that Δn is proportional to concentration and that the proportionality factor is independent of molecular weight.

SUGGESTIONS FOR FURTHER READING

BOYD, R. H., and PHILLIPS, P. J., *The Science of Polymer Molecules*, Cambridge: Cambridge University Press, **1993**.

BRAUN, J.-M., and GUILLET, J. E., "Study of Polymers by Inverse Gas Chromatography," *Advan. Polymer. Sci.*, (**1976**), *21*, 107.

CALDWELL, K. D., "Field-Flow Fractionation," *Anal. Chem.*, (**1988**), *60*(17), 959A–960A, 962A–966A, 968A, 970A–971A.

CARRAHER, C. E., JR., *Seymour/Carraher's Polymer Chemistry*, 5th ed., New York: Marcel Dekker, Inc., **2000**.

CHANDA, M., *Advanced Polymer Chemistry*, New York: Marcel Dekker, Inc., **2000**.

COLLINS, E. A., BARES, J., and BILLMEYER, F. W., JR., *Experiments in Polymer Science*, New York: Wiley-Interscience, **1973**.

DAVID, D. J., and MISRA, A., *Relating Materials Properties to Structure. Handbook and Software for Polymer Calculations and Materials Properties*. Lancaster, PA: Technomic Publishing Co., **1999**.

FLORY, P. J., *Principles of Polymer Chemistry*. Ithaca, N.Y.: Cornell University Press, **1953**, Chaps. 7, 14.

GEDDE, U. W., *Polymer Physics*. London: Chapman & Hall, **1996**.

GIDDINGS, J. C., "Characterization of Colloid-Sized and Larger Particles by Field-Flow Fractionation," *Polymer Mater. Sci. Eng.*, **1988**, *59*, 156–159.

GIDDINGS, J. C., "Polymer Characterization by Thermal Field-Flow Fractionation and Related Methods," *Polymer Mater. Sci. Eng.*, **1988**, *59*, 1–3.

HIEMENZ, P. C., *Polymer Chemistry*. New York: Dekker, **1984**, Chaps. 2, 9.

MCCAFFERY, E. L., *Laboratory Preparation for Macromolecular Chemistry*. New York: MGraw-Hill, **1970**.

MEYERHOFF, G., "Die viscosimetrische Molekulare Wichtsbestimmung von Polymeren," *Fortschr. Hochpolymer.-Forsch.*, **1961**, *3*, 59.

MOORE, W. R., "Viscosities of Dilute Polymer Solutions," *Progr. Polymer Sci.* (A. D. Jenkins, ed.), **1967**, *1*, 1.

MENCER, H. J., "Efficiency of Polymer Fractionation: A Review," *Polymer Eng. Sci.* **1988**, *28*(8), 497–505.

OUANO, A. C., "Quantitative Data Interpretation Techniques in Gel Permeation Chromatography," *J. Macromol. Sci., Rev. Macromol. Chem.*, **1973**, *C9*, 123.

PORTER, R. S., and JOHNSON, J. F., "Gel Permeation Chromatography," *Progr. Polymer Sci.* (A. D. Jenkins. ed.), **1970**, *2*, 201.

SANDLER, S. R., KARO, W., BONESTEEL, J.-A., and PEARCE, E. M. *Polymer Synthesis and Characterization. A Laboratory Manual*. San Diego: Academic Press, **1998**.

SPERLING, L. H., *Introduction to Physical Polymer Science*, 3rd ed. New York: Wiley Interscience, **2001.**

SUN, S. F., *Physical Chemistry of Macromolecules. Basic Principles and Issues*, New York: Wiley Interscience, **1994**.

TUNG, L. H., "Recent Advances in Polymer Fractionation," *J. Macromol. Sci., Rev. Macromol. Chem.*, **1971**, *C6*, 51.

16

Thermodynamics of Solutions of High Polymers, and Unperturbed Chains

INTRODUCTION

Probably the most important single physical property of a high polymer is its molecular weight and, as seen in Chapter 14, the absolute measurement of this property is based on the properties of *solutions* of high polymers. It is therefore important that the polymer chemist should have a general understanding of the thermodynamics of polymer solutions and an appreciation of how the thermodynamic properties of such solutions differ from those formed by small molecules. Because of the very large size of the polymeric solute molecules compared to solvent molecules, many of the traditional concepts of solutions must be modified. For example, even the concept of an ideal solution requires modification.

The theoretical basis for the understanding of polymer solutions was developed independently by Flory[1] and Huggins[2] some 60 years ago in essentially equivalent treatments. In this chapter the treatment and notation of the former will be followed.

[1]Flory, P. J., *J. Chem. Phys.*, **1942,** *10,* 51.
[2]Huggins, M. L., *J. Phys. Chem.*, **1942,** *46,* 151.

DEFINITION OF AN IDEAL SOLUTION

A traditional definition of an ideal solution is that it is a system in which Raoult's law (1) is obeyed.

$$a_1 = X_1 = (1 - X_2) = \frac{P_1}{P_1^o} \tag{1}$$

In this equation a_1 is the thermodynamic activity of the solvent, X_1 the mole fraction of the solvent, X_2 the mole fraction of the solute, P_1 the pressure of solvent vapor above the solution, and P_1^o the vapor pressure of the pure solvent. A thermodynamic consequence of this definition is that the chemical potential of the solvent in an ideal solution is given by (2), where μ_1^o is the chemical potential of the pure solvent, or, in other words, the Gibbs free energy per mole.

$$\mu_1 = \mu_1^o + RT \ln X_1 \tag{2}$$

It will be recalled that (1) and (2) were used in Chapter 14 in the derivation of the osmotic pressure of ideal solutions.

All solutions, including polymer solutions, obey (1) and (2) in the limit of infinite dilution where they become ideal. For solutions of small solute molecules, deviations from ideality become negligible when both the mole fractions and weight fractions of the solute are small. However, the molecular weights of high-polymer solutes are so drastically different from those of typical solvents that vanishingly small *mole* fractions of solutes (i.e., $X_2 \to 0$) are obtained even though the *weight* fraction of the polymer is very large. Under such conditions, the mole fraction and the adherence of the system to Raoult's law are not useful indicators of ideality. As a numerical example, consider a polymer of molecular weight $\overline{M}_2 = 10^6$, a solvent of molecular weight $M_1 = 10^2$ and a solution that is 91% by weight of polymer. The mole fraction of the solvent is

$$X_1 = \frac{n_1}{n_1 + n_2} = \frac{w_1/M_1}{w_2/M_2 + w_1/M_1} = \frac{M_2/M_1}{w_2/w_1 + M_2/M_1} \tag{3}$$

where n_i is the number of moles and w_i is the weight of component i. Inserting the numbers $w_2/w_1 = 10$ and $\overline{M}_2/M_1 = 100$ into (3), we find that $X_1 = 0.999$. Thus, while the solvent makes up only 9% of the solution by weight (and the solution must be expected to behave very nonideally), the mole fraction of the solvent is sufficiently close to unity to *suggest* ideal behavior. This contradiction indicates that the thermodynamic *activity* of a solvent in an ideal polymeric solution is not equal to the mole fraction, whereas the two are equal for solutions of small molecule solutes. Therefore, Raoult's law is of little use for polymer solutions. As will be shown, the ideal polymer solution is better described as one in which the activity of the solvent is equal to the *volume fraction* of the solvent. This definition can be extended to ordinary solutions, since volume fraction and mole fraction for such solutions are very nearly the same, and this definition is, therefore, of more general validity than the traditional one.

The traditional definition of an ideal solution [i.e., (1) and (2)] is based on the *interchangeability* of solvent and solute particles. This means that the replacement of a solvent molecule by a solute molecule results in no change in the net molecular attractions and repulsions. As a consequence, an equivalent traditional definition of an ideal solution is one in which the formation of the solution from n_1 moles of pure solvent and n_2 moles of pure solute meets the following thermodynamic requirements:

$$\Delta H_{\text{mix}} = 0 \tag{4}$$

$$\Delta S_{\text{mix}} = -R(n_1 \ln X_1 + n_2 \ln X_2) \tag{5}$$

Early experimental work on polymer solutions indicated that deviations from ideality depend only weakly on the temperature. In view of the thermodynamic relationship describing the temperature dependence of the free energy of mixing,

$$\frac{\partial}{\partial T}\left(\frac{\Delta G_{\text{mix}}}{T}\right)_P = -\frac{\Delta H_{\text{mix}}}{T^2} \tag{6}$$

this observation suggests that ΔH_{mix} is not generally large. Therefore, the major cause for deviations from ideality lies in the failure of (5) to describe the *entropy* of mixing in the preparation of polymer solutions. Accordingly, we shall first devote our attention to a theoretical treatment of the entropy of mixing of solvent and solute, beginning with a simple treatment applicable to small molecules. An extension will then be made to macromolecular solutes. Finally, we shall consider the enthalpy and free energy of mixing that accompany the formation of a polymer solution.

ENTROPY OF MIXING OF SOLVENT AND SOLUTE

Small-Molecule Solutes Dissolved in Small-Molecule Solvents

Let us approach the entropy of mixing of solute and solvent from the point of view of a statistical theory in which the solvent and solute particles are assigned to positions in an imaginary lattice. For the present, consider both the solute and solvent molecules to be spherical particles of the same size. Assume also that the replacement of a solvent molecule by a solute molecule results in no change in the interactions of neighboring particles. Under these conditions the entropy of mixing of the solvent and solute arises solely from the greater number of lattice arrangements (i.e., configurations) possible for the solution, as compared to the solvent.

In Figure 16.1 a finite two-dimensional representation of the imaginary lattice is shown, with open circles representing solvent molecules and closed circles denoting solute molecules. In this situation there are no restrictions on the placing of particles in the lattice positions.

Let N_0 be the number of lattice positions, N_1 be the number of solvent molecules, and N_2 the number of solute particles. The assumption is made that all the lattice positions are occupied, and this may be described by

$$N_0 = N_1 + N_2 \tag{7}$$

The problem is to calculate the number of ways that the N_0 molecules may be assigned to the N_0 positions in the lattice. If we imagine for the moment that all the N_0 molecules are distinguishable, then there are N_0 ways to choose the first molecule to drop randomly into the lattice. For each of these N_0 ways of choosing the first molecule there are $N_0 - 1$ ways to choose the second one, and for each of the $N_0(N_0 - 1)$ ways of choosing the first two molecules there are $N_0 - 2$ ways to choose the third, and so on. Therefore, for N_0 *distinguishable* particles, the number of arrangements in the lattice, Ω', is given by

$$\Omega' = N_0(N_0 - 1)(N_0 - 2)(N_0 - 3) \cdots (1) = N_0! \tag{8}$$

However, although a solvent molecule may be distinguished from a solute molecule, we cannot distinguish solvent molecules from each other nor solute mole-

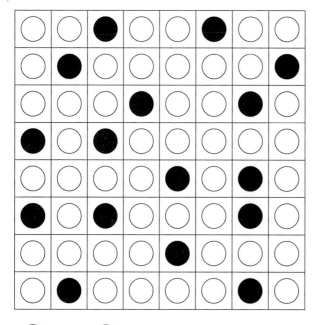

Solute   Solvent

Figure 16.1 Two-dimensional lattice representation of a solution.

Entropy of Mixing of Solvent and Solute   **495**

cules from each other. Since (8) assumes that we can, we must correct Ω' by the number of ways of permuting N_1 solvent molecules and N_2 solute molecules among themselves. Thus, the number of *distinguishable* arrangements in the lattice is

$$\Omega = \frac{N_0!}{N_1! N_2!} \tag{9}$$

For the starting materials (i.e., pure solvent and solute), the number of distinguishable arrangements is

$$\Omega_1 = \Omega_2 = \frac{N_1!}{N_1!} = \frac{N_2!}{N_2!} = 1 \tag{10}$$

According to Boltzmann the entropy of a system is given by

$$S = k \ln \Omega \tag{11}$$

where k is Boltzmann's constant (i.e., $k = 1.38 \times 10^{-23}$ J/deg-molecule) and Ω is the number of distinguishable configurations or arrangements of the system as calculated above. The entropy of mixing of the solvent and solute, in the simple case at hand, is due solely to changes in the possible number of configurations of the mixed and unmixed systems and may be written as

$$S_c = \Delta S_{\text{mix}} = S - S_1 - S_2 \tag{12}$$

or

$$S_c = \Delta S_{\text{mix}} = k \ln \Omega - k \ln \Omega_1 - k \ln \Omega_2 \tag{13}$$

where the symbol S_c denotes this configurational entropy. Following substitution of (9) and (10) into (13), equation (14) is obtained.

$$S_c = \Delta S_{\text{mix}} = k[\ln N_0! - \ln N_1! - \ln N_2!] \tag{14}$$

To proceed further, use is made of the Stirling approximation for the factorials of large numbers. This states that

$$N! = \left(\frac{N}{e}\right)^N \tag{15}$$

or

$$\ln N! = N \ln N - N \tag{16}$$

Substitution of (7) and (16) into (14) leads directly to the expression

$$S_c = \Delta S_{mix} = -k\left[N_1 \ln \frac{N_1}{N_1 + N_2} + N_2 \ln \frac{N_2}{N_1 + N_2}\right] \qquad (17)$$

Finally, from the relationships $R = N_A k$ and $N_i = N_A n_i$, where N_A is Avogadro's number and n_i represents the number of moles of the ith component, (17) may be transformed to the form shown in (18), in which X_i represents the mole fraction.

$$S_c = \Delta S_{mix} = -R[n_1 \ln X_1 + n_2 \ln X_2] \qquad (18)$$

Polymeric Solutes Dissolved in Small-Molecule Solvents

The simplicity of the treatment described above depends on the interchangeability of solute and solvent molecules. Despite its simplicity, the expression shown in (18) describes quite well the entropy of mixing of solvent and solute molecules whose ratio of sizes (i.e., molar volumes) range from unity to about 3 or 4. However, when the solute is a polymer molecule whose molar volume may be thousands of times greater than that of a solvent molecule, the concept of interchangeability of a solvent and a solute particle is absurd and must be abandoned. Yet, this simple general approach to the entropy of mixing is so attractive that it is worthwhile to retain it and modify the model to take into account the vast difference in size of solvent and solute molecules.

The model chosen[1,2] for a polymer solute is that of a long-chain molecule consisting of x chain segments, each *segment* being of the same size (i.e., volume) as a solvent molecule. Solvent *molecules* and polymer chain *segments* may now be considered interchangeable in the lattice model of the solution. A simple analogy is to regard each solvent molecule as a white pearl and the polymer molecule as a string of x black pearls. The sizes of the black and white pearls are the same and hence are interchangeable in the lattice positions. Thus, according to this model, the number of chain segments (i.e., the number of pearls in the string) is related to the size ratio by

$$x = \frac{\overline{V_2}}{\overline{V_1}} \qquad (19)$$

where $\overline{V_1}$ and $\overline{V_2}$ are the molar volumes of solvent and solute, respectively.

The assumption that solvent molecules and chain segments are interchangeable permits the derivation to proceed in an analogous manner to the simple case just described for small-molecule solutes. The only difference is that the

[1]Flory, op. cit.
[2]Huggins, op. cit.

x chain segments of the polymer solute must be connected. This means that chain segments cannot be assigned to lattice positions in a completely random manner because each segment must have at least one other polymer segment adjacent to it. The lattice model of the polymer solution may be illustrated as in Figure 16.2. The relationship between the number of lattice positions and the number of solvent and solute molecules now becomes

$$N_0 = N_1 + xN_2 \tag{20}$$

where, as before, N_0, N_1, and N_2 are the number of lattice positions, solvent molecules, and solute molecules, respectively.

To calculate the number of configurations of the mixture, first consider the number of ways in which a polymer molecule of *x* chain segments may be added to the lattice when *i* polymer molecules are already present. The number of vacant positions into which the first segment of this $(i + 1)$st molecule may be placed, and hence the number of ways in which this may be done is $(N_0 - xi)$. Having chosen one of these vacant sites in which to place the first segment of the $(i + 1)$st polymer molecule, we must now consider how many ways there are to place the second segment of the polymer. Letting Z be the coordination number of a lattice site (i.e., the number of nearest neighbor sites to any given site), the second segment must go into one of the Z sites that are nearest neighbors to the one in which the first segment was placed. However, not all of these Z sites

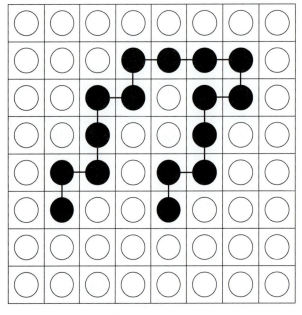

Solvent   Chain segments of the polymer

Figure 16.2 Two-dimensional lattice representation of a polymer molecule in solution.

may be available. Some may already be occupied by segments from the first i polymer molecules present in the lattice. Let the symbol f_i be the probability that a site adjacent to the one occupied by a segment of the $(i + 1)$st molecule is already occupied by a segment from one of the first i molecules. Then the number of ways in which the second segment may be added is $Z(1 - f_i)$. For the addition of the third segment, one of the sites adjacent to the second segment is already occupied by the first segment. Hence the number of ways to add the third segment, and succeeding segments, is $(Z - 1)(1 - f_i)$. The number of configurations of the $(i + 1)$st molecule in the lattice, v_{i+1}, is the product of these numbers for the individual segments, namely,

$$v_{i+1} = \underbrace{(N_0 - xi)}_{\substack{\text{1st} \\ \text{segment}}} \cdot \underbrace{Z(1 - f_i)}_{\substack{\text{2nd} \\ \text{segment}}} \cdot \underbrace{(Z - 1)(1 - f_i)}_{\substack{\text{3rd} \\ \text{segment}}} \cdot \underbrace{(Z - 1)(1 - f_i)}_{\substack{\text{4th} \\ \text{segment}}} \cdots \tag{21}$$

or

$$v_{i+1} = (N_0 - xi)Z(Z - 1)^{x-2}(1 - f_i)^{x-1} \tag{22}$$

As an approximation to f_i, it may be assumed (with a reasonably small error) that the average probability that a given site is not occupied by segments of the first i molecules is equal to the fraction of sites remaining empty after the first i molecules have been added. Thus,

$$(1 - f_i) \approx \frac{N_0 - xi}{N_0} \tag{23}$$

The use of (23) and the simplifying approximation, $Z(Z - 1)^{x-2} \approx (Z - 1)^{x-1}$, enables (22) to be reduced to the more compact form shown in (24).

$$v_{i+1} = (N_0 - xi)^x \left(\frac{Z - 1}{N_0} \right)^{x-1} \tag{24}$$

Finally, as a third and convenient approximation, it can be shown by Stirling's formula (15) that the first term of (24) can be written, with little error, in the factorial form which yields

$$v_{i+1} = \frac{(N_0 - xi)!}{[N_0 - x(i + 1)]!} \left(\frac{Z - 1}{N_0} \right)^{x-1} \tag{25}$$

Expression (25) describes the number of configurations of just one polymer molecule in the lattice. The number of ways to place the N_2 indistinguishable polymer molecules is the product of these individual numbers of configurations

Entropy of Mixing of Solvent and Solute

divided by the number of ways of permuting the N_2 molecules among themselves. Thus,

$$\Omega = \frac{1}{N_2!} \left(\prod_{i=1}^{N_2} \nu_i \right) = \frac{1}{N_2!} \left(\prod_{i=0}^{N_2-1} \nu_{i+1} \right) \tag{26}$$

Substitution of (25) into (26) and writing out the terms in the product yields

$$\Omega = \frac{1}{N_2!} \left[\frac{N_0!}{(N_0 - x)!} \cdot \frac{(N_0 - x)!}{(N_0 - 2x)!} \cdot \frac{(N_0 - 2x)!}{(N_0 - 3x)!} \right.$$

$$\left. \cdots \frac{[N_0 - (N_2 - 1)x]!}{(N_0 - N_2x)!} \right] \left(\frac{Z - 1}{N_0} \right)^{N_2(x-1)} \tag{27}$$

which on cancellation of terms simplifies to

$$\Omega = \frac{N_0!}{N_2!(N_0 - xN_2)!} \left(\frac{Z - 1}{N_0} \right)^{N_2(x-1)} = \frac{N_0!}{N_1! \, N_2!} \left(\frac{Z - 1}{N_0} \right)^{N_2(x-1)} \tag{28}$$

Because the solvent molecules can occupy the remaining lattice sites in only one way, (28) is the total number of arrangements or configurations of the solution. The reader should note that the expression in (28) is the same as that for the ordinary solution, i.e. (9), except for the factor $[(Z - 1)/N_0]^{N_2(x-1)}$. Substitution of typical numbers into this factor (i.e., $Z \sim 10$, $x = 10^3$, $N_0 \sim 10^{23}$, $N_2 \sim 10^{18}$) shows that $[(Z - 1)/N_0]^{N_2(x-1)} \ll 1$. This means that there are many fewer configurations possible for the polymer solutions compared to small-molecule solutions.

The total configurational entropy is given by (11). Substitution of (28) into (11), and the use of Stirling's approximation for the factorials, leads in a straightforward way to

$$S_c = -k \left[N_1 \ln \frac{N_1}{N_1 + xN_2} + N_2 \ln \frac{N_2}{N_1 + xN_2} - N_2(x - 1) \ln \frac{Z - 1}{e} \right] \tag{29}$$

where e is the base of natural logarithms. The configurational entropy in (29) represents the entropy of mixing of the perfectly ordered pure solid polymer, for which $S = 0$, with pure solvent. This mixing process can be broken down into two reversible steps. The first step is conversion of the perfectly ordered polymer to a randomly oriented polymer and this process corresponds, in our model, to the random placement of polymer molecules into the lattice without a solvent. The second process consists of adding solvent molecules to the empty sites in the lattice and represents the entropy of mixing of the randomly oriented polymer with the solvent. If the entropy change of the first process is designated as ΔS_{dis} and that of the second process ΔS_{mix}, expression (30) holds.

$$\Delta S_{mix} = S_c - \Delta S_{dis} \tag{30}$$

In order to use (30) to evaluate the entropy of mixing of a randomly oriented polymer with the solvent, it is important to note that S_c is given by (29) and ΔS_{dis} is given by (29) *under the special condition* that $N_1 \rightarrow 0$ (i.e., no solvent has been added to the lattice). Thus,

$$\Delta S_{dis} = \lim_{N_1 \to 0} S_c = k \left[N_2 \ln x + N_2(x-1) \ln \frac{Z-1}{e} \right] \tag{31}$$

and so, subtracting (31) from (29), we obtain

$$\Delta S_{mix} = -k \left(N_1 \ln \frac{N_1}{N_1 + xN_2} + N_2 \ln \frac{xN_2}{N_1 + xN_2} \right) \tag{32}$$

If the approximation is made that x can be replaced by the ratio of the partial molar volumes (i.e., $x = \overline{V}_2/\overline{V}_1$), the expression can be changed to a molar basis (i.e., $k = R/N_A$) and this last result may be written as

$$\Delta S_{mix} = -R(n_1 \ln \phi_1 + n_2 \ln \phi_2) \tag{33}$$

where n_i is the number of moles of ith component and ϕ_i is the volume fraction:

$$\phi_i = \frac{n_i \overline{V}_i}{\sum_i n_i \overline{V}_i} \tag{34}$$

A comparison of (33) with (18) shows that the ideal entropy of mixing of a polymeric solute with a solvent is given by an expression that is similar to the classical ideal entropy of mixing of small-molecule solute and solvent molecules. The only difference is that, for polymer solutions, the volume fraction rather than the mole fraction is the dimensionless measure of concentration. The mole fractions and volume fractions of small molecule solutes in solution are essentially the same, and it would appear that (33) is the more general expression which reduces to (18) as the molecular sizes become equal.

The expression in (33) refers to a monodisperse polymer solute in which all the molecules are the same size. For a polydisperse polymer with a distribution of molecular weights, the term $n_2 \ln \phi_2$ must be replaced by $\sum_i n_i \ln \phi_i$, where the summation goes over the solute particles only. The theory is also useful for polymer blends (in which the "solvent" is also polymeric)[1] and has even been modified to work for rod-like chains stiff enough to form liquid-crystalline phases.[2]

ENTHALPY OF MIXING OF SOLVENT AND POLYMERIC SOLUTE

When a polymeric solute is added to a solvent, an enthalpy change occurs because solvent–solvent and solute–solute interactions are replaced by solvent–solute interactions. According to the lattice theory, such interactions may be represented by the numbers and types of nearest neighbors in the lattice. A nearest-neighbor interaction may be defined as a lattice contact, so there will be three types of such contacts (i.e., $[1, 1]$, $[2, 2]$, and $[1, 2]$, respectively). The process of dissolution may then be written in terms of the change in these contacts

$$\tfrac{1}{2}[1, 1] + \tfrac{1}{2}[2, 2] \rightarrow [1, 2] \tag{35}$$

The energy change associated with the formation of one solvent–solute contact, $\Delta w_{1,2}$, is given by

$$\Delta w_{1,2} = w_{1,2} - \tfrac{1}{2}(w_{11} + w_{22}) \tag{36}$$

Now if $P_{1,2}$ is the average number of solvent–solute contacts (i.e., 1,2 contacts) over all the lattice configurations, then the enthalpy of mixing of the solvent and solute is

$$\Delta H_{\text{mix}} = \Delta w_{1,2} P_{1,2} \tag{37}$$

per solute particle. The fraction of the lattice sites that are adjacent to those which contain a polymer segment and are at the same time occupied by solvent molecules (i.e., the probability of a 1,2 contact) should be given approximately by ϕ_1, the volume fraction of solvent. The total number of all the different types of contacts of each of the $x - 2$ internal polymer segments (not counting segments to which each is chemically bound) is $Z - 2$, while the two terminal segments will each have $Z - 1$ such contacts. The total number of 1,2 contacts for each polymer molecule is then

$$P_{1,2} = [(x - 2)(Z - 2) + 2(Z - 1)]\phi_1 \tag{38}$$

For large values of Z, $P_{1,2} \approx Zx\phi_1$ and the enthalpy of mixing of N_2 polymer molecules with N_1 solvent molecules is given by

---

[1]Balsara, N. P., in *Physical Properties of Polymers Handbook*, J. E. Mark, Ed. (New York: Springer-Verlag, **1996**), p. 257.
[2]Yang, Y., Kloczkowski, A., Mark, J. E., Erman, B., and Bahar, I., *Macromolecules*, **1995**, *28*, 4920.

$$\Delta H_{mix} = Zx\phi_1 \, \Delta w_{1,2} N_2 \tag{39}$$

From the definition of volume fractions, ϕ_1 and ϕ_2, it is easily shown that $xN_2\phi_1 = N_1\phi_2$. Then, on a molar basis, the enthalpy of mixing is given by

$$\Delta H_{mix} = Z \, \Delta w_{1,2} n_1 \phi_2 N_A = Z \, \Delta W_{1,2} n_1 \phi_2 \tag{40}$$

where $\Delta W_{1,2} = N_A \, \Delta w_{1,2}$. It is convenient to describe the interaction energy per mole of solvent, $Z\Delta W_{1,2}$, in terms of a dimensionless interaction parameter multiplied by RT. Thus, defining $Z\Delta W_{1,2} = \chi_1 RT$, the enthalpy of mixing (40) becomes

$$\Delta H_{mix} = RT\chi_1 n_1 \phi_2 \tag{41}$$

The interaction parameter χ_1, given by $Z\Delta W_{1,2}/RT$, is the energy change (in units of RT) that occurs when a mole of solvent molecules is removed from the pure polymer (where $\phi_2 = 1$) and is immersed in an infinite amount of pure polymer (where $\phi_2 = 0$). Because of the approximate nature of the lattice theory, χ_1 is found to depend on the concentration of the solution. According to its definition, χ_1 depends inversely on the temperature. χ_1 is generally positive, with values at 25°C and at infinite dilution being near 0.5. According to (41), the fact that χ_1 is positive means that the dissolution of a polymeric solute in a solvent is generally an endothermic process.

FREE ENERGY OF MIXING OF POLYMERIC SOLUTE WITH SOLVENT

The Gibbs free energy change for the dissolution of a polymeric solute is easily obtained from the well-known thermodynamic expression

$$\Delta G = \Delta H - T\Delta S \tag{42}$$

because substitution of (33) and (41) into (42) leads immediately to the result

$$\Delta G_{mix} = RT(\chi_1 \, n_1 \, \phi_2 + n_1 \ln \phi_1 + n_2 \ln \phi_2) \tag{43}$$

It is now possible to answer the question of whether dissolution of a polymer in a solvent occurs with positive or negative free energy. The answer, (43), clearly depends on the concentration of the solution and on the sign and magnitude of χ_1. As the temperature is increased, χ_1 decreases and dissolution becomes thermodynamically more favorable.

CHEMICAL POTENTIAL AND ACTIVITY OF SOLVENT

Chapter 14 demonstrated that the presence of a solute lowers the chemical potential of a solvent from its value in the pure solvent. This is fundamentally important for the derivation of osmotic pressure changes. A theoretical expression for the reduction of the chemical potential of the solvent is readily obtained from the free energy of mixing since, by definition, the chemical potential of a solvent in a solution relative to that in the pure solvent is given by

$$\mu - \mu_1^\circ = \left(\frac{\partial [G_{\text{soln}} - G_1^\circ]}{\partial n_1} \right)_{T,P,n_2} = \left(\frac{\partial \Delta G_{\text{mix}}}{\partial n_1} \right)_{T,P,n_2} \tag{44}$$

Partial differentiation of ΔG_{mix}, (43), with respect to n_1 at constant T gives

$$\mu_1 - \mu_1^\circ = RT \left[\frac{n_1}{\phi_1} \left(\frac{\partial \phi_1}{\partial n_1} \right)_{n_2} + \ln \phi_1 + \frac{n_2}{\phi_2} \left(\frac{\partial \phi_2}{\partial n_1} \right)_{n_2} + \chi_1 \phi_2 + \chi_1 n_1 \left(\frac{\partial \phi_2}{\partial n_1} \right)_{n_2} \right] \tag{45}$$

The partial derivatives of the expression above may be evaluated from the definition of volume fraction. Volume fractions may be written in terms of the molar volume ratio, $x = \overline{V}_2/\overline{V}_1$, as

$$\phi_1 = \frac{n_1}{n_1 + xn_2} \tag{46}$$

and

$$\phi_2 = \frac{xn_2}{n_1 + xn_2} \tag{47}$$

The result, after some manipulation, is

$$\mu_1 - \mu_1^\circ = RT \left[\ln(1 - \phi_2) + \left(1 - \frac{1}{x} \right) \phi_2 + \chi_1 \phi_2^2 \right] \tag{48}$$

For an ideal solution, in which the solvent and solute molecules are identical in size and shape (i.e., $x = 1$), in which $\Delta H_{\text{mix}} = 0$ (i.e., $\chi_1 = 0$), and in which volume fraction and mole fraction are equal, equation (48) reduces to the classical expression shown in (49) (see also Chapter 14, equation 5)

$$\mu_1 - \mu_1^\circ = RT \ln X_1 \tag{49}$$

where X_1 is the mole fraction of solvent. In the case of a heterogenous polymer, x in equation (48) is replaced by $\overline{x}$ (i.e., by the *average* degree of polymerization).

In classical solution theory, equation (49) is valid only for ideal solutions. However, to retain the simple form of this equation for nonideal solutions, the activity of the solvent in a solution is defined by

$$\mu_1 - \mu_1^\circ = RT \ln a_1 \tag{50}$$

Hence, the activity of the solvent in a solution of polymer is given by

$$\ln a_1 = \ln (1 - \phi_2) + \left(1 - \frac{1}{x}\right)\phi_2 + \chi_1\phi_2^2 \tag{51}$$

Experimental values of a_1 can be obtained from the vapor pressures of the solvent in the solution to its value in the pure state: $a_1 = p_1/p_1^{\circ}$. and an analogous treatment beginning with (43) yields equation (52) for the activity of the solute:

$$\ln a_2 = \ln \phi_2 + (1 - x)(1 - \phi_2) + \chi_1 x\phi_2^2 \tag{52}$$

THE OSMOTIC PRESSURE OF POLYMERIC SOLUTIONS

Recall from Chapter 14 that the fundamental precept underlying our understanding of the osmotic pressure of a solution is the reduction in the chemical potential of a solvent that occurs when a solute is added to it and the compensating increase in solvent chemical potential that accompanies an increase in the external pressure applied to the solution. A repeat of the osmotic pressure derivation of Chapter 14, but with (48) instead of (49) being used to describe the dependence of chemical potential on concentration, leads to the expression

$$\Pi = -\frac{RT}{\overline{V}_1}\left[\ln (1 - \phi_2) + \left(1 - \frac{1}{x}\right)\phi_2 + \chi_1\phi_2^2\right] \tag{53}$$

where Π is the osmotic pressure and $\overline{V}_1$ is the molar volume of solvent. If the logarithmic term in (53) is expanded in the well-known series (54),

$$\ln (1 - \phi_2) = -\phi_2 - \tfrac{1}{2}\phi_2^2 - \tfrac{1}{3}\phi_2^3 - \cdots \tag{54}$$

the result shown in (55) is obtained.

$$\Pi = \frac{RT}{\overline{V}_1}\left[\frac{\phi_2}{x} + (\tfrac{1}{2} - \chi_1)\phi_2^2 + \tfrac{1}{3}\phi_2^3 + \cdots\right] \tag{55}$$

It is sometimes more convenient to express the concentration of the solution in terms of weight per unit volume (i.e., $c = w_2/V_{\text{soln}}$) than in terms of volume fraction. This can be introduced into (55) by recognizing that the volume fraction of the polymer in the solution may be written as

$$\phi_2 = \frac{n_2\overline{V}_2}{n_1\overline{V}_1 + n_2\overline{V}_2} = \frac{w_2\overline{V}_2}{M_2 V_{\text{soln}}} = \frac{c\overline{V}_2}{M_2} \tag{56}$$

Strictly speaking, the symbols $\overline{V}_i$ in (56) are partial molar volumes. However, we may assume them to be equal to the respective molar volumes, an assumption that is compatible with the other approximations of the theory. Substitution of (56) into (55) and elimination of x by its definition, $x = \overline{V_2}/\overline{V_1}$, gives after some re-arrangement

$$\frac{\Pi}{c} = \frac{RT}{M_2}\left[1 + \frac{\overline{v}_2^2 M_2}{\overline{V}_1}\left(\frac{1}{2} - \chi_1\right)c + \frac{\overline{v}_2^3 M_2}{3\overline{V}_1}c^2 + \cdots\right] \tag{57}$$

In the expression in (57), $\overline{v}_2$ is the specific volume of polymer (actually the partial specific volume) or, in other words, $\overline{v}_2 = \overline{V}_2/M_2$. In the case of a heterogenous polymer, M_2 should be replaced by the number-average molecular weight of the polymer.

It is seen that (57) is in the form of a virial equation which was written in Chapter 14 in the form

$$\frac{\Pi}{c} = \frac{RT}{M_2}(1 + \Gamma c + g\Gamma^2 c^2 + \cdots) \tag{58}$$

and in which the second and third virial coefficients Γ and $g\Gamma^2$, respectively, are given by

$$\Gamma = \frac{\overline{v}_2^2 M_2}{\overline{V}_1}\left(\frac{1}{2} - \chi_1\right) \tag{59}$$

and

$$g\Gamma^2 = \frac{\overline{v}_2^3 M_2}{3\overline{V}_1} \tag{60}$$

From experimental studies, such as an examination of the dependence of Π/c on concentration, it is possible to derive values of χ_1 provided, of course, that the densities or specific volumes of the polymer and the solvent are known. Some values of χ_1 for several polymers in a variety of solvents at 25°C and in the limit of zero concentration are shown in Table 16.1. Also included in Table 16.1 are the interaction energies $RT\chi_1$ which represent the energy change that occurs when 1 mol of solvent is transferred from the pure solvent and is immersed in an infinite amount of polymer.

All the polymer–solvent systems in Table 16.1 show positive values of χ_1. These positive values indicate that replacement of a solvent molecule by a polymer molecule occurs with a positive enthalpy change (i.e., is an endothermic process). Negative values of χ_1 would indicate exothermic dissolution, with $\Delta H_{mix} < 0$. Such negative values of χ_1 are observed only very rarely, even though they would be more likely in systems in which either the polymer or the solvent is polar (thereby increasing the attractive interactions on mixing). It must be concluded, then, that at 25°C dissolution is an endothermic process. Dissolution will

TABLE 16.1 POLYMER-SOLVENT INTERACTION ENERGIES AT 25°C AND AT INFINITE DILUTION

Polymer	Solvent	χ_1	$RT\chi_1$ (cal/mol)
Natural rubber	Benzene	0.42	249
Poly(dimethylsiloxane)	Chlorobenzene	0.47	278
Polyisobutylene	Benzene	0.50	296
Polyisobutylene	Cyclohexane	0.43	254
Polyisobutylene	n-Pentane	0.49	290
Polystyrene	Methyl ethyl ketone	0.47	278
Polystyrene	Ethylbenzene	0.40	237
Polystyrene	Cyclohexane	0.505	299
Poly(methyl methacrylate)	Chloroform	0.377	223
Poly(methyl methacrylate)	4-Heptanone	0.509	301
Poly(methyl methacrylate)	Tetrahydrofuran	0.447	265

only be favored thermodynamically (i.e., $\Delta G < 0$) at those temperatures and compositions for which the negative terms in the free-energy expression (43) are numerically greater than the enthalpy of mixing. Thus, for thermodynamically favored dissolution, expression (61) must hold.

$$-(n_1 \ln \phi_1 + n_2 \ln \phi_2) > \chi_1 n_1 \phi_2 \tag{61}$$

SOME OTHER APPLICATIONS OF THE FLORY–HUGGINS THEORY

Another application of this theory is to the liquid–liquid equilibrium that occurs when a polymer precipitates out of solution. This would occur in a solution polymerization carried out in a poor solvent as the polymer molecular weight became sufficiently high. It would also occur as the solubility of a polymer decreases in a fractional precipitation (by lowering the temperature, or by adding a nonsolvent), as was described in Chapter 15. The precipitated phase is *not* pure polymer, but rather a polymer solution in equilibrium with the supernatant solution. Compared to the supernatant phase, it has a much smaller volume, is more concentrated in polymer, and the polymer is of higher molecular weight (this is the reason for carrying out a fractional precipitation).

Experimentally, the precipitation temperature T_p is measured as a function of composition, and the usual binodial curve separates the one-phase from the two-phase region, as is typically illustrated in physical chemistry texts for partially miscible mixtures of hexane and nitrobenzene. The features that are the same as these simple liquid mixtures are (i) the existence of the critical temperature T_c at the maximum in the curve, (ii) the locating of the compositions of the two phases when phase separation occurs by end points of the "tie lines," and (iii) the applicability of the "lever rule" to the lengths of the tie-line segments to obtain the relative amounts of the two phases in equilibrium.

The features that are unique to polymer solutions are (i) the occurrence of a binodial curve for each molecular weight, (ii) the occurrence of all of these binodals at very low polymer concentrations, and (iii) the following effects that occur with increasing the polymer molecular weight: (a) the binodials move to lower concentrations, (b) their maxima (critical temperatures) increase, and (c) they become narrower. These features are shown schematically in Figure 16.3. They are all reproduced at least semi-quantitatively by the Flory–Huggins theory. The relevant equation is

$$1/T_c = (1/\Theta)[1 + (1/\psi_1)(1/x^{1/2} + 1/2x)] \qquad (62)$$

where Θ is the Flory theta temperature already mentioned in Chapter 15. It is here seen to be the critical temperature for infinite molecular polymer, but it also is important with regard to the removal of excluded-volume effects, as discussed further below. The quantity ψ_1 is an entropy of mixing parameter, x is the degree of polymerization of the polymer, and $(1/x^{1/2} + 1/2x)$ is called the "size function." Treating experimental data in terms of this equation involves plotting $1/T_c$ against $(1/x^{1/2} + 1/2x)$, as shown in Figure 16.4. The intercept gives the theta temperature, Θ, and its value, in conjunction with the slope, gives the entropy of mixing parameter.

The Flory–Huggins theory has also been used to treat the equilibrium between a polymer solution and precipitated crystalline polymer. This is called crystal–liquid equilibria, or melting-point depression (as induced by a solvent). The origin of the decrease in melting point, T_m, is shown schematically in Figure 16.5. The presence of solvent increases the entropy change for the process since the polymer can mix with it as well as disorder itself conformationally. This decreases the denominator in $T_m = \Delta H_m / \Delta S_m$, which is the ratio of heat of melting to the entropy of melting. The equation that results from application of the Flory–Huggins theory is

$$(1/T_m - 1/T_m^\circ)/v_1 = (R/\Delta H_m)(\overline{V_2}/\overline{V_1})(1 - \chi_1 v_1) \qquad (63)$$

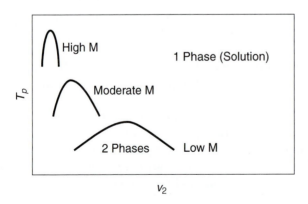

Figure 16.3   Illustrative binodials for liquid–liquid phase separations in polymer solutions.

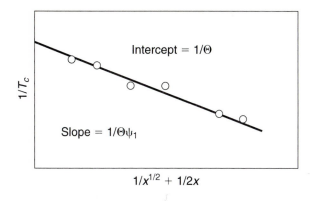

Intercept = 1/Θ

$1/T_c$

Slope = $1/\Theta\psi_1$

$1/x^{1/2} + 1/2x$

Figure 16.4 Illustrative plot of liquid–liquid phase separation data.

where T_m and T_m° are the melting points in the presence and absence of solvent, respectively, v_1 is the volume fraction of solvent present, ΔH_m is the heat of fusion per mole of repeat units, $\overline{V}_2$ and $\overline{V}_1$ are the molar volumes of polymer and solvent, respectively, and χ_1 is the usual Flory–Huggins interaction parameter. Thus, plotting $(1/T_m - 1/T_m^\circ)/v_1$ against v_1 gives a linear relationship with an intercept yielding ΔH_m, and a slope yielding χ_1.

The heat of fusion thus obtained is for the 100% crystalline polymer, since it is obtained by measurements of temperatures, which are intensive properties (since they do not depend on the number of crystallites studied in the experiment). This maximum value of the heat of fusion is much used to estimate percent crystallinity, from the ratio of the actual heat of melting (measured by calorimetry) to this maximum value obtained from melting point depression measurements. It is also used to obtain entropies of melting from $\Delta S_m = \Delta H_m/T_m^\circ$, which are useful thermodynamic measures of chain flexibility. The values of ΔS_m thus provide complementary information to the characteristic ratio of the chain dimensions, which is a structural measure of chain stiffness as described below. A

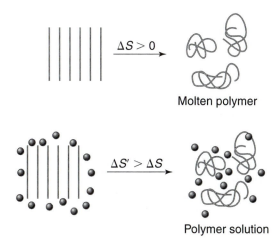

$\Delta S > 0$

Molten polymer

$\Delta S' > \Delta S$

Polymer solution

Figure 16.5 Depression of the melting point by having solvent present to mix with the chains leaving the crystalline lattice.

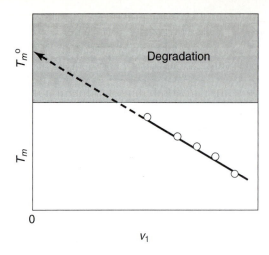

Figure 16.6 Extrapolation of solvent-depressed melting points to the value that would be exhibited by the undiluted polymer had it not degraded.

final application is the estimation of melting points of polymers that degrade before they melt, for example to get measures of chain flexibility as just described. Values of the depressed melting point are obtained at different compositions and then plotted against v_1. Extrapolation through the higher temperatures at which the polymer would degrade to $v_1 = 0$ then gives the melting point T_m^o the polymer would have exhibited had it not degraded. This is illustrated in Figure 16.6.

A final example is the use of the Flory–Huggins theory in the interpretation of network swelling. The theory is used for the thermodynamics of mixing of the swelling solvent with the network chains, and one of the molecular theories of rubberlike elasticity is used to characterize the mechanical deformation of the chains being stretched out in the dilation caused by the entering solvent. The resulting equation is given in Chapter 21.

LIMITATIONS OF THE THEORY

According to (57), a single value of χ_1 should be sufficient to describe the osmotic pressure, as well as other thermodynamic properties, over a wide range of polymer concentrations. However, experimental tests show that χ_1 depends on the *concentration* of the solution with the values usually increasing as ϕ_2 increased. Some typical results designed to test the theory are shown in Figure 16.7 for polystyrene in methyl ethyl ketone and for polyisobutylene in cyclohexane.

The failure of the theory to account for the dependence of χ_1 on the composition of the solution is caused by the approximations inherent in the theory. However, despite these shortcomings, the simple lattice theory gives us, in a relatively simple and instructive way, a semiquantitative appreciation of the factors involved in the thermodynamics of polymer solutions. Further developments of

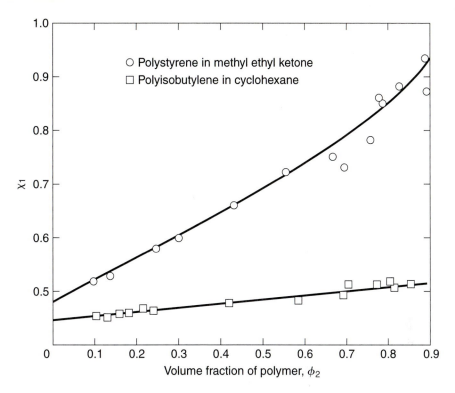

Figure 16.7 Dependence of polymer–solvent interaction parameter χ_1 on concentration at 25°C. [Reproduced from Flory, P. J., *Discussions Faraday Soc.*, **1970, 49;** with permission of the Faraday Division of the Chemical Society of London.]

the theory do account crudely for the dependence of χ_1 on composition, but these treatments are quite complex and are beyond the scope of this book.

DILUTE SOLUTIONS, AND CHAINS UNPERTURBED BY EXCLUDED VOLUME EFFECTS

There are a number of shortcomings of the Flory–Huggins theory in addition to the unanticipated concentration dependence generally shown by χ_1. Another is the fact the theory does not give reliable results for dilute polymer solutions. This problem can be traced to assumptions in the lattice calculation, such as the assumption that the site occupancy probability is uniform throughout the lattice. However, in dilute solutions this probability would be significantly higher within a polymer coil than it is between coils, as is illustrated in Figure 16.8, and an average probability will generally not suffice.

The Flory–Krigbaum theory is one approach to treating dilute polymer solutions. This theory provides molecular insights into the excluded volume effects

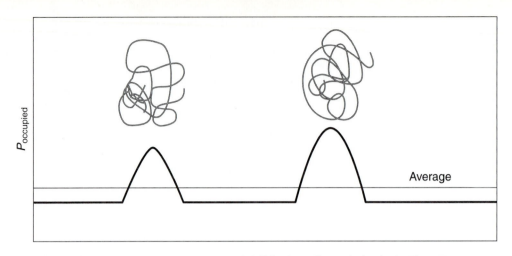

Figure 16.8 Schematic site occupancy probabilities for a dilute solution in the Flory–Huggins lattice calculation.

that have been such a problem theoretically. This is sufficiently important that some background information is needed to put it into context.[1]

The simplest example of an excluded volume effect occurs in gases, which can range in structure from monoatomic species to multiatomic molecules. This type of excluded volume is totally *inter*molecular, and thus can be suppressed by reducing the pressure to zero ($p \rightarrow 0$). As pointed out below, this makes the excluded volume b per molecule negligible relative to the now very large molar volume of the gas, i.e., $\overline{V} - b \cong \overline{V}$. Of course, there is a corresponding intermolecular-excluded volume in the case of polymer molecules; in the case of solutions, this can be made negligible by going to infinite dilution ($c \rightarrow 0$). However, this intermolecular part is greatly augmented by an *intra*molecular component because polymer molecules are so long that different parts of the same molecule can interfere with one another. Such interfering segments are close together in space, but they are generally well separated along the chain trajectory. For this reason this type of excluded volume effect is also called a "long-range interaction." There is no way to suppress this interaction because it is inherent to the long-chain structures of polymer molecules. Instead, an approach is used that parallels Boyle's treatment of non-ideal gases.[2] In this treatment, the criterion for ideality, $p\overline{V}/RT = 1$, is evaluated as a function of the pressure of the gas. As p increases, this criterion deviates from ideality, with the excluded volume effects increasing the pressure and the intermolecular attractions decreasing it. This is illustrated by the b and a corrections, respectively, in the van der Waals equation for non-ideal gases:

[1]Mark, J. E., *J. Chem. Educ.* **2002**, *79*, 1437.
[2]Atkins, P. W., *Physical Chemistry*, 4th ed. (Oxford: Oxford University Press, **1990**).

$$(p + a/\overline{V}^2)(\overline{V} - b) = RT \qquad (64)$$

where a and b are positive and specific for a particular gas. When this equation is rearranged to

$$p = RT/(\overline{V} - b) - a/\overline{V}^2 \qquad (65)$$

it becomes more obvious that a nonzero excluded volume ($b \neq 0$) increases p while nonzero intermolecular attractions ($a \neq 0$) decreases it. At high temperatures, the excluded volume effect predominates, so the temperature is lowered until the increase in pressure it causes is just offset by the pressure decrease from the intermolecular attractions. This is illustrated in Figure 16.9, for lowering the temperature from T_1 to T_2, to T_3, to T_B. At the unique temperature T_B, called the "Boyle temperature" for a particular gas, the excluded volume interactions are still there but are now nullified by being exactly offset by the intermolecular attractions. The real gas thus "masquerades" as an ideal gas, sometimes over a very large range in pressure.

This idea was extended to polymers, with the criterion for ideality now being based on the van't Hoff relationship described in Chapter 14:

$$\pi M/cRT = 1 \qquad (66)$$

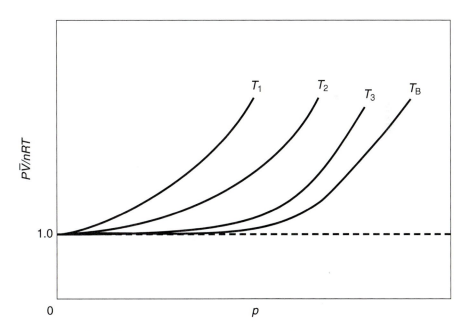

Figure 16.9 Location of the Boyle temperature T_B, at which a real gas acts ideally.

where π is the osmotic pressure of the polymer solution. Thus, this ratio is monitored as a function of concentration, c, for a range of decreasing temperatures. If the solvent chosen is a very powerful one for the polymer being characterized, then the effective excluded volume effect is very large, with a correspondingly large size of the polymer random coil. Lowering the temperature makes the solvent poorer and causes the polymer segments to prefer interacting with themselves rather than with the solvent. These enhanced attractions between polymer segments compress the random coil (and, if the temperature is decreased sufficiently, will eventually precipitate the polymer from the solution). In this case, the compressive effects are made just sufficiently large to nullify the chain expansion effects from the excluded volume interactions. This occurs at a unique temperature already described as the Θ temperature, the Flory temperature, or the ideal temperature. It was previously encountered under liquid–liquid equilibria as the critical temperature for infinite molecular weight polymer. In any case, at Θ the excluded volume interaction is there, but it is *without effect* and the chain acts as if it had zero cross-sectional area. Properties of chain molecules in this Θ state are said to be "unperturbed," and are much used to establish structure–property relationships, for example, how *para*-phenylene groups in a repeat unit increase the stiffness of the corresponding polymer.

A specific example of the above approach would be carrying out the light-scattering measurements described in Chapter 14 using a polymer in a solvent at its theta temperature. The radius of gyration $<S^2>_0$ thus obtained could be converted to the unperturbed end-to-end dimensions by the relationship $<r^2>_0 = 6<S^2>_0$. (Statistical averages in this area are designated by either overbars or bracketing fences $<\ >$). Traditionally, these dimensions are expressed as the "characteristic ratio" (CR) defined as the high molecular-weight limit of the ratio $<r^2>_0/nl^2$, where n is the number of skeletal bonds, and l is the skeletal bond length. An alternative approach to obtain values of the CR would be the intrinsic viscosity measurements described in Chapter 15. An important point is that the extent to which the CR exceeds unity is a standard (structural) measure of chain stiffness.

One complication is the fact that there is a "specific solvent" effect, in that different Θ solvents can give somewhat different values of the characteristic ratio, even at very similar Θ temperatures. The effect can be around 20% in the case of polar polymers, and some of the effect has been attributed to differences in dielectric constant that leads to changes in conformational energies and thus changes in unperturbed dimensions. Another possible effect arises because some solvents interact differently with different chain conformations and change the unperturbed dimensions in this way.

It is also possible to obtain chain dimension measurements in a thermodynamically good solvent, and then apply approximate corrections for the chain expanding away from its unperturbed value. As mentioned in Chapter 15, this expansion is characterized (linearly) by the ratio $\alpha = [<r^2>/<r^2>_0]^{1/2}$. An estimate of its magnitude in a good solvent requires some measure of the polymer-solvent interactions that contribute to this chain expansion. In the case of the Flory–Orofino equation developed for this purpose, it is the second virial coefficient A_2 described in Chapter 14:

$$A_2 M/[\eta] = (2^{5/2} \pi N_{\text{avo}}/27\Phi) \ln[1 + (\pi^{1/2}/2)(\alpha^2 - 1)] \qquad (67)$$

where π is the numerical constant 3.1416, N_{avo} is Avogadro's number, and Φ is a constant.

In conformational analyses of polymers, it is important to obtain values for the temperature coefficient of any property of interest, for example $d\ln<r^2>_0/dT$ in the case of the unperturbed dimensions. One approach is to use a series of Θ solvents where the various values of Θ are different enough to provide a reliable estimate of this coefficient. Because of the specific solvent effect, it is important that these Θ solvents should be structurally similar, for example a series of alcohols. A second approach would be to carry out the measurements in a good solvent as a function of temperature, and then use equation (67) to obtain α at each temperature of interest. Finally, there is the option of doing force-temperature measurements on elastomeric networks, as is described in Chapters 18 and 21.

The expansion of a polymer coil in a thermodynamically good solvent is used in some practical applications. A good example is the addition of a polymer to a motor oil to reduce the variation of its viscosity at different temperatures. The principle is illustrated in Figure 16.10, where T_{ambient} is the temperature of the automobile when the engine is not in operation, and $T_{\text{operating}}$ is its temperature when it is. The upper stippled region designates viscosities that are sufficiently high to interfere with starting the engine (particularly in a very cold climate). The lower stippled region specifies the viscosity range that corresponds to good lubrication. If a motor oil is formulated with hydrocarbon fractions that provide good lubrication when the engine is hot, then its viscosity may increase unacceptably during cooling, along line A. The opposite approach would be to formulate the oil so that it has a desirably low viscosity at the ambient temperature. However, its viscosity may then become too low for adequate lubrication once the temperature increases in the operating engine, as illustrated along line B. To avoid these two situations, a polymer is dissolved in the oil so that its chains expand as the oil (the "solvent") becomes thermodynamically better when the temperature rises. Because of the polymer chain expansion, the normal decrease in viscosity with increased temperature is now moderated, and follows line C into the lubrication-acceptable region at the engine's operating temperature.

MORE REFINED THEORIES

In spite of the successes of the theories mentioned, some properties of polymer solutions are beyond their reach. One example involves excess volumes, which are volume changes that occur on mixing that can correspond to either expansions or shrinkages. These departures from simple additivity of volumes require more

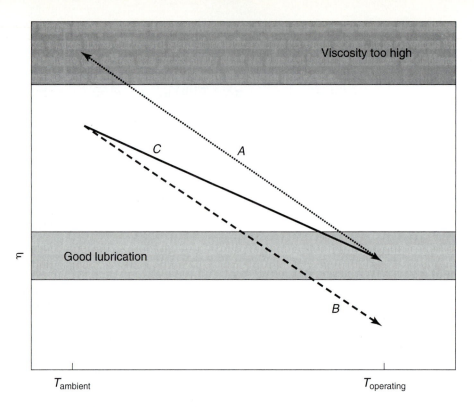

η

Viscosity too high

C

A

Good lubrication

B

$T_{ambient}$

$T_{operating}$

Figure 16.10 Dissolving a polymer in a motor oil to control its viscosity-temperature profile.

sophisticated theories. An example is the equation-of-state approach, in which equation-of-state information is needed for both the polymer and solvent. Thermal expansion coefficients are sometimes available for this purpose, but compressibilities only rarely, and this has limited the use of these theories. Scaling theories also exist that focus on obtaining universal laws to describe the properties of polymers and their solutions. The goal here is to find functional forms, but without the coefficients required to actually predict the magnitudes of the properties of interest.

These more refined approaches are described in some of the references cited in the Suggestions for Further Reading.

STUDY QUESTIONS

1. Using the simple lattice theory, evaluate the number of distinguishable arrangements, Ω, for 100 cm^3 of a solution of styrene in *m*-xylene in which the concentration of styrene is $10^{-2}M$ at 20°C(ρ_{xylene} = 0.861 g/cm^3 at 20°C). Calculate the entropy of mixing in the preparation of this solution.

2. Suppose that the dissolved styrene monomer in Problem 1 is completely converted to a dissolved polymer of $\overline{DP} = 1000$. Assuming a coordination number of 12, calculate the number of distinguishable arrangements of solute and solvent in this solution and compare with the result of Problem 1. Calculate the entropy of mixing in the preparation of 100 cm³ of this solution and the entropy of polymerization of the styrene.

3. Using the lattice theory, calculate the entropy change for the conversion of 10^{-6} mol of polymer of $\overline{DP} = 500$ from a perfectly ordered state to a randomly ordered state. (Assume a coordination number of 12.)

4. Calculate the enthalpy of mixing when 10^{-5} mol of poly(methyl methacrylate) of $\overline{M}_n = 10^5$ g/mol and of $\rho = 1.20$ g/cm³ are dissolved in 150 g of chloroform ($\rho = 1.49$ g/cm³) at 20°C. Assume that the volumes are additive.

5. Show that $x N_2 \phi_1 = N_1 \phi_2$ when the symbols are as defined in the text.

6. Calculate the Gibbs free energy change in the preparation of the solution of Problem 4.

7. Derive equation (48).

8. Derive equation (57).

9. The second virial coefficient in toluene solution of a test sample of polystyrene was found to be 219 cm³/g at 25°C. The partial specific volume of the polymer in the solution was found to be 0.91 cm³/g. Evaluate the interaction parameter χ_1.

10. Two series of solutions employ the same low molecular-weight solvent having a molar volume $\overline{V}_1 = 100$ cm³/mol; each solution has $n_1 = 1 \times 10^{23}$ solvent molecules. Series A contains low molecular-weight solute having $\overline{V}_2 = 200$ cm³/mol, and numbers of molecules $10^{-23} n_2 = 0.1, 0.3, 0.5, 0.8, 2.0,$ and 4.0. Use the Flory–Huggins equation to calculate values of the entropy of mixing $\Delta S_M/n_o k$ where n_o is the total number of molecules and k is the Boltzmann constant. Show the results as a function of solute volume fraction v_2 for this series of solutions. Do the same for Series B, which contains high molecular-weight solute having $\overline{V}_2 = 200{,}000$ cm³/mol, and $10^{-23} n_2 = 0.00004, 0.0001, 0.0004, 0.001, 0.003,$ and 0.006. Explain the marked differences between the two curves at constant v_2.

11. A solution of poly(methyl methacrylate) ($\rho = 1.20$ g/cm³, $\overline{M}_n = 3.52 \times 10^5$) in chloroform ($\rho = 1.49$ g/cm³) has been prepared by adding 50.0 mg of polymer to 150 g of the solvent. Estimate the osmotic pressure of the resulting solution.

12. Use the appropriate Flory–Huggins equation to plot the osmotic pressures (atm) that ought to be developed by solutions of high molecular-weight poly(dimethylsiloxane) in benzene, separated from pure benzene by a membrane permeable only to the benzene, as a function of polymer volume fraction $v_2 = 0.1, 0.25, 0.5, 0.8,$ and 0.9. Assume a temperature of 25°C and values of χ_1 of 0.5, 0.65, 0.75, 0.8, and 0.87, for these five volume fractions, respectively. Plot on the same diagram the theoretical osmotic pressures that should be developed by a hypothetical *athermal* solvent for this polymer having the same molar volume as benzene. Extrapolate both curves to infinite dilution, and explain why the curve for the athermal solvent lies above that for the benzene.

13. The following results for liquid–liquid equilibria have been reported for isotactic poly(isopropyl acrylate) in *n*-decane by Wessling, R. A., Mark, J. E., and Hughes, R. E. [*J. Phys. Chem.*, **1966,** *70,* 1909].

$10^{-6}\overline{M}$ (g/mol)	$T_c(°C)$
0.123	159.2
0.278	165.3
0.531	168.7
1.04	171.8
2.77	174.6

Calculate the ideal temperature Θ and entropy of mixing parameter ψ_1 for this system.

14. Quinn, F. A., Jr. and Mandelkern, L. [*J. Am. Chem. Soc.*, **1958**, *80*, 3178; **1959**, *81*, 6533] reported the following melting points for the polyethylene, α-chloronaphthalene system:

v_1	$T_m(°C)$
0.00	137.6
0.07	135.2
0.16	131.7
0.23	128.4
0.32	126.0

Using 1.25 and 0.92 cm³/g for the specific volumes of polymer and solvent, respectively, calculate the heat of fusion ΔH_u (cal per mol of CH_2CH_2 units), and the interaction parameter χ_1 for this system. Also, calculate the value of the alternative interaction parameter $B = RT\chi_1/\overline{V}_1$ at $T = 130°C$ (where R is the gas constant and $\overline{V}_1$ is the molar volume of the solvent).

15. According to Crescenzi, V. and Flory, P. J. [*J. Am. Chem. Soc.*, **1964**, *86*, 141], methyl ethyl ketone is a Θ solvent for poly(dimethylsiloxane) (PDMS) at 20°C. Under these conditions, a fraction of this polymer having a molecular weight of 1.20×10^6 had an intrinsic viscosity $[\eta]_\Theta = 0.835$ dl/g. Calculate the characteristic ratio $<r^2>_o/nl^2$ of PDMS using $\Phi = 2.5 \times 10^{21}$ dl cm^{-3}mol^{-1}, where $<r^2>_o$ is the unperturbed dimension, n is the number of skeletal bonds, and $l = 1.64$Å is the skeletal bond length.

16. A fraction of *trans*-1,4-polybutadiene $[CH_2—CH=CH—CH_2—]_x$ was studied by osmometry and viscometry in a thermodynamically good solvent. The results were $\overline{M} = 6.7 \times 10^4$ g/mol, $A_2 = 1.21 \times 10^{-3}$ cm³mol/g² and $[\eta] = 1.01$ dl/g. Calculate the chain-expansion factor α using the Flory–Orofino equation, and the characteristic ratio $CR = <r^2>_o/nl^2$ of this polymer using $\Phi = 2.1 \times 10^{21}$ dl cm^{-3}mol^{-1}, and an average bond length squared of $l^2 = (1/4)[3(1.53)^2 + 1.34^2] = 2.20$ Å².

17. Chiang, R. [*J. Phys. Chem.*, **1966**, *70*, 2348] reported the following intrinsic viscosities for a high molecular-weight fraction of polyethylene in three structurally similar Θ solvents:

Solvent	$\Theta(°C)$	$[\eta]_\Theta$ (dl/g)
Dodecanol-1	137.3	0.943
Decanol-1	153.3	0.928
Octanol-1	180.1	0.881

Calculate the temperature coefficient of the unperturbed dimensions $d\ln<r^2>_o/dT$ from these data, assuming that specific solvent interactions have no effect on this coefficient.

18. A solution of a polymer in an athermal solvent was found to have the following values of the intrinsic viscosity:

T(°C)	$[\eta]$ (dl/g)
20.	1.000
40.	1.021
60.	1.042
80.	1.064
100.	1.085

Thermal expansion coefficients for the polymer and solvent over this range of temperature are $\beta = 0.80 \times 10^{-3}$ and $1.00 \times 10^{-3}\ K^{-1}$, respectively. Calculate $d\ln<r^2>_o/dT$ at the average temperature $60°C$, using the value of the chain expansion factor $\alpha = 1.400$ independently determined at this temperature.

SUGGESTIONS FOR FURTHER READING

BILLMEYER, F. W., Jr., *Textbook of Polymer Science*, 3rd. ed. New York: Wiley, **1984**, Chap. 7.

BINDER, K. (Ed.), *Monte Carlo and Molecular Dynamics Simulations in Polymer Science*. New York: Oxford University Press, **1995.**

DES CLOIZEAUX, J., and JANNINK, G., *Polymers in Solution. Their Modelling and Structure*. New York: Oxford University Press, **1990**.

DE GENNES, P.-G., *Scaling Concepts in Polymer Physics*. Ithaca, NY: Cornell University Press, **1979**.

FLORY, P. J., *Principles of Polymer Chemistry*. Ithaca, N.Y.: Cornell University Press, **1953**, Chap. 12.

FLORY, P. J., "Thermodynamics of Polymer Solutions," *Discussions Faraday Soc.*, **(1970)**, *49*, 7.

FORSMAN, W. C. (Ed.), *Polymers in Solution. Theoretical Considerations and Newer Methods of Characterization*. New York: Plenum Press, **1986**.

FREED, K. F., *Renormalization Group Theory of Macromolecules*. New York: John Wiley & Sons, **1987**.

GROSBERG, A. Y., and KHOKHLOV, A. R., *Statistical Physics of Macromolecules*. New York: Am. Inst. Phys., **1994**.

HIEMENZ, P. C., *Polymer Chemistry*. New York: Dekker, **1984**, Chap. 8.

ISIHARA, A., and GUTH, E., "Theory of Dilute Macromolecular Solutions," *Advan. Polymer Sci.*, **1967**, *5*, 233.

KAMIDE, K., and DOBASHI, T., *Physical Chemistry of Polymer Solutions. Theoretical Background*. Amsterdam: Elsevier, **2000**.

KLENIN, V. J., *Thermodynamics of Systems Containing Flexible-Chain Polymers*. Amsterdam: Elsevier, 1999.

MORAWETZ, H., *Macromolecules in Solution*. New York: Wiley-Interscience, 1975.

PETHRICK, R. A., and DAWKINS, J. V. (Eds.), *Modern Techniques for Polymer Characterization*. New York: Wiley & Sons, Inc., **2000**.

RICHARDS, E. G., *An Introduction to Physical Properties of Large Molecules in Solution*. Cambridge: Cambridge University Press, **1980**.

SUN, S. F., *Physical Chemistry of Macromolecules, Basic Principles and Issues*. New York: Wiley Interscience, **1994**.

TERAOKA, I., *Polymer Solutions. An Introduction to Physical Properties*. New York: Wiley & Sons, Inc., **2002**.

YAMAKAWA, H., *Modern Theory of Polymer Solutions*. New York: Harper & Row, **1971**.

17

Morphology, Glass Transitions, and Polymer Crystallinity

MORPHOLOGICAL CHANGES IN POLYMERS

Most long-chain synthetic polymers show a characteristic sequence of changes as they are heated. All linear polymers are glasses at low temperatures. As the temperature is raised, a certain point is reached at which the polymer changes from a glass to a rubber. This change is known as the glass transition temperature (T_g). When heated above T_g, amorphous polymers pass successively through rubbery, gumlike, and finally liquid states with no clear demarcation between the different phases. On the other hand, crystalline polymers remain flexible and thermoplastic above T_g until the temperature is raised to the crystalline melting temperature, T_m. At this point the polymer melts to a viscous liquid at a sharply defined temperature. The crystalline melting phenomenon occurs when sections of adjacent chains are packed together in a regular array, and the melting point represents the temperature at which these microcrystallites are thermally disrupted. The different characteristics of amorphous and crystalline polymers are illustrated in Figure 17.1. Extensive cross-linking may distort this picture and mask the transitions, for example by suppressing crystallization.

When a crystalline and a noncrystalline modification of the same polymer are compared, it is sometimes found that the glass transition temperature is higher or lower in the crystalline form, and the temperature span of the rubbery phase may be truncated. In fact, in many crystalline polymers the rubbery phase of the amorphous state is replaced by a flexible, thermoplastic phase, which is less extensible than that of a conventional elastomer, but much tougher.

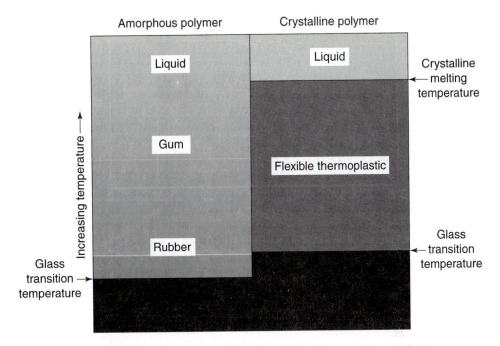

Figure 17.1 Different transition behavior exhibited by amorphous and crystalline polymers.

CHARACTERISTICS OF THE VISCOELASTIC STATE

Elementary physics divides matter into three quite distinct categories: solids, liquids, and gases. *Solids* are substances that occupy a fixed shape and volume (i.e., they do not flow). *Liquids* flow readily but occupy a fixed volume. *Gases* flow and change their volume easily. This simple view of the universe does not account for the properties of open-chain or moderately cross-linked polymers. Most polymers are neither classical solids nor liquids. They are *viscoelastic* materials. The viscoelastic state has the characteristics of *both* the solid and liquid states.

Consider a piece of lightly cross-linked natural rubber. At rest on a laboratory bench, it has all the characteristics of a solid (definite shape, fixed volume, no evidence or liquid flow.) But if we stretch the material or apply pressure to one part of it, it will change shape like a liquid. Of course, if we release the tension or pressure, it will revert to its original shape, sometimes quickly, but, for some materials, over a long period of time. These are some of the characteristics of the viscoelastic state, and these unusual properties account for the valuable properties of macromolecules. By possessing these properties, polymers can be used for many applications where conventional solids (like sodium chloride or benzoic acid) or ordinary liquids would be unsuitable.

In this section we examine some of the peculiarities of the viscoelastic state; later we will consider two critically important phenomena: the glass transition (T_g) and polymer crystallinity.

Terminology

It is relatively easy for an experimenter to describe a particular substance as a glass, elastomer, gum, or liquid. However, more precise terminology is needed if, for example, we wish to say that one elastomer is "stiffer" than another.

The following terms are commonly used. *Tensioning* is the act of attempting to stretch a material. Quantitatively, the amount of force applied per unit cross sectional area is known as the tensile *stress*, σ. The amount of stretch induced in the sample is the tensile *strain*, ε. These two factors are related by the equation, $\sigma = E\varepsilon$, where the constant, E, is known as the *modulus* or Young's modulus. This constant is a characteristic of a particular polymer. It is a measure of stiffness or rigidity, or resistance to deformation in general. A substance that has a high modulus has a high rigidity and can be deformed only by the application of appreciable stress. It is important to recognize that resistance to stretching (modulus or rigidity) is not the same as resistance to breaking. A material may be extremely rigid but may break under low tension (although this combination of properties is quite rare for polymers). Thus, another term—*tensile strength*—is needed to define the ultimate load that a material can bear without breaking. Piano wire has a tensile strength of roughly 2×10^{10} dyn/cm^2. Poly(methyl methacrylate) in the glassy state has a value of only 8×10^8 dyn/cm^2. *Tenacity* is the stress at the breaking point of the material. *Impact strength* is the resistance of a material to breakage when subjected to a sharp blow. *Toughness* is another measure of resistance to breakage, and is defined as the total energy input to the breaking point. It is thus the integral $\int f dL$, where f is the force and L the distance.

Viscoelastic behavior can also be defined in terms of *shear*. Shear occurs when, for example, a piece of polymer is deformed without a volume change by the application of force to the top of the sample in a direction parallel to the surface on which the sample is resting. The resistance to deformation is described as the *shear modulus*, G. The application of hydrostatic pressure to a material will bring about a decrease in its volume. The resistance to contraction is given by the *bulk modulus*, B.

Creep is the process by which a polymer undergoes a slow change of shape or a flowing action when subjected to a constant force such as gravity. It is the most obvious "liquid-like" feature of the viscoelastic state, and one that is obviously detrimental to most polymer applications. *Stress relaxation* measures the decrease in retractive force with time.

Some of these terms are employed in the discussion of the testing of polymers. (Chapter 21).

The Glassy State

Two typical glassy polymers are poly(methyl methacrylate) and polystyrene. They are dimensionally stable at room temperature and do not creep over reasonably short periods of time. Although they have some flexibility as thin samples, they

are not elastomeric. They have moderate strength and some impact resistance, but they can be shattered by a sharp blow with a hammer. At room temperature both are well below their glass transition temperatures ($\simeq 100°C$). Because most glassy polymers are used as structural materials, rigidity and resistance to creep, high impact strength, and a high glass transition temperature are the most desirable properties. Unfortunately, high-impact strength is more a characteristic of elastomers than of glasses. Hence, a compromise may be needed between high rigidity and high impact strength.

The phenomenon of brittleness or susceptibility to shattering is connected with the ease with which a small fracture can be propagated throughout the polymer. The application of impact force to a glassy polymer will result in a separation of chains at the point of impact and in a cleavage of skeletal bonds. The separation of individual chains takes place by overcoming the weak van der Waals attractions. More energy is required to break covalent bonds. The problem is that even a small impact-induced reentrant discontinuity at the surface of a glassy polymer will cause an increase in the local stress at that point. This focuses most of the impact stress on relatively few chemical bonds. Hence, the crack propagates rapidly as both covalent bonds and the van der Waals attractions give way. The fundamental problem with a glass is that the conversion of the impact energy into the breakage of bonds is one of the few mechanisms available for dissipation of that energy. By contrast, an elastomer *absorbs* the energy into harmless molecular motions. In the glassy state, the polymer molecules are rigidly fixed in place, and such impact-absorbing molecular mobility is not present. Hence, the material shatters.

Three ways have been developed to overcome this problem. First, the formation of covalent crosslinks between chains has the effect of increasing the amount of energy needed to propagate a crack. There are now more covalent bonds to be broken. Second, finely divided materials known as "fillers" may be added to the polymer. These serve to interrupt the propagation of cracks. Third, an impact-absorbing second phase may be incorporated into the polymer matrix. This last approach is discussed in a later section.

The Rubbery State

Elastomers are polymers that are well above their glass transition temperatures. Although a few elastomers contain microcrystallites, most are amorphous materials. The elastomeric characteristics become lost if the temperature is high enough to induce gumlike behavior. However, as discussed earlier, crosslinks between the chains maintain the elastomeric character of the material at high temperatures.

The phenomenon of elasticity has been mentioned previously, and this is discussed further in Chapter 21. Elastomeric materials have the properties of liquids—they change shape or flow readily when subjected to weak forces. But they differ from liquids in their capacity to reassume their original shape once the distorting force has been removed. This behavior is a consequence of the high mobility of the backbone bonds. The chains can readily undergo conformational

changes to yield contracted (coiled) or extended conformations in response to an external force. The ability of an elastomer to reassume its original shape after deformation can be attributed to two features. First, the highly coiled molecular conformations are preferred for entropic reasons, and a stretched polymer will contract when the tension is released to allow a maximization of the entropy. Second, cross-links prevent the chains from slipping past each other when the applied tension is excessive or prolonged. As a consequence, many cross-linked elastomers have a high modulus and high strength when stretched. The cross-links also reduce the tendency for the polymer to creep.

Elastomers have two other characteristics: they are "resilient" and they absorb "solvents" and swell to a surprising degree. *Resilience* is the ability of a material to bounce back. A rubber ball dropped from a height onto the floor rebounds. The degree to which the rebound height compares with the initial height is a measure of resilience. Resilience reflects the ability of an elastomer to absorb energy, store it, and then *rapidly* return that energy following the impact. In molecular terms, it reflects the ease with which the polymer chains can undergo a rapid conformational distortion from their preferred state, and the ease and rapidity with which they reassume their original preferred condition and release the stored free energy. In turn, this depends on the ease of torsion of the backbone bonds (Chapter 18) and on a low cohesive energy between adjacent chains (individual segments of chains must be able to slip past each other readily). The same molecular characteristics explain why elastomers have a high impact resistance. They absorb energy in conformational changes on impact and then release it. They do not absorb the energy by cleavage of chemical bonds.

The ability of an elastomer to absorb solvents and undergo a marked expansion of volume is a well-known phenomenon. The swelling of rubber in benzene or toluene is an example of this process, which is discussed further in Chapter 21. The solvent absorption is facilitated because the polymer chains are in flexural motion at room temperature, and there is a favorable entropy of mixing. Thus, the solvent molecules can readily penetrate the polymer matrix and bring about a separation of the large molecules. Eventually, the polymer will dissolve unless cross-links are present. Cross-links place a limit on the degree to which the chains can separate and, hence, on the extent of swelling. In fact, the degree of swelling in a solvent can be used as a measure of the cross-link density. Some elastomers are used in technology in a solvent-expanded form. The "solvents" are known as plasticizers or "oil extenders."

Elastomers are not generally used for structural applications because of their dimensional instability. However, they are widely employed for energy absorbing applications, tires, and as flexible, impact resistant coatings.

The Flexible, Nonelastomeric State

Many microcrystalline polymers, at temperatures above their glass transitions, exist in the form of flexible film- or fiber-forming materials. Such materials show some elastomeric-type properties (i.e., flexibility, impact resistance, some limited

elasticity, ability to swell in solvents, etc.), but their main characteristic is a *combination* of flexibility and dimensional stability.

Such materials can be viewed as amorphous elastomers in which are embedded temporary crosslink sites in the form of microcrystalline domains. The nature of these crystalline domains is discussed in a later section. Here, it is sufficient to point out that the crystalline regions impart a certain stiffness to the material (compared to the purely amorphous material), and reduce the tendency of the polymer to creep. Very high degrees of crystallinity can induce brittleness, and compromises may be necessary between the advantages of crystallinity and the advantages of impact resistance. "Tough" polymers represent a result of this compromise.

Plasticizers are often used to extend the range of properties of flexible polymers. The effect of the plasticizer is to swell the amorphous regions, lower the cohesion between the chains, and allow these regions to function as flexible elastomeric domains. Plasticizers are also employed to modify polymers that have a normal T_g value above room temperature. In such cases they serve to lower the glass transition temperature until it is below room temperature.

The Liquid State

If cross-links are absent, both amorphous and microcrystalline polymers soften at high temperatures. The melting process allows the chains to separate from each other and permits viscous flow to occur readily. Few high polymers are used as technological materials in the molten state. Decomposition reactions occur at the high temperatures required for melting. However, the molten state is used extensively for the fabrication of polymers (see Chapter 20).

Some Phenomenological Models

As already mentioned, polymers exhibit time-retarded effects, as well as equilibrium elastic effects, because of the difficulty of chains moving around in a highly viscous, entangled medium. Thus, the viscous effects associated with liquids have to be combined with the elastic effects characteristic of solids to understand the "viscoelasticity" of a polymeric material.

The simplest models used for this purpose are phenomenological, meaning that they treat a material without reference to its molecular structure, and the techniques apply equally well to a polymer film, a wooden board, or a slab of concrete. This turns out to have the advantage of great generality, but abandons the usual goal of relating properties to molecular structure, and this makes many chemists and physicists uncomfortable. In any case, the basic ideas can be illustrated using two models, one due to Maxwell (of Maxwell–Boltzmann statistics), and to Voigt, respectively. These are shown schematically in Figure 17.2.

In these models, the elastic response is modeled with a spring having a modulus, G, connected to a dashpot containing a fluid having a viscosity η. The spring models the recoverable elastic response from chains that are stretched out and

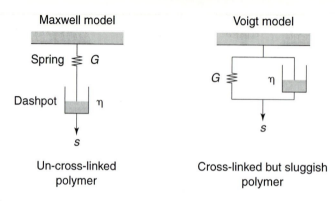

Figure 17.2 The simplest phenomenological models for viscoelasticity.

trying to return to their higher-entropy random-coil spatial configurations. By contrast, the dashpot models the time-retarded effects of the chains rearranging themselves in the polymeric medium. They can be visualized as analogues of the shock absorbers on an automobile, or the devices on doors that prevent them from closing too rapidly. The basic idea is that when a stress s is imposed as shown in the sketch, fluid has to flow into the dashpot for it to respond. Similarly, in a retraction, the flow out of the dashpot models the time-retarded recovery. In both cases, the slowness with which this occurs depends directly on the viscosity of the fluid.

In the Maxwell model, these two elements are connected in "series," and much of the terminology in this regard is reminiscent of that associated with electrical circuits. This arrangement models an un-cross-linked polymer, with the spring modeling the temporary elasticity of the stretched out chains before they can retract, and the dashpot the extent to which the elastic responses are retarded. In this case, the rate of deformation $\dot{\gamma} = d\gamma/dt$ has two components, one from the equation for the modulus $G = s/\gamma$, and the other from the viscosity equation $s = \eta\dot{\gamma}$, Specifically,

$$\dot{\gamma} = s/\eta + (1/G)(ds/dt) \tag{1}$$

This differential equation is solvable in the two cases of interest in polymer science: stress relaxation and creep. In stress relaxation, the sample is stretched to the desired length and then held at this length while the stress decreases (because of the uncross-linked chains sliding by one another). At constant length, the rate of deformation $\dot{\gamma}$ is zero, and equation (1) is then solvable by inspection. The solution is s as a specific function of t, and can be obtained by asking what function of s when added to its first derivative gives the zero now appearing on the left-hand side of the equation. This has to be an exponential in t, and continuing this argument shows it to be

$$s = s_o e^{-t/\tau} \tag{2}$$

Characteristics of the Viscoelastic State

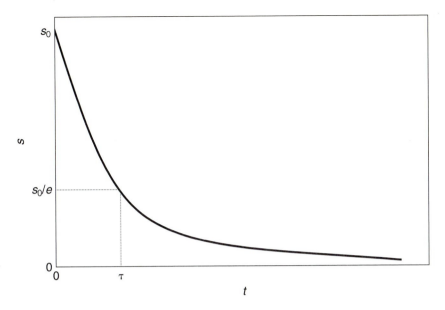

Figure 17.3 Solution to the Maxwell model for stress relaxation.

where s_o is the stress at the initial time $t = 0$ and $\tau = \eta/G$ is the relaxation time. As can be seen from equation (2), it is the time required for the stress to relax to $1/e$ of its original value s_o. Thus, it is analogous to the half-life of a nuclear species, except that the occurrence of the exponential changes the factor of 1/2 to $1/e$. The solution can be checked by substituting it into equation (1), and is shown schematically in Figure 17.3.

The other relevant experiment, creep, involves attaching the top of the un-cross-linked polymer to a fixed clamp and the other end to a container holding weights to give a constant applied (nominal) stress. The polymer sample extends until the chains flow by one another to the point of failure of the sample. Because the stress is constant, the term ds/dt in equation (1) is now zero, making the equation solvable by the separable-variables technique. The differential equation is now

$$d\gamma = (s/\eta)dt \tag{3}$$

which is immediately integratable to

$$\gamma = (s/\eta)t + I \tag{4}$$

where I is the integration constant. Since γ must be zero when t is 0, I must also be zero, making γ directly proportional to the time. This is shown by the two curves in Figure 17.4. If the proportionality factor is large (because of large stress or low viscosity), then the rate of creep is relatively rapid. If this factor is low (be-

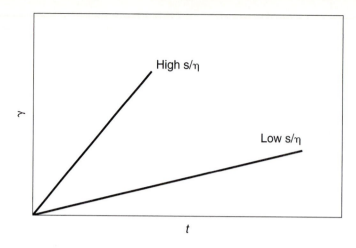

Figure 17.4 Solution to the Maxwell model for creep.

cause of low stress or high viscosity, from a higher molecular weight), then the rate of creep is lower.

The differential equation that characterizes the Voigt model is obtained from the fact that the stress, s, is now partitioned between the spring element and the dashpot element. Solving the two defining equations for s gives the two contributions:

$$s = G\gamma + \eta\dot{\gamma} \tag{5}$$

Again, the solution can be obtained by asking what function of γ when added to its first derivative gives simply s and not some function of γ. This argument gives the result

$$\gamma = (s/G)(1 - e^{-t/\tau}) \tag{6}$$

Similarly, if the stress is now removed, the left-hand side of equation becomes zero, and the solution now is analogous to that obtained earlier for the stress in stress relaxation:

$$\gamma = \gamma_o e^{-t/\tau} \tag{7}$$

where γ_o is the deformation immediately after removal of the deforming stress. Again, there is an exponential dropoff, with τ characterizing the time retardation, but now for the recovery part of the cycle. The deformation and subsequent retraction are shown in Figure 17.5. A quartz spring, without any time-dependent effect, would immediately show the maximum deformation, with immediate recovery upon removal of the stress.

Characteristics of the Viscoelastic State

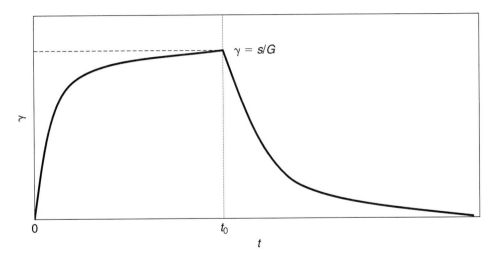

$\gamma = s/G$

Figure 17.5 Solution to the Voigt model for elongation followed by retraction.

In any real polymer sample, there would probably be a range or spectrum of spring constants, viscosities, and relaxation times. Different values of these parameters would be required for chains that differed in their molecular weights, degree of entanglement, etc.

In summary, viscoelasticity studies employ models appropriate for interpreting stress-strain-time data available for a polymer. This gives values of G, η, and τ, which are then used to predict the behavior of the material under different conditions. This phenomenological approach has the disadvantage of loss of contact with molecular information, but it has considerable advantage in the prediction of time-dependent behavior and the design of materials with optimized properties.

Considerable work is also being carried out using approaches that are molecular rather than phenomenological. One example treats the motions of a chain in terms of the "reptations" employed by snakes and some other reptiles.

An Unusual Viscoelastic Material

A novel type of polymer composite called "Silly Putty®" illustrates some interesting aspects of viscoelasticity.[1] It was first prepared at General Electric in 1950 by mixing a low molecular-weight polysiloxane with boric acid. This is essentially the same technique used today, after several hundred million samples have been sold, mostly as toys for children.

Its interest as a toy is caused by the fact that the material is a rubber-like solid over short time intervals but is a viscous liquid over more protracted intervals. Thus, the material will bounce like a rubber ball, but will slowly flow like a

[1]Thayer, A., *Chem. Eng. News*, November 27, 2000, p. 27.

viscous liquid in the earth's gravitational field. Its pliable nature has been exploited in its use to increase hand strength and to relieve stress. Its fluidity, however, can cause problems since it will flow into carpeting, etc. if left there by a child not aware that the material is really a liquid. All these aspects provide examples of the time dependence of the mechanical properties of polymeric materials in general.

Silly Putty® also has the ability to lift images from newspaper pages, and its deformability can then be used to stretch these images into humorous distortions. However, the material is not cross-linked because the usual flow processes make these images ephemeral.

Two-Phase Systems

As mentioned earlier, a major problem in the technological use of polymers is to design materials that have dimensional stability and yet have high impact strength. One way in which this can be accomplished is by the addition of "reinforcement" fillers such as glass fiber or carbon black to a polymer that has a high glass transition temperature. This method is discussed in Chapter 20. The second approach involves the preparation of a polymer system that possesses both glassy and elastomeric domains. The glassy regions provide the resistance to deformation, and the elastomeric regions function as impact-absorbing domains. For example, block copolymers can be prepared from styrene and butadiene by the use of anionic polymerization techniques. The polymers are known as SBR copolymers, and are commercially available as the Kratons™. The polystyrene end blocks have a T_g at 100°C and the polybutadiene center block at −63°C. At normal temperatures the polystyrene blocks function as "anchors" or glassy "filler" domains, preventing the polybutadiene regions from exhibiting their normal extensibility and potential for creep. Because the "cross-links" can be softened by raising the temperature above the glass transition temperature of polystyrene, these materials have the characteristic of reprocessability. It is also possible to achieve phase inversion so that the polystyrene is the continuous phase, with the polybutadiene phase dispersed in it. The rubbery phase greatly increases the impact resistance of the polystyrene to the extent that this composite is called "high-impact polystyrene" (HIPS). Acrylonitrile-butadiene-styrene (ABS) polymers have similar characteristics.

These two-phase systems present additional challenges to the polymer scientist or engineer. The interfaces between the two phases are of crucial importance with regard to physical properties, particularly in polymer-filler composites. They also raise important questions with regard to adhesion in general.

Great interest now exists in studying molecular entities that can self-assemble into higher-order structures held together by noncovalent forces, such as van der Waals forces, Coulombic attractions, or hydrogen bonding. This area of "supramolecular structures" parallels the linking of atoms into molecules, in a way much used by living organisms.

TRANSITION THERMODYNAMICS

Two kinds of transitions can be distinguished by use of the relevant thermodynamics, namely first order and second order. The formal definitions of these two types of transition will provide guidance on the corresponding experiments to be carried out to characterize them.

Specifically, a first-order transition is one in which there is a discontinuity in a first derivative of the free energy. The Gibbs free energy is the most appropriate here because of its direct relevance to processes at constant temperature and pressure. It can be obtained from the first law of thermodynamics, $dE = dq - dW$ (where E is the energy, dq the heat flow into the system, and dw is the work done by the system in the process). Combining the first law with the definition of the entropy $dS = dq_{rev}/T$, and using $dH = dE + d(pV)$, $dW = pdV$, and $dG = dH - d(TS)$, gives one of the four "fundamental" equations of thermodynamics:

$$dG = Vdp - SdT \qquad (8)$$

It describes how the Gibbs free energy changes with pressure and temperature (in the absence of any chemical reaction). Since there are three variables in this equation, the quantities relevant to our definitions are partial derivatives. One is

$$(\partial G/\partial p)_T = V \qquad (9)$$

which suggests dilatometric (volume-temperature) measurements as one method for characterizing such transitions. The other partial derivative is

$$(\partial G/\partial T)_p = -S \qquad (10)$$

This suggests the alternative of carrying out calorimetric measurements because of the close association of the entropy to the heat capacity, as will be demonstrated.

Second-order transitions, on the other hand, are those which show a discontinuity in the second derivatives of the free energy. One second derivative associated with typical polymer characterization experiments is

$$(\partial[(\partial G/\partial p)_T]/\partial T)_p = (\partial V/\partial T)_p = V\alpha \qquad (11)$$

where α is the thermal expansion coefficient $(dln\, V/dT)_p$. The calorimetric analogue to this dilatometric quantity involves

$$(\partial[(\partial G/\partial T)_p]/\partial T)_p = -(\partial S/\partial T)_p = -(dq_{rev}/T dT)_p = -C_p/T \qquad (12)$$

where $C_p = (dq/dT)_p$ is the heat capacity at constant pressure.

Another distinction between the two types of transitions is the fact that a first-order transition involves a latent heat, for example the heat required to melt a crystalline solid in contact with its molten state at its melting point, or the heat required to vaporize a liquid in contact with its vapor at its boiling point. Second-order transitions do not have such latent heats.

Figure 17.6 shows the differences to be expected in both dilatometric and calorimetric experiments. In the case of a first-order transition, an increase in temperature causes a gradual increase in volume, up to the melting point T_m, at which point the second phase (the liquid state) appears. The Gibbs phase rule

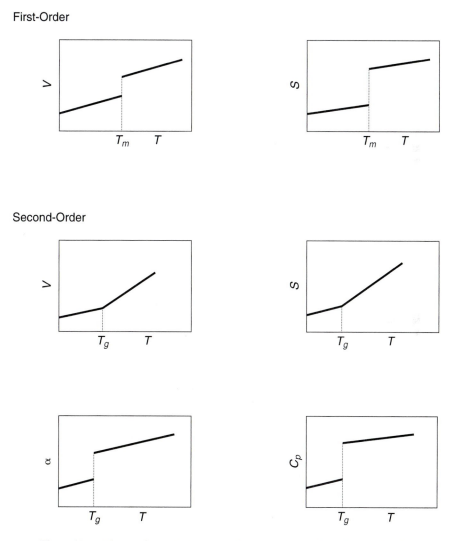

Figure 17.6 Thermodynamic distinctions between first-order and second-order transitions.

then requires that the temperature remain at $T = T_m$ until all the crystalline phase has been converted to liquid. Because the volume of the liquid phase is larger than that of the corresponding crystalline (because of less efficient packing), an abrupt increase in volume occurs at constant temperature, as shown in the figure. This portion of the curve, of infinite slope, is the discontinuity specified in the thermodynamic treatment. Temperatures above T_m simply give rise to the usual expansions associated with heating almost any liquid.

In the case of the second-order transition, the volume shows a change in slope in its dependence on temperature, but no discontinuity. The temperature at which the slope changes is generally taken to be the glass transition temperature T_g of the polymer. Because of the change in slope at T_g, the second derivative (proportional to the thermal expansion coefficient) does show a discontinuity at this temperature, which is consistent with the thermodynamics for this type of transition.

For calorimetric experiments on a first-order transition, the increase in temperature causes a gradual increase in entropy, because of increased thermal motions. This well-behaved increase continues up to T_m, at which point it becomes discontinuous, as did the volume in the dilatometric experiment. The abrupt increase in S is now caused by the liquid phase being much more disordered than the corresponding crystalline phase.

Similarly, the entropy in a second-order transition, shows a change in slope rather than a discontinuity in its dependence on temperature. As in the dilatometric experiments, the temperature at which the slope changes is generally taken to be the glass transition temperature T_g of the polymer, with C_p showing a discontinuity.

Skepticism exists that the glass transition is really a second-order transition, because of the importance of time in such measurements. Carrying out the same experiment more slowly can give a significantly lower value of T_g or perhaps none at all. One viewpoint is that T_g is simply the temperature below which the material acts like a permanent solid, given the experimenter's level of patience.

THE GLASS TRANSITION TEMPERATURE

The glass transition represents the rather sharp change that occurs from the glassy to the rubbery or flexible thermoplastic states in nearly all linear-type polymers. It is thought to be a "second-order transition." This transition is characteristic of a particular polymer in much the same way that a melting point is characteristic of ordinary low-molecular-weight compounds. In fact, the glass transition temperature varies with the types of skeletal atoms present, with the types of side groups, and even with the spatial disposition of the side groups. Table 17.1 lists values for a number of different polymers.

To a very large extent, the practical ultility of polymers and their different properties depend heavily on their glass transition temperatures. Thus, an important area of polymer research is the drive to understand how different mole-

TABLE 17.1   GLASS TRANSITION TEMPERATURES (T_g) AND CRYSTALLINE MELTING TEMPERATURES (T_m) FOR SELECTED POLYMERS[*]

Polymer	$T_g(°C)$	$T_m(°C)$
Polystyrene (isotactic)	100	240
Poly(*m*-methylstyrene) (isotactic)	70	215
Poly(methyl methacrylate) (atactic)	114	—
Poly(methyl methacrylate) (isotactic)	48	160
Poly(methyl methacrylate)(syndiotactic)	126	200
Poly(*cis*-1,4-isoprene)	−67	36
Poly(*trans*-1,4-isoprene)	−68	74
Poly(dimethylsiloxane)	−123	−29
Poly(dichlorophosphazene)	−63	−10
Poly[bis(trifluoroethoxy)phosphazene]	−66	242
Poly[bis(ethoxy)phosphazene]	−84	—
Poly(trifluoroethoxypentafluoropropoxyphosphazene) rubber	−77	—
Polyacrylontrile	85	317
Nylon 66	45	267
Poly(ethylene terephthalate)	17	—
Polyethylene	−20	141
Polyethylene	−107	95

[*] A compilation of T_g and T_m values can be found in O. G. Lewis, *Physical Constants of Linear Homopolymers* (New York: Springer-Verlag, 1968).

cular features affect the glass transition temperature. One approach to this problem is described in Chapter 18. Here it is sufficient to note that there appears to be a close connection between the T_g value and the *flexibility* of the polymer chain. The chain flexibility depends more on the rotation or torsion of skeletal bonds than on changes in bond angles or lengths. When a randomly coiled chain is pulled out into an elongated conformation, the skeletal bonds "unwind" rather than undergo angular distortion (Figure 17.7). Thus, flexibility on a macroscopic scale depends on *torsional mobility* at the molecular level. If a highly flexible chain is present, the glass transition temperature will generally be low. If the chain is rigid, the T_g value will be high. For example, poly(dimethylsiloxane) has one of

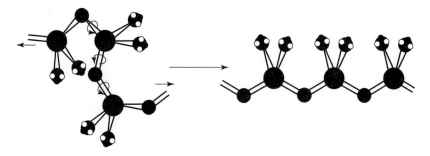

Figure 17.7   Elasticity of a polymer such as silicone rubber depends on the ease with which the chains can be stretched out from a random coil. The chain elongation is a consequence of the *unwinding* of bonds rather than a marked widening of bond angles.

the lowest T_g values known ($-123°C$), in part because the silicon–oxygen bonds have considerable torsional mobility.

However, the inherent rigidity or flexibility of the backbone structure is only one contributing factor. The torsional mobility of the skeletal bonds will also be affected by the side groups: large side groups or charged structures on the same chain will repel or attract each other and this could appreciably lower the chain mobility. Examination of Table 17.1 will provide examples of this effect. Furthermore, polar interactions *between* neighboring chains could raise the T_g. Even the chain length has an effect. For low polymers, the T_g generally rises with increasing chain length until a limiting value is reached. An alternative interpretation of the glass transition is based on the behavior of the *bulk* solid or elastomer rather than on specific interactions between individual chains. In this approach the flexibility of the bulk material depends on the "free volume" within that material. If intermolecular voids exist (as a consequence of the inefficient packing of side groups on neighboring chains), it will be easier for the polymer chains to undergo reorientation in response to an applied stress. Hence, the bulk material will be flexible or elastomeric. Of course, free volume, chain flexibility, and side-group interactions are interconnected factors and it is only rarely possible to separate one influence from the others. The prediction of glass transition temperatures on the basis of molecular structure is a complex problem that is still in the preliminary stages of development.

It should be mentioned that T_g and T_m are generally closely correlated. For example, high glass transition temperatures can be caused in part by low dynamic flexibility, which involves the ability of the chain to *change* conformations. But changes in structure that decrease (or increase) dynamic flexibility generally also decrease (or increase) equilibrium flexibility (the ability of the chain to collapse into compact, random arrangements). These correlations have been incorporated into the rule of thumb

$$T_g = (2/3)T_m \qquad\qquad (13)$$

where the temperatures are in absolute degrees (K).

DETECTION OF GLASS TRANSITIONS

It might be imagined that the most straightforward way to measure a glass transition temperature is to simply manipulate the polymer as it is cooled or heated in order to find the temperature at which it changes from a hard glass to a rubber or a flexible thermoplastic. In principle, this can be done, although the results are likely to be far less accurate than if one of the following methods is employed.

"Indentation" Techniques (Penetrometer)

Below the T_g a polymer is hard and glassy. Above T_g the material is soft and flexible. Thus, the degree to which a sharp point can penetrate the surface of the polymer at a given temperature can be used to detect the transition. In practice, the point of a weighted needle is allowed to rest on the polymer surface as the temperature is raised. As the polymer passes through its T_g the needle penetrates the surface and the movement of the needle can be monitored by means of an amplification gauge. This method is less accurate than those described in the following sections, and requires relatively large samples of polymer. However, it is a useful method for the preliminary, engineering-oriented examination of polymers.

Torsional Rigidity Methods

The resistance of a polymer sample to torsion depends on whether the polymer is in the glassy or flexible state and this principle can be used to measure T_g. Perhaps the most engineering-oriented technique based on this principle involves the application of a torsional vibration to the end of a bar of the polymer. The resistance to torsion and the energy loss of the polymer are then measured at different temperatures.

In the laboratory a more convenient technique makes use of a torsional pendulum (Figure 17.8). In one device, an inert matrix such as a braided glass

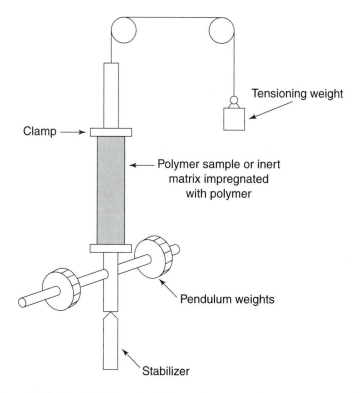

Figure 17.8 Simple device in which a polymer sample or a porous matrix (paper) impregnated with polymer forms part of a torsional pendulum. The rigidity of the polymer and its capacity to absorb torsional energy are measured from the period of the pendulum and its damping characteristics.

fiber is impregnated with a solution of the polymer and then dried. The fiber is suspended in a torsional pendulum device (see Figure 17.9) and the period of the pendulum and its damping frequency are measured (Figure 17.10). As the sample is heated through its glass transition temperature, a drastic loss in rigidity is detected, accompanied by a sharp maximum in the damping curve. This provides a very sensitive technique for the detection and measurement of the transition.

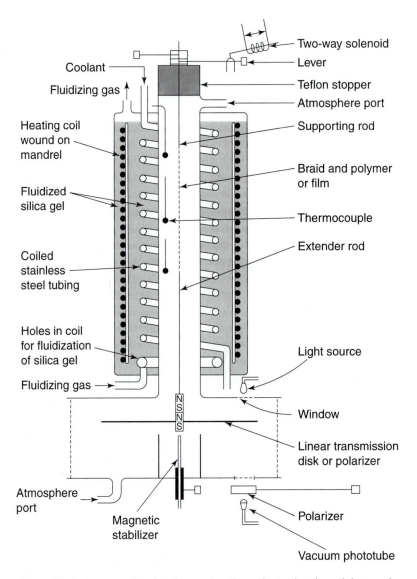

Figure 17.9 Diagram showing the constructions of a torsional pendulum, and torsional braid analyzer. [Printed with permission from Gillham, J. K., *CRC Crit. Rev. Macromol. Sci.*, **1972–73**, *1*, 83; © the Chemical Rubber Co., CRC Press, Inc.]

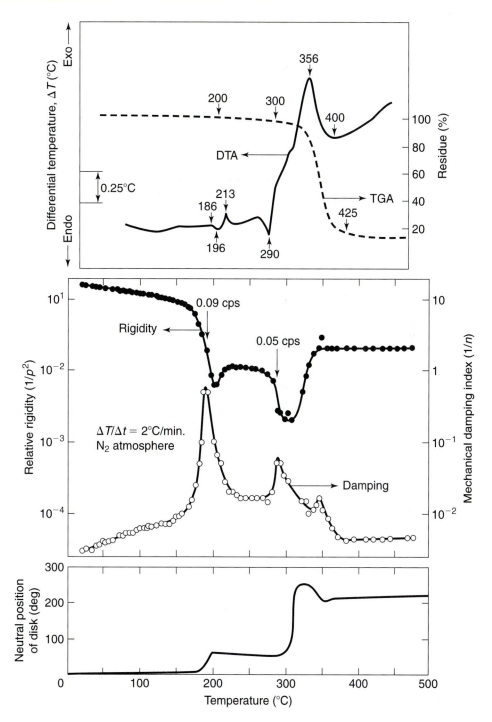

Figure 17.10 Comparison of thermomechanical (torsional braid), differential thermal analysis, and thermogravimetric analysis data for cellulose triacetate. The bottom figure shows the twisting of the sample in the absence of oscillations as a result of expansion or contraction of the sample at T_g and T_m. [Reproduced with permission from Gillham, J. K., *CRC Crit. Rev. Macromol. Sci.*, **1972–73**, *1*, 83; © the Chemical Rubber Co., CRC Press, Inc.]

Furthermore, the method is suitable for very small (0.25-g) samples of polymer and is thus ideal for exploratory work.

Broadline Nuclear Magnetic Resonance

The typical nmr signal seen for a low-molecular-weight compound in dilute solution or as a molten sample is a series of sharp peaks. Nmr spectra of solids, on the other hand, are broad, diffuse, and often difficult to detect. When a polymer is heated through the glass transition region it is, in a sense, converted from a solid to a pseudo liquid, and this transition can be detected by broadline nmr methods. In practice, several grams of the polymer are needed. Some care is required to distinguish between the glass transition and the onset of *side-group* torsional motions. Figure 17.11 illustrates the way in which nmr changes can be utilized to detect glass transitions.

Dilatometry

The coefficient of volume expansion of a polymer with temperature depends on whether the polymer occupies the glassy, rubbery, thermoplastic, or liquid states. Thus, the change of slope of a volume versus temperature plot can be used to identify the glass transition (Figure 17.12). In practice, a dilatometer (Figure 17.13) is used for these measurements, and the position of the capillary miniscus is plotted as a function of temperature.

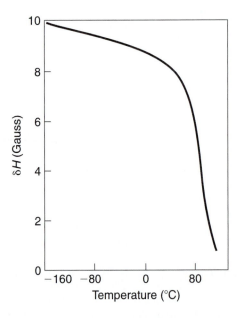

Figure 17.11 Curve showing change in ¹H nmr line width as a sample of commercial poly(vinyl chloride) is heated through the glass transition region. [From Hassan, A. M., *CRC Crit. Rev. Macromol. Sci.*, **1972–73**, *1*, 83; © the Chemical Rubber Co., CRC Press, Inc.]

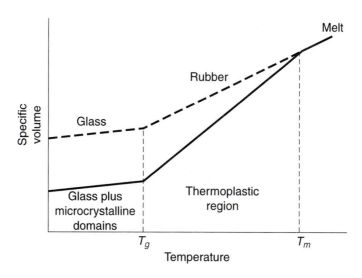

Figure 17.12 Variation of specific volume of a crystalline polymer (solid line) and an amorphous modification of the same polymer (dashed line) as a function of temperature. Glass transitions are indicated by abrupt changes in the slopes of the lines.

Figure 17.13 Dilatometer, used to monitor the change in volume of a polymer plus the surrounding liquid as a function of temperature by changes in the height of the liquid level in the capillary tube. Because the liquid does not undergo sharp transitions when heated, but the polymer does, any abrupt increase or changes in the slope of volume versus temperature may be attributed to T_m or T_g transitions, respectively, in the polymer.

Detection of Glass Transitions

Differential Thermal Analysis and Differential Scanning Calorimetry

Differential thermal analysis (DTA) and differential scanning calorimetry (DSC) are perhaps the most popular techniques for the measurement of glass transition temperatures. The DTA method requires the heating of a small polymer sample at a constant rate of temperature increase. The temperature of the polymer is compared continuously with the temperature of a control substance, such as alumina, which itself undergoes no transitions in the temperature range being scanned. The temperature difference between the polymer and the control material is then a function of the different specific heats of the two substances. The specific heat of a polymer changes rapidly as a transition region is approached with exothermic changes being characteristic of glass transitions (Figure 17.14). Endotherms indicate first-order transitions (melting temperatures). Two advantages of the DTA method are that only a small amount of the polymer is required, and the measurement is quite rapid. Also, the transition temperature can be identified to within 1 or 2°C. A disadvantage is that highly crystalline polymers may show only weak exotherms that may prove difficult to identify.

Differential scanning calorimetry operates by a slightly different principle. Two small metal containers, one containing the polymer sample and the other a control substance, are heated by individual electric heaters. The temperature of

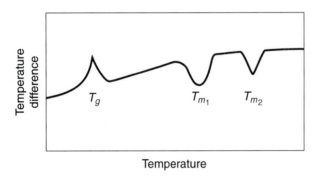

Figure 17.14 Differential thermal analysis (DTA) scan of a polymer, showing an exotherm typical of a glass transition and the endotherms that are characteristic of T_m transitions.

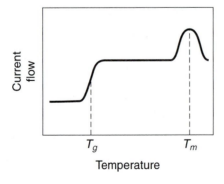

Figure 17.15 Differential scanning calorimetry scan (DSC) showing sections of the curve characteristic of T_g and T_m transitions.

each container is monitored by a heat sensor. If the sample suddenly absorbs heat during a transition, this change will be detected by the sensor, which will initiate a greater current flow through the heater to compensate for the loss. Thus, absorption of heat by the sample results in a greater current flow. Since the change in electric current can be monitored accurately, this provides a sensitive measure of transition temperatures. Figure 17.15 shows a DSC scan.

MICROCRYSTALLINITY

A distinction must be drawn between polymers that are commonly described as "crystalline" and the single crystals formed by low-molecular-weight substances. In the latter type of crystal, the crystalline order results from a regular packing of molecules or ions in a three-dimensional lattice (Figure 17.16). Crystals formed from low-molecular-weight compounds retain their integrity as the temperature is raised, until the point is reached at which the vibrational forces become more important than the intermolecular attractions. At this point the lattice breaks down over a very narrow temperature range, and the crystal melts sharply.

In microcrystalline polymers, on the other hand, the crystallinity results from the *regular packing of chains* (Figure 17.17). However, it is important to recognize that such regular packing arrangements usually exist only in small domains within the polymer. Hence, a microcrystalline polymer really consists of microcrystallites embedded in a matrix of amorphous polymer. In fact, a polymer chain may pass through several amorphous and crystalline regions as shown in Figure 17.18. It will be obvious from Figure 17.18 that a maximum degree of order can be obtained only when the crystallites themselves are lined up on the same axis, a process that is known as *orientation*. The microcrystalline regions are characterized by a more efficient use of the available space. Hence, polymers that possess microcrystalline domains have a higher density than forms of the same material that are totally amorphous. In some polymer systems, especially polyamides, the

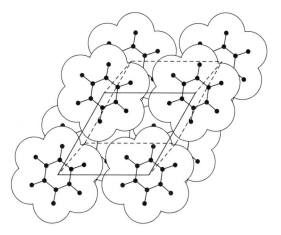

Figure 17.16 Regular packing of small molecules of hexamethylbenzene in the space lattice of a single crystal. [From Bunn, C. W., *Chemical Crystallography* (London: Oxford University Press, **1963**).]

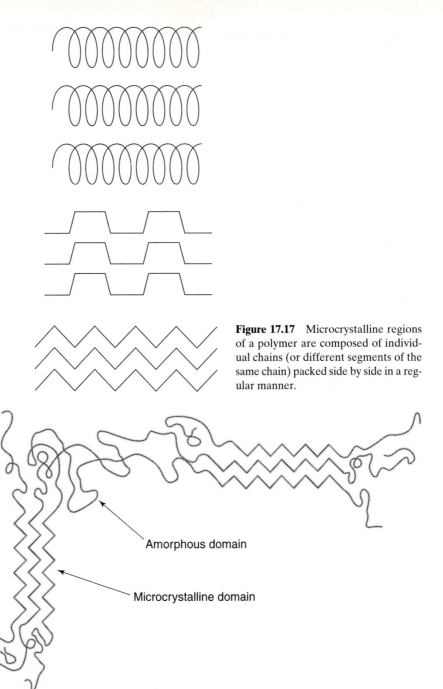

Figure 17.17 Microcrystalline regions of a polymer are composed of individual chains (or different segments of the same chain) packed side by side in a regular manner.

Amorphous domain

Microcrystalline domain

Figure 17.18 Structure of a microcrystalline polymer is generally considered to consist of microcrystalline domains separated by amorphous, random coil regions. A single polymer chain may traverse several microcrystalline and amorphous regions.

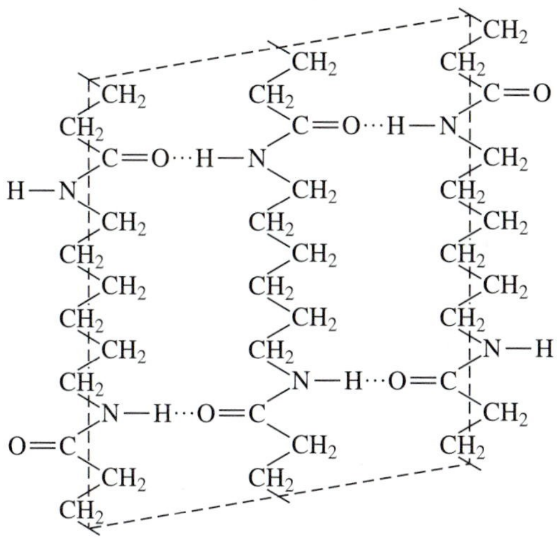

Figure 17.19 Orientation of the chains in the microcrystalline region of a polyamide (nylon 66). Individual chains are held together by hydrogen bonds.

regular packing of adjacent chains is facilitated by intermolecular hydrogen bonding as well as by the normal van der Waals attractive forces (Figure 17.19). Such polymers often have especially high crystalline melting temperatures. Apparently part of the reason is the persistence of the hydrogen bonding after melting, which decreases the entropy of fusion and increases the melting point correspondingly.

INFLUENCE OF CRYSTALLINITY ON PHYSICAL PROPERTIES

Polymers crystallize to attain a state of lower free energy. The regular packing of chains means *closer* packing and enhanced opportunities for intermolecular attractions. This is the driving force behind crystallization. However, an opposing force is the need for the system to maximize its entropy by increases in the conformational disorder. Hence, the observed structure represents a balance (sometimes unstable) between these opposing forces. Typically, 30 to 70% of the polymer may remain in the amorphous state, and the logical question must, therefore, be raised as to why the polymer does not continue to behave like an amorphous material.

In the microcrystalline regions the chains are essentially held together by dipolar, hydrogen bonding, or van der Waals forces. For this reason the crystalline domains function as *cross-links* for the amorphous regions. The cross-links are labile at the melting temperature or even during manipulation of the polymer, but cross-links they are, with all the physical property influences that would be expected from such structures. The crystalline cross-links stiffen and toughen the polymer and reduce the swelling in solvents. On a macroscopic level the introduction of microcrystallinity changes a rubbery elastomeric polymer into a tough, flexible material.

For example, at room temperature, polyethylene is roughly 100°C above its glass transition temperature, but it remains a tough plastic material. Without the

Influence of Crystallinity on Physical Properties

microcrystallites it would be a soft elastomer, as indeed it is above the crystalline melting temperature of 115°C or 135°C. This effect is strikingly demonstrated with organophosphazene high polymers. Poly[bis(trifluoroethoxy)phosphazene], $[NP(OCH_2CF_3)_2]_n$, and poly(diphenylphosphazene), $[NP(OPh)_2]_n$, are both highly crystalline, flexible, tough thermoplastics. However, if the symmetry of the structure is destroyed by the random introduction of a second substituent group, the polymers become rubbery and highly elastomeric.

Knowledge of effects such as this allows the polymer chemist to modify the materials to an even greater degree than is permitted by chemical means alone. In fact, the properties of many technologically important polymers are controlled to within fine limits by the degree of crystallinity introduced into the system.

ENHANCEMENT OF CRYSTALLINITY

Microcrystalline polymers are generally tougher than totally amorphous ones. They can be bent more without breaking, they resist impact better, and they are less affected by temperature changes or solvent penetration than are completely amorphous polymers. For these reasons, it is frequently advantageous, especially in manufacturing processes, to attempt to increase the degree of crystallinity.

The simplest technique available for this purpose is to stretch a fiber or a film of the polymer (Figure 17.20). When a polymer is cast as a film or extruded as a fiber, some microcrystallinity is often introduced. However, the microcrys-

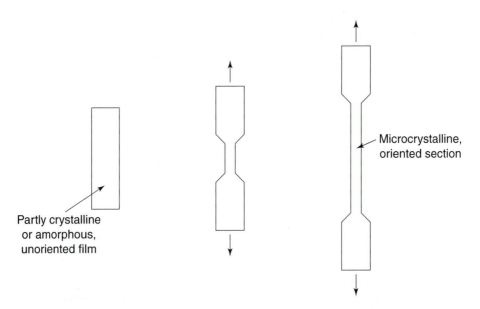

Figure 17.20 Piece of polymer film can be oriented by stretching. Often, the central section of the film elongates first with the orientation spreading toward the ends of the film as continued tension is applied.

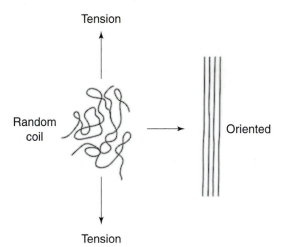

Tension

Random
coil

Oriented

Tension

Figure 17.21 On a molecular level the stretching of a polymer film or fiber results in a roughly parallel alignment of the macromolecular chains. This facilitates crystallization.

tallites tend to be few in number and randomly oriented relative to each other. The act of orienting the sample by stretching serves to pull the individual chains into a roughly parallel orientation (Figure 17.21). This enhances the chance that regular packing of adjacent chains will take place. Additional crystallization can also be introduced by heating and cooling the tensioned polymer in an "annealing" process. Altogether, a stretching of the material to four or five times its original length is not uncommon during crystallization and annealing. This process can be carried out by hand or with the use of stretching devices (see Chapter 20), but in large-scale fiber manufacture it is performed continuously by passing the fiber around heated rotating drums, as shown in Figure 17.22. The influence of the orientation process is especially evident with nylon, where unoriented fibers are

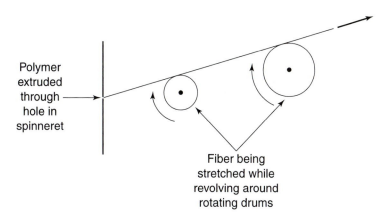

Polymer
extruded
through
hole in
spinneret

Fiber being
stretched while
revolving around
rotating drums

Figure 17.22 Stretching and orientation of fibers can be carried out in a continuous process by allowing the fiber to rotate around a revolving drum. Each drum in the sequence is of larger diameter than the preceding one, or of the same diameter but revolving at a faster speed.

Enhancement of Crystallinity

brittle, but oriented fibers are strong, tough, and somewhat elastic. It should also be mentioned that some of the strongest fibers are those that are stiff enough to self assemble into liquid-crystalline phases. These "mesophases" are typically nematic, which means the chains are parallel but sufficiently out of register that this type of ordered structure does have the fluidity required for spinning into fibers.

KINETICS OF CRYSTALLIZATION

Since crystallization can have such profound effects on the properties of a polymer, it is useful to develop some additional insights into the process. As already mentioned, the free energy change at constant temperature and pressure, $\Delta G_{T,p} = \Delta H - T\Delta S$, has to be negative for a process to occur, and this includes crystallization. When T is precisely the melting point $T_m = \Delta H/\Delta S$, the expected equilibrium result is obtained, that $\Delta G_{T,p} = 0$. In the case of crystallization, ΔH is negative and $-T\Delta S$ is positive, so the inequality gives us the obvious result that T must be less than T_m for $\Delta G_{T,p}$ to be negative and thus conducive to polymer crystallization.

The lower the temperature below T_m, the larger is the thermodynamic driving force for crystallization. However, kinetics becomes increasingly important as the temperature approaches the glass transition temperature, where the chain rearrangements required for crystallization become almost entirely suppressed and the polymer is said to be "quenched." This is illustrated in Figure 17.23, which

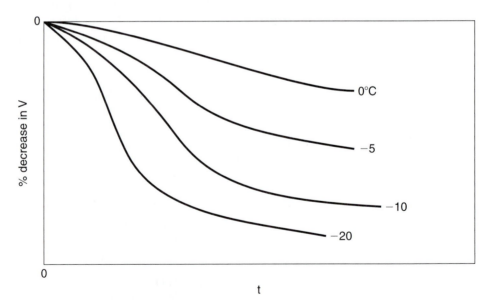

Figure 17.23 Rates of crystallization for the temperatures indicated with each curve.

shows typical rates of crystallization, as gauged by time-dependence of the volume. The larger the shrinkage, the higher the degree of crystallization, and thus slopes of such curves give rates of crystallization. The curves for different crystallization temperatures have similar shapes, in that the initial rate of crystallization is small, because most of the crystallites initially formed are unable to reach the threshold size (minimum ratio of surface area to volume) at which they overcome the destabilizing effects of the positive interfacial free energy. The rate of crystallization then goes through a maximum at the inflection point, decreasing toward zero as the number of chains able to crystallize is depleted.

The curves are drawn to illustrate the behavior for natural rubber, which has a T_m of approximately 11°C, and a T_g of approximately −72 °C. The curve for 0° thus represents an undercooling or supercooling of only 11 degrees, which explains the relatively low rate of crystallization. Larger undercoolings to −5, −10, and −20°C are seen to give increasingly larger maximum rates of crystallization. Temperatures below approximately −20 °C, however, give *lower* rates of crystallizations because of the kinetic complication already mentioned. This is illustrated in Figure 17.24, which shows the maximum rate of crystallization as a function of crystallization temperature. Experimentally, it is found that the temperature which gives the maximum rate of crystallization is two-thirds the distance from the glass transition temperature to the melting point:

$$T(\text{max rate}) = T_g + (2/3)(T_m - T_g) \tag{14}$$

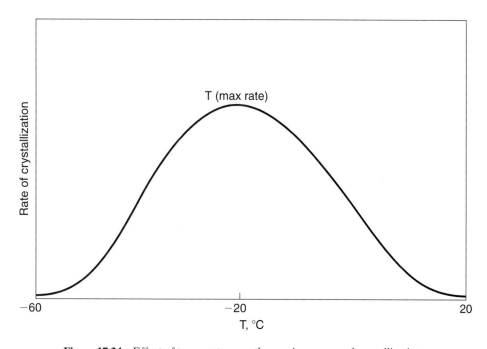

Figure 17.24 Effect of temperature on the maximum rate of crystallization.

Kinetics of Crystallization

Crystallization isotherms such as those shown schematically in Figure 17.23 are frequently interpreted in terms of the Avrami equation. It represents the fraction crystallinity as

$$\theta = 1 - \exp(-kt^n) \qquad (15)$$

where k is a constant which contains information on the crystallite growth, and n is an exponent widely used to interpret the various nucleation and growth processes.

Crystallization can be facilitated by what is called "strain-induced crystallization." The chains are stretched out in a deformation, typically elongation, and this decreases their entropies S. Only a small additional decrease in S is then required to bring about crystallization and, because the melting point T_m is given by $\Delta H / \Delta S$, this decreased ΔS elevates T_m. It is in this sense that deformation induces crystallization. The most direct experiment, illustrated in Figure 17.25, is to stretch an elastomer such as natural rubber and then hold it in the deformed state. Pass an X-ray beam through the stretched sample, note the diffraction pattern, and then raise the temperature until the diffraction spots disappear. This is the melting point as elevated by the strain. The effect can be very large, as shown schematically in Figure 17.26. As the sample is stretched (the relative length L/L_i is increased) the melting point increases monotonically until it is well above room temperature, at which point crystallization occurs. This is reversible in the sense that if the elastomer is permitted to snap back to the undeformed state its melting point will be lowered to below room temperature, and it becomes amorphous again. An interesting variation of this experiment involves wrapping a stretched rubber band around an object such as a package of leftover food, and placing

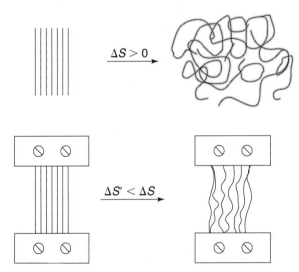

Figure 17.25 Molecular explanation of strain-induced crystallization in terms of the reduction of entropy from the chains being held in the deformed state mechanically.

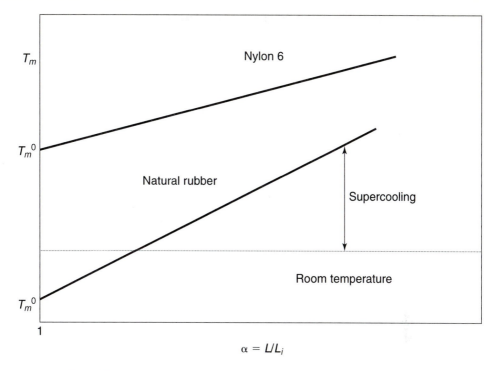

Figure 17.26 Increases in the melting point upon stretching a typical fiber and a typical elastomer.

both in the freezer. The temperature in the freezer is well below any of the melting points of the natural rubber and the elastomer will be converted to a strong fiber which will be difficult to remove until it warms to a temperature above its elevated T_m, at which point it will become elastomeric again.

For comparison, the figure also shows the behavior of a polymer whose melting point is above room temperature even in the undeformed state. In this case, illustrated by Nylon-6, the melting points are again elevated by the strain but persist when the deformation is removed. This gives rise to an interesting analogue to the law of corresponding states exhibited by gases. Different gases can vary tremendously in their properties, for example in their condensability into the liquid state. Yet if these gases are compared at their reduced temperatures $T_r = T/T_c$ and reduced pressures $p_r = p/p_c$ (where the subscripts refer to critical state values), then the behaviors of these gases would become very similar. In the case of polymers, if room temperature were around 100° lower than it is, then natural rubber would be very similar to the Nylon-6, and could compete with it in applications such as textile fibers. On the other hand, if room temperature were about 100° higher than it is, the Nylon-6 could compete with the natural rubber as an elastomeric material.

Kinetics of Crystallization

Nearly all the methods discussed earlier for the measurement of glass transition temperatures can be used to measure crystalline melting temperatures. For example, endotherms in differential thermal analysis curves (Figure 17.14) or breaks in dilatometric curves (Figure 17.11) yield T_m values. Crystallization results in a volume contraction that can be followed dilatometrically. The torsional pendulum method (Figures 17.8 to 17.10) also identifies T_m values from points of abrupt decrease in rigidity and from damping maxima. Nuclear magnetic resonance techniques can also be used. However, convincing proof that microcrystallinity exists in a particular polymer requires the use of two additional techniques: optical birefringence and X-ray crystallography.

The optical birefringence technique makes use of a polarizing microscope, preferably fitted with a heating stage for raising the temperature of the sample. When viewed through crossed polarizers, a microcrystalline polymer will show a dark background broken by bright specks which originate from the crystalline domains. Some polymers show a more complex structure in which individual microcrystallites radiate in all directions from specific points to form spherulites (Figure 17.27). The structure of each spherulite can be visualized as made up of microcrystals oriented along the radii of individual spheres. In practice, the spherulites grow until they contact each other to form the unusual patterns shown in Figure 17.27. The bright specks extinguish or reappear as the sample is rotated relative to the polarizing filter. The specular pattern should persist until the temperature is raised to the crystalline melting point, at which temperature the bright specks or spherulites should be totally extinguished.

The X-ray diffraction approach to the study of crystallinity makes use of the fact that a totally amorphous polymer will not give rise to a sharp diffraction pattern. Diffuse scattering rings only are seen. These resemble the X-ray patterns obtained from liquids. An unoriented microcrystalline polymer will yield an X-ray photograph that consists of sharp rings (see Chapter 19). Such photographs bear a striking resemblance to Debye–Scherrer "powder" photographs of powdered small-molecule crystalline materials, and for the same reason. The sharp concentric areas or rings result from diffraction by randomly oriented microcrystallites.

An *oriented* microcrystalline polymer fiber or film will yield a pattern of diffraction *spots* or arcs. The apparatus used to obtain photographs of this type is discussed in Chapter 19. Equipment is available that will allow X-ray patterns to be obtained as the temperature is raised. The appearance of amorphous-type patterns and the disappearance of a crystalline-type pattern provides a measure of the crystalline melting temperature. However, it is not uncommon that, as the temperature is raised, one crystalline pattern is replaced by another, and in this way a sequence of T_m values can be measured.

It is also important to be able to estimate the amount of crystallinity in a sample. One way to determine the fraction crystallinity, θ, of a polymer was already described in Chapter 16. It involved a measurement of the heat of melting

Figure 17.27 Polarized-light microscope photograph of a sample of low-density polyethylene showing the spherulitic structure at two different magnifications. The light polarization directions are horizontal and vertical. (Courtesy of R. S. Stein, University of Massachusetts.)

of a 100% crystalline polymer from melting point depression measurements, and then finding what fraction of this amount was actually needed to melt a known amount of the polymer being characterized.

Another important method is based on densities. The density of the 100% crystalline material, can be estimated from the characteristics of the unit cell, which is the basic building block from which the macroscopic crystallites are built. The unit cell is analogous to a brick that can be visualized as being translated in three directions to build a wall or other masonry structure. It is obtained in X-ray crystallography from the locations and relative intensities of the diffraction spots generated by the crystal structure. The dimensions of the unit cell, the angles between its faces, and the atomic contents can be determined. In the simple unit cell shown in Figure 17.28, the volume is simply the product of the three dimensions, abc. Since the contents of the cell are also known, the weight of the unit cell can be obtained by summing the weights of the atoms, and the ratio of the weight to the volume gives the "maximum" or "crystallographic" density. It is also necessary to know the density of the totally amorphous polymer, perhaps by extrapolating liquid state densities measured above the melting point down to room

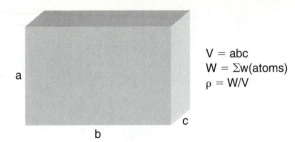

$V = abc$
$W = \Sigma w(atoms)$
$\rho = W/V$

Figure 17.28 Calculation of the density of a 100% crystalline polymer from the geometry and contents of its unit cell.

temperature or to some other temperature of interest. In terms of specific volumes $v = 1/\rho$, these two pieces of data give the maximum decrease in v that this polymer could exhibit by crystallization: v(totally amorphous) $- v$(totally crystalline). Dilatometry is then used to measure the actual decrease for the polymer being characterized: v(totally amorphous) $- v$(partially crystalline). The fraction crystallinity of this polymer is then simply

$$\theta = [v(\text{totally amorphous}) - v(\text{partially crystalline})]/$$
$$[v(\text{totally amorphous}) - v(\text{totally crystalline})] \tag{16}$$

Another example of an X-ray method is to resolve the total scattering profile into contributions from the crystalline phase, with the remaining contributions coming from the amorphous phase. This can be done by noting which peaks disappear when the crystalline phase is suppressed by melting, following either a temperature increase or the addition of a solvent. The sum of the areas under the peaks from the crystalline phase relative to the total area under the diffraction scan then gives the fraction crystallinity.

POLYMER SINGLE CRYSTALS

The preceding comments have been concerned with *micro*crystalline polymers. However, a few polymers are known to exist also as single crystals. Polyethylene is the best known example. Two main techniques are available for the preparation of polymer single crystals. There are (1) the crystallization of the polymer from an inert solvent at very high dilutions: for example, the slow cooling of linear polyethylene from hydrocarbon media at dilutions near 1:2000, and (2) the simultaneous polymerization of a monomer and crystallization of the polymer. This technique has been used to prepare single crystals of polyoxymethylene.

The crystals obtained from polyethylene are visible only under a microscope. They have a flat, lozenge shape (Figure 17.29). X-ray diffraction photographs of such crystals resemble those of low-molecular-weight materials. However, the molecular structure within the crystal consists of a folded arrangement of zigzag chains (Figure 17.30). It has been found that the folding interval

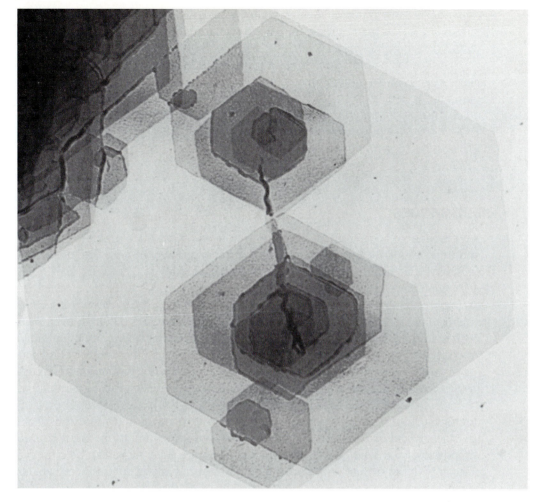

Figure 17.29 Photograph of lozenge-shaped single crystals of polyethylene. (Courtesy of I. R. Harrison, The Pennsylvania State University.)

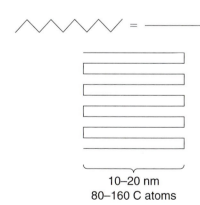

10–20 nm
80–160 C atoms

Figure 17.30 Presumed packing arrangement of polyethylene chains in a single crystal. This represents the so-called adjacent reentrant model of the structure in which a single chain bends back and forth in order to provide a parallel alignment of the chains. Another model (the so-called "switchboard" model) allows re-entry of the chain at a place some distance from the point of emergence.

Polymer Single Crystals

depends on the temperature of crystallization rather than on the polymer chain length or the nature of the solvent. The exact nature of the folding has been a subject of considerable controversy.

It should be noted that globular proteins and viruses crystallize to form single crystals. However, these molecules are arrayed individually in the lattice in the manner typified by crystals of low-molecular-weight compounds. This, in itself, is an astonishing phenomenon. The Bushy stunt virus, for instance, has a molecular weight of 13 million, with 4 molecules per cell and with a unit cell edge of 318 Å. The X-ray analysis of systems of this kind will be considered briefly in Chapter 19.

OTHER TECHNIQUES

A great deal of additional information is now being obtained about crystallinity and other aspects of polymer structure and morphology by scanning probe microscopy, which itself represents several approaches.

One approach is that of scanning tunneling microscopy (STM), in which an extremely sharp metal tip on a cantilever is passed over the surface being characterized while an electric current flowing through quantum mechanical tunneling is measured. Monitoring of the current then allows the probe to be maintained at a fixed height above the surface. A display of probe height as a function of surface coordinates then gives the desired topographic map. One limitation of this approach is the obvious requirement that the sample must be electrically conductive.

On the other hand, atomic force microscopy (AFM), does not require a conducting surface. The probe simply responds to attractions and repulsions from the surface, and its corresponding downward and upward motions are recorded directly to give a relief map of the surface structure. The probe can be either in contact with the surface, or adjacent to it and sensing only Coulombic or van der Waals forces.

Other types of scanning probe microscopy are:

- Electrochemical STM and AFM
- Frictional force microscopy
- Surface force compliance
- Magnetic force microscopy
- Electric force microscopy
- Scanning thermal microscopy
- Near-field scanning optical microscopy

Some of these various techniques not only generate topographic relief images, but they also provide the opportunity to transport separate individual atoms and molecules. Another application is to attach probes at the two ends of a polymer chain and stretch the chain out to determine its equilibrium and dynamic mechanical properties. References to some of these experiments on "single molecule elasticity" are given in Chapter 21.

STUDY QUESTIONS

1. If the physical changes that occur at the glass transition represent a sudden onset in backbone torsional motions, why should the value of T_g vary with the absence or presence of microcrystallinity?

2. Of the alternative methods mentioned in this chapter for the detection of glass transitions and melting temperatures, which methods would be more suitable to use: **(a)** in an exploratory synthetic research laboratory, **(b)** in a physical testing laboratory, or **(c)** in a manufacturing plant. Why?

3. What problems might you anticipate if, in the torsional braid method, instead of the use of a glass braid you decided to use a braid made from **(a)** nylon 66; **(b)** cellulose acetate; **(c)** paper; **(d)** copper?

4. After reference to Figure 17.19, suggest ways in which you might raise or lower the glass transition temperature of a polyamide and raise or lower the melting temperature.

5. Use some of the information in Table 17.1 to test the rule-of-thumb that $T_g \approx (2/3) \, T_m(\text{degrees } K)$.

6. Use the following dilatometric data to calculate the percentage of crystallinity of a hypothetical sample having a specific volume of 1.1 cm³/g. Assume the specific volumes of the 100% amorphous polymer and the 100% crystalline polymer are 1.2 and 1.0 cm³/g, respectively.

7. Use $T_m = 11°C$ and $T_g = -72°C$ for natural rubber to predict the temperature $T(\text{max rate})$ at which this elastomer should show the maximum rate of crystallization.

8. Suggest reasons why polyethylene is one of the few synthetic polymers known that forms single crystals.

SUGGESTIONS FOR FURTHER READING

AKLONIS, J. J., MacKNIGHT, W. J., and SHEN, M., *An Introduction to Polymer Viscoelasticity*. New York: Wiley-Interscience, **1972**.

BAER, E., "Relaxation Processes at Cryogenic Temperatures," *CRC Crit. Rev. Macromol. Sci.*, **1972–73**, *1*, 215.

BIRLEY, A. W., HAWORTH, B., and BATCHELOR, J., *Physics of Plastics, Processing, Properties and Materials Engineering*. Munich: Hanser Publishers, **1992**.

BONNELL, D. A. (Ed.), *Scanning Probe Microscopy and Spectroscopy. Theory, Techniques, and Applications*, 2nd ed. Berlin: Springer, **2001**.

BROSTOW, W., *Performance of Plastics*. Cincinnati: Hanser Gardner Publications, Inc., **2001**.

CANNON, S. L., McKENNA, G. B., and STATTON, W. O., "Hard-Elastic Fibers (A Review of a Novel State for Crystalline Polymers)," *J. Polymer Sci. (D), Macromol. Rev.*, **1976**, *11*, 209.

CASALE, A., PORTER, R. S., and JOHNSON, J. F., "Dependence of Flow Properties of Polystyrene on Molecular Weight, Temperature, and Shear," *J. Macromol. Sci., Rev. Macromol. Chem.*, **1971**, *C5*, 387.

CHANZY, H., "Nascent Morphology of Polyolefins," *CRC Crit. Rev. Macromol. Sci.*, **1972–73**, *1*, 315.

CHIU, J. (Ed.), *Polymer Characterization by Thermal Methods of Analysis* (Symposium-Selected Papers). New York: Dekker, **1974**.

DAVID, D. J., and MISRA, A., *Relating Materials Properties to Structure. Handbook and Software for Polymer Calculations and Materials Properties.* Lancaster, PA: Technonic Publishing Co., **1999**.

DEANIN, R. D., "Compatibility and Practical Properties of Polymer Blends," *Appl. Polym.*, Proc. Am. Chem. Soc. Symp. O. A. Battista, Appl. Polymer Sci., 1987 (R. Seymour and H. F. Mark, Eds.). New York: Plenum Press, **1988**, pp. 53–63.

DESPER, C. S., "Technique of Measuring Orientation in Polymers," *CRC Crit. Rev. Macromol. Sci.*, **1972–73**, *1*, 501.

DE STEFANIS, A., and TOMLINSON, A. A. G., *Scanning Probe Microscopies. From Surface Structure to Nano-Scale Engineering.* Uetikon-Zurich: Trans Tech Publications Ltd., **2001**.

DICKIE, R. A., LABANA, S. S., and BAUER R. S., (Eds.), *Cross-Linked Polymers: Chemistry, Properties, and Applications, ACS Symp. Ser.*, **1988**, *367*.

DUSEK, K., and PRINS, W., "Structure and Elasticity of Non-crystalline Polymer Networks," *Advan. Polymer Sci.*, **1969**, *6*, 1.

EISENBERG, A., "Ionomer Blends," *Polymer Mater. Sci. Eng.*, **1988**, *58*, 978–980.

ESTES, G. M., COOPER, S. L., and TOBOLSKY, A. V., "Block Polymers and Related Heterophase Elastomers," *J. Macromol. Sci., Rev. Macromol. Chem.*, **1970**, *C4*, 313.

FAN, L. T., and SHASTRY, J. S., "Polymerization Systems Engineering," *J. Polymer Sci. (D), Macromol. Rev.*, **1973**, *155*.

FAVA, R. A., "Polyethylene Crystals," *J. Polymer Sci. (D), Macromol. Rev.*, **1971**, *5*, 1.

FERRY, J. D., *Viscoelastic Properties of Polymers*, 3rd ed., New York: Wiley, **1980**.

FISA, B., "Nascent Morphology of Polyolefins," *CRC Crit. Rev. Macromol. Sci.*, **1972–73**, *1*, 315.

FRISCH, K. C., KLEMPNER, D., and FRISCH, H. L., "Recent Advances in Interpenetrating Polymer Networks," *Polymer Eng. Sci.*, **1982**, *22*(17), 1143–1152.

FRISCH, K. C., KLEMPNER, D., and FRISCH, H. L., "Recent Advances in Polymer Alloys and IPN Technology," Parts 2 and 3, *Mater. Design Eng.*, **1983**, *4*(5), 855–862.

GEIL, P. H., *Polymer Single Crystals* (*Polymer Reviews*, Vol. 5). New York: Wiley-Interscience, **1963**.

GILLHAM, J. K., "A Semimicro Thermomechanical Technique for Characterizing Polymeric Materials: Torsional Braid Analysis," *A.I.Ch.E. J.*, **1974**, *20*, 1066.

HAN, C. D. (Ed.), *Polymer Blends and Composites in Multiphase Systems, ACS Symp. Ser.*, **1984**, *206*.

HASSAN, A. M., "Application of Wide-Line NMR to Polymers," *CRC Crit. Rev. Macromol. Sci.*, **1972–73**, *1*, 399.

HAY, J. N., "Polymer Crystallization," *Macromol. Chem* (London), **1982**, *2*, 214–233.

HILTNER, A., "Relaxation Processes at Cryogenic Temperatures," *CRC Crit. Rev. Macromol. Sci.*, **1972–73**, *1*, 215.

JACOBY, M., "New Tools for Tiny Jobs," *Chem. Eng. News*, **2000**, *October 16*, 33.

KARASZ, F., "The Glass Transition of Linear Polyethylene," *J. Macromol. Sci., Rev. Macromol. Chem.*, **1979**, *C17* (1), 37.

KLEMPNER, D., and FRISCH K. C., (Eds.), *Polymer Science and Technology*, Vol. 20, *Polymer Alloys 3: Blends, Blocks, Grafts, and Interpenetrating Networks*. New York: Plenum Press, **1983**.

KLEMPNER, D., FRISCH, K. D., XIAO, H. X., CASSIDY, E., and FRISCH, H. L., "Two- and Three-Component Interpenetrating Polymer Networks," *ACS Advan. Chem. Ser.*, **1986**, *211*, 211–230.

LABANA, S. S., and DICKIE R. A., (Eds.), *Characterization of Highly Crosslinked Polymers, ACS Symp. Ser.*, **1984**, *243*.

LEE, L. H. (Ed.), *Polymer Wear and Its Control, ACS Symp. Ser.*, **1985**, *287*.

LIPATOV, Y. S., "Relaxation and Viscoelastic Properties of Heterogeneous Polymeric Compositions," *Advan. Polymer Sci.*, **1977**, *22*, 1.

LLOYD, D. R. (Ed.), *Materials Science of Synthetic Membranes, ACS Symp. Ser.*, **1984**, *269*.

MACKNIGHT, W. J., KARASZ, F. E., and FRIED, J. R., "Solid State Transition Behavior of Blends," in *Polymer Blends* (D. R. Paul and S. Newman, Eds.), pp. 185–242. New York: Academic Press, **1978**.

MACOSKO, C. W., *Rheology. Principles, Measurements, and Applications*. New York: Wiley & Sons, Inc., **1994**.

MANDELKERN, L., *Crystallization of Polymers*. New York: McGraw-Hill, **1964**.

MANDELKERN, L., "Thermodynamic and Physical Properties of Polymer Crystals Formed from Dilute Solution," *Progr. Polymer Sci.* (A. D. Jenkins, Ed.), **1970**, *2*, 163.

MANSON, J. A., and SPERLING, L. H., *Polymer Blends and Composites*. New York: Plenum Press, **1976**.

MARCHESSAULT, R. H., "Nascent Morphology of Polyolefins," *CRC Crit. Rev. Macromol. Sci.*, **1972–73**, *1*, 315.

MARK, J. E., "Thermoelastic Results on Rubberlike Networks and Their Bearing on the Foundations of Elastic Theory," *J. Polymer Sci. (D), Macromol. Rev.*, **1976**, *11*, 135.

MARTUSCELLI, E., PALUMBO, R., and KRYSZEWSKI M., (Eds.), *Polymer Blends: Processing, Morphology, and Properties*. New York: Plenum Press, **1980**.

MCCRUM, N. G., BUCKLEY, C. P., and BUCKNALL, C. B., *Principles of Polymer Engineering*. New York: Oxford University Press, **1988**.

MINNE, S. C., MANALIS, S. R., and QUATE, C. F., *Bringing Scanning Probe Microscopy Up to Speed*. Boston: Kluwer Academic Publishers, **1999**.

MIRABELLA, F. M., and JOHNSON, J. F., "Polymer Configuration and Compositional Variables as a Function of Molecular Weight," *J. Macromol. Sci., Rev. Macromol. Chem.*, **1975**, *C12*, 81.

MUNK, P., *Introduction to Macromolecular Science*. New York: John Wiley & Sons, **1989**.

PAUL, D. R., and SPERLING L. H., (eds.), *Multicomponent Polymer Materials, ACS Advan. Chem. Ser.*, **1986**, *211*.

PETRIE, S. E. B., "The Effect of Excess Thermodynamic Properties Versus Structure Formation on the Physical Properties of Glassy Polymers," *J. Macromol. Sci., Phys.*, **1976**, *B12*, 225.

Rio, A., and Cernia, E. M., "Polyblends of Cement Concrete and Organic Polymers," *J. Polymer Sci. (D), Macromol. Rev.*, **1974**, *9*, 127.

Rudolph, H., "Polymer Blends: Current State of Progress and Future Developments from an Industrial Viewpoint," *Makromol. Chem. Macromol. Symp.*, **1988**, *16*, 57–89.

Sanchez, I. C., "Modern Theories of Polymer Crystallization," *J. Macromol. Sci., Rev. Macromol. Chem.*, **1974**, *C10*, 113.

Sandman, D. J. (Ed.), *Crystallographically-Ordered Polymers, ACS Symp. Ser.*, **1987**, *337*.

Sauer, J. A., Richardson, G. C., and Morrow, D. R., "Deformation and Relaxation Behavior of Polymer Single Crystals," *J. Macromol. Sci., Rev. Macromol. Chem.*, **1973**, *C9*, 149.

Shen, M., and Croucher, M., "Contribution of Internal Energy to the Elasticity of Rubber-like Materials," *J. Macromol. Sci., Rev. Macromol. Chem.*, **1975**, *C12*, 287.

"Special Issue on Nanotechnology", *Sci. Am.*, **2001**, *285*(3).

Sperling, L. H., "Interpenetrating Polymer Networks," *ChemTech.*, **1988**, *18*(2), 104–109.

Sperling, L. H., *Interpenetrating Polymer Networks and Related Materials*. New York: Plenum Press, **1981**.

Tadokoro, H., *Structure of Crystalline Polymers*. New York: Wiley-Interscience, **1979**.

Tobolsky, A. V., and DuPré, D. B., "Macromolecular Relaxation in the Damped Torsional Oscillator and Statistical Segment Models," *Advan. Polymer Sci.*, **1969**, *6*, 103.

Vasko, M., Bleha, T., and Romanov, A. "Thermoelasticity in Open Systems," *J. Macromol. Sci., Rev. Macromol. Chem.*, **1976**, *C15*, 1.

Watts, M. P. C., Zachariades, A. E., and Porter, R. S., "New Methods of Production of Highly Oriented Polymers by Solid State Extrusion," *Contemp. Top. Polymer Sci.* (M. Shen, Ed.), **1979**, *3*, 297.

Wiesendanger, R., *Scanning Probe Microscopy and Spectroscopy*. Cambridge: Cambridge University Press, **1994**.

Wiesendanger, R. (Ed.), *Scanning Probe Microscopy. Analytical Methods*. Berlin: Springer, **1998**.

Wilkes, G. L., "The Measurement of Molecular Orientation in Polymeric Solids," *Advan. Polymer Sci.*, **1971**, *8*, 91.

Wilkes, G. L., "Rheo-Optical Methods and Their Application to Polymeric Solids," *J. Macromol. Sci., Rev. Macromol. Chem.*, **1974**, *C10*, 149.

Yannas, I. V., "Nonlinear Viscoelasticity of Solid Polymers (in Uniaxial Tensile Loading), *J. Polymer Sci. (D), Macromol. Rev.*, **1974**, *9*, 163.

Yeh, G. S. Y., "Morphology of Amorphous Polymers," *CRC Crit. Rev. Macromol. Sci.*, **1972–73**, *1*, 173.

18

Conformational Analysis of Polymers

THE ROLE OF CONFORMATIONAL ANALYSIS

As discussed elsewhere in this book, the primary motivation for most modern fundamental chemical research is the drive to relate the physical and chemical properties of materials to their molecular structure. This is true in spectroscopy, reaction mechanism studies, synthetic chemistry, and in many other areas. It is particularly true in polymer chemistry. One of the main purposes of science is to "make sense" of observable phenomena in terms of the behavior of microscopic, molecular, or atomic particles. In polymer science this need becomes manifest in attempts to relate properties, such as strength, toughness, solution viscosity, elasticity, crystallinity, or biological behavior to the composition, shape, and dynamic behavior of the component molecules.

As will be obvious from the earlier chapters, we now know a great deal about the ways in which changes in chemical composition affect the properties of polymers. However, we have only a fragmentary knowledge of the way in which the shape and flexural differences between molecules can be related to the physical properties. This, then, constitutes a considerable challenge.

Conformational analysis involves the study of the ways in which molecules can alter their geometry by torsional rotations (or "twisting" motions) of their covalent bonds. As such, it must be distinguished from *configurational* analysis which is concerned with the fixed geometric differences between closely related molecules as, for example, in the identification of *d*- or *l*-configurations in a molecule.

However, polymer physical chemists use the term *spatial* configurations to designate the overall arrangements of units in a polymer chain, and this has much precedence in the area of statistical mechanics ("configurational partition functions") and in X-ray crystallography ("configurations of atoms in a particular structure").

Conformational changes usually result in a change in the *shape* of the molecule without the cleavage of bonds having occurred. Because the torsional motions of many bonds take place readily at normal temperatures, conformational changes are responsible for many of the phenomena that we associate with changes in macroscopic physical properties. Thus, the glass and melting transitions of a polymer, the absence or presence of crystallinity, the extensibility or elasticity, and the ways in which polymers raise the viscosity of small-molecule solvents can all be related to changes in the conformations of the macromolecules. In fact, nearly all of the useful properties of polymers can be ascribed in some way to the conformational characteristics of the component molecules.

Thus, it is of vital importance that we attempt to understand the reasons why some polymers assume one conformation while others prefer another, or why a particular polymer will undergo a conformational change at a specific temperature or when the solvent or the pH is changed. Specifically, a number of good reasons exist for attempting to predict or rationalize polymer conformations. These are:

1. To explain why some polymers crystallize and others do not.
2. To explain why some polymers are glasses at room temperature while others are rubbery, thermoplastic, or even liquid.
3. To predict the properties of polymers not yet synthesized, in order to answer the question: Are they worth making?
4. To predict the solution properties of polymers, such as mean square end-to-end distance, solution viscosity, light scattering behavior, and so on.
5. To explain the biological functions of naturally occurring polymers, such as proteins and nucleic acids.

It would be misleading to imply that these objectives have been achieved on a broad scale. Nevertheless, significant progress has been made in the right direction, and the following sections summarize some of the approaches that have been taken.

Scheme 1 summarizes the relationships that exist between the various aspects of conformational analysis to be considered in this chapter. The three subjects enclosed by the circle represent the three main theoretical approaches to conformational analysis. The peripheral subjects to which they are connected by arrows are areas that provide input to and output from the theoretical treatments.

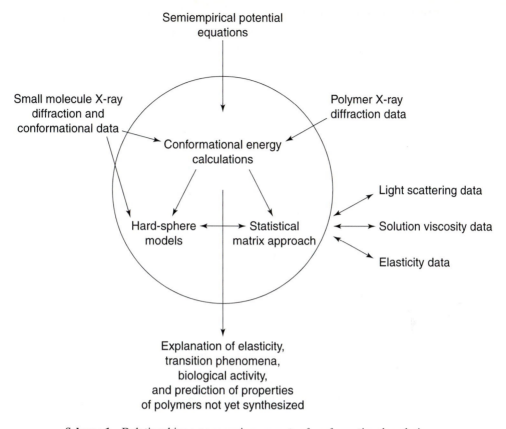

Scheme 1 — the figure text:

Semiempirical potential equations

Small molecule X-ray diffraction and conformational data

Polymer X-ray diffraction data

Conformational energy calculations

Light scattering data

Hard-sphere models

Statistical matrix approach

Solution viscosity data

Elasticity data

Explanation of elasticity, transition phenomena, biological activity, and prediction of properties of polymers not yet synthesized

Scheme 1 Relationships among various aspects of conformational analysis.

POLYMER CONFORMATIONS

A small molecule, such as ethane **(1)**, can adopt relatively few different conformations, because the detailed molecular shape is affected by the torsion of one bond only (the C—C bond). (Of course, the shape will also be modified by vibrational motions, but these have relatively little effect on physical properties.) However, propane **(2)**, with two C—C bonds, is in theory capable of generating

1

2

many more independent conformations in which the C—H bonds on different carbon atoms have different spatial relationships to each other. It is obvious that polyethylene **(3)** offers the possibility that an almost infinite number of different conformations can be generated from the same molecule.

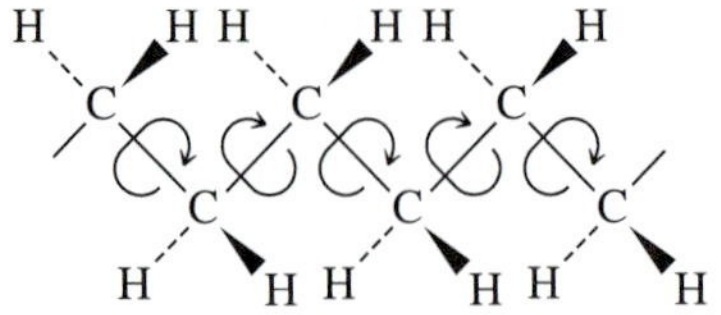

3

The superimposition of *configurational* changes onto these conformational possibilities increases the complexity of the problem even further. Thus, poly(vinyl chloride) can exist in isotactic, syndiotactic, or atactic geometries, and each of these forms will offer the prospect of an entirely new set of conformational possibilities. Consequently, at first sight, an understanding of the conformational behavior of macromolecules may appear to be too complex a problem to warrant serious discussion.

However, a considerable simplification of the problem can be introduced if polymers are divided into three categories.

1. Macromolecules that show a strong preference to adopt one distinct molecular conformation in the crystalline solid state or in solution
2. Macromolecules that assume no preferred conformation at all and adopt the overall shape of a random coil
3. Those in which an individual chain assumes either random coil or a discrete conformational state in different sections along the chain

Clearly, polymers that fall into category 3 can be treated as a composite of those in categories 1 and 2. Hence, in its simplest sense, the problem of polymer conformation can be reduced to the extreme states of the regular, repetitive conformation, on the one hand, and the random coil, on the other. As we shall see, both of these extremes can be treated satisfactorily.

SMALL-MOLECULE MODELS FOR POLYMERS

Given the complexity of macromolecular structures, it makes sense to consider first how conformational problems have been solved for small molecules and then to consider if such approaches can be extended to larger systems. The conformations of small molecules have been examined by means of two approaches, namely qualitative and quantitative methods, and these two methods will be considered in turn.

The Qualitative Approach

For many years organic and inorganic chemists have made predictions about the conformations assumed by simple, covalently bonded molecules. These predictions are based on the intuitively reasonable assumption that atoms forming part of the same molecule repel each other if they approach too closely. This is the idea behind the concept of *steric hindrance*. In fact, two guiding principles have become part of the folklore of chemistry. These are:

1. Large atoms or groups give rise to greater repulsions than small atoms for a given distance between the centers.
2. Staggered conformations are of lower energy than eclipsed conformations.

All this is responsible for the predictions that ethane should adopt a staggered conformation (4) (at least at low temperatures), while *n*-butane should occupy the conformation shown in (5).

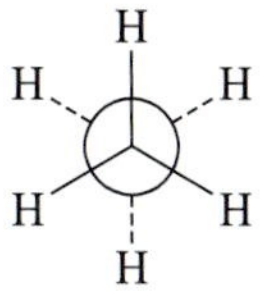

4 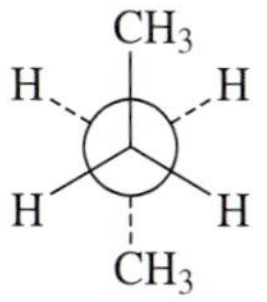   5

These arguments form the rationale behind the use of "hard-sphere" molecular models. The preferred conformation is predicted from these by the assumption that atoms and groups tend to avoid "collisions." Although these methods seem naive, they are remarkably effective for simple systems. Thus, the "hard-sphere" approach predicts that, in a molecule that can undergo conformational changes by torsion of the bond that connects two tetracoordinate atoms (such as carbon), the three "staggered" minima shown in Figure 18.1 will be accessible. As will be seen later, this leads immediately to the concept of the "three-fold rotational isomeric model" that forms the basis of some of the more complex calculations.

However, the more complex the molecule, the more difficult it is to make valid predictions from the qualitative model. For this reason, more quantitative treatments have been developed.

Quantitative Approaches

Basic principles It is assumed that a molecule will occupy that conformation which gives the lowest energy to the system. The main problem, then, is to calculate the energies of a wide variety of different conformations in order to decide which particular conformation yields the lowest energy. In practice, three factors must be taken into account:

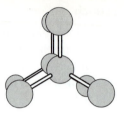

Eclipsed

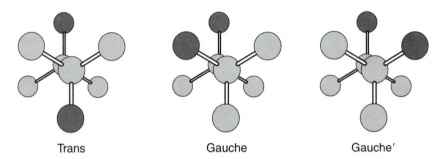

Trans   Gauche   Gauche'

Figure 18.1 The hard-sphere model predicts that eclipsed conformations can be discounted (because of intramolecular repulsions) in favor of the three staggered conformations (shown here as *trans, gauche,* and *gauche'*).

1. Short-range nonbonding intramolecular forces between the atoms.
2. Long-range nonbonding intramolecular forces (this is a particularly important factor in large flexible molecules).
3. *Inter*molecular forces between atoms on adjacent molecules. (Intermolecular forces are very difficult to predict.)

For each of these three types, it is necessary to make decisions about:

(a) The distances between the atoms
(b) The nature of the potential between the atoms

For small molecules, information about the *distances* between atoms in the same molecule can usually be obtained from single-crystal X-ray data or from infrared-Raman or microwave experiments. However, there is very little agreement between different research workers on the best nonbonding *potential* to use, especially when atoms with numerous electrons are present. This aspect will be discussed later. Most investigators also believe that an "intrinsic" threefold rotational barrier is encountered when a bond undergoes torsion, and that this effect is connected more with the characteristics of the bond itself and the orbital

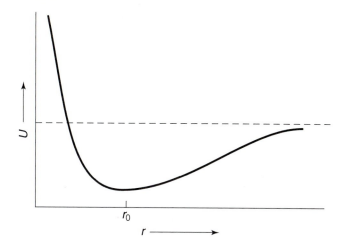

Figure 18.2 Interatomic potential between two nonbonded atoms. The energy (U) rises sharply as the nuclei approach each other at short distances (r), whereas the interaction energy is zero at very large distances. The energy minimum occurs at the point r_0.

arrangement than with the nonbonding interactions of the groups attached to the skeletal atoms.

The interatomic potential   If the assumption is made that short-range intermolecular forces dominate the energetics of conformational changes, then one of the critical problems is to identify the nature of the forces between the contributing atoms. Thus, for calculations on small molecules, use is made of quantum mechanical or semiempirical potentials such as the Lennard-Jones or Buckingham potentials (see later) which require the use of constants that have been derived from experimental data (usually from the behavior of gases).[1]

When individual nonpolar gas molecules approach each other, their energy varies in the familiar way shown in Figure 18.2. As the two molecules approach, they first experience a "London" attractive force, and eventually a repulsive force due to nuclear–nuclear repulsion. The energy minimum at r_0 represents the position at which these two opposing forces are in equilibrium. This distance often corresponds closely to the "intermolecular contact distance" or van der Waals distance in solids. It is generally assumed that the same type of potential curve describes the effects of nonbonding *intra*molecular interactions in larger molecules. The interactions are assumed to be attractive at large distances and repulsive at short distances. The main problem is to predict the exact shape of the curve for

[1]Hirschfelder, J. O., Curtiss, C. F., and Bird, R. B., *Molecular Theory of Gases and Liquids* (New York: Wiley, **1954**).

Small-Molecule Models for Polymers

different interactions, for example, for a hydrogen–hydrogen, hydrogen–carbon, or chlorine–chlorine interaction.

The Lennard–Jones potential is possibly the best known and most widely used formula. It has the form shown in (1).

$$U_{ij} = \frac{B_{ij}}{r_{ij}^{12}} - \frac{A_{ij}}{r_{ij}^{6}} \tag{1}$$

where U_{ij} is the energy at any distance, r, between the atoms i and j, and A and B are constants. The $-A/r^6$ term represents the "dispersion" attraction between the atoms, and $+B/r^{12}$ term represents the repulsion. Note that the repulsive term is steeper or "harder" than the attractive term. In fact, the "hardness" or "softness" of both the attractive and repulsive terms will depend on the types of atoms being considered. The equation given above is known as a "6–12" potential. However, different investigators may use "7–11" or other variants of the formula. The constants A and B determine the depth of the well and the distance of r_0. Thus, if the appropriate constants can be found, it should be possible to calculate the energy of any nonbonded interaction for any distance, r. The sum of all the interactions in the molecule should then represent the energy of the molecule in a particular conformation.

Other equations can be used for the energy calculation. The modified Buckingham potential, the Stockmayer potential, and others are described in standard texts[1] and in the general literature.[2]

The choice of a suitable potential is very much a matter for individual preference. However, the use of constants taken from gas data may not always be valid in calculations for larger molecules. For example, the interaction of two H_2 molecules leads to significant repulsions at distances closer than $\sim 2.5\overset{\circ}{A}$ with an r_0 minimum at distances greater than 3 Å. By contrast, it is known that when two C—H units interact, the H $\cdots$ H repulsions are significant only at distances less than 1.9 Å, with an r_0 minimum at 2.4 Å.

In addition to the use of potential functions that describe the dispersion attractions and van der Waals repulsions, it is often necessary to include a term to account for Coulombic forces. Point-charge equations can be used to simulate dipolar effects but these are not entirely satisfactory.

Model calculations on small molecules   First, let us consider *ethane*. The barrier height for the torsion of the carbon–carbon bond is known from experimental work to be about 2.8 kcal/mol. At least part of the torsional potential in this molecule can be calculated by summing the potential energies of all the in-

[1]Hirschfelder et al., op. cit.
[2]See particularly publications by Birshtein and Ptitsyn; De Santis et al.; Flory (*Statistical Mechanics of Chain Molecules*); Hopfinger; Lowe; Scheraga; Suter; and Volkenstein in the Suggestions for Further Reading at the end of this chapter.

Figure 18.3 Model for the ethane molecule, showing some of the non-bonding intramolecular interactions that must be summed to calculate the overall energy for each conformation generated by torsion of the C—C bond.

teractions shown in Figure 18.3 for all conformations generated by 360° torsion of the C—C bond in, say, 10° increments. The calculated potential curve appears as shown in Figure 18.4. It should be noted that a torsional barrier of about 2.8 kcal/mol in ethane means that at room temperature the molecule can switch from one minimum to another at a rate of about 10^{10} times per second.

As a preliminary exercise to aid in the visualization of conformational effects in polymers, it is instructive to consider also the torsional energy profile of *1,2-di-chloroethane*, as shown in Figure 18.5. Clearly, the replacement of a hydrogen atom on each carbon by a chlorine atom markedly distorts the torsional profile in such a way as to favor the *trans* conformation (Figure 18.1). The barrier designated U_1 is 3.05 kcal/mol, and U_2 is 5.58 kcal/mol.

Finally, as an additional model, we consider *n*-butane (Figure 18.6). This molecule possesses three skeletal bonds that can undergo independent torsion. Torsion of bond ψ, while bonds ϕ and χ are held in a fixed position, generates a regular, threefold profile, similar to the one shown in Figure 18.4, but with higher barriers. Torsion of bond ϕ, without torsion of ψ or χ generates a profile similar to Figure 18.5 with three stable conformers. However, if more than one bond is permitted to twist, the potential must be represented by a three-dimensional energy *surface* (based on torsional angle ψ, angle ϕ, and energy) rather than by a profile. In the case of *n*-butane, the situation is more complicated because *three* bonds can presumably undergo independent torsional motions and a four-dimensional system would be needed to represent the overall conformational energy.

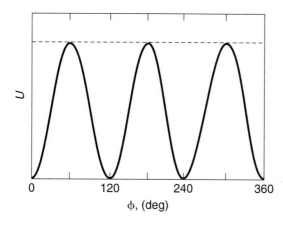

Figure 18.4 Calculated dependence of the potential energy function, U, on the angle of internal rotation in ethane. [From Volkenstein, M. V., *Configurational Statistics of Polymeric Chains* (*High Polymers*. Vol. XVII) (New York: Wiley-Interscience, **1963**).]

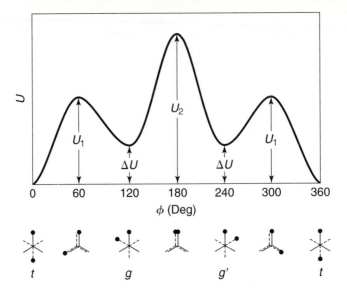

Figure 18.5 Potential as a function of torsional angle for 1,2-dichloroethane. The chlorine atoms are represented by the solid circles. [From M. V. Volkenstein, *Configurational Statistics of Polymeric Chains* (*High Polymers*, Vol. XVII) (New York: Wiley-Interscience, 1963).]

In practice, two bonds are allowed to undergo independent torsion and the third, fourth, fifth, and so on, bonds are kept fixed. Then after the energy has been minimized for the first two, successive additional bonds can be varied. Figure 18.7 shows a surface calculated for *n*-butane for independent torsion of bond ϕ, and synchronous torsion of bonds ψ and χ. Because the surface has a fourfold symmetry, only one-fourth need be shown. The calculated energies are indicated on the contours (in kcal/mol), and the *minimum energy path* or *energy pass* between two low energy sites is shown by the dashed line. When the real molecule undergoes torsional motions, it is likely to follow this path of least resistance. The preferred conformation of the molecule should in principle be represented by the ψ and ϕ values of a low-energy site on the surface.

An extension of these arguments to higher linear alkanes indicates that the number of possible stable conformers is given by $3^{(N-3)}$, where N is the number of carbon atoms in the chain. For polyethylene, the number of accessible con-

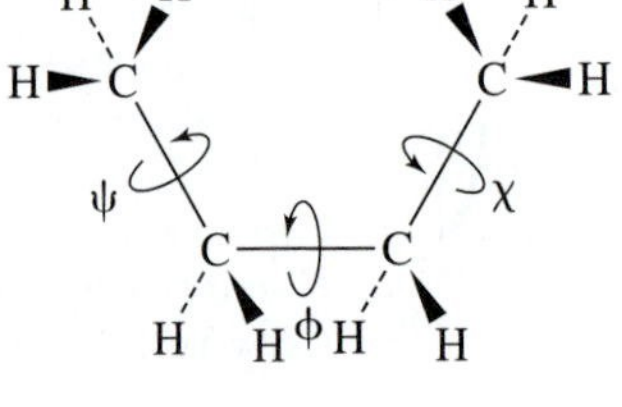

Figure 18.6 Model for exploration of the effects on the potential energy of torsions of bonds ψ, ϕ, and χ in *n*-butane.

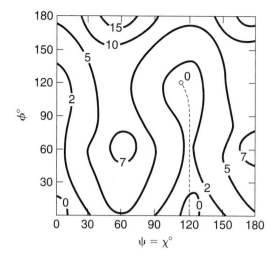

Figure 18.7 One-fourth of the energy surface calculated for *n*-butane. [From Birshtein, T. M. and Ptitsyn, O. B., *Conformations of Macromolecules* (New York: Wiley-Interscience, **1966**).]

formational states reaches an astronomical number. Hence, simplifications are needed.

SHORT-CHAIN MODELS FOR POLYMERS

The Short-Chain Concept

One simplification is to assume that the conformational preference of the whole polymer chain can be represented by the preferred conformation of a short segment of that chain. This is a key concept on which a large amount of research work is based. It depends on the assumption that the conformation of the whole polymer is determined by a composite of all the *short-range* interactions in the molecule.

This allows an enormous simplification of the problem. The question is: How valid is this assumption? It is generally believed that if the chain assumes an extended type of conformation (such as in Figure 18.8a or b), the assumption is reasonably valid. However, if a tight helix is formed (Figure 18.8c), the long-range forces become important and cannot be neglected. It is almost certain that the *inter*molecular forces exert a significant effect in the crystalline state, especially when polar side groups are present. However, relatively few calculations take this into account. Moreover, it will be clear from the assumptions made at the beginning of this chapter that the concept cannot apply accurately to a polymer random coil, although it might apply in a statistical sense to suggest relative preferences of conformational states. As will be seen later, one of the purposes of short-chain conformational analysis is to identify accessible minima and estimate the probability that specific minima are occupied. Thus, the approach discussed in the sections that follow is of value *both* in relation to discrete, repeating conformational structures and to random coils.

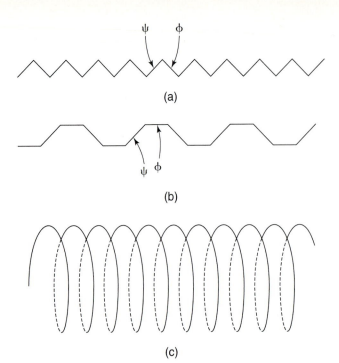

Figure 18.8 Short-range intramolecular forces (and intermolecular effects) dominate the conformational energy preferences for extended conformations, such as (a) or (b). However, longer-range forces become important if the polymer occupies a tight helical conformation (c).

Conformational Nomenclature

The conformation of the chain can be represented simply by quoting the torsional angles of the backbone bonds. Unfortunately, two quite different conventions are in use. In the first, all skeletal angles are quoted as deviations from the *trans–trans* (i.e., planar zigzag conformation of the chain). For example, the conformation shown in Figure 18.8a can be represented by the symbolism $\psi = 0°$, $\phi = 0°$. The *cis–trans*-planar conformation shown in Figure 18.8b is given the notation, $\psi = 0°$, $\phi = 180°$. Helices (Figure 18.8c) must be represented by ψ and ϕ values between $0°$ and $180°$, and these values may be positive or negative. However, some workers use the *cis–cis* conformation as the $\psi = 0$, $\phi = 0$ starting conformation, and this can lead to confusion.

Because torsional motions can be left- or right-handed, an additional convention is needed. One convention is as follows. If we look down a bond from atom 1 to atom 2 (Figure 18.9a), then ψ is *positive* if the bond (x) must be rotated *clockwise* by the angle needed to superimpose it on bond (y). Counterclockwise rotations of less than $180°$ are considered to be negative. Using this system, the *trans*-planar structure, shown in Figure 18.8a, would be described as $\psi = \phi = 180°$ rather than $\psi = \phi = 0°$. However, the *cis–trans* structure (Figure 18.8b) would continue to be described as $\psi = 0°$, $\phi = 180°$.

We prefer a system in which extended *trans*-planar structures are depicted as $\psi = \phi = 0°$, with the sign of the torsional angle determined as follows. A positive torsion angle is generated when the rear bond (y in Figure18.9) has been

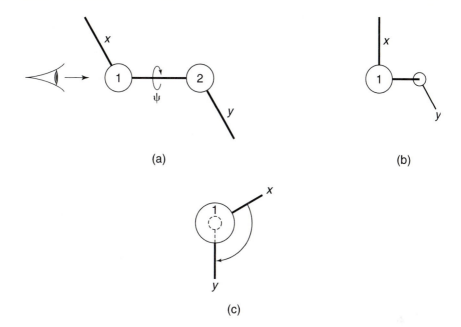

(a)   (b)

(c)

Figure 18.9   Common sign convention for torsional rotation of bonds. If bond *x* must be rotated in a clockwise direction to superimpose it on bond *y*, the torsion is given a positive value.

moved less than 180° in a clockwise direction *from* the *trans–trans* ($\psi = \phi = 0°$) position to generate the observed conformer. Positive rotations generate right-handed (clockwise) helices—a scheme that is easy to remember.

Use of the Hard-Sphere Approach

The simple hard-sphere-model approach that was described earlier for small molecules can be used for short-chain segments of polymers also. For example, polyethylene is simply a logical extension of the *n*-butane molecule, and it requires only a brief examination of models or a few sketches with pencil and paper to see that the extended zigzag conformation **(6)** is generated by this scheme. The predicted conformational profile for torsion of one bond in polyethylene would resemble Figure 18.10.

6   **7**

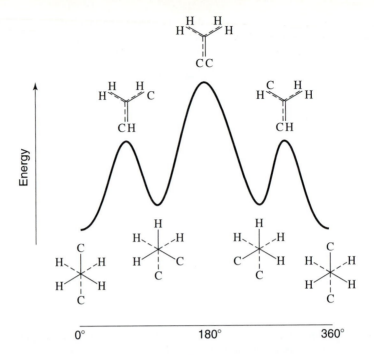

Figure 18.10 Estimated torsional profile for 360° torsion of a skeletal bond in a short-chain segment of polyethylene, based on a hard-sphere rationalization.

If it is assumed that chlorine atoms are particularly repulsive to each other, the same argument can be used to predict that syndiotactic poly(vinyl chloride) (**7**) should also form a planar zigzag arrangement (**8**), since this conformation places the longest distance between the bulky chlorine atoms. This conformation is found in the solid state, although crystal packing forces may be partly responsible for this behavior.

8

Use of Intramolecular Potentials

It will be obvious that the approach discussed earlier for the calculation of torsional energy profiles and surfaces for small molecules can be extended in a relatively straightforward manner to short segments of a long chain. The assumption must be made once again that the conformational preference is dominated by short-range forces.

Consider the short segment of a polyethylene chain shown in **9**. The *intra-molecular distances* between atoms can be calculated from bond-angle and bond-length data obtained from low-molecular-weight compounds (from single-crystal X-ray work) or from other polymers. For example, bond angles at tetrahedral carbon are often close to 109.5°, the C—H bond length in saturated organic compounds is 1.09 Å, and the C—C bond length is 1.54 Å. There are only 15 atoms in this segment, and the calculation of the distances and energies for all the non-bonding interactions (H $\cdots$ H, H $\cdots$ C, C $\cdots$ C, etc.) with the segment held in one chain conformation, is a trivial problem. Even if all the possible conformations which result from 360° incremental torsion of bonds *A* and *B* are considered, the calculations are quite manageable with the use of a modern computer.

9

Macromolecules that contain very polar skeletal bonds or side groups must be treated in a special way. This includes polymers such as poly(ethylene oxide), poly(methyl methacrylate), polyacrylonitrile, poly(vinyl alcohol), and so on. Polar or hydrogen-bonding forces are difficult to calculate.

The following sequence of operations would be followed in order to perform a preliminary calculation of the conformation of a polymer:

1. Consider a short segment of the chain and place it in a starting conformation. For example, a possible starting conformation for polyethylene would be the one shown in Figure 18.11.

2. Assign assumed bond angles and lengths. (A good starting point would be to use values obtained from X-ray single-crystal studies on smaller hydrocarbons.)

3. Set up a computer program to permit torsion of bonds C_2—C_3 and C_3—C_4 independently through increments of, say, 10°. At each conformational position, calculate the nonbonding distances ($C_1 \cdots C_5$, $H_1 \cdots H_6$, etc.).

4. Introduce a suitable nonbonding potential into the calculation and allow the computer to calculate the interaction energies that vary with torsional

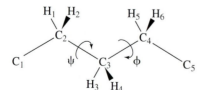

Figure 18.11 Short-chain segment used as a model for a preliminary conformational energy calculation on polyethylene.

rotations for every calculated distance. Sum all energies for each conformation until C_2—C_3 and C_3—C_4 have both undergone 360° torsion.

5. Obtain a matrix printout from the computer giving the energies for all incremental values of ψ and ϕ, and plot energy contours at preselected intervals.

6. Select the preferred conformational minima from the surface, or use the selected minimum as a starting point for additional calculations to explore the effect of the torsional motions of the adjacent skeletal bonds. In practice, this latter procedure is often incorporated into a "hunting" program to find the overall conformational minimum if four, five, or more skeletal bonds are allowed to undergo independent torsion.

7. Find the minimum energy pass between the preferred minima, and calculate the barrier heights, Boltzmann distributions, and so on.

8. Calculate the number of monomer units per repeat. This information would be used in conjunction with X-ray crystallographic studies (see Chapter 19).

Three examples of the results from this type of treatment will be mentioned.

1. A calculated potential-energy profile for a short segment of a poly(tetrafluoroethylene) chain is shown in Figure 18.12. Note that the profile shows more detail than would have been predicted from hard-sphere models alone. Several pronounced minima are present and this suggests that several crystal–crystal transitions might occur in this polymer if the system can surmount the barriers. The principal minima lie just a few degrees on each side of the main 0°:0° minimum (planar zigzag). This implies that the chain should prefer a very open helical structure. This result is compatible with X-ray diffraction data which indicate that, below 20°C in the crystalline state, the polymer assumes a helix with 13 monomer units in six turns (a 13_6 helix).

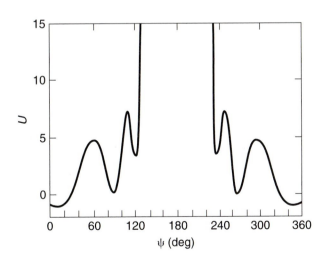

Figure 18.12 Calculated conformational potential energy profile for a short segment of a polytetrafluoroethylene chain. [Modified from De Santis, P., Giglio, E., Liquori, A. M., and Ripamonti, A., *J. Polymer Sci. (A)*, **1963**, *1*, 1383.]

2. Figure 18.13 shows an energy profile for polyisobutylene when a short segment of the molecule is allowed to find the *minimum energy path* by compensatory "avoidance" motions of adjacent units. Hard-sphere molecular models imply that this polymer is very highly hindered, an implication that is at variance with the elasticity of this material. However, the calculated torsional profile indicates that, in spite of the fact that the lowest minima correspond to a tight, helical arrangement, these minima form part of a broad, low-energy plateau that would permit torsional motions of about 200° before a high barrier was encountered. Such synchronous bond movements can also be visualized as "crankshaft motions." Conversion of one conformer to another in polyisobutylene should, therefore, be a very facile process, and it is interesting to speculate that this feature may explain the elasticity and low glass transition temperature of this polymer.

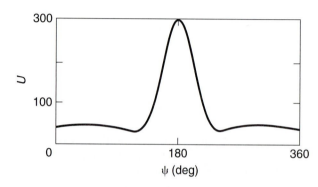

Figure 18.13 Calculated energy profile for a short segment of polyisobutylene following the minimum-energy pathway allowed by compensatory "avoidance" motions of the nearby chain units. [Modified from De Santis, P., Giglio, E., Liquori, A. M., and Ripamonti, A., *J. Polymer Sci. (A)*, **1963**, *1*, 1383.]

3. Calculations of this type can be extended to polymers that have complex side groups. An example is poly[bis(trifluoroethoxy)phosphazene], $[NP(OCH_2CF_3)_2]_n$, in which the side groups can themselves occupy a variety of different conformational orientations for each conformation of the chain. An approach to this problem is to allow the side groups to undergo energy minimization for each tested conformation of the backbone. Figure 18.14 shows an energy surface calculated in this way for the segment shown in **10**. Despite the bulkiness of the side groups, the surface shows broad areas of low potential that suggest considerable conformational mobility. This is compatible with the low T_g value of this polymer ($-66°C$).

10

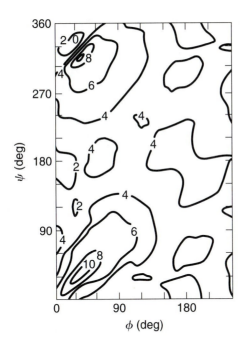

Figure 18.14 Energy surface calculated for poly[bis(trifluoroethoxy) phosphazene]. [From Allen, R. W. and Allcock, H. R., *Macromolecules*, **1976**, *9*, 956.]

Longer-Chain Models

The simplest approach based on the foregoing treatment would involve a calculation of the potential surface resulting from the independent torsion of two adjacent skeletal bonds, as shown in Figure 18.11 and **9.** Valuable data can indeed be obtained by the use of such techniques, especially when the polymer assumes an extended *trans*-planar or *cis–trans*-planar conformation. However, caution must be exercised during the choice of the actual energy minimum or minima occupied by the polymer chain. Some minima calculated for the short-chain segment may actually be eliminated by longer-range forces. For example, a model based on torsion of a short, two-bond segment of the chain might generate an energy minimum at a conformation that represents a very tight helix or even a *cis–cis* conformation. This conformation may be totally precluded by a 1:6 collision of the skeletal atoms. Thus, a more rigorous treatment would require that the torsion of a third adjacent skeletal bond should also be included in the treatment. Ideally, a segment of four or five skeletal bonds should be considered. However, the solution of a three- or four-bond problem seriously raises the complexity of the task. If the polymer possesses multiatomic side groups which can also undergo complex torsional motions, the calculations become almost prohibitively complicated.

Three approaches are possible:

1. The complete permutation of conformations for the three- or four-bond torsional segment could be calculated as an extension of the program out-

lined in the preceding section. In practice, this process could be expensive in terms of computer time.

2. A "hunting" or a "searching" procedure may be instituted. In this, the minimum for a two-bond segment may be found first, and this minimum is then used as a basis for the search for a resultant minimum when the third or fourth bonds are considered. Multiatomic side group conformations can be energy minimized for each conformation of the skeleton. This process saves computer time. However, care must be taken to ensure that minima located early in the calculation remain the principal minima at the end. In other words, a danger exists that new minima (the "real" minima) may be present in conformational space not searched by the program. Thus, judgment, insight, and intuition play an important role in this approach. Moreover, it is essential that the calculations should be tied to experimental facts at nearly every stage to prevent them from becoming unacceptably speculative. The results should be reviewed at each step in terms of actual polymer conformations, glass transition temperatures, barriers to rotation in small molecules, and so on, until the investigator is satisfied that he or she is on the right track. This back-checking procedure is essential if atoms with many more electrons than carbon or hydrogen are involved since little factual potential data are available for such atoms.

3. The process can be reduced to a search for minima in a two-, three-, four-, or higher-fold rotational isomeric model. This procedure will be discussed in a later section.

RELATIONSHIP TO THE STATISTICAL COIL CONCEPT

As mentioned earlier, the regular, repetitive, extended, or helical chain conformation is only one extreme of the conformational spectrum. This is the type of conformation that is most readily treated by the techniques discussed in the previous sections. The other extreme is the random coil, in which a wide variety of different conformations is assumed to occur more or less randomly along the length of an individual chain.

The analysis of the spatial arrangements, conformations, or "configurations" of a chain molecule invariably starts with the freely jointed chain. The main reason for this is that the statistics of such idealized chains were worked out a long time ago, and the results used successfully to interpret such phenomena as diffusion and Brownian motion. The two-dimensional version of this model is called the "random walk" and is most humorously stated in terms of the walk of a drunkard around an empty parking lot. Rather surprisingly, the same relationship holds for a three-dimensional walk, usually called the "random-flight" model. The important question is how far the person travels after n steps each of length l. This is to be contrasted to the obvious result for a directed walk in which $r = nl$. The

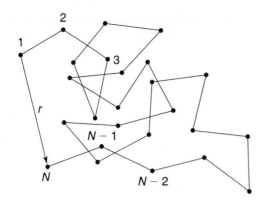

Figure 18.15 End-to-end distance, r, for a freely jointed (i.e., no fixed valence-bond angles) chain made up of N atoms.

parallel situation in polymer conformations is called the "freely jointed" chain,[1] and is illustrated schematically in Figure 18.15.

The simplest way to derive the desired relationship in three dimensions is by vector addition of the skeletal bond vectors. The mean-square distance is

$$\vec{r} = \sum_i \vec{l}_i \tag{2}$$

$$\vec{r}^2 = \vec{r} \cdot \vec{r} = \left(\sum_i \vec{l}_i\right) \cdot \left(\sum_j \vec{l}_j\right) = \left(\sum_i \sum_j (\vec{l}_i \cdot \vec{l}_j)\right) \tag{3}$$

where the dot product of two vectors is the product of their two magnitudes times the cosine of angle between them: $\vec{a} \cdot \vec{b} = |\vec{a}||\vec{b}|\cos\theta$. (Thus, such conformational problems really involve obtaining the components of each skeletal bond on every other skeletal bond in the chain.) Equation (3) can be broken into two summations

$$<r^2>_0 = \sum_{i=1} \sum <\vec{l}_j \cdot \vec{l}_j> + 2\sum_{i<j}\sum <\vec{l}_i \cdot \vec{l}_j> \tag{4}$$

where the first double sum is just nl^2, since it is n terms each of which is one skeletal bond dotted into itself.[2] The second double sum has to be zero in the present case because there is no correlations at all between the skeletal bonds in a freely jointed chain. Thus, for this very simplified model, $<r^2>_o = nl^2$.

The distribution of end-to-end distances r is also of considerable importance, being central to most theories of rubberlike elasticity and also appearing elsewhere, such as in the Flory–Krigbaum theory of dilute polymer solutions. The

---

[1] Flory, P. J., *Principles of Polymer Chemistry* (Ithaca, N.Y.: Cornell University Press, **1953**), Chap. 10.

[2] Note that $<r^2>$ means the mean square end-to-end distance which is also given as $\bar{r}^2$.

objective is to determine the probability $W(r)dr$ that r is in an infinitesimal interval, dr, about some specified value of r. Freely jointed chains show the Gaussian distribution

$$W(r)dr = (\beta/\pi^{1/2})^3[\exp(-\beta^2 r^2)]4\pi r^2 dr \tag{5}$$

where $\beta^2 = 3/2nl^2$. The Gaussian limit is also used for more realistic models and is a good approximation unless the chains are (i) very short, (ii) very stiff, or (iii) stretched close to the limits of their extensibility. A typical distribution is shown in Figure 18.16.

Although the freely jointed model is greatly oversimplified, it can be applied to real chains by defining "virtual bonds" to span a sufficient number of consecutive skeletal bonds so that these newly defined bonds are uncorrelated in direction. In polyethylene a grouping of around seven skeletal bonds is required, so the first virtual bond would span carbon atoms 1 to 8, the second would span 8 to 15, etc. A larger number of skeletal bonds would be required, for example in

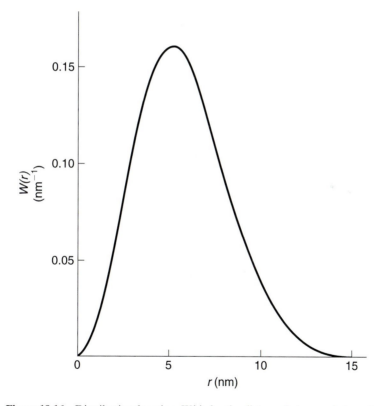

Figure 18.16 Distribution function, $W(r)$, for the distance between chain ends of a statistical coil for the case $N = 10^4$ and bond length $= 2.5\text{Å}$. [From Flory, P. J., *Principles of Polymer Chemistry* (Ithaca, N.Y.: Cornell University Press, **1953**), p. 406.]

the case of stiffer chains before the correlations between the virtual bonds became negligible.

The simplest elaborations to make the freely jointed model itself more realistic involve direct evaluations of the second double summation to determine the factors by which $<r^2>_o$ exceeds the freely jointed value nl^2. In fact, this result is the basis of the definition of the characteristic ratio (CR) as the high molecular-weight limit of the ratio $<r^2>_o/nl^2$. As already mentioned, the extent to which it exceeds unity is a measure of chain stiffness, and it typically ranges from around 4.0 to 40.

The first refinement is called the freely rotating chain, in which the angle between skeletal bonds is restricted to some value, but rotations φ about these bonds is free, meaning that all have equal probability of occurrence. It has become customary to work with bond-angle supplements θ, and in this convention the result is

$$<r^2>_o = [(1 + \cos \theta)/(1 - \cos\theta)] \, nl^2 \tag{6}$$

Frequently, the bond angles alternate between two values θ' and θ'', as they do for example in the polysiloxanes, where the bond angle about the silicon in $O—Si—O$ is approximately tetrahedral but the bond angle about O in $Si—O—Si$ is opened to around 143°. For this case, equation (6) can be generalized to

$$<r^2>_o = [(1 + \cos \theta')(1 + \cos \theta'')/(1 - \cos \theta' \cos \theta'')]nl^2 \tag{7}$$

For either case, this modification increases the CR by a factor of around 2. Because real chains show conformational preferences, this factor can be taken into account by calculating average values $<\cos \varphi>$ of the rotational angles, which introduces the additional factor shown in the equation

$$<r^2>_o = [(1 + \cos \theta)/(1 - \cos \theta)][(1 + <\cos \varphi>)/(1 - <\cos \varphi>)]nl^2 \tag{8}$$

where the additional factor $(1 + <\cos \varphi>)/(1 - <\cos \varphi>)$ suppressing the free rotation is called the "stiffening ratio." It typically increases the CR by another factor of 1.5–2.0, giving a value of around 3.4, which is still much less than the value of 6.7 for polyethylene, which is a typical random-coil polymer.

The major additional refinement still to be introduced is the knowledge that rotations about consecutive bonds are coordinated. The most important example in organic chemistry, is called the "pentane effect" since n-pentane is the simplest molecule in which it occurs. Of the four skeletal bonds in this molecule, only the central two affect the conformations, and the g^+g^- and g^-g^+ conformations are of much higher energy than the g^+g^+ and g^-g^- alternatives because they place the two CH_3 ends too close together. The same situation occurs in chains such as polyethylene, in which the two CH_3 groups are replaced by the left and right parts of the polyethylene chain. Incorporation of this effect requires the more complicated approach described in the remaining parts of this chapter. It is also necessary to note that no part of a polymer chain can place itself into a site already occupied by a preceding unit (unlike the way a diffusing particle can re-visit one of its previous locations), and account must be taken of this "excluded volume" effect as well.

As will be demonstrated, conformational analysis, plays an important role in the study of random coil polymer systems as well as regularly extended chains. Thus, connections can be made between physical properties that depend on ran-dom coil arrangements (such as elasticity or dilute solution properties) and the results of conformational calculations. These will be discussed briefly later in this chapter.

INFLUENCE OF THE BOLTZMANN DISTRIBUTION

Calculated torsional potential profiles, such as Figure 18.12 or 18.13, or energy sur-faces, such as Figure 18.14, are valuable because they suggest (1) the location of preferred (low-energy) conformations, (2) the heights of the barriers that must be surmounted by the molecule to get from one well to another, and (3) the minimum-energy pathway that the molecule probably follows in order to switch from one conformation to another.

At first sight, it might be imagined that a polymer molecule would always assume the one conformation that corresponds to the lowest-energy well. Yet the experimental facts tell us otherwise. The existence of random coil arrangements, the evidence that many extended polymer chains occupy supposed high-energy conformations, and the existence of conformational transitions all indicate that the lowest-energy well is not the only one occupied under all circumstances. Ther-modynamically, the occurrence of the higher-energy conformations randomizes the chain, and effect of this increase in entropy on the free energy more than makes up for the energy increase.

In fact, as is well known from elementary physical chemistry considerations, a molecule will "populate" different accessible energy levels to a degree that de-pends on the temperature. All other things being equal, the molecule will occupy the lowest energy well at low temperatures and will populate progressively higher levels as the temperature is raised. Thus, the probability that a particular seg-ment of the chain will occupy a well that exists for a torsional angle, ψ, will be given by the Boltzmann equation,

$$P(\psi) \sim \exp\left[\frac{-U(\psi)}{RT}\right] \qquad (9)$$

where U is the energy of that well. The higher the temperature, the greater is the probability that a chain will assume higher-energy conformations. The heights of the barriers between the wells should affect the *rates* at which bonds will switch back and forth from one conformation to another, but usually these barriers are low enough to permit many transitions per second at normal temperatures.

Thus, the important point to be noted is that conformational energy calculations can be used to estimate the *probability* that one or another conformational state will be occupied. Clearly, this has important ramifications for a consideration of the way $<r^2>_o$ varies with temperature. It is also extremely valuable for the statistical matrix approach, discussed in the next section.

THE STATISTICAL MATRIX APPROACH

The Rotational Isomeric State Model[1]

By now it is clear that any attempt to describe a polymer molecule in terms of one infinitely repeating chain conformation is not valid, except in the special case of polymer single crystals. Moreover, it will also be clear that the existence of different conformational states in the same chain adds enormously to the complexity of attempts to describe these systems, because changes in the conformation of one segment of the chain will affect the stability of the nearby conformational states. Thus, an *exact* description of the conformation of a macromolecule is beyond the capability of our present computational systems.

However, a *statistical* description is feasible if we can be satisfied with information about the *probabilities* that certain states will be occupied by the molecule. Moreover, an enormous simplification of the problem can be obtained if it is assumed that each skeletal bond has access to only a small and limited number of torsional states. For example, using the hard-sphere model mentioned earlier, these states could be the three staggered conformations (Figure 18.1) commonly used for organic compounds. Alternatively, on a more sophisticated level, they could represent the principal energy wells detected from conformational energy calculations. This approach is called the *rotational isomeric state model.* For polymers that contain aliphatic carbon atoms in the backbone, three possible states only are considered to be accessible; *trans, gauche*(+), and *gauche*(−). For polymers that contain other atoms, the number of rotational isomeric states may be less than or more than three, and could occur at very different locations.

The Statistical Weight Matrix[2]

If a threefold rotational isomeric model is used, the conformation at each chain atom can be reduced to a choice between *trans, gauche*(+), and *gauche*(−) (usually written t, g^+, g^-). Thus, torsion of two adjacent bonds (say, *A* and *B*) gener-

---

[1]Volkenstein, M. V., *Configurational Statistics of Polymer Chains* (New York: Wiley-Interscience, **1963**).

[2]Flory, P. J., *Statistical Mechanics of Chain Molecules* (New York: Wiley-Interscience, **1969**).

ates nine possibilities which can be depicted by a 3 × 3 matrix. Examination of molecular models or a consideration of torsional energy maps allows the investigator to make a judgment about the relative probabilities of each of the nine possibilities. If a hard-sphere model is to be used, conformations that generate "collisions" (serious interpenetration of van der Waals radii) are assigned a probability of zero. For example, in saturated carbon-backbone polymers, successive *gauche*(+) and *gauche*(−) conformations cause serious steric hindrance. Those interactions that appear to involve nonbonding attractions are assigned a high probability, and so on. Alternatively, the probabilities may be estimated from the calculated energies of the wells and from the Boltzmann populations of these wells.

As an example, consider a segment of a chain in which two adjacent skeletal bonds are permitted to undergo threefold rotational isomeric torsion. Assume that the *trans* conformation of each bond represents the lowest-energy state. The statistical weight matrix is

$$
\begin{array}{cccc}
 & \text{(t)} & \text{(g+)} & \text{(g−)} \\
\mathbf{U'} = \begin{matrix} \text{(t)} \\ \text{(g+)} \\ \text{(g−)} \end{matrix} & \begin{bmatrix} 1 & u_{12} & u_{13} \\ u_{21} & u_{22} & u_{23} \\ u_{31} & u_{32} & u_{33} \end{bmatrix}
\end{array}
\tag{10}
$$

The *trans–trans* combination is assigned an arbitrary statistical weight of 1. Suppose now that an examination of molecular models or of a calculated energy surface reveals that combinations of the type g(+)g(−) and g(−)g(+) are prohibitively high in energy. These can then be assigned a statistical weight of 0. If the molecule is structurally symmetric, and the energy of the *gauche–trans* combination is assumed to be one-half of the energy of the *gauche–gauche* (all energies being compared to that of the *trans–trans* combination), the matrix can be simplified to

$$
\mathbf{U'} = \begin{bmatrix} 1 & \delta^{1/2} & \delta^{1/2} \\ \delta^{1/2} & \delta & 0 \\ \delta^{1/2} & 0 & \delta \end{bmatrix}
\tag{11}
$$

where $\delta = \exp\left[-\Delta U_{g(+)\,g(+)}/kT\right]$, and $\Delta U_{g(+)\,g(+)}$ is the energy of the *gauche*(+)*gauche*(+) well above that of the *trans–trans* well.

A second 3 × 3 matrix can then be calculated for bonds B and C. The product of the two matrices will reflect the probability of the various combined conformations of the two bonds. The process can be extended for four or more bonds for more complex repeating sequences.

The partition function, Z, that is central to statistical mechanics can be generated by serial multiplications of these statistical weight matrices, and properties such as chain dimensions, dipole moments, etc. can be similarly calculated by multiplications of "generator" matrices containing these statistical weights but also bond angles, rotational angles, bond vectors, dipole moment vectors, etc.[1]

APPLICATIONS OF CONFORMATIONAL ANALYSIS

Our interest in conformational analysis is based on the theory that the physical properties of materials can be explained in terms of molecular shape or changes in molecular shape. Thus, it is worthwhile to examine briefly some of the areas in which conformational analysis has assisted in an understanding of physical phenomena.

Conformation and Crystallinity

Polymers crystallize because their chains can pack together in a regular manner. Such regular packing depends on two factors: the minimum-energy conformation of the chain and the presence or absence of tacticity.

Although globular proteins can crystallize in their native conformations, a random coil will not crystallize. Only in extended chain conformations can the conditions be met for the parallel orientation of adjacent chains. Even so, some extended chain conformations are more prone to crystallize than others. Linear polyethylene crystallizes readily because its preferred conformation is an extended zigzag (**3**). Such a molecular arrangement can easily be packed in a regular manner in a crystallite. Poly(tetrafluoroethylene) also crystallizes, but the chain conformation is a twisted zigzag arrangement which repeats only once every 13 carbon atoms. Repulsions by the fluorine atoms are responsible for the slight twist of the backbone. The fluorine atoms also maintain the extended conformation and are responsible for the chain stiffness. All this is favorable for the generation of a close-packing arrangement between neighboring chains. Syndiotactic poly(vinyl chloride) crystallizes with the chains assuming a planar zigzag conformation. Again, the chain conformation is highly favorable for crystallization.

In these cases, and in many others, conformational analysis can be used to identify low-energy wells that might be occupied in the microcrystalline or single-crystal state. However, as discussed in the earlier sections, a preferred chain conformation alone is usually not sufficient to generate a long sequence of repeating units with one discrete repetitive conformation, because the higher-energy states also will be populated at normal temperatures. Thus, *inter*molecular forces appear to play a crucial role in stabilizing one conformation at the expense of equivalent or higher-energy states. Such crystal packing forces may even favor a higher-energy single-chain conformational state over a lower one.

The case of syndiotactic poly(vinyl chloride) provides a probable example of this effect. The *trans*-planar (planar zigzag) conformation of this polymer may not be the lowest-energy state for an isolated chain. However, this conformation

---

[1]Flory, *Statistical Mechanics of Chain Molecules*, op. cit.

packs so well into a crystalline matrix (see Figure 18.17) that the overall multi-chain arrangement undoubtedly has a lower energy than other, less efficient, packing modes for other chain conformational states.

Thus, conformational energy calculations cannot be used directly and indiscriminately to predict polymer conformations in the crystalline state, and conversely the conformational results from X-ray structure studies on polymers

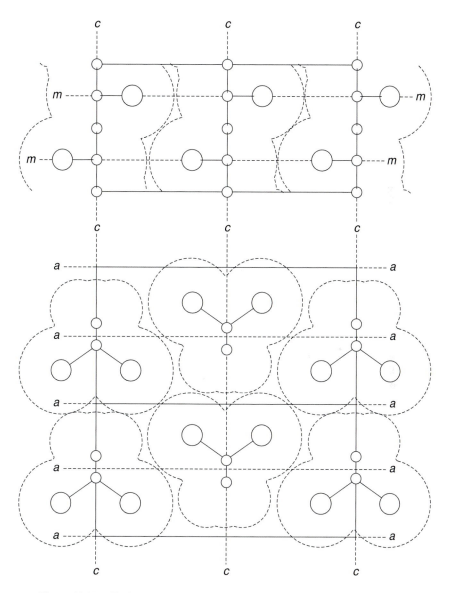

Figure 18.17 Chain packing arrangement in syndiotactic poly(vinylchloride). [From Natta, G. and Corradini, P., *J. Polymer Sci.*, **1956,** *20,* 251.]

should not be applied without question to proving the validity of single-chain conformational calculations. Of course, such comparisons would be quite valid if the *inter*chain forces could be incorporated into the conformational energy calculations. To do this would require not only an exploration of the possible single-chain conformational states, but also an investigation of a wide variety of packing arrangements for each individual chain conformation. Thus, the *prediction* of crystal structures from conformational calculations is a formidable problem. On the other hand, the *rationalization* of known crystal structures in terms of both single-chain calculations and intermolecular forces is more manageable.

The use of single-chain calculations to understand crystal conformation is probably least questionable if two conditions are met. First, if the calculated energy well occupied by a particular conformation is much deeper than competing wells, and the barriers to conformational changes are very high, the interchain packing forces may be insufficient to tip the balance in favor of another conformation. Second, if the cross-sectional profile of a chain in a low-energy conformation is close to circular, individual chains may be capable of undergoing independent rotations within the matrix without appreciable perturbation of the interchain packing forces (as if an individual cylindrically shaped pencil were to be turned in the middle of a stack of similar pencils). Some helices or extended conformations [e.g., the *cis–trans*-planar (180°:0°) conformation of poly (dichlorophosphazene)] may fall into this category.

Chain Flexibility

It is generally assumed that a correlation exists between the bulk flexibility of a polymer and the flexibility of the molecular chains. The flexibility of a polymer chain can be defined as the ease with which a randomly coiled chain can be unraveled or stretched out. Because the elongation of a chain takes place by the "unwinding" or rotation of backbone bonds (see Figure 17.2), any factor that inhibits the torsion of the skeletal bonds should decrease the polymer flexibility. If this hypothesis is correct, it should be possible to predict the relative flexibilities of polymers from their torsional energy profiles.

Two approaches to this problem exist. First, it can be assumed that the torsional flexibility of a skeletal bond depends on the "span" of torsional oscillations *within* one particular energy well. Thus, the ease of torsion depends on the *steepness* of the walls of the energy well, and flexibility is only an indirect function of the barrier height. This viewpoint is illustrated in Figure 18.18.

The width of the well is also important because broad energy wells introduce more entropy from vibrations about these minima and thus increase the probability of the occurrence of these rotational states. This is taken into account in refined approaches by multiplying the usual Boltzmann factors by a pre-exponential entropic factor.[1]

---

[1]Honeycutt, J. D. in *Physical Properties of Polymers Handbook*, J. E. Mark, ed. (New York: Springer-Verlag, **1996**).

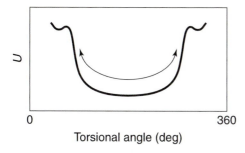

Figure 18.18 Torsional energy profile, demonstrating the concept that a broad energy well will allow appreciable torsional freedom in the skeletal bonds. This may be responsible for polymer flexibility or elasticity on a microscopic level.

The second approach assumes that the backbone bonds undergo torsion by *switching* from one well to another. In other words, the internal rotation of a bond requires a jump over the barrier. If this idea is correct, then flexibility should be determined mainly by the ease with which a bond can jump from one well to another and this, in turn, will depend on the Boltzmann distribution. Furthermore, the presence of several closely spaced wells with comparable energy minima should generate a high degree of flexibility.

It is important to differentiate between two types of flexibility, because neglecting this has caused much confusion in the literature. "Equilibrium flexibility" has to do with the ability of the chain to collapse into compact conformations of high randomness (entropy). It is determined by the locations and conformational energies of the minima in the potential energy surface, and has a significant influence on melting points through its large effect on entropies of fusion. Thus, high equilibrium flexibility can lead to low melting points. On the other hand, "dynamic flexibility" is concerned with the ability of the chain to *change* shape, and is determined in part by the barriers that *separate* energy minima. This type of flexibility has a major influence on the glass transition temperature T_g discussed in the following section. High-dynamic flexibility is associated with low values of T_g. However, it must be remembered that there is a pronounced intermolecular contribution as well, and that strong intermolecular attractions tend to increase T_g. If this were not true, then low molecular-weight solvents would not be able to function as "plasticizers" to lower glass transition temperatures.

Very Inflexible Chains

None of the statistical approaches described are useful for rigid-rod polymers, because they have so little conformational flexibility. In fact serious problems are encountered in even getting such materials into solution because they have such small entropies of mixing. The typical answer (in desperation) is to dissolve them in powerful acids that convert the chains to polyelectrolytes that repel each other long enough to allow some processing to be carried out. Alternatively, it is possible to copolymerize kinks ("molecular swivels") into the chains to increase their flexibility, and thus their entropies of mixing, solubilities, and tractibilities in general. A really innovative approach is to bond long flexible side chains onto the stiff chain backbones (comb polymers) so that they can mix with the solvent. This is an interesting example of biomimicry, since this is what occurs in biology to solubilize proteins in the α-helical state. The idea is now being used to help solubilize even nanotubes.[1]

Relationship to the Glass Transition Temperature

In Chapter 17 it was stressed that a rough relationship exists between the glass transition temperature and the flexibility of the polymer chain. The more flexible the polymer, the lower may be the T_g value. Hence, an important facet of polymer conformational analysis is the need to attempt to predict glass transition temperatures. At the present time, this part of the field is in its infancy and only rough correlations can be made. For example, the low glass transition temperature ($-76°C$) of polyisobutylene can perhaps be correlated with the broad well shown in Figure 18.13. The glass transition temperatures of poly(dihalophosphazenes) increase in the following order: $(NPF_2)_n$, $-90°C$; $(NPCl_2)_n$, $-66°C$; $(NPBr_2)_n$, $-8°C$. Calculations performed on these polymers show a striking increase in barrier height along this series and a narrowing of the principal well.[2]

Elasticity Measurements

The phenomenon of elasticity represents the tendency for a flexible polymer chain to reassume a random coil conformation after having been stretched. The driving force for this contraction is the increase in entropy that is associated with the formation of the random coil.

The assumption of a purely entropic elasticity leads to the prediction that the stress should be directly proportional to the absolute temperature at constant elongation, α (and volume, V). The extent to which there are deviations from this direct proportionality may, therefore, be used as a measure of the thermodynamic non-ideality of an elastomer. In fact, the definition of ideality for an elastomer is that the energetic contribution, f_e, to the elastic force, f, be zero. This quantity is defined by

$$f_e \equiv (\partial E/\partial L)_{V,T} \qquad (12)$$

which is a definition that closely parallels the requirement that $(\partial E/\partial V)_T$ be zero for ideality in a gas. Its relevance here is that it is directly related to the temperature dependence of the chain dimensions, as will be described.[3,4]

Force-temperature ("thermoelastic") measurements may be used to obtain experimental values of the fraction f_e/f of the force which is energetic in origin.

[1]Wilson, E., *Chem. Eng. News*, **February 7, 2002,** 12.
[2]Allcock, H. R., Allen, R. W., and Meister, J. J., *Macromolecules*, **1976,** *9,* 950.
[3]Ciferri, A., *J. Polymer Sci.*, **1961,** *54,* 149.
[4]Ciferri, A., Hoeve, C. A. J., and Flory, P. J., *J. Am. Chem. Soc.*, **1961,** *83,* 1015.

Such experiments carried out at constant volume are the most direct, and can be interpreted through use of the purely thermodynamic relationship.

$$f_e/f = -T[\partial ln(f/T)/\partial T)]_{V,L} \qquad (13)$$

However, because it is very difficult to maintain constant volume in these experiments, they are usually carried out at constant pressure instead. They are then interpreted using the equation

$$f_e/f = -T[\partial ln(f/T)/\partial T]_{p,L} - \beta T/(\alpha^3 - 1) \qquad (14)$$

in which β is the thermal expansion coefficient of the network. This relationship was obtained by using the Gaussian-based elastic equation of state to correct the data to constant pressure.

These energy changes are intramolecular, and arise from transitions of the chains from one spatial configuration to another (since different configurations generally correspond to different conformational energies). They are thus related to the temperature coefficient of the unperturbed dimensions; the quantitative relationship being

$$f_e/f = T \; dln<r^2>_o/dT \qquad (15)$$

This is a very important "bridge" equation in the sense that it relates a quantity obtained from elasticity measurements with another that has generally been obtained from intrinsic viscosity measurements in dilute solution. It is interesting to note that, because this type of non-ideality is intramolecular, it is not removed by diluting the chains (swelling the network) nor by increasing the lengths of the network chains (decreasing the degree of cross-linking). In this respect, elastomers are rather different from gases, which can be made to behave ideally by decreasing the pressure to a sufficiently low value.

In these terms, it is of great interest that polyethylene and polyisobutylene show an *increase* in restoring force (equivalent to a contraction) as the temperature is raised, whereas natural rubber and poly(dimethylsiloxane) rubber show the opposite effect. For amorphous polyethylene, the following interpretation can be made.

Use of equation (13) indicates that the energetic contribution to the elastic force is large and negative. These results may be understood in terms of the known conformational characteristics of polyethylene. The preferred (lowest-energy) conformation of the chain is the all-*trans* form, because *gauche* states (at rotational angles of $\pm 120°$) cause steric repulsions between CH_2 groups. Because the *trans* conformation has the highest possible spatial extension, stretching of a polyethylene chain requires switching some of the *gauche* states (which are present in the randomly coiled form) to the alternative *trans* states. These changes decrease the conformational energy and are the origin of the negative type of ideality represented in the experimental value of f_e/f. This physical picture also explains the decrease in unperturbed dimensions following an increase in temperature. The additional thermal energy causes an increase in the number of the higher energy *gauche* states, which are more compact than the *trans* ones.

The opposite behavior is found in the case of poly(dimethylsiloxane) $[-Si(CH_3)_2O-]$. The all-*trans* form is again the preferred conformation; the relatively long $Si-O$ bonds and the unusually wide $Si-O-Si$ bond angles reduce steric repulsions in general, and the *trans* conformation places CH_3 side groups at distances of separation where they are strongly attractive. However, because of the inequality of the $Si-O-Si$ and $O-Si-O$ bond angles, this conformation is of very low spatial extension, approximating to a closed polygon. Therefore, stretching of a poly(dimethylsiloxane) chain requires an increase in the number of *gauche* states. Because these have a higher energy, this explains why the deviations from ideality for these networks are found to be positive. This also explains the increase in unperturbed dimensions during increases in temperature. The additional thermal energy causes an increase in the number of the higher energy *gauche* states, which in this case are *less* compact than the *trans* ones.

Thermoelastic results are also used to test some of the assumptions made in the development of the molecular theories described in Chapter 21. The results indicate that the ratio f_e/f is essentially independent of degree of swelling of the network, and this supports the postulate that *inter*molecular interactions do not contribute significantly to the elastic force. This assumption is further supported by results that show the values of the temperature coefficient of the unperturbed dimensions obtained from thermoelastic experiments are in good agreement with those obtained from viscosity-temperature measurements on the isolated chains in dilute solution.

Also, because intermolecular interactions do not affect the force, they must be independent of the extent of the deformation and thus independent of the spatial configurations of the chains. This, in turn, indicates that the spatial configurations must be independent of intermolecular interactions, i. e., the amorphous chains must be in random, unordered configurations, the dimensions of which should be the unperturbed values. This conclusion has now been amply verified, in particular by neutron-scattering studies on undiluted amorphous polymers by numerous research groups.

Some Other Properties Used for Characterizing Chain Conformations and Spatial Configurations

The most common way to give a physical picture of the conformational properties of a random-coil polymer is through the end-to-end distance, r. It is easy to visualize chains being compact or expanded, or responding to a macroscopic strain by increasing their dimensions in a particular direction. The radius of gyration conveys the same information, but it is somewhat more difficult to picture. Nonetheless, a variety of other properties have been used to characterize conformations and spatial configurations, and some are listed in Table 18.1.

TABLE 18.1 ADDITIONAL PROPERTIES USED TO CHARACTERIZE SPATIAL CONFIGURATIONS OF RANDOM-COIL POLYMERS

Dipole moments	Cyclization equilibria
Optical activity	Stereochemical equilibria
NMR chemical shifts	Strain birefringence
Excimers and excitation transfer	Strain dichroism

Dipole moments have two significant advantages. First, they can be interpreted in molecular terms down to very low molecular weights, which is not possible for the intrinsic viscosities sometimes used to estimate chain dimensions (the molecular viscosity theories are developed for very long chains). In fact, dipole moments have long been used to help decide between possible structures of many small organic or inorganic molecules. Second, the corrections for excluded volume effects generally tend to be much smaller in the case of dipole moments. This approach requires that the chains have some polarity, which would seem to exclude important hydrocarbon polymers, for example. However, this has not proved to be a deterrent, and experiments and calculations have been carried out of the dipole moments of such chains modified to be polar by the attachment of polar groups at the chain ends.

Optical activity measurements are generally restricted to biopolymers, almost all of which have the required asymmetric carbon atoms C*. The repeat unit of a protein or polypeptide is a good example: $[-C^*HRCONH-]$. The simplest nonbiological analogue is poly(propylene oxide) $[-C^*H(CH_3)CH_2O-]$, and a vinyl polymer with an optically active group as a side chain would also be in this category. NMR chemical shifts are also useful since they are sensitive to local spatial arrangements, and can therefore be used to obtain information on conformational preferences. Excimers and excitation transfer can also be used since, for example, the behavior of a quencher and its chromophore located on the same chain will depend on the distance separating them and thus on the conformation of the host chain.

Cyclization equilibria involve measurments of the equilibrium constant for cyclic and linear molecules either directly after a polymerization, or after a reequilibration. The physical picture focuses on the flexibility of a chain in the equilibrium sense, i.e, of being able to collapse into the compact conformations required for the chains ends to meet and form a cyclic structure. A polysiloxane is at the extreme of these systems because it generates many cyclic species because of its high flexibility. A rigid-rod polymer would be at the other extreme, because its two ends could not loop around to become close enough to bond. An example of a stereochemical equilibrium constant would be the equilibrium between meso (isotactic) and racemic (syndiotactic) placements in the presence of a suitable catalyst. It is also dependent on the conformations of the chains.

Strain birefringence studies involve deformation of an elastomer and measuring the changes in birefringence that result from orientation of the chains. It also depends on their having different polarizabilities parallel and perpendicular to the chain direction. This is important because the chains change conformation in response to the macroscopic strain. Strain dichroism is very similar but is based on infrared spectra rather than optical birefringence.

POLYPEPTIDES AND PROTEINS

The most sophisticated objective in current polymer conformational work is the drive to simulate the conformations of polypeptides and proteins by means of nonbonding energy calculations. This objective is important because the biochemical activity of many proteins is a function of chain conformation or conformational changes (see Chapter 8). Probably the most important goal in this area is understanding "protein folding," in which the chains can rearrange from relatively featureless random coils into the precisely defined conformations needed for their functioning, for example, as globular enzymes. In principle, it should be possible to calculate the conformation of a protein on the basis of nonbonding interactions. In practice, this is a very complex problem because:

1. A protein is a complex *copolymer*. Each different monomer unit in the chain will give rise to different nonbonded interactions.
2. Coulombic forces are important in these structures.
3. Hydrogen bonding exists.
4. Extensive coiling of the chains brings *long-range* forces into play.
5. The aqueous matrix inside a living system must affect the conformation.

In spite of these difficulties, steady progress has been evident in recent years toward the calculation of protein conformations. However, it must be stressed that the ultimate objective has not yet been achieved. Rather, the research in this area must be viewed as a steady progression in which successively more complicated biopolymers have been treated with increasing success by conformational analysis techniques. Because of the specialized nature of this research, the following summary is intended as an introductory outline only. The reader is encouraged to explore the subject in greater depth by reference to the sources listed at the end of this chapter.

The general strategy has involved first, conformational energy calculations on the simplest small-molecule model structures that might simulate the behavior of amino acid residues present as middle units in a protein chain.[1,2] Linear dipeptides or other simple, short-chain linear oligopeptides that contain glycine–glycine, glycine–proline, or proline–proline sequences, and related oligomers that possess end "blocked" units to eliminate end-group effects have been investigated. The purpose of these preliminary calculations is (1) to develop suitable potential functions that can be refined by comparison with the known experimental behavior of the linear oligomers, (2) to explore the preference of par-

[1]Tanaka, S., and Scheraga, H. A., *Macromolecules,* **1974,** *7,* 698.
[2]Simon, I., Nemethy, G., and Scheraga, H. A., *Macromolecules,* **1978,** *11,* 797.

ticular peptide linkage environments for *trans*-planar, *cis*-planar, or nonplanar conformations,[1] and (3) to identify sequences of residues that favor the formation of bends in the chain. For example, calculations on the tetrapeptide, Asp-Lys-Thr-Gly, provide an explanation for the existence of bending sites when this sequence appears in the protein α-chymotrypsin.[2]

Next, the calculations have been extended to synthetic linear, homopolypeptides, such as poly(L-proline) or poly(L-valine). Such calculations, and a comparison of the results with experimental data, allow the role of an aqueous matrix or of *inter*chain packing forces to be evaluated. The interchain interaction potentials thus identified can then be employed later for the calculation of long-range *intra*molecular interactions in a folded chain. The helix-coil transition behavior of the homopolymers can also be assessed by conformational calculations.

The information derived from these types of calculations has then been used to predict the conformations of naturally occurring cyclic oligopeptide copolymers, such as, for example, the cyclodecapeptide, Gramicidin S.[3] The role played by coordinated water can also be assessed at this stage.

Finally, the calculations have been extended to long-chain copolymeric synthetic polypeptides or to proteins. Such studies take into account the interaction energies of the different amino acid residues, and the hydrophilic and hydrophobic character of the amino acid side groups. Known protein structures have been analyzed to ascertain the influence of different sequences in the generation of α-helix, extended chain, other coiled and bent states, and these data have been incorporated into the predictive framework.[4] The more sophisticated recent calculations have been directed toward understanding the reasons for cooperative helix-coil transitions in proteins, the speed and mechanism of protein folding in aqueous media, and the existence of coiled-coil structures in fibrous proteins. Eventually, such calculations will be extended to cover the role played by prosthetic groups, such as metalloporphyrins, in the structure of proteins.

MOLECULAR GRAPHICS

Conformational analysis (especially of proteins) has been automated and offered as combined software/hardware units under the name of "molecular graphics." The purpose of these assemblies is to provide a means for rapid conformational analysis of polypeptides, based on specific nonbonded potential interactions, coupled with the ability to provide a visualization of the structure on a computer screen. The effects of thermal vibrations can also be simulated. This is an important technique for molecular biology research. If applied to synthetic, nonbiological polymers, it could also be a valuable aid to molecular design and synthesis. Much of the necessary software is now commercially available, from companies such as Accelrys (San Diego) and Tripos (St. Louis).

[1]Zimmerman, S. S., and Scheraga, H. A., *Macromolecules,* **1976,** *9,* 408.
[2]Howard, J. C., Ali, A., Scheraga, H. A., and Momany, F. A., *Macromolecules,* **1975,** *8,* 607.
[3]Dygert, M., Go, N., and Scheraga, H. A., *Macromolecules,* **1975,** *8,* 751.
[4]Tanaka, S., and Scheraga, H. A., *Macromolecules,* **1977,** *10,* 9.

STUDY QUESTIONS

1. Using your personal computer, set up a simple program for calculation of the energy surfaces for *n*-propane, polyethylene, and polyoxymethylene. Choose your own potential function (variation of energy with interatomic distance) and examine how the energy surface changes with the type of potential used. (This problem could form the basis of a class project with different participants using different potential functions or structural parameters.)

2. What types of polymers might be treated more effectively by the use of a *two*fold rotational isomeric model rather than by the usual threefold rotational isomeric approach?

3. Chains of sulfur atoms have rotational states at only 90° and −90°. How many spatial configurations would there be for a chain of 8 atoms?

4. Most models for conformational energy calculations make the assumption of fixed bond angles in the skeleton. What effects on the physical properties of a polymer would you expect if the skeletal bond angles were capable of undergoing considerable valence-bond angle distortion (say over ±20°) without encountering an appreciable energy penalty?

5. Is it possible in principle to calculate a satisfactory intramolecular nonbonding potential from experimental physical data for polymers? If so, how would you do it, and what would be the weaknesses of the method?

6. Calculate the characteristic ratio $CR = <r^2>_o/nl^2$ for a hypothetical freely-rotating chain having skeletal bond angles of 90°. (Here, $<r^2>_o$ is the unperturbed dimension, n is the number of skeletal bonds, and l is the skeletal bond length.)

7. Calculate the characteristic ratio for a poly(dimethylsiloxane) chain in the idealization that the chain is freely rotating. The O—Si—O bond angle is 110° and the Si—O—Si bond angle is 143°.

8. Calculate the characteristic ratio for a hypothetical tetrahedrally-bonded polymer chain having only two rotational states: *trans* ($\phi = 0°$) and *cis* ($\phi = 180°$), with the latter 500 cal/mol higher in energy than the former. Assume independent rotational states, and a temperature of 25°C.

9. Calculate the value of the "stiffening ratio" $(1 + <\cos \phi>)/(1 - <\cos \phi>)$ for a hypothetical chain having rotational states located at $\phi = 0, 72, 144, 216,$ and 288° that are of equal energy.

10. What is the value of the characteristic ratio for a rigid-rod polymer?

11. Show that the radial form of the Gaussian distribution function has a maximum (most-probable value of r) at $r = (2 nl^2/3)^{1/2}$.

12. Suppose a polymer chain has long regions that are helical. Would you expect its dimensions to increase or decrease upon increase in temperature?

13. Obtain a numerical answer for the value of Z in:

$$Z = [1 \quad 2 \quad 3] \begin{bmatrix} 2 & 0 & 0 \\ 0 & 3 & 0 \\ 0 & 0 & 4 \end{bmatrix} \begin{bmatrix} 1 \\ 1 \\ 1 \end{bmatrix}$$

14. A. Write the statistical weight matrix for a hypothetical polymer that can exist only in g^+ states. (Use the usual indexing, with rows and columns both in the order t, g^+, g^-).

15. Why is the characteristic ratio for the dipole moment generally much smaller than the characteristic ratio for the unperturbed dimensions?

16. Would the characteristic ratio for a hypothetical hydrocarbon polymer increase, decrease, or stay the same if the *gauche*$^\pm$ states were moved from $\pm 120°$ to $\pm 100°$?

17. Describe the conformations of a polymer that would be described by the statistical weight matrix:

$$\begin{bmatrix} 0 & 0 & 0 \\ 0 & 1 & 0 \\ 0 & 0 & 1 \end{bmatrix}$$

SUGGESTIONS FOR FURTHER READING

BICERANO, J. (Ed.), *Computational Modeling of Polymers*. New York: Marcel Dekker, Inc., **1992**.

BINDER, K. (Ed.), *Monte Carlo and Molecular Dynamics Simulations in Polymer Science*. New York: Oxford University Press, **1995**.

BIRSHTEIN, T. M., and PTITSYN, O. B., *Configurational Statistics of Polymer Chains*. New York: Interscience, **1996**.

BIRSHTEIN, T. M., and PTITSYN, O. B., *Conformations of Macromolecules* (*High Polymers*, Vol. XXII). New York: Wiley-Interscience, **1966**.

BOVEY, F. A., "NMR Observations of Polypeptide Conformations," *J. Polymer Sci. (D), Macromol. Rev.*, **1974**, *9*, 1.

BOVEY, F. S., *Polymer Conformation and Configuration*. New York: Academic Press, **1969**.

BOVEY, F. A., *Chain Structure and Conformation of Macromolecules*. New York: Academic Press, **1982**.

BOYD, R. H., and Phillips, P. J., *The Science of Polymer Molecules*. Cambridge: Cambridge University Press, **1993**.

CARPENTER, D. K., "Colloids," *Encycl. Polymer Sci. Technol.* (H. F. Mark, N. G. Gaylord, and N. M. Bikales, Eds.), **1966**, *4*, 16.

CIFERRI, A., "Present Status of the Rubber Elasticity Theory," *J. Polymer Sci.*, **1961**, *54*, 149.

CIFERRI, A., HOEVE, C. A. J., and FLORY, P. J., "Stress–Temperature Coefficients of Polymer Networks and the Conformational Energy of Polymer Chains," *J. Am. Chem. Soc.*, **1961**, *83*, 1015.

DE SANTIS, P., GIGLIO, E., LIQUORI, A. M., and RIPAMONTI, A., "Stability of Helical Conformations of Simple Linear Chains," *J. Polymer Sci.*, **1963**, *A1*, 1383.

FARINA, M., "The Stereochemistry of Linear Macromolecules," *Top. Stereochem*, **1987**, *17*, 1–111.

FLORY, P. J., "Foundation of Rotational Isomeric State Theory and General Methods for Generating Configurational Averages," *Macromolecules*, **1974**, *7*, 381.

FLORY, P. J., *Principles of Polymer Chemistry.* Ithaca, N.Y.: Cornell University Press, **1953**, especially Chaps. X and XIV.

FLORY, P. J., *Statistical Mechanics of Chain Molecules.* New York: Interscience, **1969**; reissued by Hanser Publishers, Munich, **1989**.

FLORY, P. J., CIFERRI, A., and CHIANG, R., "Temperature Coefficient of the Polyethylene Chain Conformation from Intrinsic Viscosity Measurements," *J. Am. Chem. Soc.,* **1961**, *83,* 1023.

FLORY, P. J., CRESCENZI, V., and MARK, J. E., "Configuration of the Poly(dimethylsiloxane) Chain, III: Correlation of Theory and Experiment," *J. Am. Chem. Soc.,* **1964**, *86,* 146.

FLORY, P. J., and MARK, J. E., "The Conformation of the Polyoxymethylene Chain," *Makromol. Chem.,* **1964**, *75*, 11.

GELIN, B. R., *Molecular Modeling of Polymer Structures and Properties.* Cincinnati: Hanser Publishers, **1994**.

GROSSBERG, A. (Ed.), *Theoretical and Mathematical Models in Polymer Research.* New York: Academic Press, **1998**.

HONEYCUTT, J. D., "Rotational Isomeric State Models and Results," in *Physical Properties of Polymers Handbook,* (J. E. Mark, Ed.). New York: Springer-Verlag, **1996**.

HOPFINGER, A. J., *Conformational Properties of Macromolecules.* New York: Academic Press, **1973**.

LOWE, J. P., "Barriers to Internal Rotation about Single Bonds," *Progr. Phys. Org. Chem.* (A. Streitwieser and R. W. Taft, Eds.), **1967**, *6*, 1.

LOWRY, G. G., *Markov Chains and Monte Carlo Calculations in Polymer Science.* New York: Marcel Dekker, Inc., **1970**.

MATTICE, W. L., and SUTER, U. W., *Conformational Theory of Large Molecules. The Rotational Isomeric State Model in Macromolecular Systems.* New York: Wiley, **1994**.

MONNERIE, L., and SUTER, U. W. (Eds.), *Atomistic Modeling of Physical Properties. Adv. Polym. Sci.* Berlin: Springer-Verlag, **1994**, *116*.

MORAWETZ, H., "Some Studies on the Rates of Conformational Transitions and of *Cis-Trans* Isomerizations in Flexible Polymer Chains," *Contemp. Top. Polymer Sci.,* **1977**, *2*, 171.

ORVILL-THOMAS, W. J. (Ed.), *Internal Rotations in Molecules.* New York: Wiley, **1974**.

PETERLIN, A., "Conformation of Polymer Molecules," in *Polymer Science and Materials* (A. V. Tobolsky and H. F. Mark, Eds.). New York: Wiley-Interscience, **1971**, Chap. 3.

POLAND, D., and SCHERAGA, H. A., *Theory of Helix-Coil Transitions in Biopolymers.* New York: Academic Press, **1970**.

REHAHN, M., MATTICE, W. L., and SUTER, U. W., "Rotational Isomeric State Models in Macromolecular Systems," *Adv. Polym. Sci.,* **1997**, *131/132,* 1.

RIANDE, E., and SAIZ, E., *Dipole Moments and Birefringence of Polymers.* Englewood Cliffs, NJ: Prentice Hall, **1992**.

ROE, R.-J. (Ed.), *Computational Simulation of Polymers*. Englewood Cliffs: Prentice Hall, **1991**.

SCHERAGA, H. A., and coworkers. A series of papers on the conformational analysis of polypeptide chains. *Macromolecules,* **1970**, *3, 178,* 188, 628; **1971**, *4,* 112; **1972**, *5*, 455;

1973, *6,* 91, 447, 525, 535, 541; **1974**, *7,* 137, 459, 468, 698, 797; **1975**, *8,* 479, 491, 494, 504, 516, 607, 623, 750; **1976**, *9,* 142, 159, 168, 395, 408, 812; **1977**, *10,* 9, 291, 305, 1049; **1978**, *11,* 9, 552, 797, 805, 812, 819, 1168.

TERAMOTO, A., and FUJITA, H., "Conformation-Dependent Properties of Synthetic Polypeptides in the Helix-Coil Transition Region," *Advan. Polymer Sci.,* **1975**, *18,* 65.

TERAMOTO, A., and FUJITA, H., "Statistical Thermodynamic Analysis of Helix-Coil Transitions in Polypeptides," *J. Macromol. Sci., Rev. Macromol. Chem.,* **1976**, *C15,* 165.

VOLKENSTEIN, M. V., *Configurational Statistics of Polymeric Chains* (*High Polymers,* Vol. XVII). New York: Wiley-Interscience, **1963**.

19

Polymer Structure by Diffraction, Scattering, and Spectroscopy

INTRODUCTION

Many methods are available for the determination of molecular structures of polymers. These include infrared and ultraviolet spectroscopy, optical rotatory dispersion (ORD), nuclear magnetic resonance, light scattering, ultracentrifugation, gel permeation chromatography, conformational analysis, and X-ray crystallography. Ideally, as many techniques as possible should be applied to the structure solution of a given polymer. In practice, two techniques have proved to be especially powerful: nuclear magnetic resonance (nmr) and X-ray crystallography. X-ray and neutron scattering, as well as nmr and some other types of spectroscopy, are also described briefly in this chapter. For detailed information, however, the reader is referred to the Suggestions for Further Reading. In this chapter we primarily discuss the use of X-ray diffraction.

X-RAY DIFFRACTION

The X-ray diffraction method is the most powerful technique available for the examination of ordered polymers in the solid state. In general, useful information can be obtained only if the polymer forms oriented fibers, is microcrystalline, or yields single crystals (polyethylene, crystalline globular proteins, viruses, etc.). The following types of information can be obtained from X-ray diffraction experiments:

1. Estimates of the degree of crystallinity of a polymer sample.
2. Determination of the extent of orientation of crystallites in a polymer.

3. Analysis of the "macrostructure" of the polymer—the way in which the crystallites or bundles of chains are packed together.

4. Determination of the *molecular structure*, including the chain conformation and the position of individual atoms. This aspect is the most involved but yields the greatest amount of useful fundamental information.

Sections of this chapter will be concerned with a brief discussion of items 1 to 3, but most of the discussion will deal with item 4.

EXPERIMENTAL ASPECTS

Sample Preparation

1. If the polymer is a naturally occurring fibrous protein (e.g., hair), a fiber or bundle of fibers can be clamped in a metal holder (Figure 19.1a). Tension is applied to the fiber, and the holder and fiber are then mounted in an X-ray camera.

2. If the sample is a synthetic polymer, a fiber is prepared by melt or solution extrusion, or a film is cast by solvent evaporation or melt techniques (see Chapter 20). The fiber or film is then stretched (oriented) by hand before being clamped in a holder (Figure 19.1b). Further stretching is effected by the slow actuation of the screw-clamp device. A 400% elongation of the sample is not unusual during this step. The sample may be alternately warmed and cooled during this process to encourage microcrystallite growth. It makes little difference whether a fiber or a film sample is used, because no overall crystalline order normally exists *between* the bundles of polymer chains. Thus, a cylindrical symmetry is found in both fiber and film samples. However, some differences do exist between the diffraction behavior of

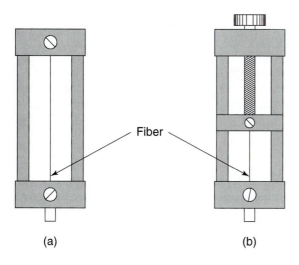

Fiber

Figure 19.1 (a) Simple brass holder for orientation of a polymer fiber or film in the X-ray beam. (b) Device for stretching a fiber or film (by turning the knurled nut and screw arrangement) and for positioning of the stretched, torsioned sample in the X-ray beam.

(a)   (b)

films and fibers. For example, a fiber or bundle of fibers may form a thicker sample than a thin film, and the intensity of X-ray diffraction may be greater with the fiber, thus permitting shorter exposures. On the other hand, very thick fibers (ca. 1 mm in diameter) sometimes yield doubled X-ray reflections because the X-rays passing through the center of the fiber are totally absorbed, but those passing through the edges are not. Biaxially oriented films may create special problems because the X-ray diffraction patterns may be dependent on the angle at which the beam enters the film. Fibers that have a cylindrical cross section yield data that can be corrected more easily for X-ray absorption effects than can thick films. Thus, unless special effects are to be studied, the sample should be a thin cylindrical fiber (ca. 0.5-mm or less in diameter) or a uniaxially oriented, very thin film.

3. When the polymer forms single crystals (globular proteins, viruses, etc.), different techniques must be used. Biological polymers must be crystallized from aqueous media, and water molecules frequently form part of the lattice. Hence, it is usually necessary to mount the crystal in contact with its mother liquor. The crystal is wedged into a capillary tube together with the mother liquor, and the tube is sealed to prevent evaporation of water. The tube is then glued to a goniometer head, and movement of the head permits orientation of the crystal relative to the X-ray beam (Figure 19.2). If the crystal is stable to air, it can be cemented to the end of a glass fiber and the latter can be attached to a goniometer head (Figure 19.2). A suitable crystal size would be less than 0.5 mm in edge length.

Generation of X-Rays

X-rays are produced when a beam of high-energy electrons, generated in an evacuated tube, strikes a metal target. The wavelength of the X-rays depends on the target material used; copper or molybdenum targets are commonly employed. The beam of X-rays emerging from the X-ray tube passes through a safety shutter arrangement and a filter, and is then collimated to a narrow beam (ca. 0.5 mm diameter) by passage through a brass tube or "collimator."

Because X-rays generated in this way are polychromatic, filtration of the beam is necessary to remove unwanted wavelengths. A thin sheet of nickel is used as a filter for copper radiation. For some applications, filtration is insufficient to produce radiation of the necessary purity. A more nearly monochromatic beam can be produced by diffraction of the main beam from the surface of a flat or curved crystal surface.

X-Ray Cameras and Diffractometers

A number of different camera or diffractometer designs are in use. The following are the most widely used instruments for polymer research:

Cylindrical cameras This type of camera is also known as a rotation camera. It is used for the preliminary investigation of all polymers and for the total

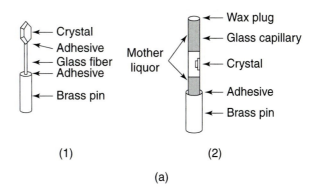

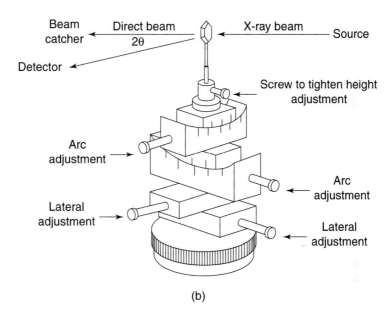

Figure 19.2 (a1) Crystal of an air-stable compound can be cemented to a glass fiber in preparation for mounting on a goniometer head; (a2) air-sensitive crystal of a protein can be wedged into a capillary tube together with the mother liquor solution; (b) goniometer head for alignment of the crystal. [Reproduced with permission from Pickworth Glusker, J. and Trueblood, K. N., *Crystal Structure Analysis* (New York: Oxford University Press, **1972**).]

molecular structure investigation of oriented fibers or microcrystalline polymers. A cylindrical camera can also be employed to measure the degree of crystallinity and the degree of crystallite orientation. The camera consists of a cylindrical brass tube (often 57.3 mm in diameter) which acts as an outer support for a curved sheet of photographic film (Figure 19.3). X-rays enter the camera via the collimator, strike the polymer sample, and the undeflected beam is trapped by a "beam

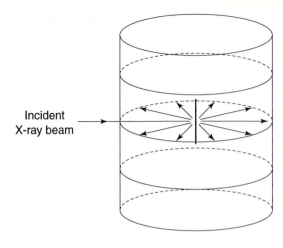

Figure 19.3 Schematic representation of a polymer fiber (vertical line) mounted in a cylindrical or "rotation" camera. Only the diffracted rays falling on the zero level (equator) of the film are shown. However, upper- and lower-layer reflections will also be generated.

Incident X-ray beam

stop." Diffracted X-rays strike the photographic film, and subsequent development reveals the positions at which the diffracted rays passed through the film. The camera is usually capped by a lid to exclude light, and the sample is mounted on a goniometer head for orientation purposes. If necessary, the sample can be rotated mechanically about its vertical axis, although in fiber crystallography this is usually unnecessary. It is sometimes advantageous to evacuate the camera or replace the air by a light gas, such as helium, to reduce the scattering of X-rays by air. A typical photograph obtained by the use of a cylindrical camera is shown in Figure 19.4.

A "powder" or Debye–Scherrer camera is simply a shallower version of a cylindrical camera. It is used mainly to obtain photographs of randomly oriented specimens. A Guinier–Wolff camera is a form of cylindrical camera in which the radiation is monochromatized and focused by a curved crystal. A Weissenberg

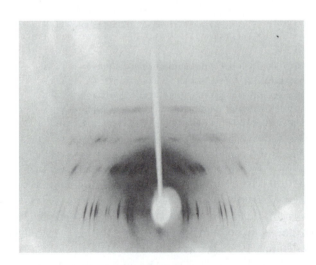

Figure 19.4 X-ray fiber photograph of poly(dichlorophosphazene), $(NPCl_2)_n$. Note the arrangement of the arclike reflections along distinct layer lines. The layer line at the bottom of the photograph is the $l = 0$ layer. (Photograph obtained by R. A. Arcus.)

camera is a cylindrical camera in which the film is mechanically translated along the camera axis in synchronization with the rotational oscillation of the crystal.

The precession camera   The Buerger precession camera[1] is sometimes used as an alternative to a cylindrical camera for molecular structural work. It has the advantage that it reveals an undistorted picture of the reciprocal lattice. Precession cameras are used extensively in the crystallography of biological macromolecules such as enzymes or heme proteins. A flat sheet of photographic film is used and this, together with the fiber or crystal, is caused to undergo a precession motion.

A simple variant (and early precursor) of the precession camera is the Laué camera, in which a fixed specimen is allowed to diffract X-rays on to a fixed, flat piece of photographic film. The interpretation of Laué photographs is more complex than the analysis of precession photographs.

Diffractometers   An alternative to the photographic detection of diffraction is the use of a Geiger, proportional, or scintillation-type counter. In practice, the use of a counter-type instrument permits the diffraction angles to be determined quickly (if only a few reflections are present), and it allows greater accuracy in the measurement of diffraction angles and intensities. An alternative detection system now commonly used is a flat "area" detector that (like film) allows large amounts of data to be collected quickly. Two main types of diffractometer instruments are available: manual and automated. With manual instruments, the operator must move the detector to the appropriate position for detection and measurement of a diffracted beam. These instruments are generally of an older design and some safety problems are associated with their use. Automated or semiautomated diffractometers operate under the control of a small computer or, indirectly, under the control of an operator. In single-crystal work, the space group of the crystal is used by the computer to orient automatically the diffractometer components and measure reflection intensities at the correct angle. These instruments cannot normally be used for small-angle work. Only the so-called "Weissenberg geometry" diffractometers offer appreciable advantages over cameras for fiber diffraction work and relatively few of these instruments are in use.

Small-angle cameras   Wide-angle instruments, such as those described in the preceding sections, are useful for the detection of crystallinity and for molecular structure determination. However, information about the *macro*structure of a polymer (i.e., the packing of fibrils or crystalites) can only be obtained from narrow-angle diffraction, from reflections that fall very close to the beam stop. Special instruments are required to detect and measure reflections such as these. The Kratky small-angle camera is one such instrument.

[1]Buerger, M. J., *The Precession Method* (New York: Wiley, **1964**).

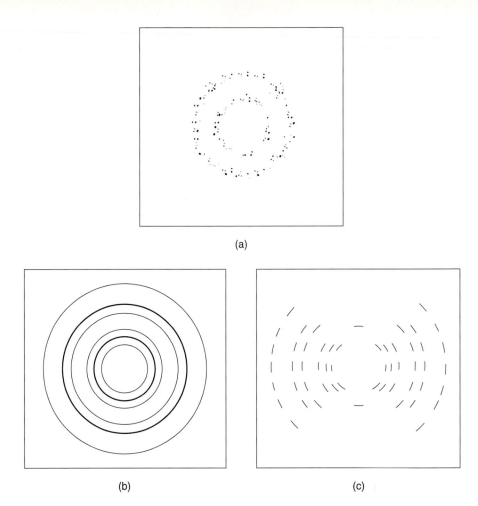

(a)

(b)   (c)

Figure 19.5   X-ray diffraction patterns typical of (a) an amorphous polymer, (b) an un-oriented microcrystalline polymer, and (c) an oriented microcrystalline polymer or an oriented helical polymer.

Measurement of the Degree of Crystallinity

A crystalline or microcrystalline material will diffract X-rays at specific angles (Figure 19.5b and c). On the other hand, amorphous materials simply scatter X-rays in all directions (Figure 19.5a). Thus, in theory, if the sharp diffraction can be compared with the general scattering from a polymer, a measure of the degree of crystallinity can be obtained. Until recently, it was common practice to do just this. However, it is now recognized that the degree of coalescence of the diffraction halos into sharp arcs or spots depends, among other things, on the presence of chain folding, lamellar crystalline growths, lattice dislocations, and on the degree of ordering in an oriented specimen, and that some of the scattering from crystalline regions is, in fact, diffuse. Moreover, crystalline-like X-ray patterns can be generated by oriented helical polymer chains even when no intermolecular crystalline packing is present. Hence, a reliable measure of the degree of crystallinity must take into account the results from nuclear magnetic resonance, density determinations, and infrared spectroscopy, as well as X-ray measurements.

Degree of Crystallite Orientation

The orientation of the crystallites in a microcrystalline polymer has a profound influence on polymer properties. Good orientation parallel to the direction of external stress can generate a high resistance to breaking and a correspondingly high tensile strength. The strength of a nylon fiber may in part be attributed to this effect. X-ray crystallography provides a convenient way for the estimation of the efficiency of orientation.

It was mentioned earlier that one of the main differences between single-crystal X-ray photographs and those derived from oriented polymers is the diffuseness and arclike qualities of the reflections often derived from polymers. This difference is illustrated by a comparison of a typical polymer photograph (Figure 19.4) with the diffraction pattern obtained from a small-molecule single crystal (Figure 19.6). Note also that in Figures 19.4 and 19.5c prominent, diffuse reflections are apparent above and below the beam stop position. These are known as meridianal reflections.

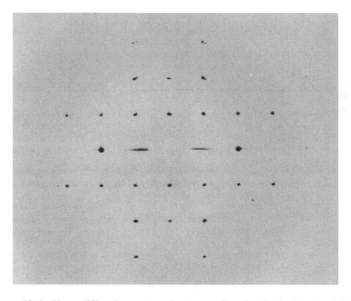

Figure 19.6   X-ray diffraction pattern from a small-molecule single crystal. Note the sharpness of the "reflections" or diffraction spots and the variations in intensities.

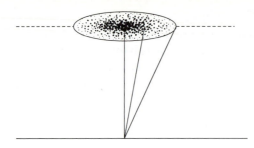

Figure 19.7 Variation of intensity across a meridianal reflection provides information about the relative numbers of microcrystallites that are oriented along specific reflections, for example along the radial lines shown.

The diffuseness and arclike character of the reflections in polymer X-ray photographs is partly a consequence of nonideal orientation of the crystallites. Totally random crystallite orientation results in an extension of the arcs into complete circles, as shown in Figure 19.5b. Thus, some measure of the degree to which the crystallites are oriented parallel to each other can be derived from the shape and size of the individual reflections. A second clue to crystallite orientation is provided by the meridianal reflections. Perfect parallel orientation of the crystallites would result in a total disappearance of the meridianal reflections if the polymer specimen were exactly at right angles to the incident X-ray beam. However, vertical tilting of a perfectly oriented specimen, toward or away from the beam, would cause the meridianal reflections to appear on the photograph. This phenomenon provides a rather sensitive measure of the degree of orientation. In addition, the way in which the intensity varies *across* a meridianal reflection corresponds to the number of crystallites oriented along a particular azimuthal angle (see Figure 19.7).

DETERMINATION OF MOLECULAR STRUCTURE AND CONFORMATION

Different Approaches

It is necessary for the reader to recognize that two closely related but quite distinct approaches are used for the structure solution of fibrous or microcrystalline polymers, on the one hand, and single-crystal materials, on the other. The single-crystal techniques used for the solution of relatively simple molecules can be expanded with some modification and an increased complexity to the structure solution of polymer single crystals. On the other hand, microcrystalline or fibrous polymers present special problems because of disorder in the packing of the chains. Preliminary structural information can be obtained quickly and easily, but a detailed analysis can be an exceedingly difficult task. The techniques used for single crystal analysis are widely known and are described in numerous textbooks (see Suggestions for Further Reading). Here, we will touch briefly on these techniques but will concentrate mainly on the study of microcrystalline materials and noncrystalline fibrous polymers.

Interpretation of X-Ray Photographs
from Microcrystalline Polymers

An oriented microcrystalline polymer can yield photographs that are superficially similar to those obtained from rotated single crystals, and for similar reasons. Figure 19.5c depicts the features of a photograph of a typical oriented microcrystalline material. The two most striking features are the existence of the meridianal reflections (discussed in an earlier section), and the presence of arcs arrayed along parallel horizontal lines. These horizontal lines are known as *layer lines*, and they are numbered according to the notation, $l = 0, 1, 2, 3, \ldots, -1, -2, -3, \ldots$. The "zero layer" always includes the point where the direct beam passes through the film or detector. This layer represents diffraction arcs or "reflections" that emerge from the sample at right angles to the fiber direction (see Figure 19.3). Arcs that lie on the higher and lower layer lines are formed from diffracted X-ray beams which emerge from the sample at an elevated or declined angle to the plane of the camera.

Individual reflections on the zero layer are formed by the following mechanism. Consider a section of a microcrystallite, as shown in Figure 19.8, with regularly packed chains oriented in a vertical direction (along the axis of orientation). Note that lines or planes can be drawn through individual atoms to generate a lattice.

Consider now the path of an X-ray beam, incident to one set of planes in Figure 19.9a. According to the Bragg approach to diffraction, X-rays will be strongly "reflected" from the layers if the angle of incidence, θ, is such that an integral number of wavelengths exists in the "path difference" $x + y$. If this condition is met, the beam emerging from the crystal will be reinforced by superimposition of waves. At other angles the waves will interfere destructively

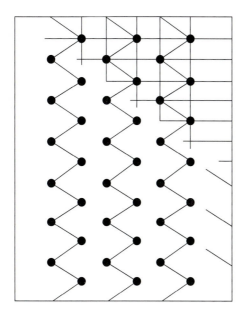

Figure 19.8 Section through a microcrystalline domain, showing laterally packed polymer chains and the way in which various interplanar spacings can be drawn through individual atoms.

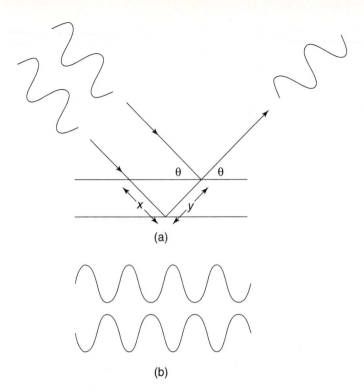

Figure 19.9 (a) Diffraction of X-rays from parallel planes of atoms within a single crystal or polymer microcrystallite. The path-length difference $(x + y)$ must correspond to an integral number of wavelengths if destructive interference is to be avoided. (b) Two wave systems that are 180° out of phase. Total destructive interference occurs.

with each other (Figure 19.9b). Thus, reflections emerge only at discrete angles, and these angles are determined by the path-length difference, and hence by the separation between the layers. The relationship between diffraction angle and layer spacing is given by the well-known Bragg equation

$$n\lambda = 2d \sin \theta \qquad (1)$$

where n is an integral number $(1, 2, 3, 4, \ldots)$, λ the X-ray wavelength, and d the layer spacing.

It is obvious that the different reflections that appear on the zero layer of the photograph represent different separations between different vertical planes. Such planes can be visualized as formed in the way shown in Figure 19.10, which represents a view *down* the chain axes.

Those planes with the largest d-spacing will generate a reflection closest to the primary beam (small value of 2θ). Those with the smallest spacings will yield reflections at wider angles.

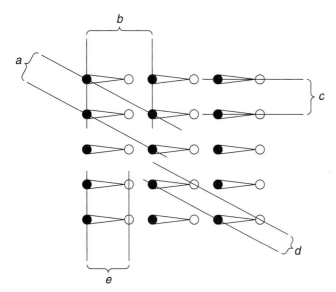

Figure 19.10 View *down* the long axis of chains packed in a microcrystallite. This view is at 90° to the one shown in Figure 19.8. The solid circles represent atoms that occupy a position closer to the viewer than those represented by the open circles. Note how sets of *vertical* planes can be drawn through individual atoms. These vertical planes are responsible for the reflections that appear on the zero layer (equator) of the X-ray photograph. The larger lattice spacings (*b*, *c*) will yield reflections at a narrow angle (close to the beam stop). The smaller spacings, such as *d*, will yield diffraction spots at wider angles (nearer the edge of the photograph).

The reflections which fall on the upper and lower layer lines ($l = 1, 2, 3, \ldots, -1, -2, -3, \ldots$) represent planes in a microcrystallite that are inclined to the fiber axis. The inclined planes in Figure 19.8 are an example. Finally, the meridianal reflections originate from planes that are at right angles to the fiber axis (see Figure 19.8). Because of the geometry of reflection (see Figure 19.9), it is clear that these planes can diffract only if the fiber is inclined at an angle to the beam. Hence, as mentioned previously, the presence or absence of a meridianal reflection can be used as a measure of crystallite orientation.

Preliminary Clues about the Structure

A simple analysis of the *d*-spacings can often provide valuable clues about the conformation of the chain or the packing of adjacent chains. For example, separation between the layer lines is inversely related to the repeating distance along the *c*-axis (fiber axis) of the sample. Hence, one measurement from the film can yield the polymer repeating distance. Second, meridianal reflections arise from planes that are perpendicular to the fiber axis. They represent reflections from planes of monomer residues. Hence, the number of the layer line on which a meridianal reflection falls is the number of monomer residues in the conformational repeat distance. This is illustrated in Figure 19.11.

Third, reflections on the zero layer of the photograph provide clues about the "width" of the zigzag, the helix, or other conformation of the chain. They may also indicate the distances between side-group atoms and the skeletal atoms, or the distances between atoms in neighboring chains. A careful analysis of these interplanar distances can suggest the dimensions and shape of the unit cell. Then,

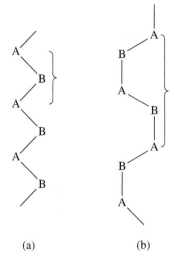

Figure 19.11 Polymers containing repeating units A–B. (a) One monomer residue per repeat: meridianal reflection expected on the $l = 1$ layer line. (b) Two monomer residues per repeat: meridianal reflection expected on $l = 2$ layer line.

(a)　　　　　(b)

if the density of the crystalline regions can be estimated, the number of monomer residues per cell, Z, can be calculated from the formula

$$\text{density} = \frac{(Z)(\text{MW})}{V \times 6.023 \times 10^{23}} \tag{2}$$

where MW is the "molecular weight" (in g/mol) of a monomer residue and V the volume in cm^3 of the unit cell.

Some X-ray studies of microcrystalline materials are terminated at this point. The information obtained is sometimes considered sufficient to define the conformation of the chain if the bond angles and bond lengths of the skeleton are known. However, it is possible for an investigator to be misled by the preliminary data. To confirm a structure, a careful analysis of the *intensities* of the reflection is required.

Preliminary Analysis of Intensity Data

The most reliable clues about the details of the molecular structure are obtained by an analysis of the relative intensity of each reflection. By "intensity" we mean the density of silver deposited on the developed film, or the number of counts recorded in a given time by a counter device. Although the *positions* of the arcs on the film indicate the crude, overall orientation of the molecules in the unit cell, the intensities allow the positions of individual atoms to be identified accurately.

The preliminary intensity analysis can be carried out with the use of two approaches. First, a simple inspection of the photograph may reveal that some reflections are much more intense than others. Because each reflection represents a series of identical parallel planes and, because the intensity of scattering is related to the electron density in those planes, a strong reflection may indicate that many atoms lie on that plane.

Thus, crystallite planes that contain planar polymer chains may reflect X-rays with a particularly high intensity, especially if heavy elements, such as phosphorus or sulfur, form part of the backbone. Planes that contain aromatic rings may be revealed in the same way.

Second, certain expected reflections may be totally absent. These "systematic absences" indicate specific orientations of the contents of the unit cell, such as the presence of glide planes, screw axes, lattice centering, and so on. Such information can provide valuable evidence in favor of one structure or another.

Often, at the end of this phase of the investigation, the investigator will be cautiously optimistic that the correct conformation and the disposition of the chains in the unit cell have been found. The next step is to analyze the intensity data in a more systematic manner.

Structure Factor Analysis

Two main factors affect the intensities of the reflections from a crystal or a microcrystallite. First, as discussed above, a high intensity in a particular reflection may result from the presence of many atoms or heavy atoms in a particular set of planes. Second, the intensities may represent the effects of destructive or constructive interference between X-ray beams reflected from different parts of the same molecule.

Consider the repeating motif shown in Figure 19.12. Although each motif occupies a lattice point (heavy lines, A), parts of the motif lie above or below or to the left or right of the lattice point. Those other parts of the motif or molecule form additional planes (which may be depicted by the dashed lines, B). Reflections of X-rays will occur at the same angle, θ, from all layers that have the same

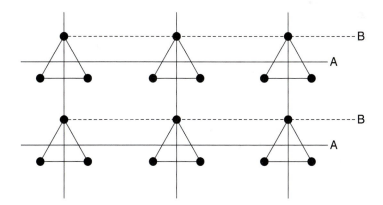

Figure 19.12 Motif of atoms centered around lattice points (intersection of solid lines). All atoms in the motif cannot lie on the same lattice points; hence the reflection of X-rays from the different resultant planes will cause destructive wave interference unless the separation permits exact phase matching.

Determination of Molecular Structure and Conformation

interplanar spacings, but the reflection from one set of planes will interfere with that from the other. The interference can be completely destructive, in which case the intensity will be zero. Alternatively, it could be partially destructive due to a partially offset phase relationship. It will be clear that the intensity is a function of the distance between the different types of planes. Because each subsidiary plane represents a different set of *atoms*, the intensity of a given reflection will be a function of the distance between different atoms in the unit cell.

The intensities are readily measured, but the "phase difference" is not known. However, by postulating models for the structure, the investigator can predict what the intensities would be if the model were correct. If the calculated intensities match the experimental ones, the model is presumed to be correct. If they do not match, the model is altered until a closer correspondence is obtained. This process is known as the *trial-and-error-technique*, and the overall method is known as a *structure-factor calculation*.

This method is the classical approach to the solution of single-crystal structures. Globular proteins, although large and complex in shape, can be studied by single crystal X-ray methods. Each protein molecule occupies a place in an ordered lattice in the same way as the scheme shown in Figure 19.12. The two main differences from small-molecule crystallography are (1) that many more reflections are generated from the complex structure and (2) that globular proteins are more sensitive to X-ray damage than typical small molecules. For this reason, the normal sequential (reflection by reflection) measurement of intensities is often replaced by a simultaneous data collection using either photographic film or an electronic "area detector." For biological macromolecules in single crystals, the structure analysis is usually facilitated by the introduction of heavy atoms such as mercury into the structure. These serve as reference points or "phasing atoms" for the structure analysis.

However, it should be recognized that this method cannot be applied satisfactorily to most oriented fibers or microcrystalline polymers. The lack of perfect crystallite orientation, the resultant diffuseness of the arcs, the cylindrical symmetry of the polymer structure, and the concentration of information in relatively few reflections generate difficulties not normally found in single crystal analysis. Moreover, a tendency is often found for the available data to become superimposed and concentrated in a few low angle reflections, on the $l = 0$ or $l = 1$ layer lines. These problems virtually eliminate the possibility that a microcrystalline or noncrystalline fiber structure can be solved with the high accuracy now common in single crystal analyses. Nevertheless, structures of microcrystalline polymers have been, and are now being, solved by this method.

Diffraction by Helices

Most synthetic polymers and some biological polymers (such as DNA, fibrous proteins, and viruses) assume helical conformations in the oriented state. These helices may or may not be packed together in microcrystallites. Even though the helices may not form part of a crystalline array, they can generate a characteris-

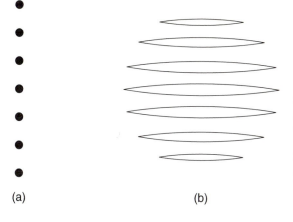

Figure 19.13 (a) Single linear array of regularly spaced diffraction centers (atoms, motifs, or repeating units) that corresponds in a very crude way to a low-resolution representation of a polymer chain. (b) Diffraction pattern from a structure of type (a) consists of a series of parallel "envelopes."

(a)   (b)

tic X-ray pattern. The study of such patterns is known as *helical transform analysis*. The method was first introduced in 1952 by Cochran, Crick, and Vand[1,2] and by Stokes. The identification of the structure of poly(α-methyl-L-glutamate) and the double helix structure of DNA constitute two of the first successes of this approach.

The analysis of diffraction effects produced by isolated helical polymer molecules can be considered at three levels of sophistication: (1) in terms of diffraction by a one- or two-dimensional regular repeating arrangement along the axis of the chain; (2) in terms of the diffraction effects expected from a continuous helix; and (3) as diffraction by a discontinuous helix made up of individual atoms and side groups. These three approaches will be considered in turn.

Diffraction by one- or two-dimensional arrays In many oriented polymer samples, the chains are extended in such a way that the macromolecular skeleton generates a regular repeating sequence along the fiber axis. This is true even if the side groups attached to the chain are disordered. A diffraction experiment will identify only what is ordered in the system and will ignore that which is disordered. If we cease to worry for a moment about the finer details of the structure, we can imagine that a polymer might be depicted in a low-resolution fashion by a one-dimensional sequence of "units" as shown in Figure 19.13a. Such a situation might arise if a marked periodicity existed *along* the chain axis, but with little or no periodicity in the directions at right angles to that axis. The Fourier transform (and hence the X-ray pattern) of such a one-dimensional array is a series of "envelopes," as shown in Figure 19.13b. These envelopes represent rudimentary layer lines that are related to those depicted earlier in Figure 19.5c. Such patterns are actually encountered in practice from some polymer systems.

[1]Cochran, W., and Crick, F. H. C., *Nature*, **1952**, *169*, 234.
[2]Cochran, W., Crick, F. H. C., and Vand, V., *Acta Cryst.*, **1952**, *5*, 581.

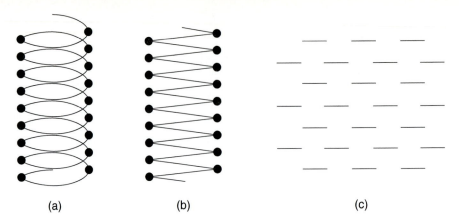

Figure 19.14 Projection of the diffracting centers (atoms or monomer motifs) on a helix (a) onto a plane surface (b), and the diffraction pattern expected from such a projection (c).

Imagine now that we can view the polymer at a slightly higher resolution, in such a way that the individual monomer residues or atoms can be recognized at all points along the chain, and the helical conformation of the chain can be crudely discerned. The projection of the atoms or monomer residues along the helix onto a surface generates a zigzag pattern (Figure 19.14a and b). Thus, the diffracting sites on the polymer, when viewed from the side, are arrayed in a two-dimensional pattern (Figure 19.14b) and the diffraction pattern generated by this arrangement is shown in Figure 19.14c. Note that the layer-line envelopes have now separated into broad but discrete "reflections." This pattern has been generated by a "convolution" of the original one-dimensional atomic array with the ordered arrangement at right angles to it. Of course, the diffraction pattern depicted is only a very crude approximation of the expected Fourier transform because the three-dimensional character of the polymer structure has been ignored. The next section partially corrects this deficiency.

Diffraction by a continuous helix   Consider the diffraction pattern that would be generated by an isolated, continuous, three-dimensional helix (e.g., by an infinitely thin wire wound helically around the surface of a cylinder) (Figure 19.15a). A cross-sectional view down the helix is a circle. Therefore, the diffraction phenomenon can be considered in terms of diffraction at the surface of a cylinder, but a cylinder that possesses a regular periodicity along its axis. Diffraction by such systems can be analyzed most easily in terms of cylindrical waves rather than by the diffraction of plane waves as used in conventional X-ray analyses. Cylindrical waves can be visualized as resembling the concentric wave fronts that radiate out when a pebble is tossed into a pond, or the circular waves that form when ocean wave fronts strike an obstruction.

Diffraction by a continuous helix gives rise to a diffraction pattern similar to that shown in Figure 19.15b. Such patterns can be generated on a macroscopic

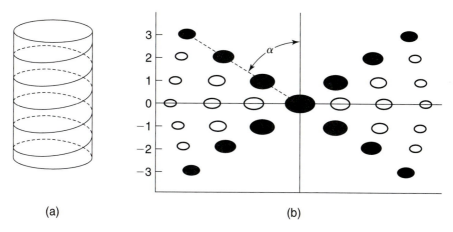

(a)   (b)

Figure 19.15 Diffraction pattern of a continuous helix. The solid ellipses represent the largest peak in the Bessel function that contributes to each layer line. The open ellipses denote the subsidiary (attenuating) peaks of each Bessel function. A zero-order Bessel function is the principal contributor to the zero-layer line; a first-order Bessel function comprises the first layer line, and so on. The angle α on the diffraction diagram equals the pitch of the helix.

scale by the use of an optical diffractometer (see a later section). The main features of this pattern are (1) the existence of layer lines (indicated by $n = 0, 1, 2, \ldots$), (2) the characteristic cross pattern formed by the most intense reflections [which results from (3) the fact that the strongest reflection on each layer line is found progressively farther and farther from the meridian for higher and higher layer lines].

Such behavior can be rationalized and simulated by the use of Bessel functions. Bessel functions are special integrals that are used in physics to describe the behavior of wave motions generated by cylindrical objects. A cylindrical wave of order n is made up from a Bessel function of order n (written as J_n) multiplied by a cosine wave of period n

$$C_n(r, \phi) = J_n(2\pi Rr) \cos n(\phi - \theta + \pi/2) \tag{3}$$

where r and ϕ are the radial and azimuthal coordinates in real space, and R and θ are the appropriate coordinates in reciprocal space.

The Fourier transform of a continuous helix (i.e., one that does not possess discontinuities from the presence of atoms or repeating residues) for each layer line, n, is given by

$$G(R, \theta, z) = J_n(2\pi Rr)e^{in(\theta + \pi/2)} \tag{4}$$

The form of the $n = 0, 1$, and 2 Bessel functions is shown in Figure 19.16.

A Bessel function of order $n = 0$ generates an attenuated wave that has its maximum peak located at $X = 0$. Note that this behavior simulates the situa-

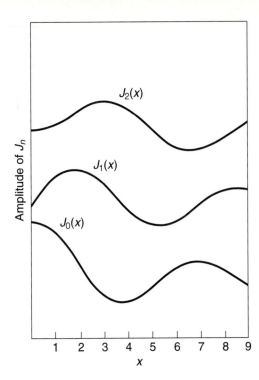

Figure 19.16 Bessel functions J_0, J_1, and J_2, showing how the strongest peak moves farther from the origin for higher and higher values of n in J_n. A comparison of these profiles with Figure 19.15 indicates how the behavior of a Bessel function of order, n, corresponds to the diffraction density profile on the nth-layer line.

tion depicted along the zero layer line of the diffraction pattern in Figure 19.15b. Similarly, the Bessel function of order $n = 1$ has a maximum farther from the origin, and the position of this peak corresponds to the location of the strongest reflection on the $n = 1$ layer line. In fact, the construction of the whole diffraction pattern can be understood if the origin of each Bessel function occupies the meridian of the photograph, with higher layer lines corresponding to higher-order Bessel functions.

A real polymer molecule differs from the two situations discussed above because it possesses both discrete atoms or monomer residues *and* a three-dimensional helical arrangement. Thus, the picture presented above for diffraction by a continuous helix must be modified to take into account the discontinuous structures. This is accomplished by combining or "convoluting" the two types of results. Descriptions of the mathematical procedures are beyond the scope of this book (for details, see the Suggestions for Further Reading), but pictorially the result is shown in Figure 19.17.

Thus, the actual diffraction pattern will consist of multiple layer lines and multiple crosses. If the spacing between the origins of the crosses is small, the cross patterns will overlap to give an extremely complex pattern (Figure 19.18). The diffraction pattern may be even more complex if several turns of the helix are needed to accomplish one repeat. Straightforward techniques (such as the "*n-l* plot," "the radial projection," and the "helical projection") have been de-

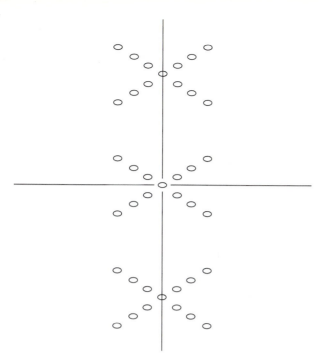

Figure 19.17 Discontinuous helix (i.e., a helix made up of atoms or monomer repeat units) gives a diffraction pattern made up of multiple crosses arrayed along the meridian.

veloped to assist in the interpretation of such complex patterns.[1-4] The actual process of structure solution for oriented, noncrystalline helical polymers involves a trial-and-error matching of the geometric relationship on the diffraction photograph and the measured intensities with plausible trial structures.

Helices within microcrystallites   If individual helices are packed together in microcrystalline domains, additional reflections will appear from the *inter*molecular spacings. At the same time, the diffraction disks will shrink in size to resemble the spots found in single-crystal photographs. In such cases, a combination of space group information (i.e., the packing symmetry within the unit cell), standard structure factor, and helical transform calculations may be needed to solve the structure. This can be an exceedingly complex undertaking and, for this reason, the structures of many polymers have not yet been worked out in detail.

[1]Cochran and Crick, op cit.
[2]Cochran et al., op. cit.
[3]Klug, A., Crick, F. H. C., and Wyckoff, H. W., *Acta Cryst.*, **1958,** *11,* 199.
[4]Holmes, K. C., and Blow, D. M., *The Use of X-Ray Diffraction in the Study of Protein and Nucleic Acid Structure* (New York: Wiley-Interscience, **1966**).

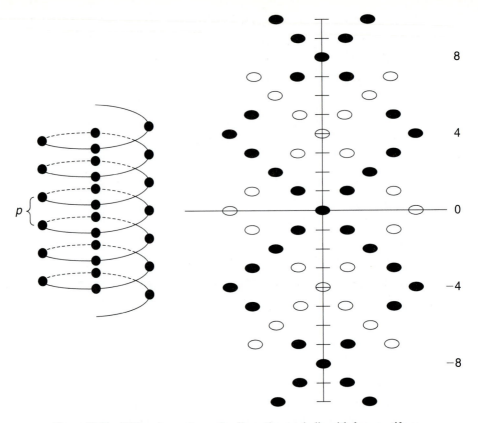

Figure 19.18 Diffraction pattern of a discontinuous helix with four motifs per turn of the helix. The cross patterns are centered on the 0, 4, 8, −4, −8, ..., layer lines. Hence, the number of monomer units per repeat (in this case four) can be determined by inspection of the photograph to ascertain the lowest layer lines above 0 that bear meridianal reflections (in this case, layer line 4). Open and closed ellipses simply indicate reflections from the different cross patterns and do not represent intensity differences.

Optical Diffraction

We cannot see the structure of molecules simply by looking at them through a very powerful optical microscope because the distances between atoms are shorter than the wavelength of light. X-rays have wavelengths similar to the distances between atoms, but no lenses exist that are capable of focusing the diffracted X-ray beams. Hence, in X-ray crystallography, we are forced to analyze the diffraction pattern mathematically rather than to reconstruct the image directly.

However, devices exist that enable the optical diffraction patterns of scaled up "models" of the polymer to be studied. If the *optical* diffraction pattern from the model resembles the X-ray diffraction pattern from the polymer, the model

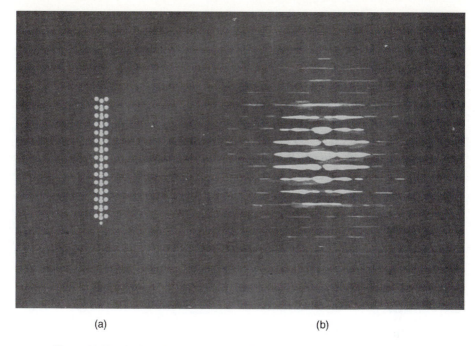

(a)	(b)

Figure 19.19   Optical diffraction pattern (b) obtained from the polymeric motif shown in (a). The motif represents one conformation of poly(dichlorophosphazene). (Motif and diffraction photograph prepared by R. A. Arcus.)

is assumed to be a true representation of the polymer structure. If not, the model can be altered in a trial-and-error procedure until a good correspondence is obtained. The method is used more commonly with fibrous polymers than with single crystals because, although Fourier calculations are now so easy to perform for single-crystal materials, they are less applicable to fiber diagrams. The optical diffraction method is really a visualization procedure for helical transform calculations.

In practice, the "model" consists of holes punched in a piece of opaque photographic film, or a photographic transparency derived from a molecular model. The holes represent the atoms, with the size of the hole being roughly related to the size of the atom. Diffraction patterns generated by single chains can be studied, or the effects of chain packing can be simulated by a model which contains a number of motifs arranged side by side (Figure 19.19).

An optical diffractometer of the Taylor and Lipson[1] design is illustrated in Figure 19.20. The light source is a low-power laser unit. The diffracted image can be examined through a small telescope or by means of a camera and photographic film. The advantage of the method is the speed with which a number of alterna-

[1]Taylor C. A., and Lipson, H., *Optical Transforms* (Ithaca, N.Y.: Cornell University Press, **1964**).

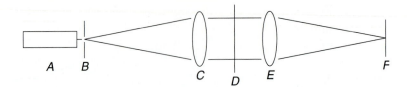

Figure 19.20 Schematic view of an optical diffractometer. Coherent light from a laser (A) is converted into a point source by a pinhole (B) and a parallel beam is formed by lens C. After passage through the perforated mask (D), the beam is focused by lens E onto a piece of photographic film (F). For the best results, lenses C and E should be of long focal length. In a typical instrument the effective diameter of the lenses is about 10 cm.

tive model structures can be evaluated. Once a reasonable correspondence between the optical and X-ray diffraction patterns has been achieved, helical transform calculations can be performed to optimize the structure.

SCATTERING

Some General Aspects

The use of scattered X-rays and beams of neutrons for investigating polymer structure has advantages over the use of photons in visible light microscopy or light scattering. Both X-rays and neutrons have much shorter wavelengths, λ, and can thus be used to investigate much smaller dimensions. X-rays scatter from electrons, and neutrons from nuclei, so the two methods can complement each other with regard to the types of information they provide. The wavelengths of the neutrons can be varied through adjustments of their velocities, v, as shown in the de Broglie equation

$$\lambda = h/p \tag{5}$$

Here, h is Planck's constant, and the momentum $p = mv$ (where m is the mass of the neutron).

The x-ray approach has the advantages of high fluxes (intensities), and the fact that the equipment is not excessively expensive and is therefore widely available. Neutron scattering, on the other hand, has the disadvantages of lower fluxes, and equipment that is very expensive and not yet widely accessible. The main advantage is the contrast between the molecule being investigated and a counterpart that is obtained by isotopic substitution. In particular, deuterium scatters very differently from hydrogen, yet the incorporation of deuterium atoms into a polymer has relatively little effect on physical properties. Thus, a deuterated polymer within the undeuterated form of the same polymer is readily distinguishable. Scattering from it can be separated from that from the surrounding host matrix (all the other chains in its environment). This unique situation can be used to answer questions such as, "What are the dimensions of a polymer chain in the undiluted amorphous state?"

Measurements of such structural features are based on "elastic" scattering, where the frequency of the radiation is unchanged in the process. "Quasi-elastic" scattering involves small changes in frequency and can give dynamic information about chain motions and diffusion. "Inelastic" scattering occurs when there are large shifts in frequency, and this gives information about local vibrations of the chains.

Structural Applications

As already mentioned, the dimensions of a chain in the bulk (undiluted) amorphous state can be determined from elastic neutron scattering.[1] The dimensions turn out to be those of the chain in the theta state, where it is unperturbed by long-range interactions (excluded volume effects). This was predicted by Flory many years before the measurements became possible, on the grounds that there was no advantage to the chain in increasing its dimensions, since this would only replace intramolecular excluded volume interactions by intermolecular ones.

Similar experiments can be used to document how the chain dimensions change as solvent is added and the system changes progressively from the bulk state to a concentrated solution, to the semi-dilute state, and finally to a dilute solution. Related work was used to investigate how a polymer chain relaxes, and specifically how a stretched-out chain reptates through an elastomer (in which the network chains and cross-links act as obstacles to its motions). Also, for elastomers, neutron scattering can be used to determine how closely the dimensions of a chain follow changes in the macroscopic dimensions of the sample during deformations. This permits distinguishing between affine deformations, where the changes do parallel one another, and non-affine (phantom) deformations, where they do not.

This can be accomplished for mechanical deformations such as simple elongation or swelling by an absorbed solvent. Another application is to networks that have bimodal distributions of network chain lengths, in particular helping to explain why such networks undergo strain-induced crystallization so readily. If the short chains are labelled but not the long chains, for example, it can be shown how these two types of chains orient differently in response to the macroscopic strain.

Crystallization in general has been investigated extensively using neutron scattering. For example, it is possible to determine how the dimensions of a chain change when the material crystallizes. Another application involves attempts to understand the nature of chain folding in thin lamellar single crystals.

The structure of phase-separated systems has also been studied by these techniques. Two important examples are block copolymers and hybrid organic/in-

[1]Ullman, R., *Polym. News*, **1987**, *13*, 42.

organic composites (such as silica-reinforced polysiloxanes). These examples have great relevance to the general area of polymer miscibility. Self-assembled systems such as core-shell structures, micelles, and latexes have also been of interest.

Dynamic Properties

Quasi-elastic neutron scattering has been used to study the motions of large segments of polymers chains. It can even give translational motion information for entire polymer molecules if the molecular weights are low enough. The method has been used to study the diffusion of cyclic oligo(dimethylsiloxane) molecules. Diffusion in general is another aspect of this important area of chain dynamics.

Inelastic scattering can also be used to characterize very local motions of a polymer chain. A specific example of this type of investigation is the determination of the local modes of motions in polyethylene crystallites.

SPECTROSCOPY

Table 19.1 lists three important types of spectroscopy, namely infrared and ultraviolet, nuclear magnetic resonance, and electron paramagnetic resonance, and mentions some of their polymer applications.

Most applications of infrared and ultraviolet spectroscopy to polymer characterization are not very different from analogous applications that involve small molecules. For example, the identification of unknown polymers is not inherently different from identifying small-molecule unknowns. The same is true for estimates of the amounts of comonomers, end groups, unsaturation, and impurities

TABLE 19.1 SOME TYPICAL APPLICATIONS OF SPECTROSCOPY TO POLYMERS

Infrared and ultraviolet	Identification of unknown polymers
	Estimation of amounts of comonomers, end groups, unsaturation, and impurities
	Study of chemical reactions of polymers
	Determination of stereochemistry
	Measurement of degree of crystallinity, and degree of orientation (by polarized IR)
Nuclear magnetic resonance	Determination of stereochemistry
	Conformational analysis (*trans* vs. *gauche*, etc.)
	Determination of chemical composition and chemical sequence distribution in chemical copolymers
	Study of relaxation processes and molecular motions in general
Electron paramagnetic resonance	Detection of free radicals in:
	Polymerization
	Cross-linking
	High-energy irradiation
	Photochemical degradation and oxidation
	Mechanical fracture

in a polymer. Another application is in the study of chemical reactions of polymers (e.g., oxidation, by measurements of the $C{=}O$ infrared absorption band in partially degraded polyethylene). Other examples include the determination of the degree of crystallinity, and of chain orientation (by polarized infrared). Polymer stereochemical structures can also be estimated from such spectroscopic measurements, but this approach has now been essentially replaced by nuclear magnetic resonance (NMR) methods.

High resolution solution-state NMR has a variety of applications in polymer chemistry. It can be used in conformational analysis (e.g., to estimate the relative amounts of *trans* and *gauche* states in a polymer), and to determine chemical compositions and sequence distributions in copolymers. The more recent use of solid state NMR spectroscopy has opened up many opportunities for the study of polymer structure and morphology. An example is the study of relaxation processes and polymer chain motions. It is also possible to determine values of the glass transition temperature, T_g, and melting point, T_m, of solid polymers from the narrowing of resonance lines that accompany the increased mobilities that occur when these temperatures are exceeded. The imaging of polymer-based composites is also feasible. Molecular structural information about polymers in the solid state is possible using solid-state NMR, although the level of detail still lags behind that of solution-state studies.

Electron paramagnetic resonance is also widely used to characterize polymers. This method involves the detection and study of free radicals, and is therefore useful in reactions that involve such species in chain-growth polymerizations and the cross-linking of elastomers. Other applications include studies of the effects of high-energy radiation on a polymer, and photochemical degradation and oxidation. A final application deals with the fracture and healing process in the deformation of polymers. For example, the bending of a strip of a high-performance thermoplastic such as Nylon-6 generates free radicals where chains are broken but before catastrophic fracture occurs. These radicals are remarkably long-lived in their high-viscosity environments, because they cannot immediately find partners to form covalent bonds.

STUDY QUESTIONS

1. Discuss the reasons why the molecular-structure determination of a synthetic high polymer is nearly always a more involved process than the structure solution of a small molecule that forms part of a single crystal.

2. What supplementary (i.e., non-X-ray) information about a new polymer would you seek before you undertook an attempted X-ray structure analysis? Give reasons for your choices.

3. For X-rays of wave length $\lambda = 1.54$ Å, what is the second-order ($n = 2$) Bragg angle for an interplanar spacing of 5 Å?

4. By an examination of molecular models, convince yourself that a three-dimensional repeating structure of two or more polymer chains can be interpreted in terms of *planes*

of atoms. Attempt to correlate different sets of planes with the directions in which diffracted X-rays might be formed. Look for atomic planes that might generate reflections on the "upper" layer lines of an X-ray photograph.

5. Suppose that you know the repeating distance of a polymer chain, but you do not know the bond angles, bond lengths, or the chain conformation, what steps would you take in order to attempt to solve the molecular structure? What serious errors could you make?

6. After reading this chapter, consult several of the books and articles mentioned in the Suggestions for Further Reading. Then write a research proposal on the subject of a proposed structure solution for any polymer of your choice. Stress experimental and structure analysis methods that you would plan to use.

7. Consider possible ways in which the projection of a polymer molecule might be depicted in two dimensions on a sheet of paper or a piece of photographic film. If this projection is used as a mask in an optical diffractometer, what effects on the optical transform would you expect from the following changes: 10 chains oriented side by side instead of one chain; increasing the separation between the chains; the use of one repeating unit instead of a long chain; use of a continuous line instead of individual atoms to represent the chain; elimination of the side groups from the structure; decreasing the wavelength of the incident light.

SUGGESTIONS FOR FURTHER READING

Polymer Crystallography and Scattering

ALEXANDER, L. E., *X-Ray Diffraction Methods in Polymer Science.* New York: Wiley-Interscience, **1969**.

ATKINS, E. D. T., "Problem Areas in Structure Analysis of Fibrous Polymers," in *Fiber Diffraction Methods*, ACS Symp. Ser., **1980**, *141*, 31–41.

AZAROFF, L. V., "X-ray Diffraction by Liquid-Crystalline Polymers," *Mol. Cryst. Liq. Cryst.*, **1987**, *145*, 31–58.

BRUMBERGER, H. (Ed.), *Small-Angle X-Ray Scattering.* London: Gordon and Breach, **1967**.

FRENCH, A. D., and GARDNER, K. H., *Fiber Diffraction Methods*, ACS Symp. Ser., **1980**, *141*.

HARBURN, G., TAYLOR, C. A., and WELBERRY, T. R., *Atlas of Optical Transforms.* Ithaca, N.Y.: Cornell University Press, **1975**.

HIGGINS, J. S., and BENOIT, H., *Neutron Scattering from Polymers*, Oxford: Clarendon Press, **1994**.

HOLMES, K. C., and BLOW, D. M., *The Use of X-ray Diffraction in the Study of Protein and Nuclei Acid Structure.* New York: Wiley Interscience, **1966**.

HOSEMANN, R., "The Paracrystalline State of Synthetic Polymers," *CRC Crit. Rev. Macromol. Sci.*, **1972–73**, *1*, 351.

KAKUDO, M., *X-Ray Diffraction by Polymers.* New York: Elsevier, **1972**.

LIPSON, H. (ed)., *Optical Transforms.* London: Academic Press, **1972**.

LOHSE, D. J., "The Use of Small Angle Neutron Scattering in Polymer Research," *Polym. News,* **1986**, *112*, 8.

RICHARDS, R. W., "Small-Angle Neutron Scattering Studies of Polymers: Selected Aspects," *J. Macromol. Sci.-Chem.*, **1989**, *A26*, 787.

ROE, R.-J., *Methods of X-Ray and Neutron Scattering in Polymer Science.* Oxford: Oxford University Press, **2000**.

SCHULTZ, J. M., *Polymer Crystallization. The Development of Crystalline Order in Thermoplastic Polymers.* Washington, DC: American Chemical Society, **2001**.

SHERWOOD, D., *Crystals, X-rays, and Proteins.* New York: Halsted Press (Wiley), **1976**.

SHIBAYAMA, M., JINNAI, H., and HASHIMOTO, T., "Neutron Scattering," in *Experimental Methods in Polymer Science. Modern Methods in Polymer Research and Technology*, (T. Tanaka, Ed.). San Diego: Academic Press, **2000**, p. 57.

SPERLING, L. H., "Characterization of Polymer Conformation and Morphology Through Small-Angle Neutron Scattering—A Literature Review," *Polym. Eng. Sci.*, **1984**, *24*, 1.

STEIN, R. S., "Studies of Polymers with Radiation," *MRS Bull.*, **2000**, *25*(10), 19.

TADOKORO, H., *Structure of Crystalline Polymers.* New York: Wiley Interscience, **1979**.

TADOKORO, H., "Structure of Crystalline Polyethers," *J. Polymer Sci. (D), Macromol. Rev.*, **1967**, *1*, 119.

TADOKORO, H., "Recent Developments in Structure Analysis of Fibrous Polymers," in *Fiber Diffraction Methods*, ACS Symp. Ser. **1980**, *141*, 43–60.

TAYLOR, C. A., and LIPSON, H., *Optical Transforms.* Ithaca, N.Y.: Cornell University Press, **1964**.

TONELLI, A. E., GOMEZ, M. A., TANAKA, H., SHILLING, F. C., COZINE, M. H., LOVINGER, A. J., and BOVEY, F. A., "Solid State NMR, DSC, and X-ray Diffraction Studies of Polymer Structures, Conformations, Dynamics, and Phase Transitions," *Polym. Mater. Sci. Eng.*, **1988**, *59*, 295–301.

ULLMAN, R., "Small Angle Neutron Scattering of Polymers," *Polym. News*, **1987**, *13*, 42.

VAINSHTEIN, B. K., *Diffraction of X-Rays by Chain Molecules.* New York: Elsevier, **1966**.

WIGNALL, G. D., "Scattering Techniques with Particular Reference to Small-Angle Neutron Scattering," in *Physical Properties of Polymers*, 2nd ed., (J. E. Mark, A. Eisenberg, W. W. Graessley, L. Mandelkern, E. T. Samulski, J. L. Koenig, and G. D. Wignall, Eds.). Washington, DC: American Chemical Society, **1993**, p. 313.

WIGNALL, G. D., "Neutron and X-Ray Scattering," in *Physical Properties of Polymers Handbook*, (J. E. Mark, Ed.). New York: Springer-Verlag New York, Inc., **1996**, p. 299.

WILSON, H. R., *Diffraction of X-Rays by Proteins, Nucleic Acids, and Viruses.* London: Arnold, **1966**.

WUNDERLICH, B., *Macromolecular Physics. Vol. 1. Crystal Structure, Morphology, Defects.* New York: Academic Press, **1973**.

NMR Spectroscopy of Polymers

AXELSON, D. E., and RUSSELL, K. E., "Characterization of Polymers by Means of Carbon-13 NMR Spectroscopy," *Progr. Polymer Sci.*, **1985**, *11*(3), 221–282.

BOVEY, F. A., *High Resolution Nuclear Magnetic Resonance of Macromolecules.* New York: Academic Press, **1972**.

BOVEY, F. A., "The High Resolution Nuclear Magnetic Resonance Spectroscopy of Polymers," *Progr. Polymer Sci.* (A. D. Jenkins, Ed.), **1971**, *3*, 1.

Bunn, A., "Synthetic Macromolecules," *Nucl. Magn. Reson.*, **1988**, *17*, 225–269.

Cunliffe, A. V., "Synthetic Macromolecules," *Nucl. Magn. Reson.*, **1986**, *15*, 216–263.

Cunliffe, A. V., "Synthetic Macromolecules," *Nucl. Magn. Reson.*, **1987**, *16*, 223–289.

Gerstein, G., "High Resolution NMR Spectrometry of Solids, Part II," *Anal. Chem.*, **1983**, *55*(8), 899A-900A, 902A-904A, 906A-907A.

Havens, J. R., and Koenig, J. L., "Applications of High-Resolution Carbon-13 Nuclear Magnetic Resonance Spectroscopy to Solid Polymers," *Appl. Spectroscopy*, **1983**, *37*(3), 226–249.

Heatley, F., "Synthetic Macromolecules," *Nucl. Magn. Reson.*, **1981**, *10*, 205–221.

Heatley, F., "Nuclear Magnetic Resonance of Solid Polymers: An Introduction" in *NATO Advan. Study Inst. Ser., Ser. C*, *94* (*Static Dyn. Prop. Polymer Solid State*), **1982**, 251–270.

Jelinski, L. W., "Solid State Deuterium NMR Studies of Polymer Chain Dynamics," *Ann. Rev. Mater. Sci.*, **1985**, *15*, 359–377.

Kausch, H. H., and Zachmann H. G. (eds.), *Advances in Polymer Science*, Vol. 66: *Characterization of Polymers in the Solid State I*: Part A; *NMR and Other Spectroscopic Methods*: Part B, *Mechanical Methods*. Berlin: Springer-Verlag, **1985**.

Komoroski, R. A. (ed.), *High Resolution NMR Spectroscopy of Synthetic Polymers in Bulk*. Deerfield Beach, Fla.: VCH Publishers, **1986**.

Liu, K.-J., and Anderson, J. E., "Proton Magnetic Resonance Studies of Molecular Interactions in Polymer Solutions." *J. Macromol. Sci., Rev. Macromol. Chem.*, **1970**, *C5*, 1.

McBrierty, V. J., "NMR of Synthetic Polymers," *Magn. Reson. Rev.*, **1983**, *8*(3), 165–242.

Ramharach, R., "NMR Spectroscopy of Polymers," *Polymer News*, **1988**, *13*(6), 174–180.

Sillescu, H., "Solid State NMR: A Tool for Studying Molecular Motion in Polymers," *Makromol. Chem., Macromol. Symp.*, **1986**, *1*(1), 39–49.

Tonelli, A. E., "NMR Spectroscopy of Polymers," in *Physical Properties of Polymers Handbook*, (J. E. Mark, Ed.). New York: Springer-Verlag New York, Inc., **1996**, p. 271.

Various authors, a series of articles on NMR spectroscopy of polymers (including solid state) in *Methods Stereochem. Anal.*, **1986**, *7*, 1 334.

Single-crystal X-ray Crystallography

Bunn, C. W., *Chemical Crystallography*, 2nd ed. London: Oxford University Press, **1961**.

Dunitz, J. D., *X-Ray Analysis and the Structure of Organic Molecules*. Ithaca, N.Y.: Cornell University Press, **1979**.

Glusker, J., and Trueblood, K. N., *Crystal Structure Analysis: A Primer*. New York: Oxford University Press, **1972**.

Kitaigorodsky, A. I., *Molecular Crystals and Molecules*. New York: Academic Press, **1973**.

Lipson, H., and Cochran W. *The Determination of Crystal Structures*, 3rd ed. Ithaca, N.Y.: Cornell University Press, **1966**.

Milburn, G. H. W., *X-Ray Crystallography*. London: Butterworth, **1973**.

Sherwood, D., *Crystals, X-rays and Proteins*. New York: Halsted Press (Wiley), **1976**.

Stout, G. H., and Jensen, L. H. *X-Ray Structure Determination*. New York: Macmillan, **1968**.

Spectroscopy of Polymers (general)

CAMPBELL, D., "Electron Spin Resonance of Polymers," *J. Polymer Sci. (D), Macromol. Rev.*, **1970**, *4*, 91.

CLARK, D. T., "ESCA Applied to Polymers," *Advan. Polymer Sci.*, **1977**, *24*, 125.

CLARK, D. T., and FEAST, W. J., "Application of Electron Spin Spectroscopy for Chemical Applications (ESCA) to Studies of Structure and Bonding in Polymer Systems," *J. Macromol. Sci., Rev. Macromol. Chem.*, **1975**, *C12*, 191.

CRAVER, C. D. (Ed). *Polymer Characterization: Spectroscopic, Chromatographic, and Physical Instrumental Methods, ACS Advan. Chem. Ser.*, **1983**, *203*.

FISCHER, E. W., "Neutron Scattering Studies of the Structure of Amorphous and Crystalline Polymers and of Polymer Blends," *Makromol. Chem., Macromol. Symp.*, **1987**, *12*, 123–144.

GRAYA, A., "IR Spectroscopy of the Low-Dimensional Organic Crystals and Polymers," *Mater. Sci. Forum*, **1987**, *21*, 69–93.

HUMMEL, D. O., *Infrared Analysis of Polymers, Resins, and Additives: An Atlas*. New York: Wiley-Interscience, **1969**.

IVIN, K. J. (ed.), *Structural Studies of Macromolecules by Spectroscopic Methods*. New York: Wiley, **1976**.

JENNINGS, B. R., "Electro-Optic Methods for Characterizing Macromolecules in Dilute Solution," *Advan. Polymer Sci.*, **1977**, *22*, 61.

KITAGAWA, T., and MIYAZAWA, T., "Neutron Scattering and Normal Vibrations of Polymers," *Advan. Polymer Sci.*, **1972**, *9*, 335.

KOENIG, J. L., *Spectroscopy of Polymers*. Washington, DC: American Chemical Society, **1992**.

KOENIG, J. L., "Spectroscopic Characterization of Polymers," in *Physical Properties of Polymers*, 2nd ed., (J. E. Mark, A. Eisenberg, W. W. Graessley, L. Mandelkern, E. T. Samulski, J. L. Koenig, and G. D. Wignall, Eds.). Washington, DC: American Chemical Society, **1993**, p. 263.

KRIMM, S., "Infrared Spectra of High Polymers," *Fortschr. Hochpolymer.-Forsch.*, **1960**, *2*, 52.

MORAWETZ, H., "Studies of Synthetic Polymers by Nonradiative Energy Transfer," *Science*, **1988**, *240*, 172–176.

NODA, I., DOWREY, A. E. and MARCOTT, C., "Group Frequency Assignments for Major Infrared Bands Observed in Common Synthetic Polymers," in *Physical Properties of Polymers Handbook*, (J. E. Mark, Ed.). New York: Springer-Verlag New York, Inc., **1996**, p. 291.

O'KONSKI, C. T. (Ed.), *Molecular Electro-optics*. New York: Dekker, **1976–77**.

RANDY, H., and NAMA, I. I., *Spectroscopy in Polymer Research*. New York: Springer Verlag, **1997.**

RUBINSON, K. A., and RUBINSON, J. F., *Contemporary Instrumental Analysis*. Englewood Cliffs, NJ: Prentice Hall, **2000**.

Safford, G. J., and Naumann, A. W., "Low Frequency Motions in Polymers as Measured by Neutron Inelastic Scattering." *Advan. Polymer Sci.*, **1967**, *5,* 1.

SNYDER, R. G. "Infrared and Raman Spectra of Polymers," *Methods Exp. Phys.,* **16A**, (Polymers), **1980,** 3–148.

SUN, S. F., *Physical Chemistry of Macromolecules. Basic Principles and Issues.* New York: Wiley Interscience, **1994**.

TOSI, C., and CIAMPBELL, F., "Applications of Infrared Spectroscopy to Ethylene Propylene Copolymers," *Advan. Polymer Sci.*, **1973**, *12*, 87.

WILLIAMS, J. L. R., and DALY R. C. "Photochemical Probes in Polymers," *Progr. Polymer Sci.* (A. D. Jenkins, Ed.). **1977**, *5*, 21.

WOODY, R. W., "Optical Rotary Properties of Biopolymers," *J. Polymer Sci. (D). Macromol. Rev.*, **1977**, *12*, 181.

20

Fabrication
of Polymers

INTRODUCTION

Throughout this book, an attempt has been made to emphasize the practical utility of polymers and the reasons for their importance to modern society. In this chapter, an emphasis is placed on the ways in which polymers are converted into useful products. To a very large extent, most polymer chemists and chemical engineers gain considerable pleasure from using polymers to make things—fibers, films, and molded objects of all kinds. These activities generate the kind of fundamental satisfaction that other scientists derive from growing crystals or observing color changes in a chemical reaction. Thus, the fabrication of polymers is as much a part of laboratory work as it is of the manufacturing process. In the following sections, when possible, both small-scale laboratory and large-scale industrial fabrication methods are mentioned.

Fabrication methods can be divided roughly into those which yield films, fibers, or bulk-molded objects. Areas of more specialized importance are elastomer technology, the formation of expanded polymers (foams), and surface coatings. Closely related to nearly all these topics are the problems of polymer compounding, blending, and curing, and the question of polymer reinforcement. Each of these topics will be considered briefly in turn.

PREPARATION OF FILMS

Polymer films can be made by two fundamentally different techniques—by solution-casting or by melt- or sinter-fabrication methods.

Solution Casting

The basic idea in solution casting is to dissolve the polymer in a suitable solvent to make a viscous solution. The solution is then poured onto a flat, nonadhesive surface, and the solvent is allowed to evaporate. The dry film can then be peeled ("stripped") from the flat surface.

One of the principal differences between high polymers and low-molecular-weight compounds is that polymers will form films, whereas small molecules are deposited as crystals or weak conglomerates. Thus, it is sometimes possible to confirm the formation of a high polymer by the fabrication of a strong, cohesive film.

On a laboratory scale, the solution casting of films is quite easy. Usually, the first problem is to find a suitable solvent for the polymer. An preferred solvent is one which is sufficiently volatile that it will evaporate at a reasonable rate at room temperature or slightly above, but not so volatile that it vaporizes rapidly and forms bubbles or semicrystalline precipitates. Rapid volatilization also causes cooling of the film, which could cause crazing or condensation of water from the atmosphere. A solvent that has a boiling point between about 60°C and 100°C will usually give good films. If the casting procedure is to be carried out on an open laboratory bench, some consideration should be given to the potential toxicity or flammability of the solvent.

High polymers dissolve only slowly in most solvents. Swelling of the polymer occurs first, and this is followed by dissolution from the edges of the polymer particles. Stirring—particularly high-speed stirring—accelerates the dissolution process. In the laboratory, a high-speed, shear disk stirrer is often used. This consists of a spindle to which is attached a metal disk, as shown in Figure 20.1. Stirred polymer solutions have a tendency to climb the stirrer spindle, and additional disks may be needed above the surface of the solution to prevent this action. The final polymer solution should be quite viscous—sufficiently fluid that slow liquid flow is possible, but not so fluid that the solution spreads out quickly on the casting surface. Solutions containing about 20 wt % of polymer often give a suitable viscosity.

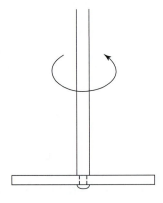

Figure 20.1 High-speed metal disk stirrer to aid the dissolution of polymers in suitable solvents.

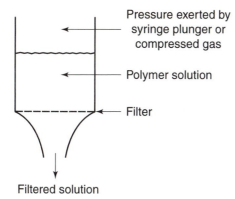

Figure 20.2 Pressure filtration unit for polymer solutions.

If the polymer solution needs to be filtered before casting, filtration must be accomplished by *pressure* techniques rather than by gravity or vacuum filtration. Small quantities of polymer solutions can often be filtered in a hypodermic syringe-filtration unit. Larger quantities must use a filter unit operated by compressed air (Figure 20.2).

The simplest casting surface for laboratory work is a sheet of plate glass. The polymer solution is simply poured onto the glass and a uniform solution thickness is ensured by spreading of the film with a glass rod (Figure 20.3). Of course, the thickness of the solution film determines the thickness of the final film once shrinkage has occurred from the loss of solvent. Thus, for a 20% polymer solution, the initial film thickness should be about five times greater than that required for the final dry film.

A number of refinements are possible to this simple technique. First, the sheet of glass may be replaced by a film of poly(tetrafluoroethylene), which has a lower tendency to stick to polymer films, or a chromium-plated, heated "casting bench" may be employed. Second, the film thickness may be determined by the use of a "Gardner knife," a device with a micrometer adjustment of the blade height above the casting surface. Third, the casting surface may be covered by a removable lid to slow the rate of solvent evaporation and prevent dust particles

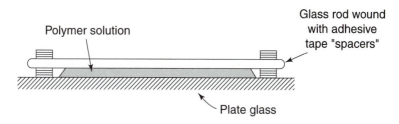

Figure 20.3 Laboratory spreading device for the solution casting of films.

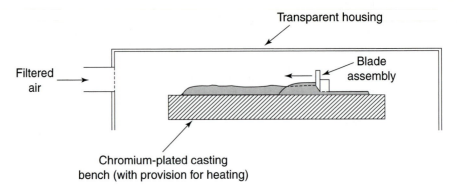

Figure 20.4   Polymer casting bench.

from settling on the film. In the most sophisticated devices a transparent poly(methyl methacrylate) cover encloses the whole unit and only dry, filtered air is admitted to the system (Figure 20.4). Such devices often resemble large glove boxes. In the laboratory, a lid fabricated from aluminum foil, or an inverted cooking pan, may serve nearly as well.

On an industrial scale, the polymer solution is fed continuously through a slit die onto a large rotating drum (Figure 20.5), or onto a moving metal belt (Figure 20.6). A hood assembly may be used to remove organic solvents from the work area. Films of poly(vinyl alcohol), poly(vinyl chloride), or vinyl chloride copolymers are often manufactured by the use of these techniques.

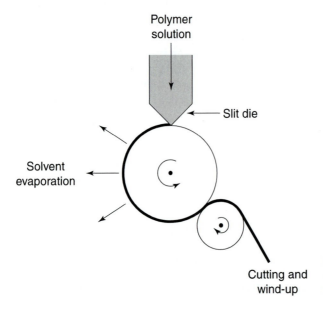

Figure 20.5   Solution casting of films on an industrial scale, with the use of rotating metal drums.

Fabrication of Polymers   Chapter 20

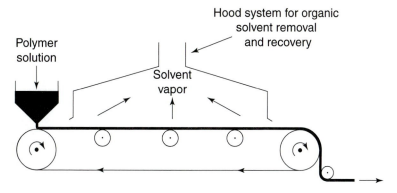

Figure 20.6 Use of a moving-belt system for the continuous solution casting of polymer films.

Melt Pressing of Film

Polymers that are thermally stable above their melting or softening temperatures can be fabricated into films by a combination of heat and pressure. The melt-pressing technique is more often used in the laboratory than in the manufacturing plant because large films are difficult to prepare by this method, and the process is discontinuous rather than continuous. The apparatus shown in Figure 20.7 is employed. It consists of two electrically heated platens. One platen can be forced against the other by means of a hydraulic unit (usually hand-pumped). The powdered or subdivided polymer is placed between two sheets of aluminum or copper foil, and this sandwich is placed between the two heated platens. Pressure is then applied (ca. 2000 to 5000 psi for about 30 s), whereupon the sandwich is removed and cooled, and the film is separated from the foil. In practice, the temperature and pressure must be determined by trial and error. If the temperature is too high, the polymer will simply flow out of the sandwich. If the tem-

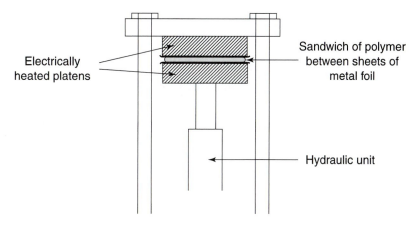

Figure 20.7 Hydraulic press for the melt pressing of polymer films.

Preparation of Films

635

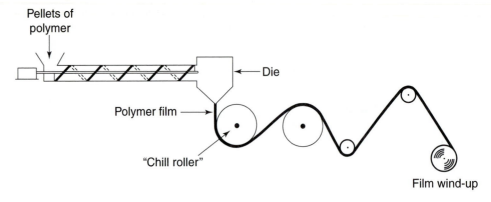

Pellets of
polymer

Die

Polymer film

"Chill roller"

Film wind-up

Figure 20.8 Sequence of operations for the melt extrusion of polymer films.

perature is too low, the film may be opaque or weak because of inadequate fusion. Metal shims or gaskets can be used in the sandwich to define the thickness of the film.

Sinter-fabrication of film   Some polymers, such as poly(tetrafluoroethylene) (Teflon[1]), have melting points that are so high that melt-fabrication techniques at high pressures are not feasible. However, powders of such polymers can be "preformed" into weak films at high pressures. Subsequent heating above the melting point completes the sintering process. This technique is reminiscent of those used for the fabrication of certain metals or ceramics. For poly(tetrafluoroethylene), the initial preforming is carried out at 1500 to 6000 psi, and the subsequent sintering takes place at 365 to 385°C during a brief exposure to the high-temperature conditions.

Melt-extrusion of films   Preferred manufacturing processes are those which are continuous. Melt-extrusion processes have this advantage. The overall sequence of operations is illustrated in Figure 20.8. Polymer pellets or powder are fed into a screw extruder. This is a device that heats the polymer and, by means of a rotating screw spindle, forces it under pressure into the die. The molten polymer is then extruded through the die slit. The flat sheet of molten polymer is collected by a rotating drum, which cools the film to below the softening temperature. Subsequent rollers complete the cooling and orientation processes. Typically, the sheet of molten polymer emerging from the die is 10 to 40 times thicker than the final film because the speed of the rotating drum exceeds the speed at which the polymer is extruded from the die. The final film may be 0.5 to 4 mils (ca. 0.01 to 0.1 mm) thick. It will be oriented uniaxially, in the direction of the extrusion. It is possible to make the orientation biaxial, for example by holding the

[1]DuPont's registered trademark for its fluorocarbon resins.

sides of the uniaxially oriented film in clamps attached to the two sides of a conveyor belt that diverges. This can also be done, more elegantly, in the blowing operation described in the next section. The effects of both types of orientation on mechanical properties are discussed following the discussion of fibers.

Bubble-Blown Films

An alternative method for the melt extrusion of films involves the extrusion of a *tube* of polymer, which is then expanded by compressed gas to form a tube of thin film. The process is shown schematically in Figure 20.9. Molten polymer from a screw extruder is forced through an annular die. Gas pressure inside the extruded tube blows the tube into a cylindrical bubble. The bubble is flattened by rollers, slit length-ways to form a continuous film, and then wound into a roll. Film made in this way has a high degree of biaxial orientation. Copolymers of vinyl chloride and vinylidine chloride (Saran) can be fabricated into films by this technique.

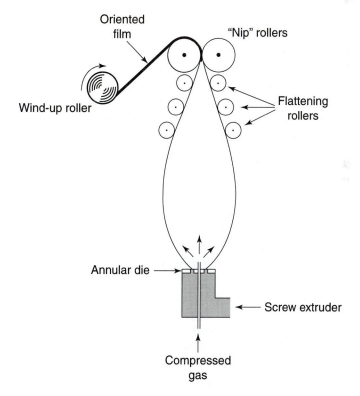

Figure 20.9 "Bubble" blowing of films.

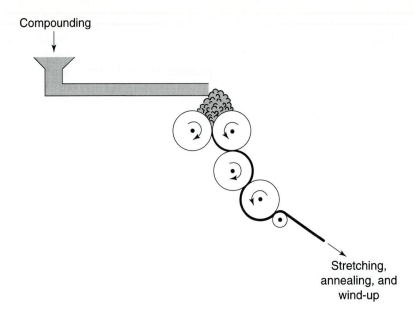

Figure 20.10   Film manufacture by calendering.

Films by Calendering

The process of calendering consists of squeezing molten polymer into a thin sheet between heated rollers. This is a method normally used for the manufacture of thick films [2 mils (0.05 mm) to more than 10 mils (0.25 mm) thick]. The process is shown schematically in Figure 20.10. Calendering is a process much used for poly(vinyl chloride) and related copolymers.

FIBERS

The production of synthetic fibers for textile manufacture comprises one of the largest and most important branches of polymer technology. Even in the research laboratory, the preparation of fibers from new polymers is an important step in the physical evaluation of a polymer by mechanical or X-ray diffraction techniques.

Two fundamental techniques exist for the production of fibers: solution-spinning and melt-spinning techniques. In solution spinning, a solution of the polymer is extruded as a filament either into a nonsolvent (*wet* spinning), or solvent is removed from the filament by a stream of hot air or inert gas (*dry* spinning). In melt spinning, the molten polymer is extruded directly into filaments. After spinning, the fiber is usually *drawn* or *oriented* by stretching to improve its strength. Conditioners are usually applied also.

Filaments are manufactured in three main forms. *Filament yarn* consists of bundles of tens to hundreds of roughly parallel, continuous individual polymer filaments of great length. *Staple* is formed from a very large number (perhaps thousands) of shorter, randomly oriented fibers. *Monofilament*, as the name suggests, consists of individual fibers of great length. Monofilaments are much thicker (0.1 to 2 mm diameter) than the other types of fiber.

Solution Spinning

Wet spinning   Solution-spinning methods comprise the oldest processes used for the preparation of synthetic fibers. The wet-spinning modification requires the coagulation of a filament of the viscous polymer solution in a nonsolvent for the polymer.

On a laboratory scale, filaments can be wet-spun with the use of the apparatus shown in Figure 20.11. A viscous solution of the polymer is extruded into a continuous filament by means of a hypodermic syringe and needle. Coagulation takes place in a trough of nonsolvent, and the solid filament is wound continuously on to a spool. In practice, the hypodermic needle diameter should be larger than the diameter of the filament needed. Moreover, the choice of the nonsolvent and the temperature of the coagulation bath are critical. If coagulation is too abrupt, a weak or "granular" fiber will be formed. Hence, the coagulation bath often consists of a mixture of solvent and nonsolvent to effect a slower, more controlled precipitation. Refinements to the laboratory apparatus include the use of a motor-driven syringe (Figure 20.12) for a steady extrusion rate, and the addition of a motor-driven take-up spool for the filament. Fiber stretching spools may also be included in the sequence.

On an industrial scale, wet spinning is used to manufacture filament yarn from viscose rayon, proteins, poly(vinyl alcohol), polyacrylonitrile, poly(vinyl chloride), and other polymers. The industrial equipment is illustrated in Figure 20.13. The polymer solution is first pressure-filtered and then forced into filaments by passage through a spinneret. A spinneret is a metal plate with holes

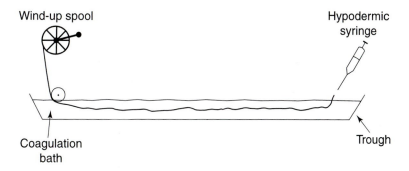

Figure 20.11   Apparatus for the laboratory wet spinning of fibers.

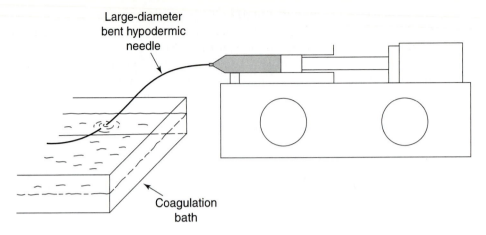

Figure 20.12 Use of a motor-driven syringe pump for the laboratory preparation of wet-spun fibers.

drilled in it. Tens to hundreds of holes may be present if the product is to be used for filament yarn; thousands of holes are present if staple is to be the product. The size of the holes does not determine the final thickness of the fiber; this factor is determined by the rate of fiber wind-up, the shrinkage during coagulation, degree of orientation, and so on.

As in the laboratory process, the viscosity of the solution, the nature of the nonsolvent, and the temperature all affect the properties of the fibers. Very viscous solutions are needed to prevent the filament from separating into droplets

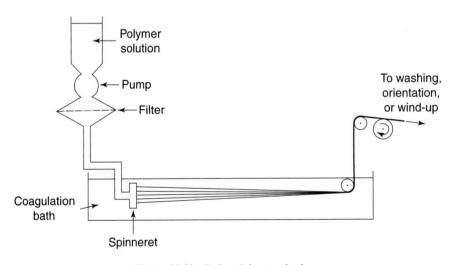

Figure 20.13 Industrial wet spinning.

at the extrusion step. Some typical coagulation systems are as follows. Dilute aqueous sulfuric acid, sodium sulfate, and zinc sulfate form the coagulation bath for the spinning of viscose rayon xanthate solutions (see page 197). Cuprammonium rayon is spun into water. Polyacrylonitrile is spun from dimethylformamide into aqueous dimethylacetamide, or from aqueous 50% sodium thiocyanate into aqueous 10% sodium thiocyanate. Spandex (polyurethane-elastomer) fibers can be spun from dimethylformamide into water.

Although wet-spun fibers account for a moderate percentage of synthetic fiber production (mainly viscous rayon), they have certain disadvantages. Fibers with a uniform cross section are very difficult to produce by the wet-spinning process. The outer surface of each filament coagulates first to form a skin. When the core coagulates at a slower rate, the outer skin shrinks and becomes convoluted. Subsequent orientation by stretching may generate greater order in the sheath than in the core. A further disadvantage of the process is that it is slow. Low extrusion speeds are needed to permit precipitation in long coagulation baths. Some of these disadvantages can be overcome by the use of dry-spinning or melt-spinning techniques.

Dry spinning   The dry-spinning process involves the extrusion of a polymer solution through a spinneret into a hot air stream which volatilizes the solvent and leaves a dry polymer fiber. The technique can be carried out on a laboratory scale, but it is difficult. As shown in Figure 20.14, a polymer solution is extruded in the laboratory from a hypodermic syringe into a glass or metal

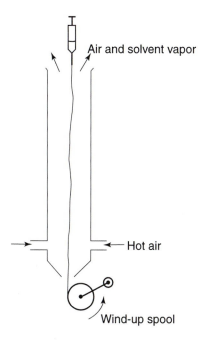

Air and solvent vapor

Hot air

Wind-up spool

Figure 20.14   Laboratory assembly for the dry spinning of fibers.

chimney containing a stream of hot air (possibly from a laboratory air blower). Problems encountered include the formation of droplets instead of fibers (if the solution is not viscous enough), and adhesion of the fiber to the wall of the chimney due to turbulence of the hot air stream. Moreover, inflammable or toxic solvent vapors must be removed effectively.

However, the process is carried out effectively on a large scale in industry. The sequence of operations is illustrated in Figure 20.15. A relatively concentrated polymer solution is filtered and pumped through a spinneret. The fibers pass down a vertical tube (which may be up to 25 ft long) countercurrent to a stream of hot gas. Often the polymer solution is heated before extrusion to lower the viscosity sufficient for passage through the spinneret holes. This enables smaller quantities of solvents to be used. Obviously, it is advantageous if the solvent used is quite volatile, such as acetone or carbon disulfide; but water has been used as a solvent for poly(vinyl alcohol), and dimethylformamide or dimethylacetamide are employed as solvents for polyacrylonitrile or Spandex.

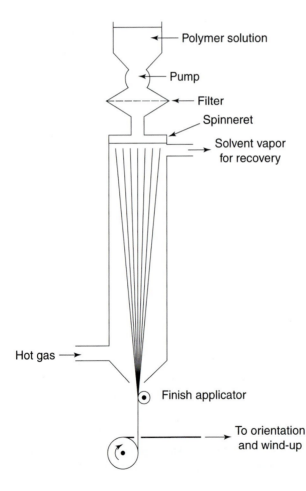

Figure 20.15 Manufacturing equipment for the dry spinning of fibers.

Melt Spinning of Fibers

In this process, *molten* polymer is extruded through spinnerets. Immediate cooling causes solidification of the fibers, which can then be stretched or collected immediately on a bobbin. The advantages of the melt-spinning technique are that (1) the spinning process is extremely rapid, and (2) the fibers have a uniform, circular cross section. The disadvantage of melt spinning is that some polymers are not sufficiently stable above their melting temperature to survive the spinning process intact.

The simplest laboratory technique for the formation of fibers from a molten polymer is to insert a glass rod into the melt and pull out a long fiber. More sophisticated laboratory or pilot-plant methods make use of forced extrusion of the melt through a small orifice under pressure generated by nitrogen gas or a hydraulically operated piston (Figure 20.16).

On a large industrial scale, the equipment illustrated schematically in Figure 20.17 is employed. Solid polymer pellets or chips are melted by a heated grid, and air is removed. A nitrogen atmosphere is often maintained over the melt. After pressure filtration, the molten polymer is forced through spinnerets and the molten fibers solidify in a stream of cold air. The fibers are then collected on rollers and bobbins. Using these techniques, filament yarn can be produced at the rate of several thousand feet per minute. The melt-spinning process is used for the preparation of fibers from nylons 6 and 66, poly(ethylene terephthalate), polyethylene, vinylidene chloride copolymers, polyurethane, and polyacryloni-

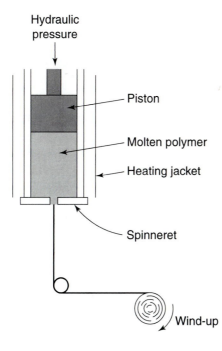

Figure 20.16 Laboratory or pilot-plant equipment for the melt extrusion of fibers.

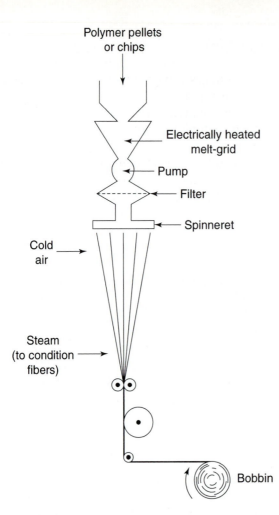

Polymer pellets
or chips

Electrically heated
melt-grid

Pump

Filter

Spinneret

Cold
air

Steam
(to condition
fibers)

Bobbin

Figure 20.17 Equipment for the melt spinning of fibers on an industrial scale.

trile. The last polymer must be plasticized with 30 to 40% dimethylformamide before it can be spun.

Glass fibers are produced in the same way—both to form filament and staple. In this case the starting material is molten glass from a manufacturing furnace, or marbles that are melted before extrusion. The extrusion temperature for glass fibers is in the region of 1250 to 1450°C.

Monofilaments are often made by melt-spinning techniques. Nylon 610, nylon 11, polyethylene, Saran, and polypropylene monofilaments are well known.

Fiber Orientation and After-Treatment

The maximum strength of a fiber is not realized until it has been drawn to orient the polymer molecules (see Chapter 17). On a large scale, orientation of the fibers is achieved by a continuous process in which the fiber passes round successive

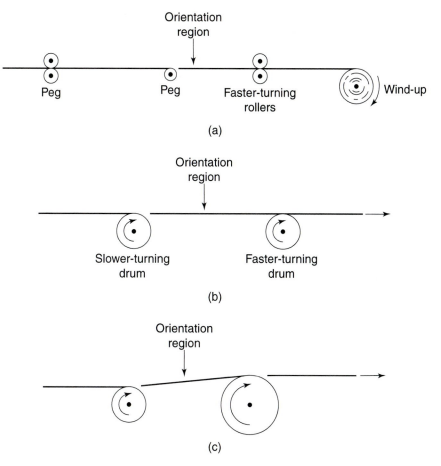

Figure 20.18 Three methods for the continuous orientation of synthetic fibers. (a) The fiber passes round a "peg" which stabilizes the stretch orientation, being induced by the faster-turning set of rollers. (b) The faster-turning drum stretches the polymer in the orientation zone. With this technique some difficulty may be experienced with stabilization of this zone. (c) The two drums turn at the same speed, but the fiber is stretched because of the greater circumference of the second drum.

drums which either rotate at increasing speeds, or which rotate at the same speed but have increasing diameters (Figure 20.18). Polymers that have a high glass transition temperature (such as some polyesters) may need to be heated during the stretching process. In the orientation zone, the fiber diameter is markedly reduced as the polymer "necks down." Some polymers are oriented immediately after spinning. Others, such as Nylon-6, may be stored first to permit crystallization before orientation is carried out.

Many textile fibers are treated with lubricants and antistatic agents immediately after the spinning process. Some fibers are dyed after spinning, but it is usually much more convenient to incorporate the dyestuff into the polymer solution or melt before spinning takes place.

CHAIN ORIENTATION AND REINFORCEMENT

Molecular Origin of the Reinforcement

Processing involves forcing polymers to change from one orientation to another. The resultant materials flow tends to align the chains and make the system anisotropic. This alignment or orientation can be exploited, and some processing steps are designed to maximize orientation, or at least to optimize its effects on properties. The anisotropy is illustrated very simply in Figure 20.19, in which two aligned chains are clamped both at their centers and at their ends. Pulling the chains apart from their centers requires relatively little energy, because the chains are held together in this direction by only van der Waals attractions, and possibly by polar interactions and hydrogen bonds. The energies involved might amount to only a few kcal/mol. Extending the material in the other direction is an entirely different matter. It would now be necessary to *break* the chains, and bond dissociation energies can be well over a hundred kcal/mol. Thus, this rudimentary system shows very anisotropic mechanical properties, and many processing techniques are designed to exploit these differences. This is accomplished by inducing orientation that improves the properties in the directions most important in a particular application, accepting as a trade-off any losses in directions that are much less important. The principles will be illustrated with respect to films, fibers, and some other examples of interest.

Films

As discussed in the preceding section, polymer chains in films are generally *uniaxially* aligned in the direction in which they were extruded. The orientation strengthens the film in this direction, but weakens it in the perpendicular direction. This is illustrated most easily by strapping tape, which is strong along its length, but pulls apart easily along its width.

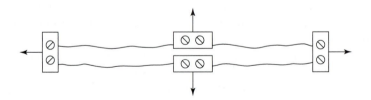

Figure 20.19 Two approximately parallel chains exhibiting very different mechanical properties parallel and perpendicular to the chains.

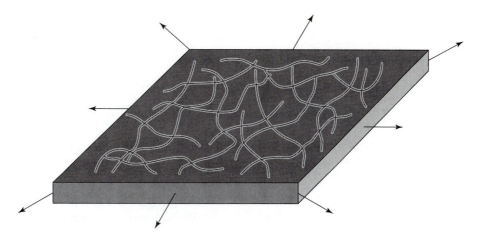

Figure 20.20 Sketch of a biaxially oriented film, showing the stretched-out chains isotropically arranged so as to give good mechanical properties in all directions parallel to the plane of the film.

This type of anisotropy is avoided in *biaxially* oriented films, as illustrated in Figure 20.20. The chains are not only stretched out, but their end-to-end vectors are randomized in the plane of the film. There is now strength in any direction in this plane, because there will always be some subset of chains aligned in the direction of any applied force, and they will resist being broken. This film is weak in the perpendicular direction and, in principle, it could be easily split in a way reminiscent of the splitting of some minerals such as mica into thinner sheets.

Fibers

When a molten polymer or a solution is forced through a spinneret the process aligns the chain in the direction of the extrusion, with a similar alignment of the crystallites if the solid polymer is partially crystalline. The additional orientation that is typically imposed after the spinning causes additional alignment of the crystallites, and the stretching and alignment of the amorphous chains separating the crystallites. This is shown schematically in Figure 20.21. The fiber can be tremendously strong in its axial direction, but weak in the perpendicular directions, in a way that could lead to "splitting hairs." This anisotropy is particularly marked in the case of the rigid-rod chains in high-performance fibers such as Kevlar®. When placed under compression, these fibers can undergo extensive

Draw direction

Figure 20.21 The drawing of a fiber so as to orient the crystallites, and to stretch out and orient the amorphous chains between them.

Chain Orientation and Reinforcement

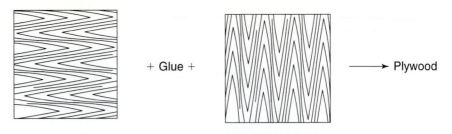

Figure 20.22 Making plywood by gluing together two or more thin sheets of wood so that their fiber directions are sufficiently different to give good mechanical properties in all directions parallel to the plane of the board.

separation into smaller fibrils. This fibrillation may not seem important in typical applications, but when a fiber is wrapped around an object such as a pressure vessel or aerospace nose cone for reinforcement, this puts the top of the fiber in tension but the bottom part in compression.

Some unpublished experiments carried out many years ago by H. F. Mark demonstrated that that silkworms have learned to appreciate such trade-offs. When silkworms were forced into spinning and drawing silks to extents different from those developed through evolution, the resulting fibers did not have as good a balance of properties as those generated by the silkworm when left to itself. On the other hand, fibers artificially reeled from immobilized silkworms can be superior to the naturally spun ones.[1]

Some Other Examples

Plywood can be thought of as an example of biomimicry or bio-inspired design, as illustrated in Figure 20.22. A thin sheet of wood is strong in the direction of the wood fibers but weak in the perpendicular direction. In this application, the fibers are analogous to the stretched-out polymer chains. In any case, the gluing of two or more sheets together in such a way that the fibers in the various layers point in several different directions gives a composite that is almost equally strong in all directions (except the one perpendicular to its surface).

In some applications, the orientation is intentionally only temporary. An example is one form of heat-shrinkable tubing. A thin tube of a partially crystalline polymer, such as polyethylene, is cross linked by radiation, heated above its melting point, expanded by introducing hot water or air under pressure, and then cooled in the expanded state. This expanded tubing has a "memory" of a higher entropy in the collapsed state, but it cannot access the original state because the chains are now locked into position by the crystallites. The tubing can then be slid over an electrical connection, and then collapsed onto it by raising the temperature above the melting point of the polymer. The resultant collapse is analogous

[1]Shao, Z., and Vollrath, F., *Nature*, **2002,** *418,* 741.

to the retraction of a stretched rubber band following its release, or the deflation of a rubber balloon when the air escapes. This solderless type of electrical connection is resistant to vibrations, water vapor, corrosive liquids, etc. Therefore, it is much preferred for crucial applications such as the wiring in space vehicles and in other applications where human lives can be at stake.

FABRICATION OF SHAPED OBJECTS

A wide variety of different techniques are available for the conversion of polymers or polymerization systems into shaped objects. With only a few exceptions these techniques require the use of complicated and expensive machinery. Hence, they are not suitable for laboratory fabrications. The types of shaped objects that can be made from synthetic polymers covers a vast range, from the plastic housings of ball-point pens or plastic caps for bottles to the interiors of refrigerators or the nose cones of missiles or supersonic aircraft. The following sections outline some of the techniques available.

Casting

Casting is a process in which a liquid monomer or prepolymer is polymerized inside a suitable mold. Initiation of polymerization is usually effected by the use of chemical reagents, although photochemical or high-energy radiation techniques have occasionally also been used. With chemical initiation, the "curing" or polymerization step usually takes place when the mold is heated in an oven. The main advantage of the casting process is that intricately shaped objects can readily be made. Furthermore, the process is inexpensive and can easily be adapted to laboratory or small-scale production procedures.

The simplest and most straightforward application of casting is in the preparation of rigid polymer sheets. This type of procedure was mentioned earlier in Chapter 3 for the preparation of clear sheets of poly(methyl methacrylate). The mold consists of two sheets of plate glass, separated by a gasket (Tygon tubing), and held together by spring clips (Figure 20.23).

The polymer–catalyst mixture is poured into the mold (with careful removal of bubbles), the gasket is adjusted to seal the inlet, and the assembly is placed in an oven to cure. Care must be taken to avoid overheating, to allow for shrinkage, and to permit annealing to occur. Large sheets of window-type material can be made in this way. Similar sheets can also be made from polystyrene, or epoxy resins. Polymer sheets of this kind are lighter than glass, but they are scratched more readily.

More intricately shaped objects can easily be made with the use of suitable molds. Plaster, clay, or wooden molds are often used if the surface of the mold is sealed with a nonadhesive coating. Such molds have a relatively short lifetime, but they are often used by hobbyists or artisans. Glass or metal molds are more durable, but molds made from Teflon or elastomers (such as silicone rubber) are

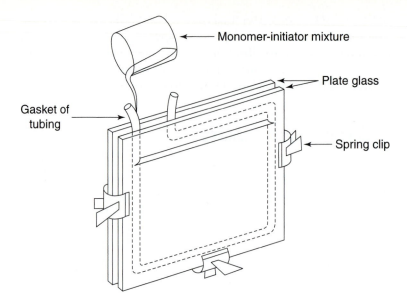

Figure 20.23 Procedure for casting sheets of polymer.

preferred when a facile release of the polymer from the mold is needed. "Release agents," such as waxes, greases, or silicones, may be used to prevent adhesion of the polymer to any mold material. The casting process can also be used to encapsulate articles such as electronic components, or to produce lamp stands or paperweights which contain shells, pictures, and so on.

Ethyl or methyl methacrylate polymers are commonly fabricated by casting techniques. However, the method is also used to make objects from polystyrene, silicones, epoxy and phenol–formaldehyde resins, and from polyurethanes. Cast epoxy resins, in particular, yield very tough, durable castings.

Compression Molding

Compression-molding techniques are normally used for the fabrication of thermo-*setting* polymers, such as phenol–, urea–, or melamine–formaldehyde, alkyd, diallyl phthalate, or silicone resins. A charge of the molding powder (resin, fillers, pigments, curing agents, and mold release agents) or a tablet of the mixture is placed in the lower half of a heated metal mold (Figure 20.24). The mold is closed, air and excess resin are forced out, and the mold is held shut until the resin has cured (30 s to several minutes). The mold is then opened and the object is released or ejected. Typically, the mold is made from chromium plated metal. Temperatures above 150°C, and pressures of 2000 psi or more, are usually employed. A modification of the compression molding technique is *transfer* molding. In this, the resin is melted outside the mold and is then rapidly injected into the mold by a plunger. This procedure causes less wear on the mold than does conventional

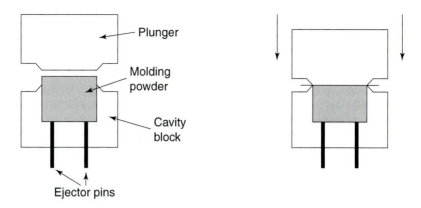

Figure 20.24   Compression molding.

compression molding. A disadvantage of the compression molding method is the time required for the curing step.

Injection Molding

Injection molding is a high-speed method that is used for the fabrication of both thermoplastic and thermosetting polymers. The equipment is illustrated in Figure 20.25. The powdered polymer or resin is heated above the melting or softening point, and the liquid is then forced by a plunger into a closed two-piece mold. The polymer cools or cures, solidifies, the mold opens, the product is ejected, and the cycle is repeated. The sequence may take only 10 to 30 s, which makes it particularly suitable for mass production. Often, the liquid is heated to temperatures above 250°C, and the injection pressure may be as high as 10,000 to 30,000 psi.

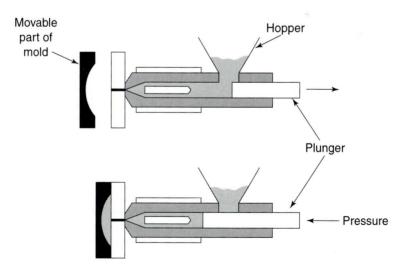

Figure 20.25   Injection-molding cycle.

Fabrication of Shaped Objects

Objects as large or larger than television cabinets can be made by injection molding.

Blow Molding

This technique is used to make bottles, toys, tanks, or other hollow objects from thermoplastic polymers. As shown in Figure 20.26, a thermally softened tube of polymer (known as "parison") is delivered from an extruder into an opened, two-piece mold. The mold closes around the parison, pinching it down at one or both ends. Compressed gas (25 to 100 psi) is then injected into the parison, which expands to line the inside of the cooled mold. The thermoplastic hardens, the mold opens, and the object is removed. The process is then repeated. In practice, all degrees of automation are possible from the manual transfer of individual parisons into the mold to a continuous but carefully synchronized extrusion of parison material into successive different molds. Blow molding is an inexpensive process. It allows facile changes to be made in the wall thickness of the product simply by making changes in the wall thickness of the parison. Moreover, the polymers used can have higher molecular weights than those used in injection molding, because high fluid viscosity is an advantage rather than a disadvantage. Polyethylene, poly(vinyl chloride), polycarbonates, methacrylate polymers, polyformaldehyde, and polystyrene are commonly blow-molded.

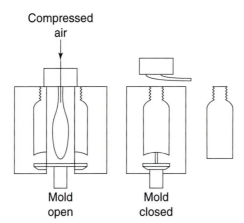

Figure 20.26 Blow-molding operation.

Thermofusion and Thermoforming

Thermofusion is a process used to make large objects, such as boats, barrels, and other large containers from a finely divided thermoplastic, usually polyethylene. The powder is placed in large sheet-metal molds, and the polymer is melted in an oven.

In thermoforming, a *sheet* of a thermoplastic is heated above the softening temperature, and is then pressed into a mold, often by the application of pressure

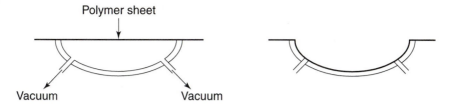

Polymer sheet

Vacuum   Vacuum

Figure 20.27 Thermoforming of polymer sheets.

or vacuum (Figure 20.27). The technique is used to make shallow trays, transparent skylight roof "blisters," or raised topographical relief maps. Polyethylene, poly(vinylchloride), poly(methyl methacrylate), polystyrene, polycarbonate, or ABS polymers are suitable for this fabrication method.

Rotational Molding

This is a technique for the fabrication of hollow objects from thermoplastic polymers. The solid polymer is melted inside a closed mold. The mold is then rotated simultaneously around its axial and equatorial axes so that the polymer coats the inside surface uniformly as it cools. The technique is used to make water tanks, fuel tanks, glove boxes, luggage, battery cases, and housing or ductwork.

Bag Molding

This technique is used for the preparation of large objects at a low production volume, especially when high-strength, high-performance structures are required. Prototype nose randomes for aircraft and missiles are made in this way. A thermosetting resin and a reinforcement material (glass cloth) are placed in an open mold and covered by a flexible diaphragm. The surrounding air chamber is then closed and air pressure is introduced. The pressure forces the diaphragm and resin into the mold. Curing then takes place as the temperature is raised.

Tube Fabrication

Tubes of polymer are normally prepared by extrusion from a heated screw extruder through an annular die. Thermoplastics that are suitable for tube extrusion include polyethylene, poly(vinyl chloride), polypropylene, polyformaldehyde, nylon, ABS terpolymers, and blends of rubber and polystyrene. Most of the tube-extruded thermoplastics are used for water pipes, drains, irrigation pipes, gas lines, and electrical or telephone conduits. Tygon tubing used in the laboratory is poly(vinyl chloride) plasticized by phthalate esters. Polymer rods and channels are also formed by extrusion techniques.

Such "tubing" is used as an insulating, protective coating for electrical wire. The polymer, typically polyethylene, is extruded as a thin small-diameter tube, which is then fed over a mandrel that holds the tube in position until it is permitted to collapse onto the wire (which is also being fed through the extruder). The polyethylene sheath is then cooled to form the protective coating around the wire. In some cases, the polyethylene coating is then cross-linked, for example by high-energy radiation, to prevent polymer "creep" at elevated temperatures, which might expose the wire. In fact, radiation cross-linking is used increasingly to stabilize polymers for high-temperature applications such as in tubes used for under-floor hot-water heating.

Extrude Swell

In polymer extrusions, a serious complication exists as the polymer exists the die. Specifically it expands, sometimes close to doubling its dimensions perpendicular to the flow direction. The misnomer "die swell" was frequently applied to this effect, which is now known by the more appropriate label "extrude swell" or "melt swell." Figure 20.28 characterizes it further and describes its molecular origins. Consider a cylindrical rod obtained by extruding a polymer through a die that has a circular opening. The magnitude of the effect is given by the *swell ratio*, or ratio of diameters D/d, as shown in the left portion of the figure. The ratio depends on the molecular weight and molecular-weight distribution of the polymer being extruded, and also on the temperature, flow rate, etc., but it can be in the vicinity of 1.8 under actual processing conditions. The expansion occurs because of the nature of the flow process, as shown in the right portion in the figure. The flow is at a minimum at the inside walls of the die (because of friction with the wall), and is a maximum at the center. The resultant shear gradient stretches the polymer molecules from the approximately spherical shapes shown to cigar-shaped ones aligned in the direction of the flow. This decreases the entropy of the polymer in the same way as when the chains are elongated when a rubber band is stretched. When the chains flow out of the die they retract to their higher entropy spatial arrangements. This is another example of a "memory" effect. The contraction in the axial direction causes an expansion in the direction perpendicular to the polymer flow. Increases in the flow rate cause a decrease in the viscosity because of the greater ease with which elongated chains can flow, but the swell ratio increases correspondingly. This is shown in Figure 20.29.

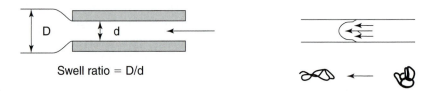

Figure 20.28 Definition of extrude swell ratio in the case of extruding a cylindrical rod, and its molecular origin.

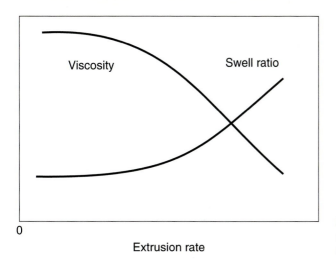

Figure 20.29 Decreases in viscosity and increases in extrudate swell accompanying increases in flow rate.

The situation worsens when the die opening is noncircular because the cross section of the extrudate now changes *shape* as well as dimensions. For example, as shown in the left part of Figure 20.30, extrusion through a square-shaped opening yields a rod with a bowed-out cross section. Thus, to obtain a rod with a square cross section it would be necessay to use a die with concave sides, and its exact shape would be extremely hard to anticipate. The right hand part of the figure shows the origin of the change in shape. The flow has to decrease from its maximum value at the center of the opening (point z) to its zero value both along the perpendicular (zy) and the diagonal (zx). Because the perpendicular is shorter than the diagonal, the shear gradient is larger along this direction and a polymer chain lying in that direction is more distorted. Thus, the sides bow out more than the corners.

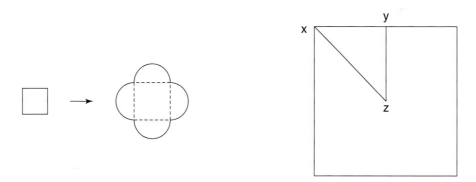

Figure 20.30 Illustrative change in shape upon extruding an object that does not have a circular cross section, and the molecular explanation of the non-uniformity.

Fabrication of Shaped Objects

655

EXPANDED POLYMERS

Foam rubber and cellular insulation material are made by generating gas bubbles in a polymer and then stabilizing the expanded structure. Such processes are used to make polyurethane foams, polystyrene foam beverage cups and furniture, poly(vinyl chloride) fabric coatings, ordinary sponge rubber, and epoxy flotation devices. The cellular expansion process is illustrated in Figure 20.31. Three methods are available for the formation of gas bubbles:

1. Latex foam rubber is made by mechanically induced frothing of a latex or a liquid rubber, followed by crosslinking the polymer in the expanded state.
2. The liquid polymer or monomer is mixed with a chemical "blowing agent," which liberates a gas when heated. Azobisisobutyronitrile (AIBN) evolves nitrogen gas when heated. Sodium bicarbonate liberates carbon dioxide.
3. A low-boiling liquid or gas is dissolved in the liquid polymer under pressure. Heating of the polymer causes boiling of the liquid to generate bubbles. Pentanes, hexane, or halocarbons are commonly used expansion agents.

Once the polymer has been expanded or "blown," the cellular structure must be stabilized rapidly; otherwise it would collapse. Two stabilization methods are used. First, if the polymer is a thermoplastic, expansion is carried out above the softening or melting point, and the form is then immediately cooled to below the melting temperature. This is called *physical stabilization*. The second method—*chemical stabilization*—requires the polymer to be crosslinked immediately following the expansion step.

Polystyrene or poly(vinyl chloride) foams are usually stabilized simply by cooling. Such polymer-blowing agent mixtures are often extruded through a slit. Expansion and simultaneous cooling occur as the polymer is extruded. Polyurethane foams are expanded by carbon dioxide bubbles generated from the reaction of excess isocyanate with water or carboxylic acids in the system, and by

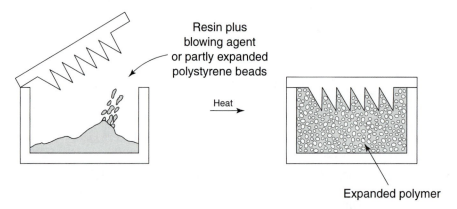

Figure 20.31 Preparation of molded objects from expanded cellular polymers.

the expansion of volatile organic expansion agents. Crosslinking occurs during foaming. Epoxy foams are stabilized by crosslinking, as are those from phenolic resins. Silicone foams are expanded by chemical blowing agents, and are stabilized by the reaction of a cross-linking catalyst (peroxide).

The individual cells within the foam may be separated from each other (closed-cell foams), or they may be interconnected (open-cell foams). The closed-cell variety are obviously better suited for thermal insulation purposes. Whether a closed- or open-cell structure is formed depends on a critical balance between the viscosity of the liquid polymer and the rate of decomposition of the chemical blowing agent.

A closed-cell system used in some flotation devices makes use of a cellular arrangement of glass or silica microballoons incorporated into the polymer.

REINFORCED POLYMERS

Many polymers must be reinforced before they can be used. For example, although most raw elastomers have an advantageous flexibility and impact strength, they are often too soft and delicate for use in rubber tires. The introduction of reinforcement introduces toughness and allows the basic shape of the object to be retained even under conditions of high stress. Reinforcement is used for solid, glassy polymers as well as for flexible or elastic polymers.

Two basic types of reinforcement are in common use: reinforcement by fabrics or cords, and reinforcement by finely divided fillers, such as carbon black, short glass fibers, carbon fibers, mica, wood flour, calcium carbonate, and so on. For some types of application (e.g., rubber tires or Fiberglas boats), both types of reinforcement are used together.

Woven fabrics of glass fiber, nylon, polyester, or rayon are used as reinforcement in automobile tires. The fabric itself is coated with a rubber compound in a calendering operation. The tread and sidewalls are made by extrusion. A tire is fabricated first from a calendered rubber liner wrapped around a drum. The fabric reinforcement is then added and the tread and sidewalls are positioned outside the reinforcement. The unit is then heated to cure the elastomer.

Glass fabrics are also used to reinforce thermosetting polymers in laminates. A *laminate* is a sandwich of fabric layers bound together by the polymer. The laminate is built up layer by layer with each layer of fabric impregnated by and separated from the next one by a layer of resin. The sandwich is then compressed in a press and heated in an oven until the resin has cured. Laminates are characterized by toughness, dimensional stability, and a resistance to cracking or shattering when placed under stress.

Finely divided fillers in a polymer matrix serve the same purpose. Most fillers are bound to the polymer by van der Waals forces; hence the binding strength increases with surface area and with the degree of subdivision of the filler. However, carbon black is a particularly effective filler for rubber and other polymers, apparently because it participates in chemical grafting to the polymer. Chopped glass fibers are extensively used as filler materials for rigid thermosetting resins in the manufacture of boats, automobile bodies, building panels, crash helmets, and so on. A commonly used resin formulation is a styrene solution of the condensation product from ethylene glycol and maleic or phthalic anhydrides. Benzoyl peroxide is a suitable curing accelerator. Other resins include phenol-formaldehyde, epoxy, silicone, alkyd, and melamine-formaldehyde formulations. The polymer may constitute 10 to 60% by weight of the material. Often a mixture of the resin and the chopped glass fiber is sprayed onto an open mold, with further reinforcement sometimes introduced by the application of a woven fabric. Boats, large tanks and housings, and crash helmets are made in this way.

Thermoplastics, such as nylons, polyacrylates, polycarbonates, polypropylene, polyformaldehyde, or ABS terpolymers can also be strengthened by the incorporation of fillers. These materials are being used increasingly in the automobile industry.

Other forms of reinforcement include the use of metal honeycombs and filament-wound products. The filament-winding technique uses a continuous filament wound around a mandrel and impregnated with the resin. The cylindrical- or spherical-shaped products have a very high strength.

ELASTOMER TECHNOLOGY

Raw rubber, either natural or synthetic, is only rarely suitable for use in most applications. The pure elastomer is usually too soft and extensible, or too readily attacked by oxygen or ozone. For these reasons, a substantial number of additives are compounded into the polymer to improve its performance. These additives include:

1. *A vulcanizing agent*, such as sulfur, and related additives, such as mercaptothiazole (a vulcanizing accelerator), activators, or retarders. Metal oxides (ZnO) are used as activators. The vulcanization process is a thermally induced cross-linking step which imparts strength, resistance to viscous flow, and elasticity to the polymer. Natural rubber, SBR, butyl, or nitrile rubber can be vulcanized with sulfur. Silicone rubber is crosslinked by means of peroxides.

2. *Fillers*, such as carbon black, to increase the tensile strength and elasticity (the rapidity of retraction). Apparently, carbon black functions by forming weak, covalent crosslinks between the chains. This effect stiffens the elastomer and generates toughness. Other fillers, such as clay, are sometimes added to rubber to improve its handling qualities before vulcanization. Silica may be used as a reinforcing filler for silicone elastomers.

3. *Pigments*, to modify the color of the elastomer. Pigments cannot be used if carbon black is employed as a filler.

4. *Plasticizers*. These may be added to soften an elastomer.

5. *Antioxidants*. Most organic elastomers react slowly with oxygen or ozone. In this process they become either soft or hard and brittle. This effect is especially serious with elastomers made from dienes. An antioxidant is a compound added to protect the elastomers from this effect.

Some of the properties of elastomers are discussed in Chapter 21.

SURFACE COATINGS

Polymers are widely used as materials for coating metals, wood, paper, fabric, or even, in rare cases, for coating stone and masonry. Polymers are used as coatings in order to confer waterproofing, flameproofing, fungus-resistance, or corrosion-protection properties to other substrates. The polymer coating may be an adhesive, insulating, decorative, or reflective coating. Such surface-coating properties are ultilized in the painting of metal, the coating of fabrics (to make rainwear or artificial leather), the production of magnetic recording tape, pressure-sensitive tape, book bindings, paper, wire enamels, floor protection, and photographic emulsions. Some of the methods employed are listed below.

Dipping

A mold is dipped first into a solution or emulsion of a polymer and then into a coagulation bath. The procedure is repeated until a thick-enough layer of polymer has been built up. In the manufacture of, for example, a rubber glove, a fabric glove placed over the mold actually serves as the reinforcing substrate on which the polymer is coated. Rubber and polyethylene coatings are often applied in this way.

Calender Coating

A film of the coating polymer is prepared by calendering. It is then applied in a continuous process to the film of substrate. The two films are then squeezed together by rollers.

Extrusion Coating

A thermoplastic is extruded through a slit to form a thin film. This film coats one side of the moving substrate film, and the two layers are forced together by passage through rollers.

Electrostatic Coating

This is essentially a spray gun method in which the coating material is given an electric charge before spraying. The object to be coated is either grounded or bears an opposite charge. The method facilitates a uniform coating procedure. It is used for coating automobile parts or even textiles or paper products.

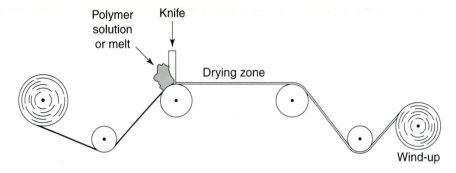

Figure 20.32 Knife coating of a polymer onto a film.

Knife Coating

This method is a large-scale continuous process development of the solution film-casting technique discussed earlier. A film is cast continuously on a moving belt formed by the substrate film material. A schematic diagram of the process is shown in Figure 20.32. The same type of process can also be used for melt casting.

Roll Coating

A continuous sheet or film of the material to be coated passes over a roller. The roller picks up a film of the molten or dissolved polymer and applies it to the film in a continuous process. The method closely resembles the techniques used in mass-production newspaper printing.

Fluidized-Bed Coating and Powder Molding

A preheated metal object is dipped into a fluidized bed of powdered polymer and a compressed gas. Polymer particles adhere to the hot object. The object is then passed through an oven to fuse the polymer and form a film. The coating may be left in place to protect the metal, or it may be removed from the metal to yield a hollow article.

Radiation-Cured Coatings

Although heat is the principal method used for the curing of thermosetting materials applied as coatings, experimental units have been tested that irradiate a coating with gamma rays or ultraviolet light in order to crosslink the polymer. Of course, an ideal process would bring about a chemical grafting of the coating to the substrate.

STUDY QUESTIONS

1. From what you know of the properties of the following polymers, suggest the types of fabrication methods that might be suitable for these materials on a large scale: poly(vinyl alcohol), polybenzimidazoles, polycarbonates, polymeric sulfur, borosilicate glass, polybutadiene, polystyrene.

2. Why does a volume contraction take place when methyl methacrylate polymerizes in a mold?

3. What precautions would be necessary if you were to carry out a solution-spinning experiment in an attempt to prepare fibers of polyethylene from hot xylene?

4. Suppose that you wished to prepare thin films of a polymer to test its ability to function as a barrier to gaseous diffusion. What fabrication problems and solutions to these problems do you foresee?

5. You have been given the task of planning the fabrication of the nose segment of a new rocket-propelled aircraft. Extreme thermal and dimensional stability of the unit will be needed. Prepare a written proposal for this work, justifying your decisions at each step.

6. Thermoplastics that are reinforced with glass fibers are being used increasingly for the fabrication of components for automobiles. What problems would you be likely to encounter if you wished to use conventional, high-volume fabrication techniques for the manufacture of such components?

7. Polymers are now used extensively for the manufacture of the hulls of motor boats, yachts, and so on. What reasons can you think of that might prevent an extension of this technology to the manufacture of ocean liners or freighters, or the manufacture of the outer shells of civil aircraft?

8. What fundamental physicochemical characteristics of a polymer are needed if it is to be used as (**a**) a textile fiber; (**b**) a film for use as a packaging material; (**c**) an artificial leather; (**d**) a surface coating for outdoor use; (**e**) a liner for a home refrigerator?

9. One fabrication technique that is not discussed in detail in this chapter is the use of polymers as wire coatings. Suggest ways in which a metal wire might be coated with (**a**) a thermoplastic polymer, and (**b**) a thermosetting resin. What special properties of the polymer would you be looking for in this application?

10. Why do inorganic substances, such as glass fiber, asbestos, carbon black, boron fibers, carbon fibers, and so on, constitute materials that are added to polymers as reinforcement fillers? Why, for example, are nylon fibers not normally used to reinforce epoxy resins?

11. Why is processing a rigid-rod polymer so difficult, for example spinning Kevlar® into high-strength fibers?

SUGGESTIONS FOR FURTHER READING

CORISH, P. J. (Ed.), *Concise Encylopedia of Polymer Processing & Applications*. Oxford: Pergamon Press, **1992**.

Encyclopedia of Polymer Science and Engineering (H. F. Mark, N. M. Bikales, C. G. Overberger, G. Menges, and J. I. Kroschwitz, eds.). New York: Wiley, **1986**, including sections on extrusion, fibers, films, laminates, and so on.

FRIED, J. R., *Polymer Science and Technology*. Englewood Cliffs, NJ: Prentice Hall, **1995**.

GRULKE, E. A., *Polymer Process Engineering*, 2nd ed. Englewood Cliffs, NJ: Prentice Hall, **1994**.

KRASSIG, H., "Film to Fiber Technology," *J. Polymer Sci. (D), Macromol. Rev.*, **1977**, *12*, 321.

KRAUS, G., "Reinforcement of Elastomers by Carbon Black," *Advan. Polymer Sci.*, **1971**, *8*, 155.

MANSON, J. A., and SPERLING, L. H. *Polymer Blends and Composites*. New York: Plenum Press, **1976**.

McCRUM, N. G., BUCKLEY, C. P., and BUCKNALL, C. B., *Principles of Polymer Engineering*. New York: Oxford University Press, **1988**.

PEARSON, J. R. A., *Mechanical Principles of Polymer Melt Processing*. New York: Pergamon Press, **1966**.

PENN, W. S., *PVC Technology*. New York: Wiley, **1971**, **1972**.

RASTOGI, A. K., "Fiber-Reinforced Plastics Today," *ChemTech.*, **June 1975**, 349.

RODRIGUEZ, F., *Principles of Polymer Engineering*, 2nd ed. New York: McGraw-Hill, **1982**.

SORENSON, W. R., and CAMPBELL, T. W. *Preparative Methods of Polymer Chemistry*, 2nd ed. New York: Wiley-Interscience, **1968**.

WILLIAMS, H. L., *Polymer Engineering*. New York: Elsevier, **1975**.

21

The Testing
of Polymers

INTRODUCTION

Polymers are used in an enormous variety of different applications. However, each different use normally requires a polymer with very specific properties. Hence, an important aspect of polymer technology is the testing of new polymers to determine their advantages and disadvantages for different applications. The ultimate test of any polymer is for it to be fabricated into a suitable object and tested under the same operating conditions that it would encounter in normal use. This is an expensive and time-consuming procedure. Thus, considerable emphasis is placed on the generalized *laboratory* testing of new polymers in the hope that the results will provide guidelines to the types of applications for which the polymer is best suited.

Many of the tests that are conducted on a new polymer involve *engineering* evaluations. A detailed discussion of such procedures and the underlying theory is beyond the scope of this book. However, the following sections outline a few of the laboratory tests that are commonly applied to any new polymer. For a more detailed discussion of this topic, the reader is referred to the specialized references at the end of this chapter. Here we will review the role of fundamental physicochemical tests, mechanical evaluations, thermal properties, electrical tests, and tests designed to predict the stability of a polymer to weathering, solar radiation, and other environmental influences.

FUNDAMENTAL PHYSICOCHEMICAL TESTS

Two characteristics of a polymer form the foundation of any use-oriented evaluation. These are the *glass transition temperature* (T_g) and the presence or absence of *crystalline melting transitions* (T_m). If a polymer has a high T_g (say, above 30°C), it will generally be unsuitable for use in applications that require flexibility and rubbery properties. If the material is contemplated for use as an elastomer in a low-temperature environment (e.g., in the arctic or in aircraft) it must have a very low T_g and also a low degree of crystallinity. For a polymer to be useful as a textile fiber, it should normally have a T_g that is below its typical operating temperature but a T_m that is above this temperature. A polymer that is to be used as a rigid structural material should have a high T_g (100°C or above).

It was once thought that a polymer designed to serve as a fiber or film should have a glass transition temperature, T_g, well below its normal exposure temperatures. Now we recognize that this is not necessary because an object will become brittle below T_g only if all of its dimensions are relatively large. A film has one dimension that is small (its thickness), and this permits polystyrene films ($T_g = 100°C$) to be used for packaging at room temperature and below. A fiber has two dimensions that are small, and this permits glass wool to be used at temperatures hundreds of degrees below its T_g ($\approx 800°C$). For polystyrene domains in a thermoplastic elastomer, all three dimensions are small, and this makes those domains deformable at room temperature!

Also note that the glass transition temperature depends on chain structure (an *intra*molecular effect) with stiff chains having higher values of T_g than flexible ones. It also depends on *inter*molecular effects as well, with strong attractions leading to higher values of T_g than weak ones. This last dependence explains how plasticizers work, with the diluent separating and "lubricating" the chains.

A complication connected with melting points is that large crystallites have significantly higher T_m values than smaller ones. This is because of the positive interfacial free energy that destabilizes small particles because of their large ratios of surface area, S, to volume, V: $S/V \propto r^2/r^3 \propto 1/r$, where r is an effective radius of the particle. Thus, the enlargement of crystallites by annealing generally increases T_m significantly, but this also means that such a polymer, once melted and recrystallized, may subsequently melt at a lower temperature.

Thus, the measurement of T_g and T_m transitions by the methods discussed earlier in Chapters 17 and 19 is the logical first step in any polymer evaluation program. This information must then be viewed in the light of the *molecular weight* of the polymer (Chapters 14 and 15), because physical properties usually depend on the average chain length for the lower-molecular-weight species.

MECHANICAL TESTS

Chemists normally seek to understand the physical properties of materials in terms of *molecular* features, for example, in terms of molecular shape, conformational mobility, crystal packing forces, and bond energies. On the other hand, the engineer or technologist, while being aware of such fundamental matters, is usually much more interested in the question: What are the special properties of this material that would favor its use in this particular application? Although marked advances have been made in recent years in the physics and chemistry of polymers, mechanical properties cannot yet be predicted from fundamental molecular structural principles, except in a few very favorable cases. For example, the actual strength of a polymer may be only $\frac{1}{10}$ to $\frac{1}{100}$ of the value calculated on the basis of bond strengths and intermolecular forces. Hence, for the present, mechanical tests provide the only method to obtain engineering-type evaluations of new polymers. Such tests are performed routinely in materials laboratories and in industrial research laboratories. Because most of the tests result in destruction of the sample, and because such tests do not have a high degree of reproducibility, multiple tests on similar samples are needed before valid results can be obtained.

Four different groups of questions about the mechanical properties form the basis of an applications evaluation. These are:

1. What is the response of the material to stress and strain? Does the material distort or elongate easily when stretched? Does it remain elongated or does it snap back to its original form when the stretching force is released? How readily will it break if subjected to stretching (tensile) forces? Is the polymer rigid or does it flow under pressure?

2. What is the resistance of the material to impact? Does the polymer shatter like glass, or does it absorb the force and remain intact, for example, like rubber?

3. What is the strength of the material to flexural distortions? Does a rod or a plate of the material break when bent, does it remain bent, or does it spring back when the stress is removed? What is the response to continued long-term flexural distortion (fatigue)?

4. What is the hardness and abrasion resistance of the material? Is the surface of the polymer readily deformed when pressed into contact with sharp objects? Does the material abrade when used in bearings? Is the surface scratched easily when abraded by metals, other plastics, fabrics, grit, and so on?

It will be clear that answers to most of these questions must be obtained before a polymer can even be considered for use in any important or large-scale application, whether that application happens to be as a textile fiber or in the nose cone of a rocket.

Stress–Strain Curves

The tensile strength of a material provides a measure of its resistance to elongation or breaking when stretching forces are applied to it. The terminology used in this field relates the stress (the force per unit cross-sectional area applied to the sample) to the strain (the elongation of the sample under a given stress) (see Chapter 17). Because the stress–strain behavior of most materials is time-dependent, the speed at which the stress is applied must also be taken into account. For instance, a sudden, abrupt pull on a fiber may cause it to break. The same

Figure 21.1 Typical shape of a flat polymer sample used for stress–strain tests.

force applied slowly may result in slight elongation of the fiber and in a higher resistance to breakage.

In practice, stress–strain experiments are often carried out on a flat sample that has been shaped into the form shown in Figure 21.1. The ends of the sample are clamped into the jaws of a testing machine and the jaws are separated by the application of a known mechanical force. The test material usually elongates or breaks in the narrower, central region of the specimen. Both the stress and the strain can be read from the dials on the machine, or the data may be provided directly on chart paper.

The ratio of stress to strain is a measure of the *stiffness* (i.e., a stiff polymer will yield very little as the stress is applied). A soft polymer, such as an elastomer, will yield considerably under the same circumstances. The data are normally plotted in the form shown in Figure 21.2, which typifies the behavior of a thermoplastic material, such as polyethylene. Note that *toughness* is a measure of resistance to breaking. This property can be crudely estimated from stress–strain curves, as well as from impact tests. The stress–strain behavior of five different types of materials is illustrated in Figure 21.3. For example, an un-cross-linked elastomer might behave in the manner shown in curve (a) of Figure 21.3. A polymer for use as a tough, structural plastic (for use in housings, gear wheels, etc.)

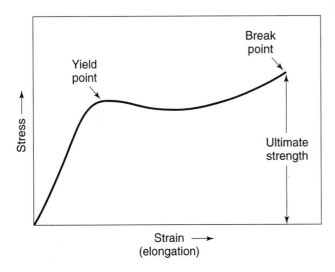

Figure 21.2 Stress–strain curve for a thermoplastic material such as polyethylene. Note the initial section of the curve in which increased stress causes a moderate, but noncatastrophic elongation. However, further applied stress causes appreciable elongation or "yield," without the application of comparable additional stress.

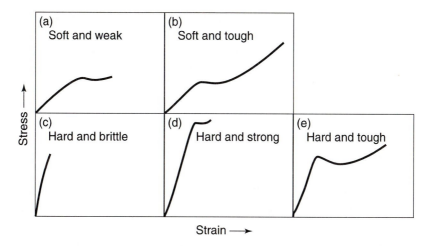

Figure 21.3 Characteristic stress–strain curves for five different types of polymeric materials.

might conform to curve (d). Polystyrene is a hard, brittle polymer with stress–strain characteristics of type (c). One of the most important pieces of information derived from stress–strain curves is the value of the tensile strength (stress at the breaking point). Some typical values for a number of different materials are given in Table 21.1.

TABLE 21.1 TENSILE STRENGTHS OF VARIOUS POLYMERS AND STEELS

Polymer	Tensile Strength (psi)
Polyethylene (low to medium density)	1,000–2,400
Poly(tetrafluoroethylene)	3,500
Polyethylene (high density)	4,400
Poly(dimethylsiloxane)	5,000
Polypropylene	5,000
Poly(vinylidene chloride)	8,000
Polystyrene	8,000
ABS terpolymer	8,500
Polyamides	9,000–12,500
Polycarbonate	9,500
Polyesters (cast)	~10,000
Polysulfone	10,200–12,000
Poly(phenylene oxide)	10,500
Epoxy resin	
Cast	12,000
Molded	16,000
Glass-filled nylon	31,000
Fabric-reinforced epoxy resin	60,000–85,000
Carbon steel	80,000
Filament-wound epoxy resin	100,000–250,000
Type 420 stainless steel	250,000
Steel wire	500,000

Thermoplastic materials generally become less rigid as the temperature is raised, although the change in rigidity is not continuous if T_g or T_m transitions are encountered. Rigidity is clearly a favorable property for materials that will be used as structural plastics. Perhaps more important is the long-term ability of a rigid polymer to withstand *cold-flow* or *creep*. Tests of this property can be conducted by measuring the long-term elongation of a test sample, one end of which has been attached to a weight.

Impact Resistance

Impact resistance is normally associated with *toughness*. A polymer that is prone to shatter on impact cannot be used in many applications (although it may be a perfectly satisfactory structural material). Polymers in their glassy state (i.e., below their T_g) are particularly prone to shatter on impact. The standard impact test involves two types of experiments. Either a swinging pendulum is allowed to strike the sample from different displacements or a falling weight is dropped on to the sample from various heights. The impact force is increased until the sample breaks. Impact tests provide only rough guidelines at best to comparative toughness. Some polymers that are quite tough under actual operating conditions (nylon, for example) perform poorly in standard impact tests. Moreover, thinner samples are relatively tougher than are thicker specimens.

Flexural Strength

The ability of a material to undergo flexural distortions without weakening or shattering is a key property for the use of polymers in gear wheels, vibrating components, structural parts of automobiles, boats, or aircraft. A commonly used criterion of flexural strength is the force needed to cause a beam of the polymer to be deflected (i.e., bent) by a known amount. However, a more meaningful criterion is the ability of the material to withstand *fatigue* (i.e., multiple cyclic flexural motions). A graphical plot of the stress versus the number of flexural cycles needed to bring about failure of the sample provides an estimate of the fatigue characteristics of that particular polymer.

Hardness and Abrasion Resistance

Hardness, abrasion resistance, scratch resistance, and friction are properties that are related but not necessarily directly related to each other. *Hardness* is measured by the distance of indentation and recovery that occurs when a steel ball (the indentor) is pressed into the surface under constant load, and is then released. In these terms, polyolefins are soft but polyimides and poly(ethylene terephthalate) are hard polymers.

Abrasion resistance and scratch resistance are subtly different properties. *Abrasion resistance* represents the ability of a polymer to retain a smooth surface while moving constantly in contact with another (smooth) surface, for example, when the polymer is used as a spindle bearing material, as sliding surfaces in reciprocating devices, in hinges, or in gears. A relationship appears to exist between abrasion resistance and the coefficient of friction. Those polymers that are both hard and have a low coefficient of friction (i.e., a slippery surface) in general have a high abrasion resistance. Abrasion tests may take the form of a measurement of the weight loss at the abrasion surface. Nylons, polyacetals, and poly(ethylene terephthalate) perform well in these tests.

The susceptibility of a smooth polymer surface to scratching may be critical in determining the use of a polymer in windows, lenses, or automobile windshields. *Scratch resistance* tests are highly subjective evaluations that are conducted differently in various laboratories. Some tests require the pressing and twisting of a piece of sandpaper on the surface. Others require scratching of the surface with pencils of differing hardness. Scratch resistance is frequently associated with rubbery, rather than hard or brittle surface properties, or with the ability of a surface to "heal" by cold-flow of the polymer.

RUBBERLIKE ELASTICITY

The preceding discussion of mechanical properties has focused on the elastic properties of thermoplastics and thermosets. Very different behavior is characteristic of another class of polymeric materials—the elastomers. They are defined by their very large deformability with essentially complete recoverability when the stress is released. In order for a material to show this type of elasticity, three molecular requirements must be met: (i) the material must consist of polymeric chains, (ii) the chains must have a high degree of flexibility and mobility, and (iii) the chains must be joined into a network structure or have physical chain linkages through microcrystallites or knots.

The first requirement results from the fact that the molecules in a rubber or elastomeric material must be able to alter their arrangements and extensions in space dramatically in response to an imposed stress, and only a long-chain molecule has the required very large number of conformational arrangements at different extensions. The second characteristic required for rubberlike elasticity specifies that the different conformational arrangements be *accessible*, i.e., changes in these arrangements should not be hindered by constraints that might, for example, result from inherent rigidity of the chains, extensive chain crystallization, or the very high viscosity characteristic of the glassy state. A network structure is needed for the elastomeric recoverability. It is obtained by joining together or "cross-linking" pairs of segments on different polymer molecules, approximately one out of a hundred, thereby preventing stretched polymer chains from irreversibly sliding past on another. The network structures thus obtained is illustrated in Figure 21.4, in which the cross-links are generally chemical bonds (as would occur in sulfur-vulcanized natural rubber) but could be physical rather than chemical connections. These elastomers are frequently included in the category of "thermosets," which are polymers that have a network structure which is generated or "set" by thermally induced chemical cross-linking reactions. How-

Figure 21.4 Sketch of a typical elastomeric network, with interchain entanglements also depicted.

ever, this has now frequently taken on the more specific meaning of networks that are very heavily cross-linked and below their glass transition temperatures. Such materials, exemplified by phenol-formaldehyde and epoxy resins, are very hard materials with none of the high extensibility associated with typical elastomers. The cross links in an elastomeric network can also be temporary or physical aggregates, for example the small crystallites in a partially crystalline polymer or the glassy domains in a multiphase triblock copolymer.

The earliest elasticity experiments involved stress–strain–temperature relationships, or network "thermoelasticity." They were first carried by J. Gough in 1805. The discovery of vulcanization or curing of rubber into network structures by C. Goodyear and N. Hayward in 1839 was important because it permitted the preparation of samples which could be investigated with much greater reliability. Quantitative experiments were carried out by J. P. Joule, in 1859. This was only a few years after entropy was introduced as a general concept in thermodynamics. Another important experimental result which contributed to the understanding of polymers was that mechanical deformations of rubberlike materials generally occur essentially at constant volume, so long as crystallization is not induced. (In this sense, the deformation of an elastomer and a gas are very different).

A molecular interpretation of the concept that rubberlike elasticity is primarily entropic in origin had to await H. Staudinger's demonstration, in the 1920s, that polymers are covalently bonded molecules, and not aggregates of small molecules of the type studied by colloid chemists. In 1932, W. Kuhn used this observed constancy in volume to point out that the changes in entropy must involve changes in orientation or spatial configuration of the network chains.

Later in the 1930s, W. Kuhn, E. Guth, and H. F. Mark first began to develop quantitative theories based on this idea that network chains undergo conformational changes by skeletal bond rotations in response to an imposed stress. More rigorous theories began with the development of the "Phantom Network" theory by H. M. James and E. Guth in 1941, and the "Affine Model" theory by F. T. Wall,

P. J. Flory, and J. Rehner, Jr. in 1942 and 1943. Modern theories generally begin with the phantom model and extend it, for example, by taking into account interchain interactions.

Because high flexibility and mobility are required for rubberlike elasticity, elastomers generally do not contain stiffening groups such as ring structures or bulky side chains. This is the reason for the low glass transition temperatures, T_g, typically found for elastomers. Such polymers also tend to have low melting points, or no melting transition, but some do undergo crystallization following sufficiently large deformations. Examples of typical elastomers include natural rubber and butyl rubber (which undergo strain-induced crystallization), and poly(dimethylsiloxane), several polyphosphazenes, poly(ethyl acrylate), styrene-butadiene copolymer, and ethylene-propylene copolymer (which generally do not).

Some polymers are not elastomeric under normal conditions but can be made so by raising the temperature or adding a diluent ("plasticizer"). Polyethylene is in this category because of its high degree of crystallinity. Polystyrene, poly(vinyl chloride), and the biopolymer elastin are also of this type but, because of their relatively high glass transition temperatures, they require elevated temperatures or the addition of a diluent to make them elastomeric.

A final class of polymers is inherently non-elastomeric. These include polymeric sulfur, because of its chains are too unstable, poly(p-phenylene) because its chains are too rigid, and thermosetting resins because the distances between crosslinks are too small.

Preparation of Networks

One of the simplest ways to introduce the cross links required for rubberlike elasticity is to carry out a copolymerization in which one of the comonomers has a functionality of three or higher. However, this method has been used primarily to prepare materials so heavily cross linked that they are in the category of hard thermosets rather than elastomeric networks. The more common techniques include vulcanization (addition of sulfur atoms to unsaturated sites), peroxide thermolysis (covalent bonding through free-radical generation), and end-linking of functional-terminated chains (isocyanates to hydroxyl-terminated polyethers, organosilicates to hydroxyl-terminated polysiloxanes, and silanes to vinyl-terminated polysiloxanes).

A sufficiently stable network structure can also be obtained by physical aggregation of some of the chain segments onto filler particles, by formation of microcrystallites, by condensation of ionic side chains onto metal ions, by chelation of ligand side chains to metal ions, and by microphase separation of glassy or crystalline end blocks in a triblock copolymer. The possibility that some polymers, under certain processing conditions, can become entangled into knots may also explain rubbery elasticity in un-cross-linked macromolecules. The main advantage of these materials is that the cross-links are generally only temporary, which means that such materials frequently exhibit reprocessability. Of course, this temporary nature of the cross-linking can also be a disadvantage since the materials are rubberlike only so long as the aggregates are not broken up by high temperatures or by the presence of diluents or plasticizers, etc.

Typical Applications of Elastomers

Typical non-biological applications of elastomers are tires, gaskets, conveyor belts, drive belts, rubber bands, stretch clothing, hoses, balloons and other inflatable devices, membranes, insulators, and encapsulants. Biological applications include parts of living organisms (skin, arteries, veins, heart and lung tissue, etc.), and various biomedical devices (contact lenses, prostheses, catheters, drug-deliver systems, etc.). It is interesting to note that most of these application require only small deformations. Relatively few take advantage of the very high extensibility that is characteristic of most elastomeric materials.

Specific applications usually require a particular type of elastomer. For example, a hose should have as large as possible mismatch of solubility parameters with the fluid it will be transporting. Thus polar elastomers such as polychloroprene would be best for hoses used with hydrocarbon fluids such as gasoline, jet fuel, greases, oils, lubricants, etc.

Experimental Details for Elastomer Studies

Most of the studies of mechanical properties of elastomers have been carried out based on elongation experiments, because of the simplicity of this type of deformation. The results are typically expressed in terms of the nominal stress $f^* \equiv f/A^*$ which, in the simplest molecular theories, is given by

$$f^* = (vkT/V)(\alpha - \alpha^{-2}) \tag{1}$$

where v/V is the density of network chains, i.e., their number per unit volume V, k is the Boltzmann constant, T is the absolute temperature, and α is the elongation or relative length of the stretched elastomer. Also frequently employed is the modulus, defined by

$$[f^*] \equiv f^* v_2^{1/3}/(\alpha - \alpha^{-2}) = vkT/V \tag{2}$$

Where v_2 is the volume fraction of polymer in the (possibly swollen) elastomer. A smaller number of studies have been carried out using types of deformation other than elongation. For example, biaxial extension or compression, shear, and torsion have been used. Some typical studies of this type are mentioned below.

Swelling is a nonmechanical property, but is also much used to characterize elastomeric materials. It is an unusual deformation in the sense that *volume changes* are of central importance, rather than being negligible. Swelling is a three-dimensional dilation in which the network absorbs solvent, and reaches an equilibrium degree of swelling, at which point the free energy decrease, because of the mixing of the solvent with the network chains, is balanced by the free energy increase that accompanies the stretching of the chains. In this type of experiment, the network is typically placed in an excess of solvent, which it imbibes until the dilational stretching of the chains prevents further absorption. This equilibrium extent of swelling can be interpreted to estimate the degree of cross linking of the network provided the polymer–solvent interactions parameter is known. Conversely, if the degree of cross-linking is known from an independent experiment, then the interaction parameter can be determined. The equilibrium degree of swelling and its dependence on various parameters and conditions provide important tests of theory.

A number of optical and spectroscopic properties are also used. An example is the optical birefringence of a deformed polymer network. This strain-induced birefringence can be used to characterize segmental orientation, both Gaussian and non-Gaussian elasticity, crystallization and other types of chain ordering, and short-range correlation. Other optical and spectroscopic techniques are also important, particularly with regard to segmental orientation. Examples are fluorescence polarization, deuterium NMR, and polarized infrared spectroscopy.

A relatively new technique that is being applied to elastomers is small-angle neutron scattering when for example, deuterated chains are present in a non-deuterated host. One application has been the determination of the degree of randomness of the chain configurations in the undeformed state, an issue that is important to the basic postulates of elasticity theory. Of even greater importance is a determination of the manner in which the dimensions of the chains follow the macroscopic dimensions of the sample, i.e., the degree of "affineness" of the deformation. This relationship between the microscopic and macroscopic levels in an elastomer is one the central problems in the study of rubberlike elasticity.

Some small-angle X-ray scattering techniques have also been applied to elastomers. Examples are the characterization of fillers precipitated into elastomers, and the corresponding incorporation of elastomers into ceramic matrices, in both cases to improve mechanical properties.

Typical Stress–Strain Behavior of Elastomers

A typical stress–strain isotherm obtained on a strip of cross-linked elastomer such as natural rubber is shown schematically in Figure 21.5. The units for the force are generally Newtons, and the curves obtained are usually checked for reversibility. In this type of representation, the area under the curve is frequently of considerable interest since it is proportional to the work of deformation. Its value up to the rupture point is thus a measure of the *toughness* of the material.

The upturn in modulus at high elongations is important because it corresponds to an increase in toughness. This is generally caused by strain-induced crystallization, which results from an increase in the melting point of the network chains. In turn, this is caused by the decreased entropy of the stretched chains and the fact that the melting point is inversely proportional to the entropy of

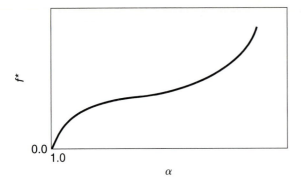

f^*

0.0

1.0

α

Figure 21.5 Stress-elongation curve for an elastomer showing an upturn in modulus at high elongations (α = elongation, f^* = stress).

melting. However, in some cases, the upturns can be caused by the limited extensibility of the chains. These instances are easy to identify, because these upturns will not be diminished by decreasing the amount of crystallization, by an increase in temperature, or by the addition of a diluent. It is in this sense that the stretching "induces" the crystallization of some of the network chains.

The initial part of the stress–strain isotherm shown in Figure 21.5 has the expected form, where f^* approaches linearity as α becomes sufficiently large to make the α^{-2} term in equation (1) negligibly small. The large increase in f^* at high deformation in the particular case of natural rubber is caused largely if not entirely by strain-induced crystallization.

Other deviations from theory are found in the region of moderate deformation identified by examination of the usual plots of modulus against reciprocal elongations. Although equation (2) predicts the modulus to be independent of elongation, it generally decreases significantly with increases in α. The intercepts and slopes of such linear plots are generally called the Mooney-Rivlin constants $2C_1$ and $2C_2$, respectively, in the semi-empirical relationship $[f^*] = 2C_1 + 2C_2\alpha^{-1}$. As described above, the more refined molecular theories of rubberlike elasticity explain this decrease by the gradual increase in the non-affiness of the deformation as the elongation increases toward the phantom limit.

Control of Network Structure

Until recently, relatively little reliable quantitative information existed about the relationship of stress to structure, primarily because of the uncontrolled manner in which elastomeric networks were prepared. Segments close together in space were linked irrespective of their locations along the chain. This resulted in a highly random network structure in which the number and locations of the crosslinks were essentially unknown. Such a structure was shown earlier in Figure 21.4. However, new synthetic techniques are now available for the preparation of "model" polymer networks of known structure. An example is the reaction shown in Figure 21.6, in which hydroxyl-terminated chains of poly(dimethylsiloxane) (PDMS) are end-linked using tetraethyl orthosilicate. Characterization of the un-cross-linked chains with respect to molecular weight, M_n, and molecular-weight

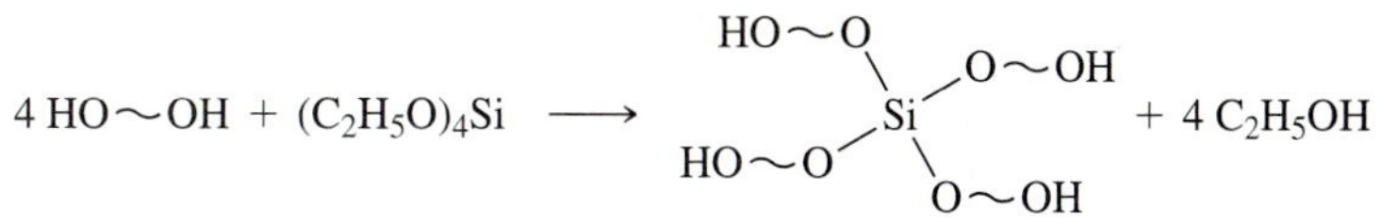

$$4\ HO\sim OH\ +\ (C_2H_5O)_4Si\ \longrightarrow\ \begin{matrix} HO\sim O & & O\sim OH \\ & \diagdown\ \diagup & \\ & Si & \\ & \diagup\ \diagdown & \\ HO\sim O & & O\sim OH \end{matrix}\ +\ 4\ C_2H_5OH$$

Figure 21.6 A typical end-linking scheme for preparing an elastomeric network of known structure.

distribution, followed by taking the reaction to completion give elastomers in which the network chains have the following characteristics. First, the molecular weight, M_c, between cross-links is equal to M_n, and second, cross-links have the functionality of the end-linking agent.

The end-linking reactions described above can also be used to make networks with unusual chain-length distributions. Those that have a bimodal distribution are of particular interest with regard to their ultimate properties, as will be described below.

Networks at Very High Deformations

As already described in Figure 21.5, some (unfilled) networks show a large and rather abrupt increase in modulus at high elongations. This increase is important because it corresponds to a significant toughening of the elastomer. However, the molecular origins of this phenomenon have been the source of considerable controversy. It had been widely attributed to the "limited extensibility" of the network chains, i.e., to an inadequacy in the Gaussian distribution function. The issue has now been resolved by the use of end-linked, non-crystallizable model poly(dimethylsiloxane) networks. The results showed that the anomalous upturn in modulus detected for crystallizable polymers such as natural rubber is due to strain-induced crystallization.

Several so-called "ultimate properties" are of interest. They are the tensile strength, maximum extensibility, and toughness (energy to rupture), and all are affected by strain-induced crystallization. The higher the temperature, the lower the extent of crystallization and, correspondingly, the poorer the ultimate properties.

In the case of such non-crystallizable, unfilled elastomers, the mechanism for *network rupture* has been elucidated by studies of model networks similar to those already described. For example, values of the modulus of bimodal networks formed by end-linking mixtures of very short and relatively long chains as illustrated in Figure 21.7, were used to test the "weakest-link" theory in which rupture was thought to be initiated at the shortest chains (because of their very limited extensibility). However, it was found that increases in the number of very short chains did *not* significantly decrease the ultimate properties. The reason is the very non-affine nature of the deformation at such high elongations. The network simply reapportions the increasing strain among the polymer chains until no further reapportioning is possible. Only after reaching this point does chain scission begin and this leads to rupture of the bulk elastomer. The weakest-link

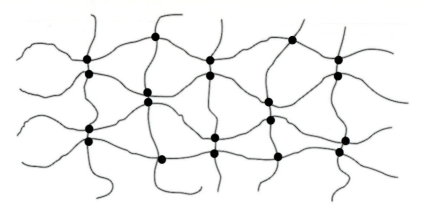

Figure 21.7 Sketch of a network having a *bimodal* distribution of network chain lengths (in other words, a mixture of very short and long chains).

theory implicitly assumes an affine deformation, which leads to the prediction that the elongation at which the modulus increases should be independent of the number of short chains in the network.

In fact, there is a bonus if a multimodal distribution of network chain lengths is formed by end linking a very large number of short chains into a long-chain network. The ultimate properties are then actually improved. Bimodal networks prepared by these end-linking techniques have very good ultimate properties, and there is currently much interest in preparing and characterizing such networks and developing theoretical interpretations for their properties. The types of improvements obtained are shown schematically in Figure 21.8.

The results are represented in such a way that the area under a stress–strain curve corresponds to the energy required to rupture the network. If the network contains all short chains it is brittle, which means that the maximum extensibility is very small. If the network contains all long chains, the extensibility is high but ultimate strength is very low. In neither case is the material a tough elastomer because the areas under the curves are relatively small. The curves in the figure illustrate why the bimodal networks are much improved elastomers. They can have a high ultimate strength without the usual decrease in maximum extensibility.

A number of experiments have been carried out in an attempt to determine if this reinforcing effect in bimodal PDMS networks could possibly be from some intermolecular effect such as strain-induced crystallization. They involved, for example, carrying out the measurements at an elevated temperature or in the swollen state. All of the results strongly argued against the presence of any crystallization or other type of intermolecular ordering. Apparently, the observed increases in modulus are due to the limited chain extensibility of the short chains, with the long chains serving to retard the rupture process. This can be thought of in terms of what executives like to call a "delegation of responsibilities."

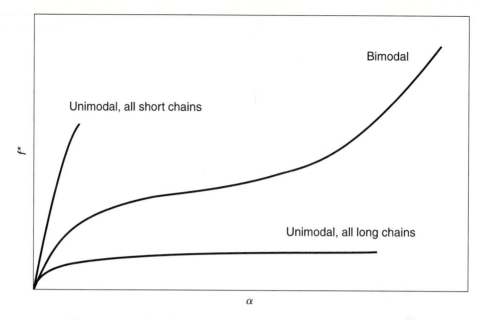

Figure 21.8 Typical plots of nominal stress against elongation for unimodal and bimodal networks obtained by end-linking relatively long chains and very short chains. The area under each curve represents the rupture energy (a measure of the "toughness" of the elastomer). (α = elongation, f^* = stress)

Another advantage to such bimodality when the network can undergo strain-induced crystallization, is that it may provide an additional toughening effect. A decrease in temperature increases the ultimate strength of at least some bimodal networks compared to those of the corresponding unimodal ones. This suggests that bimodality facilitates strain-induced crystallization.

In practical terms, the above results demonstrate that short chains of limited extensibility may be bonded into a long-chain network to improve its toughness. It is also possible to achieve the opposite effect. Thus, bonding a small number of relatively long elastomeric chains into a relatively hard short-chain PDMS thermoset greatly improves its impact resistance.

Other Types of Deformation

Numerous other deformations are of interest, including compression, biaxial extension, shear, and torsion. Unfortunately, some of these deformations are considerably more difficult to study experimentally than simple elongation and have not been investigated as extensively.

Measurements of biaxial extension are of practical importance in packaging applications. This deformation can be imposed by the direct stretching of a sample sheet in two perpendicular directions within its plane, by two independently-variable amounts. In the equi-biaxial case, the deformation is

equivalent to compression. Biaxial extension studies can also be carried out by the inflation of sheets of the elastomer (bubble-blowing). Upturns in the modulus occur at high biaxial extensions, as expected. Experimental results on natural rubber networks in shear have also been of considerable interest. The upturns in modulus in shear were found to be very similar to those obtained in elongation. Very little work has been done on elastomers in torsion. More results are presumably forthcoming, particularly on the unusual bimodal networks and on networks containing some of the unusual in-situ generated fillers described below.

Swelling and Gels

Swelling is a nonmechanical property, but it is widely used to characterize elastomeric materials. Swelling is an unusual deformation because volume changes are of central importance, rather than being negligible as they are in mechanical deformations. Swelling is a three-dimensional dilation in which the network absorbs solvent until it reaches an equilibrium degree of swelling. At this point the free energy decrease caused by the mixing of the solvent with the network chains is balanced by the free energy increase that accompanyies the stretching of the chains. In this type of experiment, the network is placed in an excess of solvent, which it imbibes until the dilational stretching of the chains prevents further absorption.

The equilibrium extent of swelling can be use to obtain structural information about the network, provided the polymer–solvent interaction parameter χ_1 is known. The relevant relationship is called the Flory–Rehner equation and the result for the molecular weight between cross-links is

$$M_c = -[\rho V_1(v_{2m}^{1/3} - v_{2m}/2)]/[ln(1 - v_{2m}) + v_{2m} + \chi_1 v_{2m}^2] \qquad (3)$$

where ρ is the density of the polymer, V_1 is the molar volume of the solvent, and v_{2m} is the (maximum) volume fraction of polymer in the network at swelling equilibrium. Conversely, if the degree of cross-linking is known from an independent experiment, the interaction parameter can be determined. The equilibrium degree of swelling and its dependence on various parameters and conditions provide additional tests of rubberlike elasticity theory. Most studies of networks in swelling equilibrium give values for the cross-link density or related quantities that are in satisfactory agreement with those obtained from mechanical property measurements.

Gels are formed when elastomers absorb liquids, for example in attempts to control oil spills on water. When organic liquids are absorbed, the resultant materials are called organogels. If water is absorbed they are called hydrogels. Hydrogels are of considerable interest in biomedicine because they resemble soft living tissues in their physical properties. They can also be used to immobilize enzymes and mammalian cells.

An important property of some hydrogels is their abrupt collapse (decrease in volume) following a relatively minor change in temperature, pH, solvent com-

position, etc. Although the collapse is quite slow in large, monolithic pieces of gel, it is rapid enough in fibers and films to make the phenomenon useful with regard to the construction of switches, actuators, responsive membranes and related mechanical devices, or artificial muscles.

Filler-Reinforced Elastomers

As mentioned above, one class of multiphased elastomers is capable of undergoing strain-induced crystallization. In this case, the second phase is made up of the crystallites, which provide considerable reinforcement. However, such reinforcement is only temporary, because it may disappear following removal of the strain, addition of a plasticizer, or an increase in temperature. For this reason, many elastomers (particularly those that cannot undergo strain-induced crystallization) are generally compounded with a *permanent* reinforcing filler. The two most important examples are the addition of carbon black to natural rubber and to some synthetic elastomers, and the addition of silica to poly(organosiloxane) elastomers. In fact, the reinforcement of natural rubber and related materials is one of the most important processes in elastomer technology. It leads to an increase in modulus at a given strain, and improvements of various technologically important properties such as tear and abrasion resistance, resilience, extensibility, and tensile strength. There are also disadvantages, including increases in hysteresis (and thus of heat build-up), and compression set (permanent deformation). The mechanism of the reinforcement is not well understood and one of the most important unsolved problems in this area is the nature of the bonding between the filler particles and the polymer chains.

Some understanding might be obtained by the *precipitation* of reinforcing fillers into network structures rather than blending badly agglomerated fillers into the polymers before cross-linking. This has been done for a variety of fillers, for example silica formed by hydrolysis of organosilicates, titania from titanates, alumina from aluminates, etc. A typical, and important, reaction is the acid- or base-catalyzed hydrolysis of tetraethylorthosilicate:

$$Si(OC_2H_5)_4 + 2H_2O \longrightarrow SiO_2 + 4C_2H_5OH \qquad (4)$$

Reactions of this type are widely used by the ceramists in the sol-gel chemical route to high-performance ceramics (see Chapter 9). The advantages for ceramic synthesis are the possibility of using low temperatures, the high purity of the products, the control of ultrastructure (at the nanometer level), and the relative ease of forming ceramic alloys. For elastomer reinforcement, the advantage include avoidance of the difficult, time-consuming, and energy-intensive process of blending agglomerated filler into high molecular-weight and high-viscosity polymers, and the ease of obtaining extremely good dispersions.

The simplest approach for elastomer reinforcement is when some of the organosilicate material is absorbed into the cross-linked network, and the swollen sample is then place in water containing a catalyst, which is typically a volatile base such as ammonia or ethylamine. Hydrolysis to form the silica-like particles pro-

ceeds rapidly at room temperature to yield roughly 50 wt% filler in less then an hour. Impressive levels of reinforcement can be obtained by this *in-situ* technique. The modulus $[f^*]$ generally increases substantially, and some stress–strain isotherms show the upturns at high elongation that are the signature of good reinforcement. As generally occurs in filled elastomers, there can be considerable irreversibility in the stretching behavior, which is thought to be caused by an irrecoverable sliding of the chains over the surfaces of the filler particles.

Elastomer-Modified Ceramics

If the hydrolyses in organosilicate-polymer systems are carried out with larger amounts of the silicate, bicontinuous phases can be obtained (with the silica and polymer phases interpenetrating on another). At still-higher concentrations of the silicate, the silica generated becomes the continuous phase, with the polymer dispersed in it. The result is a polymer-modified ceramic, variously called an "ORMOCER," "CERAMER," or "POLYCERAM." It is important to determine how the polymeric phase, often elastomeric, improves the mechanical properties of the ceramic in which it is dispersed.

Current Problems and Futures Trends

There is a real need for more high-performance elastomers, which are materials that remain elastomeric to very low temperatures and are relatively stable at very high temperatures. Some phosphazene polymers, such as $[NP(OCH_2CF_3) - (OCH_2(CF_2)_xCF_2H]_n$ are in this category (see Chapter 9). These polymers have rather low glass transition temperatures ($-66°C$) in spite of the fact that the skeletal bonds in the phosphorus–nitrogen chains have an unusual type of bonding. Thus, there are a number of interesting problems related to the elastomeric behavior of these unusual semi-inorganic polymers. There is also increasing interest in the study of elastomers that also show mesomorphic behavior.

A particularly challenging problem is the development of a more quantitative molecular understanding of the effects of filler particles, in particular carbon black in natural rubber and silica in siloxane polymers. Such fillers provide impressive reinforcement in elastomers in general, and the way in which they do this is still only poorly understood. A related but even more complex problem involved systems of the type described above in which one or both components are generated *in situ*. An almost unlimited variety of structures and morphologies can be generated by this technique. An understanding of how physical properties such as elastomeric behavior depend on these variables is a challenging but very important problem.

An example of an important objective for future work is the study of *single* polymer chains, particularly with regard to their stress–strain isotherms. Some rather sophisticated equipment is required, such as "optical tweezers," and sensitive force-measuring devices. Most of the effort in this area thus far has involved biopolymers, including examination of mechanically induced transitions between

their various conformations. Although such studies are not necessarily relevant to the many unresolved issues about the interactions among chains within an elastomeric network, they are certainly of interest in their own right.

THERMAL PROPERTIES

Perhaps the main reason why synthetic polymers have not yet replaced metals and ceramics in many applications is the inability of most polymers to maintain their advantageous physical properties at temperatures above 150 to 200°C. Other reasons include their high thermal expansion characteristics, their brittleness at low temperatures, their flammability, and poor chemical stability at high temperatures. One of the main thrusts in polymer research and technology is the drive to use synthetic polymers in ever more thermally hostile environments. Hence, thermal tests are vitally important in any polymer evaluation program.

Thermal Expansion

The thermal expansion coefficient of a polymer is measured by means of a dilatometer (see page 540) or by direct mechanical measurement of a length of a bar of polymer at different temperatures. The thermal expansion coefficients of most synthetic polymers may be as much as ten times greater than those of common metals. Hence, severe thermal distortions can arise when polymers are bonded to metals.

Mechanical Changes

The brittleness, rigidity, and strength of a polymer can be measured as a function of temperature change by the use of stress–strain or impact tests carried out on heated or cooled samples. Thermal softening can be examined with the use of penetration-type measuring devices.

Flammability

The flammability of a polymer is now a critical factor that determines its potential uses. Government flammability regulations increasingly control the types of materials that may be used in textiles, household furnishings, the interior components of civil aircraft, electrical insulation, and thermal insulation. Many organic polymers burn, and questions must be answered about the conditions under which burning can be initiated, the ability of the material to continue burning after ignition, and the generation of toxic fumes from a burning polymer.

Flammability tests in the laboratory sometimes bear little relationship to the possible burning behavior of the polymer in normal use. For example, a common flammability test for textile fabrics requires that a test sample of the fabric should be dried in an oven, cooled in a desiccator, and then ignited with a bunsen burner within a short time of exposure to the atmosphere. Another test of the polymer flammability involves an estimate of the minimum oxygen concentration

needed for sample ignition. Toxicity tests on the products from polymer combustion require an involved laboratory procedure. Combustion products may be detected by vapor-phase chromatography or mass spectrometry. Animal toxicity tests are often needed to evaluate the possible physiological effects on humans following smoke inhalation from burning polymers.

Thermal Decomposition

The chemical decomposition of a polymer at elevated temperatures usually becomes evident in a practical sense by a deterioration of the physical properties. The chemical aspects of thermal degradation were discussed in Chapter 7. Such chemical information can often be used to predict the property changes expected when a polymer is used at high temperatures. For example, depolymerization of the polymer would be expected to result in a loss of strength, increasing brittleness, and perhaps even liquefaction. However, a more meaningful test of technological thermal stability is to examine the actual mechanical properties of the material after it has been heated ("aged") for various periods of time at elevated temperatures. Stress–strain experiments and impact tests may reveal more information about thermal stability than can be estimated from chemical facts alone.

ELECTRICAL TESTS

Polymers are used as electrical insulators, electric wire coatings, as dielectric materials, as electrets, and even as semiconducting or superconducting materials. Thus, an examination of the electrical properties of a polymer forms an important part of the evaluation procedure. The following properties are usually measured.

Resistivity

The resistance of a material to the flow of an electric current can be measured from the potential gradient developed between two electrodes applied to a polymer specimen. The volume resistivity is defined as the measured resistance times the distance between the electrodes, divided by the cross-sectional area. Clearly, a material that is a candidate for use as an electrical insulator should show a high volume and surface resistivity. The following factors are also important.

Dielectric Strength and Arc Resistance

As the voltage is increased across a polymer sample, a point is reached at which catastrophic electrical breakdown occurs. This point is determined by increases in the voltage applied to electrodes placed on opposite faces of a thin sheet of polymer film. Polymers frequently show electrical "fatigue" in which the repeated application of relatively low voltages eventually causes electrical breakdown. Or-

ganic polymers are prone to undergo surface arcing by the formation of carbonized spark pathways.

Dielectric Constant and Power Factor

A knowledge of the dielectric constant of a polymer is important if the material is to be used either as an insulator or as the dielectric material in an electrical condenser. In practice, the dielectric constant is measured from the capacitance of a condenser that contains the polymer as an insulating dielectric compared to the capacitance of the same condenser containing only air as the separation medium. A high dielectric constant is associated with the polarization and polarizability of the electrons that form individual bonds in the polymer matrix. This, in turn, depends on the *orientation* of polar groups in the matrix. The orientation motion of individual segments or component parts of each polymer chain depends on conformational changes and thermal motions. Hence the ability of the polymer groups to switch orientations in phase with an alternating current may be limited, especially at high frequencies. This phase lag results in an absorption of energy by the polymer—the *loss factor*, which is a measure of the energy absorbed per cycle by the polymer from the field. The sine of the phase difference or loss angle is called the *power factor*, and this value multiplied by the dielectric constant yields a value for the loss factor.

A polymer that has a high power factor generates a considerable amount of heat from the alternating electrical field. Hence, such polymers may soften or melt and lose their insulation capability. Polyethylene has a low power factor and thus is suitable for use as an insulator. Another term, the *dissipation factor*, is the tangent of the loss angle. It measures the ratio of the in-phase to out-of-phase power.

ENVIRONMENTAL STABILITY

Most synthetic polymers are more stable than steel, copper, sandstone, or limestone in moist environments that contain dilute aqueous acid or inorganic salts. However, compared to most forms of stone and structural metals, synthetic polymers are quite unstable when exposed to solar radiation or to an ozone-containing urban atmosphere. For example, rubber automobile tires degrade quite rapidly in an atmosphere that contains photochemical smog. Polymeric surface coatings have only a limited outdoor life, especially in regions where exposure to intense sunlight is common. Even some of the most stable polymers, such as low-density polyethylene, crack and degrade after long exposure to the atmosphere, especially in sunlight. Many polymers are also affected adversely by contact with organic solvents, detergents, strong acids, or oxidizing agents. As polymers become more expensive and as the labor costs rise for the reinstallation of degraded polymeric materials, more and more emphasis will probably be placed on monitoring and improving environmental stability.

Weathering Tests

The resistance of a polymer to weathering is often tested experimentally by long-term outdoor exposure of polymer samples to the atmosphere and sunlight followed by evaluation of the changes in mechanical or optical properties. Such tests may take years to complete. Preliminary evaluations can be carried out more conveniently in the laboratory with the use of an "accelerated" weathering unit. Such an apparatus contains high-intensity lamps to simulate the effects of solar radiation. Polymers that are used for aircraft or space applications are often tested by irradiation with mercury vapor lamps that simulate the high ultraviolet-light content of unfiltered sunlight.

Many of the chemical reactions that take place in a polymer during exposure to sunlight are oxidation reactions. These reaction pathways were discussed earlier in Chapter 7.

Solvent Resistance

Some polymers are used throughout their working life in contact with organic fluids or hydrocarbon greases. O-rings in hydraulic systems must withstand the action of such fluids for long periods of time. The testing of polymers for their solvent resistance takes two different forms. First, it is necessary to establish if the polymer swells (or even dissolves) in a particular fluid. Second, it is essential to determine if the polymer cracks or crazes in contact with organic fluids. Many do, and such effects lower the strength and flexibility of these materials.

ADDITIONAL PROPERTIES

Optical Properties

Transparent polymers are used in lenses, prisms, bottles, or as a base for photographic film. The tests conducted on these materials are often designed to determine if yellowing of the polymer takes place over a long period of time, especially after exposure to sunlight. Other optical tests are designed to measure the gloss on the surface of a polymer.

Moisture and Gas Permeability

Films of hydrophilic polymers are often permeable to water vapor, whereas hydrophobic polymers form films that are impervious to water. Such considerations must be taken into account when choosing a polymer for packaging or building applications. Similarly, the use of polymer films to protect packaged items (food, oxidizable chemicals, etc.) must take into account the permeability of the polymer to gases such as oxygen. The water or oxygen permeability of specific polymers is also important when polymers are being considered for use in biomedical devices, for example, in dialysis membranes or heart–lung machines. This topic is considered further in Chapter 24.

STUDY QUESTIONS

1. What types of stress–strain behavior would you expect to be shown by materials that would be suitable for use as **(a)** automobile shock housings; **(b)** ball-point pen housings; **(c)** a plastic basin for the kitchen sink; **(d)** a cushioning material for delicate instruments; **(e)** the outer casing of a football; **(f)** a decorative paperweight; **(g)** an automobile bumper?

2. Why is the actual strength of a polymer nearly always far less than calculated on the basis of the skeletal bond strengths and intermolecular forces?

3. How would you explain the stress–strain behavior of a "soft and weak" polymer (Figure 21.3a) in terms of molecular phenomena? How might the same polymer be modified to change it to the "hard and tough" category?

4. In molecular terms, why should a polymer such as polyisobutylene or silicone rubber be more impact-resistant than, say, polystyrene?

5. The catastrophic failure of metals following multiple, apparently benign flexural motions has led to several engineering disasters in the past. Could polymers fail in the same way? If so, what might the mechanism be that gives rise to the ultimate failure? How would you set up a test procedure to evaluate polymers for possible uses **(a)** as materials for the construction of ships; **(b)** as structural materials in skyscrapers; **(c)** as the structural material in aircraft wings?

6. Suggest ways in which rigid polymers might be bonded to steel in such a way that the sandwich would be unaffected by large temperature fluctuations.

7. A strip of cross-linked natural rubber having an undistorted width of 0.5 cm and a thickness of 0.1 cm gave the following stress–strain data at 25°C:

L (cm)	w (g)
5.0	0.0
6.0	30.0
6.5	41.0
7.0	50.0
7.5	58.0

   a. Plot $[f^*] \equiv f/[A^*(\alpha - \alpha^{-2})]$ vs. α^{-1} to obtain the constants (N mm^{-2}) in the linear equation $[f^*] = 2C_1 + 2C_2\alpha^{-1}$.
   b. Calculate the cross-link density μ/V (mols cm^{-3}) using the phantom theory result $2C_1 = \nu kT/2V$ and the relationship for tetrafunctional cross links $\mu = \nu/2$.
   c. Calculate the molecular weight between cross-links $M_c = \rho/(\nu/V)$, using 0.915 g cm^{-3} for the density ρ of natural rubber.

8. Obtain the elastic equation of state for biaxial extension, in which $\alpha_x = \alpha_y$.

9. The following thermoelastic data have been obtained from constant length measurements on *cis*-1,4-polybutadiene:

$T(°C)$	f, Newtons (N)
50	0.838
60	0.857
70	0.866
80	0.884
90	0.909

Calculate f_e/f and $d\ln <r^2>_o/dT$ at the average temperature 70°C from these data, using $\alpha = 1.25$ and $\beta = 0.68 \times 10^{-3}$ deg^{-1}.

10. A natural rubber network which had been prepared in the undiluted state swells to ten times its original volume when placed into benzene at 25°C. Calculate the degree of cross-linking $(v/2V^*)$ in mols of cross-links per cm^3 of the dry network, using 1.14 cm^3 g^{-1} as the specific volume of benzene, and $\chi = 0.40$ as the polymer-solvent interaction parameter at the specified temperature.

11. Suppose there is an energy change on deforming an elastomer such that the energetic part of the force f_e is 25% of the total force, i.e., $f_e/f = 0.25$. How much would this increase the force $f = f_e + f_S$ beyond that required when only the entropy is involved?

12. Suppose that you are engaged in a search for polymers that conduct electricity. What kind of apparatus would you construct to perform the tests? What problems might you encounter?

SUGGESTIONS FOR FURTHER READING

ALLCOCK, H. R., and DUDLEY, G. K. "Lower Critical Solubility Temperature Study of Alkyl Ether Based Polyphosphazenes." *Macromolecules*, **1996**, *29*, 1313–1319.

ALLCOCK, H. R., and AMBROSIO, A. M. A. "Synthesis and Characterization of pH-Sensitive Poly(organophosphazene) Hydrogels." *Biomaterials*, **1996**, *17*, 2295–2302.

ALLCOCK, H. R., KELLAM, E. C. and MORFORD, R. V. "Gel electrolytes from co-substituted oligoethyleneoxy/trifluoroethoxy linear polyphosphazenes." *Solid State Ionics* **2001**, *143*, 297–308.

BIKERMAN, J. J., "Sliding Friction of Polymers," *J. Macromol. Sci., Rev. Macromol. Chem.*, **1974**, *C11*, 1.

BLOCK, H., "The Nature and Application of Electrical Phenomena in Polymers," *Advan. Polymer Sci.*, **1979**, *33*, 93.

BRINKER, C. J., GIANNELIS, E. P. LAINE, R. M. and SANCHEZ, C. (Eds.), *Better Ceramics Through Chemistry VIII: Hybrid Materials*. Warrendale, PA: Materials Research Society, **1998**, Vol. *519*.

BUCHHOLZ, F. L., "Superabsorbent Polyacrylates," *Trends Polym. Sci.*, **1994**, *2*, 277.

BUCHHOLZ, F. W., and GRAHAM, A. T. (Eds.), *Modern Superabsorbent Polymer Technology*. New York: Wiley & Sons, Inc., **1997**.

CHU, S., "Laser Manipulation of Atoms and Particles," *Science*, **1994**, *253*, 861.

Encyclopedia of Polymer Science and Engineering (H. F. Mark, N. M. Bikales, C. G. Overberger, G. Menges, and J. I. Kroschwitz, eds.). New York: Wiley, **1986**.

ERMAN, B., and MARK, J. E. *Structures and Properties of Rubberlike Networks*. Oxford University Press, New York: **1997**.

FRISCH, K. C., and PATSIS, A. V., *Electrical Properties of Polymers*. Lancaster, Pa.: Technomic Publishing Co., **1972**.

GENT, A. N. (Ed.), *Engineering with Rubber. How to Design Rubber Components*. New York: Hanser Publishers, **1992**.

GRINBERG, V. Y., DUBOVIK, A. S., KUZNETSOV, D. V., GRINBERG, N. V., GROSBERG, A. Y., and TANAKA, T. "Studies of the Thermal Volume Transition of Poly(N-Isopropylacrylamide) Hydrogels by High-Sensitivity Differential Scanning Microcalorimetry. 2. Thermodynamic Functions," *Macromolecules*, **2000**, *33*, 8685.

HAYAKAWA, R., and WADA, Y., "Piezoelectricity and Related Properties of Polymer Films," *Advan. Polymer Sci.*, **1973**, *11*, 1.

HERTZBERG, R. W., "Fatigue Failure in Polymers," *CRC Crit. Rev. Macromol. Sci.*, **1972–73**, *1*, 433.

JANSHOFF, A., NEITZERT, M., OBERDORFER, Y., and FUCHS, H., "Force Spectroscopy of Molecular Systems—Single Molecule Spectroscopy of Polymers and Biomolecules," *Angew. Chem. Int. Ed.*, **2000**, *39*, 3213–3237.

KAELBLE, D. H., "Rheology of Adhesion," *J. Macromol. Sci., Rev. Macromol. Chem.*, **1971**, *C6*, 85.

KAMBOUR, R. P., "A Review of Crazing and Fracture in Thermoplastics," *J. Polymer Sci. (D), Macromol. Rev.*, **1973**, *7*, 1.

KOYAMA, T., and STEINBUCHEL, A. (Eds.), *Biopolymers, Vol. 2: Polyisoprenoids*. New York: Wiley-VCH, **2001**.

LEE, L.-H., *ACS International Symposium on Advances in Polymer Friction and Wear, Los Angeles, 1974*. New York: Plenum Press, **1974**.

LI, H., RIEF, M., OESTERHELT, F., and GAUB, H. E., "Single-Molecule Force Spectroscopy on Xanthan by AFM," *Adv. Mater.*, **1998**, *3*, 316.

LI, Y., and TANAKA, T., "Phase Transitions in Gels," *Annu. Rev. Mater. Sci.*, **1992**, *22*, 243.

LOCKETT, F. J., *Non-linear Viscoelastic Solids*. New York: Academic Press, **1972**.

MANSON, J. A., "Fatigue Failure in Polymers," *CRC Crit. Rev. Macromol. Sci.*, **1972–73**, *1*, 433.

MARK, J. E., "Rubber Elasticity," *J. Chem. Educ.*, **1981**, *58*, 898.

MARK, J. E., "Molecular Aspects of Rubberlike Elasticity," *Acc. Chem. Res.*, **1985**, *18*, 202.

MARK, J. E., and ERMAN, B., *Rubberlike Elasticity. A Molecular Primer*, Wiley-Interscience. New York: **1988**.

MARK, J. E., "The Rubber Elastic State," in *Physical Properties of Polymers*, 2nd ed., J. E. MARK, A. EISENBERG, W. W. GRAESSLEY, L. MANDELKERN, E. T. SAMULSKI, J. L. KOENIG and G. D. WIGNALL (Eds.), Washington, DC: American Chemical Society, **1993**, p. 3.

MARK, J. E., and CALVERT, P. D., "Biomimetic, Hybrid, and In-Situ Composites," *J. Mats. Sci., Part C*, **1994**, *1*, 159.

MARK, J. E., LEE, C. Y-C, and BIANCONI, P. A. (Eds.), *Hybrid Organic-Inorganic Composites*. Washington: American Chemical Society, **1995**, *585*.

MARK, J. E., "The Sol-Gel Route to Inorganic-Organic Composites," *Hetero. Chem. Rev.*, **1996**, *3*, 307–326.

MARK, J. E., "Ceramic-Reinforced Polymers and Polymer-Modified Ceramics," *Polym. Eng. Sci.*, **(1996)**, *36*, 2905.

MARK, J. E., and ERMAN, B., "Elastomers and Rubber-like Elasticity," in Performance of Plastics, W. BROSTOW (Ed.), Cincinnati: Hanser, **1999**.

MARK, J. E., "Thermoset Elastomers," in Applied Polymer Science—21st Century, Craver, C. and Carraher, C. E., Jr. (Eds.), Washington: American Chemical Society, **2000**.

MARK, J. E., "Some Recent Theory, Experiments, and Simulations on Rubberlike Elasticity," *J. Phys. Chem.*, **2003,** *107,* 000.

MARK, J. E., "Some Aspects of Rubberlike Elasticity Useful in Teaching Basic Concepts in Physical Chemistry," *J. Chem. Educ.*, **2002,** *79,* 1437.

MASON, P., *Cauchu. The Weeping Wood*. Australian Broadcasting Commission, Sydney: **1979**.

ORTIZ, C., and HADZIIOANNOU, G., "Entropic Elasticity of Single Polymer Chains of Poly(Methacrylic Acid) Measured by Atomic Force Microscopy," *Macromolecules*, **(1999)**, *32*, 780.

REBENFELD, L., MAKAREWICZ, P. J., WEIGMANN, H. D., and WILKES, G. L., "Interactions between Solvents and Polymers in the Solid State," *J. Macromol. Sci., Rev. Macromol. Chem.*, **(1976)**, *C15*, 279.

SEYMOUR, R. B., *Modern Physics Technology*, Reston, Va.: Reston, **1975**.

SMITH, S. B., CUI, Y., and BUSTAMANTE, C., "Overstretching B-DNA: The Elastic Response of Individual Double-Stranded and Single-Stranded DNA Molecules," *Science*, **(1996)**, *271*, 795.

TANAKA, T., "Gels," *Sci. Am.*, **(1981)**, *244*, 124.

WILLIAMS, D. J., *Polymer Science and Engineering*. Englewood Cliffs, N.J.: Prentice-Hall, **1971**.

WILLIAMS, J. G., *Stress Analysis of Polymers*. London: Longmans, **1973**.

WRASIDLO, W., "Thermal Analysis of Polymers," *Adran. Polymer Sci.*, **(1974)**, *13*, 1.

22

Relationship between Macromolecular Structure and Properties

INTRODUCTION

Why are various polymers different? This question underlies nearly all the thinking and research in polymer chemistry and technology. There are two different but complementary approaches to answering this question. One approach seeks to understand the relationship between the structure of *individual* molecules and the properties of the substances that contain these molecules. The other approach views the properties of a substance in terms of forces *between* macromolecules as they interact in, for example, the solid state. This second viewpoint places a heavy emphasis on the role of domains and domain boundaries, reinforcement of polymers by fillers, solid-state defects, alloying of different macromolecules, and the presence or absence of ordered regions in a solid.

Both approaches are equally important, but their strengths lie in understanding different types of properties. The individual molecule method has more relevance to the properties of polymers in solution or in the melt, or to understanding chemical stability. The materials approach is needed for polymers in the solid state. Both methods are important for an understanding of surface properties.

The connection between molecular structure and properties has been mentioned throughout this book. This chapter contains a summary of the main effects gathered into two categories. First, the subject is considered from the viewpoint of different units in the polymer backbone. Second, the influence of different side groups will be reviewed.

In each case the properties of interest are related to the environment in which each macromolecule finds itself. In solution, the environment will be mainly molecules of the solvent. In the solid state, the environment will consist of other polymer molecules. At the surface, polymer molecules will interact with each other and with molecules of vapor or liquid that are in contact with the surface. Thus, the materials properties generated by the presence of *many* macromolecules in close proximity to each other may change or, in some cases, overwhelm the properties expected for individual macromolecules.

For each macromolecular feature discussed, we consider its influence on polymer solubility, bulk properties such as rigidity or flexibility, surface character, chemical stability, and in a few cases, biological compatibility. In Chapters 23 and 24 we deal with electrical and biomedical properties in some detail. The approach to be followed here is not based on theory but reflects the kind of experience-based intuition used by experts in polymer synthesis. It is the starting point for the design and tailored synthesis of new polymers.

INFLUENCE BY THE MACROMOLECULAR SKELETON

First, it must be stressed that the polymer backbone serves one main purpose— to maintain a *linearity* of molecular connections and thereby to generate materials flexibility, strength, or high viscosity. For most polymers it is the side groups that determine properties such as solubility, crystallinity, surface chemistry, and so on. The backbone structure defines the properties to a greater extent for condensation-type polymers than for polyolefins or vinyl polymers.

The two most important molecular characteristics of the skeleton are (1) its *flexibility*, which depends on the ease of torsion of backbone bonds and, for some elements, on the degree to which backbone bond angles can be widened or narrowed in response to mechanical stress or thermal activation, and (2) the *stability* of the skeletal bonds to chemical reagents and elevated temperatures. In the following paragraphs we review ways in which different skeletal structures affect these properties.

The Aliphatic Carbon–Carbon Bond

The carbon–carbon single bond is a unit that confers appreciable flexibility to a polymer chain. This flexibility is a consequence of the inherently low barrier to torsion of the carbon–carbon bond, and it becomes evident in polymers such as polyethylene, polyisobutylene, poly(methylvinyl ether), or *cis*-polypentenamer. The main weakness of the aliphatic carbon–carbon chain unit is its sensitivity to thermooxidative cleavage. This property is a consequence of the ability of the bond to undergo free-radical cleavage reactions, and it is exacerbated when the side groups are hydrogen atoms. Other side groups such as fluorine or aryl units may protect the chain against degradation.

The Aliphatic Carbon–Carbon Double Bond

The aliphatic carbon–carbon double bond has a higher barrier to internal rotation than does a carbon–carbon single bond. Hence, double bonds in a hydrocarbon chain might be expected to generate chain stiffness and high glass transition temperatures. This is not always the case. *cis*-Polypentenamer **(1)** has isolated double bonds in the chain but one of the lowest glass transition temperatures known for hydrocarbon polymers ($-114°C$). Natural rubber [poly(*cis*-1,4-isoprene] **(2)** ($T_g = -70°C$) and its *trans* analogue (gutta percha) ($T_g = -68°C$) are other examples. In fact, the presence of only one side group at the olefinic carbon sites may lower intramolecular steric hindrance sufficiently to enhance the torsional mobility of the adjacent single bonds.

$$\left[CH=CH-CH_2-CH_2-CH_2 \right]_n \qquad \left[CH_2-\underset{}{\overset{CH_3\ H}{C}}=CH-CH_2 \right]_n \qquad \left[\bigwedge\!\!\bigwedge\!\!\bigwedge \right]_n$$

$$\textbf{1} \qquad\qquad\qquad\qquad \textbf{2} \qquad\qquad\qquad\qquad \textbf{3}$$

However, *alternating* single and double bonds, as in polyacetylene **(3)** confer considerable skeletal rigidity because maximum π-orbital overlap can be achieved only if the chain is planar. Conjugated arrays of this type generate color and electrical conductivity (see Chapter 23).

Double bonds in a hydrocarbon skeleton are sites that are sensitive to chemical attack, especially by ozone or by oxygen in the presence of visible or ultraviolet radiation. Hence, polymers of this type are prone to slow oxidation in the atmosphere.

Aromatic Rings as Skeletal Units

Aromatic rings in a polymer skeleton confer rigidity and extended chain (rigid rod) character to the macromolecule. Polymers such as aromatic polyamides, polyesters, or polyarylenes can also show main-chain liquid crystallinity. In other words, even in the liquid state, the polymer molecules can become oriented with respect to each other (like logs in a river). In the solid state such orientation generates high strength. Aromatic rings are generally resistant to thermooxidative attack. Hence, many polymers designed for high temperature use contain aromatic units as an integral part of the chain. Polymers that contain strings of conjugated aromatic rings (or even aryl rings separated by heteroelements such as sulfur) can function as electronic conductors when doped (Chapter 23).

Aromatic Ladder Structures

As discussed earlier (in Chapter 2), the ultimate in macromolecular rigidity, rigid-rod character, and thermooxidative stability is found in double-strand polymers that contain aromatic rings. However, these properties are achieved at the expense of processability. Thus, the design of high-temperature polymers requires a subtle balancing of rigid segments, flexible chain segments, and solubilizing units. Species of this type are more dependent on the chain structure and less dependent on the side groups for their properties than most of the other classes of polymers.

The Etheric Carbon–Oxygen Bond

Oxygen atoms in a hydrocarbon chain (**4**) usually behave as "swivel groups," allowing skeletal torsional mobility and (unless overpowered by solid-state influences) materials flexibility. The flexibility associated with a skeletal oxygen atom may be a consequence of the absence of side groups attached to that atom and the resultant diminution of intramolecular interactions in the vicinity of that site. (However, each oxygen atom bears two lone-pair orbitals and these will exert a small but significant influence.) The bond angles at oxygen may vary in response to conformational changes and this effect may also add to the flexibility of the chain.

$$-\overset{|}{\underset{|}{C}}-O-\overset{|}{\underset{|}{C}}-$$

4

The chemical stability of the etheric carbon–oxygen bond in polymers depends on the nature of the nearby skeletal units. Ether linkages are generally stable to hydrolysis and to thermooxidation. However, polyaldehydes depolymerize readily at moderate temperatures, and this reflects an instability of the $C-O-C$ linkage, especially in the presence of Lewis acids. Poly(oxymethylene) is unstable to strong mineral acids. Chemical stability is generated if the oxygen atom is flanked by aryl groups.

If the nearby side groups are sufficiently small (hydrogen) to allow an interaction between the skeletal oxygen atoms and water, the polymers are hydrophilic or water soluble. Poly(ethylene oxides), $(OCH_2CH_2)_n$ are soluble in water. Poly(oxymethylene) is only sparingly soluble in most solvents (and is insoluble in water) because of its high crystallinity. For this reason, it is often used in applications that require solvent resistance.

The Carbon–Oxygen Bond in Polyesters

The $C-O-C$ bond in polyesters (**5**) again provides chain flexibility. However, the properties of many polyesters are determined by their solid-state structure, particularly by crystallite formation. The chemical stability depends on the other chain units present. Nearby aromatic groups provide some steric and hydrophobic protection against hydrolysis, but lacking this protection the linkage may be cleaved by acids or bases in solution or in the melt. Poly(glycolic acid) and poly(lactic acid) and their copolymers are used in biomedicine *because* the ester linkage is hydrolyzed at pH 7.5 at body temperature. Solid polyesters with high ratios of hydrocarbon to ester units are usually chemically stable because of surface effects and the protection of chain units within crystalline domains.

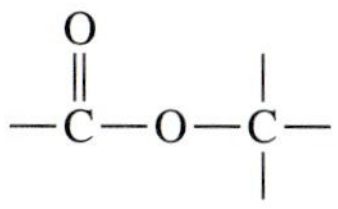

$$\overset{\displaystyle O}{\underset{\displaystyle |}{\overset{\displaystyle \|}{-C}}}-O-\overset{\displaystyle |}{\underset{\displaystyle |}{C}}-$$

5

The Anhydride Linkage

As discussed in Chapter 24, the anhydride linkage **(6)** is inherently hydrolytically unstable. Thus polyanhydrides are not used for general applications but are employed in biomedicine. However, the sensitivity to hydrolysis is greater if aliphatic rather than aromatic units are present in the chain. The hydrolytic behavior also depends on the ratio of hydrocarbon to anhydride units in the structure.

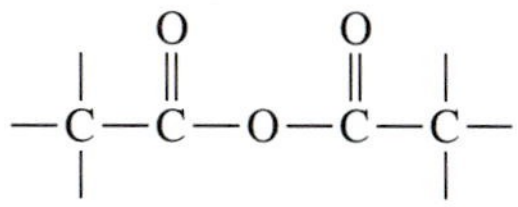

$$-\overset{|}{\underset{|}{C}}-\overset{\overset{\displaystyle O}{\|}}{C}-O-\overset{\overset{\displaystyle O}{\|}}{C}-\overset{|}{\underset{|}{C}}-$$

6

Schiff Base Linkages

Polymers prepared by the reactions of aldehydes with amines contain the Schiff base linkage (7). This is also hydrolytically unstable, although aromatic Schiff base polymers are reported to be stable under normal conditions and to form main-chain liquid crystalline polymers.

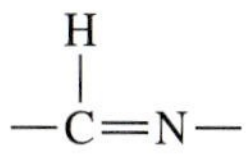

$$-\overset{\overset{\displaystyle H}{|}}{C}=N-$$

7

The Amide Linkage

It is generally believed that the amide linkage **(8)** is a "chain-stiffening" unit because of the partial double-bond character of the N—C bond and the opportunities for internal and external hydrogen bonding that exist in polyamides. However, as in other skeletal systems, the properties of a specific polyamide will depend on the ratio of flexible methylene units or inflexible aryl groups to amide linkages in the chain.

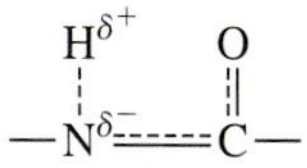

$$-\overset{\overset{\displaystyle H^{\delta+}}{|}}{N^{\delta-}}\text{------}\overset{\overset{\displaystyle O}{\|}}{C}-$$

8

The amide linkage is moderately sensitive to hydrolysis, but the behavior of a synthetic polymer under hydrolytic conditions will depend more on materials and surface effects than on the chemistry of the linkage itself. For example, many commercial polyamides are highly crystalline, and repeating units within each crystalline region are inaccessible to water or other reagents. The maximum sensitivity of the amide linkage to hydrolysis is seen in globular proteins, where individual repeating units are accessible to water, acids, or enzymes.

The Urethane Linkage

Polyurethanes contain the linkage group shown in **9**, flanked by oligo-ether or oligo-ester skeletal segments. In theory, the urethane linkages in these polymers should be hydrolytically sensitive sites. However, the chemical properties are dominated by the other components of the chain, and hydrolytic chain cleavage is encountered only after prolonged exposure of the polymer to a hydrolytic environment. The urethane linkage is considered to be a source of molecular flexibility, and this property underlies the use of many polyurethanes as elastomers. However, the flanking segments have an equal or greater influence on the materials properties since ether, ester, aliphatic hydrocarbon, or aromatic ring units impose their own characteristics on the polymer.

$$
\begin{array}{ccc}
& \text{H} & \text{O} \\
& | & || \\
\text{(C)} - \text{N} - \text{C} - \text{O} - \text{(C)}
\end{array}
$$

9

The Siloxane Linkage

The silicon–oxygen bond (**10**) has one of the highest torsional mobilities in any polymer backbone. Hence, provided that the side groups are small enough to allow that flexibility to be retained, the glass transition temperatures will be very low [-125 to $-130°C$ for poly(dimethylsiloxane)] and will favor the emergence of elastomeric character. The flexibility provided by the silicon-oxygen bond probably owes more to the oxygen atom than to the silicon. The Si—O—Si bond angle can vary over a very wide range (from $120°$ to at least $140°$) in response to external forces, and this undoubtedly reduces any intramolecular restrictions to bond torsion that might otherwise exist.

$$
\begin{array}{c}
| \\
- \text{Si} - \text{O} - \\
|
\end{array}
$$

10

The siloxane bond is apparently more stable to thermooxidative attack than is the aliphatic carbon–oxygen bond. However, in solution the Si—O bond is quite sensitive to a variety of reagents, including acids and bases. Hence the widely respected chemical stability of polysiloxanes may be more a consequence of surface effects and hydrophobicity than of intrinsic chemical inertness.

Polysiloxanes have an unusual ability to absorb and transmit oxygen: hence, they are of interest as membrane materials. This property appears to be connected with the presence of the Si—O skeletal bonds in a way that is not fully understood.

The Phosphazene Linkage

Contrary to what might be expected from the presence of a double bond in each repeating unit (**11**), the phosphazene backbone has one of the highest skeletal flexibilities known. This molecular flexibility is attributed to (1) a skeletal bond length near 1.6 Å, which is slightly longer than the C—C value of 1.54 Å, a factor that may reduce short-range intramolecular interactions; (2) a variable bond angle at nitrogen (120 to 135°); and (3) a skeletal electronic structure that imposes virtually no π-bond barrier to torsion of the skeletal bonds (see Chapter 9). Thus, the flexibility or stiffness of the chains in these polymers depends more on the structure of the side groups than on the skeleton.

$$-\overset{\displaystyle |}{\underset{\displaystyle |}{P}}=N-$$

11

The phosphorus–nitrogen skeleton has a high chemical stability—high enough to withstand a wide variety of macromolecular side-group substitution reactions. However, any reaction that places a hydroxyl side group on phosphorus sensitizes the system to eventual chain cleavage by the process shown in equation (1).

$$-\overset{\displaystyle OH}{\underset{\displaystyle R}{P}}=N- \longrightarrow -\overset{\displaystyle O}{\underset{\displaystyle R}{P}}-\overset{\displaystyle H}{N}- \overset{H_2O}{\longrightarrow} -\overset{\displaystyle O}{\underset{\displaystyle R}{P}}-OH + H_2N- \tag{1}$$

The phosphazene backbone has a high photolytic and thermooxidative stability and appears to be especially resistant to ozone and a variety of free-radical reagents. The skeletal nitrogen atoms in polyphosphazenes are basic (especially if electron-supplying side groups are present). This means that the nitrogen atoms may be protonated and can complex to metal ions.

Sulfur–Sulfur, Sulfur–Carbon, and Sulfur–Nitrogen Linkages

The sulfur–sulfur bond has a high degree of inherent flexibility, but it is thermally labile and sensitive to thermooxidative attack. Sulfur–carbon bonds in sulfone structures (**12**) are resistant to oxidation because sulfur is in a high oxidative state. Hence, this linkage is used in some high-temperature polymer systems to connect aromatic rings. The sulfur–nitrogen skeleton **(13)** is of interest mainly because of its high electrical conductivity. However, the linkage is oxygen sensitive, and as oxidation proceeds, the conductivity declines.

$$-C-\overset{\overset{\displaystyle O}{\|}}{\underset{\underset{\displaystyle O}{\|}}{S}}-C- \qquad -S{=}N-$$

12   **13**

INFLUENCE OF DIFFERENT SIDE GROUPS

The side groups attached to a polymer chain can have a more profound effect on the polymer properties than the skeleton itself. The side groups may protect the skeleton against chain cleavage reactions, they are responsible for most of the solubility properties of polymers, and the steric and polar interactions between side groups on the same chain or different chains largely determine the glass transition temperature, crystallinity, and surface properties of the material. Even when the side groups are small (as in the case of polytetrafluoroethylene, for example), it is the side groups that define the properties.

Thus, molecular design in polymer chemistry involves the choice of a backbone that will provide a certain set of "default" properties (strength, flexibility, oxidation stability, chain length and molecular-weight distribution, and synthetic accessibility), while the choice of side groups will control nearly all the other properties.

Hydrogen as a Side Group

Side-group hydrogen atoms provide properties that depend on the skeletal atom to which that element is attached. In hydrocarbon polymers the behavior of C—H bonds dominates the macromolecular properties. Thus, hydrocarbon polymers are hydrophobic, soluble in organic (especially hydrocarbon) media, sensitive to free-radical attack (aliphatic units), but relatively insensitive to a wide variety of chemicals, especially when the bulk polymer is protected by surface effects. The C—H bond imposes only a minimal barrier to torsion of the C—C backbone bonds (because of the small size of the hydrogen atoms and the low polarity of C—H bonds). Hence, polymers with hydrogen atoms as side groups often display the lowest glass transition temperatures for a particular backbone. However, the small size of the hydrogen atom may favor ordered packing of the chains, and this can lead to high degrees of crystallinity.

These attributes of hydrogen atoms do not apply if the atom is connected to a skeletal element other than carbon. As we have seen, hydrogen attached to nitrogen introduces the likelihood of hydrogen bonding. The Si—H bond is highly reactive, as is the P—H bond.

Alkyl Groups as Side Units

The "hydrocarbon" character of alkyl groups (hydrophobicity, chemical unreactivity, etc.) is nearly always retained when such groups are connected to a macromolecular backbone.

Methyl side groups are present in many polymers. Examples are shown in **14** to **17.** In polypropylene **(14)** and polyisobutylene **(15)** the methyl groups reinforce the hydrocarbon character of the skeleton, providing hydrophobicity and solubility in organic solvents. In polypropylene the methyl group is sufficiently bulky that the tacticity of the polymer has a powerful effect on the properties. Isotactic polypropylene is a crystalline thermoplastic. The atactic modification is a gum. Perhaps surprisingly, the two geminal methyl groups in polyisobutylene *enhance* the materials flexibility (the polymer flows slowly under the influence of gravity). The solid-state explanation of this is that the alternation of two methyl side groups and two hydrogens along the chain creates "free volume" that facilitates molecular reorientation. The methyl side groups in poly(dimethylsiloxane) **(16)** also generate hydrophobicity and allow macromolecular flexibility. As discussed in Chapter 24, many of the biomedical uses of polysiloxanes depend on this combination of properties.

$$\begin{array}{cccc}
CH_3 & CH_3 & CH_3 & CH_3 \\
| & | & | & | \\
-C-CH_2- & -C-CH_2- & -Si-O- & -P=N- \\
| & | & | & | \\
H & CH_3 & CH_3 & CH_3 \\
\mathbf{14} & \mathbf{15} & \mathbf{16} & \mathbf{17}
\end{array}$$

However, poly(dimethylphosphazene) **(17)** is hydrophilic. This anomaly may be due to a strong influence by the backbone. Perhaps the methyl groups are not sufficiently large to shield the skeletal nitrogen atoms against protonation or hydrogen bonding to water molecules. Polymer **17** is a microcrystalline material rather than an elastomer, perhaps a consequence of the influence by skeletal dipolar forces.

Ethyl-, propyl-, butyl-, or higher alkyl side chains generally maintain or enhance the hydrocarbon character of a polymer. The conformational disorder introduced by oligo-methylene chains may discourage crystallization, although examples exist where longer-chain side units undergo polyethylene-type crystallization. The presence of olefinic unsaturation in a side group raises the reactivity of a polymer, especially to free-radical-type reactions that form the basis of crosslinking processes.

Aryl Side Groups

Phenyl rings are hydrophobic, rigid, and relatively bulky. As side groups, they impose stiffness and steric hindrance on to any chain to which they are attached. For example, the glass transition temperature of polystyrene **(18)** of 100°C is roughly 100°C higher than that of polypropylene **(19),** and 200°C higher than that of polyethylene **(20)**. The larger the number of aromatic side rings present, the more pronounced is the chain stiffening effect.

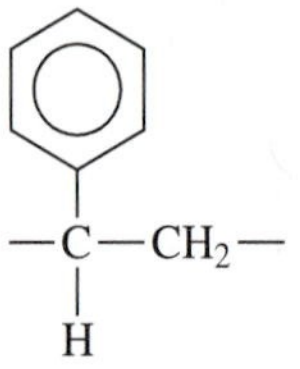

18	19	20

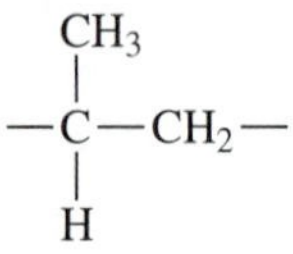

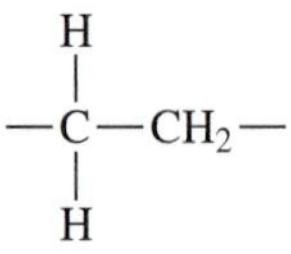

Aromatic rings absorb energy strongly in the near-ultraviolet region of the spectrum. Thus, the presence of such side groups provides a mechanism for photolytic reactions that eventually lead to discoloration and decomposition of a polymer.

Mesogenic Side Groups

Rigid, flat units, such as biphenyl **(21)**, aromatic azo **(22)**, or cholesteryl **(23)** units in a side group, may generate "side-chain liquid crystallinity." Side groups of this type have the capacity to form ordered arrays in the melt or in solution. The side-group ordering may be in the form of stacking or collinear orientation of the mesogenic units (Figure 22.1). The result is a polymer that is microcrystalline in the normal solid state but liquid crystalline over a specific temperature range in the melt or in solution.

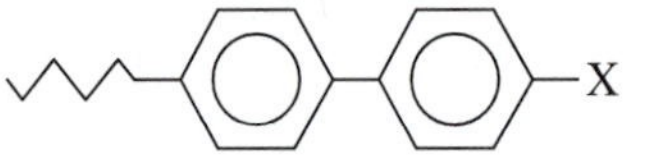

21	22

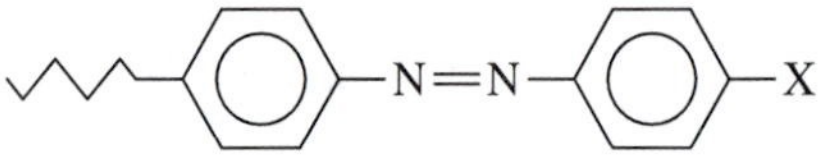

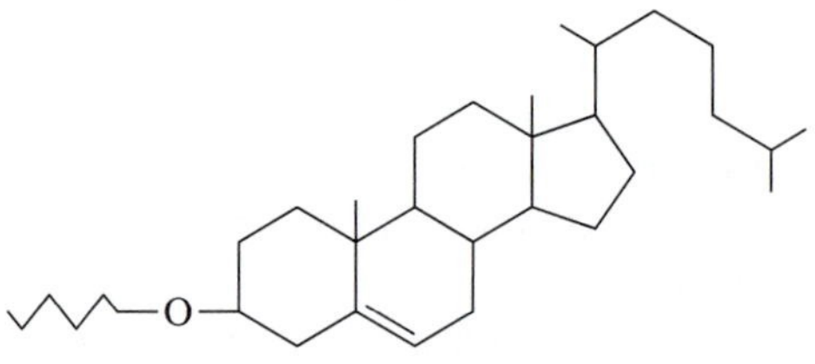

23

The existence of side-chain liquid crystallinity requires the presence of a special combination of molecular features. First, the main polymer chain must be highly flexible to ensure that the conformational preference of the chain does not unduly influence the interactions of the side groups. Polysiloxane or polyphosphazene chains have the necessary flexibility. So do some polyacrylate chains. Second, the rigid units (the mesogens) must be connected to the skeleton by flexible spacer groups. The length and flexibility of the spacer are critical for ensur-

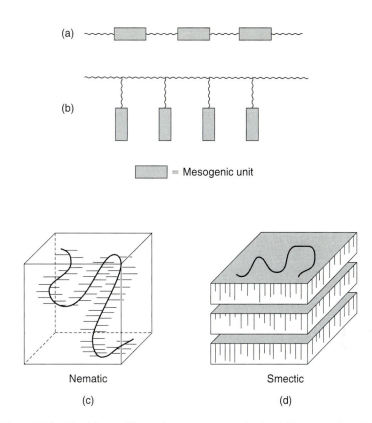

Figure 22.1 Liquid crystalline polymers may contain the rigid mesogenic units in the main chain (a) or as part of the side chains (b). In side-chain liquid crystalline systems, the mesogenic units may become organized in a number of different ways. The nematic (c) and smectic (d) arrangements are two possibilities. (Drawing by M. S. Connolly.)

ing that the motions of the skeleton are decoupled from those of the mesogenic units. Oligo-ethyleneoxy spacer units are often employed. The packing forces between mesogenic units are relatively weak, and the collinear orientation is easily disrupted if the mesogens respond to every movement of the skeleton. Finally, the nature of the terminal unit (X in structures **21** and **22**) exerts a surprisingly large effect. Polar groups, such as OCH_3, CN, or NO_2, favor the appearance of liquid crystallinity.

Fluorine as a Side-Group Unit

The presence of fluorine in the side groups of a polymer has a profound effect on nearly all the properties of that material. This effect is so striking that "fluoropolymers" are often treated as a separate, remarkable class of polymers with properties unmatched in any other system.

Fluorine confers extreme hydrophobicity and water insolubility onto a polymer. It raises the thermal and oxidative stability, and confers solvent, fuel, and oil resistance. In some instances, the element generates interchain interactions that are so strong that the polymer will not dissolve in any solvent. It is the C—F bond that confers these properties, rather than the element itself. Fluorine attached to other elements (e.g., in Si—F or P—F units) is highly reactive and sensitizes the polymer to hydrolytic attack.

Examples of fluoropolymers include poly(tetrafluoroethylene) (Teflon®) **(24),** poly(vinylidene fluoride) **(25),** a variety of fluoro- and fluorochlorocarbon elastomers, fluoroalkylsiloxane elastomers **(26),** and fluoroalkyloxyphosphazene elastomers **(27).** All are solvent-resistant, oxidatively, thermally, and hydrolytically stable materials. They are biocompatible (see Chapter 24) mainly because of their hydrophobicity. Poly(tetrafluoroethylene) is highly crystalline and is used in some applications because of its surface lubricity—again a consequence of the presence of C—F bonds. Poly(vinylidene fluoride) is used as an electret polymer—a material that can be dipole-aligned in an electric field. Electrets are widely used in microphones, speakers, and sound detection devices that make use of the piezoelectric character of this polymer.

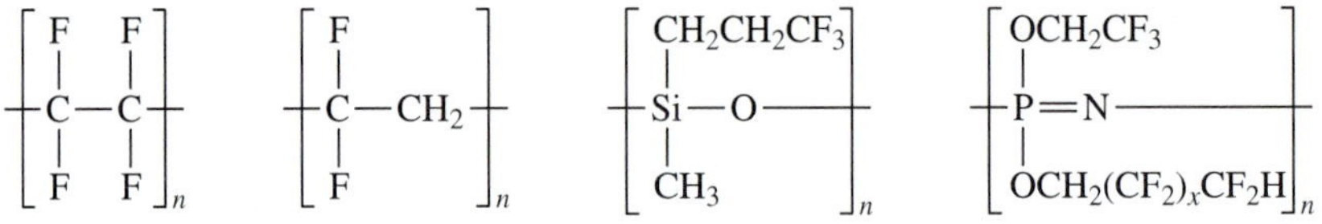

$$\begin{array}{cccc} \text{24} & \text{25} & \text{26} & \text{27} \end{array}$$

Chlorine

Chlorine attached to a hydrocarbon polymer chain or to an alkyl or aryl side group is relatively resistant to chemical attack, particularly if the polymer is in the solid state. Thus, poly(vinyl chloride) **(28)** and poly(vinylidine chloride) **(29)** are stable polymers that are widely used in technology. However, benzyl chloride units $(C_6H_4CH_2Cl)$ in a side chain are reactive to a variety of reagents. One or more chlorine atoms per repeating unit, connected directly to a carbon skeleton, increase chain stiffness.

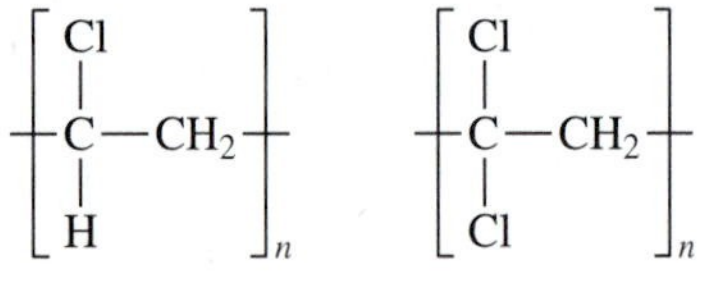

$$\begin{array}{cc} \text{28} & \text{29} \end{array}$$

Carbon–chlorine bonds in a polymer sensitize the system to photochemical reactions in sunlight. Although chlorine confers some degree of fire resistance to a polymer, the decomposition products at high temperatures (including hydrogen chloride) are toxic. The introduction of chlorine into a hydrocarbon polymer usually reduces the solubility in hydrocarbon solvents, fuels, and so on, and this is important for many applications. Polymers that contain chlorine are often incompatible with mammalian cells and are thus unsuitable for use as biomaterials.

As with fluorine, the stability of chlorine-containing polymers applies only if the halogen is linked to carbon. The Si—Cl and P—Cl bonds, for example, are highly reactive to a wide range of nucleophiles (see Chapter 9).

The Cyano Side Group

The cyano unit is a polar, hydrophilic group that imposes a special set of properties on carbon backbone polymers. The classical example is polyacrylonitrile (30). The cyano group dramatically changes the solubility behavior of a hydrocarbon polymer reducing the solubility in nonpolar organic media and increasing the solubility in solvents such as dimethylformamide, dimethyl sulfoxide, or dimethylacetamide. Thus, copolymers that contain the acrylonitrile monomer unit are used in solvent- and oil-resistant elastomers. The glass transition temperature of polyacrylonitrile is 85°C, which suggests that the polarity of the side group imposes restrictions on chain mobility, probably through dipole–dipole interactions. The polymer is also crystalline for similar reasons.

$$
\begin{bmatrix} \begin{array}{c} C\equiv N \\ | \\ C-CH_2 \\ | \\ H \end{array} \end{bmatrix}_n \qquad \begin{bmatrix} \begin{array}{c} C\equiv N \\ | \\ C-CH_2 \\ | \\ C\equiv N \end{array} \end{bmatrix}_n
$$

30   31

The thermal stability of cyano polymers is only modest. At elevated temperatures they darken and undergo internal addition reactions (see Chapter 9). However, polyacrylonitrile has excellent stability to sunlight, chemical reagents, and microorganisms. Poly(vinylidene cyanide) (31) is hydrolytically unstable.

Hydroxyl Groups

Hydroxy groups are hydrophilic and water-solubilizing substituents. They form strong hydrogen bonds. Poly(vinyl alcohol) (32), although hydrophilic, is not soluble in water at room temperature. However it dissolves in hot water: These conditions presumably disrupt the intermolecular hydrogen bonds. Polymers that contain 88% vinyl alcohol and 12% vinyl acetate repeating units are soluble in water at 25°C, apparently because the acetate side groups disrupt the intermolecular associations. As the hydroxyl-to-acetate ratio falls below 70% hydroxyl, the polymer becomes insoluble in water, but still soluble in ethanol–water media. The effect of the hydroxyl group in 32 on skeletal mobility is indicated by the glass transition temperature of 70 to 85°C. Intramolecular hydrogen bonding probably serves to restrict skeletal torsional motions.

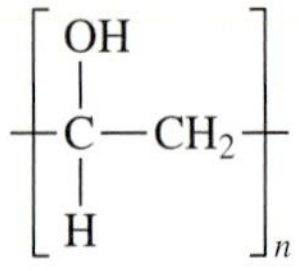

$$
\begin{bmatrix} \overset{\displaystyle OH}{\underset{\displaystyle H}{\overset{|}{\underset{|}{C}}} - CH_2} \end{bmatrix}_n
$$

32

Similar influences by side hydroxyl group can be seen in polysaccharides, proteins, and in a wide variety of synthetic polymers. For example, polyphosphazenes with glyceryl or glucosyl side groups are soluble in water. Hydroxyl groups at the terminus of an aliphatic or aromatic side group also generate hydrophilic or water-solubilizing properties. However, the greater the ratio of hydrocarbon units to hydroxyl groups, the lower will be the tendency for the polymer to dissolve in aqueous media.

Amide Side Groups

The amido side group in polyacrylamide **(33)** illustrates the characteristics of this unit. Polyacrylamide is very soluble in water (215 g per 100 mL at 30°C), presumably because both the NH_2 and $C=O$ groups form hydrogen bonds with water molecules. The solid polymer is a brittle glass (T_g = 153 to 204°), with molecular reorientation probably restricted by hydrogen bonding. The solid is very hygroscopic. Alkaline hydrolysis to give carboxylate groups proceeds rapidly, but acidic hydrolysis is slow. Acrylamide monomer is a severe neurotoxin, but the polymer appears to be less toxic.

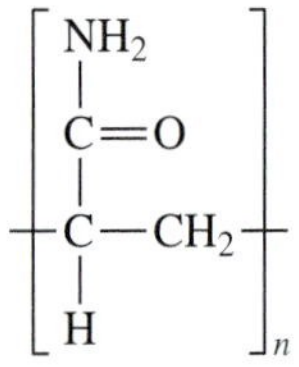

$$
\begin{bmatrix} NH_2 \\ | \\ C{=}O \\ | \\ \overset{|}{\underset{|}{C}} - CH_2 \\ H \end{bmatrix}_n
$$

33

Alkyl Ether Side Group

Poly(methylvinyl ether) **(34)** illustrates some of the characteristics of ether-type side groups. This polymer is soluble in water at 25°C, but the solubility *decreases* as the temperature is raised. The solubility and solid-state properties depend heavily on the tacticity of the polymer and on the terminal alkyl group. For example, atactic poly(methylvinyl ether) is a viscous liquid at room temperature, with a T_g at −31°C. The ethyl ether analogue has a T_g at −42°C, the *n*-propyl derivative at −49°C, and the *n*-butyl derivative at −55°C. These values illustrate how the inherently high flexibility of a hydrocarbon chain can become manifest if the side group is sufficiently flexible to allow facile internal reorientational motions. Stereoregular polymers of this type are crystalline, waxy solids. Water solubility is maintained if side-chain oligo ethyleneoxy units are present, for example, if side groups such as $-OCH_2CH_2OCH_3$ or $-OCH_2CH_2OCH_2CH_2OCH_3$ are attached to a polyphosphazene.

$$\left[\begin{array}{c} OCH_3 \\ | \\ -C-CH_2- \\ | \\ H \end{array}\right]_n$$

34

Ester Side Groups

Two types of polymers with ester side groups are of interest—esters of polyacrylic acid **(35)** and polymethacrylic acid **(36)** or esters of poly(vinyl alcohol) **(37)**. The ester groups in all three polymer types favor solubility in organic solvents. However, species of type **35** and **36** have relatively polar side groups if the group R is small (methyl or ethyl). Under these circumstances, the polymers are soluble in polar solvents such as ketones, esters, or ether-alcohols such as diglyme. As the group R becomes longer and more hydrophobic, the solubility becomes greater in nonpolar solvents, such as aliphatic or aromatic hydrocarbons. Poly(vinyl acetate) **(37)** is soluble in aromatic, ketonic, ester, or chlorinated solvents, but is insoluble in anhydrous ethanol or aliphatic hydrocarbons. All three polymers are insoluble in water.

$$\left[\begin{array}{c} O \\ \| \\ C-OR \\ | \\ -C-CH_2- \\ | \\ H \end{array}\right]_n \qquad \left[\begin{array}{c} O \\ \| \\ C-OR \\ | \\ -C-CH_2- \\ | \\ CH_3 \end{array}\right]_n \qquad \left[\begin{array}{c} O \\ \| \\ O-C-R \\ | \\ -C-CH_2- \\ | \\ H \end{array}\right]_n$$

35   **36**   **37**

The glass transition temperatures of carbon-backbone polymers with ester side groups vary widely with the tacticity and the nature of the R groups. Atactic poly(methyl methacrylate) has a T_g of 105°C. Poly(vinyl acetate) has a T_g of 28 to 31°C.

The stability of the polymers to hydrolysis declines in the order **36** > **35** > **37**. Methacrylic ester polymers are more stable to acidic and basic hydrolysis than acrylic ester polymers. Both are more resistant to hydrolysis than is poly(vinyl acetate). Indeed, hydrolysis of this latter polymer provides the main method for the preparation of poly(vinyl alcohol). The three different classes behave differently on thermolysis. Acrylate ester and vinyl ester polymers decompose by fragmentation. Methacrylic ester polymers depolymerize to the monomer. All three classes of polymers appear to be nontoxic.

The Carboxylic Acid Side Group

Poly(acrylic acid) **(38)** and poly(methacrylic acid) **(39)** are two prototype polymers that demonstrate the influence of a carboxylic acid unit on the properties of a polymer. Both polymers are hygroscopic in the solid state and are readily soluble in water, methanol, or ethanol. However, they are insoluble in aromatic or aliphatic hydrocarbons. The hydrophilic character is a direct consequence of the ability of a carboxylic acid unit to become highly solvated in water or alcohol. Solubility in aqueous base is also favored by the formation of ionic units of the type $-COO^-Na^+$, although the presence of divalent cations may generate ionic crosslinks and make the polymer insoluble. Polymers that bear carboxylic acid side groups are often insoluble in acidic aqueous media. These characteristics are maintained even when aromatic or aliphatic units separate the carboxylic acid group from the chain. However, the hydrophilic character diminishes as larger and larger hydrocarbon spacer units are introduced. Sulfonic acid (SO_3H) groups in a side chain generate properties that are similar to those induced by carboxylic acid groups.

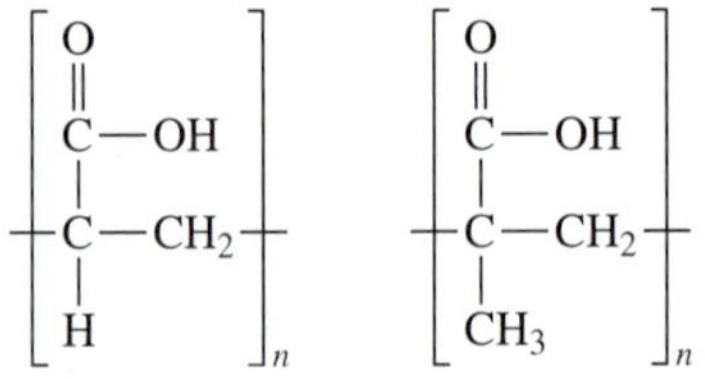

38   39

Pyridino and Other Amino Groups

Poly(vinylpyridine) **(40)** is soluble in aqueous acid or in ethanol but is insoluble in neutral or basic aqueous media. The solubility in aqueous acid follows protonation of the most basic sites in the molecule—the nitrogen atoms **(41)**. Quaternization of the same sites with an alkyl halide **(42)** also generates water solubility, and the quaternized polymers are polyelectrolytes. The unquaternized polymer **(40)** forms complexes with metal ions. Other polymers with pendent amino sites behave in the same way.

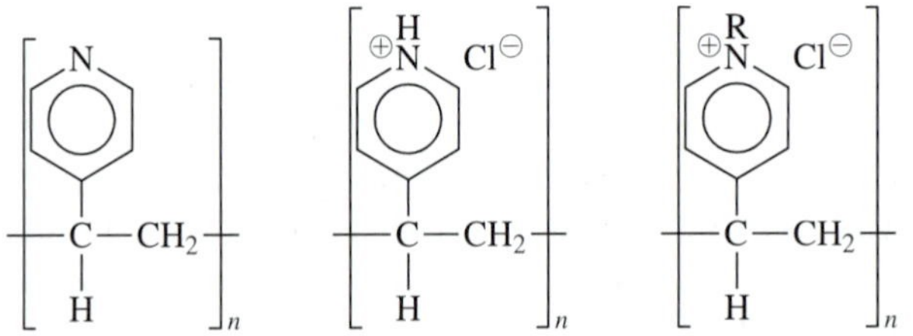

40   41   42

RELATIONSHIP TO SURFACE PROPERTIES

The influence of molecular structural features on the solubility and hydrophilicity or hydrophobicity of a polymer have been mentioned in earlier paragraphs. The surface character of a polymer is a critical factor in the choice of a material for many applications. The surface not only determines how rapidly a material may be affected by organic solvents, water, or chemical reagents, but it also plays a role in adhesion, biocompatibility, and friction.

It is possible to estimate surface behavior from a knowledge of the skeletal units and side groups present, but this is likely to give an oversimplified picture. Certain side groups may be concentrated at the surface and others buried beneath it in response to the influence of the interface at which the surface was formed (air, water, organic media, mercury, glass, etc.). Moreover, macromolecular motions may form a new surface by burying some units and exposing new segments. It is quite common, for example, to detect a change in surface hydrophobicity over a period of time as one type of side group withdraws from the surface and is replaced by another.

STRUCTURAL INFLUENCES ON SOLID STATE PROPERTIES

As discussed earlier in this book, two key factors determine the physical state of a polymer at a given temperature—the glass transition temperature, T_g, and the microcrystalline melting temperature, T_m. Intermediate transitions, such as liquid crystalline changes, may also affect the bulk properties. Thus, a major goal in polymer design is to be able to alter the various transition temperatures by control of the basic chemical structure, the stereochemistry, copolymer sequence distribution, degree of branching, and cross-linking. The approach followed in this section is to examine the extent to which solid-state structure–property relationships can be developed for homopolymers, and then to consider how these principles might be extended to copolymers.

Equilibrium and Dynamic Flexibility

Both glass transitions and melting transitions depend on chain flexibility. There are two types of flexibility. "Equilibrium" flexibility is a measure of a chain's ability to collapse into compact, random conformations of high entropy. Because a large conformational entropy leads to a large entropy of melting, ΔS_m, this leads to low melting points because $T_m = \Delta H_m/\Delta S_m$, where ΔH_m is the heat of fusion. By contrast, "dynamic" flexibility measures the ability of a chain to *change* conformation or shape. High flexibility in this sense can give rise to low values of the glass transition temperature T_g, because the chains will have to be cooled to lower temperatures before they lose the long-range motions that are suppressed in the glassy state.

Intermolecular Attractions

The melting point of a polymer is directly proportional to the heat of melting. Melting almost always requires an increase in the distance between the polymer chains, because all but a few materials expand on melting. This means that a lot of energy will have to be expended to separate chains that are held together by strong attractions, and this would lead to large values of ΔH_m, and thus high values of T_m. Other things being equal, melting points would be expected to increase in the order: nonpolar chains $<$ polar chains $<$ hydrogen-bonded chains.

It is difficult to change only one of these features of a chain at a time. For example, changes in structure that decrease the equilibrium flexibility generally also decrease the dynamic flexibility. Such changes would include the linkage of bulky side groups to a chain, and incorporation of ring structures in the polymer backbone. Similarly, changes in structure which alter flexibility of either type often change the *inter*molecular interactions as well. An example is the introduction of an oxygen atom into the chain backbone to increase flexibility, a change that might also make the chain more polar.

The oxygen "swivel unit" just cited is actually the most flexibility-inducing atom that can be introduced into a chain. It is divalent, has a small radius, and bears no substituents. Dicoordinate nitrogen in some polymers, such as phosphazenes, behaves in the same way. Sulfur is also very good in this regard, probably because the C—S bond is considerably longer than the C—O bond and this helps to offset steric effects. The ester linkage, O—C(=O)— generally increases chain flexibility. The C—O bond has some double bond character, which tends to stiffen a chain, but the small size of the oxygen atom is more important in most cases. The result depends on the type of chain being modified. Structural changes on a chain that is already very stiff will almost certainly make it more flexible, but the same changes on a very flexible chain will most likely make it stiffer.

Illustrative Examples

Equilibrium and dynamic flexibility   Some of the relationships that control flexibility are illustrated in Table 22.1. The reference material is polyethylene (PE), and other polymers are compared with it. The incorporation of one oxygen atom between every two CH_2–CH_2 units decreases the melting point by 75°C. However, for this polymer the glass transition temperature does not decrease. Instead it *rises* by 65°C. This increase is presumably caused by the increase in polarity. Indeed, the polarity is so high that it is sufficient to make poly(ethylene oxide) soluble in water. Both transition temperatures decrease markedly in the case of poly(dimethylsiloxane). This is one of the most flexible polymers known, for several reasons: (i) the Si—O bond is longer than the C—C bond in hydrocarbon polymers, (ii) the Si—O—Si bond angle is much wider than the usual tetrahedral value, (iii) the rotational barrier around the Si—O bonds is much lower than that about C—C bonds, and (iv) the Si—O—Si bond angle is highly deformable.

TABLE 22.1 SOME EFFECTS OF CHANGES IN FLEXIBILITY ON MELTING POINTS
AND GLASS TRANSITION TEMPERATURES

Change in Flexibility	Polymer and Repeat Unit	T_m, °C	T_g, °C
None—(Reference material)	Polyethylene $[CH_2-CH_2-]$	135	−125
Increase	Poly(ethylene oxide) $[CH_2-CH_2-O-]$	65	−60
Increase	Poly(dimethylsiloxane) $[Si(CH_3)_2-O-]$	−45	−125
Decrease	Poly(ethylene terephthalate) $[O-(CH_2)_2-O(C=O)-C_6H_4-C=O-]$	265	70

The effects of *decreases* in flexibility are illustrated by poly(ethylene tereph-thalate). The presence of the ester linkage increases flexibility, but this effect is overwhelmed by the stiffness of the *para*-phenylene groups in the repeat unit, and both transition temperatures are increased significantly. This is one of the few polyesters that have a high enough melting point to compete with aliphatic polyamides as textile fiber materials.

This stiffening approach is also used in living organisms to increase the melting points of fibrous proteins, but it operates through choices of α-amino acid repeat units. Fibrous proteins require some crystallinity to generate good mechanical properties. The repeat peptide unit is shown in the left portion of Figure 22.2, in which variation of the side group R provides the twenty amino acid repeat units available for making a protein. The desired stiffening effect is introduced by the presence of either proline (PRO) or hydroxyproline (HPRO), which have the atypical structures shown schematically in the right portion of the figure. The ring structures in the backbone of both units are formed by having the R groups on the carbon atoms linked to the nitrogen atoms as well. This parallels the use of *para*-phenylene groups in poly(ethylene terephthalate), or in some rigid-rod polymers such as Kevlar®. The effects of this stiffness are evident from the amino acid compositions of some forms of collagen, which is a fibrous, partially crystalline protein found in the connective tissues of a variety of animals. Some intriguing data are given in Table 22.2. The collagen in the codfish has a combined amount of PRO and HPRO of 16%. This content is enough to give a melting point of 16°C, which is above the operating temperature of this type of fish. Thus,

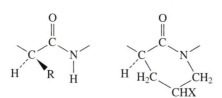

Figure 22.2 The left portion of the figure shows the general structure of the protein repeat unit, and the right portion the modification that gives a stiffening of the chain. The X is H in the case of proline and OH in the case of hydroxyproline, but this is unimportant in the present context.

TABLE 22.2 EFFECTS OF RIGIDIFYING α-AMINO ACID REPEAT UNITS ON THE MELTING POINTS OF DIFFERENT TYPES OF COLLAGEN

Source	% (PRO and HPRO)	T_m, °C	T_{Body}, °C
Cod	16	16	12
Shark	19	29	26
Calf	23	39	37

the protein remains in the crystalline state. The collagen in shark has a larger amount of PRO and HPRO, as shown, because it has to withstand a body temperature of 26°C. Mammals, such as calves, need a still higher content of these amino acids to keep their collagen melting point above their body temperature of 37°C (98.6°F). The melting points given in the table are really lower limits. The collagen is under tension in the body, and this increases its melting point above the values cited. Otherwise, a person would have to worry about the melting of his or her collagen at the onset of any significant fever!

Intermolecular attractions   The effects of intermolecular attractions on transition temperatures are illustrated in Table 22.3. Poly(vinyl chloride) can be viewed as a polyethylene in which a hydrogen on every other carbon atom is replaced by chlorine. The resultant Coulombic attractions bring about an increase in both T_m and T_g. The increase in melting point is due to the increased heat of melting. The increase in glass transition temperature, on the other hand, is due to the increased attractions between chains, which raise the temperature required to thaw the long-range motions that had been locked into the structure in the glassy state.

In the remaining part of the table, only approximate, average values for T_m and T_g are given, for the general classes of polymers cited. Thus, the dots represent aliphatic sequences $(CH_2)_m$ of unspecified lengths m. Polyesters have relatively low values of these transition temperatures, presumably because of the relatively high flexibility of the ester linkage. Polyamides have significantly higher values because of hydrogen bonding. In this case, the increase in T_m is not caused

TABLE 22.3 EFFECTS OF INTERMOLECULAR INTERACTIONS ON MELTING POINTS AND GLASS TRANSITION TEMPERATURES

Interactions	Polymers	T_m, °C	T_g, °C
Coulombic	Poly(vinyl chloride) $[CHCl—CH_2—]$	270	85
Coulombic	Polyesters $[C(=O)—O—\ldots]$	100	30
H Bonding	Polyamides $[C(=O)—NH—\ldots]$	200	50
H Bonding	Polyurethanes $[O—C(=O)—NH—\ldots]$	170	30
H Bonding	Polyureas $[NH—C(=O)—NH—\ldots]$	300	70
H Bonding	Cellulose and its derivatives	Very high	100

by increased ΔH_m, but to the value of ΔS_m being decreased by hydrogen bonding that persists into the molten state. Polyurethanes can be thought of as hybrids of polyesters and polyamides because they have both oxygen atoms and NH groups bracketing the carbonyl groups. Not surprisingly, their transition temperatures tend to fall between these two limits. Polyureas can be considered to be "double" polyamides, since they have two sites for H bonding per repeat unit. Cellulose and its derivatives have all three features that give high transition temperatures: (i) polarity, (ii) hydrogen-bonding, and (iii) stiffening by cyclic units in the chain backbone.

The Case of Elastin

The properties of bioelastomers can be used to illustrate several important structure–property relationships. Good examples are the proteins elastin in mammals and resilin in insects. The human body has to synthesize elastomers whenever it needs high deformability with recoverability. Obvious examples are in blood vessels, skin, and heart, and lung tissue. It is instructive to note how the structure of elastin is optimized for its use in mammals. Elastin is a random copolymer that does not crystallize, irrespective of the temperature or degree of stretching. The use of *random* chemical sequences is reminiscent of commercial ethylene-propylene copolymers, in which the propylene "irregularities" prevent the ethylene units from crystallizing, thus generating the amorphous chains that are necessary for rubberlike elasticity. The total absence of crystallites could present a problem with regard to strength, but this is solved in the living system by the elastin being organized around a mesh of strong collagen fibers for reinforcement. The structure is similar to laboratory pressure-tubing reinforced by a mesh of Nylon fibers, but the elastin-collagen biocomposite is much more sophisticated. The collagen fibers are slack when they are encased in elastin, which permits an artery, for example, to expand radially during a pressure pulse from the heart. This smooths out the flow of blood during the systolic and diastolic parts of the pumping cycle. However, in the case of an expanding aneurism, the collage fibers become taut and then strenuously resist the further expansions that could lead to a fatal bursting of the blood vessel.

It is also impressive to note the choices that living systems make with regard to the R side chains shown in the left portion of Figure 22.2. Some results are given in Table 22.4. The glycine, alanine, and valine have the smallest side chains and are used in an astonishingly high 65% of the choices. Glycine is particulary compact and is widely employed for avoiding steric congestion. This amino acid is employed extensively in the three helices that make up the cable-like structure of collagen, or between chain segments that are compactly folded back onto themselves in globular proteins such as enzymes. Polar side chains are avoided because they could lead to strong interchain attractions that would make the elastomer sluggish, and generate a very high T_g. Actually, the glass transition temperature of elastin *is* high (around 200°C), because of the unavoidable polarity

TABLE 22.4 SOME PEPTIDE UNITS, THEIR SIDE CHAINS, AND THEIR PREVALENCE IN ELASTIN

Peptide unit	R side group	Approximate mol %
Glycine	H	30
Alanine	CH_3	20
Valine	$CH(CH_3)_2$	15
Polar units	Various	<5

of the protein backbone. For this reason, elastin in the body is always swollen with aqueous body fluids which act as plasticizers to bring T_g below the operating temperature of the body.

Another impressive feat in living systems is connected with the fact that fish that swim into colder waters have significantly lowered body temperatures. It might be expected that the aqueous body fluids that plasticize the elastin would be less "good" in the thermodynamic sense. The elastin would therefore deswell, and the lowered plasticizer concentration would cause an increase in T_g (and thus a fatal brittleness in the connective tissues). However, elastin does exactly the opposite. It does what is required for the fish to survive: the elastin swells *more* as the temperature is decreased. It has been conjectured that this is caused by all the elastin nonpolar side chains forming an oily sheath around the polar backbone, analogous to the myelin sheath around nerve fibers. This is thought to be the origin of the very unusual thermodynamics of mixing.

Random and Block Copolymers

The use of homopolymers restricts the discussion to the stippled area in Figure 22.3, which corresponds to changes in melting points and glass transition temperatures which parallel one another. However, it is important to be able to achieve independent control of T_g and T_m. For example, it might be desirable to lower the T_m of a partially crystalline polymer without making the glassy regions rubbery (region A), or to increase T_m without making the rubbery regions glassy (region B). Access into region A can be accomplished with the use of *random* copolymers, and entry into region B can be brought about with *block* copolymers.

In random copolymers, the melting point changes much more than the glass transition temperature as the composition is changed, because melting is a cooperative process, whereas the glass transition is less so. An example is shown in Figure 22.4, where A might represent isotactic "Iso" vinyl units and B syndiotactic "Syn" units. If Syn units are added to a crystalline Iso polymer, then these "irregular" units first make the Iso crystallites smaller and decrease their melting points. Further additions of Syn units forces the Iso sequences to be so small that none are long enough to crystallize at all. The same thing would happen to a crystalline Syn polymer following the insertion of "irregular" Iso units. However, the

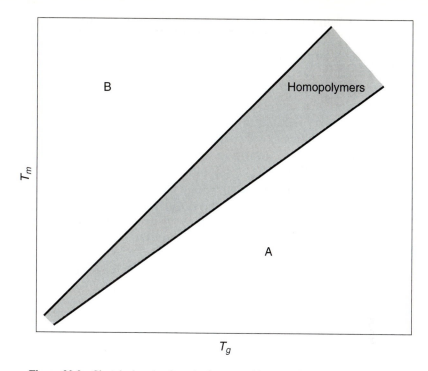

Figure 22.3 Sketch showing how, in the case of homopolymers, T_m and T_g tend to be strongly correlated.

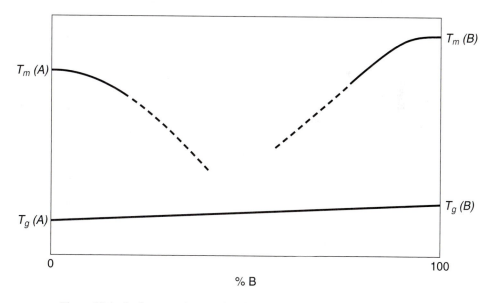

Figure 22.4 In the case of stereochemical or chemical copolymers, the melting point is much more sensitive to composition than is the glass transition temperature.

glass transition temperature would not show such a strong dependence on composition. The same arguments can be made for the ethylene sequences that are disputed by propylene units in an ethylene–propylene elastomer.

The ability to increase T_m while decreasing T_g requires the use of a block copolymer. One block, called the hard block "H," consists of a very stiff structure, and therefore has high values of both T_m and T_g. The other block, called the soft block "S," is very flexible, and therefore has low values of T_m and T_g. In a block copolymer, the blocks phase-separate into domains, and each of the two phases exhibits its respective transition temperatures. This is another case of "delegation of responsibilities." The hard blocks are there because of their high melting points (and their correspondingly high glass transitions are irrelevant). The soft blocks are there for their low glass transition temperatures (and their correspondingly low melting points are likewise irrelevant). The situation is illustrated in Figure 22.5. The material will not flow until the higher of the two melting points is exceeded, and it will not become glassy until the temperature drops below the lower of the two glass transition temperatures. This effectively increases T_m while decreasing T_g, and the material would have useful properties over a wide range of temperatures, as illustrated in the figure.

"Segmented polyurethanes" are good examples of this type of block copolymer. If the hard blocks are present in small amounts they act as crosslinks within the soft phase. Goodrich's Estane® rubbers are in this category. On the other

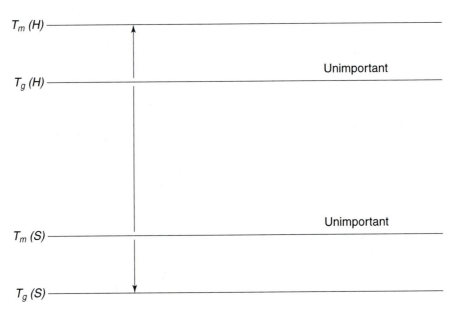

Figure 22.5 Since a block copolymer can exhibit two melting points and two glass transition temperatures, it can have useful properties from the lower of the two T_g's to the higher of the two T_m's.

hand, if the hard blocks predominate, the result is a fiber in which the soft phase gives good extensibility. The Lycra® or Spandex® fibers produced by DuPont are in this category.

Branching and Cross-Linking

Small amounts of branching will decrease crystallinity and melting points. An example is low-density polyethylene. An increase in branching completely suppresses the crystallinity.

Small amounts of cross-linking, typically one cross-link in a hundred repeat units, will convert an amorphous polymer into an elastomer. However, further increases in crosslink density, to approximately one crosslink per ten repeating units, will convert it into a hard thermoset, with no rubberlike behavior at all.

HOW SCIENTISTS DESIGN NEW POLYMERS AND POLYMERIC MATERIALS

The production of new polymers is the driving force that continues to replace metals and ceramics by synthetic polymers, and fuels the expansion of polymeric materials into fields as diverse as fuel cells, batteries, light-emitting devices, biomedical materials, and components of aircraft and automobiles. New polymeric materials arise through a combination of molecular design and the development of new polymer synthesis techniques. The principles outlined in the preceding sections of this chapter form the basis of macromolecular design, and this in turn stimulates the development of new synthesis methods and new ways to tune the solid state and surface properties of polymers. In this sense, modern polymer chemistry, with its broad portfolio of known structure-property relationships, has an advantage over the earlier approach in which the (sometimes accidental) synthesis of a new polymer often yielded unexpected combinations of properties. However, it should be recognized that the element of surprise still exists in scientific work. Indeed, it is sometimes said that the role of the scientist is to do things that cannot be predicted, and cannot always be justified on the basis of common knowledge. Only in this way can science escape from accepted practice and generate real breakthroughs into new areas. Work that follows an absolute adherence to well-established knowledge and protocol may not yield any new insights.

Thus, useful new polymers are produced by three different routes—first, by a chance discovery; second, by a scientist who sees hitherto unrecognized connections between two different types of chemistry and combines them to produce a new system; and third, by logical methodical design based on structure–property relationships. The third method has to be preceded by one of the first two. Developments that result from combining ideas from two different areas are often based on a subconscious knowledge of the general relationships between structure and properties in other systems.

The discovery of poly(dimethylsiloxane) and poly(tetrafluoroethylene) (Teflon®), and the synthesis of high-density polyethylene arose from accidental reactions (by Kipping, Plunkett, and Ziegler, respectively). The poly(organophosphazenes) arose through an integration of ideas from organic polymer chemistry and inorganic chemistry.

Usually, accidental discoveries or cross-disciplinary discoveries of this type stimulate a burst of exploratory research which then leads to the synthesis of many different derivatives and the development of structure–property relationships, sometimes by researchers other than the discoverer. By some accounts, Kipping actually threw away the first poly(organosiloxanes) made in his laboratory because they were not the compounds he was looking for. It was left to others to recognize their value and develop them in detail. There is a lesson here. On the other hand, most of the discoverers of new polymers are instantly aware of the significance of the discovery and set to work, sometimes in spite of widespread scepticism and opposition, to develop the field and establish the primary structure–property relationships needed to enable the new field to develop. After that, useful polymers are usually produced by direct design.

The developmental design approach often starts from a knowledge of the *defects* of existing polymeric materials, information that comes from engineers, physicists, physicians, and other users. Thus, the design process starts with a knowledge of the combination of properties needed to make a better material, and then proceeds to consider the alternative ways in which these properties can be generated. For example, if there is a need for fire-resistant clothing for firefighters or race-car drivers, this would require a polymer with a high melting point, and this in turn suggests the use of polymers with relatively stiff structures, such as the aromatic polyamides. The key structure–property relationship involved here would be chain stiffness to give a high melting point. The choice of this type of polymer might then require exploratory synthesis work using condensation polymerization. Other decisions would be needed about how to generate crystallinity in fibers of the polymer, because crystallinity would yield high tensile strength. Crystallinity would depend on monomer sequencing and on the types of side groups present. Resistance to fire might be achieved by the incorporation of phosphorus into the molecular structure, and so on.

Polymer chemists spend a great deal of their time designing new molecules and pondering their structure–property relationships, and the reader is encouraged to practice this process routinely to gain experience. Some example exercises are given in the Study Questions.

STUDY QUESTIONS

1. A need is identified for a polymer that exists as a hydrogel that can remove metal ions from solution. Based on the principles discussed in this chapter, design such a polymer, and outline how it would be synthesized.

2. Polymers that consist of aromatic rings linked end to end are often insoluble in most solvents. Which side units might you attach to the aromatic rings to enhance their solubility in **(a)** aliphatic hydrocarbons, **(b)** ethers, and **(c)** aromatic hydrocarbons?

3. Which side groups confer water solubility on a polymer, and why? Which groups would you choose if you wished to design a polymer that might be **(a)** insoluble in aqueous base but soluble in aqueous acid, **(b)** soluble in aqueous base but insoluble in aqueous acid, and **(c)** soluble in solutions of monovalent cations (e.g., Na^+) but insoluble in solutions of divalent cations (e.g., Ca^{2+}).

4. The Si—O bond in silicate minerals is often cleaved by strong aqueous base. Why is poly(dimethylsiloxane) relatively inert to this reagent?

5. Liquid crystallinity in polymers occurs when the mesogen is part of the chain and when it forms part of the side-group structure. In an essay-type answer discuss the possibility that flat, rigid units other than those mentioned in this chapter might generate liquid crystallinity.

6. How does the addition of a relatively small number of propylene comonomeric units to polyethylene convert it into a commercially important elastomer?

7. If one unit out of a hundred in a polymer is at a cross-link, the result is an elastomer such as natural rubber. However, if one unit out of *ten* is at a cross-link the result is a relatively hard thermoset such as an epoxy resin. What are the approximate values of the molecular weight, M_c, between cross-links for these two cases if the molecular weight of the repeat unit is 50 g/mol?

SUGGESTIONS FOR FURTHER READING

ALFREY, T., and GURNEE, E. F., *Organic Polymers*. Englewood Cliffs, NJ: Prentice Hall, **1967**.

ALLEN, G., and BEVINGTON, J. C., *Comprehensive Polymer Science* (7 volumes). Oxford: Pergamon Press, **1988**.

ASKADSKII, A. A., *Physical Properties of Polymers. Prediction and Control*. Amsterdam: Gordon and Breach Publishers, **1996**.

BERRY, G. C., "Rheological Properties of Nematic Solutions of Rodlike Polymers," *Mol. Cryst. Liq. Cryst.*, **1988**, *165*, 333.

CALUNDANN, G. W., and JAFFE, M., "Anisotropic Polymers, Their Synthesis and Properties," in *Synthetic Polymers* (Proc. of the Robert A. Welch Foundation XXVI, Houston), **1983**, p. 247.

CIFERRI, A., KRIGBAUM, W. R., and MEYER, R. B., (Eds.), *Polymer Liquid Crystals*. New York: Academic Press, **1982**.

CRAVER, C., and CARRAHER, C. E., Jr. (Eds.), *Applied Polymer Science—21st Century*. Washington: American Chemical Society, **2000**.

ECONOMY, J., "Ordering Processes in the Aromatic Polyesters," *Polymer Mater. Sci. Eng.* **1985**, *52*, 1.

EISENBERG, A., and KING, M., *Ion Containing Polymers*. New York: Academic Press, **1977**.

ENGLAND, D. C., USCHOLD, R. E., STARKWEATHER, H., and PARISER, R., "Fluoropolymers: Perspectives of Research, in *Synthetic Polymers* (Proc. of the Robert A. Welch Foundation XXVI, Houston), **1983**, p. 193.

ERMAN, B., and MARK, J. E., *Structures and Properties of Rubberlike Networks*. New York: Oxford University Press, **1997**.

FELDMAN, D., and BARBALATA, A., *Synthetic Polymers: Technology, Properties, Applications*. London: Chapman & Hall, **1996**.

GROSBERG, A. Y., and KHOKHLOV, A. R., *Giant Molecules. Here, There, and Everywhere* ... San Diego: Academic Press, **1997**.

HOLLIDAY, L. (Ed.), *Ionic Polymers*. London: Applied Science Publishers, **1975**.

LENZ, R. W., "Structure–Order Relationships in Liquid Crystalline Polyesters," *Pure Appl. Chem.*, **1985**, *57*, 1537.

MARK, H. F., BIKALES, N. M., OVERBERGER, G. C., and MENGES, G., *Encyclopedia of Science and Engineering* (16 volumes). New York: Wiley-Interscience, **1985–89**.

MORGAN, P. W., KWOLEK, S. L., and PLETCHER, T. C., "Aromatic Azomethine Polymers and Fibers," *Macromolecules*, **1987**, *20*, 729.

PORTER, D., *Group Interaction Modelling of Polymer Properties*. New York: Marcel Dekker, Inc., **1995**.

RICHARDS, R. W., and PEACE, S. K. (Eds.), *Polymer Surfaces and Interfaces III*. Chichester: John Wiley and Sons, **1999**.

VAN KREVELEN, D. W., *Properties of Polymers*. Amsterdam: Elsevier, **1997**.

23

Electroactive and Electro-Optical Polymers

INTRODUCTION

In Chapter 9 it was pointed out that a relationship exists between polymers, ceramics, semiconductors, and metals, and that an important component of modern research and technology is the preparation of new materials that have the properties of *both* polymers and ceramics or polymers and metals. Here we examine materials that combine the properties of polymers *and* metals, particularly with respect to their electrical conductivity.

Most classical organic polymers are electrical insulators (Figure 23.1), while many metals are good conductors of electricity. Polymers are useful because of their ease of fabrication, flexibility or strength, lightness of weight, and chemical inertness. Metals are valuable not only because they are good conductors of electricity, but also because of their strength and (often) their ductility. However, many metals are heavy and are prone to corrosion. Metalloid semiconductors are brittle and are exceedingly difficult to purify and fabricate.

Polymers that conduct electricity are candidates for a wide variety of uses, ranging from easily fabricated semiconductor chips and integrated circuits to electrodes, lightweight battery components, sensors, electrochromic displays, and static-free packaging materials. This is one reason why the subject of conducting polymers is the focus of intense research activity. It is a subject that will become increasingly important in the future.

Three different types of electrically conducting polymers are known—ion–polymer solid electrolyte systems, composites of electronic conducting ma-

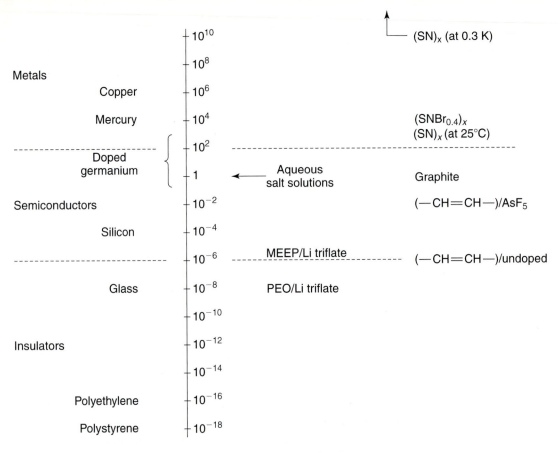

Figure 23.1 Electrical conductivity (in $ohm^{-1}cm^{-1}$) of various materials. The values shown are for measurements at 25 to 30°C except where specific temperatures are given. MEEP is methoxyethoxyethoxypolyphosphazene.

terials in nonconducting polymers, and polymers that conduct electricity by electronic transport. Electro-optical polymers are materials in which the passage of an electric current changes some optical property such as refractive index or the color, or leads to the emission of light. Alternatively, they may generate electricity when irradiated with light.

IONIC CONDUCTION IN SOLID POLYMERS

The Phenomenon

A salt such as sodium chloride dissolved in a liquid such as water conducts electricity by the migration of Na^+ and Cl^- ions to opposite electrodes under the pressure of an electrical potential (Figure 23.2). This behavior depends on the ability of water molecules to solvate the two ions and facilitate their separation from

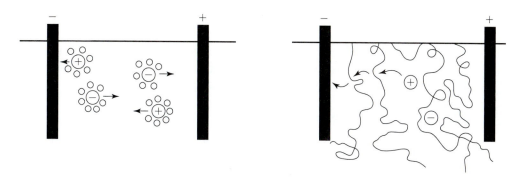

Figure 23.2 Comparison of ionic conductivity in a solution of a salt dissolved in water with the behavior of a salt dissolved in a solid, solvating polymer. In the aqueous solution, ions, accompanied by solvation sheaths of water molecules, can move freely toward the oppositely charged electrodes. In the solid polymer solution, one (or both) of the types of ions are passed from one polymer molecule to another as the macromolecules undergo thermally induced motion.

each other. In general, the greater the concentration of salt in the water, the higher is the electrical conductivity of the solution, because more charge carriers are present in the system.

Some polymer molecules can function as *solid* solvents for salts. In general these are macromolecules that possess electron donor (basic) coordination sites that can form weak linkages to the cation component of the salt. In so doing, the polymer solvates one or both ions and facilitates ion separation. Unless the ions are well separated, they will not serve as charge carriers. Instead, they would prefer to remain tightly bound as an (electrically neutral) ion pair.

However, even a system in which the ions are well separated by solvation will be a poor conductor of electricity unless the ions are sufficiently mobile to migrate readily toward the appropriate electrode. Thus, the host polymer must be sufficiently flexible and provide enough "free volume" to allow ionic migration. In practice, this means that the polymer must have a low glass transition temperature and a low degree of crystallinity.

The preparation of solid solutions of salts in suitable polymers is quite straightforward. A common solvent for the polymer and the salt is used to make separate solutions of the two. The solutions are then mixed, and the solvent is evaporated to leave a film of the polymer–salt conjugate.

Although the exact mechanism of ionic conduction in polymers is not fully understood, a plausible explanation is as follows. If one or both of the ions are *weakly* coordinated to sites along the polymer chain, they will move with that coordination site as the polymer goes through its normal thermally activated bending, twisting, and reptation motions. This movement will allow an ion to transfer from a coordination site on one polymer to a similar site on another (Figure 23.2). Ion diffusion will be mainly in one direction if an electronic potential is applied.

Ionic Conduction in Solid Polymers

Hence, a cation, for example, will be passed toward the cathode from one polymer molecule to another in a hand-to-hand fashion. Free-volume effects must also be important in ion transport to provide a pathway through which ions can diffuse. This mechanism explains why amorphous polymers are the best matrices and why the conductivity increases with rising temperature.

Host Polymers for Solid Electrolytes

A number of different host polymers have the characteristics described above, and several that have been studied in detail are shown in structures **1** to **4**.

$$\text{+CH}_2\text{—CH}_2\text{—O}\text{+}_n \qquad \text{+(CH}_2\text{—CH}_2\text{—O)}_x\text{(CH}_2\text{—CH(CH}_3)\text{—O}\text{—)}_y\text{+}_n$$

1   **2**

$$\left[\begin{array}{c} \text{OCH}_2\text{CH}_2\text{OCH}_2\text{CH}_2\text{OCH}_3 \\ | \\ \text{—N=P—} \\ | \\ \text{OCH}_2\text{CH}_2\text{OCH}_2\text{CH}_2\text{OCH}_3 \end{array}\right]_n \qquad \left[\begin{array}{c} \text{O(CH}_2\text{CH}_2\text{O)}_x\text{CH}_3 \\ | \\ \text{—O—Si—} \\ | \\ \text{CH}_3 \end{array}\right]_n$$

3   **4**

Poly(ethylene oxide)

Poly(ethylene oxide)(**1**) is the prototype solid polymeric electrolyte host, against which all the other systems are compared. It was the first polymer studied in detail (in the 1970s). The skeletal oxygen atoms are the donor sites that coordinate to alkali metal cations, and the presence of these sites explains why this polymer takes up high concentrations of salts such as lithium triflate, $LiOS(O)OCF_3$. Thus, coordination of the polymer to one (or both) of the ions brings about ion separation, and the solid becomes an ionic conductor. At 25°C the maximum conductivity that can be obtained for this system is below 10^{-6} ohm^{-1} cm^{-1} (also given as 10^{-6} S/cm). However, even when heated, the maximum conductivity that can be achieved with a poly(ethylene oxide)–salt system (about 10^{-4} ohm^{-1} cm^{-1} at 100°C) is only one-hundredth of those found for comparable aqueous solutions of salts at the same concentration. This has been attributed to residual cation–anion interactions that persist in the polymeric matrix.

A serious defect of poly(ethylene oxide) as a solid electrolyte host is its partial crystallinity. Ion transport occurs in the amorphous regions only, and systems based on this polymer must be heated to 70 to 100°C before crystallite melting occurs and the maximum conductivity can be obtained. This problem can be minimized by the use of a noncrystalline copolymer that contains both ethyleneoxy and propyleneoxy chain units, as shown in structure **2**.

Comb polymers An alternative approach to the design of solid electrolyte host polymers is illustrated in structures **3** and **4.** Here the side groups provide weak coordination sites for the cations, while the highly flexible polymer backbone assists in the conformational transitions that permit movement of the ions.

Poly[bis(methoxyethoxyethoxy)phosphazene] (**3**) (also known as "MEEP") is a comb-type macromolecule designed and synthesized to optimize the required features of a solid electrolyte system. It has a low glass transition temperature ($-84°C$), is completely amorphous, dissolves in organic solvents such as tetrahydrofuran that are also solvents for triflate salts, and it bears a high concentration of alkyl ether units in the side chains. The conductivity of MEEP–lithium triflate systems at room temperature ($2.2 \times 10^{-5} \, \omega^{-1} \, cm^{-1}$) is 1000 times higher than the value for a comparable system based on poly(ethylene oxide). Moreover, the dimensional stability of this system (resistance to flow) can be increased by chemical or radiation crosslinking without a significant decrease in the conductivity. The polysiloxane host polymer shown as structure **4** has similar attributes. The use of branched ethyleneoxy side chains in place of the linear side groups in MEEP raises the conductivity into the $10^{-4} \, \Omega^{-1} \, cm^{-1}$ range and converts the material to a solid rather than a gum.

Salts   The salts used in solid polymeric electrolytes must have a low crystal lattice energy. Otherwise, ion pairing and crystal forces will predominate over solvation by the polymer and ion-pair separation. This is the reason why bulky anions, such as triflate, are often used. The concentration of salt in the polymer has a profound effect on the conductivity behavior. At low salt concentrations the conductivity is low because, although the ion pairs are well separated, they are few in number. As the salt concentration is raised, the conductivity rises to a maximum beyond which it begins to fall off. This decline may be partly a consequence of ion pairing, but is mainly a result of the ions forming many transient crosslinks to the polymer. This has the effect of raising the glass transition temperature and reducing ion mobility.

In spite of the obvious advance represented by MEEP-type solid polymer electrolytes, their room-temperature conductivities are still an order of magnitude less than the values required for general purpose batteries. However, the addition of small amounts of coordinative organic solvents, such as propylene carbonate, raises the conductivity to the required $10^{-3} \Omega^{-1}$ level.

Uses for Solid Polymeric Electrolytes

Widespread interest exists in the use of these systems in lightweight, high-energy-density batteries. Rechargeable lithium cells based on polymeric systems should be able to store two to five times as much energy per unit weight and volume as a conventional lead-acid car battery. The design for a battery of this type is shown in Figure 23.3. Other proposed applications include electrochromic displays, electrochromic windows, solid-state photoelectrochromic cells, sensors, electrochemical transistors, and supercapacitors.

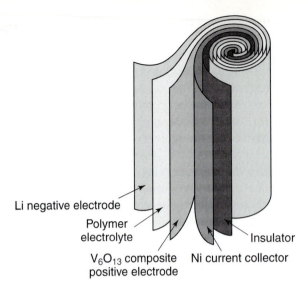

Figure 23.3 Design for a high-energy-density rechargeable lithium battery that makes use of a polymeric electrolyte. During discharge, lithium is oxidized at the lithium–polymer interface and is transported as Li^+ across the polymeric electrolyte. Simultaneously, lithium ions are inserted from the lithium triflate electrolyte into the positive electrode. During charging of the battery the process is reversed. [From D. F. Shriver and G. C. Farrington, *Chem. Eng. News*, May 20, 42 (1985). Reprinted with permission from *Chemical & Engineering News*. Copyright 1985 American Chemical Society.]

Li negative electrode

Polymer electrolyte

Insulator

V_6O_{13} composite positive electrode

Ni current collector

Proton Conductors

A second class of solid ionic conductors are those that transport protons. Membranes fabricated from these polymers are a key component of an important class of fuel cells that convert chemicals such as hydrogen, methanol, or methane into electrical energy. Figure 23.4 illustrates the structure of a polymer membrane-based fuel cell. Each cell consists of a proton-conducting membrane that separates two electrodes. The anode contains finely divided platinum-ruthenium alloy, and the cathode contains powdered platinum. Hydrogen as a fuel enters the device at the anode and is converted into protons and electrons. The electrons move through the external circuit to the cathode while the protons migrate through the polymer membrane to the cathode. At the cathode, the hydrogen is oxidized by atmospheric oxygen to water. A direct methanol fuel cell operates on the same principle except that the methanol is converted to carbon dioxide and water. Direct methanol fuel cells require a membrane that does not allow methanol to diffuse through to the cathode. Hydrogen-oxygen fuel cells have been developed for uses in buses and small stationary power plants, but direct methanol fuel cells offer some serious advantages over hydrogen for use in automobiles, cellular phones, and laptop computers. Several hundred individual cells are needed to generate the power needed for an automobile.

Typical polymers employed in this application are macromolecules with acidic side groups. These allow the protons to move from acid site to acid site across the membrane, usually assisted by the presence of water. The classical proton-conductive membrane is a material known as Nafion-117, which is a fluorocarbon polymer that bears sulfonic acid ($-SO_3H$) functional groups. However, Nafion cannot be used at temperatures higher than 80°C because it

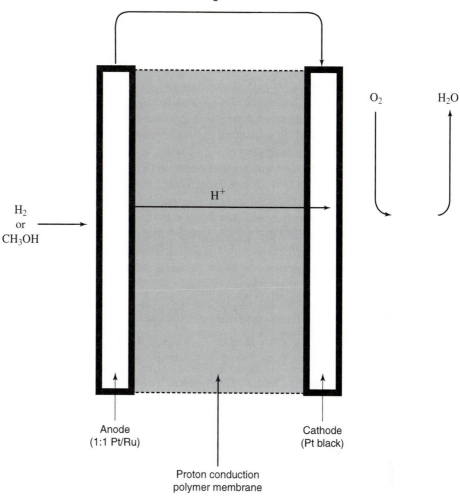

e^-

O_2 H_2O

H^+

H_2
or
CH_3OH

Anode
(1:1 Pt/Ru)

Cathode
(Pt black)

Proton conduction
polymer membrane

Figure 23.4 Schematic diagram of a polymer membrane fuel cell. The protons are transported through the membrane via acidic sites on the polymer molecules. Electrons passing through the external circuit provide the power to propel an automobile or power a laptop computer or cellular phone.

dehydrates. It is also highly permeable to methanol. Thus, considerable research is being carried out to find replacements for this material. Sulfonated polystyrene was used in the hydrogen-oxygen fuel cells on the Gemini space mission, but these had relatively short lifetimes in the aggressive reaction conditions of a fuel cell. Recent work has focused on sulfonated polyimides and other high-temperature polymers, and promising results have been obtained from aryloxyphosphazenes that bear sulfonic acid, phosphonic acid, or sulfonimide functional groups.

PYROLYSATE POLYMERS AND COMPOSITE POLYMERS WITH CONDUCTING DOMAINS

The conversion of electrically insulating polymers to conducting materials by pyrolysis to graphite or by the incorporation of graphite powder or metal particles into a composite structure has been an aspect of polymer technology for many years. Materials such as pyrolyzed polyacrylonitrile or pyrolyzed polyesters have electrical conductivities in the semiconductor range. This property is generally attributed to the presence of conducting graphitic domains embedded in an amorphous carbon matrix. Polyaromatic polymers may also be present and these may also provide conduction pathways.

The incorporation of particles or metals into a polymer matrix provides access to composite materials that sometimes have surprisingly high electrical conductivities (up to $1 \ \Omega^{-1} \ cm^{-1}$). The electronic conductivity depends on contact between the conducting particles. Hence, high loadings of the conducting material are usually needed unless it is in the form of rods or disks. Thus, the preparation of useful conducting materials by this method requires a balancing of the need for high conductivity against the deterioration of strength and other mechanical properties that may accompany the incorporation of large amounts of particulate material into the polymers. Electrically conducting composite materials are used as antistatic coating and packaging materials, electrodes, and in a variety of devices used in the electrical engineering industry.

ELECTRONICALLY CONDUCTING POLYMERS

General Description and Purpose

The third class of electrically conducting polymers includes macromolecules that in the solid state provide pathways for *electronic* conduction. Specifically, these are polymers that provide a pathway for electrons (or their counterparts, positive *holes*) to migrate along a polymer chain and to jump from chain to chain. This process superficially resembles electronic conduction in metals or metalloid semiconductors.

In polymers, the process generally depends on the presence of arrays of "conjugated," delocalized double bonds, as found in poly(sulfur nitride) or polyacetylene. "Doping" of such polymers is often required to inject electrons into the delocalized framework or to remove electrons and leave positive holes. Many systems of this type appear to depend as much on the ordered, crystalline nature of the solid-state arrangement as on the molecular structure of the polymer. A number of the polymers to be described in later sections of this chapter—poly(sulfur nitride), polyacetylene, poly(phenylene vinylene) polyphenylene, poly(phenylene sulfide), polypyrrol, polythiophene, polyaniline, and transition metal-bound polymers—provide an insight into the conduction process, although many questions about the conduction mechanisms remain. The widespread interest in electronically conductive polymers is a consequence of the possibility that these materials can be used as lightweight and easily fabricated replacements for semiconductor chips and integrated circuits, lightweight wiring, electrode materials in fuel cells and batteries, and components in a wide variety of optoelectronic devices.

Band Theory

Electronic conduction in polymers takes place by the facile migration of electrons through the molecular and solid-state structure. This process occurs because large molecules that have a delocalized unsaturated, electronic structure (many double or triple bonds in communication with each other) generate properties that are not found in simple small molecules. The electrical properties are different because the electronic structures are different.

Consider a simple diatomic molecule derived from two identical atoms that have only one electron available for bonding. The energy levels of these two electrons before and after the formation of the bond are shown in Figure 23.5. Bond formation involves the generation of two new energy levels—the bonding level, occupied by the two electrons, and an unoccupied antibonding level. This higher

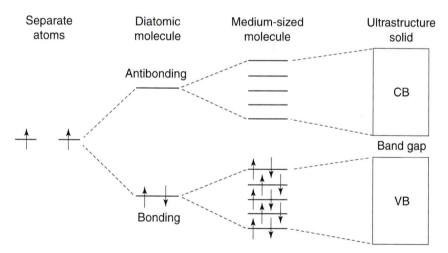

Figure 23.5   Progression from atomic orbitals in two isolated atoms, through the formation of a bonding orbital in a simple diatomic molecule, and the generation of several bonding orbitals in a medium-sized molecule, to the coalescence of orbital energy levels into bands. This last situation occurs when thousands or millions of atoms are connected to each other by a three-dimensional network of bonds. As in the case of small molecules, the energy levels are separated into a bonding set (valence band) and an antibonding set (conduction band).

level can be populated temporarily if bonding-level electrons acquire sufficient energy from heat or light to make the transition.

Bond formation in more complex molecules follows the same principle. As each bond is formed an additional bonding and antibonding level is added to the overall electronic structure. Thus, the larger the number of atoms that are linked together in a molecule or ultrastructure, the greater will be the number of bonding and antibonding orbitals. In a high-polymer molecule, or a solid-state system, thousands or millions of atoms may be involved and the number of molecular orbitals becomes correspondingly large.

Beyond a certain size of "molecule" the bonding orbitals become crowded together on the energy scale to form a set of closely spaced energy levels. This set is known as a bonding *band* or, in physics or solid-state terminology, as the *valence band*. The term "band" is used because the energy levels are so closely spaced that for all practical purposes they form an energy continuum within which electrons can exchange places and are free to wander throughout the material. A similar crowding and band formation occurs with the antibonding orbitals, which are described as the *conduction band*. The two bands are usually separated by an energy gap known as the *band gap*. The height of the band gap and the degree to which the valence band is filled with electrons determine if the material is an insulator, a semiconductor, a metal, or a "semimetal." These possibilities are shown schematically in Figure 23.6.

It should be noted that although the electrons in a band are free to wander within the energy confines of the band, there will be no overall movement of electrons unless they have access to unoccupied energy levels and unless they can function as unpaired electrons or some derivative state.

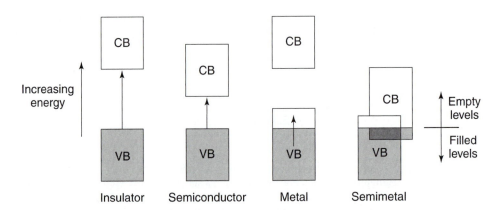

Figure 23.6 Simplified representation of the band structure found in an insulator, a semiconductor, a metal, and a semimetal. The shaded boxes indicate regions filled with electrons. Unshaded areas are regions that could accept electrons promoted from the filled levels.

Insulators

Electrical conductivity is attributed to the presence of unpaired electrons or derivative states in the conduction band, the valence band, or at new energy levels in the band gap. An electron that is unpaired (i.e., is the sole occupant of an energy level) is unconstrained by the presence of a spin-paired partner, and is free to flow in the direction dictated by an applied electrical potential.

If all levels of the valence band are full, the (coherent) flow of electrons in one direction would be difficult if not impossible. However, one way in which unpaired electrons can be generated is by electrons jumping the band gap from the top of the valence band to the lower levels of the conduction band after being activated by thermal energy or light. Insulators are materials that have band gaps that are so wide that electron jumps of this magnitude are virtually prohibited. Most polymers fall into this category.

It should be noted that no sharp demarcation line exists between insulators and semiconductors. The difference between these two types of materials is one of degree rather than kind. It is also worthwhile to note that the concept of insulators being materials with a high band gap is also an oversimplification. The electrical behavior of some insulators varies unpredictably with temperature or voltage increases in a manner that cannot be explained by band gap theory.

Semiconductors

Materials which, for one reason or another, have a narrower band gap (Figure 23.6) are called semiconductors. They conduct small but significant electric currents and show conductivities between 10^{-7} and 10^2 Ω^{-1} cm^{-1}. Most of the polymers that fall into this category have skeletal systems with long sequences of formally alternating single and double bonds—highly conjugated systems that might be expected to favor the broad delocalization of electrons. Conduction generally is via unpaired electrons or related states that are generated in one of several ways.

First, if the band gap is small, heat or light will provide enough energy for an electron to jump from the highest energy levels of the valence band to the lowest levels of the conduction band (Figure 23.6). The promoted electron will be free to move in the conduction band and will be a current carrier as it migrates along the chain toward the positive electrode. Meanwhile, the vacancy left in the valence band constitutes a positively charged hole. This, too, can migrate along the polymer chain, but in the opposite direction. Hence, an electric current flows. Such materials are known as *intrinsic semiconductors*. For obvious reasons their conductivity rises as the temperature is raised or as the light intensity is increased.

Second, the electrical conductivity of semiconductor polymers can be much improved by the injection or removal of electrons from an external source. Electrons may be injected by the addition of small-molecule electron donor chemicals (such as radical anions like sodium naphthalenide) or by reduction of a polymer in an electrochemical cell. Alternatively, electrons can be removed by treatment of the polymer with electron acceptors (such as AsF_5, SbF_5, or iodine) or by electrochemical oxidation. This is known as *doping*. Doping not only increases the number of charge carriers present, it may also lower the band gap if the injected electrons occupy a new energy state within the gap.

This explanation of semiconductivity in polymers is something of an over-simplification since the free-electron charge carriers may be modified into spin-paired species that also carry the current (see later). However, as a working description to aid molecular design, this simple explanation has some value.

Metals

Metals are materials with an unfilled conduction band and a partly filled valence band (Figure 23.6). This situation arises because many metallic *atoms* contain un-paired electrons that do not associate with their neighbors to form covalent bonds. Instead, the electrons occupy the lower levels of the valence band and have ready access to the unfilled higher energy levels of that same band. Hence, electrons need not surmount a (forbidden) band gap in order to find an energy level appropriate for migration. They simply remain in the same band. For example, sodium has *one* 3s valence shell electron per atom. This means that because no *molecular* orbitals are formed in the metallic state, the available 3s electrons fill only half of the available vacancies in the valence band, and the unoccupied higher levels of that same band are freely available for electron migration.

Thus, metallic conductors have more energy states available in the valence band than electrons to fill those states. Intrinsic semiconductors and insulators have the same number of electrons as energy states. This is a key difference.

As the temperature of a metal is raised, the conductivity falls. This is because temperature increases magnify the thermal vibrations of the lattice and this increases the scattering of the moving electrons. At low temperatures the conductivity of a metal is limited by the scattering of electrons from imperfections in the lattice. At least one of the known electronically conducting polymers has all the characteristics of a metal.

Semimetals

A semimetal is a material in which the top of the valence band overlaps the bottom of the conduction band. The valence band is formally filled, but the overlap causes a redistribution of electrons so that, like a metal, empty higher levels are accessible. Thus, metallic-type conduction can occur.

Poly(sulfur nitride), $(SN)_x$

The synthesis of poly(sulfur nitride) was described in Chapter 9. This polymer has remarkable properties, including its metallic (gold) appearance and its metal-level electrical conductivity at room temperature. This conductivity ($3 \times 10^3 \ \Omega^{-1} \ cm^{-1}$) is only slightly lower than that of mercury, nichrome, or bis-muth ($1 \times 10^4 \ \Omega^{-1} \ cm^{-1}$). The conductivity of $(SN)_x$ is more pronounced along

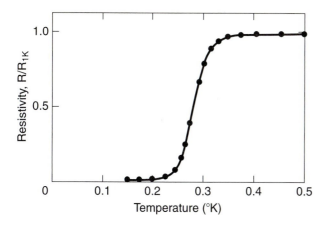

Figure 23.7 $(SN)_x$ undergoes a change from a metallic conductor to a superconductor as it is cooled below 0.3 K. [From Greene, R. L., Street, G. B., and Suter, L. J., *Phys. Rev. Lett.*, **1975**, *34*, 577.]

the direction of the polymer chains than at right angles to them. However, when cooled to 0.3 K, the material undergoes a change to a *superconductor*, at which point there is no resistance to electrical flow (Figure 23.7). The superconductivity is more isotropic (i.e., in three dimensions). If the polymer chains were totally "one-dimensional," the inevitable defects would presumably limit both the one-dimensional metallic properties and the superconductivity. Indeed, defects in the structure are known to exist; hence, conductivity *between* chains probably occurs.

The electrical properties can be explained in terms of the polymer structure. The packing of the chains is illustrated in Figure 23.8. Individual chains occupy a *cis–trans*-planar conformation (see Chapter 18), with S—N bond lengths intermediate between those of single and double bonds. This suggests that a delocalized bonding arrangement exists along the chain via delocalized, half-filled π-orbitals, and this would permit the existence of a metallic conduction band. The interchain packing is such that electronic transmission could occur via orbital overlaps between S—S, N—N, or S—N pairs on adjacent chains. This would produce a system of pockets of electrons and holes (as in a semimetal such as bismuth).

The exact reasons for the metal-like conductivity and superconductivity of $(SN)_x$ are still subject for debate. The structure of each individual molecule, with its delocalized unsaturated skeleton, clearly provides a pathway for electron migrations. Moreover, a band structure can easily be visualized, since each unit cell of the solid contains 44 electrons. Each S—N unit formally provides one unpaired electron—a plausible source of charge carriers. In a sense, this arrangement resembles that in a metal such as sodium (discussed earlier) and would lead to a half-filled valence band. Hence the electrons in this band could readily function as charge carriers by moving to infinitesimally higher energies within the band.

An alternative explanation of the room-temperature conductivity is that the material behaves as a semimetal, in which formally filled and unfilled bands overlap in energy and redistribute the electrons into two partially filled bands.

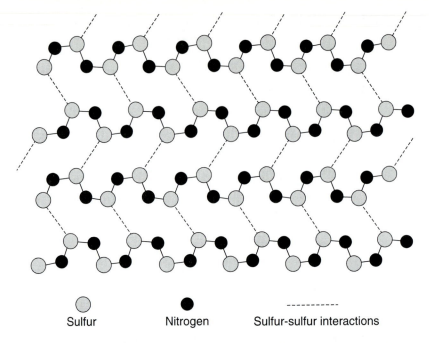

Sulfur   Nitrogen   Sulfur-sulfur interactions

Figure 23.8 Packing of the $(SN)_n$ chains in a crystalline matrix. (From Mac-Diarmid, A. G., Heeger, A. J., and Garito, A. F., *McGraw-Hill Yearbook of Science and Technology: Polymer* **1977**.)

These explanations neglect a complication normally expected for polymers with delocalized skeletal systems—the so-called Peierl's distortion that would lead to a segregation of skeletal bonds into (short) double bonds and (longer) single bonds with a resultant change from metallic properties to those of a medium-band-gap semiconductor. This does not happen with $(SN)_x$, and it is speculated that bonding linkages *between* chains prevent the distortion from occurring. The reasons for the transition to a superconductor at 0.3 K are not really understood, although many theoretical viewpoints have been presented.

Despite its unusual behavior, $(SN)_x$ is of interest more as a model for electronic conductivity in polymers than as a useful material for practical applications. The polymer is sensitive to oxidation. It also has a reported tendency to explode if heated in air or subjected to high pressures.

Brominated derivatives of $(SN)_x$ are formed when the polymer is exposed to bromine vapor. A black polymer of composition $(SNBr_{0.4})_x$ is generated. This derivative retains the fibrous structure of $(SN)_x$ and also exhibits increased electrical conductivity. Heating of $(SNBr_{0.4})_x$ in vacuum causes a partial loss of bromine and the formation of copper-colored crystals of $(SNBr_{0.25})_x$. In these structures at least part of the bromine is incorporated between the polymer chains (and parallel to them) as Br_2 and Br_3^- units. The increased conductivity may reflect the removal of electrons from the conduction band of $(SN)_x$ to generate holes.

Polyacetylene

Synthesis, structure, and morphology
Polyacetylene (**5** and **6**) has received more attention than any other electronically conducting polymer because (1) the monomer is inexpensive and readily available, (2) the synthesis of polyacetylene is a relatively straightforward process, (3) the conductivity can be varied over a very wide range by the addition of dopants, and (4) it has been the focus of considerable debate over the reasons for its properties.

5   **6**

A number of methods are available for the synthesis of electroactive quality polyacetylene, and several of these have been described in earlier chapters. However, the most widely used method is the Shirakawa technique, which involves the polymerization of acetylene in contact with a concentrated Ziegler-Natta catalyst (typically, a 4:1 mixture of triethylaluminum and titanium tetra-*n*-butoxide). Under these conditions polyacetylene is formed as a microfibrilar mat, each fibril being about 200 Å in diameter. The polymer is metallic in appearance despite the porous structure. If the polymerization is carried out at $-78°C$, the gold colored polymer occupies a *cis-trans*-planar conformation (**5**). If the reaction temperature is 150°C, the polymer is generated as the silver *trans-trans* form (**6**). The latter conformer is favored thermodynamically. Conversion of *cis-trans* to *trans-trans* occurs slowly at ambient temperature or more quickly at 200°C. The polyacetylene produced by this method is believed to be about 85% crystalline.

Conductivity and doping
The electrical conductivity of the pure polymer is quite low (10^{-9} Ω^{-1} cm^{-1} for the *cis-trans*, and 10^{-5} Ω^{-1} cm^{-1} for the *trans-trans* form), both values being near the insulator–semiconductor boundary. However, a major advance occurred in 1977 when it was discovered that pure polyacetylene can be reduced by alkali metals, radical anions, or electrochemically, or oxidized by electron acceptors such as AsF_5, SbF_5, iodine, and again, by electrochemical methods, to give materials with markedly increased conductivities. In some cases the conductivity rises by several orders of magnitude (up to the 10^6 Ω^{-1} cm^{-1} region) following doping. Hence by controlling the degree of doping, the material can be converted from a semiconductor to a metal or stabilized at any intermediate state. More recently, it has been found that strong protonic acids (perchloric, sulfuric, or trifluoromethanesulfonic acids) function as "proton dopants," generating conductivities in the range of 10^3 Ω^{-1} cm^{-1}.

Any change in structure or morphology that lowers the planarity of the chains, the length of the delocalized sequences, or the degree of crystallinity also lowers the conductivity. For example, polyacetylenes that bear alkyl or aryl side groups that force the chain into nonplanar conformations, show markedly reduced conductivities. Polymers with very high degrees of crystallinity and low degrees of cross-linking have conductivities, after doping, that are as high as $1.5 \times 10^5 \ \Omega^{-1} \ cm^{-1}$—about one-third that of copper. Moreover, these polymers are reported to be more stable in the atmosphere than their less carefully prepared counterparts.

Theory   Why is pristine *trans*-polyacetylene a poor conductor? Why does doping have such a dramatic effect on the conductivity? What are the charge carriers in a system of this kind? The answers to these questions, once thought to be straightforward, are now the subject of much debate.

First consider pure polyacetylene. Classical organic chemistry favors a structure in which π-electron delocalization would generate equal bond lengths along the chain (**7**) (as, for example, in benzene), with a continuous (valence band) molecular orbital providing a pathway for facile electron migration. This would correspond roughly to the structure of a metal and would yield a half-filled valence band. Conductivity along the chains would be high because, as in a metal, unpaired electrons would have ready access to the unoccupied levels of the valence band.

7

8   **9**

However, the Peierl's distortion, mentioned earlier, would preclude the existence of such a structure. Instead, the system would separate into a sequence of alternating single and double bonds as shown in **8** and **9**. Such a system would no longer be metallic because the energy levels would segregate into a filled valence band and an unfilled conduction band, the two being separated by a significant band gap. Hence, electrical conduction could occur only after thermal or photolytic activation of electrons to give them sufficient energy to jump the gap into the lower levels of the conduction band. This may explain why pure polyacetylene is a near insulator (*cis-trans* form) or only a modest semiconductor (*trans-trans* form) rather than a metal.

The simplest explanation for the effect of dopants on polyacetylene is that chemical or electrochemical reduction injects unpaired electrons into the lowest levels of the conduction band and that these serve as current carriers. Conversely, oxidation would remove electrons from the top of the valence band to give positive holes and unpaired residual electrons. These would be current carriers. However, experimental evidence exists that contradicts this simple explanation. For example, the conductivity of the doped polymer is much higher than is predicted from the number of unpaired electrons present. The number of unpaired electrons was measured by electron spin resonance experiments. Moreover, it was also found that undoped polyacetylene contained more unpaired electrons than can be accounted for by thermal activation into the conduction band. These observations have led to the development of a controversial interpretation based on the idea of "spinless conductivity" and electron transport via structural defects in the polymer chain.

In this interpretation it is postulated that an electron added to polyacetylene by doping goes *not* into the conduction band but into an intermediate electronic state within the band gap. This is illustrated in Figure 23.9. The product of the reduction is a radical anion, with the intergap energy states occupied by the two electrons from one π-bond and the electron added by reduction. This state is known as *polaron*. Addition of a second electron to the same site yields a di-anion, called a *bipolaron* (Figure 23.9). The bipolaron contains no unpaired electrons, but its energy levels in the band gap would allow facile jumps of electrons into the conduction band. Conduction pathways would thus be generated without the presence of semipermanent free-electron states. Oxidation of polyacetylene would lead to removal of one electron from the valence band and formation of a radical cation—a positive polaron. Further oxidation would generate a positive bipolaron.

Polyacetylene is almost unique among unsaturated backbone polymers because of the existence of two different, equal-energy resonance states, **8** and **9**.

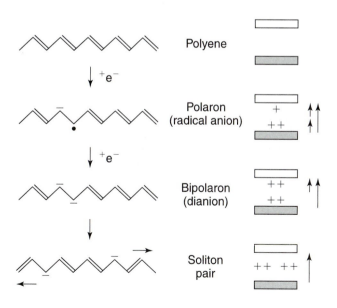

Figure 23.9 Reduction (doping) of polyacetylene to give a polaron, a bipolaron, and a soliton pair. [From Baker, G. L. in *Electronic and Photonic Applications of Polymers* (Bowden, M. J. and Turner, S. R., eds.), *ACS Advan. Chem. Ser.*, **1986**, *218*. Reprinted with permission from *American Chemical Society Advances in Chemistry Series No. 218*. Copyright 1986 American Chemical Society.]

These provide access to an alternative mechanism of conduction via defects known as solitons. Undoped polyacetylene contains approximately one unpaired electron for each 3000 carbon atoms. This is a larger number than can be accounted for by thermally induced electron transitions from the valence band to the conduction band. It has been proposed that these unpaired electrons are located at defect sites formed at the boundaries in the chain between the two types of structure shown as **8** and **9.** A defect of this type might be represented by structure **10.** Such a defect structure is known as a *soliton.*

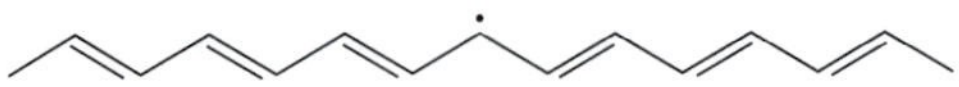

10

A "neutral" soliton (free radical) could arise from conformational changes such as the isomerization of *cis-trans* to *trans-trans* polyacetylene. But irrespective of its origin, it should be able to move readily in either direction along the chain as a current carrier. It is calculated that it costs the system only 0.4 eV to create a soliton, while it costs 0.7 eV for an electron to jump from the valence band to the conduction band. This is because the energy state for a soliton is in the middle of the band gap.

Solitons can (in theory) function as current carriers in doped polyacetylene also. As shown in Figure 23.9, a bipolaron could lower its energy by segregating into two negative solitons at the midgap energy level. Current could then be carried as the charged solitons and the defect sites move along the chain.

These proposals to explain the conductivity of polyacetylene may seem to be unduly complex. However, they do explain many of the puzzling experimental facts, and until new experimental evidence or simpler theories are available, they provide a good working hypothesis.

Uses of polyacetylene   The use of fibers of polyacetylene as lightweight "wires" to replace copper or other metals in electrical circuits is a long-range possibility. Devices based on *p-n* junctions formed from oxidized and reduced regions on a polyacetylene film, or by pressing two films together, is a more realistic possibility. However, polyacetylene appear to be particularly suitable for use as lightweight electrodes in batteries or fuel cells because of its highly porous structure. Practical devices of this type depend on the development of methods for the fabrication of the polymer into films, fibers, or intricately shaped devices.

Although polyacetylene is no longer considered to be the best candidate for electronic or electro-optical applications, the study of this polymer provided the springboard for most of the later work on conductive polymers. It was for this reason that Heeger, MacDiarmid, and Shirakawa were awarded the Nobel Prize in chemistry in 2000 for their pioneering investigation of this polymer.

Phenylene Polymers

Polyacetylene is only one of a number of polymers that have long sequences of conjugated double bonds in the skeleton. Aryl rings in a polymer chain can also serve to generate a delocalized skeletal structure. Two examples are shown in structures **11** and **12,** and a related, nonconjugated polymer is illustrated in **13.**

11   **12**

13

Poly(para-phenylene) (11)   Poly(*para*-phenylene) has long been considered a good candidate for electrical conductivity because of its extensively delocalized π-electron structure. However, the development of electroactive polymers of this type has been held up by synthesis and fabrication problems. Short-chain polyphenylenes are insoluble in most solvents and have very high melting points. Thus, during a synthesis process, oligomers precipitate from solution and chain growth is terminated. For this reason, oligomers rather than high polymers have been used for doping and conductivity experiments. However, some evidence exists that polyphenylene oligomer molecules are linked together to form higher polymers when treated with oxidizing dopants such as AsF_5. The conductivity of undoped oligophenylenes is low $(10^{-14} \ \Omega^{-1} \ cm^{-1})$, but doping with AsF_5 raises the conductivity into the range of $5 \times 10^2 \ \Omega^{-1} \ cm^{-1}$. Reduction with the usual dopants also raises the conductivity dramatically.

Theoretical explanations of the conductivity of oligo- and polyphenylenes are more speculative than in the case of polyacetylene because the available experimental data are derived from short-chain material that has been characterized in only a limited manner. Nevertheless, it has been proposed that reduction or oxidation generates a radical anion or radical cation (polaron), and that further reduction or oxidation gives a bipolaron (Figure 23.10). In contrast to polyacetylene, the conversion of a bipolaron to a soliton pair is considered to be unlikely in polyphenylenes because the quinonoid structure required for the soliton would be less stable than the aromatic form. As with polyacetylene, the polaron and bipolaron states would occupy two energy levels in the band gap, states that would provide easy access to the unfilled levels of the conduction band.

Poly(phenylene vinylenes) (12)   By far the greatest practical advances in electronically conductive polymers have come from the synthesis and study of

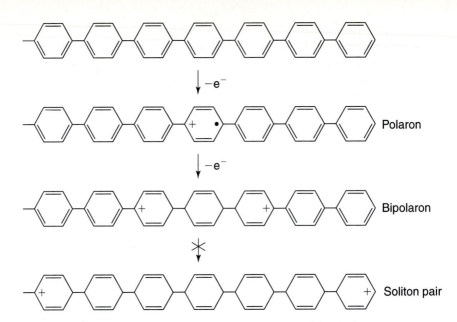

Figure 23.10 Oxidative doping of an oligo-*para*-phenylene. Removal of one electron would generate first a radical cation (polaron) and then a bipolaron. The separation of a bipolaron into a soliton pair is unlikely in this system because of the relatively high energy of the quinonoid structure. [From G. L. Baker in *Electronic and Photonic Applications of Polymers* (M. J. Bowden and S. R. Turner, eds.), *ACS Advan. Chem. Ser.*, **218** (1986). Reprinted with permission from *American Chemical Society Advances in Chemistry Series No. 218*. Copyright 1986 American Chemical Society.]

poly(phenylene vinylenes). These are polymers with the general structure shown in **12**, but with aliphatic side groups linked to the aryl rings to improve the solubility in organic solvents. Without the solubilizing side groups these species are insoluble short chain oligomers.

However, even these materials show conductivities of 3 ohm^{-1} cm^{-1} when doped with AsF$_5$. Later studies showed that high polymeric modifications can be prepared by elimination reactions from high polymeric precursors. These species show conductivities in the range of 5×10^2 to 3×10^3 Ω^{-1} cm^{-1} when doped with AsF$_5$ or iodine. The iodine-doped species are stable in the atmosphere.

These polymers can be viewed as structural hybrids of polyacetylene and poly(*para*-phenylenes) and, as such, their conductivity behavior can be visualized in similar terms. However, their stability to air is better than that of polyacetylene and this is one reason for the widespread interest in these polymers as nanowires and for their current use in commercial electronic light-emitting devices. This topic is discussed later in this chapter.

Poly(phenylene sulfide) (13) This polymer is commercially available, has excellent chemical and thermal stability, and can be fabricated readily into films and fibers. When oxidized with AsF_5 its conductivity rises into the 1- to $10\text{-}\Omega^{-1}\,cm^{-1}$ region. The behavior of this material on doping is puzzling since the primary structure **(13)** is nonconjugated beyond each aromatic ring. Presumably resonance forms such as **14** could participate in long-range electron transport. However, the situation is complicated by the fact that the polymer *reacts* with AsF_5 in ways that form bonds between the *ortho*-carbon atoms of the aromatic rings. Thus, a clear understanding of the electroactivity of this polymer may be hard to obtain.

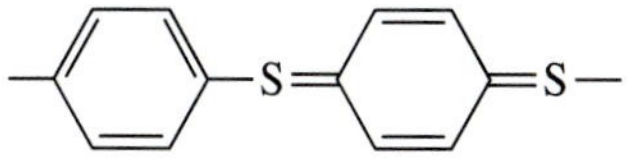

14

Polypyrrole and Polythiophene The electrochemical polymerization of pyrrole was discussed in Chapter 5. The polypyrrole formed in this way **(15)** is a smooth, continuous, black material that coats the anode. Polymerization yields a polymer that bears positive charges which are balanced to overall electrical neutrality by the presence of anions (X^-) at a high loading of one anion for every three or four pyrrole residues. The anion (from the supporting electrolyte) is typically the toluenesulfonate $(CH_3C_6H_4SO_3^-)$ or perchlorate (ClO_4^-) ion.

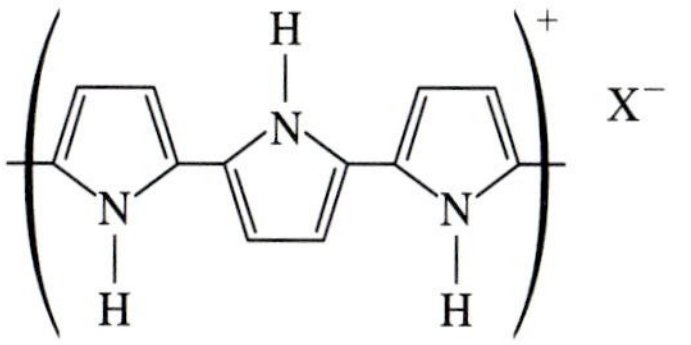

15

Such a polymer can be reduced electrochemically to neutralize the positive charges, a process that changes the color from black to transparent green. This same reduction converts the material from an electrical conductor to an insulator: Reoxidation reverses the process and regenerates the conductivity.

The conductivity of the black (oxidized) polymer can be quite high, up to $100\ \omega^{-1}\,cm^{-1}$. Despite the high conductivity, the polymer is a semiconductor rather than a metal: the conductivity rises with increased temperature.

The mechanism of electronic conduction is believed to involve polarons and bipolarons. Removal of an electron (electrochemically) generates a delocalized radical cation (polaron). Further oxidation converts this to a bipolaron which has charge carriers but no free-radical (spin) component. These two states are located within the band gap, providing electron migration within easy access of electrons in the valence band.

Polypyrrole has been the focus of much technological as well as scientific interest. Films of this polymer have good mechanical strength, the conductive polymer is stable in the atmosphere, and the color changes on oxidation and reduction could lead to uses as electrochromic switches in display devices. The polymer is reported to be in commercial use in batteries.

Polythiophene (**16**) is formed from thiophene under electrochemical oxidation conditions similar to those used for the preparation of polypyrrole. However, unlike polypyrrole, polythiophene gives relatively low conductivities (10^{-3} to 10^{-4} ohm^{-1} cm^{-1}) and it is unstable in air. As in the case of polypyrrole, a range of derivative polymers have been prepared by polymerization of substituted thiophenes.

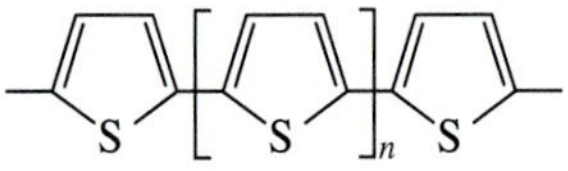

16

Polyaniline

"Polyaniline" is a black, dark green, or blue-violet material of uncertain composition known for 100 years as "aniline black" or "emeraldine." It is formed by the oxidation of aniline. Various idealized structures have been proposed, including the ones shown in **17** and **18**.

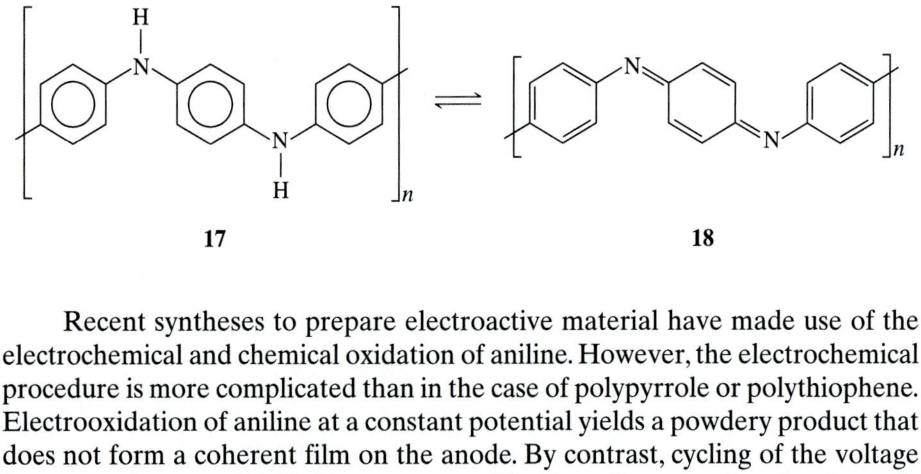

17   **18**

Recent syntheses to prepare electroactive material have made use of the electrochemical and chemical oxidation of aniline. However, the electrochemical procedure is more complicated than in the case of polypyrrole or polythiophene. Electrooxidation of aniline at a constant potential yields a powdery product that does not form a coherent film on the anode. By contrast, cycling of the voltage between -0.2 and $+0.8$ V gives a coherent film that adheres well to the electrode surface.

Paradoxically, although the exact characterization of polyaniline is more difficult than that of nearly all the other polymers mentioned in this chapter, it may prove to be one of the most useful conductive polymeric materials. It is suitable as an electrode for aqueous electrochemistry, and this opens up a range of possible applications that are closed for many of the other systems.

Stacked Phthalocyanine Polymers The synthesis and structure of poly(elementophthalocyanine) polymers was mentioned in Chapter 9. These polymers consist of a saturated elemento-oxy backbone (e.g., a siloxane backbone) with each nonoxygen heteroelement forming part of a phthalocyanine structure. The arrangement can be described as a shishkebab structure (**19**). Polymers of this type can be dissolved in sulfuric acid and converted to fibers. Treatment with iodine (the dopant) or electrochemical oxidation generates electrical conductivity along the chain axis, with conductivities of nearly 1 ohm^{-1} cm^{-1} having been reported.

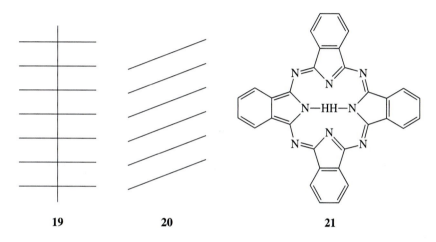

19	**20**	**21**

The mode of conduction differs markedly from that in the examples discussed in the preceding sections. Electron movement does not take place down the polymer skeleton. Indeed, high conductivities are detected if the skeleton is not present at all—when the flat phthalocyanine molecules are simply allowed to form stacks within a crystal (**20**). Even the central element is not necessary for conduction: The compound shown in **21** forms stacks that generate a room-temperature conductivity of approximately $7 \times 10^2\ \Omega^{-1}$ cm^{-1}, which rises to $3.5 \times 10^3\ \Omega^{-1}$ cm^{-1} when the system is cooled to 1.5 K. Thus, the material behaves like a metal.

The conclusion from this is that the electrical conductivity involves the transmission of electrons or holes by jumps *between* the phthalocyanine structures, which are separated by about 3.3 Å. The iodine used as the dopant is present as I_3^- and is located between neighboring stacks of the phthalocyanine molecules.

Polymers with Transition Metals in the Side-Group Structure The last group of electroactive polymers to be mentioned in this chapter are those in

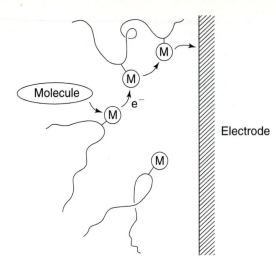

Figure 23.11 Metal atoms attached to a polymer that itself adheres to an electrode surface. The immobilized metal atoms can transmit electrons from molecules in solution to the electrode. The electrons jump from one metal atom to another without encountering a significant barrier.

which the polymer skeleton is electronically inactive but bears side groups that contain transition metals. The transition metals are capable of oxidation or reduction at an electrode surface, and may serve a catalytic function as they facilitate the migration of electrons from electrode to reagent, or vice versa. The linkage of small-molecule transition metal compounds to a polymer, which is itself attached to an electrode, restricts their mobility to regions close to the electrode surface. In addition, electron transfer can occur between different metal atoms attached to the polymer. Hence, the metallo polymer can function as a type of quasi-metal, serving as an intermediary between the electrode and reactions taking place in the solution phase (Figure 23.11).

Examples of polymers that can function in this way are given in Chapter 9. In general, polymers that bear metallocenyl side groups, such as ferrocene units, metallo porphyrins, or amine–transition metal complexes, are employed in this way. Poly(vinylpyridinium) salts of transition metal anions have also been used. In some cases, the preformed metallo polymers are solvent-cast on the surface of an electrode. In other instances a vinyl-ligand or vinylmetallo monomer is electropolymerized on the electrode surface.

Integrated Circuits Based on Polymeric Conductors

The technological imperative toward miniaturization of electronic circuits beyond those currently possible by microlithography on silicon, and the high cost of presently available devices, has focused attention on ways that polymeric electronic conductors might be used. In particular, the use of conductive polymers as "nanowires" and as molecular-level transistors has been the focus of much attention. However, significant questions and hurdles remain. For example, how can thin films of conductive polymers be subjected to microlithography and then exposed to selective doping? Can the dopants be immobilized within the polymers to prevent their migration? Can wires only nanometers thick be fabricated on a chip, and how can they be connected to polymeric transistors? How long-lived might such devices be, what is their stability to air, and could such small-sized features carry enough current without burn-out to sustain reliable devices?

Considerable effort is being expended to answer these questions and develop useable devices. The use of classical microlithography techniques (Chapter 5) employing polymer resists or electron-beam writing is feasible, and lightweight, flexible, physically robust integrated circuits made in this way would find some specialized uses. Nevertheless, the semiconductor industry is heavily invested in silicon- and, to a lesser extent, gallium arsenide-technology, and it seems unlikely that polymers will displace these materials any time soon.

However, significant advantages can be foreseen for the use of polymeric conductors in conjunction with the newer and still experimental methods of "soft lithography." In this process, a master pattern that represents a circuit etched into a disk of silicon or another material is used as a template for producing a rubber stamp (Figure 23.12). The stamp is usually made by cross-linking poly(dimethylsiloxane) pressed in contact with the master template. After the elastomer has cross-linked, it is peeled from the template, is "inked" with a solution of a polymer, and then printed on a prepared surface. Continuous roller printing, as in multicolor newspaper production, can be envisaged. In this way, successive layers of polymer and connecting wires could be assembled on a surface, with the limits of miniaturization set mainly by the accuracy with which the silicone rubber stamp can be fabricated.

This type of process could lead to mass production of miniature circuits and would allow polymers to be used as conductors or semiconductors in ways that are not possible with silicon or gallium arsenide. Although ultra-miniaturization (nano-circuits) may be difficult to achieve by this method, its use in flat panel display devices (computer and television screens, automobile dashboard panels, cellular telephones, etc.) seems feasible.

From a scientific viewpoint, this process provides opportunities for the development of other elastomers for the stamp, and for controlling its surface character. A need also exists for improving the solubility and morphological character of the conductive polymers themselves.

OPTICAL AND ELECTRO-OPTICAL DEVICES

Different Optical Properties

Lenses, optical waveguides, nonlinear optical devices, and light-emitting diodes have traditionally been made from totally inorganic materials. Yet the lightness of weight and ease of fabrication of polymers makes then attractive candidates to replace traditional materials in these applications. Optical uses for polymers can be viewed in terms of three types of properties. First, clear, colorless, glassy polymers are useful as lenses, prisms, optical waveguides, or windows. These are materials in which the color or refractive index does not change with variations in light intensity or the presence of an electric field. Second are polymers in which the color or optical density can be altered by the application of an electric field. Third are poly-

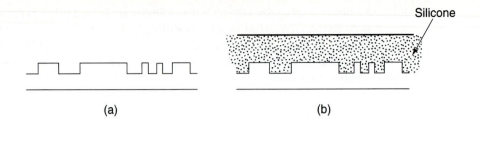

Silicone

(a)   (b)

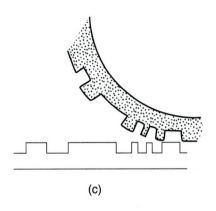

(c)

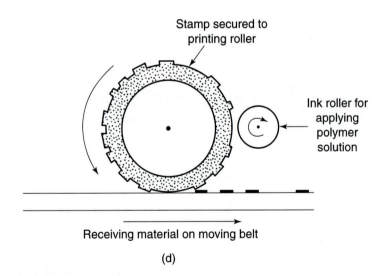

Stamp secured to
printing roller

Ink roller for
applying
polymer
solution

Receiving material on moving belt

(d)

Figure 23.12   "Soft lithography" carried out by: (a) preparation of a negative master with the features of the circuit etched via photolithography in a suitable rigid substrate such as silicon; (b) replication of the surface features by crosslinking poly(dimethylsiloxane) in contact with the surface; (c) the silicone rubber stamp is peeled from the template surface; and (d) the silicone stamp secured to a printing roller is used to transfer conductive polymer solutions to the final substrate.

meric glasses in which the refractive index varies with changes in light intensity or with the strength of an applied electric field—so-called nonlinear optical (NLO) materials. Finally, there are polymers which emit light when an electric current is applied, known as electroluminescent materials or polymer light emitting diodes. Each of these is considered in turn.

"Linear" Optical Materials

Classical inorganic glasses, such as silicates, used in windows, lenses, prisms, and optical waveguides, are characterized by the absence of absorption in the visible region of the spectrum and a refractive index that depends only on composition and temperature. These materials are heavy, brittle, and must be fabricated at high temperatures. Limits exist to the complexity of shapes that can be produced quickly, routinely, and inexpensively. Thus, polymers are used increasingly to replace inorganic glasses in camera lenses (particularly aspherical lenses) and prisms, impact-resistant eyeglasses, and bulletproof windows. Their use as optical waveguides for control systems or communications in automobile or aircraft is an important objective for the future because of the savings in weight that may be achieved. The key factors required for optical applications are optical clarity (which results from the absence of microcrystallites or other scattering centers), absence of color, high melting point, and the ability of the producer to control the refractive index.

Table 23.1 gives the refractive indices of a number of inorganic glasses and polymers. One disadvantage of classical polymers compared to inorganic glasses is that different polymers cannot be mixed to generate intermediate refractive indices. As discussed in earlier chapters, different polymers do not mix well—the mixture tends to separate into opaque, heterogeneous phases. Thus, the refractive index must be controlled at the *molecular* level rather than at the materials level.

TABLE 23.1 REFRACTIVE INDICES OF INORGANIC MATERIALS, SMALL-MOLECULE ORGANIC COMPOUNDS, AND POLYMERS

Material	Refractive Index (at 589nm–623nm)
Water	1.33
Quartz	1.50
Crown glass	1.52
Flint glass	1.65
Benzene	1.49
Nitrobenzene	1.55
Aniline	1.58
Carbon disulfide	1.63
Poly(methyl methacrylate)	1.49
Polycarbonate (Lexan)	1.57
Polystyrene	1.59
Poly(2-chlorostyrene)	1.61
Poly(ethylene terephthalate)	1.64
Polyphosphazene with 2- and 4-phenylphenoxy side groups	1.69
Polyphosphazene with iodonaphthoxy side groups	1.75

This is accomplished either by random copolymerization of two or more different monomers or by macromolecular substitution.

Refractive index depends on the number of electrons per unit volume of material. This is why lead glass ("Flint glass") has a higher refractive index than most borosilicate glasses. In polymer chemistry, high electron densities are achieved in two ways—by ensuring a high concentration of π-electrons (aryl rings) or by the presence of halogen atoms, especially bromine or iodine. Thus, some of the most widely used optical polymers contain aryl rings in the main chain, which has the added advantage that the glass transition temperatures are high. The macromolecular substitution approach has been investigated for polyphosphazenes where biphenyloxy and naphthaleneoxy side groups have been used, especially with bromine or iodine atoms linked to the aromatic rings.

Optical switches for telephone switchboards make use of the refractive index change that occurs when a polymer is heated. Thus, application of heat to a waveguide junction can switch a signal into one of several optical circuits. This effect is most pronounced for polymers that are above their T_g.

Note that it is relatively easy to prepare colored polymeric glasses for optical filters or decorative effects by dissolving an organic dye in the polymer or by linking a dye molecule to the polymer by copolymerization or macromolecular substitution.

Polymers with Controllable Opacity

Photochromic sunglasses are well-known. These use materials in which the opacity of the lens increases when the light intensity increases. Windows that become less transmissive as the light intensity increases are also available. These materials undergo some molecular change from a colorless to a colored state when irradiated with strong light, especially ultraviolet light. Typically polymers of this type contain a photochromic organic molecule dissolved in the polymer or coupled to the polymer chains. Spiropyrans (**22**), which are converted to colored merocyanines (**23**), are examples of such compounds.

Spiropyran (Sp)
(pale yellow)

22

UV
or heat

Merocyanine
(reddish-violet)

23

Polyaromatic compounds, which undergo a π-π^*-transition to a colored state when exposed to intense light have been investigated for eye protection applications.

A second form of controlled opacity is based on the behavior of liquid crystalline molecules in an electric field. Small-molecule liquid crystals serve a well-known role in liquid crystalline displays. However, limits exist to the temperature range over which these small-molecule systems will switch their orientation. Side chain liquid crystalline polymers have been proposed as a means to extend the temperature range over which the effect can be used. The principles are illustrated in a simplified form in Figure 23.13.

A large number of polymers that bear liquid crystalline (so-called *mesogenic*) side groups have been synthesized. The mesogens are typically polar biphenyl or aromatic azo units that respond to the electric field. They must be connected to the main polymer chain through flexible spacer groups so that the orientation of the mesogen is not constrained by the conformation of the polymer. High flexibility in the main chain is also preferred. Polymers that have been investigated for this purpose include polyacrylates, polysiloxanes, and polyphosphazenes.

Nonlinear Optical Polymers

Unlike the transparent polymeric glasses described above, nonlinear optical polymers are designed to undergo a change in refractive index following exposure to an electric field or when the light intensity is changed. Two types of nonlinear optical (NLO) polymers are known—designed as χ^2 and χ^3 (second- and third-order) materials.

χ^2-Polymers are species in which the *side groups* are large, polarizable organic units such as aromatic azo groups or cinnamate moieties. Normally, they bear an electron polarizing group at each end and are connected to the main chain through a flexible spacer group. Typical NLO units are shown as **24** and **25**.

| 24 | 25 |

A requirement for χ^2 activity is that the overall materials structure must be non-centrosymmetric. In other words, all the side groups must point in the same direction. This is usually accomplished by heating a film or fiber of the polymer above its glass transition temperature, subjecting the polymer to a strong, kilovolt-level electric field (poling) then cooling of the polymer below T_g, while the electric field in maintained. Below T_g the orientation of the NLO group should be retained because of the rigidity of the polymer matrix. However, unless the T_g is well above room temperature, thermal motion within the material will slowly randomize the side group orientation over a period of days or weeks, and the

A. Small-molecule liquid crystalline system

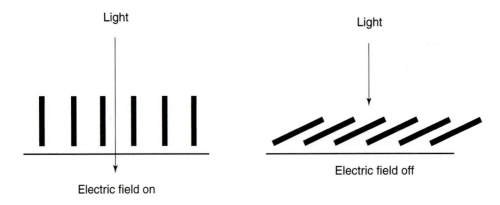

Light

Light

Electric field on

Electric field off

B. Side-chain liquid crystalline polymer

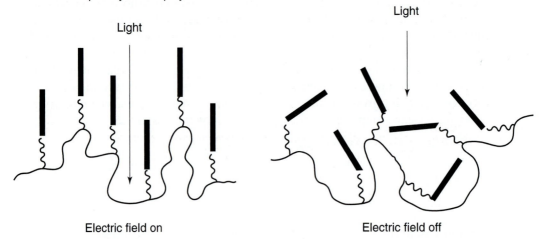

Light

Light

Electric field on

Electric field off

Figure 23.13 Principles of operation of a small-molecule liquid crystalline system compared to a side-chain liquid crystalline polymer. The thick lines represent the mesogens and the wavy lines the flexible spacer groups.

NLO effect will be lost. Crosslinking of the polymer during poling improves the chances that the ordering will be retained. Use of an NLO polymer in a simple device is depicted in Figure 23.14.

Thus, a light beam entering the device can be switched into waveguides A or B depending on whether the electric field applied to the electrodes is on or off because the refractive index of the device will change under these two sets of conditions. Such a switch could be the key feature of devices in which an electronic signal is converted into an optical signal for transmission into one of several optical waveguides in, for example, a telephone switchboard. Other devices (known as Mach-Zender interferometers) exist in which a beam of light passing through the NLO material is slowed by half a wavelength by exposure to an electric field (because the refractive index is changed) and this is used to produce destructive inteference when combined with an unmodified beam. This is the basis of an on-off optical switch controlled by the electric field.

The second class of NLO polymers are χ^3 materials that contain broadly delocalized electrons in the main chain. Examples include polyacetylene, poly(phenylene vinylene), poly-p-phenylene, etc. The χ^3 effect is weaker than the χ^2, but the molecules do not have to be aligned by poling. Thus, χ^3 materials hold the promise of longer lifetimes before the NLO effect is lost.

One of the problems with polymeric NLO materials is that they may be unstable over long periods of time to the laser light used in communications devices. Moreover, highly electron-delocalized polymers tend to be colored. Hence, the balancing of NLO structure and wavelength of operation is a major challenge.

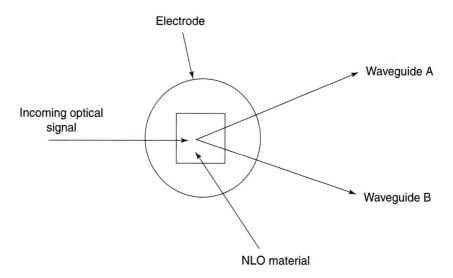

Figure 23.14   A simple switching device based on a change in refractive index following the application of an electric field to an NLO polymer.

Photovoltaic Cells

A photovoltaic cell is a device that generates direct-current electricity from light. The potential advantage for the use of polymeric semiconductors for this application is that the existing technology uses either semiconductor-grade silicon, which is expensive, or amorphous semiconductor silicon, which requires specialized equipment for its production. Photovoltaic cells based on thin polymer films cast from organic solvents would provide a serious alternative to the existing technology. A simplified explanation of their operation is as follows.

A hypothetical photovoltaic cell is shown in Figure 23.15. Absorption of a photon would cause excitation of an electron at the *p-n* junction from the valence band to the conduction band. If the polymer is doped, the electron jump will be to or from an energy level within the band. Because the bands are bent at a *p-n* junction, the promoted electron will move "downhill" toward the *p*-doped side. The hole left in the valence band will move in the opposite direction. Hence, an electric current flows as long as the junction is exposed to light.

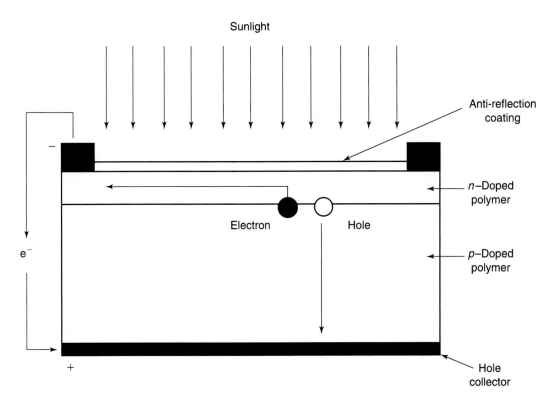

Figure 23.15 Possible design for a photovoltaic cell based on electronically conductive polymers. Electrons are promoted from the valence band to the conduction band at the *p-n* junction. Electrons move into the *p*-type layer while holes migrate through the *n*-type material, the overall result of which would be the generation of an electric current.

Electroluminescent Polymers

As mentioned earlier in this chapter, poly(phenylene vinylenes) have been commercialized for uses in light-emitting flat panel devices. These devices work on the principle shown in Figure 23.16. Electrons and holes are injected at opposite sides of a polymeric semiconductor. When the two meet, they annihilate each other and the energy is released in the form of photons. In other words, light is emitted when an electric current is passed through the polymeric semiconductor. The color of the emitted light depends on the polymer structure and its band gap. For example, a polymer with a high band gap will emit blue light; one with a smaller band gap will emit in the green, and yet a smaller band gap will give rise to red. Thus a combination of red, green, and blue emitting pixels allows any color in the visible spectrum to be generated. The development of blue-emitters has been a challenge that has now been overcome.

The initial devices being manufactured from poly(phenylene vinylenes) are small displays for use in cellular phones but, coupled with microprinting techniques of the type described earlier, it is reasonable to assume that conductive polymers may one day be used in full color screens for laptop computers and television screens.

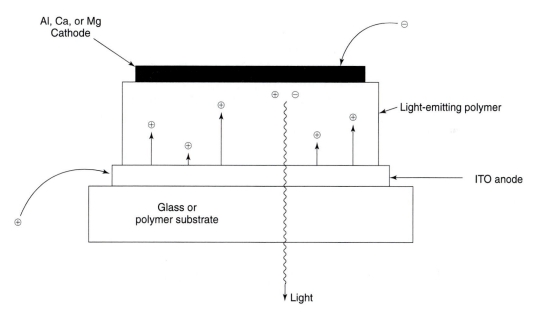

Figure 23.16 Cross section of an electroluminescent device based on an electronically-conductive polymer. Electrons are injected into the polymer via a reactive metal. Holes are injected via a transparent indium tin oxide (ITO) electrode. Electrons and holes annihilate each other within the polymer and light is emitted.

COMMENT

The study and development of electroactive and electro-optical polymers provide an illustration of a growing trend in contemporary polymer research. This is the use of polymers for highly specialized purposes rather than as bulk, commodity materials. In the field of conductive or optical materials, the polymers are being used not because they are cheap or readily available on a large scale, but because they promise to provide a *combination* of properties not found in any other materials. The same is true for the polymers that are being developed for use in bio-medicine, as discussed in Chapter 24.

STUDY QUESTIONS

1. Design a lightweight battery for use in aerospace applications. What materials would you select for the anode, cathode, solid solvent, and electrolyte?

2. Discuss the likelihood that conducting polymers might one day be used for high-voltage electric power lines. What are the advantages and disadvantages of specific polymers for this application?

3. You have been presented with the task of choosing a polymer for use as a semiconductor for integrated circuit fabrication. Which polymer would you choose? How would the polymer be *n*-doped and *p*-doped in various regions? What level of miniaturization seems possible? What problems would you have to overcome?

4. What are the potential disadvantages in the design of the photovoltaic device shown in Figure 23.15 and how would you correct any problems that might exist? How might such devices be mass-produced in order to cover a large area of barren land with a pollution-free source of electric power?

5. Through discussions with other students in your class, consider the advantages and disadvantages in the use of polymer-based electroluminescent flat panel screens in computer monitors or televisions, compared to the existing technology. What others uses might be found for the polymer-based devices especially if they could be produced as large-area units?

6. Discuss the problems that might be encountered in the synthesis of side-chain liquid crystalline, NLO, and high-refractive index polymers. How would you overcome these hurdles? Can you think of an application in which two of these three features might be incorporated into the same polymer?

7. If you had a choice to use a liquid-crystalline device based on a small-molecule system or one based on a polymeric system, under what circumstances would you use one or the other?

8. What are the advantages and disadvantages of polymer membrane based fuel cells compared to devices that use molten salt electrolytes? Based on a review of Chapter 22 discuss how you would design an improved fuel cell membrane (a) for a hydrogen-oxygen fuel cell and (b) for a direct methanol cell. What advantages can you identify for the use of fuel cells rather than rechargeable lithium batteries in satellites, automobiles, laptop computers, and cellular phones?

9. Suppose one fiber is made from an electrically conducting polymer of high molecular weight, and a second fiber from the same polymer of lower molecular weight. Why would their conductivities be expected to differ?

10. Some polymers can have nonlinear optical (NLO) properties that are much more pronounced than those of competing inorganic materials such as lithium niobate. What other advantages might polymers have in this application?

11. In the area of NLO materials, what would be a possible advantage of frequency doubling?

SUGGESTIONS FOR FURTHER READING

Polymer Electrolytes

ABRAHAM, K. M., JIANG, Z., and CARROLL, B., "Highly Conductive PEO-like Polymer Electrolytes," *Chem. Mater.*, **1997**, *9,* 1978.

ALLCOCK, H. R., NAPIERALA M. E., OLMEIJER, D. L., CAMERON, C. G., KUHARCIK, S. E., REED, C. S., and O'CONNOR, S. J. M., "Macromolecules for Solid Polymer Electrolytes," *Electrochimica Acta*, **1998**, *43*, 1145.

BLONSKY, P. M., SHRIVER, D. F., AUSTIN, P. E., and ALLCOCK, H. R., "Polyphosphazene Solid Electrolytes," *J. Am. Chem. Soc.*, **1984**, *106*, 6854.

GOTTESFELD, S., HALPERT, G., and LANDGREBE, A., (Eds.) *Proton Conducting Membrane Fuel Cell*. Pennington, N.J.: Electrochem. Proc. Series, **1995**.

GRAY, F. M. *Solid Polymer Electrolytes*. New York: VCH Publishers, **1991**.

KOPPEL, T. *Powering the Future*, Canada: John Wiley & Sons, Ltd., **1999**.

RATNER, M. A., and SHRIVER, D. F., "Ion Transport on Solvent-Free Polymers," *Chem. Rev.*, **1988**, *88*, 109.

SHRIVER, D. F., and FARRINGTON, G. C., "Solid Ionic Conductors," *Chem. Eng. News*, May 20, **1985**, p. 42.

SKOTHEIM, T. A. (ed), *Handbook of Conducting Polymers*, Vols. 1 and 2. New York: Dekker, **1986**.

Electronically-Conductive Polymers

BAKER, I. B., "Progress toward Processable, Environmentally Stable Conducting Polymers," in *Electronic and Photonic Applications of Polymers* (M. J. Bowden and R. S. Turner, eds.) *ACS, Advan. Chem. Ser.*, **1988**, *218*.

BOWDEN, M. J., and TURNER, S. R., (Eds.), *Electronic and Photonic Applications of Polymers, ACS Advan. Chem. Ser.*, **1988**, *218*.

BREDAS, J. I., and Street, G. B., "Polarons, Bipolarons, and Solitons in Conducting Polymers," *Acc. Chem. Res.*, **1985**, *18*, 309.

COWAN, D. O., and WIYGUL, F. M., "The Organic Solid State," *Chem. Eng. News*, July 21, **1986**, p. 28.

DIAZ, A. F., and BARGON, J., "Electrochemical Synthesis of Conducting Polymers," in *Handbook of Conducting Polymers*, Vol. I, (T. A. Skotheim, ed.). New York: Dekker, **1986**, p. 81.

EPSTEIN, A. J., and MILLER, J. S., "Linear-Chain Conductors," *Sci. Am.*, October, **1979**, p. 52.

EPSTEIN, A. J., "Electrically-Conducting Polymers: Science and Technology," *Mater. Res. Bull.*, **June 1997**, 16 .

FAULKNER, L. R., "Chemical Microstructures on Electrodes," *Chem. Eng. News*, Feb. 27, **1984**, p. 28.

FROMMER, J. E., and CHANCE, R. R., "Electrically Conductive Polymers," in *Encyclopedia of Polymer Science and Engineering*, 2nd ed., Vol. 5 (H. F. Mark, N. Bikales, C. G. Overberger, G. Menges, and J. I. Kroschwitz, Eds.). New York: Wiley, **1986**, p. 462.

LABES, M. M., LOVE, P., and NICHOLS, L. F., "Polysulfur-Nitride: A Metallic, Superconducting Polymer," *Chem. Rev.*, **1979**, *79*, 1.

MACDIARMID, A. G., and KANER, R. B., "Electrochemistry of Polyacetylene: Application to Rechargeable Batteries," in *Handbook of Conducting Polymers*, Vol. 1 (T. A. Skotheim, Ed.). New York: Dekker, **1986**, p. 689.

MACDIARMID, A. G., "Polyaniline and Polypyrrole: Where are We Headed?" *Synth. Mater.*, **1997**, *84*, 27.

MACDIARMID, A. D., and ZHENG, W., "Electrochemistry of Conjugated Polymers and Electrochemical Applications," *Mater. Res. Soc. Bull.*, **June 1997**, 24 .

MARKS, T. J., "Electrically Conductive Metallomacrocyclic Assemblies," *Science*, **1985**, *227*, 181.

MCCULLOUGH, R. D. "The Chemistry of Conducting Polythiophenes," *Adv. Mater.*, **1998**, *10*, 93.

REYNOLDS, J. R., "Advances in the Chemistry of Conducting Organic Polymers: A Review," *J. Mol. Electron.*, **1986**, *2*, 1.

REYNOLDS, J. R., "Electrically Conductive Polymers," *ChemTech.*, **1988**, *18*, 440.

SEANOR, D. A., *Electrical Properties of Polymers*. New York: Academic Press, **1982**.

TOUR, J. M., "Soluble Oligo- and Polyphenylenes," *Adv. Mater.* **1994**, *6*, 190.

General and Electro-Optical

CIFERRI, A. (Ed.), *Supramolecular Polymers*. New York: Marcel Dekker, Inc., **2000**.

CRAVER, C., and CARRAHER, C. E., Jr. (Eds.), *Applied Polymer Science—21st Century*, Washington: American Chemical Society, **2000**.

HAMLEY, I. W., *Introduction to Soft Matter. Polymers, Colloids, Amphiphiles and Liquid Crystals*, Chichester: John Wiley and Sons, **2001**.

KIM, D. Y., CHO, H. N., KIM, C. Y., "Blue Light Emitting Polymers," *Prog. Polymer Sci.*, **2000**, *25*, 1089.

LEE, L.-H. (ed.), *New Trends in Physics and Physical Chemistry of Polymers*, New York: Plenum, **1989**.

PRASAD, P. N., and NIGAM, J. K. (Eds.), *Frontiers of Polymer Research*, New York: Plenum Press, **1991**.

PRASAD, P. N., and WILLIAMS, D. J., *Introduction to Nonlinear Optical Effects in Molecules and Polymers*, New York: John Wiley and Sons, **1991**.

PRASAD, P. N. (Ed.), *Frontiers of Polymers and Advanced Materials*, New York: Plenum, **1994**.

SAEGUSA, T., HIGASHIMURA, T., and ABE, A. (Eds.), *Frontiers of Macromolecular Science*, Oxford: Blackwell Scientific Publishers, **1989**.

TAKEMOTO, K., OTTENBRITE, R. M., and KAMACHI, M. (Eds.), *Functional Monomers and Polymers*, 2nd ed., New York: Marcel Dekker, **1997**.

TANABE, Y. (Ed.), *Macromolecular Science and Engineering. New Aspects*, New York: Springer, **1999**.

VOGTLE, F., *Supramolecular Chemistry. An Introduction*, Chichester: John Wiley and Sons, **1991**.

Devices

BREDAS, J.-L., CORNIL, J., and HEEGER, A. J., "The Exciton Binding Energy in Luminescent Conjugated Polymers," *Adv. Mater.*, **1996**, *8*, 447.

GAO, J., YU, G. and HEEGER, A. J., "Polymer p-i-n Junction Photovoltaic Cells," *Adv. Mater.* **1998**, *10*, 692.

MAO, S. S., RA, Y., GUO, L., DALTON, R. L., CHEN, A., GARNER, S., and STEIER, W., "Progress Toward Device-Quality Second-Order Nonlinear Optical Materials," *Chem. Mater.* **1998**, *10*, 146.

RENAK, M. L., BAZAN, G. C., and ROITMAN, D., "Microlithographic Process for Patterning Conjugated Emissive Polymers," *Adv. Mater.* **1997**, *9*, 392.

WILBUR, J. L., JACKMAN, R. J., WHITESIDES, G. M., CHEUNG, E. L., LEE, L. K., and PRENTISS, M. G., "Elastomeric Optics," *Chem. Mater.*, **1996**, *8*, 1380.

WOHRLE, D., and MEISSNER, D., "Organic Solar Cells," *Adv. Mater.* **1991**, *3*, 129.

ZHAO, X.-M., STODDART, A., SMITH, S. P., KIM, E., XIA, Y., PRENTISS, M., and WHITESIDES, G. M., "Fabrication of Single-Mode Polymeric Waveguides Using Micromolding in Capillaries," *Adv. Mater.*, **1996**, *8*, 420.

YANG, Y., "Electroluminscent Devices," *Mater. Res. Soc. Bull.*, **(June 1997)**, 31.

24

Biomedical Applications of Synthetic Polymers

USES FOR POLYMERS IN BIOMEDICINE

The widespread use of synthetic polymers in technology and in everyday life is an accepted feature of modern civilization. Polymers are now being used for almost every conceivable application, and there is every indication that these uses will continue to increase in future years. However, there exists one important area in which the use of synthetic polymers has generally been cautious and limited—the area of medicine. There are a number of scientific reasons for this, and these will be discussed below. The important point is that profound changes are expected in medical techniques as new synthetic polymers are developed. In fact, the application of polymers to medicine has become one of the principal challenges facing the polymer scientist.

The types of synthetic polymers needed for biomedical applications can be grouped roughly into three categories: (1) polymers that are sufficiently biostable to allow their long-term use in artificial organs—blood pumps, blood vessel prostheses, heart valves, skeletal joints, kidney prostheses, and so on; (2) polymers that are bioerodable—materials that will serve a short-term purpose in the body and then decompose to small molecules that can be metabolized or excreted, sometimes with the concurrent release of drug molecules; and (3) water-soluble polymers (usually bioerodable) that form part of plasma or whole blood substitute solutions or which function as macromolecular drugs. In general, the largest amount of effort has been evident for polymers of types 1 and 2, but materials in category 3 are of great interest as improved pharmaceuticals.

Importance of Surface Character of Biomaterials

Although the bulk properties of biomaterials (elasticity, flexibility, or rigidity) are important, surface properties play a more critical role in this field than in almost any other. This is because living tissues are exceedingly sensitive to "foreign" surfaces. This sensitivity ranges from the tendency of blood to clot in contact with many polymer surfaces, to the ability of microorganisms to colonize nonliving surfaces and to release toxins that kill nearby tissue cells.

The biological responses to polymer surfaces are exceedingly complex. They depend on the hydrophobicity or hydrophilicity of the surface; the smoothness, roughness, or porosity; the presence or absence of ionic groups; the types of elements exposed at the surface, bioerodability or biostability; and whether the surface is that of a coherent solid or a hydrogel. The synthetic polymer chemist has a critical role to play in this field as he or she attempts to design new materials with an exact combination of bulk and surface properties. Many of the principles mentioned in Chapter 22 have a strong bearing on the development of new biomaterials.

BIOSTABLE MATERIALS

General Principles

Everyone is familiar with the idea that body organs can be transplanted from one person to another. Heart, kidney, and corneal transplants are now performed frequently. One of the problems with organ-transplant procedures is that there are rarely enough donor organs to meet the need. Another problem is that the antibodies of the recipient often reject the donor tissues and attempt to destroy them. Although this effect can be suppressed by immunosuppressor drugs, the transplant may eventually be rejected. Moreover, immunosuppressor drugs reduce the body's ability to combat microorganisms or to destroy abnormal cells. Hence, a risk of serious infection or even cancer is associated with their use.

For these reasons, there is an increasing drive to bypass the need for live organ transplants by the use of artificial organs made from synthetic polymers. Polymeric devices that can fulfill the functions of the heart valves, blood vessels, lungs, or kidneys have been under development for many years, but polymers are now being used in experimental heart pumps, bone or socket replacements, intraocular lenses, artificial corneas, and permanently implanted artificial teeth, as well as in tooth reconstruction materials.

In all these uses, many synthetic polymers offer a broad range of advantages over metals, glass, or ceramics. Prominent among these advantages are their low density, chemical inertness, flexibility, elasticity, or rigidity according to need, and ease of fabrication into intricate shapes. Moreover, the texture, hardness, or softness of the original tissue can be mimicked by the choice of a suitable polymer. Almost all the major classes of polymers have been investigated for possible biomedical uses. However, a polymer must fulfill certain critical requirements if it is to be used in an artificial organ.

First, it must be physiologically inert. Nearly all synthetic polymers suffer from one common disadvantage—their ability to trigger rejection mechanisms by the body. These rejection processes become manifest in the coagulation of blood in contact with polymers or the inflammation or even tumor formation which occurs when some polymers remain in contact with internal tissues for long periods of time. The overcoming of these deleterious interactions is one of the most urgent problems faced by the synthetic polymer chemist. Some of the approaches that have been tried will be outlined in the following sections.

Second, the polymer itself should be stable during many years of exposure to hydrolytic or oxidative conditions at body temperature. It must be resistant to enzyme attack, and it must not change dimensions, disintegrate, or dissolve in aqueous media or in contact with lipids or other fatty materials.

Third, if it is to be used as a structural material to replace bone, it must be strong and resistant to impact.

Fourth, the polymer must be sufficiently stable chemically or thermally that it can be sterilized by chemicals, heat, or radiation.

Stability of Polymers in Living Systems

It is important to recognize that the use of synthetic polymers in living systems revolves around one of two requirements: that the polymer should be totally inert, or that the polymer should be totally biodegradable. Unfortunately, most polymers fall between these two extremes.

Biological inertness is difficult to predict on the basis of intuitive chemical knowledge. Almost certainly, hydrolysis reactions form the first line of attack by the body on a polymer but, because most synthetic polymers are totally insoluble in aqueous media, the conventional reactivity relationships familiar to chemists are irrelevant. Nevertheless, it is a fact that some insoluble synthetic polymers more than others initiate inflammation of the surrounding tissues, blood clots, and so on, and even tumor formation.

The body has three basic responses to the implantation of a foreign body. First, it responds to the physical characteristics of the object (shape, roughness, presence of sharp edges, etc.). These responses may take the form of epithelial encapsulation of the foreign body, keratinization of the surrounding tissue, thickening of the connective tissue, or generation of giant cells. Second, the body reacts to the chemical toxicity (if any) of the polymer by the appearance of issue inflammation, inhibition of epithelial growth, and other effects. Finally, there is a possibility of bacterial, fungal, or viral infection originating at the surface of the implant or of the direct generation of an antigenic reaction by some chemical component of the polymer surface.

Thus, the comparison of different polymers for biomedical uses is not a straightforward process. Research workers disagree about the relative significance of implant design and the chemical properties of the material. They also disagree on the question of whether demonstrated tumor formation in rodents means that some polymers will initiate tumor growth in human beings. Human metabolism is much slower than that of rodents, although human beings live much longer. Add to this the complication that many commercial polymers contain potentially toxic or carcinogenic monomers or additives that can be leached out easily in the body, and it will be seen that enormous difficulties face the researcher who wishes to answer the question: Which polymer is the best for a particular biomedical application?

With these uncertainties it will be recognized that the following observations are tentative indeed. Tissue culture experiments suggest that the following order represents an *increasing* degree of toxicity of various polymers.[1,2]

Silicone rubber $\approx$ polyethylene $<$ poly(tetrafluoroethylene)
$\approx$ fluorinated poly(ethylene−propylene)
$\approx$ poly(phenylene oxide)
$\approx$ poly(methyl methacrylate) $<$ poly(vinylidene fluoride)
$\approx$ nylon $\approx$ polystyrene $<$ polyurethane
$\approx$ poly(vinyl chloride)
$\approx$ ABS polymer.

Those toward the end of the list totally inhibited the growth of tissue culture cells. The preeminence of the "silicone" (polysiloxane) rubber that heads this list was challenged recently with regard to possible migration of oligosiloxane molecules from the gels in breast implants. These molecules were thought to have been transported, both passively (by diffusion) or actively (by cells), to various organs in the body. However, the claims that systemic problems were caused by these transported molecules were not supported by reports from the medical community.

Of course, the interaction between a polymer and the body may also lead to a weakening of the polymer itself. Polyurethanes disintegrate after only 16 months in the body. Nylon apparently loses 80% of its tensile strength after being implanted in the body for 3 years, polyacrylonitrile loses 24% of its strength in 2 years, and poly(tetrafluoroethylene) 6% in a year. Silicone rubber, on the other hand, is hardly affected at all in a year and a half in the body.

Four polymers account for most of the materials currently used in biostable medical applications. These are: poly(tetrafluoroethylene), polyesters, poly(di-

[1]Lee, H., and Neville, K., *Handbook of Biomedical Plastics* (Pasadena, Calif.: Pasadena Technology Press, **1971**), Chap. 14, p. 4.

[2]Homsy, C. A., K. Ansevin, D., O'Bannon, W., Thomson, S. A., Hodge, R., and Estrella, M. E., in *Biomedical Polymers*, A. Rembaum and M. Shen, eds. (New York: Dekker, **1971**), p. 132.

methylsiloxane), and polyurethanes. Although polyurethanes are nowhere near the top of the list of biocompatible polymers, they are one of the most widely used biomaterials, mainly because of their elasticity and flex strength. Compromises are nearly always needed in biomedical engineering.

CARDIOVASCULAR APPLICATIONS

Heart Valves and Vascular Prostheses

Damaged heart valves, weakened arterial walls, and blocked arteries constitute some of the commonest cardiovascular disorders, and polymers have been used extensively to correct such problems. Defective heart valves can be replaced by mechanical valves based on various designs. In one design, a ball of silicone rubber is retained inside a stainless steel cage (Figure 24.1). Silicone rubber is used because of its inertness, elasticity, and low capacity to cause blood clotting. The failure of such valves is usually due to "wedging" of the ball (i.e., the formation of a trough at the points where the ball is forced against the cage) or surface breakdown from abrasion or lipid absorption. Valves of this type are still being used. A more recent design makes use of a small, circular plate as a flap valve, with the flap made from pyrolytic carbon or poly(oxymethylene). Another surgical practice is to implant modified (crosslinked) tissue heart valves from pigs ("porcine valves"). Devices fabricated from synthetic hydrogels may eventually replace porcine valves.

Aneurisms (balloon-like expansions of the arterial wall) can be repaired by reinforcement of the artery with a tube of woven polyester or poly(tetrafluoroethylene) fabric. Completely blocked arterial sections are removed and re-

Figure 24.1 Starr-Edwards ball-type heart valves constructed from a silicone rubber ball, a chrome-cobalt cage, and a Teflon® ring for suturing to the heart tissue. [Photograph courtesy of Edwards Laboratories, American Hospital Supply Corporation.]

placed by a tube of porous poly(tetrafluoroethylene) (Gore Tex). The polymer is relatively noninteractive with blood, and its porosity favors the growth and anchoring of a lining of endothelial cells that insulate the blood from the polymer. Failure of such blood vessel prostheses usually occurs by thrombus buildup at the points of connection to the natural vessel. Evidence exists that this is partly due to the different elasticity of the natural vessel and the polymer, which induces fluid turbulence and deposition of blood platelets and thrombin. A clear need exists for new biopolymers that have an elasticity comparable to that of the living tissue.

The Artificial Heart

Heart disease and circulatory disorders are responsible for more deaths in North America and western Europe than any other ailment. The most serious problems arise from arteriosclerosis and from the progressive narrowing of the cardiac arteries. A blockage of one of these arteries can precipitate a "heart attack." For patients with an irreversibly damaged heart, two prospects exist. First, a heart transplant may be possible, but this procedure is limited for the reasons discussed earlier. Second, the functions of the damaged heart may be taken over permanently or temporarily by an artificial pump. A considerable amount of research has been devoted to the design and testing of artificial heart pumps. Synthetic elastomers and rigid polymers have been used extensively for the construction of these devices. Unfortunately, most synthetic polymers accelerate the clotting of blood. This problem is so serious that animals on which the pumps are tested sometimes die within hours from the massive, gelatinous blood clots that form in the pumps. Avoidance of the clotting process is a complex problem because it depends on the design of the pump and the presence or absence of turbulence as well as on the materials used for construction. In the following sections we will consider briefly first the design problem, and then the polymer problem.

Heart pump designs   Two types of pumps have been developed on an experimental basis: (1) auxiliary blood pumps to bypass or supplement the action of a damaged heart until it can repair itself, and (2) total artificial heart pumps that can completely replace the living organ. Many of the booster pumps have used a rigid housing, often made of reinforced epoxy resin, with an internal tube of silicone rubber (Figure 24.2). Compressed air applied inside the rigid casing compresses the silicone or polyurethane rubber inner tube and this forces blood from the pump. Valves may be used to prevent backflow, or the compression cycle may be synchronized with the pumping motion of the heart. A related device is the intraaortic balloon, a 25 cm × 2 cm polyurethane balloon inserted into the aorta which expands as compressed helium or carbon dioxide is pulsed in or out. Other devices use hemispheres of titanium, polycarbonate, or poly(methyl methacrylate) containing a polyurethane diaphragm. Pulses of compressed air or carbon dioxide actuate the diaphragm and cause the pumping of the blood (Figure 24.3).

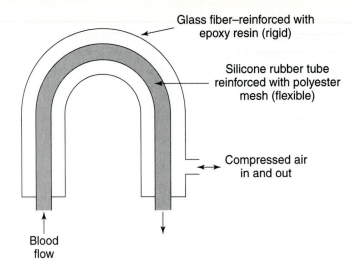

Figure 24.2 Relatively simple "artificial heart" device designed for implantation in the body. Pulses of compressed air compress the silicone-rubber inner tube, which is connected to the aorta. The phase of the pumping cycle is synchronized with that of the patient's heart.

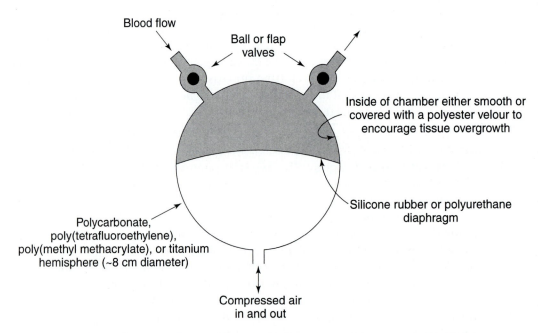

Figure 24.3 Schematic design of a hemispherical "artificial heart" pump, designed to operate outside the patient's body. The polyurethane rubber diaphragm is actuated by compressed air. Turbulence of the blood as it passes through the valves is a major cause of clotting.

Total artifical hearts have been constructed which resemble the general structure of a human heart but which are actuated by compressed gas or oil, or by electrical induction through the skin. Centrifugal pump designs are also under development.

Polymers for heart pumps   A wide variety of different polymers have been used for the fabrication of heart pumps. These include silicone rubber, polyurethane rubber, Dacron® polyester, Teflon®, polycarbonate, poly(methyl methacrylate), poly(vinyl chloride), and pyrolytic carbon. Most of these materials cause blood clotting, destruction of red cells, or alteration of the blood proteins, although some are markedly better than others.

As mentioned earlier, polyurethanes are among the most commonly used flexible biomaterials. They have excellent flexing strength. (The diaphragm in a heart pump would have to withstand about 90 million flexing motions without breaking over a 10-year period.) However, they are chemically unstable during long-term exposure to aqueous media. Some authorities believe that this could be an advantage if decomposition of the polymer is paralleled by endothelial tissue ingrowth. Calcification of polyurethane membranes is a problem during long-term use.

In theory, silicone rubber is an ideal biomaterial. Its chemical inertness is impressive, and it is soft and flexible. However, it can promote blood clotting if the blood is flowing slowly, and it can fail after continuous flexing. Another problem is the tendency of silicone rubber to absorb fats from the blood, to swell, and eventually to weaken. Fluoroalkylsiloxane polymers or polyphosphazenes may prove to be more suitable for artificial heart applications.

The ability of a synthetic polymer to initiate the clotting of blood depends on the nature of the surface (smooth surfaces are better than rough) and on the chemical and physical properties of the polymer. For example, highly water repellent polymers appear to be among the best materials for contact with blood. Because the inside lining of blood vessels is negatively charged, it has also been speculated that polymers with a surface charge might be more effective than polymers with no change.

Two additional approaches to the problem of blood incompatibility have been examined. In the first, an anticoagulant, such as heparin, is bonded to the surface or absorbed into the polymer. In the second, an attempt is made to use a polymer that will facilitate tissue overgrowth to insulate the polymer from the blood. A velvety velour of polyester fibers has been tested and has been found to function in this way. Tissue overgrowth takes more than 16 weeks. "Seeding" of the velour with living cells before implantation of the device speeds up this process. Webs of fine polypropylene fibers function in the same way.

Bacterial or fungal colonization of the surface of an implant may be the eventual cause of the failure of most artificial cardiovascular devices to sustain a patient.

TISSUE ADHESIVES AND ARTIFICIAL SKIN

It has been recognized for many years that a need exists for synthetic polymers that can be used to glue tissues together. The use of an adhesive would be much more rapid and effective than the sewing of a wound with a suture. A group of polymers based on the poly(α-cyanoacrylate) structure have proved to be effective for this purpose. α-Cyanoacrylates have the general formula shown in **1**, where the group, R, can be methyl, butyl, hexyl, octyl, and so on. These monomers polymerize by an anionic mechanism in the presence of water. Higher alkyl derivatives polymerize more rapidly on biological substrates and are less irritating to tissues than are the lower alkyl derivatives. However, their curing characteristics are somewhat unpredictable. In addition to their use as skin adhesives, they have been tested as adhesives in corneal and retinal surgery, and as an adjunct to suturing in internal surgery.

$$CH_2 = \overset{\displaystyle C \equiv N}{\underset{\displaystyle \underset{\displaystyle \underset{\displaystyle R}{|}}{\overset{\displaystyle |}{O}}}{\underset{\displaystyle \underset{\displaystyle \underset{\displaystyle |}{C = O}}{|}}{\overset{\displaystyle |}{C}}}$$

1

The search for polymeric materials that can be used as synthetic skin to cover large burns has led to the use of synthetic poly(amino acid) films for this purpose. Velours of nylon fiber have also been tested for this use, as have films of poly(α-cyanoacrylates).

BONES, JOINTS, AND TEETH

Bone fractures are occasionally repaired with the use of polyurethanes, epoxy resins, and rapid-curing vinyl resins. Silicone rubber rods and closed-cell sponges have been used as replacement finger and wrist joints, and vinyl polymers and nylon have been investigated as replacement wrist bones or elbow joints. Furthermore, cellophane and, more recently, silicone rubber have been used in knee joints to prevent fusion of the bones. Dramatic advances have been made in hip-joint surgery with the use of stainless steel or polyethylene ball joints attached to the femur by means of a poly(methyl methacrylate) filler and binder. Teflon fabric and silicone rubber have been used to make synthetic ligaments and tendons.

Synthetic polymers have been utilized for many years in the fabrication of dentures. Poly(methyl methacrylate) is the principal polymer used both for acrylic teeth and for the base material. Acrylic resins are also used for dental crowns, and epoxy resins are sometimes employed to cement crowns to the tooth post. Considerable interest has been generated in the implantation of replacement teeth directly into the mandible or maxilla. A fused carbon base for the tooth is used to delay rejection.

More recent work and anticipated developments include the use of polymeric coatings or paint to prevent the decay of teeth and the development of thermo- or photo-setting polymers to replace silver amalgam or gold as tooth-filling materials. Some progress along these lines has been made with the use of inorganic powders bound together by means of a rapid-curing poly(acrylic acid) cement.

CONTACT LENSES AND INTRAOCULAR LENSES

The correction of vision by the use of contact lenses is a very challenging problem. One constraint arises from the fact that the eye needs oxygen to carry out its metabolic processes, and obviously cannot tolerate a complicated network of arteries and veins that would interfere with its optical functions. As a result, the oxygen required is absorbed directly from the atmosphere by diffusion through the surface of the eye. This severely limits the type of material that can be used as a contact lens that covers a large part of this surface. Early materials made from a glassy polymer such as poly(methyl methacrylate) had superb transparency, but were unsatisfactory because the oxygen-permeability of a glassy polymer is typically only 1/1000 of that of a rubbery counterpart. One remedy was to fabricate a small bulge on such a "hard" lens so that, when a person blinked, the lens was moved around sufficiently for oxygen to diffuse through the newly exposed area. This approach was helped by the fact that a person blinks roughly 20,000 time a day, but this remedy was not very effective or elegant.

The modern tendency is toward "soft" contact lenses. The approach is to use a rubbery polymer such as poly(dimethyl siloxane) which has excellent transparency and almost unrivalled oxygen-permeability. However, this polymer has the disadvantage that it is hydrophobic (repels water). This creates discomfort for the user because the device is not wet by the tears that lubricate the surface of the eye. In extreme cases, the siloxane can repel the water phase to the extent that the lens becomes fused to the eye, and surgery may be required to remove it. A solution to this problem is to graft a thin coating of a hydrophilic polymer onto the inner surface of the lens. This changes the surface character without any discernible loss of oxygen permeability.

A soft contact lens can also be made from a lightly crosslinked, water-soluble polymer. Such polymers swell in aqueous media, but do not dissolve. Instead, they form soft *hydrogels,* the expanded shape of which is defined at the point of cross-linking. Hydrogel research is an important area in many fields of biomedicine because hydrogels can be designed to mimic the physical character of many tissues (cartilage, skin, blood vessel linings, etc.). The design of hydrogels for intraocular lenses (i.e., for lenses to replace the natural lens following eye injury or removal of cataract-damaged lenses) is a special challenge since the replacement lens must be folded without damage into a small cross section before insertion through a small incision into the eye.

ARTIFICIAL KIDNEY AND HEMODIALYSIS MATERIALS

The function of a kidney is to remove low-molecular-weight waste products from the bloodstream. Artificial kidneys have been available for several years. They function by passage of the blood between the walls of a dialysis cell which is immersed in a circulating fluid. Because conventional hospital hemodialysis equipment is bulky and expensive, a continuing need exists to construct smaller and cheaper units. Synthetic polymers form the basis of these new developments.

Cellophane (regenerated cellulose) has been used for semipermeable dialysis membranes in conventional kidney machines. However, the need for miniaturization has been responsible for the use of bundles of hollow fibers as a dialysis cell. In one particular development, a bundle of 2000 to 11,000 hollow fibers of modified polyacrylonitrile (17 cm long and 300 μm diameter) are used. The polymer is "heparinized" to prevent blood clotting. Hollow rayon fibers or polycarbonate or cellulose acetate fibers have also been used for the same purpose.

OXYGEN-TRANSPORT MEMBRANES

Surgical work on the heart frequently requires the use of a heart-lung machine to circulate and oxygenate the blood. A variety of devices have been developed, but many make use of a membrane through which oxygen and carbon dioxide must pass. Poly(dimethylsiloxane) membranes are highly efficient gas transporters. They are made by dip-coating a Dacron® or Teflon® screen in a xylene dispersion of silicone rubber. When dried, a film of 0.075 mm or more in thickness can be obtained, and this can be incorporated into the oxygenator. Silicone rubber membranes have also been tested in "artificial gills" for underwater breathing. It is of interest that silicone rubber has approximately six times the oxygen permeability of fluorosilicones, nearly 80 times the value for polyethylene, and 150,000 times the permeability of Teflon.[1] This could be connected with the high torsional mobility of the siloxane chains.

BIOERODIBLE POLYMERS

Most polymer research is organized to design and synthesize new polymers that are more stable than their predecessors. However, an important area of biomedical research involves a search for polymers that will decompose to nonpolymeric molecules when implanted in the body. The decomposition reactions usually involve hydrolysis (either enzymatically induced or by nonenzymatic

---

[1]DuPont's registered trademark for its fluorocarbon resins.

mechanisms) to nontoxic small molecules that can be metabolized by or excreted from the body. Three medical applications exist for polymers of this type—for use as surgical sutures, tissue ingrowth materials, or controlled drug-release devices.

SURGICAL SUTURES

The use of sutures to close an internal or external wound is well known. Catgut was used for all sutures until recently. However, catgut is relatively inert, and postoperative procedures were usually necessary for the removal of the suture after the normal 15-day healing of the tissues. A replacement for catgut is synthetic poly(glycolic acid), or condensation copolymers of glycolic acid with lactic acid (2). Poly(glycolic/lactic acid) has a high tensile strength and is compatible with human tissue. However, it differs from catgut in being totally absorbable by many patients within 15 days, thus removing the need for a suture-removal operation. The polymer degrades by hydrolysis to nontoxic glycolic acid. Poly(glycolic/lactic acid) is also used as a substrate for tissue regeneration, as described in the next section.

$$\left[\left(\begin{matrix} CH_3 & O \\ | & \| \\ -CH-C-O \end{matrix}\right)_{\!x}\!\left(\begin{matrix} & O \\ & \| \\ CH_2-C-O \end{matrix}\right)_{\!y}\right]_n$$

2

In the glycolide-lactide copolymers, the glycolide sequences tend to be much more crystalline than those of the lactide, because the lactide unit $[-CH(CH_3)C(O)O-]$ has an asymmetric carbon atom and these sequences may be stereochemically irregular. The crystallites in the glycolide sequences would make them more resistant to degradative hydrolysis. This is exploited by increasing the amount of the glycolide in the copolymer when a more slowly degrading suture is desired for the closure of more serious wounds.

TISSUE ENGINEERING POLYMERS

Tissue engineering is a process whereby polymers are used as a matrix for human cell regeneration. A polymeric object in the shape of the tissue to be regenerated is seeded with the patient's own cells. These undergo cell division and spread as the polymer hydrolyzes and the hydrolysis products diffuse from the site. Eventually the new tissue will occupy the volume and shape originally occupied by the polymer. Regenerated skin, bone, and other tissues are the subject of intensive research. Most of this work is being carried out with poly(glycolide/lactide) as the bioerodible matrix material in the form of fiber mats or other porous templates. However, newer materials such as polyphosphazenes are being investigated for bone and nerve regeneration because they hydrolyze to neutral pH products. For example, polymer 3 degrades slowly at pH 7 to give amino acid, phosphate, and ethanol, which are metabolized, and traces of ammonia, which are excreted. Hydrogels are of interest for use in the regrowth of functional organ tissue.

$$\left[-N=P-\begin{array}{c} NHCH_2COOC_2H_5 \\ | \\ | \\ NHCH_2COOC_2H_5 \end{array}\right]_n$$

3

CONTROLLED RELEASE OF DRUGS

The treatment of many diseases requires the introduction of drugs into the body. Frequently, much larger quantities of the drug are used than are needed because a large fraction of the drug will be excreted before serving its function. As a result of this procedure, the patient is subjected to an alternating overdose and then a deficiency of the chemotherapeutic agent.

This undesirable situation is illustrated by the dashed line in Figure 24.4. The optimal delivery rate would correspond to "zero-order" kinetics, in which the same amount of drug per hour is released independent of time. Also, the amount should be within the desired therapeutic range, and the delivery should continue for as long as needed. This goal would correspond to a horizontal line within the stippled area of the figure.

Thus, it has been recognized for some years that a system for the slow, continuous release of drugs would be a decided advantage for the treatment of many ailments. For example, the controlled release of insulin into the bloodstream would markedly improve the well-being of diabetics. Similarly, the targeted release of anticancer agents within the body might permit smaller doses of these highly toxic agents to be used. Advantages can also be foreseen for the slow, continuous release of birth control agents into the bloodstream over a period of months or years. On a more speculative level, the possibility exists that polymeric antioxidants introduced into the body could protect individuals against the effects of high-energy radiation.

Three approaches are being investigated in an attempt to use polymers to bring about a slow release of drugs. These are (1) the use of polymer membranes as diffusion-controlling barriers, (2) the employment of solid but biodegradable polymers to effect the controlled release of an encapsulated drug, and (3) the use of drugs that are chemically bound to water-soluble polymers.

Diffusion-Controlling Membranes or Matrices

Many chemotherapeutic drugs are relatively small molecules that can diffuse slowly through polymer membranes. Thus, if an aqueous solution of a drug is enclosed by a polymer membrane, the drug will escape through the membrane at a rate that can be controlled by membrane thickness and composition. A device that

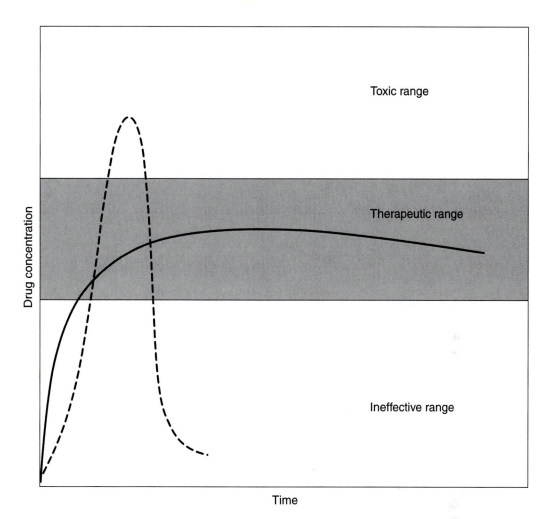

Figure 24.4 Time dependence of drug concentrations in the body. The dashed line shows the uncontrolled dissolution of a pill, with two regions so low that the concentration would be ineffective, and another region possibly high enough to be toxic. The solid line shows the goal of controlled delivery, specifically a relatively constant delivery rate in the therapeutic range for an extended period of time.

employs this principle has been used for the slow, controlled release of the antiglaucoma drug, pilocarpine, from a polymer capsule placed beneath the eyelid (Figure 24.5).

The same principle applies if a film, rod, or bead of a polymer is impregnated with a drug and is then implanted in the body at a site where the drug can have the maximum beneficial effect. Alternatively, diffusion of the drug from a polymer that is in contact with the skin can be the basis of a transdermal drug delivery system. Diffusion of the drug from the polymer matrix permits a continuous controlled release to be achieved over a period of weeks or months. Regular in-

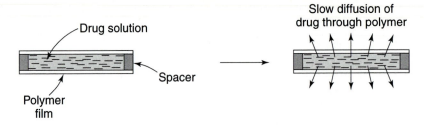

Figure 24.5 Slow, controlled release of a drug by diffusion through a polymer membrane.

jection or oral ingestion of the drug is no longer needed. This technique has been used for the slow release of birth control drugs and it has been suggested for possible use in cancer chemotherapy.

One important medical advantage in controlled-release devices of this kind is that the drug delivery system can be removed at any time when the therapy is no longer needed.

Solid, Biodegradable Matrices

An excellent way to achieve the slow, controlled release of a drug from a solid matrix is to use a biodegradable polymer as the matrix. As the polymer degrades slowly (usually by hydrolysis), the chemotherapeutic molecules are released (Figure 24.6). An important requirement is that the hydrolysis products from the polymer should be nontoxic and readily excreted.

A second requirement is that the rate of release should follow a predetermined protocol. For example, for treatment of some diseases a burst of drug in high local concentration might be followed by no release, with the cycle repeated at precise intervals. Alternatively (and more usually), a so-called "zero-order" protocol may be preferred, in which the rate of release of the drug remains constant over a period of days, weeks, or months. These requirements provide many

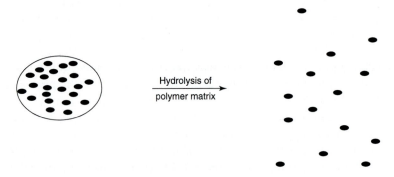

Figure 24.6 Controlled release of a chemotherapeutic drug by dissolution ("erosion") of an encapsulating polymer matrix.

opportunities for ingenuity in both the design of the matrix polymer and the shape of the device.

A number of the polymers mentioned earlier are of interest for this application. Poly(glycolic acid), poly(amino acid ester phosphazenes) such as **3,** polyphosphazenes with imidazolyl sugar- or glyceryl residues as side groups, and aliphatic polyanhydrides are all under investigation. The use of polyanhydrides that contain both aromatic (hydrophobic) and aliphatic copolymer components allows the rate of bioerosion to be controlled over a wide range, since the aliphatic anhydride linkages are more susceptible to hydrolysis. An erosion matrix of this type has reached an advanced state of development for the controlled release of an antitumor drug with special applications in brain tumor therapy. The same principle, of controlling the release rate by variations in hydrophobic and hydrophilic components of the matrix polymer, underlies much of the emerging work with polyphosphazene drug delivery systems. One polyphosphazene with 4-carboxyphenoxy side groups is used as a microencapsulant to permit oral delivery and slow, controlled release of vaccines.

Water-Soluble Polymer-Bound Drugs

Water-soluble biodegradable polymers are of interest for two types of applications. First, water-soluble polymers that are bound to drug molecules could bring about a marked improvement in the behavior of most pharmaceuticals. Second, there is the prospect that such polymers can be used in synthetic blood substitutes as viscosity enhancers or as oxygen-transport macromolecules.

As discussed earlier, two major reasons for the inefficiency of conventional drug therapy are the rapid dilution of the drug as it diffuses to and from the target site, and the ease of excretion of small-molecule drugs through the kidneys. Thus, much larger quantities of the drug must be introduced into the body than should be needed to correct the medical problem.

Water-soluble polymers diffuse only slowly through the tissues and, moreover, will not be excreted as rapidly as small molecules because macromolecules cannot normally pass through semipermeable membranes. Thus, a polymer-bound drug should offer considerable advantages over a small molecule drug.

Three possibilites exist. First, the drug could be linked to a relatively stable molecule, in which case the activity of the drug and its entry into the cell may be modified by the presence of the polymer. Second, if the polymer degrades in the body and concurrently releases the drug, the chemotherapeutic activity of the drug will be unchanged. A third possibility is that the water-soluble polymer itself is bioactive. An example of this third class is the polyanion derived from divinyl ether–maleic anhydride cyclopolymer (DIVEMA) mentioned in Chapter 2. This polymer stimulates interferon production. It will be clear that the design and synthesis of polymers that have the correct water solubility, lack of toxicity, and an appropriate rate of hydrolytic decomposition at body temperature is one of the most demanding challenges faced by the polymer chemist during the coming years.

POLYMERIC BLOOD SUBSTITUTES

Blood serves many functions, including the transport of oxygen and carbon dioxide, nutrients, minerals, and white cells. To fulfill these functions it must maintain a suitable viscosity to prevent turbulence. Synthetic polymers have been investigated for use in plasma substitutes and as volume expanders to reduce the amount of whole blood needed, for example, during the use of a heart-lung machine. Furthermore, the transmission of hepatitis and other diseases through the use of pooled plasma provides a continuing incentive for the development of a synthetic substitute for this fluid. Poly(vinyl pyrrolidone) (**4**) was used extensively by the Germans in World War II as a colloidal plasma substitute for the treatment of casualties. Its disadvantages for this application are connected with its poor biodegradability. In fact, it is retained indefinitely in the spleen, lymph nodes, liver, and bone marrow, and it may initiate carcinogenic changes. Hence, there is a serious need for the development of a water-soluble or hydrophilic polymer that is nontoxic and biodegradable. Some water-soluble polyphosphazenes may be of value for this application.

4

On an even more ambitious level, certain polymers have been investigated as oxygen-transport compounds for use in blood. For example, an emulsion of poly(tetrafluoroethylene) particles (less than 0.001 mm in diameter) or liquid fluorocarbons in water, together with glucose, salts, and surfactants, has been used to replace the blood of rats. The rats remained alive and active for periods from 5 h to several days. The long-range possibility also exists that biodegradable, water-soluble macromolecules can be synthesized that possess oxygen-carrying side groups such as metalloporphyrins. Solutions of such polymers could be used as blood replacement fluids, but would be degraded and excreted as the body produced new blood cells. Prototypes of such polymers are under development.

STUDY QUESTIONS

1. Of all the polymers discussed in this book, which ones appear to you to be the most suitable for the construction of (**a**) artificial heart pumps; (**b**) surgical tapes for covering superficial wounds; (**c**) contact lenses for the eye; (**d**) replacement tubes for varicose veins; (**e**) a tympanic membrane for the ear; (**f**) replacement aqueous humor of the eye; (**g**) a replacement for nose cartilage? In each case give the reasons for your choice.
2. How would you undertake the task of purifying the following commercial polymers to prepare "surgical-grade" material for use in implanted devices: poly(vinyl chloride), poly(dimethylsiloxane), a fluoroalkoxyphosphazene rubber, polystyrene, a polyester? What effect might the purification have on the properties in each case?

3. Which synthetic polymers might be suitable for use in replacement nerve fibers, and why?

4. Design a miniature artificial kidney using synthetic polymers for all the components.

5. What water-soluble polymers are known that might be used as carrier molecules for chemotherapeutic drugs? Suggest ways in which drugs such as steroids, antibiotics, or anticancer agents might be linked to the polymer.

6. Speculate on the advantages and disadvantages of controlled drug delivery devices that use the solid bioerodable matrix method. Are there any legal or ethical reasons why this method might be better or worse than, say, the use of a drug molecule covalently bound to a water-soluble polymer?

7. Design a new polymer to be used for the manufacture of surgical catheters, with the proviso that the material must be resistant to colonization by bacteria and fungi.

8. Immunization and vaccination of humans is normally accomplished by the injection of living or killed microorganisms into a patient to stimulate an antibody response. Design an implantable device in which microorganisms are immobilized on a polymer surface in order to stimulate antibody production (see also Chapters 7 and 8).

9. Discuss how ^{29}Si NMR might be used to test for possible transport of low molecular-weight polysiloxanes from the gels inside breast implants to other parts of the body.

SUGGESTIONS FOR FURTHER READING

ALBERTSSON, A. C., DONARUMA, L. G., and VOGL, O., Functional Polymers. Part XXXVIII. Synthetic Polymers as Drugs, *Ann. N. Y. Acad. Sci.*, **1985**, *446*, 105.

ALLCOCK, H. R., "Polyphosphazenes as New Biomedical and Bioactive Materials," in *Biodegradable Polymers as Drug Delivery Systems* (R. Langer and M. Chasin, Eds.), New York: Dekker, **1989**.

ALLCOCK, H. R., GEBURA, M., KWON, S., and NEENAN, T. X., "Amphiphilic Polyphosphazenes as Membrane Materials," *Biomaterials*, **1988**, *19*, 500.

ALLCOCK, H. R., and KWON, S., "Glyceryl Polyphosphazenes: Synthesis, Properties, and Hydrolysis," *Macromolecules*, **1988**, *21,* 1980.

ALLCOCK, H. R., "Inorganic-Organic Polymers as a Route to Biodegradable Materials," *Macromol. Symp.*, **2000**, *144*, 33.

ANDERSON, J. M., "Selected Examples of Pathologic Processes Associated with Human Polymeric Implants," *Polymer Sci. Technol.*, **1981**, *14*, 11.

ANDERSON, J. M., and KIM, S. W., (Eds.), *Recent Advances in Drug Delivery Systems*. New York: Plenum Press, **1984**.

ATALA, A., and LANZA, R. P. (Eds.), "Methods of Tissue Engineering", San Diego, Academic Press, **2002**.

BANO, M. C., COHEN, S., VISSCHER, K. B., ALLCOCK, H. R., and LANGER, R., "A Novel Synthetic Method for Hybridoma Cell Encapsulation," *Biotechnology*, **1991**, *9*, 468.

BATZ, H. G., "Polymeric Drugs," *Advan. Polym. Sci.*, **1977**, *23*, 25.

Biomaterials, Medical Devices, and Artificial Organs, **1973**, *1*, and subsequent issues.

BRALEY, S. A., "Acceptable Plastic Implants," in *Modern Trends in Biomechanics* (D. C. Simpson, Ed.). London: Butterworth, **1970,** 25–51.

BRALEY, S. A., "The Chemistry and Properties of the Medical-Grade Silicones," *J. Macromol. Sci. Chem.*, **1970**, *A4*, 529.

BRESLOW, D. S., "Biologically Active Synthetic Polymers," *Pure Appl. Chem.*, **1976**, *46*, 103.

BRUCK, S. D., "Polymeric Materials in the Physiological Environment," *Pure Appl. Chem.*, **1976**, *46*, 221.

BRUCK, S. D., *Properties of Biomaterials in the Physiological Environment*. Boca Raton, Fla.: CRC Press, **1980**.

BUTLER, G. B., "Synthesis and Properties of Novel Polyanions of Potential Antitumor Activity," *J. Macromol. Sci. Chem.*, **1979**, *A13*, 351.

CHANG, T. M. S., *Artificial Cells*. Springfield, Ill.: Charles C. Thomas, **1972**.

CHASIN, M., LEWIS, D., and LANGER, R., "Polyanhydrides for Controlled Drug Delivery," *Biopharm. Manuf.*, **1988**, *1*(2), 33–35, 38–40, 46.

CHIELLINI, E., SOLARO, R., "Biodegradable Polymeric Materials," *Adv. Materials*, **1996**, *8*, 305.

COHEN, S., BANO, M. C., CIMA, L. G., ALLCOCK, H. R., VACANTI, J. P., VACANTI, C. A., and LANGER, R., *Clinical Materials*, **1993**, *13*, 3.

COLE, G., HOGAN, J., and AULTON, M., *Pharmaceutical Coating Technology*, 2nd ed., Bristol, PA: Taylor & Francis Inc., **1995**.

COOPER, S. L., PEPPAS, N. A., HOFFMAN, A. S., and RATNER B. D. (Eds.), *Biomaterials: Interfacial Phenomena and Applications, Advan. Chem. Ser.*, **1982**, *199*.

CRAVER, C., and CARRAHER, C. E., Jr. (Eds.), *Applied Polymer Science—21st Century*. Washington: American Chemical Society, **2000**.

DONARUMA, L. G., "Synthetic Biologically Active Polymers," *Progr. Polymer Sci.* (A. D. Jenkins. Ed.), **1974**, *4*, 1.

DONARUMA, L. G., and VOGL, O. (Eds.), *Polymeric Drugs*. New York: Academic Press, **1978**.

FENDLER, J. H., "Membrane Mimetic Chemistry," *Chem. Eng. News*, **Jan. 2 1984**, *25*.

GOEDEMOED, J. H., and DE GROOT, K., "Development of Implantable Antitumor Devices Based on Polyphosphazenes," *Makromol. Chem., Macromol. Symp.*, **1988**, *19*, 341.

HASTINGS, G. W., and DUCHEYNE P. (Eds.), *Macromolecular Biomaterials*. Boca Raton, Fla.: CRC Press, **1984**.

KAMMERMEYER, K., "Biomaterials: Developments and Applications," *ChemTech.*, **1971**, *1*, 719.

LANGER, R., "Controlled Release of Macromolecules," *ChemTech.*, **1982**, *12*, 98.

LANGER, R., and CHASIN M. (Eds.), *Biodegradable Polymers as Drug Delivery Systems*. New York: Dekker, **1989**.

LANGER, R., SIEGEL, R., BROWN, L., LEONG, K., KOST, J., and EDELMAN, E., "Controlled Release: Three Mechanisms," *ChemTech.*, **1986**, *16*, 108.

LARSSON, R., LARM, O., and OLSSON, P., "The Search for Thromboresistance Using Immobilized Heparin," in *Blood in Contact with Natural and Artificial Surfaces, Ann. N. Y. Acad. Sci.*, **1987**, *516*, 102.

LEE, H., and NEVILLE, K., *Handbook of Biomedical Plastics*. Pasadena: Pasadena Technology Press, **1971**.

LINDSEY, A. S., "Polymeric Enzymes and Enzyme Analogs," *J. Macromol. Sci., Rev. Macromol. Chem.*, **1969**, *C3*, 1.

LUBICK, N., "Contact Lenses. The Hard and the Soft," *Sci. Am.*, **2000**, *284*(10), 88.

LYMAN, D. J., "Biomedical Polymers," *Rev. Macromol. Chem.*, **1966**, *1*, 355.

MCGINITY, J. W., *Aqueous Polymeric Coatings for Pharmaceutical Dosage Forms*, 2nd ed. New York: Marcel Dekker, Inc., **1997**.

OKAWARA, M., and MATSUNGA T. (Eds.), *Advanced Materials*. Tokyo: CMC Co., **1987**.

PAUL, D. R., and HARRIS F. W. (Eds.), *Controlled Release Polymeric Formulations, ACS Symp.* Ser., **1976**, *33*.

PRASAD, P. N., MARK, J. E., and TING, F. J. (Eds.), *Polymers and Other Advanced Materials. Emerging Technologies and Business Opportunities*, New York: Plenum, **1995**.

PRASAD, P. N., MARK, J. E., KANDIL, S. H., and KAFAFI, Z. H. (Eds.), *Science and Technology of Polymers and Advanced Materials*, New York: Plenum Press, **1998**.

RATNER, B. D., "Surface Structure of Polymers for Biomedical Applications," *Makromol. Chem., Macromol. Symp.*, **1988**, *19*, 163.

RHODES, C. T., and PORTER, S. C., "Coatings for Controlled-Release Drug Delivery Systems", *Drug Dev. Ind. Pharmacy*, **1998**, *24*, 1139.

RINGSDORF, H., "Structure and Properties of Pharmacologically Active Polymers," *J. Polymer Sci. Polymer Symp.*, **1975**, *51*, 135.

ROUHI, A. M. "Contemporary Biomaterials," *Chem. & Eng. News* (**Jan. 18, 1999**).

RUPP, F., GEIS-GERSTORFER, J., and GECKELER, K. E., "Dental Implant Materials: Surface Modification and Interface Phenomena," *Adv. Materials*, **1996**, *8*, 254.

SEDLACEK, B., OVERBERGER, C. G., and MARK H. F. (Eds.), "Medical Polymers: Chemical Problems," *J. Polymer Sci. Polymer Symp.*, **1979**, *66*.

SHALABY, S. W., and PEARCE, E. M., "The Role of Polymers in Medicine and Surgery," *Chemistry*, **1978**, *51*, 20.

TIRRELL, D. A., DONARUMA, L. G., and TUREK, A. B. (Eds.), *Macromolecules as Drugs and as Carriers for Biologically Active Materials, Ann. N. Y. Acad., Sci.*, **1985**, *446*.

TIRRELL, J. G., FOURNIER, M. J., MASON, T. J., and TIRRELL, D. A., "Biomolecular Materials," *Chem. & Eng. News*, (**December 19, 1994**).

VISSCHER, G. E., ROBINSON, R. L., MAULDING, H. V., FONG, J. W., PEARSON, J. E., and ARGENTIERI, G. T., "Biodegradation of and Tissue Reaction to 50:50 Poly(DL-lactide-co-glycolide) Microcapsules," *J. Biomed. Mater. Res.*, **1985**, *19*, 349.

Appendix **I**

Polymer Nomenclature

INTRODUCTION

Polymer nomenclature is not yet completely systematic and some aspects of the subject are in a state of flux. Different naming systems are often used in research and in technology, and occasionally some of the well-established names for polymers are confusing or even ambiguous. Some aspects of nomenclature were dealt with in Chapter 1 and 4. This appendix provides a summary of the main principles that are involved.[1]

STRUCTURAL AND SOURCE NAMES

Two fundamentally different systems of polymer nomenclature are in widespread use—the "structural" and "source" (or "derivative") methods. A structural name emphasizes the *actual* structure of the polymer without reference to the monomer from which it was derived. The source name is based on the name of the original monomer and is obtained simply by addition of the prefix, *poly*. For example, structures **1** and **2** represent the same polymer. The name "polymethylene" is

$$-(CH_2)_{\overline{n}}- \qquad -(CH_2-CH_2)_{\overline{n}}$$

Polymethylene   Polyethylene

1   **2**

[1]Further details will be found in *Macromolecules*, **1968**, *I*, 193; *Macromolecules*, **1973**, 6, 149; *J. Polymer Sci., Polymer Lett.*, **1973**, *11*, 389.

a fundamental structural description.[1] The term "polyethylene" tells how the polymer was made. In general, source names are more common, but structural names are preferred in the newer nomenclature systems. However, both types of names are used interchangeably at the present time. For example, the names "polyoxymethylene" and "polyformaldehyde" are both in widespread use.

PARENTHESES

Parentheses and square brackets are used only to prevent confusion. The name "polyethyl acrylate" is ambiguous. "Poly(ethyl acrylate)" is not. Neither is "polyacrylonitrile." In general, parentheses or brackets are needed if the monomer name consists of two or more words or if the root name itself contains numbers. An extreme example is the name poly[bis(2,2,2-trifluoroethoxy)phosphazene], which, because of the placing of the brackets and parentheses, is unambiguous.

END GROUPS

End groups are not specified for high polymers. In low polymers (telomers) the end groups may be depicted by the symbols α and ω as in α-chloro-ω-(trichloromethyl) polymethylene, $Cl-(CH_2-)_n CCl_3$.

POLYMERS PREPARED FROM UNSATURATED MONOMERS

Addition polymers are usually described by the source name. Examples are poly(vinyl chloride) (*not* polyvinyl chloride), poly(methyl methacrylate), polystyrene, polypropylene, and so on.

Two complications may be encountered. First, if the monomer is a diene, the polymerization may involve different addition pathways (1,2 or 1,4, for example). The name poly(1,2-butadiene) is ambiguous as a source name, since the polymer was not derived from 1,2-butadiene. A more correct name is 1,2-polybutadiene. The polymer with the formula $-(CH=CHCH_2CH_2-)_n$ is named poly(1-butenylene) to avoid ambiguity.

The second problem is concerned with the use of an unambiguous description of side groups attached to the main chain. Consider the polymer repeating structures shown in **3** and **4**. The name poly(methylstyrene) would apply to both structures. The name poly(2-methylstyrene) is also ambiguous. Structure **3** is best described by the name poly(α-methylstyrene), and structure **4** by poly(*o*-methylstyrene). Note that conventional organic nomenclature describes compound **5** as 1,1-dichloroethylene (also known as vinylidine chloride), whereas compound **6** is

---

[1] Of course, the name "polymethylene" would be both a structural *and* a source name if the polymer were to be synthesized from carbene, CH_2 or from a condensation polymerization of a molecule such as $I-CH_2-I$.

4-chloro-1-butene. These designations should be included in the source name of the polymer.

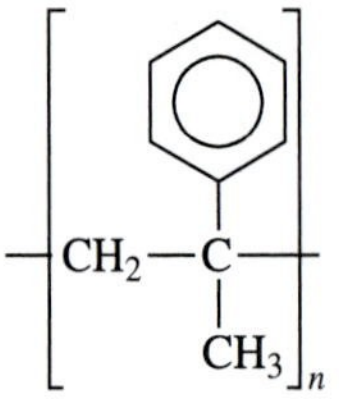

3

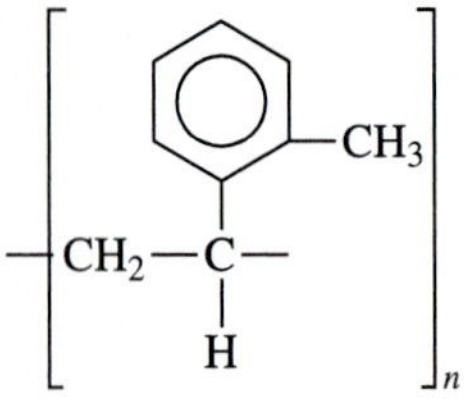

4

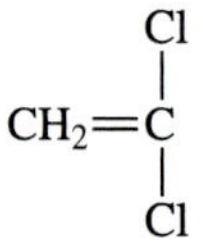

5

$CH_2{=}CH{-}CH_2{-}CH_2{-}Cl$

6

POLYMERS PREPARED BY STEP-TYPE PROCESSES

The nomenclature used for these systems can be confusing. Polyamides and polyesters are named by both structural and derivative methods. The polymer depicted in **7** could be called poly[imino(1-oxohexamethylene)], poly(6-hexanoamide), poly(6-aminocaproic acid), or poly(ε-caprolactam). In practice, the source name polycaprolactam is in more common use. The polyester **8** is usually called poly(ethylene terephthalate)—a name that is based mainly on structural considerations. Particular ambiguity is evident with the polymer of structure **9**, which is called poly(ethylene oxide), poly(oxyethylene), or poly(ethylene glycol). Compounds of structure **10** are called poly(*p*-phenylenes), and those of structure **11** are known as poly(phenylene oxide) or poly(oxy-1,4-phenylene).

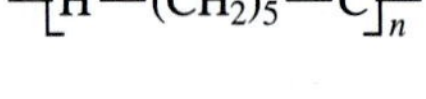

7

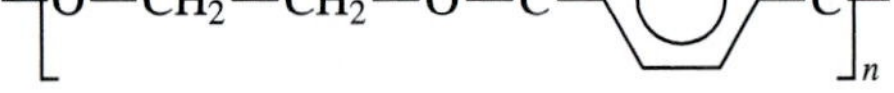

8

$\text{-}(CH_2{-}CH_2{-}O)_n\text{-}$

9

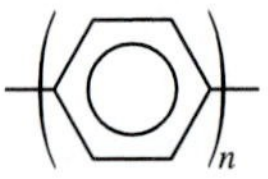

10

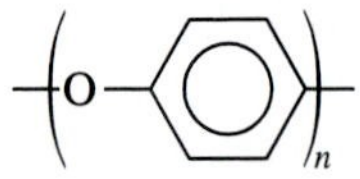

11

INORGANIC POLYMERS

The field of inorganic polymers is still in its infancy and the nomenclature is in a "trial-and-error" phase. The problem arises because developments in the inorganic macromolecular field have involved contributions by researchers trained in mineralogical, small-molecule inorganic, and organic disciplines, each making use of a different nomenclature system. However, an attempt is being made to establish a uniform system,[1] and some of the suggestions are mentioned below.

First, a few one-dimensional inorganic macromolecules are known that are constructed from covalent bonds and have well-established structural names derived from the conventions of organic polymer nomenclature. Examples are polysiloxanes [e.g., poly(dimethylsiloxane)]; polyphosphazenes [e.g., poly(diphenoxyphosphazene)]; polythiazene or polythiazyl [poly(sulfur nitride)]; and polysulfur.

Second are those inorganic polymers that possess coordinative binding either alone or together with covalent binding. The multiplicity of coordination numbers and structures that is theoretically possible within this class of compounds is almost too large for comprehension. Hence, the nomenclature problems are staggering.

The names of one-dimensional polymers are preceded by the prefix *catena-*. Two-dimensional polymers are designated by the prefix *phyllo-*, and three-dimensional macromolecules are indicated by the prefix *tecto-*. The fundamental nomenclature is based on the name of the *central atom* and all groups attached to the central atom are named as ligands. Bridging groups are indicated by the symbol μ. Points of attachment of a bridging group are indicated by the addition of the italicized symbols for the atom or atoms through which attachment to each center occurs. The symbols for atoms attached to different centers are separated by colons. Two examples are:

catena-Poly(difluorosilicon)   *catena*-Poly[bis[μ-diphenylphosphinato(1-)*O:O′*]beryllium(II)]

[1]Block, B. P., and Donaruma, G., Committee on Nomenclature, American Chemical Society Division of Polymer Chemistry.

ABBREVIATIONS, ACRONYMS, AND TRADE NAMES

Polymers are often named by abbreviations. For example, poly(methyl methacrylate) is known as PMMA, polyacrylonitrile as PAN, polyoxymethylene as POM, and so on. If these represented a universal "shorthand" notation, they would be useful as a nomenclature system. Unfortunately, no agreement exists on the preferred acronym for every polymer, and the same abbreviations are often used to represent different polymers. Communication in acronyms across a language barrier (e.g., in scientific journals) can lead to confusion. Hence, our advice is to avoid such abbreviations except in casual conversations between close colleagues.

Some trade names for polymers are now part of the language (nylon and rayon, for example). The problem with trade names is that different names are often coined for the same polymer. For this reason, trade names are generally not used in the fundamental scientific journals, although they are employed in the technological literature.

SUGGESTIONS FOR FUTHER READING

LENZ, R. W., *Organic Chemistry of Synthetic High Polymers*. New York: Interscience Publishers, **1967**.

ODIAN, G., *Principles of Polymerization*, 3rd ed. New York: Wiley-Interscience, **1991**.

SPERLING, L. H., METANOMSKI, W. V., and CARRAHER, C. E., Jr., "Polymer Nomenclature," in *Applied Polymer Science—21st Century*, (C. Craver and Carraher, J. C. E. (Eds.), Washington: American Chemical Society, **2000**, p. 49.

Appendix **II**

Properties and Uses of Selected Polymers

A problem encountered by many newcomers to polymer chemistry is the need to remember the structures, properties, and technological uses of a broad range of synthetic polymers. The data in this Appendix have been compiled as a reference source to which the reader can refer as he or she encounters new polymer names or structures in the main text or in the study questions. It is hoped that this compilation will also stimulate the reader to think about the relationship between polymer structure and uses in order to propose new polymer structures that might have more favorable properties for specific applications.

Polyamides

Name	Repeating Unit	$T_g(°C)$	$T_m(°C)$	Properties and Uses
Poly(decamethylene carboxamide) or poly(11-aminoundecanoic acid) (nylon 11, Rislan)	$-\text{N}(\text{H})-(\text{CH}_2)_{10}-\overset{\text{O}}{\text{C}}-$	46	198	Manufacture of fishing lines, bristles, gunstocks, and gasoline lines.
Poly(hexamethylene adipamide) (nylon 66, Bri-Nylon)	$-\text{N}(\text{H})-(\text{CH}_2)_6-\text{N}(\text{H})-\overset{\text{O}}{\text{C}}-(\text{CH}_2)_4-\overset{\text{O}}{\text{C}}-$	45	267	Fibers used in textiles, tire cords, rope, thread, belting, and fiber cloth. Polymer also used in molded objects such as high-impact gear wheels and electrical insulators.
Poly(hexamethylene sebacamide) (nylon 610)	$-\text{N}(\text{H})-(\text{CH}_2)_6-\text{N}(\text{H})-\overset{\text{O}}{\text{C}}-(\text{CH}_2)_8-\overset{\text{O}}{\text{C}}-$	50	165,226	Used in sports equipment and bristles for brushes.
Poly(nonamethylene urea) (Urylon)	$-\text{N}(\text{H})-(\text{CH}_2)_9-\text{N}(\text{H})-\overset{\text{O}}{\text{C}}-$		236	Fibers.
Polycaprolactam, poly(pentamethylene carboxamide), or poly(6-aminohexanoic acid) (nylon 6, Perlon, Caprolan)	$-\text{N}(\text{H})-(\text{CH}_2)_5-\overset{\text{O}}{\text{C}}-$		223	Fibers used in textiles and tire cords. Polymer also molded into gears, cams, and shoe heels.
Poly(m-phenylene isophthalamide) (Nomex)	$-\text{N}(\text{H})-\text{C}_6\text{H}_4-\text{N}(\text{H})-\overset{\text{O}}{\text{C}}-\text{C}_6\text{H}_4-\overset{\text{O}}{\text{C}}-$		390	Heat-resistant polymer: retains dimensional stability and mechanical properties up to or above 250°C; decomposes above 370°C. Used as fiber in manufacture of heat-resistant textiles for use in space suits, filter fabrics for high-temperature filtration, conveyer belts, parachute cables, and aircraft tire cords. Also used in electrical insulation and aircraft panels.

Polyesters and Polycarbonates

Name	Repeating Unit	$T_g(°C)$	$T_m(°C)$	Properties and Uses
Poly(cyclohexane-1,4-dimethylene terephthalate) (Kodel)	$-\text{O}-\text{CH}_2-\text{C}_6\text{H}_{10}-\text{CH}_2-\text{O}-\overset{\text{O}}{\text{C}}-\text{C}_6\text{H}_4-\overset{\text{O}}{\text{C}}-$	92(cis)	318(cis), 256(trans)	Fabricated into textile fibers. As isophthalate copolymer, used in "blister" packaging material.
Poly(ethylene terephthalate) (Dacron, Terylene, Fortrel, Mylar)	$-\text{O}-(\text{CH}_2)_2-\text{O}-\overset{\text{O}}{\text{C}}-\text{C}_6\text{H}_4-\overset{\text{O}}{\text{C}}-$	69	270	Strong, tough, thermoplastic with surface lubricity and resistance to wear. As textile fibers, used in tire cords, yacht sails, and electrical insulation. Films used as base for photographic film or magnetic tape. Bulk polymer used to make gear wheels and structural objects.

780

Polymer	Structure	T_m, °C	T_g, °C	Uses
Poly(butylene terephthalate)	$-O-(CH_2)_4-O-\overset{O}{\overset{\|}{C}}-$[benzene ring]$-\overset{O}{\overset{\|}{C}}-$			Tough, solvent resistant, thermoplastic with good fatigue resistance, and low moisture absorption. Used in automobile ignition systems.
Poly(4,4′-isopropylidine-diphenyl carbonate) or poly(4,4′-carbonato-2,2-diphenylpropane) (Lexan)	$-O-$[ring]$-\underset{CH_3}{\overset{CH_3}{\overset{\|}{\underset{\|}{C}}}}-$[ring]$-O-\overset{O}{\overset{\|}{C}}-$	267	150	Tough, transparent polymer with high impact strength and tensile strength near 5500 psi; fire resistant. Used as safety glass, bullet-proof windows, skylights, bathroom fixtures, plumbing, automobile components, lighting fixtures, food containers, and expanded foams. Also used in automobile doors and as expanded foam in automobile roofs.

Polymer	Structure	T_m, °C	T_g, °C	Uses
Poly(butylene glycol) (Polyglycol B)	$-O-\underset{C_2H_5}{\overset{\|}{CH}}-CH_2-O-\underset{CH_3}{\overset{\|}{CH}}-\underset{CH_3}{\overset{\|}{CH}}-$			Used as an additive to gasoline, oils, greases, antifoaming agents, and detergents.
Poly(epichlorohydrin) (Polyglycol 166)	$-O-CH_2-\underset{CH_2Cl}{\overset{\|}{CH}}-$	121 (isotactic)		Used in manufacture of urethanes, coatings, resins, and surfactants, specialty elastomers, fuel-pump diaphragms, oil and fuel hoses, and oil-well equipment.
Poly(epichlorohydrin-ethylene oxide) copolymers (ECO, Hydrin)	$-CH_2-\underset{CH_2Cl}{\overset{\|}{CH}}-O-CH_2-CH_2-O-$			Used in seals, gaskets, and hoses for automobiles.
Poly(ethylene oxide) (Carbowax)	$-O-CH_2-CH_2-$	66.2	−67	Used as a thickening agent; textile sizing agent, polymer intermediate water-soluble films, and pharmaceutical binder.
Polyformaldehyde (Delrin, Celcon)	$-O-CH_2-$	(182.5) (60)	(−30) (−82)	Tough plastic used for fabrication of gears, brushes, pipes, molded articles, pens, and carburetor components.
(Nitroso rubber)	$-O-CF_2-CF_2-\underset{CF_3}{\overset{\|}{N}}-$			Solvent-resistant, low-temperature elastomer used for the manufacture of nonburning molded objects.
Poly(tetramethylene oxide), poly(tetrahydrofuran)	$-(CH_2)_4-O-$			Used as a plasticizer for cellulose or chlorinated rubber. Used in artificial leather. Soft block in segmented polyurethanes.

(Continued)

Name	Repeating Unit	$T_g(°C)$	$T_m(°C)$	Properties and Uses
Poly(2,6-xylenol) or poly(2,6-dimethyl-1,4-phenylene oxide) (Parlene)	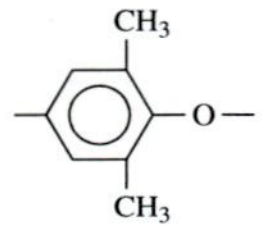		338	Structural plastic used in appliances, business machines, automobile components, water distribution equipment, and high-temperature applications.
Poly(phenylene sulfide)	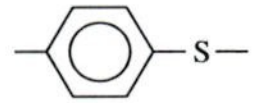	85	288	Solvent-resistant polymer below 200°C, with high thermal stability. Used in protective coatings for valves, pumps, pipes, and tanks, and in the manufacture of injection-molded articles.
Polyimides				
Poly(pyromellitimide) (Kapton)	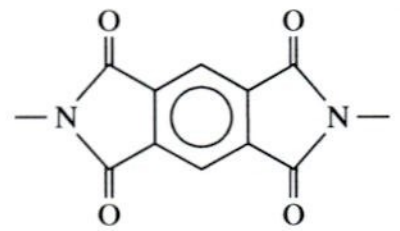			Used in high-temperature applications.
Polyimide	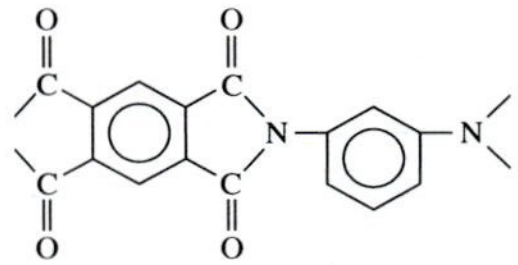			Used in wire enamels, laminates, and high-temperature applications.
Polymines				
Poly(ethylene imine)				Wet-strength improver for paper.
Inorganic				
Poly[bis(aryloxy)phosphazenes]	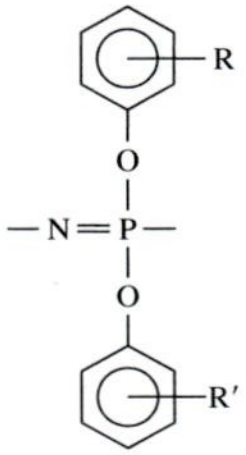			Nonburning elastomers

Name	Structure			Applications and remarks
Poly[bis(methylamino) phosphazene]	$-N=P-$ with NHCH$_3$, NHCH$_3$	14	—	Water-soluble polymer. Experimental carrier polymer for chemotherapeutic agents.
Poly[bis(trifluoroethoxy) phosphazene]	$-N=P-$ with OCH$_2$CF$_3$, OCH$_2$CF$_3$	242	−66	Highly water-repellent, ultraviolet stable, nonburning, film- and fiber-forming polymer. Projected uses in biomedical implantation devices.
Poly[bis(fluoroalkoxy)phosphazene] mixed substituent polymers (PNF or Eypel-F rubber)	$-N=P-$ with OR$_F$, OR$_F$	—	~−80	Low-temperature, solvent and oil-resistant, nonburning elastomers, used on O-rings, gaskets, seals, fuel lines, pipes, carburetor components and other automotive and aircraft applications. Possible biomedical applications.
Poly(carborane-siloxanes)(Dexsil)	$\left[-m\text{-carborane}- \begin{array}{c} CH_3 \\ \vert \\ Si-O \\ \vert \\ CH_3 \end{array} \right]_x$ (Others contain phenylsiloxane units)			Thermal stability up to 300–500°C. Used in heat-stable elastomers, O-rings, etc., and as a stationary phase in chromatography.
Poly(dimethylsiloxane)(silicone rubber)	$-O-Si-$ with CH$_3$, CH$_3$	−29	−123	Oxidation-resistant elastomer, used in seals, hoses, biomedical devices, mold releases, and water-proofing agents.
Carbon fibers				Strengths of up to 100×10^6 lb/in^2 in extension. Used as reinforcing agent in high-strength, heat-resistant composites. Some use in heat-resistant fabrics.
Poly(sulfur nitride), polythiazyl	$-S=N-$			Purple-gold fibrous polymer that is an electrical conductor at room temperature and a superconductor at very low temperatures.

(Continued)

Phenol- and amine-formaldehyde

Name	Repeating Unit	T_g(°C)	T_m(°C)	Properties and Uses
Polyphenol formaldehyde resins (Bakelite)	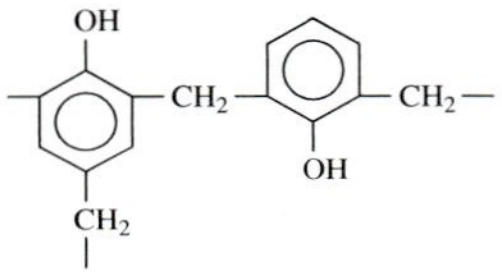 (Three-dimensional network)			Hard thermosetting polymer with high resistance to deformation under load. Used for the manufacture of cast and molded articles such as telephones, electrical insulators, buttons, laminates, and heat-resistant objects. Resist layer for printed circuits.
Poly(melamine-formaldehyde) resins	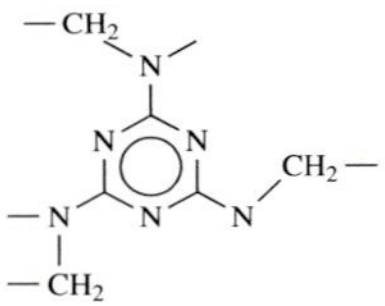			Thermosetting polymer fabricated into molded objects (e.g., dinnerware), laminates, surface coatings, table and countertops: used in textile and paper-treatment agents, adhesives, and wall paneling.
Poly(urea-formaldehyde) resins	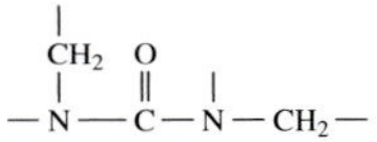 (Three-dimensional network)			Used in the manufacture of molded objects, thermal insulation, adhesives, lighting fixtures, and plywood.

Polysaccharides

Name	Repeating Unit	T_g(°C)	T_m(°C)	Properties and Uses
Cellulose (R = H)	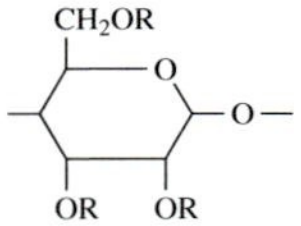		>270	Used in paper, textiles, and wood products; employed as starting material for rayon manufacture; as rayon, used in textile fiber and tire cord; as Cellophane, used as packaging film.
Carboxymethylcellulose	R = H, CH$_2$COOH			Used as adhesive and emulsifying agent; utilized in pharmaceuticals.
Ethylcellulose	R = H, C$_2$H$_5$			Transparent film used for packaging, molded articles, laquers, and as a printing ink stabilizer.
Cellulose acetate	R = H, and COCH$_3$	157	306	Clear, transparent material used as fibers, or as injection-molded, extruded, and sheet plastics.
Cellulose nitrate	R = H, NO$_2$			Used in lacquers, adhesives, and molded objects.
Ethylcellulose Methylcellulose	R = H, CH$_3$	43	165	Employed as textile finish; as an adhesive or sizing agent, and as a thicking agent.

(Continued)

Polysulfones

Polymer	Structure	Number	Description
Poly(diphenylether sulfone) (polyether sulfone)	aromatic ether–sulfone repeat unit	230	Tough polymer with good electrical insulation and self-extinguishing flame behavior. Used for the manufacture of heat-stable, injection-molded articles.
Poly(diphenyl sulfone-diphenylene oxide sulfone) copolymer (Astrel 360) (polyether sulfone)	aromatic sulfone–ether copolymer repeat unit	250–285, (depending on copolymer composition)	Tough polymer with good electrical insulation and self-extinguishing flame behavior. Used for the fabrication of heat-stable, injection-molded articles.
(Udel polysulfone)	bisphenol-A sulfone repeat unit $\left[-C(CH_3)_2-\right]$	190	Tough impact-resistant polymer with good electrical insulation and fire-resistant properties. Manufactured into injection-molded articles, housings for power tools, computer parts, circuit breakers, meter housings, battery cases, etc.

Polyurethanes

Polymer	Structure	Number	Description
Polyurethane	$-(CH_2)_3N-\overset{\text{H}}{}-\overset{\text{O}}{C}-O-$	148	Foam rubber, synthetic leather. Used as segmented copolymers with poly(tetramethylene oxide) in Spandex elastomers.
Polyurethane	$-N-\overset{\text{O}}{C}-N-\langle\text{aromatic}\rangle-CH_2-\langle\text{aromatic}\rangle-N-\overset{\text{O}}{C}-O-[-(CH_2)_4-]_8\,C-$	317	Expanded foam rubber; textile laminates; carpet underlays; thermal insulation.

Polyvinyl and Polyolefin Compounds

Polymer	Structure	Number	Description
Polyacrylamide	$-CH_2-CH-$ with $-C(=O)NH_2$	165	Water-soluble polymer used in paper treatment or as a thickening agent.
Poly(acrylic acid)	$-CH_2-CH-$ with $-COOH$	106	Water-soluble polymer used as an adhesive or thickening agent.
Polyacrylonitrile (Orlon, Acrilan, Creslan)	$-CH_2-CH-$ with $-C\equiv N$	85	Often copolymerized with small amounts of acrylamide. A fiber-forming polymer with extensive use in textiles, netting, and as a precursor for the pyrolytic preparation of carbon fibers.

Name	Repeating Unit	T_g(°C)	T_m(°C)	Properties and Uses
Poly(acrylonitrile-butadiene) copolymers (nitrile rubber)	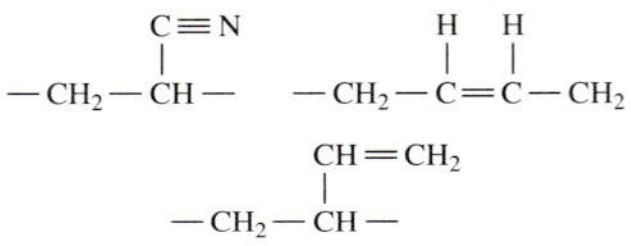 —CH₂—CH— (C≡N); —CH₂—C=C—CH₂— (H, H)			Solvent-resistant elastomer used in gaskets, oil hoses, oil seals, fan belts, oil-well components, adhesives, and in tank linings.
Poly(acrylonitrile-butadiene-styrene) copolymers (ABS polymers)	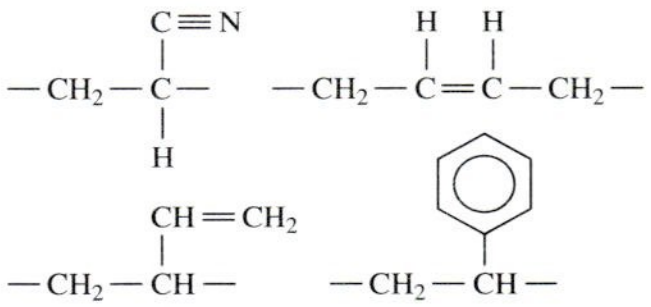 —CH₂—CH— (C≡N); —CH₂—C=C—CH₂— (H, H); —CH₂—CH— (C₆H₅)			Tough structural plastic or rubber used in the manufacture of telephones, pipes, and a wide variety of molded articles.
Poly(acrylonitrile-vinyl chloride) copolymer (Dynel)	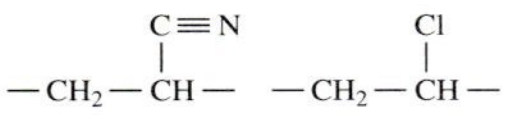 —CH₂—CH— (C≡N); —CH₂—CH— (Cl)			Used as a textile fiber material.
Polybutadiene (butadiene rubber)	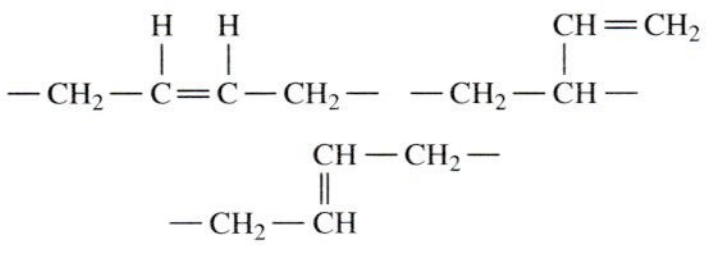 —CH₂—C=C—CH₂— (H, H); —CH₂—CH— (CH=CH₂) etc.	−58(1,3); −65(1,2-isotactic); −102(1,4-cis); −10, −48 (1,4-trans); −85(20% 1.2)	125(1,2-isotactic); 154(1,2-syndiotactic); 6.3(1,4-cis); 148, 109(1,4-trans)	Rubber polymer used as an alternative to natural rubber or SBR in the manufacture of footwear, belting, hoses, pneumatic tires, and toys.
Butadiene-acrylonitrile copolymers		−56(80–20 copolymer); −41(70–30 copolymer)		Used as adhesives.
Poly(1-butene)	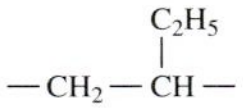 —CH₂—CH— (C₂H₅)	−45, −24(isotactic)	142, 126, 106, 65 (isotactic)	Rubbery polymer used in heavy-duty plastic sheet, as a base for pressure-sensitive tapes, and in pipes and tubes.
Poly(butyl-α-cyanoacrylate)	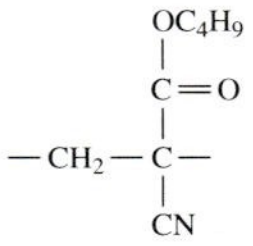 —CH₂—C— (OC₄H₉, C=O, CN)	85		Used as adhesive.
Polychloroprene (neoprene)	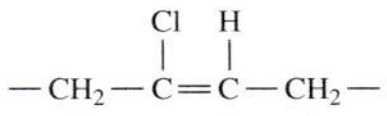 —CH₂—C=C—CH₂— (Cl, H)	−45(85% trans-1,4); −20(cis-1,4)	43	Solvent-resistant elastomer used in adhesives and cable jackets, seals, and golf-ball covers.
Poly(chlorotrifluoroethylene-vinylidene fluoride) copolymers (Kel-F)	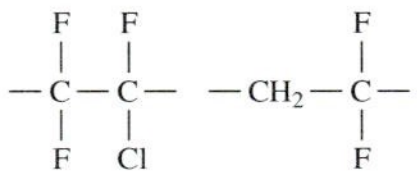 —C—C— (F,F / F,Cl); —CH₂—C— (F, F)		70(cis-1,4)	Solvent-resistant, high-temperature elastomer used in rocket motors, molded articles, O-rings, and pipe lining.

786

Polyvinyl and Polyolefin Compounds

Polymer	Structure	T	T	Uses
Poly(ethyl acrylate)	$-CH_2-CH-$, side: $C=O$, OC_2H_5	-22		Used in varnishes and printing inks.
Poly(ethyl vinyl ether)	$-CH_2-CH-$, side: OC_2H_5	-42		Elastomeric polymer used in adhesives and as a nonmigratory plasticizer.
Polyethylene or polymethylene	$-CH_2-CH_2-$	$-125, -20$	$\sim-140, 95$	Tough plastic with extensive uses in monofilament fibers, films, extrusion-molded objects, electrical insulation, bottles, and toys.
Poly(ethylene-vinyl acetate) copolymers	$-CH_2-CH_2-$ $-CH_2-CH-$, side: O, $C=O$, CH_3			Used in medical tubing and syringes, toys, and cable insulation.
Poly(ethylene-propylene) copolymers (Noedel) (EPR)	$-CH_2-CH_2-$ $-CH_2-CH-$, side: CH_3	-60 (50–50 copolymer)		Used in high-pressure steam hoses, automobile parts, appliances, and seals.
Fluorinated ethylene-propylene copolymers (Teflon FEP)				Used for electrical cable insulation.
Polyisobutylene (butyl rubber)	$-CH_2-C-$, side: CH_3, CH_3 (often copolymerized with small amounts of isoprene)	-70	1.5	Rubbery, elastomer used in adhesives, tire inner tubes, caulking compounds, dairy hoses, raincoats, and seals.
Poly(cis-1,4-isoprene) (natural rubber)	$-CH_2-C=C-CH_2-$, side: CH_3, H	-70	36	Extensive uses in automobile tires and in a wide range of other industrial products.
Poly(trans-1,4-isoprene) (gutta percha)	$-CH_2-C=C-CH_2-$, side: CH_3, H	-68	74	Used in toys, balloons, golf-ball covers, and automobile equipment.
Poly(methacrylic acid)	$-CH_2-C-$, side: OH, $C=O$, CH_3			Adhesive and thickening agent.

(Continued)

Name	Repeating Unit	$T_g(°C)$	$T_m(°C)$	Properties and Uses
Poly(methyl acrylate)	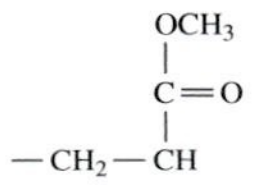 $-CH_2-CH-$ / $C=O$ / OCH_3	5–9 (atactic)		Used in surface coatings.
Poly(methyl-2-cyanoacrylate)	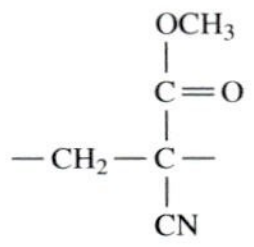 $-CH_2-C-$ / CN / $C=O$ / OCH_3			Adhesive, especially for metals.
Poly(methyl methacrylate) (Plexiglas, Lucite, Perspex, PMMA)	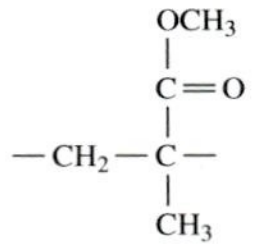 $-CH_2-C-$ / CH_3 / $C=O$ / OCH_3	105 (114); (60) (−7); 48 (isotactic); 128 (syndio)	160 (isotactic); 200 (syndio)	Clear, transparent, glassy polymer used extensively in castings, lenses, roof "bubbles," windows, dentures, fiber optics, and illuminated signs.
Poly(styrene butadiene) copolymers (SBR and GRS elastomers)	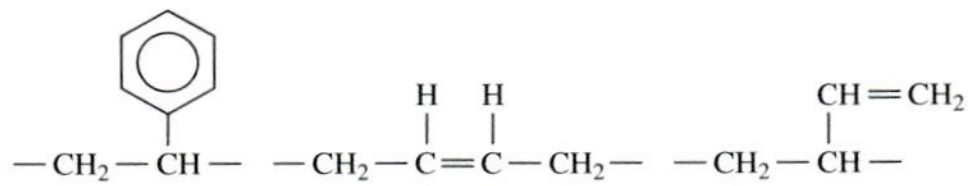 $-CH_2-CH-$ (phenyl) $-CH_2-C=C-CH_2-$ with H H and $CH=CH_2$ $-CH_2-CH-$	−56, (23–77 copolymers); −41 (30–70 copolymers)		Elastomeric polymers used as a replacement for natural rubber: for example in footwear, latex paints, and tire treads.
Poly(styrene-α-methylstyrene) copolymer	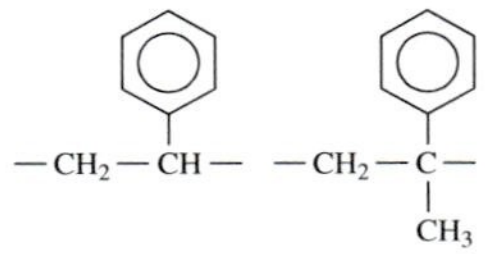 $-CH_2-CH-$ (phenyl) $-CH_2-C-$ / CH_3 (phenyl)			Used in electrical appliances and refrigerator linings.
Poly(tetrafluoroethylene) (Teflon)	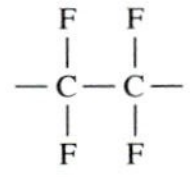 $-CF_2-CF_2-$	130; −113	327; 30	Highly water repellent polymer with a surface lubricity. Extensively utilized in bearings and other sliding surfaces, nonsticking cooking utensils, seals, machined parts, and protective liners.
Poly(tetrafluoroethylene-hexafluoropropylene) copolymers (Teflon FEP)	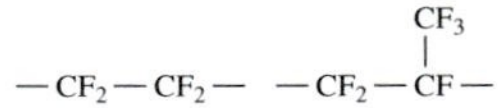 $-CF_2-CF_2-$ $-CF_2-CF-$ / CF_3			Used in capacitors, printed circuits, mold liners, textile finishes, wire insulation, tubing, and in fluid power transmission.
Poly(vinyl acetate)	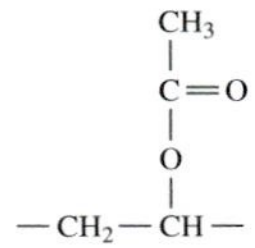 $-CH_2-CH-$ / O / $C=O$ / CH_3	30		Used in emulsion paints and as a precursor for poly(vinyl alcohol) manufacture: a component of chewing gum, drinking straws, and adhesives.

Polyvinyl and Polyolefin Compounds

Name	Repeating Unit			Uses
Poly(vinyl alcohol) (Vinylon)	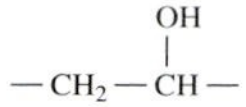 $-CH_2-CH-$ with OH	99	258	Water-soluble or hydrophilic polymer used in fibers, aqueous adhesives, sizing agents for textile fibers, as a binder for the fluorescent layer in TV tubes, as a thickening agent, as wet-strength adhesives, and films.
Poly(vinyl butyral)*	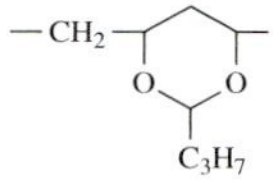 $-CH_2-CH-$ (cyclic acetal with C_3H_7)			Adhesive used in laminated safety glass.
Poly(N-vinylcarbazole) (Luvican, Polectron)	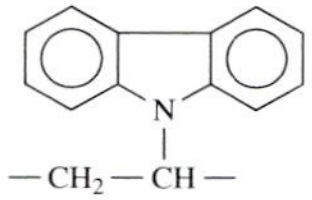 $-CH_2-CH-$ with carbazole N	200		Used for high-temperature electrical insulation, injection-molded products, and as an asbestos substitute.
Poly(vinyl chloride) (PVC)	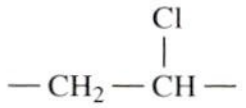 $-CH_2-CH-$ with Cl	+78–81 (atactic)	285 (extrapolated)	Hard, inflexible polymer in the unplasticized state. When plasticized (usually with phthalate esters), used in Tygon tubing and films, automobile seat covering, electrical insulation, floor tiles, molded and extruded objects, coated fabrics, and plumbing pipes.
Poly(vinyl chloride vinyl acetate) (Vinylite)	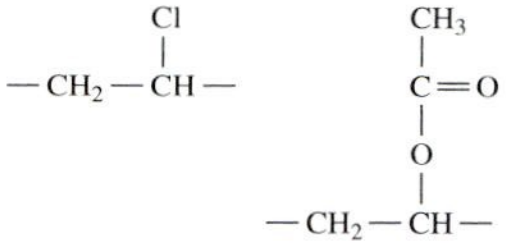 $-CH_2-CH-$ with Cl; $-CH_2-CH-$ with $O-C=O-CH_3$			Used for the manufacture of phonograph records, and in coatings for cans and other metal containers.
Poly(vinyl cinnamate)	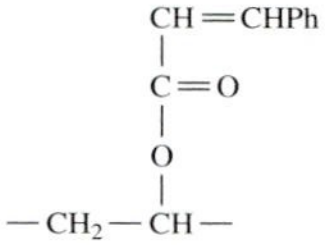 $-CH_2-CH-$ with $O-C=O-CH=CHPh$			Photocrosslinking polymer for use in the preparation of photoresist printing plates and printed circuit boards.
Poly(vinyl fluoride)	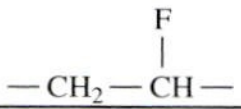 $-CH_2-CH-$ with F		200	Used in films, window glazing, and in coatings for aluminum and wood.

*Prepared from poly(vinyl alcohol) and butyraldehyde.

(*Continued*)

Name	Repeating Unit	T_g(°C)	T_m(°C)	Properties and Uses
Poly(vinyl pyrrolidone) (Kollidon, Periston)	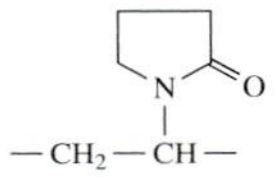			Used formerly as a blood plasma extender. Used as a protective colloid, thickening agent, emulsion stabilizer in cosmetics, surfactant in dyeing, clearing agent in beer and other beverages, and in hair sprays.
Poly(vinylidine chloride)	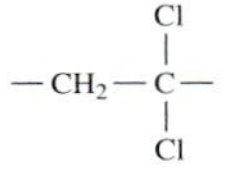	−18	210	Polymer with low-gas permeability. A component of Saran copolymer films [with poly(vinyl chloride)]: used in blow-molded bottles, pipes, and tape.
Poly(vinylidine fluoride)	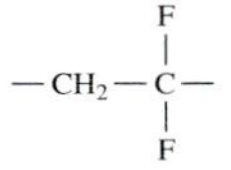	−39(13)	171	Piezoelectric polymer used in microphone diaphragms, molded and extruded objects.
Poly(vinylidine fluoride-hexafluoropropylene) copolymer (Viton)	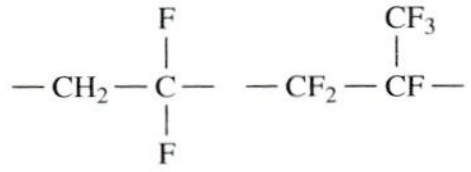	−55		Solvent-resistant elastomer used for the manufacture of O-rings, seals, hose, tubing, and fuel-resistant diaphragms.
Poly(methyl vinyl ether)	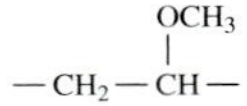	−13(−31) (atactic); −21 (isotactic)	150 (isotactic)	Water-soluble polymer used in adhesives and nonmigratory plasticizers.
Polypropylene (Herculon)	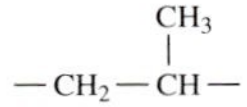	26, −35 (isotactic)	183, 130; 150 (isotactic)	Tough plastic widely used as fibers for ropes, seat covers and carpets; as films for packaging, and in injection-molded articles, especially for automotive applications.
Polystyrene	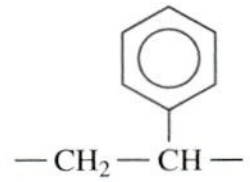	100 (atactic and isotactic)	240 (isotactic)	Clear, transparent, glasslike polymer used widely for the fabrication of molded objects and foamed insulation. Also used as a substrate (when crosslinked) for polymer-bound transition metal catalysts or as a gel permeation chromatography substrate.

Polyalkyne

Name	Repeating Unit	T_g(°C)	T_m(°C)	Properties and Uses
Polyacetylene	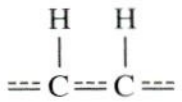			Silver- or gold-colored, metallic-type polymer that conducts electricity.

SUGGESTIONS FOR FUTHER READING

BRANDRUP, J., IMMERGUT, E. H., and GRULKE, E. A. (Eds.), *Polymer Handbook*, 4th ed. New York: Wiley, **1999**.

ELIAS, H.-G., *New Commercial Polymers, 1969–1975*. New York: Gordon and Breach Science Publishers, **1977**.

ELIAS, H.-G., *New Commercial Polymers 2*. New York: Gordon and Breach Science Publishers, **1986**.

MARK, J. E. (Ed.), *Physical Properties of Polymers Handbook*, New York: Springer-Verlag, **1996**.

MARK, J. E. (Ed.), *Polymer Data Handbook*, New York: Oxford University Press, **1999**.

RODRIGUEZ, F., *Principles of Polymer Engineering*, 2nd ed. New York: McGraw-Hill, **1982**.

ULRICH, H., *Introduction to Industrial Polymer*. Munich: Hanser Publishers, **1982**.

Author Index

Hiemenz, P.C., 332, 371, 391, 449, 491, 520
Higashimura, T., 115, 753
Higgins, J.S., 449, 626
Hildebrand, J.H., 398n
Hiltner, A., 558
Hirschfelder, J.O., 567n, 568n
Hjelm, R.J., 275
Hodge, P., 190
Hoeve, C.A.J., 590n, 598
Hofer, D., 272
Hoffman, A.S., 772
Hogan, J., 772
Holliday, L., 716
Holmes, K.C., 619n, 626
Holmes-Farley, S.R., 190
Homsy, C.A., 757n
Honeycutt, J.D., 588n, 598
Hopfinger, A.J., 568n, 598
Hosemann, R., 626
Houwink, R., 459n
Howard, J.C., 595n
Hower, J., 275
Hsieh, H.L., 114
Hu, Y., 114
Huggins, M.L., 458n, 492n
Huglin, M.B., 449
Hummel, D.O., 629

Immergut, B., 364
Immergut, E.H., 364, 463n
Inori, T., 243n, 272
Inoue, K., 190
Inoue, S., 166
Interrante, L.V., 272, 275
Irving, E., 142
Isihara, A., 520
Ittel, S.D., 371
Ivin, K.J., 114, 167, 304, 629

Jackman, R.J., 753
Jacobson, H.W., 294n, 302n
Jacoby, M., 558
Jaffe, M., 715
James, W.J., 167
Janca, J., 484n, 486n
Jannink, G., 519
Janshoff, A., 687
Jelinski, L.W., 227, 628
Jellinek, H.H.G., 305
Jenkins, A.D., 102n
Jennings, B.R., 440, 449, 629
Jensen, L.H., 628
Jerrard, H.G., 440, 449
Jiang, Z., 751
Jinnai, H., 627

Johnson, B.F.G., 142
Johnson, J.F., 491, 557, 559
Jordan, D.O., 91

Kaelble, D.H., 687
Kakudo, M., 626
Kamachi, K., 76
Kamachi, M., 753
Kambour, R.P., 687
Kamide, K., 396n, 520
Kamigaito, M., 371
Kammermeyer, K., 772
Kandil, S.H., 773
Kaner, R.B., 752
Kang, A., 190
Karasz, F.E., 559
Karo, W., 450, 491
Kauffmann, H.F., 66n
Kausch, H.H., 628
Kawagoe, Y., 228
Kazmeier, P.M., 77
Kellam, E.C., 686
Kennedy, J.P., 109n, 115, 385
Khokhlov, A.R., 26, 519, 716
Khotimskii, V.S., 271
Kim, S.W., 771
Kim, Y-B., 225n
Kimura, Y., 167
King, M., 716
King, T.A., 449
Kinsinger, J.B., 294n
Kinter, M., 228
Kitagawa, T., 629
Kitaigorodsky, A.I., 628
Klabunde, U., 114
Klavetter, F.L., 114
Klempner, D., 558, 559
Klenin, V.J., 520
Kloczkowski, A., 502n
Klotz, I.M., 404n
Klug, A., 221, 619n
Klumperman, B.J., 77
Knox, A., 449
Kobayashi, S., 167
Koch, K.A., 273, 274
Koenig, J.L., 628, 629
Koide, N., 272
Kolinsky, M., 484n, 486n
Komoroski, R.A., 628
Koppel, T., 751
Koyama, T., 687
Krassig, H., 661
Kraus, G., 662
Krigbaum, W.R., 458n, 715
Krimm, S., 629

Subject Index

Abrasion resistance, 668
Absolute molecular weights, 393–444
  colligative properties, 403–404
  end-group analysis, 402–403
  light scattering measurement, 414–431
  mass spectrometry, 432
  osmotic pressure measurement, 404–414
  ultracentrifugation, 432–444
  weight average, 419–420
Absolute viscosity, 454, 455
Acrylamide, 132–133
Acrylonitrile, 133
Acrylonitrile-butadiene-styrene (ABS), 531
Acyclic diene metathesis polymerization (ADMET), 100
Addition reactions, 55
Additives used to improve polymer performance, 658–659
  antioxidants, 659
  fillers, 658
  pigments, 658
  plasticizers, 658
  vulcanizing agent, 658
Aldehydes, 133–134
Aliphatic carbon-carbon bond, 690
Aliphatic carbon-carbon double bond, 691
Alkyl ether side group, 702–703
Alkyl groups as side groups, 697
Amide linkage, 693–694
Amide side groups, 702
Amino acids, 199–201
Amorphous polypropylene, 397
Amplification, 117
Amylopectin, 193
Amylose, 103
Anhydride linkage, 693
Anionic polymerization, 78, 81–102, 252, 373–382
  average degree of, 375
  copolymerization, 88, 381–382
  distribution degree of, 376–379
  incomplete dissociation of initiator, 380–381
  initiators for, 81–82
  kinetic chain length, 374–375
  mechanism of, 84–86
  monomers for, 82
  propagation, 379–380
  rate of, 374
Anti-freeze proteins, 204
Aromaticity, 298
Aromatic ladder polymers, 45–47

Aromatic ladder structures, 691–692
Aromatic polyamides, 41–42
Aromatic rings as skeletal units, 691
Arrhenius formulation, 344
Artificial kidney and hemodialysis materials, 764
Aryl side groups, 697–698
Atomic force microscopy (AFM), 552
Atom transfer radical polymerization, 73–74
Average kinetic chain length, 345
Avrami equation, 550
Azide decomposition route, 252–253

Bag molding, 653
Band gap, 726
Bifunctional monomers, 325
Binodial curve, 508
Bioelastomers, 709
Bioerodable polymers, 764–765
Biological condensation polymers, 18
Biological polymers, 191–226
Biomedical applications of synthetic polymers, 754–770
  artificial kidney and hemodialysis materials, 764
  bioerodable polymers, 764–765
  biostable materials, 755–758
  bones, joints, and teeth, 762–763
  cardiovascular applications, 758–761
  contact lenses and intraocular lenses, 763
  controlled release of drugs, 766–769
  oxygen-transport membranes, 764
  polymeric blood substitutes, 770
  surgical sutures, 765–766
  tissue adhesives and artificial skin, 762
  uses for polymers in biomedicine, 754–755
Biostable materials, 755–758
  synthetic polymers in living systems, stability of, 756–758
Bipolaron, 733
Block copolymers, 88
Blow molding, 652
Boltzmann distribution, 583–584
Boron nitride (BN), 268
Boyle's treatment of gases, 512
Branched polymers, 5
Branching and cross-linking, 713
Bubble-blown films, 637
Butadiene, 95

Calender coating, 659
Calendering, process of, 638

Helices (*cont.*)
    diffraction by a continuous helix, 616–619
    within microcrystallites, 619–620
Hexachlorocyclotriphosphazene, 135
Hexmethyldisiloxane, 237
High-impact polystyrene (HIPS), 531
High-molecular-weight polymers, 31–32
    scrambling reactions, 32
Homopolymer synthesis, 205
Hydrogels, 763
Hydrogen as a side group, 696–697
Hydrogen chloride, 169
Hydrolysis
    of chains, 170
    of polymethacrylamide, 178
    of side-group structures, 177–179
Hydroxyl groups, 701–702
Hydroxyproline (HPRO), 707

Ideal solution, 493–494
Imines(cyclic amines), 145
Immobilization of enzymes, 214–216
Impact resistance, 668
Impact strength, 523
Inflexible chains, 589–590
Initiation
    direct thermal and photolytic, 66–67
    free-radical chain, 128–130
    ionic chain, 130–131
    by ionizing radiation, 67
    by redox reactions, 65–66
Initiators, 21
    for anionic polymerization, 81–82
    for cationic polymerization, 104
    free-radical, 62–67
    thermal decomposition of, 62–65
Injection molding, 651–652
Inorganic condensation polymers, 18
Inorganic polymers, 229–268
Interchangeability, 494
Interfacial polymerization, 40–41
Intrinsic semiconductors, 727–728
Intrinsic viscosity, 457
    accuracy, 466–468
    and molecular size, 464
Ionic conduction in solid polymers, 718–723
    comb polymers, 720–721
    host polymers for solid electrolytes, 720
    poly(ethylene oxide), 720
    proton conductors, 722–723
    salts, 721
    solid polymeric electrolytes, uses for, 721
Ionic polymerization, 78, 372–389
    and free-radical kinetics, differences between,
        372–373

"living" polymers, 86–87
stereoregular polymers, 79–81

Kevlar, 41, 647, 707
Kinematic viscosity, 454
Kinetics
    of condensation polymerization, 306–328
    of crystallization, 548–551
    of free radical polymerization, 333–368
    of ionic polymerization, 372–389
    of polyesterification, 310–314
Knife coating, 660

Lactams, 157–159
Lactams (cyclic amides), 145
Lactides, 155
Lactones (cyclic esters), 145, 155
Ladder polymers, 7–8
Lambert law, 416
Laminate, 657
Light scattering measurement, 414–431
    accuracy of, 429–431
    experimental apparatus and technique,
        423–425
    polymer dimensions, 420–423
    Rayleigh ratio, 414, 417
    refractive index increment, 428–429
    turbidity, 415–418
Light-scattering photometry, 425–428
Linear condensation polymers, 325
Linear polyester, 34–35
Linear polymers, 4
Liquid-crystalline polymer, 14
Liquid state, 526
Lithioaryl derivatives, 180
Lithiopolystyrene, 180
Loss factor, 683
Lysozyme, 204

Macromolecular hypothesis, 21–25
    structural work, 22
    synthetic polymers, 22–25
Macromolecular skeleton, influence by, 690–696
    aliphatic carbon–carbon bond, 690
    aliphatic carbon–carbon double bond, 691
    amide linkage, 693–694
    anhydride linkage, 693
    aromatic ladder structures, 691–692
    aromatic rings as skeletal units, 691
    carbon–oxygen bond in polyesters, 692–693
    etheric carbon–oxygen bond, 692
    phosphazene linkage, 695
    Schiff base linkages, 693
    siloxane linkage, 694–695
    urethane linkage, 694

Macromolecular structure and properties, 689–714
  branching and cross-linking, 713
  elastin, 709–710
  illustrative examples, 706–709
  influence by different side groups, 696–704
  influence by the macromolecular skeleton,
    690–696
  new polymers, how scientists design, 713–714
  random and block copolymers, 710–713
  structural influences on solid state properties,
    705–706
  surface properties, 705
Main chain, reactions, 168–175
  addition, 168–169
  electron beam depolymerization, 175
  high temperature degradation, 172–175
  oxidation, 171–172
Mark–Houwink equation, 459, 462
Mass spectrometry, 432
Materials science, 25
Maxwell model, 527–529
Mechanical changes, thermal properties, 681
Mechanical tests, 664–669
  flexural strength, 668
  hardness and abrasion resistance, 668–669
  impact resistance, 668
  stress-strain curves, 665–668
Melamine–formaldehyde, 49
Melt extrusion of films, 636–637
Melt polymerization, 38–39
Melt pressing of film, 635–636
Merrifield synthesizer, 208
Mesogenic side groups, 698–699
Messenger (RNA), 220
Meta-carborane, 240
Metallocarbene, 98
Metallocenes, 96
Methyl iodide, 197
Microcrystalline polymer fiber, 554
Microcrystalline polymers, 525
  analysis of intensity data, 612–613
  clues about the structure, 611–612
  interpretation of X-ray photographs, 609–611
  structure factor analysis, 613–614
Microcrystallinity, 543–545
Microlithography and polymer reactions,
  184–185
Mineralogical and preceramic polymers, 261–268
  carbon fibers, 265–266
  sol-gel process, 264–265
Modulus, 523
Molecular colloids, 452
Molecular graphics, 595
Molecular size from intrinsic viscosity, 464
Molecular structural effects, 296

Molecular structure, 601
  and conformation, determination of, 608–622
Molecular weight, 442–444
  absolute, 393–444
  distributions, 471–477
  illustrative calculations, 471–472
  and intrinsic viscosity, 458–463
  representation of distributions, 472–473
  secondary methods for determining molecular
    weight and distributions, 451–487
Molten polymer, 643
Molybdenum catalyst, 99
Monofilament, 639
Monomer, 2
Monomer-polymer equilibria, 278
  specific, 284–285
Monomers
  for anionic polymerization, 82
  for cationic polymerization, 104
Mooney-Rivlin constants, 674
Morphological changes in polymers, 521
Myoglobin, 203

Network rupture, 675
New polymers, how scientists design, 713–714
No-catalyst copolymerization, 161–162
Nuclear magnetic resonance (NMR), 474,
  600
Number average molecular weight, 408
Nylon, 17
Nylon 610, 41

Octamethylcyclotetrasiloxane, 235
Olefin metathesis, 97
Oligomer, 3
Optical birefringence technique, 552
Optical diffraction, 620–622
Optical rotatory dispersion (ORD), 600
Organic polymers with inorganic elements
  in the main chain, 235
  polymerization of unsaturated monomers, 232–234
  in the side group, 232–234
Organosilicate-polymer systems, 680
Organosiloxane ladder polymers, 239
Orientation of polar groups, 683
Osmometry, practical, 409–412
Osmotic pressure
  accuracy of molecular weights determined from,
    412–414
  measurement, 404–414
  of polymeric solutions, 505–507
  semipermeable membranes, 414
Ostwald viscometer, 455
Oxazoline polymerization, 160–161
Oxepanes, 151–152

Random coil polymer, 422
Raoult's law, 403, 493
Rate of polymerization, 339–345
    average kinetic chain length, 345
    experimental measurement of, 339–343
    theoretical, 343–345
Rates
    of anionic polymerization, 374
    of cationic polymerization, 382–384
Rayleigh ratio, 414, 417
Rayon, 197
Reactivity ratios
    copolymerization, 363–365
    experimental determination of, 363
Redox reactions, 65
Reduced specific viscosity, 457
Reduced viscosity, 457
Reinforced polymers, 657–658
Relative viscosity, 456
Repeating unit, 316
Ribonucleic acid (RNA), 216
    messenger, 220
    ribosomal, 221
    transfer, 220–221
Ribosomal (RNA), 221
Rigid-rod polymer, 422
Ring-chain equilibria, 286–289
    statistical point of view, 292–296
Ring-opening metathesis polymerization (ROMP),
    97–100
Ring-opening polymerization, 20–21, 144–165
    cyclic anhydrides, 156
    cyclic carbonates, 156
    of cyclic compounds, 144–165
    cyclic ethers, 147–155
    cycloalkenes, 160
    cyclopolymerization, 163–164
    epoxides, 152–155
    ethylenimine, 159
    free radical, 162–163
    glycolides, 155–156
    lactams, 157–159
    lactides, 155–156
    lactones, 155
    mechanisms of, 147
    monomers, catalysts, and polymerizability in, *table*,
        146
    no-catalyst copolymerizations, 161–162
    oxepanes, 151–152
    oxetanes, 151–152
    tetrahydrofuran, 150–151
    tetrathiane, 149
    trioxane, 147–149
    trithiane, 149
Ring-polymer interconversions, examples of, 278–279

Roll coating, 660
Rotational isomeric state model, 584
Rotational molding, 653
Rubberlike elasticity, 669–681
    control of network structure, 674–675
    experimental details for elastomer studies, 672–673
    filler-reinforced elastomers, 679–680
    networks at very high deformations, 675–677
    other types of deformation, 677–678
    preparation of networks, 671–672
    swelling and gels, 678–679
    typical applications, 672
    typical stress-strain behavior, 673–674

SBR copolymers, 531
Scanning tunneling microscopy (STM), 552
Scattering, 622–624
    dynamic properties, 624
    structural applications, 623–624
Schiff base linkages, 693
Schlieren optics, 438
Schotten-Baumann condensation, 38
Scratch resistance, 669
Second-order transition, 532
Sedimentation coefficient, 434
Sedimentation constant, 437
Sedimentation equilibrium method, 435
Sedimentation velocity, 432, 433
Semipermeable membranes, 414
Shear modulus, 523
Short-chain diols, 52
Short-chain models for polymers, 571–579
    conformational nomenclature, 572–573
    hard-sphere approach, 573–574
    short-chain concept, 571
    use of intramolecular potentials, 574–577
Side groups
    alkyl ether, 702–703
    alkyl groups as, 697
    amide, 702
    aryl, 697–698
    carboxylic acid, 704
    cyano, 701
    ester, 703
    fluorine, 699–700
    hydrogen, 696–697
    hydroxyl groups, 701–702
    interactions, 298–302
    mesogenic, 698–699
    reactions of, 176–188
Silicone ladder polymer, 239
Siloxane-arylene and siloxane-carborane polymers,
    239–240
Siloxane linkage, 694–695
Sinter-fabrication of film, 636